FUNCTIONAL ANALYSIS

FUNCTIONAL ANALYSIS

Useful for

Post Graduate students of
all Indian Universities and Various Competitive Examinations

DR. SUDHIR KUMAR PUNDIR
M.Sc., M.Phil., NET, Ph.D.
Associate Professor
Department of Mathematics
S.D. (P.G.) College, Muzaffarnagar (U.P.)

CBS Publishers & Distributors Pvt Ltd

New Delhi • Bengaluru • Chennai • Kochi • Kolkata • Lucknow • Mumbai
Hyderabad • Jharkhand • Nagpur • Patna • Pune • Uttarakhand

Disclaimer

Science and technology are constantly changing fields. New research and experience broaden the scope of information and knowledge. The author has tried his best in giving information available to him while preparing the material for this book. Although, all efforts have been made to ensure optimum accuracy of the material, yet it is quite possible some errors might have been left uncorrected. The publisher, the printer and the author will not be held responsible for any inadvertent errors, omissions or inaccuracies.

FUNCTIONAL ANALYSIS

ISBN: 978-81-239-2976-7

First Edition: 2016

Reprint: 2024

Published by Satish Kumar Jain and produced by Varun Jain for

CBS Publishers & Distributors Pvt Ltd

4819/XI Prahlad Street, 24 Ansari Road, Daryaganj, New Delhi 110 002, India

Ph: 011-23289259, 23266861 Website: www.cbspd.com

e-mail: delhi@cbspd.com

Corporate Office: 204 FIE, Industrial Area, Patparganj, Delhi 110 092

Ph: 011-4934 4934 Fax: 011-4934 4935 e-mail: publishing@cbspd.com

publicity@cbspd.com

Branches

- **Bengaluru:** Seema House 2975, 17th Cross, KR Road, Banasankari 2nd Stage, Bengaluru 560 070, Karnataka, India
  Ph: +91-80-26771678/79 Fax: +91-80-26771680 e-mail: bangalore@cbspd.com
- **Chennai:** 7, Subbaraya Street, Shenoy Nagar, Chennai 600 030, Tamil Nadu, India
  Ph: +91-44-26680620, 26681266 Fax: +91-44-42032115 e-mail: chennai@cbspd.com
- **Kochi:** 42/1325, 1326, Power House Road, Opp KSEB, Power House, Ernakulam 682 018, India
  Ph: +91-484-4059061–65 Fax: +91-484-4059065 e-mail: kochi@cbspd.com
- **Kolkata:** 147, Hind Ceramics Compound, 1st Floor, Nilgunj Road, Belghoria, Kolkata 700 056, West Bengal, India
  Ph: +91-9096713055/56 e-mail: kolkata@cbspd.com
- **Lucknow:** Basement, Khushnuma Complex, 7-Meerabai Marg (behind Jawahar Bhawan), Lucknow 226 001, India
  Ph: +91-522-4000032 e-mail: tiwari.lucknow@cbspd.com
- **Mumbai:** PWD Shed. Gala no. 25/26, Ramchandra Bhatt Marg, Next to JJ Hospital Gate no. 2, Opp. Union Bank of India Noorbaug Mumbai 400 009, Maharashtra, India
  Ph: +91-22-66661880/89 e-mail: mumbai@cbspd.com

Representatives

- **Hyderabad** 0-9885175004 • **Jharkhand** 0-9811541605 • **Nagpur** 0-8692091830
- **Patna** 0-9334159340 • **Pune** 0-9664372571 • **Uttarakhand** 0-9716462459

Printed at SRK Graphics, Shahdara, Delhi, India

Preface

The book entitled 'FUNCTIONAL ANALYSIS' is meant for PG students of all Indian Universities. Besides, it will also be very useful for those students preparing for various competitive examinations like NET and GATE.

This book has evolved the lectures that I have giving to the students for the last seventeen years. Special and conscious efforts have been made to keep the writing style simple. Students who are tired of complex concepts and abstract presentations styles, will find this book simple and straight forward. It is a collection and compilation work from various sources and has been endeavoured to include as much as information could be possible. The book's objective is to provide a conceptual understanding of the fundamentals of functional analysis. Different concepts have been explained with the help of examples. A large number of problems with solutions have been provided to assist one get a firm grip on the ideas developed. There is plenty of scope in the form of exercise for the reader to try and solve the problem on his own. To make the book self-contained and competition oriented a chapter review of basic terms, results and questions has been given at the end of each chapter. Also, at the end of each section, graded examples have been given with the name of self assessment test, which help the students to grasp the thing better. These include problems on the entire section and are carefully selected to represent variety.

I express my gratitude to the authors and publishers of various books I consulted.

I wish to sincerely thank Sh S.K. Jain, Managing Director, CBS Publishers and Distributors, New Delhi for his encouragement and help in bringing out this publication in a present nice form.

My special thanks to Sh. Y.N. Arjuna, Sh. B.M. Singh, Sh. Sunil Dutt and entire team of CBS Publishers and Distributors, New Delhi whose encouragement and unstinted support enabled me to complete my book. Mr. Peeyush Goel, M/s Dreamshapers also deserve special mention for nice type setting.

I must also record my appreciation due to my wife Dr. Rimple, daughter Rijuta and son Shrish for their understanding and love during the long period that I have taken to complete this book.

Above all I am thankful to The Almighty God, without whose grace nothing is possible for any one.

Readers are welcomed to point out errors, if any and send their valuable suggestions for improving the quality of the book.

Dr. SUDHIR KUMAR PUNDIR

email : skpundir05@yahoo.co.in

Contents

Chapter 1 Introduction

1.1 INTRODUCTION

In previous classes, we have studied about the concept of sets, their types and properties. We are also familiar about the Venn diagrams and its uses. In this chapter we shall discuss about the relation and functions, concept of topological spaces and review of measure and integration, which are widely used in the further study.

1.2 RELATION

Let us take two sets of natural numbers N_1 and N_2. We define R as a relation between them such that N_1 is a square of N_2. Then we can write $1R1$, $2R4$, $3R9$, ...

In terms of ordered pair, we can write

$$R = \{(1, 1), (2, 4), (3, 9), (4, 16), \ldots\} = \{(x, y : x, y \in N \text{ and } y = x^2\}$$

The relation from set N to N is a subset of $N \times N$ such that $y = x^2$.

Definition. *Let A and B be two sets. Then a relation R from A to B is a subset of A×B.*

Symbolically: R is a relation from A to $B \Leftrightarrow R \subseteq A \times B$.

REMARKS

- If R is a relation from A to B, then A is called the domain and B the range of R.
- If R is a relation from a non-empty set A to a non-empty set B and if $(a, b) \in R$, then we write aRb, read as "a is related to b by the relation R." On the other hand, if $(a, b) \notin R$, we write $a\not{R}b$ and say that 'a is not related to b by the relation R'.
- In particular, any subset $A \times A$ defined a relation in A, known as Binary relation.

ILLUSTRATIONS

(1) If $a, b \in N$ and R is defined as "a is divisor of b" then R is relation on N.
The subset $N \times N$, which corresponds to the relation R is $S = \{(n, r): n \in N, r \in N\}$
Here, it is clear that (1, 3), (2, 4), (3, 9) (4, 8), (4, 4), are in S, whereas (2, 3), (4, 5), (5, 6) are not in S.

(2) If R is a relation from set $A = \{1,2,3\}$ to the set $B = \{-1, -2\}$ defined by $x + y = 0$, then $R = \{(1, -1), (2, -2)\}$
Here, domain of R is $\{1, 2\}$ and Range $= \{-1, -2\}$.

(3) If $A = \{a, b, c, d, e\}$ and $B = \{f, g, h, i\}$ and let $R = \{(a, g), (a, i), (d, h), (e, f)\}$ by a relation from A to B then
Domain of $R = \{a, d, e\}$ and Range of $R = \{g, i, h, f\}$

(4) If $a, b \in R$, the set of real numbers and R is "$|a - b|$ is a rational number" then R is a relation on R. The subset S of $R \times R$ which corresponds to the relation is

$$S = \{(a, b + a): a \in R, b \in Q\}$$

It is observed that $\left(1, 2\frac{1}{2}\right), \left(\pi, \pi - \frac{1}{2}\right)$ belongs to S, while $(\sqrt{2}, \pi + \sqrt{2}) \notin S$.

(5) If $A = \{2, 3, 4\}$and $B = \{a, b, c\}$, then $R = \{(2, b), (3, c), (2, a), (4, a)\}$ being a subset of $A\times B$, is a relation from $A\times B$. Here $(2, b), (3, c), (2, a), (4, a) \in R$, so we may write $2Rb$, $3Rc$, $2Ra$, $4Ra$. But $(3, b) \notin R$ therefore, $3 \not R b$.

(6) If $a, b \in N$ and R is defined by "$a - b$ is divisible by a number $n \in N$", then R is a relation on N. The subset S of $N \times N$ corresponding to the relation by

$$S = \{n, n + rm : n \in N, r \in N\}$$

Here, $m = 3$, $(2, 8), (5, 11) \in S$ [$\because 2 - 8 = 6$, which is divisible by 3]

While $(3, 8) \in S$ [$\because 3 - 8 = 5$, which is not divisible by 3]

1.2.1 TOTAL NUMBER OF RELATIONS

Let A and B be two non-empty finite sets consisting p and q elements respectively, then $A \times B$ consists of $p\,q$ ordered pairs. Therefore, total number of subset of $A \times B$ is 2^{pq}.

REMARKS

- For a non-empty set A, $\phi \in A \times A$, therefore it is a relation on A, called void or empty relation on A.
- The void relation ϕ and the universal relation $A \times B$ are called trivial relations from A to B.
- The void and universal relation on set A respectively the smallest and the largest relation on A.

1.2.2 IDENTITY RELATION

Let A be a set. The identity relation on A is the relation $I_A = \{(x, x) : x \in A\}$ on A.

For example : If $A = \{a, b, c\}$ then the relation $I_A = \{(a, a), (b, b), (c, c)\}$is the identity relation. $R=\{(a, a), (b, b)\}$ is not an identity relation as $(c, c) \notin R$.

1.2.3 INVERSE OF A RELATION

Let A, B be two non-empty sets and R be a relation from a set A to B and let (x,y), number of the subset D of $A \times B$ corresponding to the relation R from A to B.

To the relation R from the set A to the set B, there corresponds a relation from the set B to the set A called the inverse of the relation, denoted by R^{-1} such that the subset $B \times A$ corresponding to the relation R^{-1} is $= \{(y, x): (x, y) \in D\}$.

i.e., $$yR^{-1}x \Leftrightarrow xRy$$

☛ ILLUSTRATIONS

(1) Let $A = \{a, b, c\}$ and $B = \{1,2,3\}$ be two sets and let $R = \{(a, 1), (a, 2), (b, 1), (b, 2)\}$ be a relation from A to B then $R^{-1} = \{(1, a), (2, a), (1, b), (2, b)\}$

(2) If $A = \{1, 2, 3\}$, $B = \{5, 6, 7\}$ and let $R = \{(1, 5), (2, 5), (2, 7)\}$ be a relation from A to B.

Then $R^{-1} = \{(5, 1), (5, 2), (7, 2)\}$ which is a relation from B to A.

Also, Domain $(R) = \{1, 2\} =$ Range (R^{-1})

And, Range $(R) = \{5, 7\} =$ Domain (R^{-1})

(3) The inverse of the relation "*is less than*" In R "*is greater than*".

REMARK

- Sometimes, the inverse of a relation coincides with the relation itself.

  For example, the inverse of the relation "perpendicular to" in the set of straight lines coincides with itself.

1.3 CLASSIFICATION OF RELATIONS

(a) Reflexive Relation: Let R be a relation on a set A.

"A relation R is said to be reflexive if $(x, x) \in R \ \forall \ x \in A$*", i.e.,* $x R x \ \forall \ x \in A$

ILLUSTRATIONS

(1) In a set of integers, a relation R defined by $x R y$ iff $x - y$ is divisible by 4, then R is a reflexive relation because $x - x = 0$ which is a divisible by 4.

(2) The universal relation on a non-empty set A is reflexive.

(3) The relation "is less than," *i.e.*, '<' in the set of rational number is not reflexive, because no member have the relation is less than to itself.

(4) The relation "is a factor of" in the set of rational number is reflexive, since every rational number is a factor of itself.

(5) The relation "is less than or equal to." *i.e.*, $\leq$ in the set of natural number is reflexive.

$$n \leq n \ \forall \ n \in N$$

(b) Symmetric Relation. *A relation R on a set A is said to be symmetric if*

$$(y, x) \in R \text{ whenever } (x, y) \in R \ \forall \ x, y \in R$$

i.e., $$x R y \Leftrightarrow y R x \ \forall \ x, y \in R$$

ILLUSTRATIONS

(1) Let l_1, l_2 be two lines such that l_1 is perpendicular to l_2,
i.e., $l_1 \perp l_2$. Then $l_1 \perp l_2 \Rightarrow l_2 \perp l_1$. Therefore the relation $\perp$ is symmetric.

(2) The identity and the universal relation on a non-empty set are symmetric relations.

(3) Consider the set N of natural numbers and the relation 'is less than'. This relation is not symmetric. Since if $2 < 3$ then $3 \nless 2$.

Let $A = \{1, 2, 3\}$ and relations R_1 and R_2 defined by

$R_1 = \{(1, 2), (1, 3), (3, 1), (2, 1)\}$ and $R_2 = \{(1, 2), (2, 3), (3, 1)\}$

Then R_1 is a symmetric relation, but R_2 is not symmetric.

(c) Transitive Relation: *A relation R on a set A is said to be transitive iff* $(x,y) \in R$ *and* $(y, z) \in R \Rightarrow (x, z) \in R \ \forall \ x, y, z \in A$, *i.e.,* $x R y, y R z \Rightarrow xRz$.

ILLUSTRATIONS

(1) Let a, b, c be three numbers such that a is a factor of b and b is a factor of c, then obviously a is a factor of c. Therefore, 'is a factor of' is a transitive relation.

(2) If l_1, l_2, l_3 are three lines such that $l_1 \perp l_2$ and $l_2 \perp l_3$ then it is obvious that l_1 is parallel to l_3. Therefore the relation "$\perp$" is not transitive.

(3) The identity and universal relation on a non-empty set are transitive.

(4) Let l_1, l_2, l_3 be three straight lines, such that l_1 is parallel to l_2 and l_2 is parallel to l_3 then it is clear that l_1 is parallel to l_3. Therefore, 'is parallel to' is a transitive relation.

(d) Anti-symmetric Relation. *A relation R on a non-empty set A is said to be an anti-symmetric relation iff* $(x, y) \in R$ *and* $(y, x) \in R \Rightarrow x = y \ \forall \ x, y \in R$

REMARKS

- The identity relation R on a set A is an anti – symmetric relation.
- If $(x, y) \in R$ and $(y, x) \notin R$, then it may be noted that $x = y$.
- The universal relation on a set A containing at least two elements is not anti – symmetric.

1.3.1 EQUIVALENCE RELATIONS

A relation R on a set E is said to be equivalence if it is
(i) Reflexive, (ii) Symmetric and (iii) Tansitive

ILLUSTRATIONS

(1) In a set of integers, a relation R is defined by $x R y$ if and only if $x - y$ is divisible by 4. Then R is an equivalence relation. Since

(a) For $x R x$, $x - x = 0$ is divisible by 4. Therefore, it is reflexive.

(b) For $x R y$. Let $x - y = 4m$ so $y - x = 4m$, which is also divisible by 4. Therefore, it is symmetric.

(c) For $x R y$, let $x - y = 4m$; for $y R z$, let $y - z = 4n$. By adding these two equations, we get $x - z = -4(m + n)$,
which is divisible by 4. Therefore it is transitive.

(2) Let R be a relation on the set of all lines in a plane L defined by $(l_1, l_2) \in R$ if and only if line l_1 is parallel to l_2, then R is an equivalence relation because

(a) For each line $l \in L$, we have l is parallel to l.
$\Rightarrow lRl \Rightarrow R$ is reflexive.

(b) Let $l_1, l_2 \in L$ such that $(l_1, l_2) \in R$, then
$\Rightarrow (l_1, l_2) \in R \Rightarrow l_1$ is parallel to $l_2 \Rightarrow l$ is symmetric.

(c) Let $l_1, l_2, l_3 \in L$ such that (l_1, l_2) and $(l_2, l_3) \in R$, then obviously $(l_1, l_3) \in R$ because if l_1 is parallel to l_2 and l_2 is parallel to l_3, then l_3 should be parallel to l_1.

1.3.2 CONGRUENCE MODULO 'm'

Let m be an arbitrary but fixed integer. If $x - y$ is divisible by m, then two integers x and y are said to be congruence modulo m of one another.

Symbolically : $x \equiv y \pmod{m}$ if $x - y$ divisible by m.

For example : $32 \equiv 2$ (mol 3), as $32 - 2 = 30$ which is divisible by 3.

1.3.3 COMPOSITION OF RELATIONS

Let R_1 and R_2 be two relations from set A to B and B to C respectively, then we can define a relation $R_1 \, o \, R_2$ from A to C, such that $(x, z) \in R_1 \, o \, R_2$ if and only if there exist $y \in Y$ such that $(x, y) \in R_1$ and $(y, z) \in R_2$.

This relation is called composition of R_1 and R_2.

REMARKS

- $R_1 o R_2 \neq R_2 o R_1$
- $(R_2 o R_1)^{-1} = R_1^{-1} o R_2^{-1}$
- The intersection of two equivalence relations on a set is an equivalence relation.
- The union of two equivalence relations on a set is not necessarily an equivalence relation.
- If R is an equivalence relation, then R^{-1} is also an equivalence relation.

SOLVED EXAMPLES

EXAMPLE 1. *Let Z be the set of integers. Define a relation R on Z such that x R y holds if and only if x – y is divisible by 5, x ∈ Z, y ∈ Z. Show that it is an equivalence relation.*

SOLUTION. (i) For each $x \in Z$, $x - x$ *i.e.*, 0 is divisible by 5.
Therefore, for all $x \in Z$, $x R x \Rightarrow x$ is reflexive.

(ii) Let $x\,R\,y \Rightarrow x - y$ is divisible by 5.

$\Rightarrow y - x$ is divisible by 5.

Thus $xRy = yRx$

Therefore R is symmetric.

(iii) Let us suppose xRy and yRz, then $(x - y)$ and $(y - z)$ are both divisible by 5. Hence, 5 is also a divisor of $(x - y) + (y - z)$.

5 is a divisor of $(x - z)$.

Therefore, $xRy,\ yRz \Rightarrow xRz \Rightarrow R$ is transitive.

From (i), (ii) and (iii), we conclude that R is an equivalence relation.

EXAMPLE 2. *Let N×N be the set of ordered pairs of natural numbers. Also, let R be the relation in N×N, defined by (a, b) R (c, d) if and only if a+d = b+c. Show that R is an equivalence relation.*

SOLUTION. (i) For all $(a, b) \in N \times N$, we have $a+b = b+a$, *i.e.*, $(a, b)\ R\ (b, a)$.

Therefore, R is reflexive.

(ii) Let $(a, b)\ R\ (c, d)$, then, by definition of R

$$(a+d) = (b+c) \text{ or } (c+b) = (d+a)$$

$$(c, d)\ R\ (a, b) \Rightarrow R \text{ is symmetric.}$$

(iii) Let us suppose $(a, b)\ R\ (c, d)$ and $(c, d)\ R\ (e, f)$, then

$$a + d = b + c \text{ and } c + f = d + e$$

$$\Rightarrow \quad (a + d) + (c + f) = (b + c) + (d + e) \quad \Rightarrow \quad a + f = b + e$$

$$\Rightarrow \quad (a, b) R (e, f)$$

Therefore, R is transitive.

Hence, from (i), (ii) and (iii), we conclude that R is an equivalence relation.

EXAMPLE 3. *If R is the relation for natural number defined by x+4y = 20. Find the domain and range of the relation R.*

SOLUTION. Let $x + 4y = 20 \quad \Rightarrow \quad y = \dfrac{20 - x}{4}$

For $x = 4, y = 4$ and for $x = 8, y = 3$.

For $x = 16, y = 1$ and for $x = 12, y = 2$

Therefore, Domain = {4, 8, 12, 16} and range = {4, 3, 2, 1}

EXAMPLE 4. *A relation R defined on the set of integers Z, as follows*

$$(x, y) \in R \Rightarrow x^2 + y^2 = 25$$

Express R and R^{-1} as the sets of ordered pairs and hence find their respective domains.

SOLUTION. Since $(x, y) \in R \Leftrightarrow x2 + y2 = 25 \quad \Rightarrow \quad y = \pm\sqrt{25 - x^2}$

If $x = 0 \quad \Rightarrow \quad y = 5$.

Therefore, $(0, 5) \in R$ and $(0, -5) \in R$

Now, $x = 3 \quad \Rightarrow \quad y = \sqrt{25 - 9} = \pm 4$

$(3, 4) \in R$, $(-3, 4) \in R$, $(3, -4) \in R$ and $(-3, -4) \in R$

$x = \pm 4 \quad \Rightarrow \quad y = \pm 3$

Therefore, $(4, 3) \in R$, $(-4, 3) \in R$, $(4, -3) \in R$ and $(-4, -3) \in R$

$x = \pm 5 \quad \Rightarrow \quad y = \sqrt{25 - 25} = 0 \quad \therefore \ (5, 0) \in R$ and $(-5, 0) \in R$

Here, it is clear that for any other integral value of x, y is not an integer. Therefore,

$$R = \{(0, 5), (0, -5), (3, 4), (-3, 4), (3, -4), (-3, -4), (4, 3), (-4, 3), (4, -3), (-4, -3), (5, 0), (-5, 0)\}$$

and $R^{-1} = \{(5, 0), (-5, 0), (4, 3), (4, -3), (-4, 3), (-4, -3), (3, 4), (3, -4), (-3, 4), (-3, -4), (0, 5), (0, -5)\}$

Also, domain $(R) = \{(0, 3, -3, 4, -4, 5, -5\}$ = domain of (R^{-1}).

1.3.4 RELATIONS OTHER THAT EQUIVALENCE

Let R be a given relation on the set X. Then R is

(1) non-reflexive if $\exists x$, such that $(x, x) \notin R$.

(2) anti-reflexive or reflexive if $i_x \cap R = \phi$ (where i_x is the identity relation on X or $\forall x \in X : (x, x) \notin R$

(3) non-symmetrical if for some $(x, y) \in R$, we have $(y, x) \notin R$

(4) anti-symmetric if $R \cap R^{-1} = i$, *i.e.*, $(x, y) \in R$ and $(y, x) \in R \Rightarrow x = y$

(5) asymmetric if $R \cap R^{-1} = \phi$, *i.e.*, $(x, y) \in R \Rightarrow (y, x) \notin R$

(6) non-transitive if $R \circ R \not\subset R$

(7) anti-transitive if $(R \circ R) \cap R = \phi$

(8) A reflexive and symmetric, but not transitive relation is called a tolerance relation.

(9) A non-symmetric transitive relation is called an ordered relation.

(10) A reflexive, anti-symmetric and transitive relation is called partial-ordered relation.

EXERCISE 1.1

1. If R is the relation 'is less than' from $A = \{1, 2, 3, 4, 5\}$ to $B = \{1, 4, 5\}$, find the set of ordered pairs corresponding to R. Also find R^{-1}.

2. A relation R defined from a set $A = \{2, 3, 4, 5\}$ to a set $B = \{3, 6, 7, 10\}$ as follows :

$(x, y) \in R \Rightarrow x$ divides y. Write R as a set of ordered pairs and determine the domain and range of R. Also find R^{-1}.

3. Find the domain and range of $A = \{1, 2, 3, 4, 5, 6\}$ when the relation are defined as

(i) xR_1y if and only if $x - y > 0$

(ii) xR_2y if and only if $x + y < 0$

4. Two sets A and B are given by $A = \{1, 2, 8, 9\}$ and $B = \{2, 3, 4, 6, 7\}$ and if R is the relation form A to B given by $\{(1,2), (1,3), (2,4), (2,6)\}$, then which of the following statement is true?

(i) Domain (R) = Range (R^{-1}) and Range (R) = Domain (R^{-1})

(ii) Domain (R) = Domain (R^{-1}) and Range (R) = Range (R^{-1})

(iii) Domain (R) = Range (R^{-1}) and Range (R) = Domain (R^{-1})

(iv) Domain (R) = Range (R)

5. If R is a relation on a set A, then which of the following statement is not true?

(i) If R is reflexive then R^{-1} is reflexive.

(ii) If R is symmetric then R^{-1} is symmetric.

(iii) If R is transitive, then R^{-1} is transitive.

(iv) None of these

6. Find the domain and range of the following relations:

(i) $R = \{(x + 1, x + 5)\} : x \in \{0, 1, 2, 3, 4, 5\}$

(ii) $R = \{(x, x^3) : x$ is a prime number, less than $10\}$

(iii) $R = \{(a, b) : a \in \mathrm{N}, a < 5, b = 4\}$

(iv) $R = \{(a, b) : b = |a - l|, a \in Z$, and $|a| \leq 3\}$

7. Let R_1 be the relation defined on the set of reals R such as $(a, b) \in R_1$ if and only if $1 + ab > 0$ for all $a, b \in R$. Show that R_1 is reflexive, symmetric but not transitive.

8. Let R be relation on $N \times N$, defined by $(a, b) R (c, d)$ if and only if ad $(b + c) = bc(a + d)$. Show that R is an equivalence relation.

9. Show that the relation 'congruence modulo m' on the set of integers is an equivalence relation.

10. Let R_1 be a relation on the set of reals defined by $R_1 = \{(a, b) \in R \times R : a^2 + b^2 = 1\}$
Show that R_1 is not an equivalence relation on R.

11. In a set L of all straight lines in a plane, discuss which of the following two relations are equivalence relations L.
(i) $R_1 = \{(x, y): x, y \in L$ and x is parallel to $y\}$
(ii) $R_2 = \{(x, y): x, y \in L$ and x is perpendicular to $y\}$.

12. Show that the relation
$R = \{(a, b): a - b$ = even integer $\forall$ $a, b \in Z\}$, *i.e.*, $aRb \Leftrightarrow a - b$ = even integer, is an equivalence relation.

13. Show that the relation R in N, the set of natural numbers, defined by xRy if $x^2 - 4xy + 3y^2 = 0$, ($x, y \in N$) is reflexive, not symmetric and not transitive.

14. For the given relation R on a set S, determine which are equivalence relations:
(i) S is the set of all rational numbers, aRb if and only if $a = b$
(ii) S is the set of all real numbers iff
(a) $|a| = |b|$ (b) $a \geq b$
(iii) S is the set of all triangles in a plane, aRb iff a is congruent to b.
(iv) S is the set of all triangles in a plane, aRb iff a and b have equal perimeters.

15. An integer m is said to be related to another integer n if m is a multiple of n. Show that this relation is reflexive and transitive but not symmetric.

16. Let R be a relation defined on the set of natural number N as $R = \{(x, y): x, y \in N, 2x + y = 41\}$. Find the domain and range of R.

17. Let O be the origin. Define a relation between two points P and Q in a plane if $PO = OQ$. Show that the relation is an equivalence relation.

18. Given the relation $R = \{(1, 2), (2, 3)\}$ on the set of natural number N, add a minimum of ordered pairs so that the enlarged relation is symmetric, transitive and reflexive.

19. Let N denote the set of all natural numbers and R be the relation on $N \times N$ defined by $(a, b) R (c, d) \Leftrightarrow ad(b + c) = bc(a + d)$. Show that R is an equivalence relation.

20. Show that the relation, which is symmetric and transitive, is not necessarily reflexive.

ANSWERS

1. $aRb = \{(1, 4), (1, 5), (2, 4), (3, 4), (2, 5), (3, 5), (4, 5)\}$,
$R^{-1} = \{(4, 1), (5, 1), (4, 2), (5, 2), (4, 3), (5, 3), (5, 4)\}$
2. Domain (R) = $\{2, 3, 5\}$, Range (R) = $\{3, 6, 10\}$, $R^{-1} = \{(6, 2), (10, 2), (3, 3), (6, 3), (10, 5)\}$
3. (i) $\{2, 3, 4, 5, 6\}$, $\{1, 2, 3, 4, 5\}$, (ii) ϕ, ϕ **4.** (iii) **5.** (iv) **6.** (i) Domain (R) = $\{1, 2, 3, 4, 5, 6\}$, Range (R) = $\{5, 6, 7, 8, 9, 10\}$ (ii) Domain (R) = $\{2, 3, 5, 7\}$, Range (R) = $\{8, 27, 125, 243\}$ (iii) Domain (R) = $\{1, 2, 3, 4\}$, Range (R) = $\{4\}$ (iv) Domain (R) = $\{0, -1, -2, -3, 1, 2, 3\}$, Range ($R$) = $\{1, 2, 3, 4, 0, 1, 2\}$ **11.** R_1 = Equivalence relation, R_2 = Not equivalence **14.** (i), (ii) **16.** Domain (R) = $\{1, 2, ..., 19, 20\}$, Range (R) = $\{39, 37, 35, ..., 5, 3, 1\}$ **18.** $\{(1, 2), (2, 1), (2, 3), (3, 2), (1, 3), (3, 1), (1, 1), (2, 2), (3, 3), (4, 4), ...\}$

1.4 FUNCTIONS

Definition: *Let A and B be two sets, then a rule or correspondence, which associates each element of A to a unique element of B, is called a function from set A to set B.*

If a general element of set A is denoted by x, and of set B is denoted by y, then we say that y is a function of x if, for every $x \in A$, one and only one value of $y \in B$ can be determined.

Symbolically: If f is a function from a set A to a set B, then we write $f: A \to B$, read as f is a function from A to B or f maps A to B.

1.4.1 RANGE AND DOMAIN OF A FUNCTION

Let an element $y \in B$ be corresponded by an element $x \in A$, then y is called the image of x and is denoted by $f(x)$. Here, x is defined as the pre-image of y.

The set A is called the domain and the set B is called the co-domain of the function f.

The set of all f-images of the elements of A, is called image set or the range of f and is denoted by

$$f(A) \quad \text{or} \quad \{f(x) : x \in A\}$$

Evidently, $f(A) \subseteq B$.

Thus, a mapping $f : A \to B$ is the set of ordered pairs $\{(a, b) : a \in A, b \in B\}$, so that no two ordered pairs have the same finite element.

$$f = \{(a, b): a \in A, b \in B, b = f(x) \ \forall \ a \in A\}$$

For example : Let $A = \{-2, -1, 0, 1, 2\}$ and B is the set of natural numbers for every $x \in A$, $f(x) \in B$ and $f(x) = x^2$.

Here, A is the domain and B is the co-domain.

$f(a)$ is the value of the function $f(x)$, when x takes the value a, *i.e.*, when x is replaced by a.

The elements of the co-domain which is equal to $f(x)$ form the range.

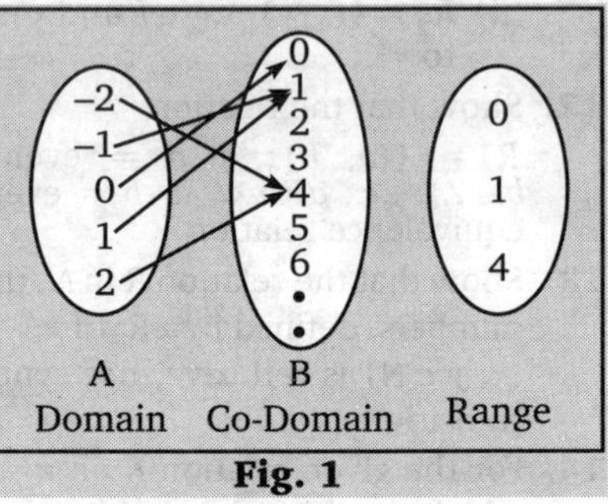

Fig. 1

When $x = -2$, $f(-2) = (-2)^2 = 4$
When $x = -1$, $f(-1) = 1$
When $x = 0$, $f(0) = 0$
When $x = 1$, $f(1) = 1$
When $x = 2$, $f(2) = 4$.
Which can be illustrated in the figure 1.

REMARKS

- If $f : A \to B$ then a single element in A cannot have more than one image in B. However, two or more elements in A may have the same image in B.
- Every element in A must have its image in B, but every element in B may not have it pre-image in A.
- To each element x in A, there exists a unique element y in B such that $y = f(x)$.
- The unique element y of B is called the value of f at x (the image of f under x), and written as $y = f(x)$.
- The range of f consist of those elements in B which appear as the image of at least one element in A.
- Range of a function is the image of its domain.
- Range is a subset of co-domain.

1.5 TYPE OF FUNCTIONS

(a) One-One function: A function f from A to B, *i.e.*, $f : A \to B$ is said to be one-one (or injective) iff distinct elements of A have distinct images.

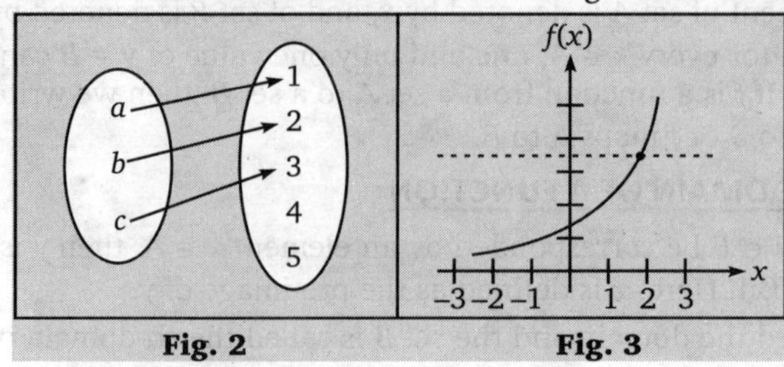

Fig. 2 **Fig. 3**

Symbolically: f is one-one if for $x_1, x_2 \in A$, we have

$$x_1 \neq x_2 \quad \Rightarrow \quad f(x_1) \neq f(x_2) \ \forall \ x_1, x_2 \in A$$

or

$$f(x_1) = f(x_2) \quad \Rightarrow \quad x_1 = x_2 \ \forall \ x_1, x_2 \in A$$

It is also called Univalent function.

Graphically, a function is one-one if and only if no line parallel to x-axis meets the graph of the function at more than one point.

(b) Many-One Function: A function $f : A \to B$ is called many-one, if at least one element of co-domain B has two or more than two pre-images in domain A.

Symbolically: f is many-one if for $x_1, x_2 \in A$, we have $x_1 \neq x_2 \Rightarrow f(x_1) = f(x_2)$

This can be illustrated in the following figures.

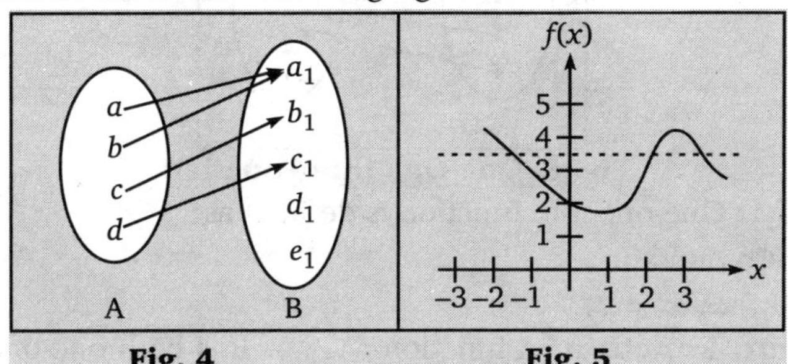

Fig. 4 **Fig. 5**

Graphically, a function is many-one if and only if a line parallel to x-axis meets the graph of the function at more than one point.

REMARK

- One-many function does not exist.

(c) Onto function: A function $f : A \to B$ is called an onto function, if there is no element of B which is not an image of some element of A, *i.e.*, every element of B appears as the image of at least one element of A. This is illustrated in Figure 6.

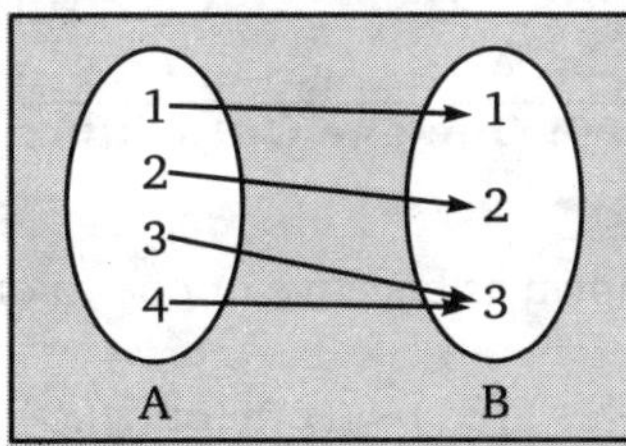

Fig. 6: Onto Function

REMARKS

- In an onto function, Range = Co-domain
- Onto function is also called surjective.

(d) Into function: A function $f : A \to B$ is called an into function, *i.e.*, if there is at least one element of set B which has no pre-image in the set A. This is illustrated in Figure 7.

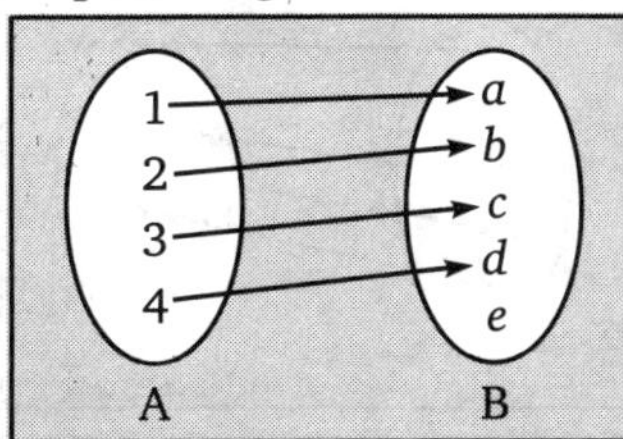

Fig. 7: Into Function

REMARK

- In an into function, Range $\subset$ Co-domain.

(e) One-One Into Function: A function $f : A \to B$ is called a one-one into function, if it is both one-one and into, *i.e.,* the different points in A are joined to different points in B and there are some points in B which are not joined to any point in A. This is illustrated in Figure 8.

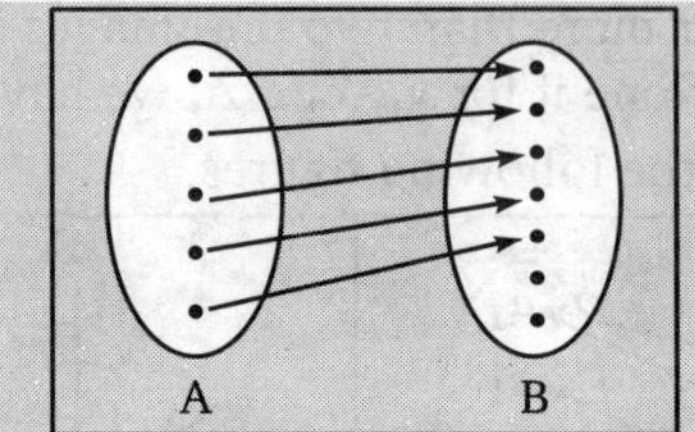

Fig. 8: One-One Into Function

Symbolically : One-one into function is defined as

(i) Range $\subset$ Co-domain.

(ii) $f(x_1) \neq f(x_2) \Rightarrow x_1 \neq x_2$.

(f) One-One Onto Function: A function $f : A \to B$ is both one-one and onto, *i.e.,* the different points in A are joined to different points in B and no point in B is left vacant. This is illustrated in Figure 9.

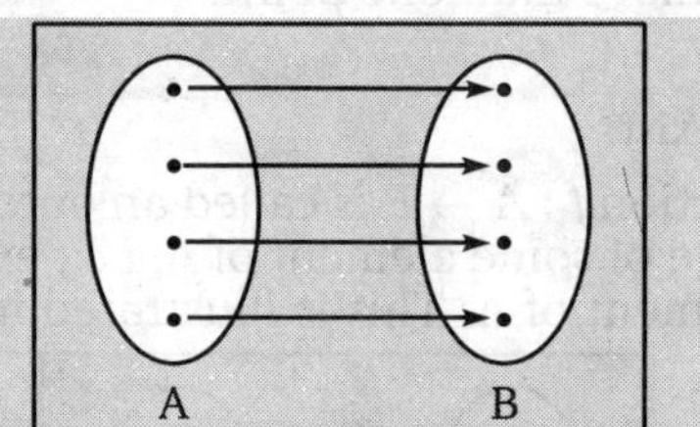

Fig. 9: One-one Onto Function

REMARKS

- One-one onto mapping is also known as bijective or one-to-one.
- For a one-one onto function
  Range = Co-domain, and $x_1 \neq x_2 \Rightarrow f(x_1) \neq f(x_2)$ or $f(x_1) = f(x_2) \Rightarrow x_1 = x_2$

(g) Many-One Into Function: A function $f : A \to B$ which is both many-one and into function is called a many-one into function, *i.e.,* two or more points in A are joined to some points in B and there are some point in B which are not joined to any point in A. Therefore, for many-one into function.

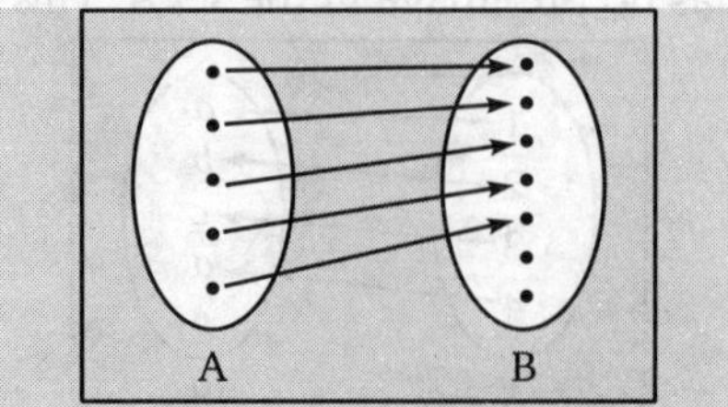

Fig. 10: Many-One Into Function

(i) Range $\subset$ Co-domain.
(ii) $x_1 \neq x_2$
$\Rightarrow f(x_1) = f(x_2)$

(h) Many-One Onto Function: If function $f : A \to B$ is both many-one and onto function is called a many one onto function, *i.e.,* in B one point is joined to at least one point in A and two or more points in A are joined to some points in B. Therefore, for many-one onto function.

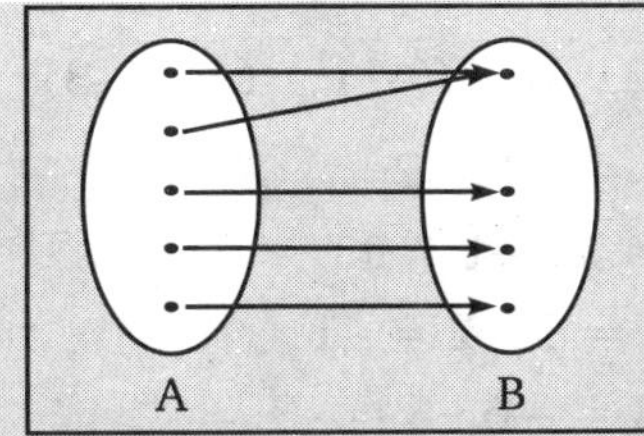

Fig. 11: Many-One Onto Function

(i) Range = Co-domain.
(ii) $x_1 \neq x_2 \Rightarrow f(x_1) = f(x_2)$

WORKING PROCEDURE

1. For checking the Injectivity (One-One) of the function Let x and y be two arbitrary elements in the domain of f.	
Step 1.	Take $f(x) = f(y)$
Step 2.	If we get $x = y$, after solving $f(x) = f(y)$. Then, $f : A \to B$ is one-one.
2. For checking the surjectivity (Onto) of a function	
Step 1.	Take an arbitrary element y in the co-domain.
Step 2.	Put $f(x) = f(y)$
Step 3.	Solve $f(x) = y$ for x and obtain x in terms of y.
Step 4.	Get the equation of the form $x = g(y)$
Step 5.	If $x = g(y)$ belongs to domain f, for all values of y, then f is onto.

RECAPITULATIONS

- For a function $f : A \to B$, A = domain, B = co-domain.
- **For one-one function:** $x_1 \neq x_2 \Rightarrow f(x_1) \neq f(x_2)\ \forall\ x_1, x_2 \in A$ or $f(x_1) = f(x_2) \Rightarrow x_1 = x_2\ \forall\ x_1, x_2 \in A$
- **For many-one function:** $x_1 \neq x_2 \Rightarrow f(x_1) = f(x_2),\ x_1, x_2 \in A$
- **For onto function:** Range = co-domain
- **For into function:** Range $\subseteq$ co-domain
- **For one-one into function:**
  (i) Range $\subseteq$ co-domain
  (ii) $f(x_1) \neq f(x_2) \Rightarrow x_1 \neq x_2$
- **For one-one onto function:**
  (i) Range = codomain
  (ii) $x_1 \neq x_2 \Rightarrow f(x_1) \neq f(x_2)$ or $f(x_1) = f(x_2) \Rightarrow x_1 = x_2$
- **For many-one into function:**
  (i) Range = co-domain
  (ii) $x_1 = x_2 \Rightarrow f(x_1) = f(x_2)$
- **For many-one onto function:**
  (i) Range = co-domain
  (ii) $x_1 \neq x_2 \Rightarrow f(x_1) = f(x_2)$

SOLVED EXAMPLES

EXAMPLE 1. *Let $f : R \to R$ be a function defined by*

$$f(x) = \begin{cases} 3x - 1 \text{ when } & x > 3 \\ x^2 - 1 \text{ when } & -2 \le x \le 3 \\ x + 3 \text{ when } & x < -2 \end{cases}$$

Find (i) $f(2)$, (ii) $f(4)$, (iii) $f(-1)$, (iv) $f(-3)$

SOLUTION. (i) $f(2) = (2)^2 - 2 = 4 - 2 = 2$

(ii) $f(4) = 3(4) - 1 = 12 - 1 = 11$

(iii) $f(-1) = (-1)^2 - 2 = 1 - 2 = -1$

(iv) $f(-3) = 2(-3) + 3 = -6 + 3 = -3$

EXAMPLE 2. *For $y = +\sqrt{x}$, say whether it is a function or not. If it is a function, find its domain and range.*

SOLUTION. Here we have $y = +\sqrt{x}$ ...(1)

Since y is real if $x \ge 0$ and is unique and finite for each $x \ge 0$.

Therefore, (1) is a function with domain $[0, \infty[$.

Again from (1), $y \ge 0 \ \forall \ x \ge 0$

Hence, range $= [0, \infty[$

EXAMPLE 3. *Find the domain of* $f(x) = \dfrac{x^3 - x^2 + 4x + 2}{3x + 11}$

SOLUTION. Since f is defined for all real values of x except when $3x + 11 = 0$

i.e., when, $x = -\dfrac{11}{3}$

Hence, domain of $f = R - \left\{-\dfrac{11}{3}\right\}$

EXAMPLE 4. *Let $f : N - \{1\} \to N$ be defined by $f(n)$ = the highest prime factor of n. Show that f is neither one-one nor onto. Also, find the range f.*

SOLUTION. Since we have

$f(6)$ = the highest prime factor of 6 = 3

$f(9)$ = the highest prime factor of 9 = 3

$f(12)$ = the highest prime factor of 12 = 3

Therefore, f is a many-one function.

Clearly, image of any $n \in N - \{1\}$ is the largest prime number that divides n. So the range of f consists of prime number only. Consequently, range of $f \ne N$ (Co-domain)

$\Rightarrow$ f is not onto function.

Hence, f is neither one-one nor onto. The range of f is the set of all prime numbers.

EXAMPLE 5. *Let $A = \{1, 2\}$. Find all one-to-one function from A to A.*

SOLUTION. Let $f : A \to A$ be a one-one function.

Then, for $f(1)$, there are two choices, *i.e.*, 1 or 2.

Let us first suppose $f(1) = 1$.

As $f : A \to A$ is one-one, $f(2) = 2$

Therefore, we have $f(1) = 1, f(2) = 2$

Now, let $f(1) = 2$

Since, $f : A \to A$ is one-one, therefore $f(2) = 1$.

Therefore, we have $f(1) = 2$ and $f(2) = 1$.

Hence, we have two one-one function say f and g form A and A given by

$$f(1) = 1, f(2) = 2 \text{ and } f(2) = 1 \text{ and } f(1) = 2.$$

EXAMPLE 6. *Let* $\{x \in R : -1 \le x \le 1\} = B$. *Show that* $f : A \to B$ *given by* $f(x) = x\,|x|$ *is one-one and onto.*

SOLUTION. Let x, y be any two elements in A, then

$$x \ne y \Rightarrow x|x| \ne y|y| \Rightarrow f(x) \ne f(y).$$

Therefore, f is one-one.

Since, range of $f = f(A) = B$ so $f : A \to B$ is onto mapping. Hence f is one-one and onto.

EXAMPLE 7. *Find the domain and range of the function.*

$$f(x) = -\sqrt{-5 - 6x - x^2}$$

SOLUTION. Given that, $f(x) = -\sqrt{-5 - 6x - x^2}$

For f to be real, $-5 - 6x - x^2 \ge 0 \quad \Rightarrow \quad x^2 + 6x + 5 \le 0$

$\Rightarrow \quad x^2 + 6x \le -5 \quad \Rightarrow \quad x^2 + 6x + 9 \le -5 + 9$

$\Rightarrow \quad (x + 3)^2 \le 4 \quad \Rightarrow \quad |x + 3|^2 \le 4$

$\Rightarrow \quad |x + 3| \le 2 \quad \Rightarrow \quad -2 \le x + 3 \le 2$

$\Rightarrow \quad -2 - 3 \le x \le 2 - 3 \quad \Rightarrow \quad -5 \le x \le -1$

Therefore, domain of $f(x) = [-5, -1]$

To find the range of $f(x)$, put $y = f(x)$

Therefore, $f(x) = -\sqrt{-5 - 6x - x^2}$, $y \le 0$

$\Rightarrow \quad y^2 = -5 - 6x - x^2 \quad \Rightarrow \quad x^2 + 6x + (y^2 + 5) = 0$

For real x, discriminant ≥ 0, *i.e.*, $(6)^2 - 4 \times 1 \times (y^2 + 5) \ge 0$

$\Rightarrow \quad 36 - 4y^2 - 20 \ge 0 \quad \Rightarrow \quad -4y^2 \ge -16$

$\Rightarrow \quad y^2 \le 4 \quad \Rightarrow \quad |y|^2 \le 4$

$\Rightarrow \quad |y| \le 2$ *i.e.*, $-2 \le y \le 2$

But $y \le 0$ therefore, $-2 \le y \le 0$.

Hence, Range of $f = [-2, 0]$

1.6 BINARY OPERATION

Let S be a non-empty set. Then any function from $S \times S$ to S is called binary operation. It is usually denoted by any of the following symbols : *, *o*, *O*, $\oplus$, +, . etc.

That is, any function $* : S \times S \to S$ is said to be binary operation, where $S \times S = \{(a, b) : a, b \in S\}$.

Instead of writting $*(a, b)$, we write $a * b$ for $a, b \in S$. Thus a binary operation * on S is a rule which assign to a pair $a, b \in S$ another element $a * b \in S$.

ILLUSTRATIONS

(1) The addition $+: Z \times Z \to Z$ is a binary operation, because for all $a, b \in Z$, we have $a+b \in Z$.

(2) The multiplication $'.': Z \times Z \to Z$ is binary operation, as $a.b \in Z$ for all $a, b \in Z$.

(3) The subtraction $'-': N \times N \to N$ is not a binary operation, as $3 \in N$, $5 \in N$ but $3 - 5 = -2 \notin N$.

(4) The subtraction $'-': Z \times Z \to Z$ is binary operation, as $a - b \in Z$ for all $a, b \in Z$.

(5) Let S be a non-empty set and $P(S)$ be its power set. Then the union opeartion on $P(S)$ is a binary opaeration as $A \cup B \in P(S)$ for all $A, B \in P(S)$.

(6) Addition of the set S of all irrationals is not a binary operation as $2+\sqrt{3} \in S$ and $2-\sqrt{3} \in S$ but $2+\sqrt{3}+2-\sqrt{3} \notin S$.

(7) Multiplication on the set S of all irrationals is not a binary operation as

$$\sqrt{2} \in S, -\sqrt{2} \in S \text{ but } \left(\sqrt{2}\right)\left(-\sqrt{2}\right) = -2 \notin S.$$

REMARKS

- A binary operation combines any two elements a and b of a set to give another element of the same set, this fact is also known as closure property or we can say that the set is closed with respect to binary opeartion.
- Binary operation is never one-one.

1.7 NUMBER OF BINARY OPERATIONS

Let $f: A \to B$ be a function, where $n(A) = p$ and $n(B) = q$. Then, the total number of functions from A to B is given by $[n(B)]^{n(A)}$.

Let S be a finite set containing n elements, then $S \times S$ will have n^2 elements. Since the binary operation is a function form $S \times S$ to S, so that total number of binary operations on S is $(n)^{n^2}$.

ILLUSTRATION

(1) If $S = \{1, 2\}$, then the total number of binary operations on $S = (2)^{2^2} = 2^4 = 16$.

1.8 PROPERTIES OF BINARY OPERATION

Let '*' : $S \times S \to S$ and 'o' : $S \times S \to S$ be any two binary operations, where S is any non-empty set.

Then we define the following properties.

(i) Associative property : A binary operation * on S is said to be associative, if

$$(a*b)*c = a*(b*c) \text{ for all } a, b, c \in S.$$

If '*' denotes addition (+) and multiplication (×), then '*' is always associative on S but if '*' denotes subtraction (–), then '*' is not associative on S.

For example : Addition and multiplication on Z, the set of all integers are always associative, *i.e.*,

$$(a+b)+c = a+(b+c)$$

and $$(a \times b) \times c = a \times (b \times c) \text{ for all } a, b, c \in Z.$$

But subtraction is not associative on Z, *i.e.*,

$$3-(5-4) \neq (3-5)-4$$

(ii) Commutative property : A binary opeartion '*' on a set S is said to be *commutative* if

$$a*b = b*a \text{ for all } a, b \in S.$$

If '*' denotes additon (+) and multiplication (×), then '*' is always *commutative* on S but if * denotes subtraction (–), then '*' is not commutative on S.

For example : If $S = Z$, the set of all integers, then

$$a+b = b+a$$

and $$a \times b = b \times a \text{ for all } a, b \in Z.$$

But subtraction is not commutative as $3-5 \neq 5-3$ for $3, 5 \in Z$.

(iii) Distributive property : A binary opeartion '*' is said to be *distributive* over another binary operation ' o' on S if

$$a*(b \text{ o } c) = (a*b) \text{ o } (a*c) \text{ for all } a, b, c \in S$$

and $$(b \text{ o } c)*a = (b*a) \text{ o } (c*a) \text{ for all } a, b, c \in S$$

Here the first distribution is known as *left distribution* whereas the second is known as *right distribution.* That is,

Left distributive property :

$$a*(b \text{ o } c) = (a*b) \text{ o } (a*c) \text{ for all } a, b, c \in S$$

Right distributive property :

$$(b \text{ o } c)*a = (b*a) \text{ o } (c*a) \text{ for all } a, b, c \in S$$

For example : If $S = Z$, the set of all integers and '*' denotes multiplication, 'o' denotes addition, then '*' is distributive over 'o' on Z, *i.e.*,

$$a \times (b+c) = a \times b + a \times c$$

and $$(b+c) \times a = b \times a + c \times a$$

But if '*' denotes addition and 'o' denotes multiplication, then '*' is not distributive over 'o' on Z, *i.e.*,

$$5+(3 \times 2) \neq (5+3) \times (5+2)$$

1.9 ALGEBRAIC STRUCTURE

Let S be a non-empty set and '*' be a binary operation on S, then $(S, *)$ is known as algebraic structure.

SOLVED EXAMPLES

EXAMPLE 1. *Determine whether the following operations are binary on the given set or not:*

(i) '*' *on N defined by* $a*b = a^b$ *for all* $a, b \in N$

(ii) '*' *on N defined by* $a*b = a+b-2$ *for all* $a, b \in N$

(iii) '*' *on N defined by* $a*b = a^b+b^a$ *for all* $a, b \in N$

(iv) '*' *on Z defined by* $a*b = \sqrt{|ab|}$ *for all* $a, b \in Z$

(v) '*' *on Z defined by* $a*b = \sqrt{a^2+b^2}$ *for all* $a, b \in Z$

(vi) '*' *on Z defined by* $a*b = \dfrac{1}{a-b}$ *for all* $a, b \in Z$

SOLUTION. (i) For all $a, b \in N$, $a^b \in N$.

$\therefore \qquad a * b \in N$

Hence, * is binary operation on N.

(ii) For all $a, b \in N$, $a + b \in N$

But $a+b-2 \notin N$ for all $a, b \in N$,

For example if $a = b = 1$ then $a + b - 2 = 0 \notin N$.

Hence, * is not binary on N, where $a * b = a + b - 2$.

(iii) For all $a, b \in N$, $a^b \in N$ and $b^a \in N$, then

$$a^b + b^a \in N \Rightarrow a * b \in N$$

Thus $a * b \in N$ for all $a, b \in N$. Hence, * is binary operation on N.

(iv) If $a = 1$ and $b = 2$, then $|ab| = |1 \times 2| = 2$.

Now $\qquad \sqrt{|ab|} = \sqrt{2}$

$\therefore \qquad \sqrt{|ab|} \notin Z$ for all $a, b \in Z$

$\Rightarrow \qquad a * b \notin Z$ for all $a, b \in Z$

Hence * is not binary operation on Z.

(v) Since $a * b = \sqrt{a^2 + b^2}$, for all $a, b \in Z$

If $a = b = 1$, then $\sqrt{a^2 + b^2} = \sqrt{1^2 + 2^2} = \sqrt{2} \notin Z$

$\therefore \qquad \sqrt{a^2 + b^2} \notin Z$ for all $a, b \in Z$

$\Rightarrow \qquad a * b \notin Z$ for all $a, b \in Z$

Hence * is not binary operation on Z.

(vi) Since $\qquad a * b = \dfrac{1}{a-b}$ for all $a, b \in Z$.

If $a = 3$ and $b = 1$, then $\dfrac{1}{a-b} = \dfrac{1}{3-1} = \dfrac{1}{2} \notin Z$

$\therefore \qquad \dfrac{1}{a-b} \notin Z$ for all $a, b \in Z$

$\Rightarrow \qquad a * b = \dfrac{1}{a-b} \notin Z$ for all $a, b \in Z$

Hence * is not binary operation on Z.

EXAMPLE 2. *Let S be a set having more than one element. Let '*' $S \times S \to S$ be binary operation defined by $a * b = a$ for all $a, b \in S$. Is $(S, *)$ associative or commutative?*

SOLUTION. Since S has more than one elenment, so if $a, b \in S$, then $a \neq b$.

Now, $\qquad a * b = a$ and $b * a = b$

$\therefore \qquad a + b \neq b * a \qquad$ for all a, b

Thus * is not commutative on S.

Again, $\qquad a * (b * c) = a * b = a$

and $\qquad (a * b) * c = a * c = a$

$\therefore \qquad a * (b * c) = (a * b) * c \quad$ for all $a, b, c \in S$

Thus * is associative on S.

EXAMPLE 3. *Let '*' be a binary opearation on Q the set of all rational numbers. Find which of the binary opeartions are commutative.*

(i) $a * b = a - b$ for all $a, b \in Q$

(ii) $a * b = a^2+b^2$ for all $a, b \in Q$.

SOLUTION. (i) $a * b = a - b = -(b - a) = - b * a$

$\therefore$ $a * b \neq b * a$ for all $a, b \in Q$.

Thus * is not commutative on Q.

(ii) $a^2 \in Q$, $b^2 \in Q$ for all $a, b \in Q$. Then

$$a * b = a^2 + b^2$$
$$= b^2 + a^2 \; [\because \text{Addition is always commutative on } Q]$$
$$= b * a$$

$\therefore$ $a * b = b * a$ for all $a, b \in Q$

Thus * is commutative on Q.

1.10 IDENTITY ELEMENT

Let $(S, *)$ be an algebraic structure. If there exists an element $e \in S$ such that $a*e = a = e * a$ for all $a \in S$ then e is called on identity element of S with respect to binary operation '*'.

ILLUSTRATIONS

(1) Let $S = Z$, the set of all integers and addition (+) is a binary operation on Z. Then we know that $0 \in Z$ such that

$$0 + a = a = a + 0 \text{ for all } a \in Z.$$

Thus, 0 is the identity element for addition on Z.

(2) If $S=N$, the set of all natural numbers and multiplication '*' is binary operation on N. Then we know that $1 \in N$ such that

$$1 \cdot a = a = a \cdot 1 \text{ for all } a \in N.$$

Thus, 1 is the identity element for multiplication on N.

(3) Let $P(S)$ denote the power set of a non-empty set S. Then we know that

$$A \cup \phi = A = \phi \cup A \quad \text{for all } A \in P(S)$$

Thus, ϕ is the identity element for the union of sets on $P(S)$.

(4) Let $P(S)$ denotes the power set of a non-empty set S. Then we know that

$$A \cap S = A = S \cap A \quad \text{for all } A \in P(S)$$

REMARK

- '0' is not an identity element for addition on N.

1.11 INVERSE OF AN ELEMENT

Let $(S, *)$ be an algebraic structure, and e be the identity element of S. Then an element $a \in S$ is said to be invertible element, if there exists an element $b \in S$ such that

$$a*b = e = b*a$$

The element b is called an inverse of a.

ILLUSTRATIONS

(1) Let $(Z, +)$ be an algebraic structure. Then $0 \in Z$ is the identity element for addition. Now corresponding to each element $a \in Z$, there exists $-a \in Z$ such that

$$a + (-a) = 0 = (-a) + a$$

Thus, $-a$ is the additive inverse of a.

(2) Let $(R,*)$ be an algebraic structure. Then $1 \in R$ is the multiplicative identity. If $a \neq 0 \in R$, then correspondig to each non-zero element $a \in R$, then exists an element $\frac{1}{a} \in R$, such that

$$a.\frac{1}{a} = 1 = \frac{1}{a}.a$$

Thus $\frac{1}{a}$ is the multiplicative inverse of a.

REMARK

- The inverse of an element a (if it exists) with respect to the addition (or multiplication) binary operations is generally is denoted by $a^{-1}\left(\text{or } \frac{1}{a}\right)$

THEOREM 1. ***Let (S,*) be an algebraic structure. If S has the identity element for '*', then it is unique.***

PROOF. Let e_1 and e_2 be two identities for '*' on S. Then by definition of identity, we have,

$$e_1 * e_2 = e_2, \text{ as } e_1 \text{ is the identity} \quad \ldots (1)$$

and $$e_1 * e_2 = e_1, \text{ as } e_2 \text{ is the identity} \quad \ldots (2)$$

From equations (1) and (2), we get

$$e_1 = e_2$$

Hence, the identity, if exists, is unique.

THEOREM 2. ***Let (S, *) be algebraic structure with identity element e. Then every element $a \in S$ has unique inverse in S, if '*' is associative.***

PROOF. Let b and c be two inverses of an element $a \in S$ in S. Then we have

$$a*b = e = b*a$$

and $$a*c = e = c*a$$

Now, $$(b*a)*c = e*c \qquad (\because b*a = e)$$

$$= c \qquad \text{[By the definition of identity]}$$

and $$b*(a*c) = b*e \qquad (\because a*c = e)$$

$$= b \qquad \text{[By the definition of identity]}$$

Since '*' is associative on S, so that $(b*a)*c = b*(a*c)$ for all $a, b, c \in S$

$\Rightarrow$ $$c = b$$

Hence, $a \in S$ has unique inverse in S.

THEOREM 3. ***Let (S,*) be an algebraic structure and '*' be an associative binary operation of S. If a be an invertible element of S, then***

$$\boldsymbol{(a^{-1})^{-1} = a}$$

PROOF. Let e be the identity element of S for '*' and a be an invertible element of S, then

$$a * a^{-1} = e = a^{-1} * a$$

$\Rightarrow$ $$a^{-1} * a = e = a * a^{-1}$$

$\Rightarrow$ a is the inverse of a^{-1}

$\Rightarrow$ $$(a^{-1})^{-1} = a$$

REMARK

- For identity element e, we have $e = e^{-1}$

THEOREM 4. ***Let (S, *) be an algebraic structure and * be an associative binary operation on S. If each element of S has inverse in S. Then $(a * b)^{-1} = b^{-1} * a^{-1}$.***

PROOF. Let e be the identity element of S for '*'. Now for $a, b \in S$, we have

$$(a * b)*(b^{-1} * a^{-1}) = a* (b * b^{-1})*a^{-1} \quad \text{(By associativity)}$$
$$= a* e * a^{-1} \quad (\because b*b^{-1} = e)$$
$$= a* a^{-1} \quad (\because a*e = a)$$
$$= e \quad (\because a * a^{-1} = e)$$

Similarly, $(b^{-1} * a^{-1}) * (a * b) = e$

$\therefore$ $(a *b)*(b^{-1} *a^{-1}) = e = (b^{-1} *a^{-1})*(a*b)$

$\Rightarrow$ $b^{-1} * a^{-1}$ is the inverse of $a*b$

$\Rightarrow$ $(a * b)^{-1} = b^{-1} * a^{-1}$

1.12 COMPOSITION TABLE FOR BINARY OPERATION ON FINITE SETS

Let $S = \{a_1, a_2, \ldots a_n\}$ be a finite set and '*' be a binary operation on S. Then we construct a composition table by means of following instructions:

First we write down the elements $a_1, a_2, \ldots, a_n$ of S in a top horizontal row and in a left vertical column as shown below. Now we write down the element $a_i * a_j$ at the intersection of row headed by a_i $(1 \leq i \leq n)$ and the column headed by a_i $(1 \leq j \leq n)$ to get the following table:

*	a_1	a_2	...	a_j	...	a_n
a_1	$a_1 * a_1$	$a_1 * a_2$	...	$a_1 * a_j$	...	$a_1 * a_n$
a_2	$a_2 * a_1$	$a_2 * a_2$	...	$a_2 * a_j$	...	$a_2 * a_n$
⋮	⋮	⋮	⋮	⋮	⋮	⋮
a_i	$a_i * a_1$	$a_i * a_2$	...	$a_i * a_j$	...	$a_i * a_n$
⋮	⋮	⋮	⋮	⋮	⋮	⋮
a_n	$a_n * a_1$	$a_n * a_2$	...	$a_n * a_j$	...	$a_n * a_n$

From above table, we conclude the following properties:

(i) Closure property: If all the entries in the table are the elements of S, then S is closed under binary operation '*'.

(ii) Commutative property: If the entries in every row coincide with the corresponding entries in the coerresponding column, we say that the composition is commutative, otherwise it is said to be non-commutative.

(iii) Existence of identity: If the row headed by an element a_i of S just coincides with the top row of the table and the column headed by a_i coincides with column on extreme left of the table, then a_j is the identity element of S for the composition '*'.

(iv) Existence of inverse: Look at the position of the identity element e anywhere in the table except in the top row and in the extreme column. If e is placed at the intersection of the row headed by a_i and the column headed by a_j, then a_i and a_j are the inverse of each other.

SOLVED EXAMPLES

EXAMPLE 1. *Let S= {1,2,3,4} and '*' be a binary operation in S defined by a*b = r, where r is the least non-negative remainder when ab is divided by 5. Construct the composition table for '*' on S.*

SOLUTION. We have

1*1 = least non-negative remainder, when 1 is divided by 5 = 1

2*3 = least non-negative remainder, when 6 is divided by 5 = 1

Similarly,

1*2 = 2, 1*3=3, 1*4 = 4,

2*1 = 2, 2*2 =4, 2*3 =1, 2*4 =3,

3*1 = 3, 3*2 =1, 3*3 =4, 3*4 =2,

4 *1 =4, 4*2 =3, 4*3=2, 4*4 =1

Now, we construct the composition tabel for '*' on S as follows:

*	1	2	3	4
1	1	2	3	4
2	2	4	1	3
3	3	1	4	2
4	4	3	2	1

From above composition table, we observe that

(i) All the entries in the table are the elements of S, so S is closed under '*'.

(ii) The binary operation '*' is commutative on S, as the composition table is symmetrical about the diagonal starting at the upper left corner and ending at the lower right corner.

(iii) 1 is the identity element for '*' because the row headed by 1 coincides with the top row and the column headed by 1 coincides with extreme left column of the table.

(iv) Every element of S is invertible with respect to '*' because the identity element 1 appears in each row headed by 2 and column headed by 3, so 2 and 3 are the inverse of each other. Similarly 4 is the inverse of itself.

EXAMPLE 2. *Construct the composition table for the composition of functions, defined on the set $S = \{f_1, f_2, f_3, f_4\}$ of four functions from C, the set of all complex numbers to itself, given by*

$$f_1(z) = z, f_2(z) = -z, f_3(z) = \frac{1}{z} \text{ and } f_4(z) = -\frac{1}{z} \text{ for all } z \in C.$$

SOLUTION. Since each function is defined from C to C, so $f_1 \circ f_1, f_1 \circ f_2, f_1 \circ f_3, f_1 \circ f_4$ etc, exist.

So,

$$(f_1 \circ f_1)(z) = f_1(f_1(z)) = f_1(z)$$

$$(f_1 \circ f_2)(z) = f_1(f_2(z)) = f_1(-z) = -z = f_2(z)$$

$$(f_1 \circ f_3)(z) = f_1(f_3(z)) = f_1\left(\frac{1}{z}\right) = \frac{1}{z} = f_3(z)$$

$$(f_1 \circ f_4)(z) = f_1(f_4(z)) = f_1\left(-\frac{1}{z}\right) = -\frac{1}{z} = f_4(z)$$

$$(f_2 o f_1)(z) = f_2(f_1(z)) = f_2(z) = -z = f_2(z)$$
$$(f_2 o f_2)(z) = f_2(f_2(z)) = f_2(-z) = -(-z) = z = f_1(z)$$
$$(f_2 o f_3)(z) = f_2(f_3(z)) = f_2(1/z) = -1/z = f_4(z)$$

Similarly, other compositions can be obtained.

Now we construct the composition table for the composition of functions 'o' as follows:

o	f_1	f_2	f_3	f_4
f_1	$f_1 o f_1 = f_1$	$f_1 o f_2 = f_2$	$f_1 o f_3 = f_3$	$f_1 o f_4 = f_4$
f_2	$f_2 o f_1 = f_2$	$f_2 o f_2 = f_1$	$f_2 o f_3 = f_4$	$f_2 o f_4 = f_3$
f_3	$f_3 o f_1 = f_3$	$f_3 o f_2 = f_4$	$f_3 o f_3 = f_1$	$f_3 o f_4 = f_2$
f_4	$f_4 o f_1 = f_4$	$f_4 o f_2 = f_3$	$f_4 o f_3 = f_2$	$f_4 o f_4 = f_1$

EXERCISE 1.2

1. Let $A = \{-2, -1, 0, 1, 2\}$ and $f : A \to Z$ given by $f(x) = x^2 - 2x - 3$. Find :

(i) the range of f,

(ii) pre-image of 6, –3 and 5.

2. Find the domain and range of the following function $f(x) = \sqrt{(x-1)(3-x)}$

3. Find the range of the following function $f(x) = \dfrac{1}{(2x-3)(x+1)}$

4. Find the domain and range of the following functions :

(i) $f(x) = \dfrac{x^2-1}{x-1}$ (ii) $y = -|x|$

(iii) $f(x) = \dfrac{|x-1|}{x-1}$ (iv) $y = \sqrt{x-3}$

5. If $A = \{-1, 0, 2, 5, 6, 11\}$, $B = \{-2, -1, 0, 18, 25, 108\}$ and $f(x) = x^2 - x - 2$, find $f(A)$.

6. Let A be the set of two positive integers. Let $f : A \to Z^+$, set of positive integers be defined by $f(n) = p$, where p is the highest prime factor of n. If range of $f = \{3\}$, find A.

7. Find the domain for which the function $f(x) = 2x^2 - 1$ and $g(x) = 1 - 3x$ are equal.

8. Let $f_1 : R \to R$ and $f_2 : C \to C$ be two functions defined as $f_1(x) = x^3$ and $f_2(x) = x^3$. Show that they are not equal.

9. Let $A = \{p, q, r, s\}$ and $B = \{1, 2, 3\}$. Which of the following relations from A to B not a function?

(i) $R_1 = \{(p, 1), (q, 2), (r, 1), (s, 2)\}$

(ii) $R_2 = \{(p, 1), (q, 1), (r, 1), (s, 1)\}$

(iii) $R_3 = \{(p, 1), (q, 2), (r, 2), (s, 3)\}$

(iv) $R_4 = \{(p, 2), (q, 3), (r, 2), (s, 2)\}$

10. Write the following relations as sets of ordered pairs and find which of them are functions :

(i) $\{(x, y) : y = 3x, x \in (1, 2, 3), y \in (3, 6, 9, 12)\}$

(ii) $\{(x, y) : y > x + 1, x = 1, 2$ and $y = 2, 4, 6\}$

(iii) $\{(x, y) : x + y = 3x, y \in (0, 1, 2, 3)\}$

11. Express the following functions as sets of ordered pairs, and find their range :

(i) $f_1 : A \to R : f_1(x) = x^2 + 1$ where $A = \{-1, 0, 2, 4\}$

(ii) $f_2 : A \to N : f_2(x) = 2x$ where $A = \{x : x \in N, x \leq 10\}$

12. Let $f : R \to R$ be a function such that $f(x) = 2^x$. Determine :

(i) range of f (ii) $\{x : f(x) = 1\}$

(iii) whether $f(x + y) = f(x) \cdot f(y)$ holds

13. Let $f : R^+ \to R$, be a function such that $f(x) = \log x$. Determine :

(i) the image set of domain of f

(ii) $\{x : f(x) = -2\}$

(iii) whether $f(xy) = f(x) + f(y)$ holds

14. Give an example of a map which is :

(i) one-to-one but not onto

(ii) not one to one, but onto

(iii) neither one-to-one nor onto

ANSWERS

1. (i) $f(A) = \{-4, -3, 0, 5\}$, (ii) ϕ, $\{1, 2\}$, -2 **2.** Domain = $[1, 3]$, Range = $[-1, 1]$

3. $\left]-\infty, \frac{-8}{25}\right] \cup [0, \infty[$

4. (i) $R - \{1\}$, $R - \{2\}$, (ii) $R : R - R^+$, (iii) $R - \{1\}$, $\{-1, 1\}$, (iv) $[3, \infty[$, $[0, \infty]$

5. $f(A) = \{1, -2, 18, 28, 108\}$ **6.** $A = \{3, 6\}$ or $(3, 9)$ or $[3, 12]$ etc. **7.** $(-2, 1/2)$ **9.** (iii)

10. (i) $\{(1, 3), (2, 6), (3, 9)\}$, function, (ii) $\{(1, 4), (1, 6), (3, 4), (3, 6)\}$, not function (iii) $\{(0, 3), (1, 2), (2, 1), (3, 0)\}$, function

11. (i) $f_1 = \{x, f(x) : x \in A\} = \{(-1, 2), (0, 1), (2, 5), (4, 17)\}$ (ii) $f_2 = \{(x, g(x)) : x \in A\} = \{(1,2),(2,4), (3, 6), \ldots, (10, 20)\}$

12. (i) Range of $f = R^+$, the set of positive real numbers, (ii) $(x : f(x) = 1) = \{0\}$, (iii) $f(x+y) = f(x).f(y)$ holds for all $x, y \in R$

14. (i) $n \to n^2 : N \to N$ (ii) $n \to |n| : Z \to N \cup \{0\}$ (iii) $n \to |n|^2 : Z \to N \cup \{0\}$

1.13 SEQUENCES

Let N be the set of natural numbers and S be any set of real numbers. A function, whose domain is the set of natural numbers and range is a subset of S, is called a sequence in S.

Symbolically. If we define a function $f : N \to S$, then f is a sequence. We shall denote a sequence in a number of ways as follows :

(i) Usually, a sequence is denoted by its images. For a sequence f, the image corresponding to $n \in N$ is denoted by f_n or $<f(n)>$ and $f(n)$ is called the nth term of the given sequence. For example $<1,4,9,\ldots>$ is the sequence whose n^{th} term is n^2.

(ii) Using in order, the first few element of a sequence, here the rule for writing down different elements becomes clear. For example, $<1, 2, 3, \ldots>$ is the sequence whose n^{th} term is n.

(iii) Defining a sequence by a recurrence formula, *i.e.*, by a rule which express the nth term by $(n-1)^{th}$ term. For example, let $a_1 = 1, a_{n+1} = 2a_n \quad \forall n \geq 1$.

These above relations define a sequence whose n^{th} term is 2^{n-1}.

REMARKS

- A sequence is represented as $<s_n>$ or $\{s_n\}$, when s_n is the nth term of the sequence.
- The set of all distinct terms of a sequence is called the range set of that sequence.
- A sequence, whose range is a subset of real numbers R is called a real sequence or a sequence of real numbers.

1.13.1 CONSTANT SEQUENCE

A sequence $< s_n >$ defined by $s_n = a, \ \forall n \in N$, is called a constant sequence.

1.13.2 EQUALITY ON SEQUENCE

Two sequence $< s_n >$ and $< t_n >$ are said to be equal, if $s_n = t_n \quad \forall n \in N$.

1.13.3 OPERATION ON SEQUENCE

Since, the sequence are real valued functions, therefore, the sum, difference, product etc. of two sequence are defined as follows :

(i) If $< s_n >$ and $< t_n >$ be any two sequence, then the sequence, whose n^{th} terms are $s_n + t_n$, $s_n - t_n$ and $s_n . t_n$ are respectively known as the sum, difference and product of the sequence $< s_n >$ and $< t_n >$ and are denoted by $< s_n + t_n >, < s_n - t_n >$ and $< s_n . t_n >$ respectively.

(ii) If $s_n \neq 0 \quad \forall n \in N$, then the sequence, whose n^{th} term is $\frac{1}{s_n}$ is called reciprocal of the sequence $< s_n >$ and is denoted by $< \frac{1}{s_n} >$.

(iii) The sequence, whose n^{th} term $s_n / t_n (t_n \neq 0 \quad \forall n \in N)$ is known as the quotient of the sequence $< s_n >$ by the sequence $< t_n >$ and is denoted by $< \frac{s_n}{t_n} >$.

(iv) The sequence, whose n^{th} term is ks_n, where $k \in R$ is known as the scalar multiple of the sequence $< s_n >$ by k and is denoted by $< ks_n >$.

1.13.4 BOUNDED SEQUENCE

(i) Bounded below sequence. A sequence $< s_n >$ is said to be bounded below if there exists a real number l such that $s_n \geq l \quad \forall n \in N$.
The number l is known as the lower bound of the sequence $< s_n >$.

(ii) Bounded above sequence. A sequence $< s_n >$ is said to be bounded above if there exists a real number u such that

$$s_n \leq u \quad \forall n \in N.$$

The number u is said to be upper bound of the sequence $< s_n >$.

(iii) Bounded sequence. A sequence $< s_n >$ is said to be bounded if it is bounded above as well as bounded below.

(iv) Unbounded sequence. A sequence $< s_n >$ is said to be unbounded if it is not bounded.

(v) Least upper bound. If a sequence $< s_n >$ is bounded above, then there exists a number u_1 such that

$$s_n \leq u_1 \quad \forall n \in N. \qquad \ldots(1)$$

This number u_1 is called an upper bound of the sequence $< s_n >$. If $u_1 < u_2$. Then from (1), we find that

$$s_n < u_2 \quad \forall n \in N$$

which implies, u_2 is also an upper bound of the sequence $< s_n >$. Hence, we can say that any number greater than u_1 is an upper bound of $< s_n >$.
Hence, a sequence has an infinite number of upper bounds, if it is bounded above. Let u be the least of all the upper bound of the sequence $< s_n >$. Then u is defined as the least upper bound $(l.u.b)$ or supremum of the sequence $< s_n >$.

(vi) Greatest lower bound. If a sequence $< s_n >$ is bounded below, then there exists a number $l_1 \in R$ such that

$$l_1 \leq s_n \quad \forall n \in N \qquad \ldots(2)$$

This number l_1 is known as the lower bound of $< s_n >$. If $l_2 < l_1$, then from (2)

$$l_2 \leq s_n \quad \forall n \in N$$

which implies, l_2 is also a lower bound of the sequence $< s_n >$. Hence, we can say that any number less than l_1 is a lower bound of $< s_n >$.
Hence, a sequence has infinite number of lower bounds, if it is bounded below. Let l be the greatest of all the lower bounds of the sequence $< s_n >$. Then l is known as greatest lower bound $(g.l.b)$ or infimum of the sequence $< s_n >$.

☛ **ILLUSTRATIONS**

(1) The sequence $< n^2 >$ is bounded below by 1 but not bounded above.

(2) The sequence $< \frac{n}{n+1} >$ is bounded as $\frac{1}{2} \leq \frac{n}{n+1} < 1 \quad \forall n \in N$.

(3) The sequence $< \frac{1}{n} >$ is bounded since $\left|\frac{1}{n}\right| \leq 1 \quad \forall n \in N$.

(4) The sequence $< 2^n >$ is bounded below and has smallest term as 2. Every member of $]-\infty\ 2]$ is a lower bound of the sequence and the sequence is not bounded above.

1.13.5 LIMIT POINT OF THE SEQUENCE

A real number l is called a limit point of a sequence $< s_n >$ if every nbd of l contains infinite number of terms of the sequence.

Thus, $l \in R$ is a limit point of the sequence $< s_n >$ if for given $\varepsilon > 0, s_n \in]\, l-\varepsilon, l+\varepsilon]$ for infinitely many points.

Here it must be noted that

(i) limit point of a sequence need not be a member of the sequence.

(ii) A limit point of a sequence may or may not be a limit point of the range of the sequence but the limit point of the range of a sequence is always a limit point of the sequence.

(iii) In case of real numbers, limit points of a sequence may also be called accumulation, cluster or condensation points.

☛ **ILLUSTRATIONS**

(1) The sequence $< \frac{1}{n} >$ has one limit point, *i.e.*, 0.

(2) The sequence $< (-1)^n >$ has two limit points 1 and –1.

(3) The sequence $< n >$ has no limit point.

(4) The sequence $< 1 + \frac{(-1)^n}{n} >$ has one limit point, *i.e.*, 1.

Sufficient Conditions for number *l* to be or not be a limit point of the sequence $<s_n>$:

(i) If for every $\varepsilon > 0, \exists m \in N$ such that $s_n \in]l-\varepsilon, l+\varepsilon[\ \forall n \geq m$ or equivalently $|s_n - l| < \varepsilon\ \forall n \geq m$, then l is the limit point of the sequence $< s_n >$.

(ii) If for any $\varepsilon = 0, s_n \in]\, l-\varepsilon, l+\varepsilon[$ for only a finite number of values of n, then l is not a limit point of the sequence $< s_n >$. Such a condition is also necessary for a number l not to be limit point of the sequence $< s_n >$.

1.13.5 BOLZANO-WEIRSTRASS THEOREM FOR SEQUENCE

(KANPUR–2000)

Statement. ***Every bounded sequence has at least one limit point.***

Proof. Let $S = \{s_n : n \in N\}$ be the range set of the bounded sequence $< s_n >$.

Then, S is a bounded set. Now, there may be two cases :

(i) Let S be a finite set. Then $s_n = p$ for infinitely many indices n. Here $p \in R$. Obviously p is a limit point of $< s_n >$.

(ii) Let S be an infinite set. Since, S is bounded, then by Bolzano-Weirstrass theorem for set of real numbers, S has a limit point say p. Therefore, every nbd of p contains infinity many distinct points of S, *i.e.*, infinitely many term of $< s_n >$ and hence p is a limit point of the sequence $< s_n >$.

1.13.6 LIMIT SUPERIOR AND LIMIT INFERIOR

The greatest limit point of a bounded sequence is called the upper limit or limit superior and is denoted by $\overline{\lim}\, s_n$ and the smallest limit point of a bounded sequence is called the lower limit or limit inferior and is denoted by $\underline{\lim}\, s_n$

- By definition, it is obvious that $\underline{\lim}\, s_n \le \overline{\lim}\, s_n$.
- A bounded sequence $< s_n >$ for which the upper limit and lower limit coincide with real number l is said to converge to l.

1.13.7 LIMIT OF A SEQUENCE

A sequence $< s_n >$ is said to have a limit l if for a given $\in > 0\ \exists$ a positive integer m such that

$$|s_n - l| < \in \quad \forall n \ge m$$

1.13.8 CONVERGENT SEQUENCE

A sequence $< s_n >$ is said to converge to a number l, if for a given $\in > 0$ there exists a positive integer m such that

$$|s_n - l| < \in \quad \forall n \ge m$$

REMARK

- A sequence $<s_n>$ is said to be convergent iff it is bounded and has exactly one limit point.

1.13.9 DIVERGENT SEQUENCE

A sequence, which is not convergent, is known as divergent sequence.

1.13.10 OSCILLATORY SEQUENCE

A sequence $< s_n >$ is said to be an oscillatory sequence if it is neither convergent nor divergent.

An oscillatory sequence is said to be oscillate finitely or infinitely according as it is bounded or unbounded.

In other words, we can say

(i) a bounded sequence, which is not convergent is said to be oscillate finitely.

(ii) an unbounded sequence, which does not diverge, is said to be oscillate infinitely.

(iii) a bounded sequence, which does not converge and has at least two limit points is said to be oscillate finitely.

ILLUSTRATIONS

(1) The sequence $\langle 1 + (-1)^n \rangle$ oscillate finitely.

(2) The sequence $\langle (-1)^n \rangle$ oscillate finitely.

(3) The sequence $< (-1)^n \left(1 + \frac{1}{n}\right) >$ oscillate finitely.

(4) The sequence $< n(-1)^n >$ oscillate infinitely.

1.14 BASIC CONCEPTS OF TOPOLOGY

1.14.1 TOPOLOGICAL SPACES

Let $X \neq \phi$ and ζ be the collection of all those subsets of X which satisfy the following three conditions :

(i) $\phi \in \zeta,\ X \in \zeta$

(ii) If $G_1 \in \zeta$ and $G_2 \in \zeta$, then $G_1 \cap G_2 \in \zeta$.

(iii) If $\{G_\lambda : \lambda \in \Delta\} \in \zeta$, then $\bigcup_{\lambda \in \Delta} [G_\lambda] \in \zeta$

Then ζ is called a topology on X and (X, ζ) is called a topological space.

1.14.2 TYPES OF TOPOLOGIES

(i) Stronger and Weaker Topologies. Let ζ_1 and ζ_2 be two topologies defined on X. If $\zeta_1 \subset \zeta_2$, then ζ_1 is said to be coarser or weaker or smaller than ζ_2 and ζ_2 is said to be finer or stronger or larger than ζ_1.

(ii) Discrete Topology. Given a non-empty set X and D be a family of all subset of X so that (X, D) is a topological space, D is called the discrete topology for X.

(iii) Indiscrete Topology. For $X \neq \phi$ and $I = \{\phi, X\}$, I is said to be the indiscrete topology for X and (X, I) is called indiscrete topological space.

(iv) Non-trivial Topology. A topology defined on X other than the two trivial topologies is said to be non-trivial topology.

(v) Usual Topology. If U be the family of subsets of R (set of real numbers) defined such that a subset $a < r < b$ and $[\, x : x \in R, a < x < b] \subset A$, then U is called a usual topology.

(vi) Cofinite or Finite Compliment Topology. Given a non-empty set X, if ζ is the family of subset of X defined such that a subset A of $X \in \zeta \Leftrightarrow A = \phi$ or $X - A = A'$ is finite, then the topology is known as cofinite topology.

1.14.3 PRODUCT AND QUOTIENT SPACES

(1) If (X_1, ζ_1) and (X_2, ζ_2) be two topological spaces and we define

$$X = X_1 \times X_2$$

and
$$\beta = (G_1 \times G_2 : G_1 \in \zeta_1, G_2 \in \zeta_2)$$

such that β is a base for same topology ζ on X, then ζ is said to be the product topology and (X, ζ) is known as product of topological spaces.

(2) If (X, ζ) be a topological space, Y is a set and f is a mapping of X and Y, then the largest topology say ζ^* for Y such that $\zeta \to \zeta^*$ is continuous is said to be quotient topology for Y relative to f and T.

1.15 COMPACTNESS

Cover. Let (X, ζ) be a topological space and $A \subset X$, then a class $C^* = [G_i : i \in \Delta]$ of subset of X is said to be the cover or covering of A if and only if $A \subset \bigcup_i G_i$ or $A \subset \bigcup\, [G_i : i \in X]$

Open Cover. If each G_i is open, then the cover defined above is called an open cover, *i.e.*, X is said to be an open cover if each $x \in A \subset X$ there exists at least one G_i, *i.e.*, $\bigcup G_i = A$ such that $x \in G_i$.

Sub Cover. A subclass of an open cover which is itself an open cover is said to be a subcover.

1.16 CONNECTEDNESS

Separated Sets. Let (X, ζ) be a topological space. Two non-empty subsets A and B of X are said to be separated if and only if

$$A \cap \bar{B} = \phi \quad \text{and} \quad \bar{A} \cap B = \phi$$

or equivalently

$$(A \cap \bar{B}) \cup (\bar{A} \cap B) = \phi$$

Therefore, A and B are separated if and only if A and B are disjoint and neither of them contains limit point of the other.

REMARK

- Any two separated sets are disjoint. But two disjoint sets are not necessarily separated.

1.16.1 CONNECTED AND DISCONNECTED SETS

Let (X, ζ) be a topological space. A subset A of X is said to be disconnected iff it is the union of two non-empty separated sets, *i.e.*, if there exist two non-empty sets C and D such that $C \cap \bar{D} = \phi$, $\bar{C} \cap D = \phi$ and $A = C \cup D$.

REMARK

- A is said to be connected if and only if it is not disconnected.

1.16.2 GENERAL RESULTS (TO BE USED DIRECTLY)

(1) A topological space X is disconnected if and only if there exists a non-empty proper subset of X which is both open and closed in X.

(2) A topological space X is connected if and only if every non-empty proper subset of X has a non-empty frontier.

(3) If connected subset E of X such that $E \subset A \cup B$, where A and B are separated sets. Then $E \subset A$ or $E \subset B$, *i.e.*, E can not intersect both A and B.

(4) Let E be a connected subset of X. If F is a subset of X such that $E \subset F \subset \bar{E}$, then F is connected. In particular $\bar{E}$ is connected.

(5) A subset E of R is connected if and only if it is an interval.

(6) Continuous image of a connected space is connected.

(7) A topological space X is disconnected iff there exists a continuous mapping of X onto the discrete two point space [0, 1].

(8) A maximal connected subset of a topological space X is called a component of a space.

(9) A topological space X is said to be locally connected at a point x iff every open neighbourhood of x contains a connected open neighbourhood of x.

(10) A topological space X is locally connected if and only if the components of every open subset of X are open in X.

(11) The image of a locally connected space under a mapping which is both continuous and open is locally connected.

1.16.3 SUMMARY OF RESULTS ON COMPACTNESS (TO BE USED DIRECTLY)

(1) Every compact subset A of a T_2-space X is closed.

(2) Closed subsets of compact sets are compact.

(3) A topological space X is compact if and only if every basic open cover of X has a finite subcover.

(4) A topological space X is compact iff every collection of closed subsets of X with the finite intersection property is fixed, *i.e.*, has a non-empty intersection.

(5) Every compact space has Bolzano Weirstrass property.

(6) A subset A of R is compact if and only if A is bounded and closed (Heine-Borel Theorem).

(7) A topological space X is said to be countably compact iff every countable open cover of X has a finite subcover.

(8) A topological space X is said to be locally compact if and only if every point in X has at least one neighbourhood whose closure is compact.

(9) A T_2-space is locally compact if and only if each of its points is an interior point of some compact subspace of X.

(10) In a T_2-space, any locally compact subspace A is the intersection of an open set and a closed set and that in a locally compact space X, the intersection of an open set and a closed set is a locally compact subspace of X.

(11) The intersection of two locally compact subsets of X is locally compact.

(12) Let f be a continuous mapping of a compact topological space X into a topological space Y. Then $f(X)$ is compact, *i.e.*, continuous image of a compact space is compact.

1.17 COUNTABLE SET

A set which can be put in one to one correspondence with the set of all natural numbers or with its subset is said to be a countable set. (DELHI–2011, 13, MEERUT–2003,04, KANPUR–2000)

For example: If $A = \{2, 4, 6, 8, \ldots\}$, then $\exists$ an one to one mapping $f : A \to N$ s.t. $f(x) = x/2$. Then A is a countable set.

Further if $B = \{2, 4, 6, 8, 10\}$, then this set can be put in one to one correspondence with the set $A = \{1, 2, 3, 4, 5\}$, which is a subset of N.

Obviously a countable set can be finite or infinite both.

An infinite countable set is also called as countable infinite or enumerable set or denumerable set.

Thus enumerable set always means an infinite countable set.

Further if a set A is countable then its elements can be put in one to one correspondence with set $N = \{1, 2, 3, \ldots\}$. If we denote the elements of A corresponding to the natural numbers $1, 2, 3, \ldots$, by $a_1, a_2, a_3, \ldots$ etc., then the set A can be written as $A = \{a_1, a_2, a_3, \ldots\}$.

Thus a set is countable iff its elements can be written in the form of a sequence.

A set which is not countable, is said to be uncountable.

THEOREM 1. ***Every subset of a countable set is countable.***

(DELHI–2007, MEERUT–1987, 88, 94, 2003(BP), KANPUR–2000)

PROOF. Let A be a countable set, then A can be written as a sequence. Let $A = \{a_1, a_2, a_3, \ldots\}$. If A is finite then its every subset will also be finite and so will be countable. Now consider that A is enumerable. Let B be any subset of A.

If B is finite or empty then the result is trivial, so let $B \neq \phi$ and let k_1 be the least positive integer s.t. $a_{k_1} \in B$. Again, let k_2 be the least positive integer with $k_2 > k_1$ s.t. $a_{k_2} \in B$. Dealing with the elements of B in this way, we reach to the conclusion that B can be written as

$$\{a_{k_1}, a_{k_2}, \ldots\}$$

which is a sequence and hence B is a countable set.

THEOREM 2. ***The union of enumerable collection of enumerable sets is also enumerable.*** (DELHI–2008, MEERUT–1972, 84, GARHWAL–1991, 95, KANPUR–2000, MADRAS–2005, PATNA–2004, AMRITSAR–2008)

PROOF. Let $A = \{A_1, A_2, \ldots,\}$ be an enumerable collection of enumerable sets where $A_i = \{a_{i_1}, a_{i_2}, a_{i_3}, \ldots\}$, $\forall i \in N$ denote enumerable sets. For the proof of the theorem, we shall construct a progression in which all the elements of A appear. Such a list is as follows :

$$\bigcup_{i \in N} A_i = \{\begin{array}{llll} a_{11} \rightarrow a_{12} & a_{13} \rightarrow a_{14} \ldots \\ a_{21} \quad a_{22} & a_{23} \quad a_{24} \ldots \\ a_{31} \quad a_{32} & a_{33} \quad a_{34} \ldots \\ a_{41} \quad a_{42} & a_{43} \quad a_{44} \ldots \end{array}$$

Choose the path as shown above; then every element of A lies somewhere on the path (*i.e.*, each element occupies some particular position say rth position $r \in N$) and hence a one-one correspondence between $\bigcup A_i$ and N is implied showing thereby that A is enumerable. Thus we can write $\bigcup A_i = \{a_{11}, a_{12}, a_{21}, a_{31}, a_{22} \ldots\}$ as a sequence.

$\Rightarrow$ set $\bigcup A_i$ is enumerable.

Besides the above argument, note that the mapping $f : \bigcup_{i \in N} A_i \rightarrow N$ given by

$$f(a_{pq}) = \frac{(p+q-2)(p+q-1)}{2} + p \quad (= \text{say } n \in N)$$

shows the enumeration of $\bigcup A_i = \{a_{11}, a_{21}, a_{12}, a_{13}, a_{22}, a_{31}, a_{14}, a_{23}, a_{32}, a_{41}, \ldots\}$. *i.e.*, this mapping assigns a unique place to each element of the set $\bigcup A_i$ in the above sequence showing that $\bigcup A_i$ is countable.

REMARKS

- The union of countable collection of countable sets is also countable.
- If A_i is countable infinite set, then $\bigcup_{i=1}^{n} A_i$ is also countable infinite.

THEOREM 3. ***The set $N \times N$ is enumerable.*** (MEERUT–1983, 98, GARHWAL–2003)

PROOF. We have $N \times N = \{(u, v) : u, v \in N\}$; then clearly $N \times N$ can be arranged as shown below :

$$\begin{aligned} N \times N = \{&(1,1), (1,2), (1,3), (1,4) \ldots \\ &(2,1), (2,2), (2,3), (2,4) \ldots \\ &(3,1), (3,2), (3,3), (3,4) \ldots \\ &\ldots \quad \ldots \quad \ldots \quad \ldots\} = \bigcup A_i, \quad \text{where } A_i = \{(i, n) : n \in N\}. \end{aligned}$$

Now obviously each $A_i (i = 1, 2, 3, \ldots)$ is enumerable which shows that the set $N \times N$, union of enumerable collection of enumerable sets, is also enumerable by Theorem 2.

THEOREM 4. ***The set Q of all rational numbers is enumerable.***

(MEERUT-1984,90,95,97,2001(BP), KANPUR-1996)

PROOF. Recall that a rational number is written as $p/q : p,q \in Z$ and $q \neq 0$.

Let $A_q = \{p/q : p,q \in Z\}$. Then obviously A_q is equivalent to Z and we know that I is enumerable and so A_q is enumerable. Also the collection $Q = \bigcup_{q \in I_0} A_q$ is enumerable where Z_0 is a subset of Z, because Q is expressed as enumerable union of enumerable sets.

THEOREM 5(a). ***The set of all real numbers in the open interval (0, 1) is not enumerable.***

(KERALA–2006, MEERUT–1989, 96, HIMACHAL–2000, KANPUR–2004)

(b). ***The set of all real numbers in the closed interval [0, 1] is not enumerable.***

(MEERUT–1998, 2004, GARHWAL–1995, 97, 2003, HIMACHAL–2002)

PROOF. (a) Suppose contradiction, *i.e.*, let the given interval be enumerable. Then the elements of the interval can be written as a sequence $\{x_1, x_2, x_3, ...\}$. Now using the decimal expansion of these x_m's, we can set as follows :

$$\begin{aligned} x_1 &= .a_{11}a_{12}a_{13}...a_{1n}... \\ x_2 &= .a_{21}a_{22}a_{23}...a_{2n}... \\ x_3 &= .a_{31}a_{32}a_{33}...a_{3n}... \\ &... \quad ... \quad ... \quad ... \quad ... \quad ... \\ x_n &= .a_{n1}a_{n2}a_{n3}...a_{nm}... \\ &... \quad ... \quad ... \quad ... \quad ... \quad ... \end{aligned}$$

where a_{ij}'s may be any integer form 0 to 9. Now let

$$x_n = b_1\, b_2\, b_3 ...$$

where b_1, b_2, ... are the digits from 0 to 9 s.t. $b_1 \neq a_{11}$, $b_2 \neq a_{22}$, $b_3 \neq a_{33}$ etc.

In general $b_m \neq a_{mm}, \forall m \in N$.

Hence, $\quad x \neq x_1, x \neq x_2, ..., x \neq x_m, ...$

Obviously, $x \in (0,1)$ and $x \notin \{x_1, x_2, ...\}$; thus we set that besides countable set $\{x_1, x_2, ...\}$ there exists element belonging to interval (0 ,1), showing that the set of all real numbers lying in the interval (0, 1) is not enumerable.

(b) Open interval (0, 1) is a subset of [0, 1]. Since (0, 1) is not denumerable, the interval [0, 1] is also not denumerable.

THEOREM 6. ***The set of all real numbers is not enumerable, i.e., R is not enumerable.***

(KURUKSHETRA–2011, MEERUT–1991, 97, GARHWAL–1996, KANPUR–1996, 2000, 02)

PROOF. Suppose contradiction, *i.e.*, let R be enumerable. Then its every subset must be enumerable. Consider the set of all real numbers lying in the interval (0, 1) which is not enumerable by preceding theorem 5; which is again a contradiction since it is the subset of R.

THEOREM 7. ***The set of all irrational numbers is not enumerable.***

PROOF. Let the contrary be true. Also, the set of all rational numbers is enumerable. Thus union of rational and irrational numbers would also be enumerable. But we know that R is the union of rational and irrational numbers and hence R would be enumerable which is a contradiction. Hence the set of all irrational numbers is not enumerable.

1.18 ALGEBRAIC NUMBERS

Let $P_n(x) = \alpha_n x^n + \alpha_{n-1}x^{n-1} + ... + \alpha_1 x + \alpha_0 (\alpha_n \neq 0)$

be a polynomial with α_i's as integral numbers : then we define the algebraic number as a root of the polynomial equation $P_n(x) = 0$ of the above form.

THEOREM 8. ***The set of all algebraic numbers is enumerable.***

(LUCKNOW–2005, PATNA–2004, MEERUT–1996, GARHWAL–1993, KANPUR–1972)

PROOF. Consider the algebraic equation, $\alpha_0 x^n + \alpha_1 x^{n-1} + ... + \alpha_n$ of degree n with $\alpha_0 \neq 0$. Now we define the rank of this equation :

$$|\alpha_0| + |\alpha_1| + |\alpha_2| + ... + |\alpha_n| = m.$$

Clearly, rank is a positive number. Also α_i's are integers; so rank is an integer ≥ 1. Obviously for a given rank the roots of the equation will be finite and therefore will be enumerable.

Again we can put a one-one correspondence in the set of natural numbers with the algebraic equation arranged with respect to rank and hence the set of all algebraic equations is enumerable. Now each algebraic equation has enumerable number of roots and so the set of all algebraic numbers is the enumerable collection of enumerable sets and hence enumerable.

REMARK

- The set P of all polynomials

$$P_n(x) = \alpha_n x^n + \alpha_{n-1}x^{n-1} + ... + \alpha_1 x + \alpha_0, (\alpha_n \neq 0)$$

with integral (rational) coefficients is denumerable. (GARHWAL–2001)

1.19 CARDINALLY EQUIVALENT SETS

A set P is said to be cardinally equivalent to a set Q, if there exists at least one one-to-one map from P to Q and is denoted by $P \sim Q$.

THEOREM 9. ***The relation P ~ Q in the family of sets is an equivalence relation.***

PROOF. Recall that a relation R in a set is an equivalence relation in P, iff

(i) R is reflexive, *i.e.*, $aRa, \ \forall a \in P$

(ii) R is symmetric, *i.e.*, $aRb \Rightarrow bRa, \quad \forall a, b \in P$,

(iii) R is transitive, *i.e.*, $aRb, bRc \Rightarrow aRc \quad \forall a, b, c \in P$.

Here we observe that the given relation is

(i) Reflexive. Since the identity map $I_P : P \to P$ given by $I_P(a) = a, \forall a \in P$ is one-one onto, *i.e.*, $P \sim P$ for every set P.

(ii) Symmetric. $P \sim Q \Rightarrow \exists$ a one-one map $f : P \xrightarrow{\text{onto}} Q$. Therefore,

$$f^{-1} : Q \xrightarrow{\text{onto}} P \text{ is also one-one} \Rightarrow Q \sim P.$$

(iii) Transitive. We have

$$P \sim Q \Rightarrow \exists \text{ a one-one map } f : P \xrightarrow[\text{onto}]{\text{onto}} Q$$

$$Q \sim R \Rightarrow \exists \text{ a one-one map } g : Q \xrightarrow{\text{onto}} R$$

$$\Rightarrow g \circ f : P \xrightarrow{\text{onto}} R \text{ is also one-one}$$

$$\Rightarrow P \sim R$$

Thus the given relation is an equivalence relation.

REMARK

- It is to be noted that an equivalence relation decomposes the set P into equivalence classes, any two of which are either equal or mutually disjoint. The set of these mutually disjoint equivalence classes is said to be the quotient set of the set for the given equivalence relation. The equivalence relation is also known as equipollent relation.

EXAMPLE 1. *If A and B are the sets of all the real numbers of the intervals* $[a_1, a_2]$ *and* $[b_1, b_2]$ *respectively, then show that* $A \sim B$.

SOLUTION. Consider a mapping $f : A \to B$ s.t.

If $x \in A$, then $f(x) = \dfrac{b_2 - b_1}{a_1 - a_2} x - a_1 \dfrac{b_2 - b_1}{a_2 - a_1} + b_1$

$$= \left(\frac{b_2 - b_1}{a_2 - a_1}\right)(x - a_1) + b_1.$$

Now, to show mapping f is one-one. For if $f(x) = f(y); x, y \in A$, then

$$\frac{(b_2 - b_1)}{(a_1 - a_2)}(x - a_1) + b_1 = \frac{(b_2 - b_1)}{(a_2 - a_1)}(y - a_1) + b_1$$

$$\Rightarrow \quad \left(\frac{b_2 - b_1}{a_2 - a_1}\right)(x - a_1) = \left(\frac{b_2 - b_1}{a_2 - a_1}\right)(y - a_1) \Rightarrow x = y$$

$\Rightarrow$ f is one-one.

Also, to prove mapping f in onto. Let $z \in B \Rightarrow b_1 \le z \le b_2$.

If $$z = \left(\frac{b_2 - b_1}{a_2 - a_1}\right)(x - a_1) + b_1, \text{ then } x = \frac{(z - b_1)(a_2 - a_1)}{(b_2 - b_1)} + a_1.$$

Obviously $a_1 \le x \le a_2$ and hence $x \in A$. Thus for any $z \in B$, $\exists x \in A$ s.t. $f(x) = z$ and hence f is onto

$\Rightarrow$ f is one-one onto and hence $A \sim B$.

REMARK

- If $A = (a_1, a_2)$, $B = (b_1, b_2)$ then $A \sim B$.

1.20 CARDINAL NUMBER OF A SET

We know that the relation of equivalence decomposes any collection of sets into equivalence classes containing the equivalent sets. Each equivalent class has a cardinal number which we shall use to represent the property the equivalent sets have in common. It will be in some sense a measure of the points in sets. It is denoted by the card X, *i.e.*, cardinal number of X. The basic property of cardinal number is that card X = card Y, iff $X \sim Y$.

Definition. *Any set which is equivalent to the set* $\{1, 2, 3, \ldots, n\}$ *is said to have the cardinal number n.*

REMARK

- The null set is said to have the cardinal number zero. Obviously for finite sets, cardinal number is just the number of element in the sets. (MEERUT–2003, 04, GARHWAL–2002)

The cardinal number of N (set of natural numbers) is denoted by a and hence all enumerable sets will have the cardinal number a.

The cardinal number of the set of real numbers is denoted by c and all the sets equivalent to the set R are said to have the cardinal number c. The set of all real numbers in the interval [0, 1] also has the cardinal number c. Since we have proved that all intervals open or closed are equivalent to [0, 1], hence every interval also has cardinal number c. The cardinal number of the set of all real-valued functions defined in the interval [0, 1] is denoted by f.

The cardinal number of an infinite set is called a transfinite cardinal number a is considered to be the first (smallest) transfinite cardinal.

Since every finite set has an enumerable subset which is equivalent to be the set of natural number N and so every infinite set has a subset with cardinal number a.

REMARK

- The set of cardinal number $\{0,1,2,3,...,a,c,f\}$ is a superset of set of all natural numbers.

1.20.1 SUM OF CARDINAL NUMBERS

Let A and B be any two disjoint sets; then

$$\text{card } A + \text{card } B = \text{card } A \cup B.$$

In general $\sum\limits_{\alpha\in\Delta} \text{card}A_\alpha = \text{card}\left(\bigcup\limits_{\alpha\in\Delta} A_\alpha\right)$ where $A_\alpha \cap A_\beta = \phi$, $\forall \alpha,\beta,\in \Delta$ (index set) such that $\alpha \neq \beta$.

1.20.2 PRODUCT OF CARDINAL NUMBERS

We define the product of two cardinal numbers as card $P \times$ card Q = card $(P \times Q)$ where P, Q are any sets.

In general, card $P_1 \times$ card P_2 ... = card [Cartesian product of sets P_i $(i = 1, 2, 3, ...)$]

1.20.3 COMPARISON OF CARDINAL NUMBERS

Let P and Q be any two sets, then

(i) card $P <$ card Q, if $\exists$ a set $R \subset Q$ s.t. $P \sim R$ and if $P \sim Q$ then card P = card Q.

(ii) card $P \leq$ card Q, if $\exists$ a set $R \subseteq Q$ s.t. $P \sim R$, *i.e.*, there exists a one-one map

$$f : P \xrightarrow{\text{onto}} R \subseteq Q.$$

(iii) Evidently, card $(P \cup Q) \geq$ card P or card Q.

(iv) a is the smallest infinite cardinal number.

THEOREM 1. $\textbf{card}\,(P \times Q) = \textbf{card}Q + \textbf{card}Q + ...\textbf{card}\,P$ **terms.**

PROOF. We have

$$P \times Q = \{x, y) : x \in P, y \in Q\} = \bigcup_{x\in P} (x, y) : y \in Q\}.$$

$$\therefore \quad \text{card}(P \times Q) = \text{card}\left(\bigcup_{x\in P} (x, y) : y \in Q\}\right). \qquad ...(1)$$

Let $x \in P$ be arbitrary but fixed, Consider the map

$$f : Q \to \{(x, y) : y \in Q\}$$

defined by $\quad f(y) = (x, y) \quad \forall y \in Q.$

This show that f is one-one. Therefore

$$\text{card}Q = \text{card}((x, y) : y \in Q))$$

Observing (1) we get one required result.

EXAMPLE 2. *Show that* $\alpha \le \text{card}P \le \alpha \Rightarrow \text{card } P = \alpha$

SOLUTION. If $\alpha \le \text{card}P \Rightarrow \exists$ a set Q such that card $Q \le$ card P where card $Q = \alpha$

$\Rightarrow$ Q is equivalent to a subset of P ...(i)

Again card $P \le \alpha \Rightarrow \text{card}P \le Q$

$\Rightarrow$ P is equivalent to a subset of Q ...(ii)

Combining (i) and (ii), we get

card P = card Q or card$P = \alpha$ $(\because \text{card}\, Q = \alpha)$

THEOREM 2. ***Let P and Q any two sets; then***

(i) card P + card Q is unique, (ii) card P.card Q is unique.

PROOF. Let $P \sim P_1, Q \sim Q_1, P_1 \cap Q_1 = \phi$. Now

(i) $P \sim P_1 \Rightarrow \exists$ a one-one map $f : \xrightarrow{\text{onto}} P_1$

and $Q \sim Q_1 \Rightarrow \exists$ a one-one $g : Q \xrightarrow{onto} Q_1$.

Define a function $\psi : P \cup Q \to P_1 \cup Q_1$ s.t.

$$\psi(x) = \begin{cases} f(x) & , \forall x \in P, \\ g(x) & , \forall x \in Q. \end{cases}$$

Evidently ψ is one-one onto since f and g are one-one onto. Hence

$$P \cup Q \sim P_1 \cup Q_1 \Rightarrow \text{card } (P \cup Q) = \text{card } (P_1 \cup Q_1)$$

This shows that card P + card Q is unique.

(ii) $x \in P, y \in Q \Rightarrow (x, y) \in P \times Q$

$\Rightarrow (f(x), g(y)) \in P_1 \times Q_1$.

Define a function, $\psi : P \times Q \to P_1 \times Q_1$ s.t.

$$\psi(x.y) = (f(x), g(y)) \quad \forall (x, y) \in P \times Q.$$

Again f and g are one-one onto, therefore ψ is also one-one onto

$\Rightarrow P \times Q \sim P_1 \times Q_1 \Rightarrow \text{card}(P \times Q) = \text{card}(P_1 \times Q_1)$.

Hence the result follows.

EXAMPLE 3. *Show that* $(P \times Q) \sim (Q \times P)$, *i.e.,* $(P \times Q)$ *is cardinally equivalent to* $(Q \times P)$.

SOLUTION. We have

$$(P \times Q) = \{p, q) : p \in P, q \in Q\}$$

Define a map $f : (P \times Q) \to (Q \times P)$ by the formula $f((p, q) = (q, p)$.

f is one-one. Let $f((p_1, q_1)) = f((p_2, q_2))$

$\Rightarrow$ $(q_1, p_1) = (q_2, p_2) \Rightarrow q_1 = q_2$ and $p_1 = p_2$

$\Rightarrow$ $(p_1, q_1) = (p_2, q_2)$.

Obviously f is onto also.

THEOREM 3. ***Let card P = p, card Q = q, card R = r, then :***

(i) $p + q = q + p$, ***i.e., addition of cardinal numbers is commutative.***

(ii) $p \cdot q = q \cdot p$, ***i.e., multiplication of cardinal numbers is commutative.***

(GARHWAL–2002)

(iii) $p \cdot (q + r) = p \cdot q + p \cdot r$, ***i.e., multiplication is distributive over addition.***

(iv) $p \cdot (q \cdot r) = (p \cdot q) \cdot r$, ***i.e., associative law for multiplication holds.***

(v) $p + (q + r) = (p + q) + r$, ***i.e., associative law for addition holds.***

PROOF. (i) Let $P \cap Q = \phi$. The elements of an arbitrary set may be in any order, therefore

$$P \cup Q = Q \cup P \Rightarrow \text{card}(P \cup Q) = \text{card}(Q \cup P)$$
$$\Rightarrow \text{card } P + \text{card } Q = \text{card } Q + \text{card } P.$$

(ii) We know that $P \times Q \sim Q \times P$

$$\Rightarrow \text{card } (P \times Q) = \text{card } (Q \times P)$$
$$\Rightarrow p \cdot q = q \cdot p.$$

(iii) Let Q, R be disjoint sets. Then

$$p.(q+r) = \text{card } [P \times (Q \cup R)] = \text{card } [(P \times Q) \cup (P \times R)]$$
$$= \text{card } (P \times Q) + \text{card } (P \times R) = p \cdot q + p \cdot r.$$

(iv) Since $P \times (Q \times R) \sim (P \times Q) \times R$ under the map $f(x,(y,z)) = ((x,y),z)$,
Hence card $[P \times (Q \times R)] = \text{card } [(P \times Q) \times R]$
or $p \cdot (q \cdot r) = (p \cdot q) \cdot r.$

(v) Suppose that P, Q, R are pairwise disjoint sets. Also we know that

$$(P \cup Q) \cup R = P \cup (Q \cup R)$$
$$\Rightarrow \text{card } [(P \cup Q) \cup R] = \text{card } [P \cup (Q \cup R)]$$
$$\Rightarrow (p+q)+r = p+(q+r)$$

THEOREM 4(i). ***If A_i is enumerable set for $i = 1, 2, 3, \ldots, n$ then $\bigcup_{i=1}^{n} A_i$ is enumerable and hence $n.a = a$.*** (GURUKUL KANGRI–2001,GARHWAL–1995)

(ii) ***If A_i is enumerable set for $i = 1, 2, 3, \ldots, n$, then $\bigcup_{i=1}^{n} A_i$ is enumerable and hence $a + a + a + \ldots$ to a terms $= a$***

PROOF. (i) Let $A = \bigcup_{i=1}^{n} A_i$. We know that countable union of enumerable set is enumerable, hence A is enumerable.

DEDUCTIONS. Let $A_i \cap A_j = \phi$ for $i \neq j$; then by the definition of sum of cardinal numbers,

$$\text{card } A_1 + \text{card } A_2 + \ldots + \text{card } A_n = \text{card } A \quad \left(\text{since } \bigcup_{i=1}^{n} A_i = A\right) \quad \ldots(1)$$

We know that cardinal number of an enumerable set is a and since each $A_i (i = 1, 2, 3, \ldots, n)$ is enumerable, so we have

$$\text{card } A_1 = \text{card } A_2 = \ldots = \text{card } A_n = a$$

Also A is enumerable so card $A = a$. So from (1), $n \cdot a = a$

(ii) Let $A = \bigcup_{i=1}^{\infty} A_i$; then A is enumerable being enumerable union of enumerable sets and hence card $A = a$

REMARK

- Let $A_i \cap A_j = \phi$ for $i \neq j$; then card $A = a$ gives

$$\sum_{i=1}^{\infty} \text{card } A_i = a$$

$\Rightarrow \quad a + a + a + a + \ldots$ to a terms $= a$.

THEOREM 5(i). ***If A_i is non-enumerable set for $1 \le i \le n$, then $\bigcup_{i=1}^{n} A_i$ is non-enumerable and hence***

$$c + c + c + \dots \text{ to } n \text{ terms} = c.$$

(ii) ***If A_i is non-enumerable, $\forall i \in N$, then $\bigcup_{i=1}^{\infty} A_i$ is non-enumerable and hence***

$$c + c + c + \dots \text{ to a terms} = c.$$ (GARHWAL–1991)

PROOF. (i) Each A_i is non-enumerable for $1 \le i \le n$, therefore card $A_i = c$ for $1 \le i \le n$. This implies $A_i \sim [a_i, a_{i+1})$ where $a_i, a_{i+1} \in R$ for $i = 1, 2, 3, \dots, n$. Let $a_i < a_{i+1}$ for $i = 1, 2, 3, \dots, n$. Thus, we have

$$A_1 \sim \{a_1, a_2\},$$
$$A_2 \sim \{a_2, a_3\},$$
$$\dots \dots \dots$$
$$\dots \dots \dots$$
$$A_n \sim [a_n, a_{n+1}).$$

At first suppose that $A_i \cap A_j = \phi$ for $i \ne j$; then $\bigcup_{i=1}^{n} A_i$ is equivalent to some subset of $[a_1, a_{n+1})$.

$$\therefore \quad \text{card}\left(\bigcup_{i=1}^{n} A_i\right) \le c. \qquad \dots(1)$$

Now evidently

$$A_i \subseteq \bigcup_{i=1}^{n} A_i \Rightarrow \text{card } A_i \le \text{card}\left(\bigcup_{i=1}^{n} A_i\right)$$

$$\Rightarrow \quad c \le \text{card}\left(\bigcup_{i=1}^{n} A_i\right). \qquad \dots(2)$$

Combining (1) and (2), we have

$$c \le \text{card}\left(\bigcup_{i=1}^{n} A_i\right) \le c \text{ ,i.e., card}\left(\bigcup_{i=1}^{n} A_i\right) = c.$$

$\therefore \bigcup_{i=1}^{n} A_i$ is non-enumerable.

Now suppose that $A_i \cap A_j = \phi$ for $i \ne j$, then

$$\bigcup_{i=1}^{n} A_i \sim [a_1, a_{n+1}) \Rightarrow \text{card}\left(\bigcup_{i=1}^{n} A_i\right) = \text{card}[a_1, a_{n+1})$$

$$= \text{card}\left(\bigcup_{i=1}^{n} A_i\right) = c.$$

REMARK

- If we assume that $A_i \cap A_j = \phi$ for $i \ne j$, then as proved above, we have

$$\sum_{i=1}^{n} \text{card}(A_i) = c\left(\because \text{card} \bigcup_{i=1}^{n} A_i = \sum_{i=1}^{n} \text{card } A_i \text{ when } A_i \cap A_j = \phi \text{ for } i \ne j\right)$$

i.e., $c + c + c + \dots$ to n terms $= c$.

(ii) Let $A = \bigcup_{i=1}^{\infty} A_i$, where card $A_i = c, \forall i \in N$. Now

$$\text{card } A_i = c \Rightarrow A_i \sim \left[1-\frac{1}{2^{i-1}}, 1-\frac{1}{2^i}\right).$$

Thus, we have

$$A_1 \sim \left[0,\frac{1}{2}\right), A_2 \sim \left[\frac{1}{2},\frac{3}{4}\right), A_3 \sim \left[\frac{3}{4},\frac{7}{8}\right), \ldots \quad \ldots A_i \sim \left[1-\frac{1}{2^{i-1}}, 1-\frac{1}{2^i}\right) \ldots.$$

Now assume that $A_i \cap A_j = \phi$ for $i \neq j$.

Then obviously $\bigcup_{i=1}^{\infty} A_i - [0,1)$.

$$\therefore \quad \text{card}\left(\bigcup_{i=1}^{\infty} A_i\right) = \text{card } [0,1) \text{ or card } A = c.$$

Next suppose that $A_i \cap A_j \neq \phi$ for $i = j$, then $\bigcup_{i=1}^{\infty} A_i$ is cardinally equivalent to some subset of $[0,1)$.

So, $\quad \text{card}\left(\bigcup_{i=1}^{n} A_i\right) \leq c$. ...(3)

Again $A_i \subset \bigcup_{i=1}^{n} A_i \Rightarrow \text{card } A_i \leq \text{card}\left(\bigcup_{i=1}^{n} A_i\right)$

$\Rightarrow c \leq \text{card}\left(\bigcup_{i=1}^{\infty} A_i\right)$ ...(4)

Combining (3) and (4) we get, $\text{card}\left(\bigcup_{i=1}^{\infty} A_i\right) = c$.

Hence $\bigcup_{i=1}^{\infty} A_i$ is non-enumerable.

REMRK

- Suppose that $A_i \cap A_j = \phi$ for $i \neq j$, we have

$$\text{card}\left(\bigcup_{i=1}^{\infty} A_i\right) = c \quad \Rightarrow \quad \sum_{i=1}^{\infty} \text{card } A_i = c$$

or $c + c + c + \ldots$ to a terms $= a$.

THEOREM 6. $\mathbf{a + \alpha = \alpha, \alpha}$ ***is any transfinite cardinal number.***

(GURUKUL KANGRI–2001)

PROOF. Let A be an infinite set therefore $\exists$ a subset B of A s.t. B is enumerable.

$\therefore \quad$ card $B = a$.

Now we can write $A = (A - B) \cup B$.

$\therefore \quad A \cup N = (A - B) \cup B \cup N = (A - B) \cup (B \cup N)$.

B and N are enumerable sets $\Rightarrow B \cup N$ is enumerable

$\Rightarrow \quad B \cup N \sim N$.

Now $\quad B \cup N \sim N, N \sim B \Rightarrow B \cup N \sim B$.

$\therefore \quad (A-B)\cup(B\cup N)\sim(A-B)\cup B$, *i.e.*, $A\cup N\sim A$

or $\quad \text{card}(A\cup N)=\text{card}A$, *i.e.*, $\alpha+a=\alpha$.

THEOREM 7. $\mathbf{c}\,.\,\mathbf{c}=\mathbf{c}$. (BANARAS–1997)

PROOF. Let $A=\{x: 0\le x\le 1\}$. Then card $A=c$.

Now, let $\quad B=\{(0,x): x\in A\}$

Then obviously $\quad B\subset A\times A \qquad \therefore \text{card}B\le \text{card}(A\times A)$

Again $A\sim B$ under the map f s.t. $f(x)=(0,x)$, $\forall x\in A$, $\therefore$ card $A\le$ card B.

$\Rightarrow \qquad \text{card } A\le \text{card }(A\times A)$. ...(1)

Let x, y be any two real numbers in the closed interval $[0, 1]$. Then x and y can uniquely be expanded in the form of infinite decimals which contain non-zero digits. Now define a map $g:(A\times A)\to A$ by writing

$$g(x,y)=0.\ x_1y_1x_2y_2x_3y_3\ldots.$$

Obviously g is one-one. So by definition

$$\text{card }(A\times A)\le \text{card } A. \qquad ...(2)$$

From (1) and (2), we get

$$\text{card } A\le \text{card }(A\times A)\le \text{card } A; \qquad \text{card }(A\times A)=\text{card } A$$

or $\quad c\,.\,c=c \qquad [\because \text{card } A\times \text{card } B=\text{card }(A\times A)]$

1.21 TRANSCENDENTAL NUMBER

Definition. *A real number which is not an algebraic number is called transcendental number. Thus the numbers e and π which are real but not algebraic numbers, are transcendental numbers.* (BANARAS–2014, AVADH–2011, DELHI–2012, MEERUT–2003)

All rational numbers are algebraic number hence every rational number is not transcendental, implying that every transcendently number must be irrational, for $R=$(rational numbers) $\cup$ (irrational numbers).

REMARK

- It must be noted that there are so many irrational numbers which are algebraic *e.g.*, $(n)^{1/r}$. Therefore every irrational number is not transcendental number.

THEOREM 1. ***Every monotonic function in a closed interval is discontinuous at a countable number of points of that interval.***

PROOF. Let $f(x)$ be a monotonic function in the closed interval $[a, b]$. Also let it be a monotonically increasing function and be discontinuous at an arbitrary point x. Then

$$\delta(x)=f(x+0)-f(x-0)>0 \qquad ...(1)$$

where $\qquad f(a)=f(a-0), f(b)=f(b+0)$.

Let $\xi_1,\xi_2,\ldots,\xi_{m-1}$ be numbers in the intervals $x_k<\xi_k<x_{k+1}$ where $a<x_1<x_2<\ldots x_m<b$, where $\xi_0=a$ and $\xi_m=b$.

$$\therefore\ f(\xi_k)-f(\xi_{k-1})\ge f(x_k+0)-f(x_k-0)=\delta(x_k) \qquad [\text{By }(1)] \quad ...(2)$$

Therefore $f(b)-f(a)=\sum\limits_{k=1}^{m}[f(\xi_k)-f(\xi_{k-1})]\ge \sum\limits_{k=1}^{m}\delta(x_k)$.

Let $\qquad \delta(x_k)>\dfrac{1}{n}$, $\forall k$. Then by last inequality, we have

$$f(b)-f(a)>\frac{m}{n} \text{ or } [f(b)-f(a)]n>m.$$

This shows that m which is the number of points of discontinuity x with $\delta(x) > \frac{1}{n}$ is bounded above, *i.e.*, the number of points of discontinuity x with $\delta(x) > \frac{1}{n}$ are finite in the closed interval $[a, b]$. Since $n \in N$, we see that the number of points of discontinuity x with $\delta(x) > \frac{1}{n}$ are finite in the closed interval $[a, b]$. Since every finite set is countable and for every $x \ \exists \ n \in N$ therefore the number of points of discontinuity in the closed interval $[a, b]$ will be an enumerable union of countable sets and hence countable.

THEOREM 2. **(Cantor's Theorem).** ***card A < card P(A), P(A) being power set of the set A.*** (MEERUT–1983, 91, GURUKUL KANGRI–2001)

Or

For every cardinal number n, $2^n > n$.

PROOF. Let $B^* = \{\{x\} : x \in A\}$; then obviously $B^* \subset P(A)$. Now define a map $f : A \to B^*$ s.t. $f(x) = \{x\}$. Obviously $A \sim B^*$. Hence

c card $A \leq$ card $P(A)$.

Thus we are only to show that card $A \neq$ card $P(A)$,

i.e., $A \times P(A)$ ...(1)

Suppose contradiction, *i.e.*, $A \sim P(A)$. So $\exists$ a one-one map

$$f : A \xrightarrow{\text{onto}} P(A).$$

Let $B = \{x \in A : x \notin f(x)\}$.

Clearly, $B \subset A \Rightarrow B \in P(A)$.

Since the mapping f is onto, there must exist $x \in A$ s.t. $f(x) = B$. Now if $x \in B$ then by definition of $B, x \in f(x)$ which is not possible.

Consider the second possibility that $x \notin B$, then $x \in f(x) = B$ which is again impossible. It means that our assumption is wrong. Hence (1) is true.

REMARKS

- From the last theorem, it follows that $n < 2^n$ where card $A = n$ and card$P(A) = 2^{\text{card}A} = 2^n$. (MEERUT–1988)
- $2^a > a$.

THEOREM 3. **(Equivalence Theorem).** ***If $A_1 \subset B \subset A$ and $A \sim A_1$, then $A \sim B$.***

Or

If $A_1 \subset B \subset A$ and* card *A =* card *A_1 then* card *A =* card *B.

PROOF. Since $A \sim A_1 \Rightarrow \ \exists$ a one-one map $f : A \xrightarrow{\text{onto}} A_1$.

As $B \subset A$ so f_B is a one-one onto where f_B is the restriction of f to B. This means that $B \sim B_1 \subset A_1$. Similarly, $A_1 \sim A_2 \subseteq B_1$.

Continuing in this way we get equivalent sets

$$A, A_1, A_2, \ldots \text{ and } B, B_1, B_2, \ldots$$

s.t. $A \supset B \supset A_1 \supset B_1 \supset A_2 \supset B_2 \supset A_3 \supset B_3 \supset \ldots$

Let $S = A \cap B \cap A_1 \cap B_1 \cap A_2 \cap B_2 \cap \ldots$

Then we can write

$$A = (A - B) \cup (B - A_1) \cup (A_1 - B_1) \cup \ldots \cup S,$$

$$B = (B - A) \cup (A_1 - B_1) \cup (B_1 - A_2) \cup \ldots \cup S$$

Define a map $\psi : A \to B$ s.t..

$$\psi(A - B) = A_1 - B_1,$$
$$\psi(A_1 - B_1) = A_2 - B_2,$$
$$\psi(A_2 - B_2) = A_3 - B_3,$$
$$\ldots \quad \ldots \quad \ldots \quad \ldots$$
$$\ldots \quad \ldots \quad \ldots \quad \ldots$$
$$\psi(B - A_1) = B - A_1,$$
$$\psi(B_1 - A_2) = B_1 - A_2,$$
$$\ldots \quad \ldots \quad \ldots \quad \ldots$$
$$\ldots \quad \ldots \quad \ldots \quad \ldots$$
$$\ldots \quad \ldots \quad \ldots \quad \ldots$$
$$\psi(S) = S.$$

Above definition of ψ makes the mapping ψ one-one and showing that $A \sim B$.

THEOREM 4. **(Schorder-Bernstein Theorem).** ***If* card $A \leq$ card B *and* card $B \leq$ card A, *then* card A = card B.** (MEERUT–1989)

Or

If each of the sets A and B is equivalent to a subset of other, then $A \sim B$.

PROOF. Let f and g be one-to-one mapping from A onto B and from B into A respectively. Let $f(A) = B_i \subset B$ and $g(B) = A_2$ and $g(B_1) = A_3$, then we have $A \supset A_2 \supset A_3$. Further $g(B_1) = A_3, f(A) = B_1$ implies $g(f(A)) = A_3$, giving gof is a one-to-one mapping from A to $A_3 \Rightarrow$ card A = card A_3. Hence by the above theorem card A = card A_2. Also existence of g s.t. $g(B) = A_2$ shows that card B = card A_2. Hence card B = card A.

THEOREM 5. $\mathbf{2^a = c}$.

PROOF. We know that card $[0,1] = c$. On the other hand each $x \in [0,1]$ can be written in the form of binary expansion as $x \in 0.x_1x_2x_3\ldots$, where each $x_i = 0$ or 1.

But selecting each x_i in two ways (either 0 or 1) we can form at most 2^a numbers. So the card $[0,1] = 2^a$ implying that $2^a = c$.

THEOREM 6. ***Every superset of an uncountable set is uncountable.***

PROOF. Suppose contradiction, *i.e.*, if B is the superset of an uncountable set A, then B is countable. But we know that every subset of a countable set is countable and so A must be countable which is a contradiction and so B is uncountable. That is to say that every superset of an uncountable set is uncountable.

THEOREM 7. ***Union of two enumerable sets is also enumerable.***

PROOF. Let A and B be the two enumerable sets.

Case I. When $A \cap B = \phi$.

Let $A = \{a_1, a_2, \ldots\}$ $B = \{b_1, b_2, \ldots\}$. Now establish correspondence $f : A \cup B \to N$

s.t.

$$f(a_n) \to 2n-1 \text{ (odd positive integer)},$$
$$f(b_n) \to 2n \text{ (even positive integer)}.$$

Clearly this mapping is one-one from $A \cup B$ onto N.

Case II. When $A \cap B \neq \phi$, then we can write $A \cup B = A \cup (B - A)$. Taking $B_1 = B - A$ we have $A \cap B_1 = \phi$. As already proved $A \cup B_1$ is countable where B_1 is countable where B_1 is countable being the subset of countable set B and hence $A \cup B = A \cup B_1$ is countable when B_1 is countable infinite but if B_1 is finite say

$$B_1 = \{e_1, e_2, ..., e_m\}.$$

Then $\quad A \cup B = A \cup B_1 = \{e_1, e_2, e_3, ..., e_m, a_1, a_2, a_3, ...\}$.

Now set a correspondence

$$f : N \to A \cup B_1$$

s.t. $\quad f(i) = e_i, \quad 1 \le i \le m,$

$$f(m+i) = a_i, \ \forall i$$

$\Rightarrow \quad A \cup B$ is enumerable.

REMARK

- We can generalize the result that union of finite countable sets is countable (whether each is countably infinite or finite).

THEOREM 8. ***Every infinite set is equivalent to its proper subset.***

PROOF. **Case I.** When A is countably infinite, then A can be written as a sequence. Let $A = \{a_1, a_2, a_3 ...\}$. Then the function $f(a_n) = a_{n+1}$ establishes a one to one correspondence between the set A and $A - \{a_1\}$ which is a proper subset of A.

Case II. When A is uncountably infinite, then it has an enumerable subset say B where $B = \{a_1, a_2, a_3 ...\}$. We shall show that

$$A \sim (A - \{a_i\}).$$

Write $C = A - B$; then $A = B \cup C$ and $B \cap C = \phi$. Also $A - \{a_i\} = B - \{a_i\}) \cup C$. Let $e(x)$ be the identity mapping which associates each $x \in A$ onto itself. Let f be the function: $f(a_n) = a_{n+1}$. Now define a function h:

$$h(x) = \begin{cases} e(x) : x \in C, \\ f(x) : x \in B. \end{cases}$$

Then the range of h is $(B - \{a_i\}) \cup C$ which is a proper subset of $B \cup C = A$. Thus the result follows.

THEOREM 9. ***If α and β are cardinal numbers such that $\alpha \le \beta$ and $\beta \le \alpha$, then $\alpha = \beta$.***

PROOF. Let card $A = \alpha$, card $B = \beta$

Now, $\quad \alpha \le \beta \Rightarrow$ card $A \le$ card B

$\Rightarrow \quad A \sim B$ or $A \sim$ to a subset of B, ...(1)

$\beta \le \alpha \Rightarrow B \sim A$ or $B \sim$ to a subset of A. ...(2)

(1) and (2) give the required result.

THEOREM 10. ***If an enumerable set is subtracted from an enumerable set, the remaining set will be enumerable.***

PROOF. Suppose contradiction, *i.e.*, $A - B$ is non-enumerable where A and B are enumerable sets. We can write $A = (A - B) \cup B$. Now $A - B$ is non-enumerable gives to be non-enumerable which is a contradiction. Hence the result.

THEOREM 11. ***If we subtract an enumerable set from a non-enumerable set, then the remaining set is non-enumerable.***

PROOF. Suppose contradiction, *i.e.*, $A-B$ is enumerable where A is non-enumerable and B is an enumerable set. We can write $A=(A-B)\cup B$. Since both $A-B$ and B are enumerable sets it implies A is enumerable which a contradiction as A is non-enumerable. Hence the result.

SOLVED EXAMPLES

EXAMPLE 1. *Prove that the set Z, of all integers is countable.*

(MEERUT–1993, 98, 2004, KANPUR–2001)

SOLUTION. Write

$$Z^+=\{1,2,3,\ldots\},$$
$$Z^-=\{-1,-2,-3,\ldots\}.$$

Then, we have $Z^+\sim N$ under the mapping $n\to n$.

Now $Z^+\sim N$ under the mapping $n\to n$,

$Z^-\sim N$ under the mapping $(-n)\to n$.

Also singleton set $\{0\}$ is finite, so countable. Thus Z is the countable union of countable sets and hence countable.

Alternatively, define the mapping $f:N\to Z$ s.t. $f(x)=(-1)^x\left[\frac{x}{2}\right]$,

where $\left[\frac{x}{2}\right]$ represents the integral value of $\frac{x}{2}$, *i.e.*, represents the largest integer less than or equal to $\frac{x}{2}$.

Establish that this mapping gives a one-to-one correspondence between N and Z implying that Z is countable.

Alternatively the $f:N\to Z$ given by

$$f(n)=\frac{n-1}{2} \text{ if } n \text{ is odd} \in N$$
$$=-\frac{n}{2} \text{ if } n \text{ is even} \in N.$$

This is also a one-to-one correspondence. Hence, $Z\sim N$.

REMARK

- An infinite set can be equivalent to its proper subset, e.g. $Z\sim N$.

EXAMPLE 2. *Find the power of an aggregate of numbers given by $\frac{M}{2^m}$, M and m being positive and integral* (GARHWAL–1994, 98)

(Power of a set means the cardinal number of that set).

SOLUTION. Let us suppose that

$$B=\left\{\frac{M}{2^m}: M,\ m\in N\right\}.$$

Write $$B_M=\left\{\frac{M}{2^m}: m\in N\right\}.$$

Then we have,

$$B_1 = \left\{\frac{1}{2^1}, \frac{1}{2^2}, \ldots, \frac{1}{2^n}, \ldots\right\}$$
$$B_2 = \left\{\frac{2}{2^1}, \frac{2}{2^2}, \ldots, \frac{2}{2^n}, \ldots\right\}$$
$$B_3 = \left\{\frac{3}{2^1}, \frac{3}{2^2}, \ldots, \frac{3}{2^n}, \ldots\right\}$$
$$\ldots \quad \ldots \quad \ldots \quad \ldots$$
$$B_n = \left\{\frac{n}{2^1}, \frac{n}{2^2}, \ldots, \frac{n}{2^n}, \ldots\right\}$$
$$\ldots \quad \ldots \quad \ldots \quad \ldots$$

Evidently,

(i) B_i is enumerable $\forall i \in N$ under the mapping $\frac{i}{2^n} \to n$,

(ii) B_i's are pairwise disjoint.

(iii) $B = \bigcup_{i=1}^{\infty} B_i$.

Thus B is enumerable being the enumerable union of enumerable sets. Hence card $B = a$, *i.e.*, the power of the given set is a.

EXAMPLE 3. *Prove that* $a < c$.

SOLUTION. Since $N \subset R \Rightarrow$ card $N <$ card R

$\Rightarrow \quad a < c$.

EXAMPLE 4. *Prove that* card$P(A) = 2^{\text{card}A}$ *for any finite set A.* (MEERUT–1988)

SOLUTION. Let card $A = \alpha$.

Then, card$(P(A)) = 1 + {}^{\alpha}c_1 + {}^{\alpha}c_1 + \ldots + {}^{\alpha}c_{\alpha} = (1+1)^{\alpha} = 2^{\alpha} = 2^{\alpha} = 2^{\text{card}\,A}$.

EXAMPLE 5. *Prove that* $\alpha \le \alpha$ *for any cardinal number* α.

SOLUTION. Let card $P = \alpha$.

Define an identity map $f : P \to P$ written by

$$f(x) = x, \forall x \in P.$$

Obviously f is one-one. Hence by definition,

card $P \le$ card P *i.e.*, $\alpha \le \alpha$.

EXAMPLE 6. *If each* $a_i (i = 1, 2, \ldots, n)$ *is a rational number, then the point* $a = (a_1, a_2, \ldots, a_n) \in R^n$ *is called a rational point.*

Show that the set of all rational points in R^n *is denumerable.*

SOLUTION. We know that set of rational numbers is countable. So varying $a_1, a_2, \ldots, a_n$ we can form a^n rational points.

But we know that $a \cdot a = a$ and hence $a^n = a$.

$\Rightarrow$ set of all rational points has the same cardinal number as that of N.

$\Rightarrow$ set of rational points is countable.

EXAMPLE 7. *Prove that* $c^c = 2^c$.

SOLUTION. We have $c^c = (2^a)^c = 2^{ac} = 2^c$.

EXAMPLE 8. *Let X be any non-empty set and let C be the family of functions* $f : X \to \{0,1\}$. *Then show that the family of subset of X, i.e., the power set of X is equivalent to C.*

SOLUTION. Let $A \in P(X)$ where $P(X)$ denotes the power set of X. Also let ϕ_A denote the characteristic function of A relative to X.

Now define a map $f : P(X) \to C$ by the formula

$$f(A) = \phi_A.$$

Obviously f as defined above is one-one onto. Hence, $P(X) \sim C$.

EXAMPLE 9. *Prove that* $]0,1] \sim]0,1[$.

SOLUTION. Denote the points of]0, 1] by x and of]0, 1[by y. Now define a correspondence

$$y = \frac{3}{2} - x \text{ for } \frac{1}{2} < x \le 1\text{; then } \frac{1}{2} \le y < 1,$$

$$y = \frac{3}{4} - x \text{ for } \frac{1}{4} < x \le \frac{1}{2}\text{; then } \frac{1}{4} \le y < \frac{1}{2},$$

$$y = \frac{3}{8} - x \text{ for } \frac{1}{8} < x \le \frac{1}{4}\text{; then } \frac{1}{8} \le y < \frac{1}{4},$$

and so on.

From the above correspondence, we see that for every $x \in]0,1]$, there corresponds one and only one y of]0, 1[. Hence by definition $]0,1[\sim]0,1[$.

EXAMPLE 10. *Show that for every real number x, the real number in the semi-open interval* $[x, x+1)$ *form an uncountable set.*

SOLUTION. Let x be any real number. Define a function

$$f : [x, x+1) \to [0,1) \text{ given by } f(y) = y - x.$$

Then f is well defined, for obviously,

$$f(x) = x - x = 0,\ f(x+1) = x + 1 - x = 1.$$

Again $\quad f(y_1) = f(y_2) \;\Rightarrow\; y_1 - x = y_2 - x \;\Rightarrow\; y_1 = y_2 \;\Rightarrow\; f$

Also f is a continuous map which implies that f is an onto map.

Hence $\quad [x, x+1) \sim [0,1)$

$\therefore \quad \text{card}\,[x, x+1) \sim \text{card}\,[0,1)$.

Now, since we know that the set of all real numbers in the set semi-open interval [0, 1) is uncountable and hence the set of all real numbers in [x, x + 1) which is cardinally equivalent to [0, 1) is uncountable.

EXAMPLE 11. *If* α *is any transfinite cardinal number, then* $a \le \alpha$.

SOLUTION. Let A be an infinite arbitrary set s.t. card $A = \alpha$.

Now, A is infinite set $\Rightarrow \exists$ an enumerable subset B of A $\Rightarrow$ card $B = a$

$$B \subseteq A \;\Rightarrow\; \text{card } B \le \text{card } A \;\Rightarrow\; a \le \alpha.$$

EXAMPLE 12. *Two enumerable sets are equivalent.*

$$P \sim N, Q \sim N \text{ or } P \sim N, N \sim Q \;\Rightarrow\; P \sim Q,$$

Since the relation $P \sim Q$ *in the family of sets in an equivalent relation.*

EXAMPLE 13. *Show that the set of all transcendental numbers in any interval is non-enumerable.*

(MEERUT–1995, 2003, PUNJAB–2002)

SOLUTION. We know that the set of all algebraic numbers and transcendental numbers is the set of all real numbers which is known to be uncountable. Also we know that the set of algebraic numbers in an interval is enumerable.

But we have already proved that if an enumerable set is removed from a non-enumerable set, the remaining set is non-enumerable. Therefore the complement of the set of all algebraic numbers in any interval relative to the set of all real numbers in that interval is uncountable. But this is the set of all transcendental numbers.

EXAMPLE 14. *Show that the interval* (0, 1) *is equivalent to the set R of all real numbers and hence show that* card $(0,1) =$ card R.

SOLUTION. Define a function $f:(0,1)\to R$ s.t.

$$f(x)=\begin{cases}\dfrac{2x-1}{x}, x\in\left(0,\dfrac{1}{2}\right)\\ \dfrac{2x-1}{1-x}, x\in\left[\dfrac{1}{2},1\right)\end{cases}.$$

Show that this function is one-one and onto implying that $(0,1)\sim R$ and hence

$$\text{card } (0,1) = \text{card } R = c.$$

Also since $(0,1)$ is uncountable, the set R is also uncountable.

Above property supports out idea of defining the same cardinal numbers c of the two sets $(0,1)$ and R. The c is called the Cardinal number of continuum.

EXAMPLE 15. *Prove that* $a<c<f$ *where* a, c *and f denote the cardinal numbers of set of all natural numbers, and real numbers and set of all real valued functions defined over* $[0,1]$ *respectively.*

(GARHWAL–1997, MEERUT–1983, 89)

SOLUTION. We have already proved that $a<c$. Now it remains to prove that $c<f$. Let F be set of all real valued functions defined over [0, 1].

Now consider the mapping $f_k:[0,1]\to R$ defined as $f_k(x)=k, \forall x\in[0,1]$ and k being a real number in [0, 1].

All these functions are real valued and so that set $F^* = \{f_k : 0\le k\le 1\}$ is a proper subset of the set F.

We can set up a one-to-one correspondence between $[0,1]$ and F^* (s.t. $F^*\subset F$). Hence

$$\text{card}[0,1] = \text{card}\, F^* < \text{card}\, F \text{ or } c<f.$$

EXAMPLE 16. *Show that a countable set is a Borel set.*

SOLUTION. Let $A=\{a_1, a_2, a_3, \ldots\}$ be a countable set. Now note that

$$\{x : x = a_r\} = \bigcap_{n=1}^{\infty}\left\{x : a_r \le a_r + \frac{1}{n}\right\}$$

and
$$A = \bigcup_{r\in N}\{a_r\}$$

$\Rightarrow$ A is obtained by the formation of countable union and intersection of closed and open sets and hence A is a Borel set.

EXERCISE 1.3

1. (a) Define an enumerable set. Show that the set of real numbers can not be enumerable, although the set of rationales is enumerable. (MADRAS–2008)
   (b) If $f : A \to B$ and the range of f is uncountable, prove that domain of f is also uncountable.
2. Define cardinal number of a set. Show that $n < 2^n$ for any cardinal number n. (RAJASTHAN–2006, INDORE–2011)
3. Prove the set of all real numbers in the closed interval [0, 1] is uncountable.
4. Prove that if A and B are enumerable then $A \times B$ is also enumerable.
5. Prove that $\alpha + \alpha = \alpha$ for any infinite cardinal number α.
6. Find the cardinal number of the set $\{x\}$ of those numbers in the interval $[0,1]$ whose ternary expansion does not have the digit 1.
7. Prove that
   (i) $[0,1] \sim]0,1[$, (ii) $[0,1] \sim [2,5]$,
   (iii) $[0,1[\sim]0,1[$.
8. If $\{E_n\}$ be a sequence of countable sets and $S = \bigcup_{n=1}^{\infty} E_n$, then prove that S is countable. (DELHI–2008, KANPUR–2003)
9. Let α and β be any two cardinal numbers such that $\alpha \le \beta$ and $\beta \le \alpha$, then prove that $\alpha = \beta$.
10. (i) Prove that the set of all numbers in any interval can not be enumerable.
    (ii) Show that the set of all characteristic functions on R is uncountable.
11. Set of real numbers is __________. (KANPUR–2003)
12. Every isolated set of point is __________. (KANPUR–2002)
13. By an example show that cancellation law does not hold in case of cardinal numbers ?
14. Set of integers $\{0,1,2,...\}$ is uncountable. (KANPUR–2003)
15. Show that the family of all finite subsets of the natural numbers is countably infinite.
16. Prove that
    (i) $a + c = c$ (ii) $a + a = a$
    (iii) $c + c = c$ (iv) $a.c = c$
    (v) $a \,.\, a = a$
17. State and prove Schroder-Bernsteing theorem.
18. Prove that the set of complex numbers is uncountable.
19. Exhibit a $1-1$ correspondence between the points of the closed interval $[0,1]$ of R and the points of the half closed interval $[0,1]$ of R.
20. Show that the set of all polynomial functions with integer (rational) coefficients is countable (or say has the cardinal a)
21. Prove that card $P(A) = 2^{\text{card } A}$, where A is any finite set. (GARHWAL–1992)
22. Using the mapping $f : N \times N \to N$ given by $f(x,y) = 2^x(x^y + 1) - 1$, show that set $N \times N$ is countable equivalent.
23. Show that the set of points in the closed interval $[2,4]$ and in the open interval $(1,2)$ are cardinally equivalent. (MEERUT–2004)
24. If a finite set of elements is added to an enumerable set, the executing set is also enumerable.
25. If B is a countable subset of an uncountable set A. Then $A - B$ is __________. (MEERUT–2001)
26. Show that the set of all sequence whose elements are the digit 0 and 1 is uncountable. (KANPUR–2001, PUNJAB–2003)
27. If $\{E_n\}$ be a sequence of countable sets and $S = \bigcup_{n=1}^{\infty} E_n$ then prove that S is countable.

1.22 INTRODUCTION TO MEASURE AND INTEGRATION

We know that the length of an interval I, written as $l(I)$ is defined to be the difference of the end points of the interval I. Therefore, irrespective of whether an interval I with a and b as its end points is closed, open, open-closed or closed-open, the length $l(I)$ is $b - a$ where $a < b$. The length is an example of set function, *i.e.*, a function which associates an extended real number to each set in some collection of sets.

REMARK

- If $a = b$, the interval $[a, b]$ degenerates to a point and has length zero while an infinite interval has length infinity.

1.22.1 LENGTHS OF OPEN AND CLOSED SETS

Definition. *Let G be any open subset of $[a, b]$, then length of G denoted by $l(G)$ or $|G|$ is defined as the sum of the lengths of the intervals of this family, i.e.,*

$$|G| = l(G) = \sum_n |I_n|,$$

where $G = \cup I_n$, a countable family of disjoint open intervals.

Definition. *If F is any closed subset of $[a, b]$, then the length $|F|$ of the closed set F is defined as*

$$|F| = |G| - |G - F| = (b - a) - |F'|$$

where G is an open subset of $[a, b]$ such that $F \subseteq G$.

We have extended the concept of lengths to open and closed sets. Since, the classes of these sets are too restricted, we would like to extend the concept of length to a wider class of sets in R. For this, we imagine a function m which assigns to each set E in R, a non-negative extended real number, $m(E)$ (called the measure of E) satisfying the following properties.

(1) $m(I) = l(I)$ for an interval I.

(2) $m(E)$ is defined for all sets $E \subset R$.

(3) If $<E_i>$ is a sequence of disjoint sets, then

$$m\left(\bigcup_{i=1}^{\infty} E_i\right) = \sum_{i=1}^{\infty} m(E_i) \qquad \text{[Countable additive property]}$$

(4) $m(E + y) = m(E)$, for any fixed number y. [Translation Invariant Property]

1.23 OUTER LEBESGUE MEASURE OF A SET

The outer Lebesgue measure of a set $A \subset R$ is defined as follows :

$$m^*(A) = \begin{cases} 0, & \text{if } A = \phi \\ \inf.\{y : y = \Sigma |I_i|\} \end{cases}$$

where $[I_i]$ is a countable family of open intervals such that $\cup I_i \subset A$, when $A \neq \phi$.

REMARK

- The outer Lebesgue measure of a set is also known as Lebesgue Exterior measure or outer measure and is also denoted by $m_e(A)$ or $\bar{m}(A)$.

1.23.1 SOME IMPORTANT OBSERVATIONS

(1) The outer measure $m^*(A)$ of any set A is always non-negative, *i.e.*, $m^*(A) \geq 0 \ \forall A$

(2) Infimum gives that $m^*(A)$ is the least length to cover the set A from outside.

(3) For each $\in > 0$, there exists at least one countable family of open intervals such that $A \subseteq \cup I_n$ and $m^*(A) + \in > \sum_n |I_n|$.

(4) If $A \subseteq B$, then $m^*(A) \leq m^*(B)$.

(5) For an interval I, $m^*(I)$ = length of I.

(6) If $A \subset [a, b]$, then $m^*(A) \leq (b - a)$.

(7) If A is singleton, then $m^*(A) = 0$.

(8) $m^*(A)$ is always unique for any set A.

(9) For each $\in > 0$, $\exists$ at least one open set $G \supset A$ such that $m^*(A) + \in > |G|$.

(10) The outer measure m^* is the set function from the set R to the set of all non-negative extended real numbers.

1.23.2 SOME RESULTS BASED ON OUTER MEASURE (TO BE USED DIRECTLY)

(1) The outer measure of an interval is its length.

(2) If $\{E_n\}$ be a countable collection of sets, then

$$m^*\left(\bigcup_n E_n\right) \le \sum_n m^*(E_n)$$

(3) If E is a countable set, then $m^*(E) = 0$.

(4) Any set with the outer measure different from zero is uncountable.

(5) The cantor's set C is uncountable with outer measure zero.

(6) For any subset A, B of R

$$m^*(A \cup B) \le m^*(A) + m^*(B)$$

(7) The inner measure of a set A, denoted by $m_*(A)$ or $m_i(A)$ is defined by $m_*(A) = b - a - m^*(A')$.

(8) $m^*(A) \ge m_*(A)$

1.24 LEBESGUE MEASURABLE SET

Definition. *A set E is said to be Lebesgue measurable or briefly measurable if for each set A, we have* (MEERUT–2008)

$$m^*(A) = m^*(A \cap E) + m^*(A \cap E')$$

REMARKS

- A set E is said to be measurable if $m^*(E) = m_*(E)$
- The measurable sets are those (bounded or unbounded) which split every set (measurable or not) into two pieces that are additve with respect to outer measure.

1.24.1 PROPERTIES OF MEASURABLE SETS

THEOREM 1. ***A linear set A of outer measure zero is Lebesgue measurable.***

PROOF. By definition, we have

$$m^*(A) \ge m_*(A) \qquad \ldots(1)$$

It is given that $m^*(A) = 0$.

Therefore, (1) gives

$$m_*(A) \le 0$$

but $\quad m_*(A) \ge 0$

$\therefore \quad m_*(A) = 0$

Hence, $\quad m^*(A) = m_*(A) = 0$

i.e., A is measurable.

REMARKS

- Any subset A, whose outer measure is zero is also measurable.
- The necessary and sufficient condition for a set E to be measurable with measure zero is that $m^*(E) = 0$.

THEOREM 2. ***Any set A is measurable iff an open set G containing A and a closed set H contained in A can be so determined that***

$$|G| - |H| < \in$$

where $\in$ is an arbitrary positive real number.

PROOF. We can find an open set $G \supseteq A$ and a closed set $H \subset A$ such that

$$|G| < m^*(A) + \in/2 \quad \ldots(1)$$

and $$m_*(A) < |H| + \in/2 \quad \ldots(2)$$

Since A is measurable, therefore

$$m^*(A) = m_*(A) \quad \ldots(3)$$

From (1), (2) and (3), we conclude that

$$|G| - \in/2 < |H| + \in/2 \quad \Rightarrow \quad |G| - |H| < \in$$

Conversely, let $|G| - |H| < \in$.

By definition of measure of A, we have

$$m^*(A) \le |G| \text{ and } m_*(A) \ge |H| \Rightarrow -m_*(A) \ge -|H|$$

On adding, we get

$$m^*(A) - m_*(A) \le |G| - |H| < \in \quad [\because |G| - |H| < \in]$$

$$\therefore \; m^*(A) - m_*(A) < \in$$

but $\in$ is arbitrary, hence, $m^*(A) = m_*(A)$

Hence, A is measurable.

THEOREM 3. ***The union of two measurable sets is measurable.***

(MEERUT–2000, 01, 04; ROHTAK–2002, 04; HIMACHAL–2002; KANPUR–2002)

PROOF. Let E_1 and E_2 be two measurable sets. Then, for any set A, we have

$$m^*(A) = m^*(A \cap E_1) + m^*(A \cap E_1')$$

$$= m^*(A \cap E_1) + m^*[(A \cap E_1') \cap E_2] + m^*[(A \cap E_1') \cap E_2']$$

$$= m^*(A \cap E_1) + m^*[(A \cap E_2) \cap E_1') + m^*[(A \cap E_1') \cap E_2')]$$

$$= m^*(A \cap E_1) + m^*(A \cap E_2 \cap E_1') + m^*(A \cap (E_1 \cup E_2)']$$

$$\ge m^*(A \cap (E_1 \cup E_2)) + m^*(A \cap (E_1 \cup E_2)')$$

which shows that $E_1 \cup E_2$ is measurable.

REMARKS

- The union of finite number of measurable sets is also measurable.
- The arbitrary union of measurable sets is also measurable.
- The intersection and difference of two measurable sets are measurable.
- The symmetric difference of two measurable sets is measurable.

THEOREM 4. **(First Fundamental Theorem).** ***If $E_1, E_2, \ldots$ are measurable pairwise disjoint sets and $E = E_1 \cup E_2 \cup \ldots$, then E is measurable.***

and $$m(E) = \sum_{k=1}^{\infty} m(E_k)$$ (MEERUT–2003, 11, 12, HIMACHAL–2001, 04)

PROOF. We know that if $E_i \cap E_j = \phi, \ \forall \ i \neq j$ (Pairwise disjoint), then

$$m^*\left(\bigcup_k E_k\right) \leq \sum_k m^*(E_k)$$

or $$m\left(\bigcup_{k=1}^{\infty} E_k\right) \leq \sum_{k=1}^{\infty} m(E_k) \quad \ldots(1)$$

We have

$$\bigcup_{k=1}^{\infty} E_k \supseteq \bigcup_{k=1}^{n} (E_k), \quad \forall \ n$$

$$\Rightarrow \quad m\left(\bigcup_{k=1}^{\infty} E_k\right) \geq m\left(\bigcup_{k=1}^{n} E_k\right) \quad \ldots(2)$$

Further, we have, if $E_1 \cap E_2 = \phi$

then $m(E_1 \cup E_2) = m(E_1) + m(E_2)$

Extending this result, we get

$$m\left(\bigcup_{k=1}^{n} E_k\right) = \sum_{k=1}^{n} m(E_k)$$

Using (2), we have

$$m\left(\bigcup_{k=1}^{\infty} E_k\right) \geq \sum_{k=1}^{n} m(E_k)$$

Letting $n \to \infty$, we get

$$m\left(\bigcup_{k=1}^{\infty} E_k\right) \geq \sum_{k=1}^{\infty} m(E_k) \quad \ldots(3)$$

Finally, (1) and (3) gives

$$m\left(\bigcup_{k=1}^{\infty} E_k\right) = \sum_{k=1}^{\infty} m(E_k)$$

REMARK

- If $E_1, E_2, \ldots$ are measurable sets, then the set $\bigcap_{k=1}^{\infty} E_k$ is measurable (Second fundamental theorem).

1.25 SOME MORE DEFINITIONS ON MEASURABILITY

(1) Boolean Ring (Rings of sets) : A non-empty class of sets which is closed under the formation of union and difference is called Boolean ring.

i.e., a non-empty class of ring β is said to be boolean ring if

$$A, B \in \beta \ \Rightarrow \ A - B \in \beta \quad \text{and} \quad A \cup B \in \beta$$

REMARKS

- A boolean ring is closed under the formation of symmetric differences.
- The class of sets (boolean ring) is called finitely additive class.

(2) Boolean Algebra (or algebra of sets or a field) : A non-empty family R of sets is called an algebra of sets or boolean algebra if it is closed under the formation of union and compliment, *i.e.*,

(1) $A \in R \ \Rightarrow \ A' \in R$ and (2) $A, B \in R \Rightarrow A \cup B \in R$

(3) σ-Ring : A boolean ring R is called a σ-ring if it is closed under the formation of countable unions, *i.e.*, for all $A_i \in R \Rightarrow UA_i \in R$

REMARKS

- Every σ-ring is a *B*-ring.
- A non-empty class of sets is called a σ-ring if it is closed under the formation of countable unions and differences.

(4) σ-Algebra of Sets : A non-empty class A of sets is called σ-algebra if it is closed under the formation of compliments and countable unions.

REMARKS

- Every σ-algebra is a σ-ring but converse is not true.
- This class of sets is also known as "Completely Additive Class".

1.26 MEASURABLE FUNCTIONS

Definition. *An extended real valued function f defined over a measurable set E is said to be measurable (Lebesgue) if the set* $E[f > a] = [x \in E : f(x) > a]$ *is measurable for all* $a \in]-\infty, \infty[$.

(DELHI–2011, CALCUTTA–2014, MEERUT–2008)

1.26.1 EQUIVALENT DEFINITIONS OF MEASURABLE FUNCTIONS

Let f be an extended real valued function defined on a measurable set E (of finite or infinite measure). Then, the following statements are equivalent :

(1) $E[f > a]$ is measurable for all $a \in R$.

(2) $E[f \geq a]$ is measurable for all $a \in R$.

(3) $E[f < a]$ is measurable for all $a \in R$.

(4) $E[f \leq a]$ is measurable for all $a \in R$.

REMARK

- An extended real valued function f defined over a measurable set E is measurable if and only if one of the statements (1), (2), (3) or (4) defined above, holds.

1.27 PROPERTIES OF MEASURABLE FUNCTIONS

THEOREM 1. ***Let f be a function defined on a measurable set E. Then f is measurable if and only if for any open set G in R, $f^{-1}(G)$ is a measurable set.***

PROOF. Let us first suppose that f is measurable. Since G is an open set in R. Then G can be written as a countable union of disjoint open intervals such that

$$G = \bigcup_n I_n ,$$

where $$I_n =]a_n, b_n[.$$

Therefore,

$$f^{-1}(G) = \bigcup_n [x \in E : f(x) \in I_n]$$

$$= \bigcup_n [E(f > a_n) \cap E(f < b_n)]$$

$$= \text{Finite union of measurable sets}$$

$\Rightarrow$ $f^{-1}(G)$ is measurable.

Conversely, let us assume that $f^{-1}(G)$ is a measurable set, where G is any arbitrary open set in R. Particularly, let us take

$$G =]\alpha, \infty[, \quad \alpha \in R$$

But, $\qquad f^{-1}(G) = E\,(f > \alpha)$ for $G =]\,\alpha, \infty\,[$

Hence, f is a measurable function.

THEOREM 2. ***A continuous function defined over a measurable set E is measurable.***

(MEERUT–2004, 05, 05BP; ROHTAK–2002)

PROOF. Define a set $B = E(f \geq a)$.

Let us first suppose that B is closed. For this, we shall prove that $D(B) \subset B$, where $D(B)$ is the derived set of B.

Let $x_0 \in D(B)$ be any arbitrary point $\Rightarrow$ x_0 is the limit point of B.

Then, by definition, for every nbd. G of x_0, we have $(G - \{x_0\}) \cap B \neq \phi$

Let $x \in [G - \{x_0\}] \cap B \Rightarrow x \in G, \; x \neq x_0$ and $x \in B$

$\Rightarrow \qquad f(x) \geq a \qquad [\because \; x \in B]$

Therefore, for all $x \in G$ such that $x \neq x_0 \;\Rightarrow\; f(x) \geq a$.

Now, since f is continuous, therefore $f(x_0) \geq a \;\Rightarrow\; x_0 \in B$.

i.e., $\qquad x_0 \in D(B) \Rightarrow x_0 \in B$

$\Rightarrow \qquad D(B) \subset B$

$\Rightarrow$ B is closed.

Since, we know that every closed set is measurable. Hence $B = E\,(f \geq a)$ is measurable.

1.27.1 SOME IMPORTANT OBSERVATIONS

(1) A constant function over a measurable set E is measurable over E.

(2) If f is measurable over a measurable set E, then f is also measurable over any measurable subset A of E.

(3) If $f(x)$ and $g(x)$ are measurable functions, then $\max\{f(x), g(x)\}$ and $\min\{f(x), g(x)\}$ are measurable.

(4) If f is a measurable function defined over a measurable set E, then cf, $-f$, $f + c$, $|f|$, f^2, $\frac{1}{f}$ $(f \neq 0)$ are also measurable functions.

(5) If f and g both are measurable defined over a measurable set E, then $f + g$, $f - g$, fg and $\frac{f}{g}$ $(g \neq 0$ on $E)$ are measurable functions over E.

(6) Let $< f_n >$ be a sequence of measurable functions defined over a measurable set E, then $\sup < f_1, f_2, \ldots\ldots, f_n, \ldots\ldots >$, $\inf < f_1, f_2, \ldots\ldots, f_n, \ldots\ldots >$, $\overline{\lim}\, f_n$, $\underline{\lim}\, f_n$ and $\lim f_n$ are measurable.

(7) Let f and g be the measurable functions over a measurable set E, then $f \cup g$ and $f \cap g$ are measurable functions.

(8) Every function defined on a set of measure zero is measurable.

(9) A step function is a measurable function.

(10) The inverse image of any measurable functions is measurable.

(11) If f is measurable, then any positive integral power of f is also measurable.

(12) If f is a real valued measurable functions defined over a measurable set E. Also, if g is a function defined and continuous on the range of f, then $g \circ f$ is a measurable function on E.

(13) If f is a measurable real valued function, then $|f|^p, p > 0$ is measurable.

(14) The set of all measurable function is an algebra.

1.28 CONCEPTS OF INTEGRABILITY

Generally, the process of integration is defined as the inverse of the differentiation. A function F is called an integral of a function f if $F'(x) = f(x)$ on some domain D of f. In view of this definition, we may define that "definite integration is a process of summation".

The German mathematician G.F.B. Riemann defined the process of integration on certain arithmetical concepts free from dependence on geometrical concepts.

1.28.1 RIEMANN INTEGRAL

Let f be a bounded real valued function defined on a bounded and closed interval $[a, b]$ and $P = \{a = x_0, x_1, ..., x_n = b\}$ be any partition of $[a, b]$. Also, let m_r and M_r denoted the infimum and supremum of the function f on the subinterval $[x_{r-1}, x_r]$ respectively, then the two sums

$$L(P, f) = \sum_{r=1}^{n} m_r \, \delta x_r \quad \text{and} \quad U(P, f) = \sum_{r=1}^{n} M_r \, \delta x_r$$

(where $\delta x_r = x_r - x_{r-1}$) are respectively called the lower Riemann sum and upper Riemann sum of f on $[a, b]$ with respect to partition P.

Also,
$$\begin{aligned} U(P, f) - L(P, f) &= \sum_{r=1}^{n} [M_r - m_r] \, \delta x_r \\ &= \sum_{r=1}^{n} \omega_r \, \delta x_r, \text{ where } \omega_r = M_r - m_r \end{aligned}$$

Then, sum $\sum_{r=1}^{n} \omega_r \, \delta x_r$ is called the oscillatory sum for the function f with respect to partition P on $[a, b]$.

The infimum of the set of the upper sums is called the upper integral of f over $[a, b]$ and is denoted by

$$U = \overline{\int_a^b} f(x)\, dx$$

Also, the supremum of the set of the lower sums is called the lower integral of f over $[a, b]$ and is denoted by

$$L = \underline{\int_a^b} f(x)\, dx$$

Definition. *A bounded function f is said to be Riemann integrable or simply integrable over $[a, b]$, if its upper and lower integrals are equal and their common value being called Riemann integral or simply the integral denoted by* $\int_a^b f(x)\, dx$.

1.28.2 SOME IMPORTANT OBSERVATIONS

(1) Let f be a bounded function defined on $[a, b]$ and let m and M be the infimum and supremum of $f(x)$ in $[a, b]$ respectively, then for every partition P of $[a, b]$, we have

$$m(b-a) \le L(P, f) \le U(P, f) \le M(b-a).$$

(2) If f_1 and f_2 are two real valued bounded functions defined on $[a, b]$, then

(a) $L(P, f_1 + f_2) \ge L(P, f_1) + L(P, f_2)$.

(b) $U(P, f_1 + f_2) \le U(P, f_1) + U(P, f_2)$.

(c) $L(P_1, f) \le L(P_2, f)$; $U(P_2, f) \le U(P_1, f)$ for $P_1, P_2 \in P(a, b)$.

(3) A necessary and sufficient condition for R-integrability of a bounded function $f : [a, b] \to R$ is that for every $\in > 0$, there exists a partition P of $[a, b]$ such that

$$U(P, f) - L(P, f) < \in \quad \forall \|P\| < \delta$$

(4) Every continuous function is Riemann integrable.

(5) Every monotonic function is Riemann integrable.

(6) A bounded function f is Riemann integrable in $[a, b]$ if the set of its points of discontinuity is finite.

(7) If f is Riemann integrable, then $|f|$ is also Riemann integrable.

(8) If f and g be two integrable functions, then $f + g, f - g, fg$ and $\frac{f}{g}$ $(g \neq 0)$ are Riemann integrable.

THEOREM 1. **(Fundamental Theorem of Integral Calculus).** ***Let f be a Riemann integrable function on [a, b] and F be a differentiable primitive function on [a, b] such that $F'(x) = f(x), x \in [a, b]$, then***

$$\int_a^b f(t)dt = F(b) - F(a).$$

(MEERUT–1998, 2006, 06BP, 07; GARHWAL–2001, 04)

PROOF. Let f be a continuous function on $[a, b]$.

Also, we have

$$F'(x) = f(x) \quad \forall\, x \in [a, b].$$

Since, f is R-integrable function on $[a, b]$, then $F'(x)$ is Riemann integrable on $[a, b]$, *i.e.*, for a given positive number $\in$, there exists a partition P of $[a, b]$ such that

$$\left| \sum_{r=1}^{n} F'(t_r)(x_r - x_{r-1}) - \int_a^b F'(x)\,dx \right| < \in \qquad t_r \in [x_{r-1}, x_r] \qquad \ldots(1)$$

By Lagrange's mean value theorem of differential calculus, we find that, there exists $t_r \in (x_{r-1} - x_r)$ such that,

$$F(x_r) - F(x_{r-1}) = (x_r - x_{r-1})F'(t_r)$$

$$\Rightarrow \sum_{r=1}^{n} \left[(x_r - x_{r-1})F'(t_r)\right] = \sum_{r=1}^{n} \left[F(x_r) - F(x_{r-1})\right] = F(b) - F(a)$$

Putting this value in (1), we get

$$\left| F(b) - F(a) - \int_a^b F'(x)\,dx \right| < \in$$

which gives

$$F(b) - F(a) = \int_a^b F'(x)\,dx = \int_a^b f(x)\,dx \qquad [\because F'(x) = f(x)]$$

Hence, $\int_a^b f(x)\,dx = F(b) - F(a).$

THEOREM 2. ***Let f be a Riemann integrable function on [a, b], then the integral function F of f given by***

$$F(x) = \int_a^b f(t)\,dt, \qquad a \le x \le b$$

is continuous on [a, b].

PROOF. Let f is Riemann integrable function on $[a, b]$. Then obviously, it is bounded on $[a, b]$, therefore, there exists a positive number M such that

$$|f(t)| \le M \qquad \forall\, t \in [a, b]$$

Let $x_1, x_2 \in [a, b]$ such that $x_1 < x_2$. Then, we have

$$|F(x_2) - F(x_1)| = \left| \int_a^{x_2} f(t)\,dt - \int_a^{x_1} f(t)\,dt \right|$$

$$= \left| \int_a^{x_2} f(t)\,dt + \int_{x_1}^{a} f(t)\,dt \right| = \left| \int_{x_1}^{x_2} f(t)\,dt \right|$$

$$\le M \left| \int_{x_1}^{x_2} dt \right| = M\,|x_2 - x_1|.$$

Let $|x_2 - x_1| < \in /M$, for a given positive number $\in$. Then, we have

$$|F(x_2) - F(x_1)| < M.\frac{\in}{M}$$

$\Rightarrow \qquad |F(x_2) - F(x_1)| < \in$, whenever $|x_2 - x_1| < \delta \;\; \forall\, x_1, x_2 \in [a, b]$.

Hence, F is uniformly continuous on $[a, b]$ and hence it is continuous on $[a, b]$.

1.29 LEBESGUE INTEGRAL OF A BOUNDED FUNCTION

Definition 1. *Let f be a bounded function defined on [a, b]. Then lower Lebesgue integral and upper Lebesgue integral can be defined as follows :*

$$L\underline{\int_a^b} f(x)dx = \text{supremum } \{L[f : Q] : Q \text{ is a measurable partition of } [a, b]\}$$

and $$L\overline{\int_a^b} f(x)\,dx = \text{infimum } \{U[f : P] : P \text{ is a measurable partition of } [a, b]\}$$

Here, $L\underline{\int_a^b} f(x)dx$ *and* $L\overline{\int_a^b} f(x)\,dx$ *are respectively called the lower Lebesgue integral and upper Lebesgue integral.*

Definition 2. *Let f be a bounded function defined on interval [a, b]. Then f is said to be Lebesgue integrable if and only if*

$$L\underline{\int_a^b} f(x)dx = L\overline{\int_a^b} f(x)\,dx$$

and their common value is called L-integral of f on [a, b] and is denoted by $\int_a^b f$.

REMARK

- The class of all bounded function f which are Lebesgue integrable on $[a, b]$ is denoted by $L[a, b]$. Therefore, f belongs to $L[a, b]$ if and only if f is Lebesgue integrable on $[a, b]$. Here, the numbers a and b called the lower and upper limits of integration respectively.

1.29.1 IMPORTANT CONCEPTS

(1) The concept of integrability of a function over an interval is introduced here is subject to two limitations :

(a) the function is bounded.

(b) the interval of integration is finite so that neither of the end points is infinite.

(2) Every bounded function is not necessarily integrable.

(3) The statement that $\int_a^b f$ exists means that the function f is bounded and integrable over $[a, b]$.

(4) Every simple function f is Lebesgue integrable and its Lebesgue integral is nothing but the same as the elementary integral of f.

(5) Every bounded measurable function defined on a set whose measure is finite is Lebesgue integrable.

1.29.2 RELATION BETWEEN RIEMANN AND LEBESGUE INTEGRALS

THEOREM 1. ***If f is a bounded function defined on [a, b] and f is R-integrable on [a, b] then f is also L-integrable on [a, b] and***

$$L\int_a^b f = R\int_a^b f$$

(DELHI–2011, MEERUT–2004; KANPUR–1999; HIMACHAL–2000, 02, 03; ROHTAK–2001)

PROOF. Let σ_1 and σ_2 be two Riemann subdivision of $[a, b]$ such that

$$R\underline{\int_a^b} f = L[f:\sigma_1] \quad \text{and} \quad R\overline{\int_a^b} f = U[f:\sigma_2]$$

Now, σ_1 and σ_2 will also give rise to measurable partitions P_1 and P_2 of $[a, b]$. Therefore,

$$L\overline{\int_a^b} f \le U[f:P_1] \quad \text{and} \quad L\underline{\int_a^b} f \ge L[f:P_2]$$

$$\therefore \quad L[f:P_2] \le L\underline{\int_a^b} f \le L\overline{\int_a^b} f \le U[f:P_1]$$

$$\Rightarrow \quad L[f:\sigma_1] \le L\underline{\int_a^b} f \le L\overline{\int_a^b} f \le U[f:\sigma_2]$$

$$\Rightarrow \quad R\underline{\int_a^b} f \le L\underline{\int_a^b} f \le L\overline{\int_a^b} f \le R\overline{\int_a^b} f \quad \ldots(1)$$

Since f is given to be R-integrable, therefore,

$$R\underline{\int_a^b} f = R\overline{\int_a^b} f = R\int_a^b f$$

Hence, (1) reduces to

$$R\int_a^b f = L\underline{\int_a^b} f = L\overline{\int_a^b} f = R\int_a^b f$$

$\Rightarrow$ f is also L-integrable over $[a, b]$ and $L\int_a^b f = R\int_a^b f$

REMARKS

- If both the R- and L-integrals exist then their values are equal. In this case, we may denote the integral by the symbols $\int_a^b f$ or $\int_a^b f(x)\,dx$.
- Converse of the above theorem is not necessarily true.

THEOREM 2. **(First Mean Value Theorem).** ***Let f be a bounded measurable real valued function such that*** $a \le f(x) \le b$ ***on a measurable set E [p, q]*** $\subset$ ***R. Then***

$$a.m(E) \le \int_E f(x)\,dx \le b.m(E)$$

(KANPUR-2001; ROHTAK-2002; MEERUT–2011, 14; DELHI–2010; RAJASTHAN–2013; AMRITSAR–2013)

PROOF. For any $m \in N$, we have

$$a \le f(x) \le b \Rightarrow \left(a - \frac{1}{m}\right) < f(x) < \left(b + \frac{1}{m}\right), \quad \forall\, x \in E$$

Let us take $\alpha = a - \frac{1}{m}$ and $\beta = b + \frac{1}{m}$

$\therefore \alpha < f(x) < \beta, \quad \forall\, x \in E$

Now, divide the interval $[\alpha, \beta]$ by means of points

$$\alpha = y_0 < y_1 < y_2 < ... < y_n = \beta$$

Define $E_0 = \{ x \in E : y_0 < f(x) < y_1\}$

and $E_r = \{ x \in E : y_r < f(x) < y_{r+1}\}, \quad r = 1, 2, ..., n-1$

Clearly $E = \bigcup_{r=0}^{n-1} E_r$ and $E_r \cap E_s = \phi$ for $r \neq s$

Further, since f is measurable over E and hence measurability of the sets E_r is implied and

$$m(E) = \sum_{r=0}^{n-1} m\,(E_r) \qquad ...(1)$$

Then, we have a measurable partition

$$P = \{E_0, E_1, ..., E_{n-1}\} \text{ of } E$$

Also, $\alpha \leq y_r \leq \beta \Rightarrow \alpha\,.\,m\,(E_r) \leq y_r\,.\,m\,(E_r) \leq \beta.m\,(E_r)$

$$\Rightarrow \sum_{r=0}^{n-1} \alpha.m(E_r) \leq \sum_{r=0}^{n-1} y_r\,.m(E_r) \leq \sum_{r=0}^{n-1} \beta.m(E_r)$$

$$\Rightarrow \quad \alpha\,.\,m(E) \leq \sum_{r=0}^{n-1} y_r\,.\,m(E_r) \leq \beta.m(E) \qquad ...(2)$$

Let Max $(y_{r+1} - y_r) \to 0$, we get $y_{r+1} \to y_r$ and hence for this partition

$$U[f : P] = \sum_{r=0}^{n-1} y_{r+1}\,.m(E_r) \to \sum_{r=0}^{n-1} y_r\,.m(E_r) = L[f : P]$$

$$\therefore \quad \int_E f = \Sigma\, y_r\,.\,m(E_r)$$

Now, (2) implies

$$\alpha.m(E) \leq \int_E f(x)\,dx \leq \beta.m(E)$$

$$\Rightarrow \left(\alpha - \frac{1}{m}\right).m(E) \leq \int_E f(x)\,dx \leq \left(b + \frac{1}{m}\right).m(E)$$

Making $m \to \infty$, we get

$$a.m(E) \leq \int_E f(x)\,dx \leq b.m(E)$$

1.30 LEBESGUE INTEGRAL FOR BOUNDED FUNCTIONS OVER A SUBSET OF REAL NUMBERS

Let $E \subset [a, b]$ be a measurable subset and f be a bounded function in $L[a, b]$. Then $\int_E f$ is defined as follows :

$$\int_E f = \int_a^b f\,\phi_E$$

where, ϕ_E is the characteristic function of E.

THEOREM 1. ***If A and B are disjoint measurable subsets of [a, b] and if f is a bounded L-integrable function on [a, b], then***

$$\int_{A \cup B} f = \int_A f + \int_B f$$

(MEERUT–2001, ROHTAK–2002, HIMACHAL–2002)

PROOF. Since A and B are disjoint measurable subsets of $[a, b]$, *i.e.*, $A \cap B = \phi$,

Therefore, $\phi_{A \cup B} = \phi_A + \phi_B$.

$$\Rightarrow \quad f\phi_{A \cup B} = f\phi_A + f\phi_B$$

Now, using the definition, we get

$$\int_{A \cup B} f = \int_a^b f\,\phi_{A \cup B} = \int_a^b (f\phi_A + f\phi_B)$$

$$= \int_a^b f\phi_A + \int_a^b f\phi_B = \int_A f + \int_B f$$

REMARK

- The first mean value theorem, when set E is not necessarily an interval, may be any subset of real numbers is defined as follows :

  "If f is bounded real valued measurable function defined on a measurable set E of finite measure such that

  $$a \le f(x) \le b, \text{ then } a\,.\,m(E) \le \int_E f \le b\,.\,m(E).$$

1.31 GENERAL LEBESGUE INTEGRAL

Now, we extend the definition of the Lebesgue integral to include the most general possible measurable functions that can take both positive and negative values. A measurable function can be written as the difference of two non-negative measurable functions.

Let f be a real valued function defined on a set E. Its positive and negative parts are defined as

$$f^+ = \max\{f, 0\}$$

and

$$f^- = \max\{-f, 0\}$$

such that f is measurable if and only if both f^+ and f^- are measurable.

Here, we have

$$f = f^+ - f^-$$

and

$$|f| = f^+ + f^-$$

Definition. *A measurable function f is said to be integrable over E if f^+ and f^- both are integrable over E. Then, we can define*

$$\int_E f = \int_E f^+ - \int_E f^- \qquad \text{(MEERUT–2008)}$$

1.31.1 IMPORTANT FACTS

(1) If f is a bounded function, Lebesgue integrable on a measurable subset of $[a, b]$, then $|f|$ is also L-integrable on E and

$$\left|\int_E f\right| \le \int_E |f|$$

The above equality will occur when either $f \ge 0$ a.e. or $f \le 0$ a.e. .

(2) If f is measurable on a measurable set E, then f is Lebesgue integrable if and only if $|f|$ is Lebesgue integrable.

(3) If f be a bounded function defined on a measurable set E with $m(E) < \infty$, then

ad

$$\inf_{f \le \psi} \int_E \psi(x)\, dx = \sup_{f \ge \phi} \int_E \phi(x)\, dx$$

for all simple function ϕ and ψ if and only if f is measurable.

1.32 THE LEBESGUE INTEGRAL OF UNBOUNDED FUNCTIONS

Let $f(x)$ be an unbounded, measurable and non-negative real valued function defined on $[a, b]$. Let $n \in N$ be arbitrary. Then, we can define a function $[f(x)]_n$ on $[a, b]$ such that

$$[f(x)]_n = \begin{cases} f(x), & \text{when } f(x) \le n \\ n, & \text{when } f(x) > n \end{cases}$$

Therefore,

$$[f(x)]_n = \min \{f(x), n\}$$

REMARK

- The function $[f(x)]_n$ defined above is bounded and measurable over $[a, b]$.

Definition. *A non-negative measurable function f defined on a measurable set E is said to be integrable if* $\int_E f < \infty$.

where,

$$\int_E f = \sup_{g \le f} \int_E g(x)$$

g being a bounded measurable function such that

$$m [E (x : g(x) \ne 0] < \infty$$

THEOREM 1. ***If f, g are non-negative measurable function defined on E, then***

$$\int_E (f + g) = \int_E f + \int_E g$$

PROOF. Define two functions $h(x)$ and $k(x)$ such that

$$h(x) \le f(x) \quad \text{and} \quad k(x) \le g(x)$$

then,

$$h(x) + k(x) \le f(x) + g(x) = (f + g)(x)$$

$$\therefore \quad \int_E h + \int_E k \le \int_E (f+g)$$

On taking summation, we get

$$\int_E f + \int_E g \le \int_E (f+g) \qquad \ldots(1)$$

Now, let $F(x)$ be a bounded measurable function such that

$$F(x) \le (f+g)(x)$$

and $F(x)$ vanishes outside a set of finite measure.

Let $\quad h(x) = \min \{f(x), F(x)\}$

and $\quad k(x) = F(x) - h(x)$

Then, $\quad h(x) + k(x) = F(x) \le f(x) + g(x)$

Clearly, we have

$$h(x) \leq f(x) \text{ and } k(x) \leq g(x)$$

Also, $\int_E F = \int_E h + \int_E k \leq \int_E f + \int_E g$

On taking summation, we get

$$\int_E (f+g) \leq \int_E f + \int_E g \quad \ldots(2)$$

Using (1) and (2), we conclude that

$$\int_E (f+g) = \int_E f + \int_E g$$

THEOREM 2. ***Let f and g be non-negative measurable functions defined on a measurable set E such that f > g on E and f is integrable over E, then g is also integrable on E.***

PROOF. We can write

$$\int_E f = \int_E (f-g+g) = \int_E (f-g) + \int_E g \quad \ldots(1)$$

Since, f is given to be integrable over E, therefore,

$$\int_E f(x) < \infty \quad \ldots(2)$$

Also, $f-g$, g both are non-negative measurable functions.

$$\therefore \quad \int_E (f-g) \geq 0, \quad \int_E g \geq 0$$

and $\quad \int_E (f-g) < \infty, \quad \int_E g < \infty$

Therefore, g is integrable over E.

Also, $\quad \int_E f - g = \int_E f - \int_E g$

THEOREM 3. ***If f is a non-negative integrable function over a set E, then for given $\in > 0$, $\exists\, \delta > 0$ such that for every set $A \subset E$ with $m(A) < \infty$, $\int_A f < \in$.***

(ROHTAK–1997, 98, 2000, 01, 04; HIMACHAL–2003, 04)

PROOF. **Case I :** If f is a bounded function on E such that

$$|f(x)| \leq M, \quad \forall x \in E$$

Let us choose

$$\delta = \frac{\in}{M} > 0$$

Now, for $A \subset E$, $m(A) < \infty$

We get

$$\int_A f \leq M \,.\, m(A) < M \,.\, \delta = \frac{M \,.\in}{M} = \in$$

Case II : Let f be unbounded on E. Then, we can define

$$f_n(x) = \begin{cases} f(x), & \text{if } f(x) \leq n \\ n, & \text{if } f(x) > n \end{cases}$$

Clearly $< f_n >$ is an increasing sequence of bounded functions on E such that $f_n \to f$. Then by Monotone Convergence Theorem, for given $\in > 0$, there exists an integer $n_0 \in N$ such that

$$\int_E f_{n_0} > \int_E f - \frac{\in}{2}$$

i.e., $$\int_E f - \int_E f_{n_0} < \frac{\in}{2}$$

Since $f > f_{n_0}$, therefore

$$\int_E (f - f_0) = \int_E f - \int_E f_{n_0} < \frac{\in}{2}$$

Choosing $\delta < \frac{\in}{2n_0}$, $\forall A \subset E$, such that $m(A) < \delta$, we get

$$\Rightarrow \quad \int_A f = \int_A (f - f_{n_0}) + \int_A f_{n_0}$$

$$\leq \int_E (f - f_{n_0}) + n_0 . m(A) \qquad (\because (f_{n_0}) \leq n_0)$$

$$< \frac{\in}{2} + n_0 . \delta < \frac{\in}{2} + n_0 . \frac{\in}{2n_0} = \in$$

$$\Rightarrow \quad \int_A f < \in$$

EXERCISE 1.4

1. Let f be a bounded measurable function defined on a measurable set E such that $f(x) \geq 0$ and $\int_E f(x)\,dx = 0$, show that $f(x) = 0$, a.e. on E.
2. If $\int_A f\,dx = 0$ for every measurable subset A of a measurable set E, show that $f(x) = 0$ a.e. on E.
3. If $f, g \in L[a, b]$ and if $f(x) \leq g(x)$ a.e. on $[a, b]$, then show that $\int_a^b f \leq \int_a^b g$.
4. If f and g are bounded measurable functions defined on a set E of infinite measure, show that $\int_E (pf + q.g) = p\int_E f + q\int_E g$.
5. Show that a function which is Lebesgue integrable is not necessarily R-integrable.
6. Show that the function
$$f(x) = \frac{d}{dx}\left(x^2 \sin\frac{1}{x^2}\right) = 2x \sin\frac{1}{x^2} - \frac{2}{x}\cos\frac{1}{x}$$
is not L-integrable over $[0, 1]$.
7. Show that the integral of a nowhere zero function can be zero.
8. If f is bounded and integrable in the sense of Riemann in a closed interval $[a, b]$, then show that f is measurable and also integrable in the sense of Lebesgue and R-integrable of f over $[a, b]$ is equal to the Lebesgue integrable over $[a, b]$.
9. Let f be a function defined on the interval $[a, b]$ such that
$$f(x) = \begin{cases} 0 & x \text{ is irrational} \\ 1 & x \text{ is rational} \end{cases}$$
Is f integrable in Riemann sense? Is f integrable in the Lebesgue sense ?
10. If f be a bounded function defined on a measurable set E with $m(E) < \infty$, then show that f is Lebesgue integrable if and only if f is a measurable function.

Chapter Review: A Competitive Approach

SELECTED TERMS AND RESULTS

▶ TERMS

- **Relation :** Let A and B be two sets. Then a relation R from A to B is a subset of $A \times B$.
- **Reflexive Relation :** A relation R is said to be reflexive if $(x, x) \in R \ \forall \ x \in A$.
- **Symmetric Relation :** A relation R on a set A is said to be symmetric if $(y, x) \in R$ whenever $(x, y) \in R$.
- **Transitive Relation :** A relation R on a set A is said to be transitive if $(x, y) \in R$, $(y, z) \in R \Rightarrow (x, z) \in R$.
- **Anti-symmetric Relation :** A relation R on a non-empty set A is said to be an anti-symmetric iff $(x, y) \in R$ and $(y, x) \in R \Rightarrow x = y \ \forall \ x, y \in R$.
- **Equivalence Relation :** A relation R on a set E is said to be equivalence if it is
  (i) reflexive
  (ii) anti-symmetric and
  (iii) transitive
- **Partial Ordered Relation :** A relation R on a set E is said to be partial ordered relation if it is
  (i) reflexive
  (ii) anti-symmetric and
  (iii) transitive
- **Function :** Let A and B be two sets, then the rule or correspondence which associates each element of A to a unique element of B is called a function or mapping.
- **Range and domain of a function:** Let an element $y \in B$ be corresponded by an element $x \in A$, then y is called the image of x and is denoted by $f(x)$. The set A is called the domain and the set B is called the co-domain of the function f.
- **One-One function :** A function $f : A \to B$ is said to be one-one iff distinct elements of A have distinct images.
- **Onto function :** A function $f : A \to B$ is called an onto function if there is no element of B which is not the image of some element A, *i.e.*, every element of B appears as the image of at least one element of A.
- **Into function :** A function $f : A \to B$ is called an into function, if there is at least one element of set B which has no pre-image in the set A.
- **Even function :** A function $f : A \to B$ is said to be an even function if $f(-x) = f(x) \forall x \in A$.
- **Odd function :** A function $f : A \to B$ is said to be an odd function if $f(-x) = -f(x) \forall x \in A$.

▶ RESULTS

- Total number of subsets of a set A is equal to 2^n where n is the number of elements of A.
- The number of proper subsets of a set with n elements is $2^n - 2$.
- Union of sets is commutative, associative and idempotent.
- Difference of two sets is not commutative.
- Difference of a set with the universal set is called complementation.
- The identity and universal relations on a non-empty sets are transitive.
- The intersection of two equivalence relations on a set is equivalence relation.
- The union of two equivalence relations on a set is not necessarily an equivalence relation.
- If *gof* and *fog* both exist, they may not be equal.
- The composition of function is associative but not commutative.
- The composition of any function with the identity function is the function itself.
- The inverse of bijective function is unique.
- The inverse of bijective function is again bijective.
- Every bounded sequence has at least one limit point.
- Compact subsets of a metric spaces are closed.

- Closed subsets of compact sets are compact.
- If F is closed and K is compact then $F \cap K$ is compact.
- A set which can be put in one to one correspondence with the set of all natural numbers or with its subset, if said to be a countable set.
- Every subset of a countable set is countable.
- The union of enumerable collection of enumerable sets is also enumerable.
- The set $N \times N$ is enumerable.
- The set Q of all rational numbers is enumerable.
- The set of all real numbers in the open interval (0, 1) is not enumerable.
- The set of all real numbers is not enumerable, *i.e.*, R is not enumerable.
- The set of all irrational numbers is not enumerable.
- A real number which is not an algebraic number is called Transcendental number. Thus the numbers e and π which are real but not algebraic numbers, are transcendental numbers.
- Every monotonic function in a closed interval is discontinuous at a countable number of points of that interval.
- Union of two enumerable sets is also enumerable.
- Every superset of an uncountable set is uncountable.
- Every infinite set is equivalent to its proper subset.
- If an enumerable set is subtracted from an enumerable set, the remaining set will be enumerable.
- If we subtract an enumerable set from a non-enumerable set, then the remaining set is non-enumerable.
- The outer Lebesgue measure of a set is also known as Lebesgue exterior measure.
- The measurable sets are those, which split every set into two pieces that are additive w.r.t. outer measure.
- The linear set of outer measure zero is Lebesgue measurable.
- The necessary and sufficient condition for a set E to be measurable with measure zero is that $m^*(E) = 0$.
- The union of two measurable sets is measurable.
- The intersection and difference of two measurable sets are measurable.
- A boolean ring is closed under the formation of symmetric difference.
- A continuous function defined over a measurable set E is measurable.
- Lebesgue integrable function is not necessarily R-integrable.
- The integral of a nowhere zero function can be zero.
- Every bounded function 'is not necessarily integrable.
- Any set whose outer measure is zero is also measurable.

REVIEW QUESTIONS

1. Define union, intersection, difference and symmetric difference of two sets.
2. Define the power set of a set.
3. How many element does the power set of a set S with n elements have?
4. Define what it mean for a function from the set of positive integers to the set of positive integer to be one to one.
5. Define the inverse of a function.
6. Let $f(n)$ be the function from the set of integers to the set of integers such that $f(n) = n^2 + 1$. What are the domain, co-domain and range of this function?
7. Give an example of a function from the set of positive integers to the set of positive integers that is :
   (a) both one-one and onto.
   (b) one-one but not onto.
   (c) neither one-one nor onto.
   (d) not one-one but is onto.
8. When the empty set the power set of a set?
9. (a) Define what is means for two sets to be equal?
   (b) Describe the ways to show that two sets are equal.
10. Let A and B be sets in a finite universal set U. List the following in order of increasing size :
    (a) $|A|, |A \cup B|, |A \cap B|, |U|, |\phi|$
    (b) $|A - B|, |A \oplus B|, |A| + |B|, |A \cup B|, |\phi|$
11. Research where the concept of a function first arose and describe how this concept was first used.

OBJECTIVE TYPE QUESTIONS

▶ FILL IN THE BLANKS

1. A relation R on a set A is symmetric iff R = ____________ .

2. Let R be an anti-symmetric relation on a set A such that $(a, b) \in R$ and $(b, a) \in R$. Then __________.

3. Let R be a relation on a set A such that $R = R^{-1}$. Then R is __________.

4. Let $A = \{1, 2, 3\}$, then the smallest equivalence relation on A is __________.

5. Let A be a finite set. Then the smallest equivalence relation on A is the __________ relation on A.

6. The void relation on a set is __________ and __________ but not __________.

7. Let R be a relation defined by $R = \{(4, 5), (1, 4), (4, 6), (7, 6), (3, 7)\}$ on N. Then $R \circ R^{-1} =$ __________.

8. Let $R = \{(a, a), (b, c), (a, b)\}$ be a relation on a set $A = \{a, b, c\}$. Then the minimum number of ordered pairs which when added to R make it transitive is __________.

▶ TRUE/FALSE

Write T for True and F for False statement.

1. A binary relation is a set. **(T/F)**
2. A void set defines a relation. **(T/F)**
3. The total number of relations from a set containing m elements to a finite set containing n elements is 2^{mn}. **(T/F)**
4. Every relation is a function. **(T/F)**
5. Every function is a relation. **(T/F)**
6. The total number of bijections from a set containing n elements to a set containing n elements is n^n. **(T/F)**
7. Every equivalence relation is symmetric. **(T/F)**
8. Every symmetric relation is equivalence. **(T/F)**
9. Every anti-symmetric relation is symmetric. **(T/F)**
10. The composition of functions is commutative. **(T/F)**
11. Reflexivity is redundant in the definition of an equivalence relation on a set A, because by symmetry $(a, b) \in R \Rightarrow (b, a) \in R$ and by transitivity $(a, b) \in R$ and $(b, a) \in R \Rightarrow (a, a) \in R$ **(T/F)**
12. The relation $R = \{(1, 2), (1, 3)\}$ is a transitive relation on a set $A = \{1, 2, 3\}$. **(T/F)**
13. The identity relation on a finite set A is the smallest equivalence relation on A. **(T/F)**

▶ MULTIPLE CHOICE QUESTIONS

Choose the most appropriate one.

1. Let R_1 and R_2 be two equivalence relation on a set. Consider the following assertion :
   (i) $R_1 \cup R_2$ is an equivalence relation.
   (ii) $R_1 \cap R_2$ is an equivalence relation.
   Which of the following is correct?
   (a) Both assertions are true.
   (b) Assertion (i) is true but assertion (ii) is not true.
   (c) Assertion (ii) is true but assertion (i) is not true.
   (d) Neither (i) nor (ii) is true.

2. The 'subset' relation on a set of set is :
   (a) a partial ordering
   (b) an equivalence relation
   (c) transitive and symmetric only
   (d) transitive and anti-symmetric only

3. Let R be a symmetric and transitive relation on a set A, then :
   (a) R is reflexive and hence an equivalence relation.
   (b) R is reflexive and hence a partial order.
   (c) R is not reflexive and hence is not an equivalence relation .
   (d) None of the above

4. The number of equivalence relations of the set $\{1, 2, 3, 4\}$ is :
   (a) 4 (b) 15
   (c) 16 (d) 24

5. Suppose A is a finite set with n elements. The number of elements in the large equivalence relation of A is :
   (a) 1 (b) n
   (c) $n + 1$ (d) n^2

6. The binary relation $S = \phi$ on the set $A = \{1, 2, 3\}$ is :
   (a) neither reflexive nor symmetric
   (b) symmetric and reflexive
   (c) transitive and reflexive
   (d) transitive and symmetric

7. Let $f(x) = x^2 + x$ and $g(x) = x + 1$ then fog is :
   (a) $x^2 + 3x + 2$ (b) $x^2 + x + 1$
   (c) $(x+1)^2 + (x+1)$ (d) None of these

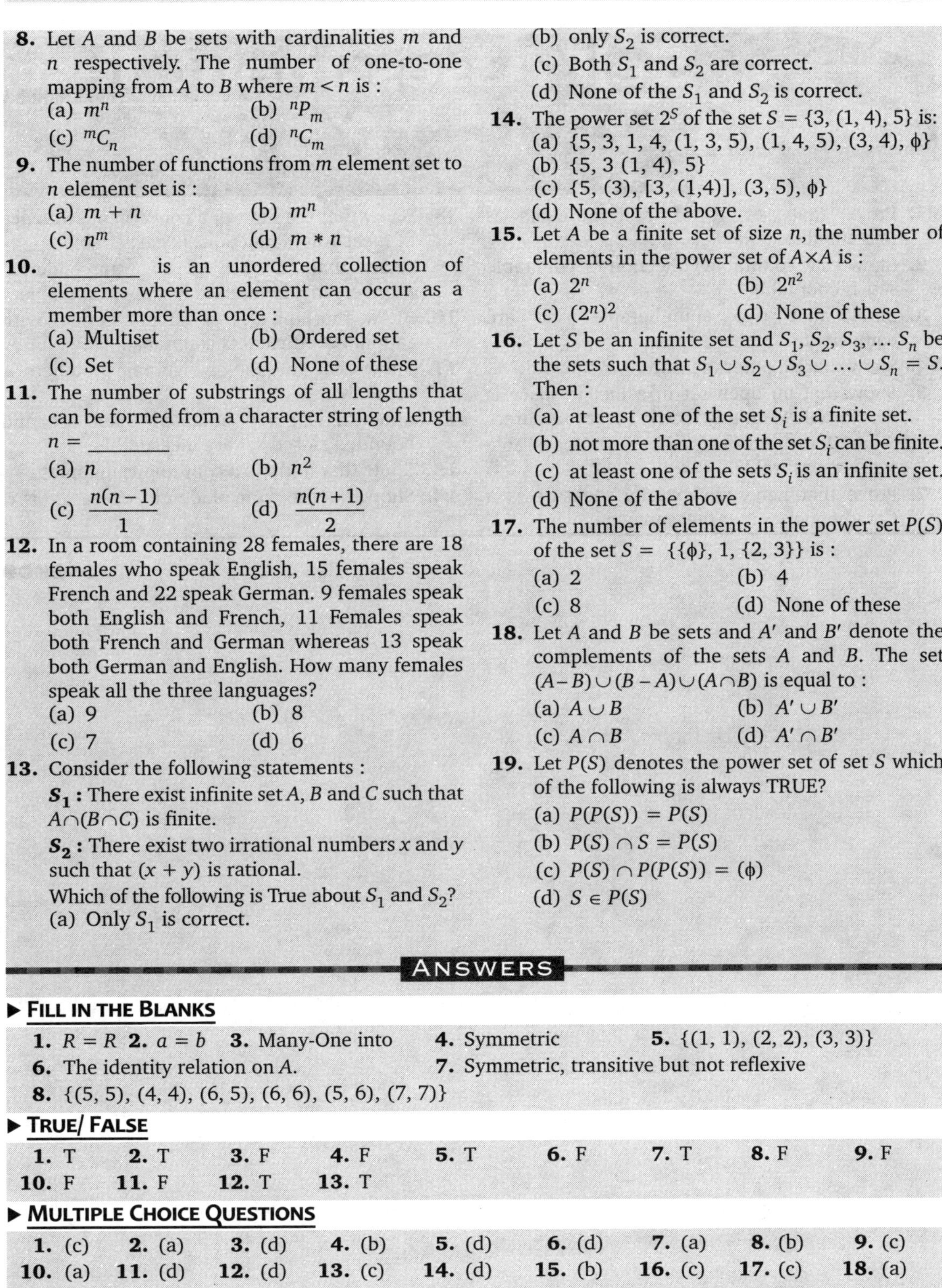

8. Let A and B be sets with cardinalities m and n respectively. The number of one-to-one mapping from A to B where $m < n$ is :
(a) m^n (b) ${}^nP_m$
(c) ${}^mC_n$ (d) ${}^nC_m$

9. The number of functions from m element set to n element set is :
(a) $m + n$ (b) m^n
(c) n^m (d) $m * n$

10. ________ is an unordered collection of elements where an element can occur as a member more than once :
(a) Multiset (b) Ordered set
(c) Set (d) None of these

11. The number of substrings of all lengths that can be formed from a character string of length n = ________
(a) n (b) n^2
(c) $\frac{n(n-1)}{1}$ (d) $\frac{n(n+1)}{2}$

12. In a room containing 28 females, there are 18 females who speak English, 15 females speak French and 22 speak German. 9 females speak both English and French, 11 Females speak both French and German whereas 13 speak both German and English. How many females speak all the three languages?
(a) 9 (b) 8
(c) 7 (d) 6

13. Consider the following statements :
S_1 : There exist infinite set A, B and C such that $A \cap (B \cap C)$ is finite.
S_2 : There exist two irrational numbers x and y such that $(x + y)$ is rational.
Which of the following is True about S_1 and S_2?
(a) Only S_1 is correct.
(b) only S_2 is correct.
(c) Both S_1 and S_2 are correct.
(d) None of the S_1 and S_2 is correct.

14. The power set 2^S of the set $S = \{3, (1, 4), 5\}$ is:
(a) $\{5, 3, 1, 4, (1, 3, 5), (1, 4, 5), (3, 4), \phi\}$
(b) $\{5, 3\ (1, 4), 5\}$
(c) $\{5, (3), [3, (1,4)], (3, 5), \phi\}$
(d) None of the above.

15. Let A be a finite set of size n, the number of elements in the power set of $A \times A$ is :
(a) 2^n (b) 2^{n^2}
(c) $(2^n)^2$ (d) None of these

16. Let S be an infinite set and $S_1, S_2, S_3, \ldots S_n$ be the sets such that $S_1 \cup S_2 \cup S_3 \cup \ldots \cup S_n = S$. Then :
(a) at least one of the set S_i is a finite set.
(b) not more than one of the set S_i can be finite.
(c) at least one of the sets S_i is an infinite set.
(d) None of the above

17. The number of elements in the power set $P(S)$ of the set $S = \{\{\phi\}, 1, \{2, 3\}\}$ is :
(a) 2 (b) 4
(c) 8 (d) None of these

18. Let A and B be sets and A' and B' denote the complements of the sets A and B. The set $(A-B) \cup (B-A) \cup (A \cap B)$ is equal to :
(a) $A \cup B$ (b) $A' \cup B'$
(c) $A \cap B$ (d) $A' \cap B'$

19. Let $P(S)$ denotes the power set of set S which of the following is always TRUE?
(a) $P(P(S)) = P(S)$
(b) $P(S) \cap S = P(S)$
(c) $P(S) \cap P(P(S)) = (\phi)$
(d) $S \in P(S)$

ANSWERS

▶ FILL IN THE BLANKS

1. $R = R$ **2.** $a = b$ **3.** Many-One into **4.** Symmetric **5.** $\{(1, 1), (2, 2), (3, 3)\}$
6. The identity relation on A. **7.** Symmetric, transitive but not reflexive
8. $\{(5, 5), (4, 4), (6, 5), (6, 6), (5, 6), (7, 7)\}$

▶ TRUE/ FALSE

1. T **2.** T **3.** F **4.** F **5.** T **6.** F **7.** T **8.** F **9.** F
10. F **11.** F **12.** T **13.** T

▶ MULTIPLE CHOICE QUESTIONS

1. (c) **2.** (a) **3.** (d) **4.** (b) **5.** (d) **6.** (d) **7.** (a) **8.** (b) **9.** (c)
10. (a) **11.** (d) **12.** (d) **13.** (c) **14.** (d) **15.** (b) **16.** (c) **17.** (c) **18.** (a)
19. (c)

1. Prove that set R of real numbers is uncountable.
2. Show that countable collection of countable sets is countable.
3. Show that two enumberable sets are equivalent.
4. Show that an outer measure is σ-subadditive.
5. Show that an open set in a metric space is measurable w.r.t. any metric outer measure.
6. Show that the difference of two measurable sets is measurable.
7. Prove that Lebesgue exterior measure is a Caratheodory outer measure.
8. Show that the limit of a convergent sequence of measurable functions is measurable.
9. Show that the set of all transcendental numbers in any interval is non-enumerable.
10. Show that the set of all polynomial with rational coefficient is countable.
11. Show that the set of algebraic numbers is countable.
12. Show that every bounded open set and bounded closed set are measurable.
13. Show that norm is a continuous mapping.
14. Show that the norm of identity operator I is 1.

●●●●●●

Chapter 2 Metric Spaces

2.1 INTRODUCTION

The set of real numbers has two type of properties. The first type consist of the algebraic which deals with addition, multiplication etc. The second type consists of properties having to do with the notion of distance between two numbers and with the concept of limit. The second type of properties are called topological or metric property. In this chapter we shall study these properties in a general space in which the notion of distance is defined.

Definition. *Let X be a non-empty set. Then a mapping* $d : X \times X \to R^+ \cup \{0\}$ *is said to be metric, if it satisfies the following conditions :*

M(1) $d(x, y) > 0 \; \forall x, y \in X$

M(2) $d(x, y) = 0$ iff $x = y \; \forall \, x, y \in X$

M(3) $d(x, y) = d(y, x) \; \forall \, x, y \in X$ *(Symmetric property)*

M(4) $d(x, y) < d(x, z) + d(z, y) \forall x, y, z \in X$ *(Triangle inequality).*

If d is a metric on X, then ordered pair (X, d) is said to be metric space.

REMARKS

- In 1905, the French mathematician Maurice Frechet thought of generalizing the notion of distance and extending it to arbitrary set, which seems the begining of metric space.
- The conditions of metric space can be defined in terms of distance as follows :
  (i) $d(x, y) \geq 0 \Rightarrow$ The distance between any two points of X is a non-negative real number.
  (ii) $d(x, y) = 0$ iff $x = y \Rightarrow$ If two points coincides then the distance is zero and if the distance is zero, then two points are same.
  (iii) $d(x, y) = d(y, x) \Rightarrow$ The distance does not depend on the order of the points x and y.
  (iv) $d(x, y) \leq d(x, z) + d(z, y) \Rightarrow$ The sum of the length of two sides of a triangle is greater than or equal to the length of the third side. The sign of equality holds when three points lie on the straight line.

2.2 SOME PARTICULAR METRIC SPACES

(1) The real line R. Let R be the set of real numbers and let $d : R \times R \to R$ be the function defined by

$$d(x, y) = |x - y| \; \forall \, x, y \in R$$

Then clearly, we have

(i) d is a non-negative real valued function on $R \times R$

(ii) $d(x, y) = 0 \Leftrightarrow |x - y| = 0 \Leftrightarrow x = y \forall x, y \in R$

(iii) $d(x, y) = |x - y| = |-(y - x)| = |y - x| = d(y, x) \; \forall \, x, y \in R$

(iv) Let x, y, z be any three elements of R.

Then $d(x, y) = |x-y| = |(x-z)+(z-y)|$

$\Rightarrow \leq |x-z| + |z-y| = d(x, z) + d(z, y)$

$\Rightarrow d(x, y) < d(x, z) + d(z, y) \ \forall \ x, y, z \in R.$

$\Rightarrow$ d is a metric on R and the ordered pair (R, d) is a metric space.

REMARK

- The metric defined above is called usual metric on R.

(2) The Euclidean plane R². Let R^2 be the set of all ordered pairs of real numbers. Define a mapping $d : R^2 \times R^2 \rightarrow R$ such that

$$d(x, y) = \sqrt{\left\{(x_1 - y_1)^2 + (x_2 - y_2)^2\right\}} \ \forall \ x, y \in R^2.$$

where $x = (x_1, x_2), y = (y_1, y_2)$.

Then, clearly we have

(i) d is a non-negative real valued function on $R^2 \times R^2$.

(ii) $d(x, y) = 0 \Leftrightarrow \sqrt{\left\{(x_1 - y_1)^2 + (x_2 - y_2)^2\right\}} = 0$

$\Leftrightarrow x_1 - y_1 = 0$ and $x_2 - y_2 = 0$

$\Leftrightarrow x_1 = y_1$ and $x_2 = y_2 \Leftrightarrow x = y$

(iii) For all $x, y \in R^2$

$$d(x, y) = \sqrt{\left\{(x_1 - y_1)^2 + (x_2 - y_2)^2\right\}}$$

$$= \sqrt{\left\{(y_1 - x_1)^2 + (y_2 - x_2)^2\right\}} = d(y, x).$$

(iv) Let $x = (x_1, x_2)$, $y = (y_1, y_2)$, $z = (z_1, z_2)$ be any three elements of R^2. Now, by triangle inequality for real numbers, we have

$$\sqrt{\left\{(a_1 + b_1)^2 + (a_2 + b_2)^2\right\}} \leq \sqrt{\left(a_1^2 + a_2^2\right)} + \sqrt{\left(b_1^2 + b_2^2\right)} \text{ for } a_1, b_1, a_2, b_2 \in R.$$

Put $a_1 = x_1 - z_1$, $a_2 = x_2 - z_2$, $b_1 = z_1 - y_1$, $b_2 = z_2 - y_2$ in above inequality, we have

$$\sqrt{\left\{(x_1 - y_1)^2 + (x_2 - y_2)^2\right\}} \leq \sqrt{\left\{(x_1 - z_1)^2 + (x_2 - z_2)^2\right\}} + \sqrt{\left\{(z_1 - y_1)^2 + (z_2 - y_2)^2\right\}}$$

$\Rightarrow d(x, y) < d(x, z) + d(y, z).$

Hence, d is a metric on R^2.

REMARK

- The metric space (R^2, d) defined above is called Eucledian plane.

(3) Discrete metric space. Let X be any non-empty set. The mapping $d : X \times X \rightarrow R$ defined by

$$d(x, y) = \begin{cases} 0 \text{ if } x = y \\ 1 \text{ if } x \neq y \end{cases}$$

is a metric on X, called the discrete metric on X.

Proof. $d(x, y)$ is a metric because it satisfy all the following properties such that

(i) $d(x, y)=0$ or $d(x, y)=1$

$\Rightarrow \quad d(x, y) \geq 0 \; \forall \; x, y \in X.$

(ii) If $x = y \Rightarrow \quad d(x, y)=0.$

Conversely, if $d(x, y)=0$ then $x = y.$

If it is not possible *i.e.,* $x \neq y$ Then $d(x, y)=1.$

It contradicts the fact that $d(x, y)=0 \Rightarrow d(x, y)=0$ iff $x=y.$

(iii) If $x = y \quad d(x, y)=0=d(y, x)$

If $x \neq y \quad d(x, y)=1=d(y, x) \; \forall \; x, y \in X.$

Hence $d(x, y)=d(y, x).$

(iv) Let $x, y, z \in X.$

If $x = y$ then $(x, y)=0$ ...(1)

$d(x, z) \geq 0 \Rightarrow d(z, y) \geq 0$

$\Rightarrow \quad d(x, z) + d(z, y) \geq 0$

$\Rightarrow \quad d(x, z) + d(z, y) \geq d(x, y)$

Similarly, if $x \neq y$, $d(x, y)= 1$ [from equation (1)]

$\Rightarrow \exists \; z \in X$ such that at least

$d(x, z)=1$ and $d(z, y)=1$

$\Rightarrow \quad d(x, y) \leq d(x, z)+d(z, y)$

Hence d is metric and (X, d) is a metric space which is called Discrete metric space.

(4) Let (X_1, d_1) and (X_2, d_2) be two metric spaces.

Define $X = X_1 \times X_2$

and $d(x, y) = d_1(x_1, y_1)+d_2(x_2, y_2) \forall x, y \in X$, when $x = (x_1, y_1)$ and $y=(x_2, y_2)$.

Then, (X, d) is a metric space. Hence, cartesian product of two metric spaces is a metric space.

REMARK

- The space (X, d) defined above is called Product metric space.

(5) The Postman metric for $\mathbf{R}^2$. Consider a well-planned city, in which the roads are either parallel or perpendicular to each other and there are rectangular blocks of housing complexes. Suppose someone wants to go from point A to point B.
How do we find the minimum distance that he has to travel.
Since he can not go as a cross flies.
Therefore, the Eucledian metric is useless.
The product metric is also useless.

Then he has to go along one road till he reaches a road on which B is situated and then move along the perpendicular road till he reaches B.

If the co-ordinates of A and B with reference to a pair of rectangular axes, one of which is parallel to one set of roads, and the other is perpendicular to it, be (x_1, x_2) and (y_1, y_2) respectively, then he will have to move a distance

$$|x_1-y_1| + |x_2-y_2|$$

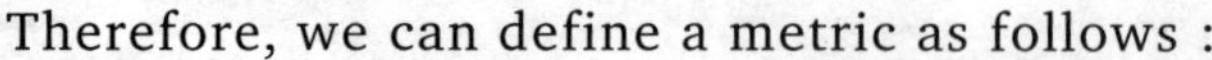
Therefore, we can define a metric as follows :

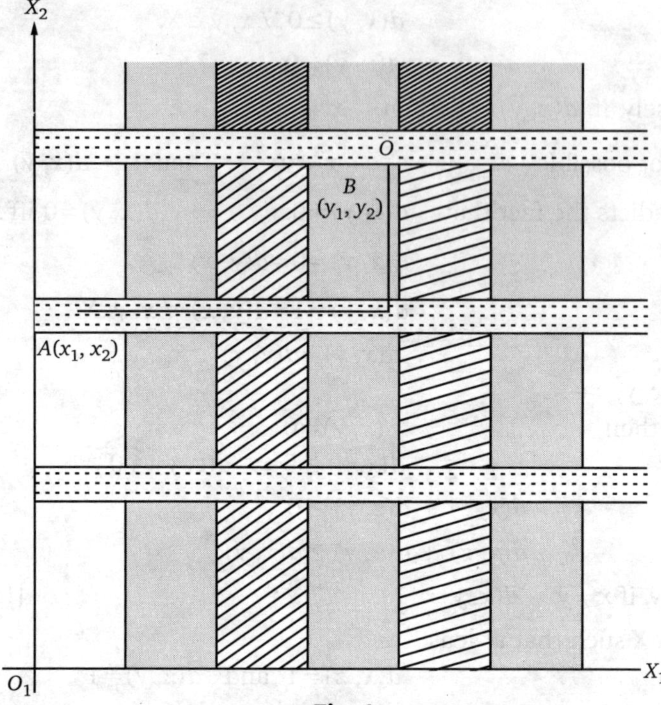

Fig. 1

Let R^2 be the set of all ordered pairs of real numbers and let $d : R^2 \times R^2 \to R$ defined by

$$d(x, y) = |x_1 - y_1| + |x_2 - y_2|$$

where

$$x = (x_1, x_2) \text{ and } y = (y_1, y_2).$$

Then, clearly we have

(i) d is a non-negative real-valued function on $R^2 \times R^2$.

(ii) $d(x, y) = 0 \quad \Leftrightarrow |x_1 - y_1| + |x_2 - y_2| = 0$

$$\Leftrightarrow |x_1 - y_1| = 0 \text{ and } |x_2 - y_2| = 0$$

$$\Leftrightarrow x_1 = y_1 \text{ and } x_2 = y_2 \Leftrightarrow x = y.$$

(iii) For all $x, y \in R^2$

$$d(x, y) = |x_1 - y_1| + |x_2 - y_2| = |y_1 - x_1| + |y_2 - x_2| = d(y, x).$$

(iv) Let $x = (x_1, x_2)$, $y = (y_1, y_2)$, $z = (z_1, z_2)$ be any three points of R^2.

Consider
$$\begin{aligned} d(x, y) &= |x_1 - y_1| + |x_2 - y_2| \\ &= |(x_1 - z_1) + (z_1 - y_1)| + |(x_2 - z_2) + (z_2 - y_2)| \\ &\leq (|x_1 - z_1| + |z_1 - y_1|) + (|x_2 - z_2 + y_2 - z_2|) \\ &= (|x_1 - z_1| + |x_2 - z_2|) + (|z_1 - y_1| + |z_2 - y_2|) \\ &= d(x, z) + d(z, y) \end{aligned}$$

$\Rightarrow \quad d(x, y) \leq d(x, z) + d(z, y).$

Hence, d is a metric on R^2.

2.2.1 SUBSPACE

Let (X, d) be a metric space and Y be a proper subset of X. Let d_1 be the restriction of d on Y *i.e.*, $d_1(x, y)=d(x, y)$, $\forall x, y \in Y \times Y$.

Then (Y, d_1) is called subspace of (X, d).

2.2.2 PSEUDO-METRIC

A mapping $d : X \times X \to R^+ \cup \{0\}$ is called a pseudo-metric or semi-metric for X if and only if

(i) $d(x, y) \geq 0$, $\forall x, y \in X$

(ii) $d(x, x)=0$, $\forall x \in X$

(iii) $d(x, y)=d(y, x)$; $\forall x, y \in X$

(iv) $d(x, y) \leq d(x, z)+d(z, y)$; $\forall$ $x, y, z \in X$.

REMARKS

- The Pseudo metric is said to be finite if $d(x, y)<\infty$; $\forall x, y \in X$.
- The Pseudo metric d differ from metric in the sense that
  (a) $d(x, y)$ may be equal to zero even if $x \neq y$ *i.e.*, distance between a pair of distinct points may be zero.
  (b) $d(x, y)=\infty$, for some $x, y \in X$ *i.e.*, ∞ is defined as a measure of a distance between a pair of points.

ILLUSTRATIONS

(1) Consider a set X of real valued functions defined over the closed interval $[-1, 1]$. Let $f(x)$ and $g(x)$ be two arbitrary real valued functions defined over $[-1, 1]$. Let us define

$$d(f, g)=\int_{-1}^{1}\left\{\left[f(x)-g(x)\right]^2 dx\right\}^{1/2}.$$

It is easy to verify that d is a pseudo metric on X.

(2) Let (X, d) be a pseudo metric and let '$\sim$' be a relation in X defined by setting

$x \sim y$ iff $d(x, y)=0$.

The relation '$\sim$' is an equivalence relation in X. In fact,

(i) Since $d(x, x)=0$, $\forall$ $x \in X$

$\Rightarrow$ $x \sim x$, $\forall$ $x \in X$.

Therefore, the relation '$\sim$' is reflexive.

(ii) $x \sim y \Leftrightarrow d(x, y)=0 \Leftrightarrow d(y, x)=0 \Leftrightarrow y \sim x$ $\Rightarrow$ '$\sim$' is symmetric.

(iii) Since $x \sim y$ and $y \sim z$ $\Rightarrow$ $d(x, y)=0$ and $d(y, z)=0$

$\Rightarrow$ $d(x, y)+d(y, z)=0$

$\Rightarrow$ $d(x, z)=0$

$\Rightarrow$ $x \sim z$

$\Rightarrow$ '$\sim$' is transitive.

2.2.3 NORM

The size of an element x is a real number denoted by $\|x\|$ and is called norm. Norm is a generalisation of the real valued functions, which satisfying the following conditions :

(i) $\|x\| \geq 0$ (ii) $\|x\|=0 \Leftrightarrow x=0$

(iii) $\|cx\|=|c|\,\|x\|$ (iv) $\|x+y\| \leq \|x\|+\|y\|$.

REMARK

- The metric defined with the help of norm as follows $d(x, y)=\|x-y\|$
This metric is known as metric induced by norm.
For Example :
Consider the set of all bounded real valued functions defined on [0, 1].
Define norm of a function as follows :
$$\|f\|=\sup \{ |f(x)| : x\in[0, 1]\}.$$
Here, it can be easily verified that
$$d(f, g)=\|f-g\|=\sup |f(x)-g(x)|$$
is a metric on [0, 1]. We denote this space by C[0, 1].

2.3 DISTANCE BETWEEN TWO SETS : DIAMETER OF A SET

Definition 1. *Let (X, d) be a metric space and let A be a non-empty subset of X. Then diameter of the set A, denoted by $\delta(A)$, is defined by*
$$\delta(A)=\sup \{d(x, y) : x, y\in A\}.$$

Definition 2. *The distance between a point $x\in X$ and a set A is denoted by d(x, A) and is defined by*
$$d(x, A)=\inf\{d(x, y):y\in A).$$

Definition 3. *The distance between two non-empty subsets A and B of a metric space X is denoted and defined by*
$$d(A, B)=\inf\{d(x, y):x\in A, y\in B\}.$$

REMARKS

- The diameter of a set is always non-negative
- The diameter of empty set is ∞.
- *A* set *A* is said to be bounded if $\delta(A)<\infty$.
- If *A* is a closed sphere of radius *r*, then $d(A)=2r$.
- The diameter of a finite set is finite and diameter for an infinite set is infinite.
- For the empty set ϕ, the distance of ϕ from point *x* is denoted by $d(x, \phi)=\infty$.
- $d(x, A)=0$ if $x\in A$.
- The distance of set *A* and empty set is ∞ *i.e.*, $d(A, \phi)=\infty$.
- $d(A, B)=0$ iff $A\cap B\neq\phi$.

THEOREM 1. ***Let (X, d) be a metric space and let x, y, z be any three points of X, then***
$$d(x, y) \geq |d(x, z)-d(z, y)|.$$

PROOF. By definition of metric space, we have
$$d(x, z) \leq d(x, y)+d(y, z)$$
$$=d(x, y)+d(z, y) \qquad \text{(By symmetric property)}$$
$$\Rightarrow \quad d(x, z)-d(z, y) \leq d(x, y). \qquad ...(1)$$
By *M* (iv), we have
$$d(z, y) \leq d(z, x)+d(x, y)$$
$$= d(x, z)+d(x, y) \qquad [d(x, z)=d(z, x)]$$
$$\Rightarrow \quad d(z, y)-d(x, z) \leq d(x, y). \qquad ...(2)$$
From (1) and (2), we conclude that
$$d(x, y) \geq |d(x, z)-d(z, y)|.$$

REMARK

- The above inequality states that the difference of the lengths of any two sides of a triangle is less than or equal to the third side. Also, the sign of equality occurs, when three points lie on the straight line.

THEOREM 1. ***Let (X, d) be a metric space, then***

$$|d(x,y) - d(x',y')| \le d(x,x') + d(y,y') \ \forall x,x',y,y' \in X$$

PROOF. By M(iv), we have

$$\begin{aligned} d(x,y) &\le d(x,x')+d(x',y) \\ &\le d(x,x')+d(x',y')+d(y',y) && [\text{By } M(\text{iv})] \\ &= d(x,x')+d(x',y')+d(y,y') && [d(y,y')=d(y',y)] \end{aligned}$$

$$\Rightarrow \quad d(x,y)-d(x',y') \le d(x,x')+d(y,y'). \qquad ...(1)$$

Also, we have

$$\begin{aligned} d(x',y') &\le d(x',x)+d(x,y') \\ &\le d(x',x)+d(x,y)+d(y,y') && [\because d(x,y') \le d(x,y)+d(y,y')] \\ &= d(x,x')+d(x,y)+d(y,y') && [\because d(x,x')=d(x',x)] \end{aligned}$$

$$\Rightarrow \quad d(x',y')-d(x,y) \le d(x,x')+d(y,y'). \qquad ...(2)$$

From (1) and (2), we conclude that

$$|d(x,y)-d(x',y')| \le d(x,x')+d(y,y')$$

THEOREM 3. ***Let $X = \phi$. Then a mapping $d : X \times X \to R$ is a metric if and only if the following conditions hold***

(i) $d(x,y) = 0 \Leftrightarrow x = y; \forall x,y \in X$

(ii) $d(x',y) \le d(x,y) + d(y,z); \forall x,y,z \in X.$

PROOF. Let us first suppose d is a metric on R.

Then by M(iv), we have

$$d(x,y) \le d(x,z)+d(z,y) \qquad ...(1)$$

Also $\quad d(z,y)=d(y,z). \qquad ...(2)$

From (1) and (2), we conclude that

$$d(x,y) \le d(x,z)+d(y,z); \ \forall \ x,y,z \in X.$$

Now, suppose condition (i) and (ii) holds : To show d is a metric.

Let x, y be any two elements of X, then by (ii), we have

$$d(x,x) \le d(x,y)+d(x,y)$$

[Replacing x, x, y for x, y, z respectively]

i.e., $\quad 2d(x,y) \ge d(x,x). \qquad ...(3)$

But from (i) $d(x,x)=0$, therefore from (3), we have

$$d(x,y) \ge 0 \ \forall \ x,y \in X \Rightarrow M \text{ (i) is satisfied.}$$

Now apply condition (ii) for the points x, y, z, we have

$$d(x,y) \le d(x,x)+d(y,x)=0+d(y,x) \qquad [\because d(x,y)=0]$$

$$\Rightarrow \quad d(x,y) \le d(y,x). \qquad ...(4)$$

Again applying condition (ii) for the points y, x, y, we get

$$d(y, x) \leq d(y, y) + d(x, y) = 0 + d(x, y) \qquad [\because d(y, y) = 0]$$
$$= d(x, y). \qquad \text{...(5)}$$

From (4) and (5), we conclude that

$$d(x, y) = d(y, x) \qquad \Rightarrow M(\text{iii}) \text{ is satisfied.}$$

Finally, for any x, y, z, we have

$$d(x, y) \leq d(x, z) + d(y, z) = d(x, z) + d(z, y)$$
$$\Rightarrow \qquad d(x, y) \leq d(x, z) + d(z, y)$$

$\Rightarrow M$ (iv) is satisfied.

Hence, d is a metric on X.

THEOREM 4. ***Let (X_1, d_1) and (X_2, d_2) be any two metric spaces defined by***

$$d[(x_1, x_2), (y_1, y_2)] = \sqrt{d_1^2(x_1, y_1) + d_2^2(x_2, y_2)}$$

where $x_1, y_1 \in X_1, x_2 y_2 \in X_2$. Show that d is a metric for $X_1 \times X_2$.

PROOF. Since (X_1, d_1) and (X_2, d_2) are two metric spaces. Therefore, by definition, we have

(i) $d_1(x_1, y_1) \geq 0$, $d_2(x_2, y_2) \geq 0$

(ii) $d_1(x_1, y_1) = 0 \Leftrightarrow x_1 = y_1$ and $d_2(x_2, y_2) = 0 \Leftrightarrow x_2 = y_2$

(iii) $d_1(x_1, y_1) = d_1(y_1, x_1)$ and $d_2(x_2, y_2) = d_2(y_2, x_2)$

(iv) $d_1(x_1, y_1) \leq d_1(x_1, z_1) + d_1(z_1, y_1)$ and $d_2(x_2, y_2) \leq d_2(x_2, z_2) + d_2(z_2, y_2)$

To show $d[(x_1, x_2), (y_1, y_2)] = \sqrt{d_1^2(x_1, y_1) + d_2^2(x_2, y_2)}$

(a) Since $d_1(x_1, y_1) \geq 0$ and $d_2(x_2, y_2) \geq 0$

Therefore $\quad d_1^2(x_1, y_1) \geq 0$ and $\quad d_2^2(x_2, y_2) \geq 0$

$$\Rightarrow \quad d_1^2(x_1, y_1) + d_2^2(x_2, y_2) \geq 0$$
$$\Rightarrow \quad \sqrt{d_1^2(x_1, y_1) + d_2^2(x_2, y_2)} \geq 0$$
$$\Rightarrow \quad d[(x_1, x_2), (y_1, y_2)] \geq 0.$$

(b) Here, we have

$$d_1(x_1, y_1) = 0 \Leftrightarrow x_1 = y_1$$

and $d_2(x_2, y_2) = 0 \Leftrightarrow x_2 = y_2$.

Now $\quad d[(x_1, x_2), (y_1, y_2)] = 0$

$$\Leftrightarrow \sqrt{d_1^2(x_1, y_1) + d_2^2(x_2, y_2)} = 0$$
$$\Leftrightarrow d_1^2(x_1, y_1) = 0,\ d_2^2(x_2, y_2) = 0$$
$$\Leftrightarrow d_1(x_1, y_1) = 0 \Leftrightarrow x_1 = y_1 \text{ and } d_2(x_2, y_2) = 0 \Leftrightarrow x_2 = y_2$$
$$\Rightarrow \quad (x_1, x_2) = (y_1, y_2)$$

(c) Here, we have

$$d_1(x_1, y_1) = d_1(y_1, x_1) \quad \text{and} \quad d_2(x_2, y_2) = d_2(y_2, x_2).$$
$$\Rightarrow \quad d_1^2(x_1, y_1) = d_1^2(y_1, x_1) \quad \text{and} \quad d_2^2(x_2, y_2) = d_2^2(y_2, x_2)$$

$$\Rightarrow \quad d_1^2(x_1, y_1)+d_2^2(x_2, y_2)= d_1^2(y_1, x_1)+d_2^2(y_2, x_2)$$

$$\Rightarrow \sqrt{d_1^2(x_1,y_1)+d_2^2(x_2,y_2)} = \sqrt{d_1^2(y_1,x_1)+d_2^2(y_2,x_2)}$$

$$\Rightarrow \quad d[(x_1, x_2), (y_1, y_2)] = d[(y_1, y_2), (x_1, x_2)].$$

(d) Let $(x_1, x_2), (y_1, y_2), (z_1, z_2) \in X_1 \times X_2$.
Then, we have

$$d(x, y)= \sqrt{d_1^2(x_1,y_1)+d_2^2(x_2,y_2)}$$

$$\leq \sqrt{\left\{\left[d_1(x_1,z_1)+d_1(z_1,y_1)\right]^2+\left[d_2(x_2,z_2)+d_2(z_2,y_2)\right]^2\right\}}$$

(By triangle inequality)

$$\leq \left[d_1^2(x_1,z_1)+d_2^2(x_2,z_2)\right]^{1/2}+\left[d_1^2(z_1,y_1)^2+d_2^2(z_2,y_2)^2\right]^{1/2}$$

$$\leq d(x, z)+d(z, y).$$

Hence, from (a), (b), (c) and (d), we conclude that

$$d(x, y)= \sqrt{\left[d_1^2(x_1,y_1)+d_2^2(x_2,y_2)\right]} \text{ is a metric on } X.$$

2.4 SOME IMPORTANT INEQUALITIES

(1) Triangle Inequality.

$$|z+w| \leq |z|+|w|, \; z, w \in C$$

(2) Holder's Inequality.

(i) If a_i, b_i ($i=1, 2, \quad , n$) are non-negative real numbers, then

$$\sum_{i=1}^{n} a_i b_i \leq \left(\sum_{i=1}^{n} a_i^p\right)^{1/p} \left(\sum_{i=1}^{n} a_i^q\right)^{1/q} \quad \text{where } p>1 \text{ and } \frac{1}{p}+\frac{1}{q}= 1.$$

(ii) Holder's Inequality for integrals. If f, g are non-negative real-valued integrable functions defined on $[a, b]$, then

$$\int_a^b f.g\, dx \leq \left\{\int_a^b (f(x))^p\, dx\right\}^{1/p} \left\{\int_a^b (g(x))^q\, dx\right\}^{1/q}.$$

(3) Cauchy-Schwarz inequality. Let $z=(z_1, z_2,\ldots, z_n)$ and $w=(w_1, w_2,\ldots, w_n)$ be two n-tuples of real or complex numbers, then

$$\sum_{i=1}^{n} |z_i w_i| \leq \left(\sum_{i=1}^{n} |z_i|^2\right)^{1/2} \left(\sum_{i=1}^{n} |w_i|^2\right)^{1/2}$$

REMARK

- In terms of norms, the above inequality can be written as $\sum_{i=1}^{n} |z_i w_i| \leq \|z\| \|w\|$.

(4) Minkowski's inequality.

(i) If $p \geq 1$ and a_i, b_i $(i=1, 2, \ldots, n)$ are non-negative real numbers then,

$$\left(\sum_{i=1}^{n}(a_i b_i)^p\right)^{1/p} \leq \left(\sum_{i=1}^{n} a_i^p\right)^{1/p} + \left(\sum_{i=1}^{n} b_i^p\right)^{1/p}$$

(ii) Minkowski's inequality for integrals. Let f, g be non-negative real-valued functions defined on $[a, b]$ then

$$\left\{\int_a^b (f+g)^p\,dx\right\}^{1/p} \leq \left\{\int_a^b (f(x))^p\,dx\right\}^{1/p} + \left\{\int_a^b (g(x))^p\,dx\right\}^{1/p}$$

(iii) Minkowski's inequality in terms of norm. Let $z = (z_1, z_2, \ldots, z_n)$ and $w=(w_1, w_2, \ldots, w_n)$ be two n-tuples of real or complex numbers. Then

$$\left(\sum_{i=1}^{n}|z_i + w_i|^2\right)^{1/2} \leq \left(\sum_{i=1}^{n}|z_i|^2\right)^{1/2} + \left(\sum_{i=1}^{n}|w_i|^2\right)^{1/2}$$

or

$$\|z+w\| \leq \|z\| + \|w\|.$$

(iv) Minkowski's inequality for complex numbers. If z_i, w_i $(i=1, 2, \ldots, n)$ are complex numbers, then

$$\left(\sum_{i=1}^{n}|z_i + w_i|^p\right)^{1/p} \leq \left(\sum_{i=1}^{n}|z_i|^p\right)^{1/p} + \left(\sum_{i=1}^{n}|w_i|^p\right)^{1/p}, (p \geq 1).$$

RECAPITULATIONS

- Let $X \neq \phi$. Then, a mapping $d: X \times X \to R^+ \cup \{0\}$ is said to be metric if following four conditions are satisfied:

  (i) $d(x, y) \geq 0 \ \forall\, x, y \in X$

  (ii) $d(x, y) = 0 \Leftrightarrow x = y \ \forall x, y \in X$

  (iii) $d(x, y) = d(y, x) \ \forall\, x, y \in X$

  (iv) $d(x, y) \leq d(x, z) + d(z, y) \ \forall x, y, z \in X$

- Diameter of a set : $\delta(A) = \sup\{d(x, y) : x, y \in A\}$
- Distance between two sets A and B : $d(A, B) = \inf\{d(x, y) : x \in A, y \in B\}$
- Triangle Inequality. $|x+y| \leq |x| + |y|$
- Holder's Inequality. $\sum_{i=1}^{n} a_i b_i \leq \left(\sum_{i=1}^{n} a_i^p\right)^{1/p} \left(\sum_{i=1}^{n} a_i^q\right)^{1/q}$
- Cauchy-Schwarz Inequality. $\sum_{i=1}^{n}|z_i w_i| \leq \left(\sum_{i=1}^{n}|z_i|^2\right)^{1/2} \left(\sum_{i=1}^{n}|w_i|^2\right)^{1/2}$
- Minkowski's Inequality. $\left(\sum_{i=1}^{n}(a_i + b_i)^p\right)^{1/p} \leq \left(\sum_{i=1}^{n} a_i^p\right)^{1/p} + \left(\sum_{i=1}^{n} b_i^p\right)^{1/p}$

SOLVED EXAMPLES

EXAMPLE 1. *Show that the mapping* $d: R^2 \times R^2 \to R^2$ *defined by* $d(x, y) = \max.\{|x_1 - y_1|, |x_2 - y_2|\}$ *where* $x = (x_1, x_2), y = (y_1, y_2) \in R$ *is metric on* R^2.

SOLUTION. (i) Since $|x_1-y_1| \geq 0$ and $|x_2-y_2| \geq 0$

which implies

$\max\{(x_1-y_1), (x_2-y_2)\} \geq 0$

$\Rightarrow \quad d(x, y) \geq 0.$

(ii) Since, we know that

$$|x_1-y_1| = 0 \text{ iff } x_1 = y_1$$

and $$|x_2-y_2| = 0 \text{ iff } x_2 = y_2.$$

Therefore,

$$|x_1-y_1| + |x_2-y_2| = 0 \text{ iff } \quad x_1 = y_1, x_2 = y_2.$$

$\Rightarrow \max\{|x_1-y_1| + |x_2-y_2|\} = 0$

iff $x_1 = y_1, x_2 = y_2$

$$d(x, y) = 0 \text{ iff } \quad x = y.$$

(iii) We know that

$$|x_1-y_1| = |y_1-x_1|$$

and $$|x_2-y_2| = |y_2-x_2|$$

$\Rightarrow |x_1-y_1| + |x_2-y_2| = |y_1-x_1| + |y_2-x_2|$

$\Rightarrow \quad d(x, y) = d(y, x)$

(iv) Consider

$$\begin{aligned} d(x, y) &= \max.\{|x_1-y_1|, |x_2-y_2|\} \\ &= \max\{|x_1-z_1+z_1-y_1|, |x_2-z_2+z_2-y_2|\} \\ &\leq \max\{|x_1-z_1| + |z_1-y_1|, |x_2-z_2| + |z_2-y_2|\} \\ &\leq \max\{|x_1-z_1|, |x_2-z_2|\} + \max[|z_1-y_1| + |z_2-y_2|] \end{aligned}$$

$\Rightarrow \quad d(x, y) \leq d(x, z) + d(z, y).$

From (i), (ii), (iii) and (iv), we conclude that d is a metric on R^2.

EXAMPLE 2. *Let X be the set of real valued bounded continuous function defined on the closed interval* [0, 1]. *We define the norm of a function* $f \in X$ *by*

$$\|f\| = \int_0^1 |f(x)|\, dx.$$

Define a mapping $d: X \times X \to R$ *by* $d(f, g) = \|f-g\| = \int_0^1 |f(x) - g(x)|\, dx \;\forall f, g \in X.$
Show that d is a metric on X.

SOLUTION. (i) Since, we have

$$|f(x) - g(x)| \geq 0.$$

Therefore, $\int_0^1 |f(x) - g(x)|\, dx \geq 0$

$\Rightarrow \quad \|f-g\| \geq 0$

$\Rightarrow \quad d(f, g) \geq 0.$

(ii) Since

$$|f(x) - g(x)| = 0 \text{ iff } f(x) = g(x)$$

therefore, $\int_0^1 |f(x)-g(x)|dx = 0$ iff $f(x)=g(x)$

$\Rightarrow$ $\|f-g\|=0$ iff $f(x)=g(x)$

$\Rightarrow$ $d(f, g)=0$ iff $f(x)=g(x)$

(iii) Since $|f(x)-g(x)| = |g(x)-f(x)|$.

Therefore, $\int_0^1 |f(x)-g(x)|dx = \int_0^1 |g(x)-f(x)|dx$

$\Rightarrow$ $\|f-g\| = \|g-f\|$

$\Rightarrow$ $d(f, g) = d(g, f)$.

(iv) Let $f, g, h \in X$. Then, we have

$$\|f-g\| = \int_0^1 |f(x)-g(x)|dx = \int_0^1 |f(x)-h(x)+h(x)-g(x)|dx$$
$$\leq \int_0^1 |f(x)-h(x)|+|h(x)-g(x)|dx$$
$$= \int_0^1 |f(x)-h(x)|dx + \int_0^1 |h(x)-g(x)|dx$$
$$= \|f-h\| + \|h-g\|$$

$\Rightarrow$ $d(f, g) \leq d(f, h) + d(h, g)$.

Hence, from (i), (ii), (iii) and (iv), we conclude that d is a metric on R.

EXAMPLE 3. *Let L_∞ denote the set of all bounded sequence. If $x=\langle x_n \rangle$ and $y=\langle y_n \rangle$ are any two parts of L_∞, we define*

$$d(x, y)=\sup\{|x_n-y_n| : n \in N\}.$$

Show that l_∞ is a metric space under d.

SOLUTION. (i) Since $|x_n-y_n| \geq 0$. Therefore, $\sup|x_n-y_n| \geq 0$

$\Rightarrow$ $d(x, y) \geq 0$.

(ii) We have $|x_n-y_n|=0$ iff $x_n=y_n$ $\Rightarrow$ $\sup.|x_n-y_n|=0$ iff $x_n=y_n$

$\Rightarrow$ $d(x, y) = 0$ iff $x=y$.

(iii) Since $|x_n-y_n| = |y_n-x_n|$. Therefore $\sup|x_n-y_n| = \sup|y_n-x_n|$

$\Rightarrow$ $d(x, y)=d(y, x)$.

(iv) Let $z=\langle z_n \rangle$ be an element of l_∞. Then for any positive integer n, we have

$$|x_n-y_n| = |x_n-z_n+z_n-y_n|$$
$$\leq |x_n-z_n| + |z_n-y_n|$$
$$\leq \sup\{|x_n-z_n| : n\in N\} + \sup\{|z_n-y_n| : n \in N\}$$
$$= d(x, z)+d(z, y)$$

$\Rightarrow$ $d(x, y) \leq d(x, z)+d(z, y)$.

EXAMPLE 4. *Let (X, d) be a metric space and let M be a positive number, then there exists a metric d_1 on X such that the metric space (X, d_1) is bounded with $\delta(x) \leq M$.*

SOLUTION. Define d_1 by

$$d_1(x, y)=\frac{Md(x, x_1)}{1+d(x, x_1)};\text{where } x, y \in X.$$

To show d_1 is a metric,

(i) Since $d(x, y) \geq 0$ and $M \geq 0$.

$$\therefore \quad \frac{Md(x,x_1)}{1+d(x,x_1)} \geq 0 \quad \Rightarrow \quad d_1(x, y) \geq 0$$

(ii)
$$d_1(x, y)=0 \quad \Leftrightarrow \quad \frac{Md(x,x_1)}{1+d(x,x_1)}=0$$
$$\Leftrightarrow \quad d(x, y) = 0 \Leftrightarrow x=y.$$

(iii) $d_1(x, y)=\dfrac{Md(x,x_1)}{1+d(x,x_1)}=\dfrac{Md(y,x)}{1+d(y,x)}=d_1(y, x).$

(iv) Let $x, y, z \in X$, therefore

$$d_1(x, y)=\frac{Md(x,y)}{1+d(x,y)}=M-\frac{M}{1+d(x,y)}$$
$$\leq M-\frac{M}{1+d(x,z)+d(z,y)}+\frac{M[d(x,z)+d(z,y)]}{1+d(x,z)+d(z,y)}$$
$$=\frac{Md(x,z)}{1+d(x,z)+d(z,y)}+\frac{Md(z,y)}{1+d(x,z)+d(z,y)}$$
$$\leq\frac{Md(x,z)}{1+d(x,z)}+\frac{Md(z,y)}{1+d(z,y)}$$
$$\leq d_1(x, z)+d_1(z, y)$$

$$\Rightarrow \quad d_1(x, y) \leq d_1(x, z)+d_1(z, y).$$

From (i), (ii), (iii) and (iv), we conclude that

$$d_1(x, y)=\frac{Md(x,x_1)}{1+d(x,x_1)} \text{ is a metric on } X$$

also since $\quad d_1(x, y)=\dfrac{Md(x,x_1)}{1+d(x,x_1)} \leq M \quad$ for every points $x, y \in X$

therefore, d_1 is a bounded metric for X with $\delta(x) \leq M$.

REMARKS

- A metric space (X, d) is said to be bounded if there exist a positive number M such that $d(x, y) \leq M$ for every pair of points x, y of X.
- A metric space which is not bounded is said to be unbounded.
- A metric space X is said to be bounded if its diameter is finite.

EXAMPLE 5. *Let (X, d) be a metric space and let $d_1(x, y)= min.\{1, d(x, y)\}$ Then show that d_1 is a metric on X.*

SOLUTION. (i) Here, we have

$$d(x, y)=1 \text{ or } d_1(x, y)=d(x, y).$$

Clearly $\quad d(x, y) \geq 0 \quad$ [d is a metric]

Therefore, in both the cases $d_1(x, y) \geq 0$.

(ii) If $d_1(x, y)=0$, then $d_1(x, y)=d(x, y)=0$.

Since d is a metric, therefore $d(x, y)=0 \Leftrightarrow x=y$.

Hence $\quad d_1(x, y)=0 \Leftrightarrow x=y$

(iii) Since either

$$d_1(x, y)=1 \text{ or } d_1(x, y)=d(x, y).$$

Therefore, if $d_1(x, y)=d(x, y)$, then $d(x, y)<1$.

Hence, $d(y, x)=d(x, y)<1$.

But $d(y, x)<1$, gives $d_1(y, x)=d(y, x)=d(x, y)=d_1(x, y)$.

If $d_1(x, y)=1$, then $d(x, y) \geq 1$ and therefore $d(y, x) \geq 1$.

But $d(y, x) \geq 1$ gives $d_1(y, x)=1$.

Hence, $d_1(x, y)=d_1(y, x)$.

Therefore, in each case $d_1(x, y)=d_1(y, x)$.

(iv) Here, we want to prove that

$$d_1(x, y) \leq d_1(x, z)+d_1(z, y). \quad \text{...(A)}$$

If $d(x, y) \leq 1$, then if either $d_1(x, z)=1$ or $d_1(z, y)=1$

$\Rightarrow$ inequality (A) holds good.

If both $d_1(x, z)\neq 1$ and $d_1(z, y)\neq 1$, then

$$d_1(x, z)=d(x, z) \text{ and } d_1(z, y)=d(z, y).$$

Then, we have

$$d(x, y) \leq d(x, z)+d(z, y)= d_1(x, z)+d_1(z, y). \quad \text{...(B)}$$

But $\quad d_1(x, y)= \min \{1, d(x, y)\} \leq d(x, y). \quad$ ...(C)

From (B) and (C), we have

$$d_1(x, y) \leq d_1(x, z)+d_1(z, y).$$

Hence from (i), (ii), (iii) and (iv), we conclude that d_1 is a metric on X.

EXAMPLE 6. *Let d be a metric for non-empty set X. Show that d_1 defined by*

$$d_1(x, y)=2d(x, y) \text{ is also metric for } X.$$

SOLUTION. (i) Since $d(x, y)$ is a metric on X, therefore $d(x, y) \geq 0$.

$\Rightarrow \quad 2d(x, y) \geq 0 \Rightarrow d_1(x, y) \geq 0$.

(ii) Since $d(x, y)$ is a metric on X, therefore

$$d(x, y)=0 \Leftrightarrow x=y$$

or $\quad 2d(x, y)=0 \Leftrightarrow x=y$

or $\quad d_1(x, y)=0 \Leftrightarrow x=y$

(iii) Since $d(x, y)$ is a metric on X, therefore $d(x, y)=d(y, x)$

$\Rightarrow \quad 2d(x, y)=2d(y, x)$

$\Rightarrow \quad d_1(x, y)=d_1(y, x)$.

(iv) Since $\quad d(x, y) \leq d(x, z) + d(z, y)$

$\Rightarrow \quad 2d(x, y) \leq 2d(x, z)+2d(z, y)$

$\Rightarrow \quad d_1(x, y) \leq d_1(x, z)+d_1(z, y)$.

Hence, from (i), (ii), (iii) and (iv), we conclude that d_1 is a metric on X.

EXAMPLE 7. *Show that the function d, defined by $d(p, q)=\|p-q\|$ where p, q are vectors in a normed vector space* V *is metric on* V.

SOLUTION. Let $p, q, r \in$ V and x is any scalar.

By definition of normed vector space, we have

(a) $\|p\|>0$ (b) $\|p\|=0 \Leftrightarrow p=0$

(c) $\|xp\|=|x|\,\|p\|$ (d) $\|p+q\| \leq \|p\|+\|q\|$.

To show d is a metric on V.

(i) Since $d(p, q)=\|p-q\|$

Then by (a) $\|p-q\| \geq 0$.

(ii) Using (b), $d(p, q)=0 \Leftrightarrow \|p-q\|=0$

$\Leftrightarrow p-q=0 \Leftrightarrow p=q$

(iii) Using (c), we have

$$d(p, q)=\|p-q\|$$

$$\Rightarrow \quad d(q, p)=\|q-p\|=\|-(p-q)\|=\|p-q\|=d(p, q).$$

(iv) Consider

$$\|p-q\|=\|p-r+r-q\| \leq \|p-r\|+\|r-q\|$$

$$\Rightarrow \quad d(p, q) \leq d(p, r)+d(r, q).$$

Hence, from (i), (ii), (iii) and (iv), we conclude that $d(p, q)$ is a metric on V.

EXAMPLE 8. *Let C be the set of all complex number then show that mapping, d:*$C \times C \rightarrow R$ *is a metric on C if d is defined as*

$$d(z_1, z_2)=|z_1-z_2|, \forall z_1, z_2 \in C.$$

SOLUTION. Since d is defined by

$$d(z_1, z_2)=|z_1-z_2|, \forall z_1, z_2 \in C.$$

We have to show that (X, d) is metric space *i, e.*, d is metric.

(i) Since $|z_1-z_2| \geq 0, \forall z_1, z_2 \in C$

$\Rightarrow \quad d(z_1, z_2) \geq 0, \forall z_1, z_2 \in C$

(ii) Since $|z_1-z_2|=0 \Leftrightarrow z_1-z_2=0 \Leftrightarrow z_1=z_2$

$\Rightarrow \quad d(z_1, z_2)=0$ iff $z_1=z_2$.

(iii) $|z_1-z_2|=|z_2-z_1|, \forall z_1, z_2 \in C$

$\Rightarrow \quad d(z_1, z_2)=d(z_2, z_1), \forall z_1, z_2 \in C.$

(iv) Let $z_1, z_2, z_3 \in C$

$$|z_1-z_2|=|(z_1-z_3)+(z_3-z_2)|$$

$$\leq |z_1-z_3|+|z_3-z_2|, \forall z_1, z_2, z_3 \in C$$

$$\Rightarrow \quad d(z_1, z_2) \leq d(z_1, z_3)+d(z_3, z_2), \forall z_1, z_2, z_3 \in C.$$

Hence by above properties d is a metric.

EXAMPLE 9. *Let* R *be the set of real numbers and the mapping* $d: R \times R \rightarrow R$ *defined by*

$$d(x, y)=|x^2-y^2|, \forall x, y \in R$$

show that it is a Pseudo-metric on R *which is not a metric on* R.

SOLUTION. Here 'd' is defined as

$$d(x, y)=|x^2-y^2|, \forall x, y \in R$$

To show it a metric.

(i) Since $|x^2-y^2| \geq 0, \forall\, x, y \in R$

$\Rightarrow \quad d(x, y) \geq 0, \forall\, x, y \in R.$

(ii) Since $\quad d(x, x) = |x^2-x^2| = 0, \forall\, x \in R$

but here $\quad d(x, y) = |x^2-y^2| = 0$

$\Rightarrow \quad x^2-y^2 = 0 \Rightarrow x = \pm y.$

$\Rightarrow \quad d(x, y) = 0$ if $x = y$ but not converse,

(iii) Since $\quad d(x, y) = |x^2-y^2| = |y^2-x^2|$

$\Rightarrow \quad d(x, y) = d(y, x), \forall\, x, y \in R.$

(iv) Let $x, y, z \in R$

$$|x^2-y^2| = |(x^2-z^2)+(z^2-y^2)|$$
$$\leq |(x^2-z^2)| + |(z^2-y^2)|$$

$\Rightarrow \quad d(x, y) \leq d(x, z) + d(z, y), \forall\, x, y, z \in R$

By above properties $d(x, y)$ is Pseudo metric space.

EXAMPLE 10. *Let d_1, d_2 be two metric for a non-empty set X. Show that the mapping d defined by*

$$d(x, y) = d_1(x, y) + d_2(x, y), \quad \forall x, y \in X$$

is also a metric for X.

SOLUTION. Let d_1, d_2 are matrices on a non-empty set X which satisfying the following properties:

(i) $d_1(x, y) \geq 0,\ d_2(x, y) \geq 0, \forall\, x, y \in X$

(ii) $d_1(x, y) = 0$ iff $x = y$ and $d_2(x, y) = 0$ iff $x = y$

(iii) $d_1(x, y) = d_1(y, x)$ and $d_2(x, y) = d_2(y, x), \forall\, x, y \in X$

(iv) $d_1(x, y) \leq d_2(x, z) + d_1(z, y), \forall\, x, y \in X$

$d_2(x, y) \leq d_2(x, z) + d_2(z, y), \forall\, x, y \in X.$

Now we have to show that

$$d(x, y) = d_1(x, y) + d_2(x, y), \quad \forall\, x, y \in X$$

is metric on X.

(i) $\quad d_1(x, y) \geq 0,\ d_2(x, y) \geq 0$

$\Rightarrow \quad d_1(x, y) + d_2(x, y) \geq 0\ \forall x, y \in X$

$\Rightarrow \quad d(x, y) \geq 0\ \forall\, x, y \in X.$

(ii) Since $\quad d_1(x, y) = 0$ iff $x = y$ and $d_2(x, y) = 0$ iff $x = y$

$\Rightarrow \quad d_1(x, y) + d_2(x, y) = 0$ iff $x = y$

$\Rightarrow \quad d(x, y) = 0$ iff $x = y$.

(iii) Since $\quad d_1(x, y) = d_1(y, x)$ and $d_2(x, y) = d_2(y, x)\ \forall\, x, y \in X$

$$d(x, y) = d_1(x, y) + d_2(x, y)$$
$$= d_1(y, x) + d_2(y, x) = d(y, x)\ \forall x, y \in X$$

(iv) Let x, y, z be elements of X such that

$$d_1(x, y) \leq d_1(x, z) + d_1(z, y)$$

and $\quad d_2(x, y) \leq d_2(x, z) + d_2(z, y)$

$\Rightarrow \quad d_1(x, y) + d_2(x, y) \leq d_1(x, z) + d_1(z, y) + d_2(x, z) + d_2(z, y)$

$$\Rightarrow \qquad d(x, y) \le [d_1(x, z) + d_2(x, z)] + [d_1(z, y) + d_2(z, y)]$$

$$\Rightarrow \qquad d(x, y) \le d(x, z) + d(z, y) \;\; \forall\, x, y, z \in X.$$

so $d(x, y)$ is a metric on X.

EXAMPLE 11. *Let (X_1, d_1) and (X_2, d_2) be two metric spaces. For any pair of points $x=(x_1, x_2)$, $y=(y_1, y_2)$ in $X=X_1 \times X_2$ and is defined as*

$$d(x, y) = d_1(x_1, y_1) + d_2(x_2, y_2)$$

then prove that d is metric on $X=X_1 \times X_2$.

SOLUTION. Since d_1 and d_2 are two metric spaces then it satisfying all properties of metric spaces. Proceeding same as above example, we have to show that d is metric for $X=X_1 \times X_2$ defined by $d(x, y)= d_1(x_1, y_1)+d_2(x_2, y_2)$.

(1) Since $\quad d_1(x_1, y_1) \ge 0, \forall\, x_1, y_1 \in X_1$ and $d_2(x_2, y_2) \ge 0, \forall\, x_2, y_2 \in X_2$

$\Rightarrow d_1(x_1, y_1)+d_2(x_2, y_2) \ge 0, \forall\, x_1, y_1 \in X_1, \forall\, x_2, y_2 \in X_2$

$\Rightarrow \quad d(x, y) \ge 0, \forall\; x, y \in X$, where $x=(x_1, x_2), y=(y_1, y_2)$.

(2) $d_1(x_1, y_1)=0$ iff $x_1 = y_1$ and $d_2(x_2, y_2)=0$ iff $x_2=y_2$

$\Rightarrow d_1(x_1, y_1)+d_2(x_2, y_2)=0$ iff $x_1= y_1$ and $x_2=y_2$.

$\Rightarrow \quad d(x, y) = 0$ iff $(x_1, x_2)=(y_1, y_2)$

$\Rightarrow \quad d(x, y) = 0$ iff $x=y$.

(3) We have $\quad d_1(x_1, y_1)= d_1(y_1, x_1), \forall\, x_1, y_1 \in X_1$

$$d_2(x_2, y_2)= d_2(y_2, x_2), \forall\, x_2, y_2 \in X_2$$

$$\begin{aligned} d(x, y) &= d_1(x_1, y_1)+d_2(x_2, y_2) \\ &= d_1(y_1, x_1)+d_2(y_2, x_2) \\ &= d_1(y, x) \;\forall\, x, y \in X. \end{aligned}$$

(4) Let$x=(x_1, x_2), y=(y_1, y_2), z=(z_1, z_2)$

and $\quad X=X_1 \times X_2, x_1, y_1, z_1 \in X_1, x_2, y_2, z_2 \in X_2$

since $\quad d_1(x_1, y_1) \le d_1(x_1, z_1)+d_1(z_1, y_1)$

and $\quad d_2(x_2, y_2) \le d_2(x_2, z_2)+d_2(z_2, y_2)$

Now $\quad d_1(x_1, y_1)+d_2(x_2, y_2) \le d_1(x_1, z_1)+d_1(z_1, y_1)+d_2(x_2, z_2)+d_2(z_2, y_2)$

$$d_1(x_1, y_1)+d_2(x_2, y_2) \le d_1(x_1, z_1)+d_2(x_2, z_2)+d_1(z_1, y_1)+d_2(z_2, y_2)$$

$$d(x, y) \le d(x, z)+d(z, y), \forall x, y, z \in X.$$

Hence, d is metric space.

EXAMPLE 12. *Let* R *be the set of all real number and let* R^2 *the set of all ordered pairs of real number. Then the function*

$$d : R^2 \times R^2 \to R$$

where d is defined as $d(x, y)=\left[(x_1-y_1)^2+(x_2-y_2)^2\right]^{1/2}$, *is a metric on* R.

SOLUTION. Here we shall show that d is a metric space defined as

$$d : R^2 \times R^2 \to R$$

such that $\quad d(x, y)=\left[(x_1-y_1)^2+(x_2-y_2)^2\right]^{1/2}$

(1) Since $(x_1-y_1)^2 \geq 0, (x_2-y_2)^2 \geq 0$

$\Rightarrow \quad (x_1-y_1)^2+(x_2-y_2)^2 \geq 0$

$\Rightarrow \quad \left[(x_1-y_1)^2+(x_2-y_2)^2\right]^{1/2} \geq 0$

$\Rightarrow \quad d(x, y) \geq 0, \forall\, x, y \in R^2.$

(2) Let $d(x, y) = 0 \Leftrightarrow \left[(x_1-y_1)^2+(x_2-y_2)^2\right]^{1/2}=0$

$\Leftrightarrow \quad (x_1-y_1)^2+(x_2-y_2)^2=0$

$\Leftrightarrow \quad (x_1-y_1)^2=0$ and $(x_2-y_2)^2=0$

$\Leftrightarrow \quad x_1-y_1=0$ and $x_2-y_2=0$

$\Leftrightarrow \quad x_1=y_1$ and $x_2=y_2 \Leftrightarrow x=y$

Hence $d(x, y)=0 \Leftrightarrow x=y \;\forall\, x, y \in R$

(3) $d(x, y)=\left[(x_1-y_1)^2+(x_2-y_2)^2\right]^{1/2}$

$=\left[(y_1-x_1)^2+(y_2-x_2)^2\right]^{1/2}=d(y, x)$

(4) $d(x, y)=\left[(x_1-y_1)^2+(x_2-y_2)^2\right]^{1/2}$

$=\left[\{(x_1-z_1)+(z_1-y_1)\}^2+\{(x_2-z_2)+(z_2-y_2)\}^2\right]^{1/2}$

$\leq\left[(x_1-z_1)^2+(x_2-z_2)^2\right]^{1/2}+\left[(z_1-y_1)^2+(z_2-y_2)^2\right]^{1/2}.$

$= d(x, z)+d(z, y)$

$\left[\because \text{By Minkowski's inequality, we have } \left[\sum_{i=1}^{n}(x_i+y_i)^p\right]^{1/p} \leq \left(\sum_{i=1}^{n} x_i^p\right)^{1/p}+\left[\sum_{i=1}^{n} y_i^p\right]^{1/p}\right]$

$\therefore \quad d(x, y) \leq d(x, z)+d(z, y), \forall\, x, y, z \in R^2.$

Hence, we conclude that d is a metric on R^2.

EXERCISE 2.1

1. Give an example of a pseudo-metric which is not metric. Is every metric a pseudo metric?
2. Let (X, d) be a metric space and let x, y, z be any three points of X, then show that
$$d(x, y) \geq |d(x, z)-d(z, y)|$$
3. Let (X, d) be a metric space and let $x, x', y, y' \in X$. Then show that
$$|d(x, y)-d(x', y')| \leq d(x, x')+d(y, y').$$
4. Let A, B be subsets of a metric space (X, d), then show that
$$\delta(A \cup B) \leq \delta(A)+\delta(B)+d(A, B).$$
5. Define the diameter of subset A of the metric space X. What is the diameter of empty set? What do you mean by the distance between two non-empty subset A and B of metric space X. If x is a point of X and A is a subset of X, write the distance of x from A.
6. Let R[0, 1] denote the classes of all Reimann integrable function f from [0, 1] into R. Consider the mapping $d : R[0, 1] \times R[0, 1] \to R$ defined by $d(f, g)=\int_0^1|f-g|(x)\,dx=\int_0^1|f(x)-g(x)|\,dx$. Show that d is a pseudo-metric but not metric on R.
7. Show that $d : R^2 \times R^2 \to R$ defined by $d(x,y)=|x_1-y_1|+|x_2-y_2|, x=(x_1,x_2), y=(y_1, y_2)$ where $x \in R^2$ is a metric on R^2.

8. Let $X=R^n$ denote the set of all ordered n-tuples of real numbers for a fixed $n \in N$. Let
$$x=(x_1, x_2,\ldots, x_n), y=(y_1, y_2,\ldots, y_n).$$
Define the mapping d_1, d_2 and d_3 of $R^n \times R^n$ into R by

(i) $d_1(x,y)=\left[\sum_{i=1}^{n}(x_i-y_i)^2\right]^{1/2}$.

(ii) $d_2(x,y)=\sum_{i=1}^{n}|x_i-y_i|$.

(iii) $d_3(x,y)=\max.\{|x_1-y_1|, |x_2-y_2|,\ldots,|x_n-y_n|\}$.

Show that d_1, d_2, d_3 are metrics on R^n.

9. Show that the set C of all complex numbers is a metric space under
$$d(z_1, z_2)=\frac{|z_1-z_2|}{\left[\left(1+|z_1|^2\right)\left(1+|z_2|^2\right)\right]^{1/2}}.$$

10. If (X, d) be a metric space, then show that d_1, defined by $d_1(x, y)=\dfrac{d(x,y)}{1+d(x,y)}$ is also a metric on X.

11. Show that sum of two metric spaces is again a metric space.

12. If $d(x, y)=2|x-y|$. Then show that d is a metric on R.

13. Let (X, d) be a metric space and let A, B be subsets of X, then show that $A \subset B \Rightarrow \delta(A) \leq \delta(B)$.

14. Let $d(x, y)=\min\{2, |x-y|\}$. Show that d is a metric on R.

15. Let $C[0, 1]$ denote the collection of all real valued bounded continuous functions defined on the closed interval $[0, 1]$, we define the norm of $f \in C[0, 1]$ by $\|f\| = \sup\{|f(x)| : x \in [0, 1]\}$ where d is defined as $d(f, g)=\|f-g\|=\sup\{|f(x)-g(x)| : x \in [0, 1]\}$ then show that d is a metric for $C[0, 1]$.

16. Let $x=\langle x_n\rangle$ be a sequence then $F=\{x : x=\langle x_n\rangle\}$ is said to be Frechet space, for which d is defined as $d(x, y)=\sum_{n=1}^{\infty}\dfrac{|x_n-y_n|}{2^n(1+|x_n-y_n|)}, (x, y \in F)$.
Show that d is metric on F.

17. Let R be the set of real numbers and let $d(x, y)=\dfrac{|x-y|}{1+|x-y|}, \forall x, y \in R$ show that d is a metric for R.

18. Let (X, d) be any metric space then show that function d^* defined by $d^*(x, y)=\min\{2, d(x, y)\}$, $\forall x, y \in X$ is a metric on X.

HINT TO SELECTED PROBLEMS

1. $d(x, y)=|x^2-y^2| \ \forall x, y \in R$.

12. Since, we know that $d(x, y)=|x-y|$ is a metric. Use this result to prove $d(x, y)=2|x-y|$ is a metric.

13. By definition of the diameter of a set, we have
$$d(A)=\sup\{d(x, y): x, y \in A\}$$
Since $A \subset B$

$x, y \in A \Rightarrow x, y \in B$.

$\therefore \{d(x, y) : x, y \in A\} \Rightarrow \{d(x, y) : x, y \in B)$

$\Rightarrow \sup\{d(x, y) : x, y \in A\}$

$\leq \sup|d(x, y) : x, y \in B\}$ $(A \subset B)$

$\delta(A) \leq \delta(B)$.

17. Do same as example 4.

18. Do same as example 5.

ANSWERS

4. $d(x, y)=|x^2-y^2|$, Yes

2.5 OPEN AND CLOSED SETS IN A METRIC SPACE

Let (X, d) be a metric space. Let $p \in X$ and $r > 0$ be given. Then, the set of all points $x \in X$ such that $d(p, x) < r$ is called the open ball or open sphere of radius r and centre p in X and is denoted by $S(p, r)$ or $\mathring{d}(p, r)$

i.e., $S(P, r)=\{x \in X : d(x, p)<r\}$.

The set of all points $x \in X$ such that $d(p, x) \leq r$ is called closed ball of radius r and centre p and is denoted by $S^*(p, r)$
$$S^*(P, r)=\{x \in X: d(x, p) \leq r\}.$$

☛ ILLUSTRATIONS

In the case of real line :

(1) $S(p, r)$ is the open interval $]p-r, p+r[$.

(2) The open interval $]a, b[$ is the open ball with centre at the points $p=\frac{1}{2}(a+b)$ and radius $r=\frac{1}{2}(b-a)$.

(3) $S(p, r)$ is the closed interval $[p-r, p+r]$.

(4) The closed interval $[a, b]$ is the closed ball $S^*(p, r)$ with centre $p=\frac{1}{2}(a+b)$ and radius $r=\frac{1}{2}(b-a)$.

REMARKS

- Open sphere is defined as spherical neighbourhood of a point p.
- An open (or closed) sphere is always non-empty, since it contains its centre at least.

2.5.1 OPEN SET

Definition. *Let (X, d) be a metric space. A subset A of X is said to be open iff to each $x \in A$, there exist $r>0$ such that $S(x, r) \subseteq A$.*

For Example :

(1) The subset $]0, 3[$ is open in $X=[0, 3]$ under the metric d given by $d(x, y)=|x-y|$.

Since d is open set in X.

2.5.2 LIMIT POINT

Let (X, d) be a metric space and $A \subset X$. A point $x \in X$ is called a limit point or limiting point or accumulation point or cluster point if every open sphere centered at x contains a point of A, other than x

i.e., $x \in X$ is called limit point of A if $[S(x, r)-\{x\}] \cap A \neq \phi$, $r \in R^+$.

2.5.3 ADHERENT POINT

Let (X, d) be a metric space and A be any subset of X. A point $x \in X$ is said to be adherent point of A if every open sphere centered at x contains at least one point of A.

i.e. if $\{S(x, r)-\{x\}\} \cap A$ is empty or non-empty for every open sphere $S(x, r)$.

There are two type of adherent points :

(a) *Limit point.* A point $x \in X$ is called limiting point of A if $[S(x, r)-\{x\}\} \cap A \neq \phi \ \forall \ r \in R^+$.

(b) *Isolated point.* A point $x \in X$ is called isolated point of A if

$$[S(x, r)-\{x\}] \cap A=\phi \ \forall \ r \in R^+.$$

i.e. if x is not the limit point of A.

2.5.4 DERIVED SET

Set of all limit points of a set A is called derived set of A and is denoted by $D(A)$.

2.5.5 CLOSED SET

Let (X, d) be a metric space and $A \subset X$. Then A is called closed set if the derived set of A contains in A *i.e.* $D(A) \subset A$. (or) if every limit point of A belongs to the set itself.

2.5.6 PERFECT SET

A closed set which has no isolated point is called perfect set.

2.5.7 DIFFERENCE BETWEEN LIMIT AND LIMITING POINT

The limit and limit point both are different terms. For example, consider the sequence $\langle 1,1,1,\ldots\rangle$ in which every element is identically equal to 1. The limit of the sequence is 1, but the set of all points of this sequence is the singleton set [1] and hence its limit point does not exists, because finite set has no limit point.

Here, it is also possible that the set of points of a sequence may have a limit point, but cannot have a limit.

THEOREM 1. ***In a metric space (X, d), the empty set ϕ and the whole space X, are open sets.***

PROOF. Let (X, d) be a metric space. To show that ϕ and X are open sets.

To prove that ϕ is open in X, it is suffices to show that

$$x \in \phi \Rightarrow \exists\ \varepsilon>0: S(x, \varepsilon) \subset \phi.$$

Since ϕ does not contain any element and hence this condition is automatically fulfilled.

Now to show X is open.

Since, corresponding to every point $x \in X$ $\exists$ an open sphere with its centre at x which is contained in X. Hence, X is an open set.

THEOREM 2. ***In a metric space (X, d), ϕ and X are closed sets.***

PROOF. Let (X, d) be a metric space. To show ϕ and X are closed sets.

We know that $\quad D(\phi)=\phi.$

Therefore $\quad D(\phi) \subset \phi.$

$\Rightarrow$ ϕ contain all its limit points.

$\Rightarrow$ ϕ is closed.

Now to show X is closed.

Since, all the limit points of X belongs to X. $\quad$ ($\because$ X is the universal set)

i.e. $\quad x \in D(X) \Rightarrow x \in X.$

$\Rightarrow$ X contains all its limit points.

$\Rightarrow$ X is a closed set.

THEOREM 3. ***In a metric space, every open sphere is an open set.***

PROOF. Let (X, d) be a metric space and let $S(p, r)$ be any open sphere.

To show that to each point $x_0 \in S(p, r)$ there exists an open sphere centred at x_0 and contained in $S(p, r)$.

Let $x_0 \in S(p, r)$ be arbitrary.

Now, $\quad x_0 \in S(p, r) \Rightarrow d(x_0, p) < r \Rightarrow r-d(x_0, p) > 0.$

Let us define $\quad \rho=r-d(x_0, p).$ $\quad$ Then clearly $\rho>0$.

Define an open sphere $S(x_0, \rho)$.

To show $\quad S(x_0, \rho) \subset S(p, r).$

Let $x \in S(x_0, \rho)$. Then by definition of open sphere, we have

$$d(x, x_0)<\rho=r-d(x_0, p).$$

Now, $\quad d(x, p) < d(x, x_0)+d(x_0, p)$ $\quad$ (By Triangle inequality)

$$< r - d(x_0, p) + d(x_0, p) = r$$

$\Rightarrow \quad d(x, p) < r \Rightarrow x \in S(p, r).$

Since x is arbitrary.

Hence, $\quad S(x_0, \rho) \subset S(p, r) \quad \Rightarrow \quad S(p, r)$ is an open set.

THEOREM 4. ***In a metric space (X, d), arbitrary union of open sets is open.***

PROOF. Let (X, d) be a metric space and let $[G_\lambda : \lambda \in \Lambda]$ be an arbitrary collection of open subset of X.

Define $\quad G = \cup [G_\lambda : \lambda \in \Lambda].$

To show G is open.

Let $x \in G \quad \Rightarrow \quad x \in G_\lambda$ for some λ.

Since G_λ is open, therefore there exist $r>0$ such that

$$S(x, r) \subset G_\lambda. \quad \text{...(1)}$$

By definition of G, we have

$$G_\lambda \subset G. \quad \text{...(2)}$$

From (1) and (2), we have

$$S(x, r) \subset G_\lambda \subset G$$

Since x is arbitrary therefore, we can say that to each $x \in G$, there exists a number $r>0$ such that $\quad S(x, r) \subset G.$

$\Rightarrow$ G is open.

Hence, the arbitrary union of open sets in a metric space, is open.

THEOREM 5. ***In a metric space (X, d), the intersection of a finite number of open sets is open.***

PROOF. Let (X, d) be a metric space and G_i $(i=1, 2, ..., n)$ be a finite collection of open subsets of X.

Define $\quad H = \cap [G_i : i = 1, 2, ..., n].$

To show H is open.

Let $x \in H$. Then by definition of H, $x \in G_i$ for each $i = 1, 2, ..., n$.

Also each G_i is open, therefore there exist $r_i > 0$ such that

$$S(x, r_i) \subset G_i, \text{ for each } i.$$

Let $r = \min\{r_1, r_2, ..., r_n\}$

Then $\quad S(x, r) \subset S(x, r_i) \subset G_i, \ \forall\, i = 1, 2, ..., n$

$\Rightarrow \quad S(x, r) \subset G_i, \quad \forall\, i = 1, 2, ..., n$

$\Rightarrow \quad S(x, r) \subset \cap [G_i : i = 1, 2, ..., n]$

Since x is arbitrary, therefore it is shown that to each $x \in H$, there exist $r>0$, such that $S(x, r) \subset H$. Hence H is open.

THEOREM 6. ***In a metric space (X, d), for every pair of distinct points x, y ∈ X, there exist disjoint open sets U and V such that x ∈ U and y ∈ V.***

PROOF. Let (X, d) be a metric space and x, y be two distinct point of X such that $x \neq y$.

Then clearly $\quad d(x, y) > 0.$

Define $\varepsilon = \dfrac{1}{3} d(x, y)$, then clearly $\varepsilon > 0$.

Also, define $U=S(x, \varepsilon)$ and $V=S(y, \varepsilon)$, then $x \in U$ and $y \in V$.

Since every open sphere is an open set, therefore U and V both are open sets.

Now to show U and V are distinct *i.e.* $U \cap V=\phi$.

Let if possible $U \cap V \neq \phi$ and $p \in U \cap V$.

Therefore, $\quad p \in U \cap V \Rightarrow p \in U$ and $p \in V$.

Now $\quad p \in U \Rightarrow d(p, x)< \varepsilon \quad$ ($\because$ U is an open sphere)

and $\quad p \in V \Rightarrow d(p, y)< \varepsilon. \quad$ ($\because$ V is an open sphere)

By triangle inequality, we have

$$\begin{aligned} d(x, y) &\le d(x, p)+d(p, y) \\ &= d(p, x)+d(p, y) \quad [\because d(p, x)=d(x, p)] \\ &<\varepsilon+\varepsilon=2\varepsilon. \end{aligned}$$

$$\Rightarrow \quad d(x, y)<2\varepsilon$$

$$\Rightarrow \quad d(x, y)<2.\frac{1}{3}\, d(x, y)$$

$$\Rightarrow \quad d(x, y)<\frac{2}{3}\, d(x, y) \text{ which is absurd.}$$

Therefore, we have a contradiction $\quad$ Hence, $\quad U \cap V=\phi$

Hence, we can find two disjoint open sets U and V such that $x \in U$ and $y \in V$.

THEOREM 7. ***A subset of a metric space is open if and only if it is the union of a family of open spheres.***

PROOF. Let (X, d) be a metric space and A be any subset of X.

Let us first suppose A is open. To show A can be written as the union of family of open spheres.

If $A=\phi$, then it can be written as the union of empty family of open sphere.

If $A\neq\phi$, let $x_1 \in A$. Now, since A is open, therefore, there exists an open sphere $S(x_1, r)$ such that $S(x_1, r) \subset A$

Similarly for $x_2 \in A$ There exists an open sphere $S(x_2, r)$ such that $S(x_2, r) \subset A$.

Proceeding in the same way, we can say that for each point x_i of A there exists an open sphere $S(x_i, r)$ such that $S(x_i, r) \subset A$

Therefore, $\quad A \subset \cup\, [S(x_i, r) : x_i \in A] \subset A$

$$\Rightarrow \quad A = \cup\, [S(x_i, r) : x_i \in A].$$

Hence A can be written as the union of a family of open spheres.

Conversely, let A can be written as the union of open spheres.

To show A is open set.

Since, we know that every open sphere in a metric space is an open set. Therefore $S(x_i, r)$ is an open set. Also, finite union of open sets is again open.

Since $\quad A=\cup\, [S(x_i, r) : x_i \in A]$.

Also the right hand side is the union of open sets. Therefore, A is open.

THEOREM 8. ***Every non-empty open set on the real line is the union of a countable collection of pairwise disjoint open intervals.***

PROOF. Let G be an open subset of R. Let $x \in G$.

Since G is open, there exists an open interval $S(x, r)$, centered at x such that

$$S(x, r) \subset G.$$

Let I_x be the union of all open intervals, which contain x and are contained in G. Then we have

(a) I_x is open interval containing x and contained in G.

(b) I_x contains each open interval which contains x and is contained in G.

(c) If y is any other point in I_x, then $I_x = I_y$.

If x and y are two distinct points of G, then either $I_x = I_y$ or $I_x \cap I_y = \phi$.

If $z \in I_x \cap I_y$. Then we have $z \in I_x$ and $z \in I_y$

$$\Rightarrow \qquad I_x = I_z \text{ and } I_y = I_z \Rightarrow I_x = I_y.$$

Now, let I be the collection of all distinct sets of the form I_x for points x belonging to G.

Then clearly, we have I is the collection of open intervals and G is the union of this collection.

Now, to show I is countable.

Let G_i be the set of all rational points in G. Then, obviously $G_i \neq \phi$.

Now, define a map f of G_i onto I such that for each $i \in G_i$, $f(i)$ be the unique interval in I to which i belongs.

$\Rightarrow G_i$ is countable ($\because G_i$ is a non-empty subset of the countable set Q of all rational numbers)

$\Rightarrow I$ is countable.

Hence, we can say that every non-empty open set on the real line is the union of a countable collection of pairwise disjoint open intervals.

THEOREM 9. ***Let (X, d) be a metric space and A, any subset of X. Then A is closed if and only if its complement (i.e. X – A) is open.***

PROOF. Let (X, d) be a metric space. Let us first suppose A is closed. To show its complement $(X–A)$ is open.

Let $x \in X–A$, then $x \notin A$. Since A is closed and $x \notin A$.

$\Rightarrow x$ is not the limit point of A.

Then by definition of limit point $\exists$ an open sphere $S(x, r)$ such that

$$S(x, r) \cap A = \phi$$

$$\Rightarrow \qquad S(x, r) \subset X–A \text{ for some } r>0.$$

Since $x \in X–A$ is arbitrary, therefore each point of $X–A$ is the centre of some open sphere which is contained in $X–A$.

$\Rightarrow X–A$ is open.

Conversely, let $X–A$ is open. Let $x \in X$ be any limit point of A

If $x \in A$, then A is closed (By definition of closed set)

If $x \notin A$, then $x \in X–A$, also since $X–A$ is open therefore there exists an open sphere $S(x, r)$ which contained in $X–A$ i.e. $S(x, r) \subset X–A$

and $\qquad S(x, r) \cap A = \phi$ for some $r>0$

$\Rightarrow x$ is not the limit point, which is a contradiction. Therefore, $x \in A$.

Hence, A is closed.

THEOREM 10. ***In a metric space, every closed sphere is a closed set.***

PROOF. Let (X, d) be a metric space. Consider a closed sphere $S(x_0, r)$ in X. To show $S(x_0, r)$ is a closed set. For this, we shall show that the complement $S'(x_0, r)$ of $S(x_0, r)$ is open.

Let $x \in S'(x_0, r)$. Then $x \notin S(x_0, r)$.

Then by definition of open sphere $d(x, x_0) > r$.

Define $\rho = d(x, x_0) - r$.

Obviously $\rho > 0$. ...(1)

Now, we take an open sphere $S(x, \rho)$ of radius ρ centered at x.

To show $S(x, \rho) \subset S'(x_0, r)$.

Let $y \in S(x, \rho) \Rightarrow d(x, y) < \rho$. ...(2)

Consider $d(x, x_0) \leq d(x, y) + d(y, x_0)$

or, $d(y, x_0) \geq d(x, x_0) - d(x, y)$

$> d(x, x_0) - r$ [using (2)]

$= d(x, x_0) - [d(x, x_0) - r]$ [using (1)]

$= r$

$\Rightarrow$ $d(y, x_0) > r$

$\Rightarrow$ $y \in S'(x_0, r)$

$\Rightarrow$ $S(x, p) \subset S'(x_0, r)$ ($\because$ y is arbitrary)

$\Rightarrow$ $S'(x_0, r)$ is void

$\Rightarrow$ $S'(x_0, r)$ is open

Hence, $S(x_0, r)$ is closed.

REMARK

- Let (X, d) be a metric space and S be any subset of X defined by

$$S = \{x \in X : d(x, x_0) = r\}$$

where $r > 0$ and $x \in X$. Then S is a closed set.

THEOREM 11. ***In a metric space (X, d), the intersection of an arbitrary family of closed sets is closed.***

PROOF. Let (X, d) be a metric space and let $(H_\lambda : \lambda \in \Lambda)$ be an arbitrary collection of closed subset of X. Then to show $\cap\, [H_\lambda : \lambda \in \Lambda]$ is also a closed set.

Since H_λ is closed for each $\lambda \in \Lambda$

$\Rightarrow$ $X - H_\lambda$ is open

$\Rightarrow$ $\cup\, (X - H_\lambda)$ is open ($\because$ Arbitrary union of open sets is open)

$\Rightarrow$ $X - \cap\, [H_\lambda : \lambda \in \Lambda]$ is open (By De-morgan's law)

$\Rightarrow$ $\cap\, [H_\lambda : \lambda \in \Lambda]$ is closed ($\because$ Complement of an open set is closed)

Hence, the arbitrary intersection of closed sets is closed.

THEOREM 12. ***In a metric space (X, d) the finite union of closed sets is closed.***

PROOF. Let (X, d) be a metric space and $H_i : i = 1, 2, \ldots, n$ be closed subsets of X. To show $\bigcup_{i=1}^{n} H_i$ is closed.

Since each H_i is closed

$\Rightarrow H_i'$ is open (∵ Complement of a closed set is open)

$\Rightarrow \bigcap_{i=1}^{n} H_i'$ is open (∵ Finite intersection of open sets is open)

$\Rightarrow \left[\bigcup_{i=1}^{n} H_i\right]'$ is open (By De-morgan's law)

Hence, $\bigcup_{i=1}^{n} H_i$ is closed.

THEOREM 13. ***In a discrete metric space, every set is open.***

PROOF. Let A be any non-empty subset of a discrete metric space (X, d).

If $A=\phi \Rightarrow A$ is open.

If $A \neq \phi$ and $x \in A$, where x is any arbitrary point of A

and $$S\left(x, \frac{1}{2}\right) = \{x\} \subseteq A$$

$\Rightarrow A$ is open.

Hence in a discrete metric space every set is open.

REMARKS

- The union of infinite number of closed sets may or may not be closed.

  For Example: Let $F_x = \left[0, \frac{n}{n+1}\right]$ Then $\bigcup_{n=1}^{n} F_n = [0, 1[$ = semi open, which is not closed.

- The intersection of an infinite number of open sets is not necessarily open.

  For Example : Consider an open interval collection $\left\{\left]-\frac{1}{n}, \frac{1}{n}\right[: n \in N\right\}$ in R with usual metric $d(x, y) = |x-y|$.

  So the intersection of this type collection will be

  $$\cap \left\{\left]-\frac{1}{n}, \frac{1}{n}\right[: n \in N\right\} = \{0\} \text{ which is not open.}$$

  Hence in a metric space the intersection of an infinite collection of open sets is not open.

2.6 NEIGHBOURHOOD

Let (X, d) be a metric space. A set N of X is said to be neighbourhood (*nbd*) of a point $p \in X$ if there exists an $\varepsilon > 0$ such that $S(x, \varepsilon) \subset N$.

For Example :

(1) *On the real line* :

(i) The open interval $]a, b[$ is a nbd of each of its points

(ii) R is a nbd of each of its points.

(iii) The closed interval $[a, b]$ is a nbd of each of its points except the end points.

(iv) The set of integers Z is not a nbd of any of its points.

(v) The set of rational numbers Q is not a nbd of any of its points.

(2) Let (X, d) be a discrete metric space and $x \in X$. Then $\{x\}$ is a nbd of x. Further every superset of $\{x\}$ is also a nbd of x.

2.7 PROPERTIES OF NEIGHBOURHOOD

THEOREM 1. ***Let (X, d) be a metric space and A be any subset of X. If N is the neighbourhood of A and $M \supset N$, then M is also a neighbourhood of A. i.e., 'every superset of a neighbourhood of A is also a neighbourhood of A'.***

PROOF. Let (X, d) be a metric space and N be a nbd of $A \subset X$.

Since N is a nbd of A, therefore, by definition there exists an open set G such that

$$A \subset G \subset N \quad \text{...(1)}$$

Given that $\quad N \subset M. \quad$ ...(2)

From (1) and (2), we conclude that

$$A \subset G \subset M$$

$\Rightarrow$ M is a nbd of A.

THEOREM 2. ***The intersection of a finite number of neighbourhood of A is also a neighbourhood of A.***

PROOF. Let (X, d) be a metric space and A is any subset of X.

Also let $N_1, N_2, ..., N_k$ are a finite number of neighbourhoods of A, then to show that

$$\cap \{N_i : i = 1, 2, ..., k] \text{ is also a neighbourhood of } A$$

Since each N_i $(i = 1, 2, ..., n)$ is a nbd of A, then by definition there exist open sets G_i $(i = 1, 2, ..., k)$ such that

$$A \subset G_i \subset N_i \,;\, i=1, 2, ..., k.$$

Since $A \subset G_i$ for each $i = 1, 2, ..., k$.

Therefore $\quad A \subset \cap \{G_i : i=1, 2, ..., k\}$

Also, $\quad G_i \subset N_i \quad \forall\, i = 1, 2, ..., k$

$\Rightarrow \quad \cap \{G_i : i = 1, 2, ..., k\} \subset \cap \{N_i : i = 1, 2, ..., k\}$

$\Rightarrow \quad A \subset \cap \{G_i : i = 1, 2, ..., k\}$

$\quad \subset \cap \{N_i : i = 1, 2, ..., k\}. \quad$ ...(1)

Since each G_i is open, therefore $\cap\, [(G_i : i = 1, 2, ..., k]$ is open.

[$\because$ Finite intersection of open sets is again open]

Hence, from (1), we conclude that $\cap \{N_i : i = 1, 2, ..., k\}$ is also a neighbourhood of A.

REMARK

- The intersection of an infinite number of neighbourhood of a set A is not necessarily the nbd of A. For example, in the metric space (R, d) where $d(x, y) = |x-y|$, (*i.e.* usual metric) $\left]-\frac{1}{n}, \frac{1}{n}\right[$ is a nbd of A for each $n \in N$ but $\cap \left\{\left]-\frac{1}{n}, \frac{1}{n}\right[: n \in N\right\} = \{0\}$, which is not a nbd of A.

THEOREM 3. ***In a metric space, every open sphere is a neighbourhood of each of its points.***

PROOF. Let (X, d) be a metric space and let $S(x_0, r)$ be an open sphere centred at x_0, and of radius r; Let $y \in S(x_0, r)$.

Now to show that $S(x_0, r)$ is a neighbourhood of y. We have to show that there exists an open sphere centred at p, which is contained in $S(x_0, r)$.

Now we have

$$y \in S(x_0, r) \Rightarrow d(y, x_0) < r$$

$\Rightarrow \quad r-d(y, x_0)>0.$

Let $\quad \varepsilon=r-d(y, x_0)$ where $\varepsilon>0$.

Now we have to show that $S(y, \varepsilon)$ contained in $S(x_0, r)$. Let $x \in S(y, \varepsilon)$

$\Rightarrow \quad d(x, y)< \varepsilon \Rightarrow d(x, y) < r-d(y, x_0)$

$$d(x, x_0) \leq d(x, y)+d(y, x_0)<r-d(y, x_0)+d(y, x_0)<r$$

$\Rightarrow \quad d(x, x_0)<r \Rightarrow x \in S(x_0, r)$

$\Rightarrow \quad S(y, \varepsilon) \subseteq S(x_0, r)$ $\quad$ ($\because$ x is arbitrary)

$\Rightarrow \quad S(x_0, r)$ is a neighbourhood of y

Since, $\quad y$ is an arbitrary point of $S(x_0, r)$,therefore,$S(x_0, r)$ is a neighbourhood of each of its points.

THEOREM 4. ***A subset in a metric space is open if and only if it is a neighbourhood of each of its points.***

PROOF. Let(X, d) be a metric space and A is any subset of X *i.e*, $A \subset X$.

Let us first suppose A is open. To show it is a nbd of each of its points.

Let x be any arbitrary point of A *i.e.* $x \in A$.

Since A is open, we can write $x \in A \subset A \quad \Rightarrow \quad A$ is a nbd of x.

Conversely, let A is a nbd of each of its points. To show A is open.

Let $x \in A$, and A is a nbd of x. Then by definition of nbd, there exists an open set G_x such that

$$x \in G_x \subset A$$

Let $\quad G=\cup \{G_x : x \in A\}$.

We claim that $\quad G=A$

If $x \in A$ then by definition $x \in \cup \{G_x : x \in A\}=G$

$\Rightarrow \quad A \subset G.$ ...(1)

Now, if $y \in G$, then $y \in G_x$ for some $x \in A$. But $G_x \subset A$ and hence $y \in A$

$\Rightarrow \quad G \subset A.$ ...(2)

From (1) and (2) we conclude that $A=G$.

Now since G is open $\quad$ (Being the union of collection of open sets)

Therefore, A is open.

THEOREM 5. ***Let (X, d) be a metric space and let $x \in X$. If $\{N_i : i= 1, 2, ..., k\}$ are finite number of neighbourhood of X, then $\cap \{N_i : i=1, 2,...,k\}$ is also a nbd of x.***

PROOF. Given that N_i ($i= 1, 2,..., k$) is a nbd of x for each i.

Therefore by definition of neighbourhood there exist open sets G_i ($i=1, 2,..., k$) such that

$$x \in G_i \subset N_i : i =1, 2,..., k$$

$\Rightarrow \quad x \in \cap \{G_i : i = 1, 2,\ldots, n] \subset \cap \{ N_i : i = 1, 2,\ldots, k\}.$ ...(1)

Since, each G_i is open, therefore $\cap \{G_i : i = 1, 2,\ldots, n\}$ is open

($\because$ Finite intersection of open sets is open)

Hence, from (1) we conclude that $\cap [N_i : i = 1, 2,\ldots, k]$ is a nbd of x.

THEOREM 6. ***Let (X, d) be a metric space and A be a subset of X. A point $x \in X$ is a limit point of A if and only if every open sphere S(x, r) centered at x contains infinitely many points of A.***

PROOF. Let us first suppose every open sphere $S(x, r)$ centered at x contains infinitely many points of A. Then clearly, we can say that every open sphere $S(x, r)$ centered at x contains at least one point of A other than x. Therefore x is the limit point of A.

Conversely, let x is the limit point of A. To show every open sphere $S(x, r)$ centered at x contains infinitely many points of A.

Let if possible there exists a sphere $S(x, r)$ which contains only a finite number of points of A. Let $x_1, x_2, ..., x_n$ be those points of $S(x, r) \cap A$ which are distinct from x.

Define $\quad r_1 = \min\{d(x, x_m): 1 \leq m \leq n\}$

Then clearly $\quad r_1 > 0$.

Then open sphere $S(x, r_1)$ contains no points of A distinct from x.

$\Rightarrow$ x is not the limit point of A; which is a contradiction.

Hence, every open sphere centered at x must contain infinitely many points of A.

THEOREM 7. ***Let (X, d) be a metric space and $A \subset X$. A point $x \in X$ is an adherent point of A if and only if d(x, A)=0.***

PROOF. Let us first suppose $d(x, A)=0$

Then $\quad d(x, A) = \inf. \{d(x, y): y \in A\} = 0.$

Let $S(x, r)$ be any sphere centered at x.

By definition of infimum, there exists a point $y_0 \in A$ such that $0 \leq d(x, y_0) \leq r$

$\Rightarrow \quad y_0 \in S(x, r)$

$\Rightarrow$ x is an adherent point of A.

Conversely, let us suppose x be an adherent point of A. To show $d(x, A)=0$.

Since x is an adherent point of A therefore, every open sphere centered at x must contain a point of A.

$\Rightarrow$ to each $r>0$ there exists a $y \in A$ such that $0 \leq d(x, y) < r$.

Since r is arbitrary, letting $r \to 0$ (By taking r very small)

$\Rightarrow \quad \inf\{d(x, y): y \in A\} = 0$

$\Rightarrow \quad d(x, A) = 0.$

THEOREM 8. ***A subset A of a metric space X is closed if and only if $D(A) \subset A$, i.e., iff A contains all its limit points.***

PROOF. Let (X, d) be a metric space and $A \subset X$.

Let us first suppose A is closed. To show A contains all its limit point.

Since A is closed $\Rightarrow A'$ is open.

Let $x \in A'$, since A' is open, therefore by definition of open set there exists a nbd N of x such that $N \subset A$.

Now, since $A \cap A' = \phi$. $\Rightarrow$ N contains no point of A.

$\Rightarrow$ x is not a limit point of A.

$\Rightarrow$ No point of A' can be a limit point of A.

$\Rightarrow$ A contains all its limit points.

$\Rightarrow \quad D(A) \subseteq A$

Conversely, let A contains all its limit point *i.e.* $D(A) \subset A$. To show A is closed.

For this, we shall show that A' is open.

Let $x \in A'$, then $x \notin A$. Since $D(A) \subset A$ and $x \notin A$ therefore $x \notin D(A)$

$\Rightarrow$ x is not the limit point of A

$\Rightarrow$ there exists a nbd N of x such that $N \cap A = \phi$

$\Rightarrow \quad N \not\subset A$

$\Rightarrow \quad N \subset A'$

$\Rightarrow$ A' contains a neighbourhood of x.

Since x is arbitrary, therefore, we can say A' is a nbd of each of its points

$\Rightarrow$ A' is open.

Hence, A is closed.

THEOREM 9. ***Let A be any subset of a metric space (X, d) then derived set D(A) of A is a closed set.***

PROOF. To show $D(A)$ is closed, we show that $D(A)$ contains all its limit point.

Let x be a limit point of $D(A)$. Then for all $r>0$, the open sphere $S(x, r)$ contains infinitely many points of $D(A)$.

We know that each point of $D(A)$ is a limit point of A

$\Rightarrow$ every open sphere $S(x, r)$ must contain infinitely many points of A

$\Rightarrow$ x is a limit point of A.

$\Rightarrow \quad x \in D(A)$

$\Rightarrow$ $D(A)$ contains all its limit points and so $D(A)$ is closed.

THEOREM 10. ***Let A and B be subset of a metric space X. Then***

(i) $A \subset B \Rightarrow D(A) \subset D(B)$ ***(ii)*** $D(A \cap B) \subset D(A) \cap D(B)$

(iii) $D(A \cup B) = D(A) \cup D(B)$.

PROOF. (i) $A \subset B$. To show $D(A) \subset D(B)$.

Let $\quad x \in D(A)$

$\Rightarrow$ x is the limit point of A

$\Rightarrow$ every nbd of x contains at least one point of A, other than x

$\Rightarrow$ every nbd of x contains at least one point of B, other than x $(\because A \subset B)$

$\Rightarrow$ x is the limit point of B

$\Rightarrow \quad x \in D(B)$

Since x is arbitrary, therefore $D(A) \subset D(B)$.

(ii) To show $D(A \cap B) \subset D(A) \cap D(B)$.

Since we know that $A \cap B \subset A \Rightarrow D(A \cap B) \subset D(A)$ ...(1)

[Using (1)]

and $\quad A \cap B \subset B \Rightarrow D(A \cap B) \subset D(A)$

From (1) and (2), we conclude that $D(A \cap B) \subset D(A) \cap D(B)$.

(iii) To show $D(A \cup B) = D(A) \cup D(B)$

Let $x \notin D(A) \cup D(B)$

$\Rightarrow$ $x \notin D(A)$ and $x \notin D(B)$

$\Rightarrow$ x is not the limit point of A and x is not the limit point of B

$\Rightarrow$ $x \notin D(A \cup B)$

Since x is arbitrary, therefore

$$D(A \cup B) \subset D(A) \cup D(B). \qquad ...(1)$$

Also $A \subset A \cup B \Rightarrow D(A) \subset D(A \cup B)$ ...(2)

and $B \subset A \cup B \Rightarrow D(B) \subset D(A \cup B)$ ...(3)

Now, (2) and (3) gives

$$D(A) \cup D(B) \subset D(A \cup B). \qquad ...(4)$$

From (1) and (4) we conclude that

$$D(A) \cup D(B) = D(A \cup B).$$

2.7.1 EQUIVALENT METRICS

Two metrices d and d^* on the same set X are said to be equivalent iff every d-open set is d^*-open and every d^*-open set is d-open.

EXAMPLE 1. *Let (X, d) be a metric space and d^* a mapping such that $d^*: X \times X \to R$*

defined by $$d^*(x, y) = \frac{Md(x,y)}{1+d(x,y)}, M > 0$$

is also a metric for X. Also show that d and d^* are equivalent.

SOLUTION. We have already proved that $d^*(x, y) = \dfrac{Md(x,x_1)}{1+d(x,x_1)}$ is metric for X as it follows the four properties of metric.

Now it remains only to show that d and d^* are equivalent.

For this we shows that d-open sphere centred at $x \in X$ contains a d^* open sphere centred at x and *vice-versa*.

Let $S(x, r)$, $r>0$ be any d open sphere centred at $x \in X$.

Let $S(x, \rho)$, $\rho>0$ be any d^* open sphere centred at $x \in X$ where $\rho = \dfrac{Mr}{1+r}$.

Now we have to show that $S^*(x, \rho) \subseteq S(x, r)$.

Here let $x_1 \in S^*(x, \rho)$ $\Rightarrow$ $d^*(x, x_1) < \rho$

$$\Rightarrow \quad \frac{Md(x,x_1)}{1+d(x,x_1)} < \frac{Mr}{1+r}$$

$$\Rightarrow \quad d(x, x_1) + rd(x, x_1) < r + rd(x, x_1)$$

$$\Rightarrow \quad d(x, x_1) < r \Rightarrow x_1 \in S(x, r)$$

$$\Rightarrow \quad S^*(x, \rho) \subseteq S(x, r)$$

Now it remains to show that $S(x, r) \subseteq S^*(x, \rho)$.

Here $r = \dfrac{\rho}{M-\rho}$.

Let $x_1 \in S(x, r) \Rightarrow d(x, x_1) < r$

$$\Rightarrow \frac{d^*(x,x_1)}{M-d^*(x,x_1)} < \frac{\rho}{M-\rho}.$$

$$\Rightarrow Md^*(x,x_1) - \rho d^*(x,x_1) < \rho M - \rho d^*(x,x_1)$$

$$\Rightarrow d^*(x,x_1) < \rho \Rightarrow x_1 \in S^*(x,\rho)$$

$$\Rightarrow S(x,r) \subseteq S^*(x,\rho)$$

Hence d and d^* are equivalent metric.

RECAPITULATIONS

➠ A subset A of a metric space (X, d) is said to be open if and only if to each, $x \in A\ \exists\ r>0$ such that $S(x, r) \subseteq A$.	➠ For x to be limit point of A : $\{S(x, r)-\{x\}\} \cap A \neq \phi$
➠ For x to be an isolated point of A : $\{S(x, r)-\{x\}\} \cap A = \phi$	➠ If $D(A) \subset A$, then A is closed.
➠ A set N of a metric space $[X\ d]$ is said to be nbd of a point $p \in X$ if there exist an $\varepsilon>0$ such that $S(x, \varepsilon) \subset N$.	➠ $A \subset B \Rightarrow D(A) \subset D(B)$
➠ $D(A \cap B) \subset D(A) \cap D(B)$	➠ $D(A \cup B) = D(A) \cup D(B)$

SOLVED EXAMPLES

EXAMPLE 1. *Find the closed and open spheres for the usual metric for* R.

SOLUTION. We know that the usual metric for R is defined by $d(x, y) = |x-y|$

Let $x_0 \in$ R. Then the open sphere $S(x_0, r)$, centered at x_0, with radius r is given by

$$S(x_0, r) = \{x \in \mathrm{R} : |x-x_0| < r\} = \{x \in \mathrm{R} : x_0-r < x < x_0+r\} =]x_0-r, x_0+r[.$$

Hence, the open spheres on the real line are open intervals.

Similarly, the closed sphere with centre x_0 and radius r is the closed interval $[x_0-r, x_0+r]$.

EXAMPLE 2. *Consider, the usual metric $d(x, y) = |x-y|$ for* $[0, 1]$

describe $S\left(\left]\frac{1}{2}, 1\right[\right)$ *and* $S\left[\frac{1}{2}, 1\right]$

SOLUTION. We have $S\left(\left]\frac{1}{2},1\right[\right) = \left\{x \in [0,1] : \left[x - \frac{1}{2}\right] < 1\right\} = \left\{x \in [0,1] : \frac{1}{2} - 1 < x < \frac{1}{2} + 1\right\}$

$$= \left[x \in [0,1] : -\frac{1}{2} < x < \frac{3}{2}\right]$$

$$= [0, 1] \qquad \text{[Don't go outside the given reason [0, 1]]}$$

Similarly, $S\left[\frac{1}{2}, 1\right] = [0, 1]$.

EXAMPLE 3. *Let R be the set of all real numbers with usual metric $d(x, y)=|x-y|$. Find whether or not the given sets are open*

(i) $A=[0, 1[$ (ii) $B=]0, 1[$ (iii) $C=]0, 1]$

(iv) $D=[0, 1]$ (v) $E=\{1\}$ (vi) $F=\{1, 2, 3\}$.

SOLUTION. Since, we know that an open sphere about $x_0 \in$ R and 'r' is the radius of open interval or sphere $]x_0-r, x_0+r[$. To show whether a set is open or not we check that for each point of given set an open interval of type described above exists or not and also contained in the given set.

(i) Here $A=[0, 1[$. Let us choose a positive number r.So the open interval $]0-r, 0+r[=]-r, r[\notin A$.

$\Rightarrow$ No open sphere with centred '0' contained in A

$\Rightarrow A$ is not open.

(ii) Let us take a point x of set $B=]0, 1[$ and let $r=\min \{x-0, 1-x\}$ so it is obvious that $]x-r, x+r[\subseteq B$

$\Rightarrow B$ is an open set.

(iii) Here $C=]0, 1]$. Let us choose a positive number r.

So the open interval $]1-r, 1+r[\notin A$.

Thus no open sphere contained in A having radius $(1+r)$.

$\Rightarrow C$ is not open set.

(iv) $D=[0, 1]$ is not an open set because D an interval $]0-r, 0+r[\not\subset D$.

(v) $E=\{1\}$ is not open set because it contains a single point 1 and so it is not possible to find $r>0$ such that $]1-r, 1+r [\subseteq F$.

(vi) F is not open because it consist elements $\{1, 2, 3\}$ but it is not possible to find $r>0$ such that $]1-r, 1+r[\subseteq F$ and other so it is not open.

EXAMPLE 4. *Describe the open spheres of unit radius about* (0, 0) *for each of the following matrices for* R^2.

(i) $d(z_1, z_2)=\sqrt{(x_1-x_2)^2+(y_1-y_2)^2}$

(ii) $d(z_1, z_2)=max\{|x_1-x_2|, |y_1-y_2|\}$

where $z_1=(x_1, y_1)$, $z_2=(x_2, y_2)$ *are any two points of* R^2.

SOLUTION. **(i)** Let d be the usual metric on R^2 and, here we are given that

$$d(z_1, z_2)=\sqrt{(x_1-x_2)^2+(y_1-y_2)^2}$$

where $z_1=(x_1, y_1)$ and $z_2=(x_2, y_2)$ be any two points of R^2.

The open space with centre z_0 and radius r is given by

$$S(z_0, r)=\left\{\sqrt{(x_1-x_0)^2+(y_1-y_0)^2}\right\}<r$$

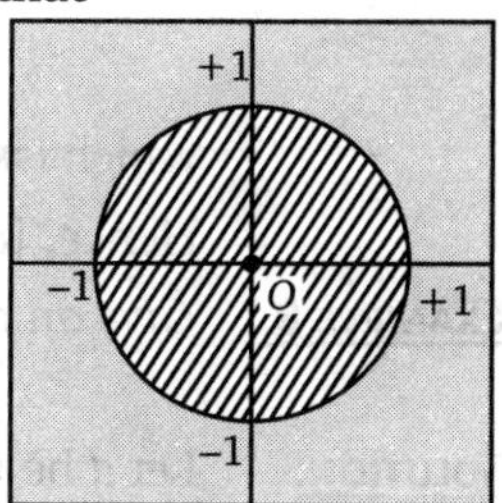

Fig. 2

Here, the open spheres $S(z_0, r)$ consists of all points of the cartesian plane which lie within the circle.

$$(x-x_0)^2+(y-y_0)^2=r^2.$$

(ii) Here, we have

$$S(z_0, 1)=\{(x, y) \in R^2 : d(z, z_0)<1\}$$
$$=\{(x, y) \in R^2 : \max[\{|x-0|, |y-0|\}<1]\}$$
$$=\{(x, y) \in R^2 : \max [\{|x|, |y|\}<1]\}.$$

$\Rightarrow$ The open spheres $S(z_0, 1)$ is the interior space of the square bounded by the lines

$$x=1, x=-1, y=1 \text{ and } y=-1.$$

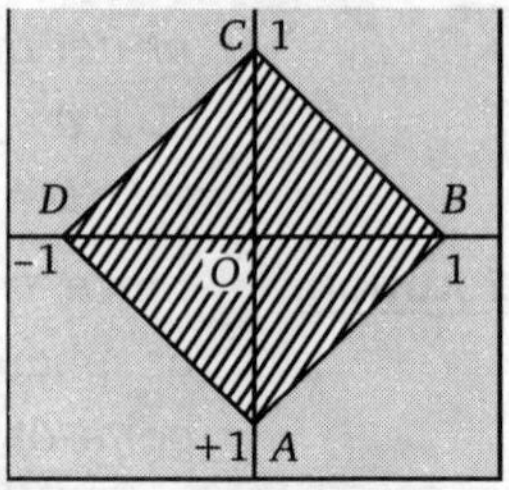

Fig. 3

EXAMPLE 5. *Let (X, d) be any discrete metric space. Describe open sphere for d.*

SOLUTION. The discrete metric space is defined by

$$d(x, y)=\begin{cases}0 & \text{if } x=y\\ 1 & \text{if } x\neq y\end{cases}$$

Let $x_0 \in X$ and $r>1$, then we have

$$S(x_0, r) = \{x \in X : d(x, x_0 < r] = X$$

[Since $d(x, x_0)=0$ or 1, each of which is less than r so that $x \in X \Rightarrow x \in S(x_0, r)$].

If $r \leq 1$ then, we have

$$S(x_0, r)= \{x \in X : d (x, x_0)<r\}=\{x_0\}$$

$$[\because d(x_0, x_0)=0<r \text{ and } d(x, x_0)=1 \ \forall \ r \text{ if } x\neq x_0]$$

EXAMPLE 6. *Show that every singleton set in* R *is closed for the usual metric d for* R.

SOLUTION. Let $a \in$ R. To show $\{a\}$ is closed.

Consider $\quad$ R$-\{a\}=]-\infty, a[\cup] a, \infty[$

= union of two open sets = open

$\Rightarrow$ R$-\{a\}$ is open

$\Rightarrow$ $\{a\}$ is closed.

EXAMPLE 7. *Show that every closed interval is a closed set for the usual metric for* R.

SOLUTION. Consider a closed interval [a, b] where $a \in$ R, $b \in$ R.

Also, consider

$$R-[a, b]=[x \in R : a<x<b]$$
$$=[x \in R : x<a, x>b]$$
$$=[x \in R : x<a] \cup [x \in R : x>b]=]-\infty, a [\cup] b, \infty[$$

($\because$ Union of two open sets is again open)

which is open.

Hence, $[a, b]$ is closed.

EXAMPLE 8. *Give an example of two closed subsets A and B of real line* R *such that*

$$d(A, B)=0, \ A \cap B= \phi.$$

SOLUTION. Let d be the usual metric on R.

Define $\quad N_1=[n+1 : n \in N]$

$$A=[n+1 : n \in N_1] \text{ and } B=\left[n+\frac{1}{n} : n \in N_1\right].$$

Now if $n \in A \Rightarrow n+\frac{1}{n} \in B \Rightarrow B–A=\frac{1}{n}$ which tends to 0 as $n \to \infty$.

Now $\quad d(A, B) = \inf \{d(x, y) : x \in A, y \in B\}$

$= \inf. \{|x-y| : x \in A, y \in B\}$

$= \inf. \left\{\frac{1}{n} : n \in A, n+\frac{1}{n} \in B\right\} = 0$

Hence $\quad d(A, B)=0$ and $A \cap B = \phi$.

EXAMPLE 9. *Give an example of a set which has*

(i) *no limit point* (ii) *exactly one limit point*

(iii) *exactly two limit points* (iv) *infinite number of limit points*

(v) *every point of the set as its limit points.*

SOLUTION. (i) The set of rational number Q, has no limit, point.

(ii) The set $S=\left[\frac{1}{n} : n \in \mathrm{N}\right]$ has exactly one limit point namely 0.

(iii) The set $S=\left\{\frac{1}{2} - \frac{1}{2}, \frac{2}{3}, -\frac{2}{3} \right\}$ has exactly two limit points 1 and –1.

(iv) The open interval]1, 2[has infinite number of limit points.

(v) The closed interval [1, 2] is a set in which every point is its limit point.

EXAMPLE 10. *Given an example to show that in a metric space the union of an infinite collection of closed sets is not necessarily closed.*

SOLUTION. Let us consider an infinite collection $F_n=\left[\frac{1}{n}, 1\right], n \in \mathrm{N}$ of closed intervals for usual metric space (*R*, *d*) and we know that in usual metric every closed interval is a closed set

$\Rightarrow F_n$ is a closed set in (*R*, *d*).

Now $\quad \cup\{f_n : n \in \mathrm{N}\} = \{1\} \cup \left[\frac{1}{2}, 1\right] \cup \left[\frac{1}{3}, 1\right] \cup \left[\frac{1}{4}, 1\right] \cup \ldots =]0, 1]$

$\Rightarrow$ which is not closed

Hence, union of an infinite collection of closed sets is not necessarily closed.

EXAMPLE 11. *If A and B are closed sets of a metric space* (*X, d*) *then show that*

(a) $A \cup B$ *is closed* (b) $A \cap B$ *is closed.*

SOLUTION. (a) Since *A* and *B* are closed set of a metric space *X*

$\Rightarrow \quad A' = X–A$ and $B'=X–B$ both are open set of *X*

$\Rightarrow \quad A' \cap B'$ is an open set of *X*

$\Rightarrow \quad (A' \cap B')'$ is closed set of *X*

$\Rightarrow \quad (A')' \cup (B')'$ is closed set of *X* (By De-Morgan's law)

($\because (A')'=A, (B')'=B$)

$\Rightarrow \quad A \cup B$ is closed set of *X*.

(b) Since *A* and *B* are closed set of *X*

$\Rightarrow \quad A'$ and B' both are open sets of *X*

$\Rightarrow$ $A' \cup B'$ is an open sets of X

$\Rightarrow$ $(A' \cup B')'$ is closed set of X

$\Rightarrow$ $(A')' \cap (B')'$ is closed set of X (By De-Morgan's law)

($\because$ $(A')'=A$, $(B')'=B$)

$\Rightarrow$ $A \cap B$ is closed set of X.

EXAMPLE 12. *Let (X, d) be a metric space and $S(x_0, r)$ the open sphere with centre x_0 and radius r. Let A be a subset of X which intersects $S(x_0, r)$ and has diameter less than r. Show that $A \subseteq S(2r, x_0)$.*

SOLUTION. Since given that $S(r, x_0)$ is open sphere with centre x_0 and radius r and A is any non-empty subset of X which intersects $S(x_0, r)$

Therefore, $A \subseteq S(x_0, r) \neq \phi$.

Let $y \in A \cap S(x_0, r)$

Then $y \in A$ and $y \in S(x_0, r) \Rightarrow d(y, x_0) < r$.

Since $x \in A$ and x is arbitrary point of A and also $y \in A$

$\Rightarrow$ $d(x, y) < r$

Now $d(x, x_0) \leq d(x, y) + d(y, x_0)$ (By triangular inequality)

$< r + r = 2r$

$\Rightarrow$ $d(x, x_0) < 2r \Rightarrow x \in S(x_0, 2r)$

$\Rightarrow$ $A \subseteq S(x_0, 2r)$.

EXAMPLE 13. *Let (X, d) be a metric space and let $p \notin S(x_0, r)$ where $x_0 \in X$ and $r > 0$ then show that $d(p, S(x_0, r)) \geq d(x_0, p) - r$.*

SOLUTION. Let $x \in S(x_0, r)$ where x is any arbitrary point of $S(x_0, r)$, then

$$d(x_0, p) \leq d(x_0, x) + d(x, p)$$

$$d(x, p) \geq d(x_0, p) - d(x_0, x) \quad \text{...(1)}$$

But we consider that $x \in S(x_0, r)$

$\Rightarrow$ $d(x_0, x) < r$ ...(2)

Using equation (2) in equation (1) we have $d(x, p) \geq d(x_0, p) - r$

This condition holds for all $x \in S(x_0, r)$

$\Rightarrow d(p, S(x_0, r)) \geq d(x_0, p) - r$

EXAMPLE 14. *Let (X, d) be a metric space and let $p \notin S(x_0, r)$, where $x_0 \in X$ and $r > 0$, then show that $d(p, S[x_0, r]) > d(x_0, p) - r$.*

SOLUTION. Let x be an arbitrary point of $S(x_0, r)$

Now, $d(x_0, p) \leq d(x_0, x) + d(x, p)$

($\because$ in a metric space, $d(x, y) \leq d(x, z) + d(z, y)$ by triangle inequality)

$d(x, p) \geq d(x_0, p) - d(x_0, x)$

But $x \in S(x_0, r) \Rightarrow d(x_0, x) < r$

So $d(x, p) \geq d(x_0, p) - r$

Since $x \in S(x_0, r) \Rightarrow d(p, S(x_0, r)) \geq d(x_0, p) - r$.

EXAMPLE 15. *Let (X, d) be a metric space and let $p \notin S[x_0, r]$ and $x_0 \in X$ and $r > 0$. Then show that $d(p, S[x_0, r]) > d(x_0, p) - r$.*

SOLUTION. We have $d(p, S[x_0, r])$ is infimum of $\{d(x, p) : x \in S[x_0, r]\}$ that is

$$d(p, S[x_0, r]) = \inf \{d(x, p): x \in S[x_0, r]\}$$

Let $x \in S[x_0, r]$ then we have

$$d(x_0, p) \leq d(x_0, x) + d(x, p) \qquad \text{(By triangular inequality)}$$

or $\quad d(x, p) \geq d(x_0, p) - d(x_0, x)$.

But $\quad x \in S[x_0, r]$

$\Rightarrow \quad d(x_0, x) \leq r \quad$ so $\quad d(x, p) \geq d(x_0, p) - r$

$\Rightarrow \quad \inf\{d(x, p): x \in S[x_0, r]\} \geq d(x_0, p) - r$

$\Rightarrow \quad d(p, S[x_0, r]) \geq d(x_0, p) - r$.

EXAMPLE 16. *In the usual metric space* (R, d) *find the derived set of the set* Q *of all rational numbers.*

SOLUTION. Here we have to show that every real number is a limit point of the set Q.

Let x be any real number and for given $\varepsilon>0$, then $x-\varepsilon$ and $x+\varepsilon$ are two distinct real numbers and exist infinitely many points between them thus for given $\varepsilon>0$ open interval $]x-\varepsilon, x+\varepsilon[$ contain at least one point of rational set Q other than x

$\Rightarrow$ x is a limit point of Q

$\Rightarrow$ every real number is a limit point of Q

$\Rightarrow$ the set of the limits of Q is the set of all real number R so $D(Q) = R$.

EXAMPLE 17. *On the real line* R, *the set of integers,* Z *has no limit point.*

SOLUTION. Let $\quad x \in R$.

Consider $\quad]x-\varepsilon, x+\varepsilon[$.

Then $\quad]x-\varepsilon, x+\varepsilon[\subseteq Z \qquad$ (By Denseness property of real numbers)

EXERCISE 2.2

1. Consider the usual metric $d(x, y) = |x-y|$, for $[0, 1]$ Describe $S(]1/4, 1/4[)$, $S[1/4, 1/4]$, $S\left(\left]0, \frac{1}{8}\right[\right)$ and $S\left(\left]\frac{1}{16}, \frac{1}{16}\right[\right)$.
2. Show that the Cantor set C is not open.
3. Show that on the real line, every open interval is an open set.
4. Show that on the real line, every closed interval is a closed set.
5. Show that in a metric space, the intersection of two open spheres need not be an open sphere but that it will always contain another open sphere.
6. If A and B are open sets of a metric space X. Then show that
   (i) $A \cap B$ is open set (ii) $A \cup B$ is also open.
7. Let x_1 and x_2 be two distinct elements in the metric space (X, d), show that two disjoint open spheres will exists, which are centred at x_1 and x_2 respectively.
8. If A_1 is open set in metric space $(X_1, d_1)$and A_2 is open set in metric space (X_2, d_2) then if $X = X_1 \times X_2$, show that $A_1 \times A_2$ is open.
9. Show that the right half open inerval $[a, b[$ is neither closed nor open (that is say clopen) for the usual metric on R.
10. In a metric space (X, d) the empty set ϕ and the whole space X are closed as well as open.
11. Show that every finite subset of R is closed with respect to the usual metric for R.
12. Show that every subset of X containing x is a neighbourhood of x, where (X, d) is discrete space defined as
$$d(x, y) = \begin{cases} 0 \text{ if } x = y \\ 1 \text{ if } x \neq y \end{cases}$$
13. Let (X, d) be any metric space and let
   (a) $d^*(x, y) = \dfrac{d(x,y)}{1+d(x,y)} \; \forall x, y \in X$.

Show that d^* is also a metric on X and d and d^* are equivalent.

(b) If $d^*(x, y)=\min\{1, d(x, y)\}$ then show that $d^*(x, y)$ is a metric on X and also d and d^* are equivalent.

14. Show that in a discrete metric space, every set is open as well as closed.

15. Show that the closed open interval $[a, b[$ is neither closed nor open for the usual metric on R.

16. Give an example of a set which (i) is both open and closed (ii) is neither open nor closed.

17. If A' denote the derived set of A, then find a set A such that

(i) $A \cap A' = \phi$ (ii) $A = A'$

(iii) $A \subseteq A$ (iv) $A \subseteq A'$.

18. On the real line, show that the set of rational numbers Q, has no limit point.

HINT TO SELECTED PROBLEMS

1. Since we know that the Cantor set is the intersection of closed sets, therefore, being the intersection of closed set, it is again closed.

4. Using the following result

$R-[a, b]=\{x \in R: x<a \text{ or } x>b\} = \{x \in R: x<a\} \cup \{x \in R: x>b\}$

$=]\infty, a[\cup]b, \infty[$, union of two open sets.

16. (i) ϕ and R (ii) Set of rational numbers.

18. Using the denseness property of real numbers.

ANSWERS

1. $\left]0, \frac{1}{2}\right[, \left[0, \frac{1}{2}\right], \left[0, \frac{1}{8}\right[, \left]0, \frac{1}{8}\right[$

8. (i) ϕ and R (ii) Q, the set of rationals

9. (i) $A=\left[1, \frac{1}{2}, \frac{1}{3}, \ldots, \frac{1}{n}\right]$ (ii) $A=$\{set of closed interval\}.

2.8 CLOSURE, INTERIOR AND EXTERIOR

Definition. *In a metric space (X, d), the set of all adherent points of a set $A \subseteq X$ is called the closure of A and is denoted by $C(A)$ or $\bar{A}$.*

ILLUSTRATIONS

(1) On the real line, every real number is an adherent point of the set R~Q of all irrational numbers. For, if p be any real number whatever any $\varepsilon>0$ be given, then $]p-\varepsilon, p+\varepsilon[$ contain infinitely many irrational numbers and hence $]p-\varepsilon, p+\varepsilon[\cap (R\sim Q) \neq \phi$. Hence, p is an adherent point of R~Q and $[R\sim Q] = R$.

(2) On the real line, the closure of each of the sets $]0, 1[$, $]0, 1]$ and $[0, 1]$ is $[0, 1]$.

(3) The closure of the set of integers is Z itself.

(4) Let $A=\left\{\frac{1}{n}: n \in Z^+\right\}$. Then closure $A = \bar{A} = \left\{\frac{1}{n}: n \in Z^+\right\} \cup \{0\}$.

THEOREM 1. ***In a metric space (X, d), the closure of a set $A \subset X$, is a closed superset of A.***

PROOF. Let p be any point of A and N be any nbd of p. Then $p \in N \cap A$ *i.e.* $N \cap A \neq \phi$. Since every nbd of p intersects A, therefore $p \in \bar{A}$. Since

$$p \in A \Rightarrow p \in \bar{A} \Rightarrow A \subset \bar{A}.$$

Now, $\bar{A}$ is a closed set $\Leftrightarrow X - \bar{A}$ is an open set.

$\Leftrightarrow X - \bar{A}$ is a nbd of each of its points.

Therefore, to show that $\bar{A}$ is a closed set, it is enough to show that $X - \bar{A}$ is a nbd of each of its points. Consider an arbitrary point $q \in X - \bar{A}$

Now, $$q \in X - \bar{A} \Rightarrow q \notin \bar{A}$$

$\Rightarrow$ there exists an $\varepsilon > 0$ such that $S(q, \varepsilon)$ contains no point of A.

It can be easily seen that no point of $S(q, \varepsilon)$ can be in $\bar{A}$. In fact if $r \in S(q, \varepsilon)$, then $S(q, \varepsilon)$ is a nbd of r containing no point of A and therefore $r \notin \bar{A}$. Since r is arbitrary, it follows that no point of $S(q, \varepsilon)$ is in $\bar{A}$. Thus $S(q, \varepsilon) \subset X - \bar{A}$, showing that $X - \bar{A}$ is a nbd of q. Again since q is arbitrary point of $X - \bar{A}$, therefore it follows that $X - \bar{A}$ is a nbd of each of its points. Hence $\bar{A}$ is closed.

THEOREM 2. ***In a metric space (X, d), the closure of a set $A \subset X$ is the smallest closed set containing A.***

PROOF. Since we have already shown that the closure of a set A is a closed superset of A, it only remains to show that $\bar{A}$ is the smallest among all closed superset of A *i.e.* if B is any closed set containing A, then $\bar{A} \subset B$.

Now, B is closed superset of $A \Rightarrow X–B$ is an open set disjoint from A

$\Rightarrow$ no point of $X–B$ is an adherent point of A

$$\Rightarrow \qquad \bar{A} \subset B$$

REMARK

- The properties of $\bar{A}$, expressed in the above two theorems is sometimes taken as definition of closure of A. It gives us a description of $\bar{A}$ without using the notion of an adherent point of A. Therefore, we can say :

  (i) The closure of A *i.e.* $\bar{A}$ is the smallest closed set containing A.

  (ii) The closure of $\bar{A}$ *i.e.,* $\bar{A}$ is the intersection of all closed sets containing A *i.e.*

  $$\bar{A} = \cap\ [F : F \text{ is closed and } A \subset F].$$

THEOREM 3. ***A subset A of a metric space is closed if and only if $\bar{A}=A$.***

PROOF. Let (X, d) be a metric space.

Let us first suppose A is closed, then A is the smallest closed set containing A

$$\Rightarrow \qquad \bar{A} = A$$

Conversely if $\bar{A}=A$, then since $\bar{A}$ is closed, therefore A is closed.

THEOREM 4. ***Let A be a subset of a metric space, then $\bar{A} = A \cup D(A)$.***

PROOF. Let p be the limit point of A then p is an adherent point of A also.

i.e. $$A' \subset \bar{A}. \qquad ...(1)$$

Now every point of A is an adherent point of A

i.e. $$A \subset \bar{A}. \qquad ...(2)$$

From (1) and (2), we conclude that

$$A \cup A' \subset \bar{A}. \qquad ...(3)$$

Now let p be any point of $\bar{A}$.

If $p \in A$, then clearly $\bar{A} \subset A \cup A'$.

Now, if $p \in \bar{A}, p \notin A$. Let N be any nbd of p. Since $p \in \bar{A}$, $N \cap A \neq \phi$ and $p \notin A$. It follows that $p \notin N \cap A$ so that $N \cap A$ is a non-empty set, containing a point other than p. Since N is any nbd of p, it follows that $p \in A'$.

Therefore $$\bar{A} \subset A \cup A'. \qquad ...(4)$$

From (3) and (4), we conclude that

$$\bar{A} = A \cup A' \Rightarrow \bar{A} = A \cup D(A)$$

THEOREM 5. ***Let (X, d) be a metric space and A, B be subsets of X. Then***

(i) $\bar{\phi} = \phi$. ***(ii)*** $A \subset \bar{A}$

(iii) $A \subset B \Rightarrow \bar{A} \subset \bar{B}$ ***(iv)*** $\overline{A \cup B} = \bar{A} \cup \bar{B}$

(v) $\overline{A \cap B} \subset \bar{A} \cap \bar{B}$ ***(vi)*** $\bar{\bar{A}} = \bar{A}$

PROOF. (i) Since ϕ is closed, therefore using theorem (3), we have $\bar{\phi} = \phi$.

(ii) By definition, $\bar{A}$ is the smallest closed set containing A *i.e.* $A \subset \bar{A}$

(iii) From (ii), we have $B \subset \bar{B}$.

Since $A \subset B$. Therefore $A \subset B \subset \bar{B}$. $\Rightarrow$ $A \subset \bar{B}$.

But $\bar{B}$ is closed $\Rightarrow$ $\bar{B}$ is the closed set containing A

Also, $\bar{A}$ is the smallest closed set containing A. Therefore $\bar{A} \subset \bar{B}$.

(iv) Since we know that

$$A \subset A \cup B \text{ and } B \subset A \cup B.$$

Therefore from (iii)

$$\bar{A} \subset \overline{A \cup B} \text{ and } \bar{B} \subset \overline{A \cup B}$$

$$\bar{A} \cup \bar{B} \subset \overline{A \cup B}. \quad \text{...(1)}$$

Now, $\bar{A}$ and $\bar{B}$ are closed sets, $\bar{A} \cup \bar{B}$ is also closed (being the union of two closed sets). Also $A \subset \bar{A}$ and $B \subset \bar{B}$

$\Rightarrow$ $A \cup B \subset \bar{A} \cup \bar{B}$

$\Rightarrow$ $\bar{A} \cup \bar{B}$ is the closed set containing $A \cup B$.

But $\overline{A \cup B}$ is the smallest closed set containing $A \cup B$.

Therefore $\overline{A \cup B} \subset \bar{A} \cup \bar{B}$. ...(2)

From (1) and (2), we conclude that $\overline{A \cup B} = \bar{A} \cup \bar{B}$.

(v) Since we know that

$$A \cap B \subset A \Rightarrow \overline{A \cap B} \subset \bar{A} \quad \text{and} \quad A \cap B \subset B \Rightarrow \overline{A \cap B} \subset \bar{B}$$

Hence $\overline{A \cap B} \subset \bar{A} \cap \bar{B}$.

(vi) Since $\bar{A}$ is a closed set, therefore using theorem (3), we have

$$\left(\bar{\bar{A}}\right) = \bar{A}.$$

REMARK

- The inclusion in part (v) of the above theorem can not be replaced by an equality.

For Example. Let $A =]0, 1[$ and $B =]1, 2[$ Then $\bar{A} = [0, 1]$ and $\bar{B} = [1, 2]$

$\Rightarrow$ $\bar{A} \cap \bar{B} = \{1\}$.

Also $A \cap B = \phi \Rightarrow \left(\overline{A \cap B}\right) = \phi$.

we find that in this case $\left(\overline{A \cap B}\right)$ is a proper subset of $\bar{A} \cap \bar{B}$.

THEOREM 6. ***A finite set in a metric space has no limit point.***

PROOF. Let (X, d) be a metric space and A be any finite subset of X.

As $p \in X$ is a limit point of any set B if every open sphere $S(x, r)$ contains an infinite number of points of B, other then p.

But A has finite number of points. Hence, A has no limit point.

THEOREM 7. ***Let (X, d) be a metric space and $A \subset X$. Then $\bar{A} = \{x \in X : d(x, A) = 0\}$.***

PROOF. Let (X, d) be a metric-space and A be any subset of X.

To show

$$\bar{A} = \{x \in X : d(x, A) = 0\} \text{ where } \bar{A} = A \cup D(A).$$

Let $d(x, A) = \varepsilon > 0$, then $x \notin A$

and $\{S(x, \varepsilon/3) - [x]\} \cap A = \phi$.

Let $x \notin D(A)$, then

$$d\,(x, A) = \varepsilon \Rightarrow x \notin A, x \notin D(A) \Rightarrow x \notin A \cup D(A)$$
$$\Rightarrow x \notin \bar{A}. \qquad [\because \bar{A} = A \cup D(A)]$$

Now $d\{x, A) = 0 \Rightarrow x \in A$ and $\{S(x, \varepsilon) - [x]\} \cap A \neq \phi$

$$\Rightarrow x \in D(A).$$

Finally, $d(x, A) = 0 \Rightarrow x \in A$ or $x \in D(A)$

$$\Rightarrow x \in A \cup D(A)$$
$$\Rightarrow x \in \bar{A}. \qquad [\because \bar{A} = A \cup D(A)]$$

and $d(x, A) = \varepsilon > 0 \Rightarrow x \notin \bar{A}$.

Further more $d(x, A) \geq 0$ and hence the result follows.

THEOREM 8. ***In a metric space (X, d), all finite sets are closed.***

PROOF. Since we know that every finite set can be written as the finite union of singletons and each singleton is closed. Also, finite union of closed sets is again closed.

Hence, finite set is always closed.

THEOREM 9. **(Bolzano-Weirstrass Theorem)** ***Every bounded sequence in a metric space has a limit point.***

PROOF. Let $<x_n>$ be a bounded sequence in a metric space (X, d). If there are only a finite number of distinct elements in $\{x_n : n \in \mathrm{N}\}$ then at least one of them must occur infinitely often. If this element be x, then x is the limit point of $<x_n>$.

If the sequence $<x_n>$ contains infinitely many distinct elements in X, then $S = \{x_n : n \in \mathrm{N}\}$ is an infinite bounded set and therefore has a limit point, say y. Since y is the limit point of S, therefore every nbd of y must contains infinitely many points of S. This implies that the sequence $<x_n>$ must be frequently in every nbd of y. Hence, y is a limit point of $<x_n>$.

2.9 INTERIOR POINT AND INTERIOR OF A SET

Definition 1. *Let (X, d) be a metric space and A be any subset of X. A point $p \in A$ is said to be an interior point of A, if A is a nbd of p.*

Definition 2. *Let (X, d) be a metric space and $A \subset X$. A point $p \in A$ is said to an interior point of A if there exist $\varepsilon > 0$ such that $S(p, \varepsilon) \subset A$.*

Definition 3. *The set of all interior points of a set $A \subset X$, is called the interior of A and is denoted by A°.*

☛ **ILLUSTRATIONS**

On the real line :

(1) Every point of $]a, b[$ is an interior point of $]a, b[$ *i.e.* $\text{int}(]a, b[)=]a, b[$.

(2) No point of Z, is an interior point of Z *i.e.* $Z^\circ=\phi$.

(3) The interior of the closed ray $[0, \infty[$ is the open ray $]0, \infty[$.

(4) 0 is not an interior point of $[0, 1[$. Every point of $]0, 1[$ is an interior point of $[0, 1[$. The interior of $[0, 1[$ is $]0, 1[$.

(5) The set of rational numbers Q has no interior point *i.e,* $Q^\circ= \phi$.

(6) Let d be the discrete metric on a non-empty set X, and let $p \in X$. Then $\text{int}\,\{p\}=\{p\}$. In fact, for each set $A \subset X$, $\text{int}\,A=A$.

2.10 PROPERTIES OF THE INTERIOR

THEOREM 1. ***Let (X, d) be a metric space and A be a subset of X then***

(i) A° is an open set.

(ii) A° is the largest open set contained in A.

(iii) A is open if and only if $A^\circ=A$.

PROOF. (i) Let $x \in A^\circ$. Then x is an interior point of A

$\Rightarrow$ A is a nbd of x (By definition of interior point)

$\Rightarrow$ $\exists$ an open set G such that $x \in G \subset A$ (By definition of neighbourhood)

Since G is open, therefore G is a nbd of each of its points. Also, Since $G \subset A$, therefore A is also a nbd of each point of G

$\Rightarrow$ Every point of G is an interior point of A

$\Rightarrow G \subset A^\circ$

$\Rightarrow$ to each $x \in A^\circ$, there exists an open set G such that $x \in G \subset A^\circ$

$\Rightarrow$ A° is the nbd of each of its points

(ii) Let G be an open subset of A and let $x \in G$. Then $x \in G \subset A$

Now, since G is open, A is a nbd of x

$\Rightarrow$ x is an interior point of A $\Rightarrow x \in A^\circ$

Therefore, $x \in G \Rightarrow x \in A^\circ$ $(\because G \subset A^\circ)$

$\Rightarrow$ A° contains every open subset of A and it is, therefore, the largest open subset of A.

(iii) Let us first suppose $A=A^\circ$.

Since A° is open, therefore A is also open.

Conversely, let A is open. To show $A=A^\circ$.

Since, G is open, therefore A is the largest open set contained in A.

Then A is surely identical with A°.

Also, A° is the largest open subset of A. Hence, $A=A^\circ$.

THEOREM 2. ***Let (X, d) be a metric space and A be a subset of X. Then A° is equals the set of all those points of A which are not limit point of A'.***

PROOF. Let x be a point of A which is not a limit point of A'. Then there exists a nbd N of x which contains no point of A'.

$\therefore\ N \subset A$

$\Rightarrow$ A is also a nbd of x.

$\Rightarrow$ $x \in A^{\circ}$.

Conversely, let $x \in A^{\circ}$.

Since A° is open (By definition)

$\Rightarrow$ A° is a nbd of x ($\because$ Every open set is a nbd of each of its points)

Also, A° contains no point of A' ($\because$ $A^{\circ} \subset A$ and A contains no point of A')

$\Rightarrow$ x is not a limit point of A'

$\Rightarrow$ no point of A° is a limit point of A'.

Hence, A' consists exactly those points of A which are not limit points of A'.

2.11 EXTERIOR POINT AND EXTERIOR OF A SET

Definition 1. *Let A be a subset of a metric space (X, d). A point $x \in X$ is said to be exterior point of A if it is an interior point of the complement of A.*

Definition 2. *The set of all exterior points of A is called the exterior of A and is denoted by ext. (A) or A^e.*

REMARKS

- Clearly ext.(A) = $(A')^{\circ}$ and ext. (A') = $(A'')^{\circ}$ = A°. Also $A \cap$ ext $(A) = \phi$
- Since ext. (A) is the interior of A' therefore ext.(A) is the largest open set containing A'.

2.12 FRONTIER POINT AND FRONTIER OF A SET

Definition 1. *Let (X, d) be a metric space. A point x of a metric space is said to be frontier point of a subset $A \subset X$ if and only if it is neither an interior nor exterior point of A.*

Definition 2. *The set of all frontier points of A is called the frontier of A and is denoted by $F_r(A)$.*

2.13 BOUNDARY POINT AND BOUNDARY OF A SET

Definition 1. *Let (X, d) be a metric space. A point $x \in X$ is said to be boundary point of a subset $A \subset X$ if it is a frontier point of A and belongs to A*

Definition 2. *The set of all boundary points of A is called the boundary of A and it is denoted by b(A).*

REMARK

- $b(A) \subset F_r(A)$

2.13.1 SOME MORE DEFINITIONS

Let (X, d) be a metric space and A, B be subsets of X. Then

(1) A is said to be dense in B if $B \subset \overline{A}$

(2) A is said to be dense in X or everywhere dense if $\overline{A} = X$.

(3) A is said to be nowhere dense in X if$(\overline{A})^{\circ} = \phi$ *i.e.* interior of the closure of A is empty.

(4) A is said to be dense in itself if $A \subset D(A)$.

(5) A is said to be perfect iff A is dense-in-itself and closed.

REMARKS

- A is said to be every where dense iff every point of X is an adherent point of A.
- A is said to be perfect if $A = D(A)$.

THEOREM 1. ***Let (X, d) be a metric space and A be any subset of X. Then***

(i) $A^\circ = \cup [G : G \text{ is open}, G \subset A]$

(ii) $\text{ext.}(A) = \cup [G : G \text{ is open}, G \subset A']$.

PROOF. (i) Let $x \in A^\circ$, then by definition of interior point.

A is a nbd of x. $\Rightarrow$ there exists an open set G such that $x \in G \subset A$.

$\Rightarrow$ $x \in \cup [G : G \text{ is open}, G \subset A]$.

Conversely, let $x \in \cup \{G : G \text{ is open}, G \subset A\}$. Then $x \in G$ for some open set G such that $G \subset A$

$\Rightarrow$ A is a neighbourhood of x

$\Rightarrow$ $x \in A^\circ$.

Therefore, $x \in A^\circ$ if and only if $x \in \cup \{G : G \text{ is open}, G \subset A\}$.

Since, x is arbitrary, therefore

$\Rightarrow$ $A^\circ = \cup \{G: G \text{ is open}, G \subset A\}$.

THEOREM 2. ***Let A be a subset of a metric space (X, d). Then a point x∈X is an exterior point of A if and only if x not an adherent point of A i.e.*** $x \in (\bar{A})'$.

PROOF. Let us first suppose x be an exterior point of A.

$\Rightarrow$ x is an interior point of A'.

$\Rightarrow$ A' is a nbd of x containing no point of A

$\Rightarrow$ x is not an adherent point of A

$\Rightarrow x \in (\bar{A})'$.

Conversely, let $x \in (\bar{A})'$. To show x is an exterior point of A

Now, $x \in (\bar{A})' \Rightarrow x$ is not an adherent point of A

$\Rightarrow$ there exists a nbd N of x which contains no points of A

$\Rightarrow x \in N \subset A'$

$\Rightarrow$ A' is a nbd of x.

$\Rightarrow$ x is an interior point of A'

$\Rightarrow$ x is an exterior point of A.

THEOREM 3. ***Let (X, d) be a metric space and A⊂X. Then, a point x∈X is a frontier point of A if and only if every nbd of x intersects both A and A′.***

PROOF. Here, we have

$$x \in F_r(A) \Leftrightarrow x \notin A^\circ \text{ and } x \notin \text{ext}(A) = (A')^\circ$$

$\Leftrightarrow$ neither A nor A' is a nbd of x

$\Leftrightarrow$ no nbd of x can be contained in A or in A'.

$\Leftrightarrow$ every nbd of x intersects both A and A'.

REMARK

- $F_r(A) = F_r(A')$

THEOREM 4. ***Let A be any subset of a metric space (X, d). Then*** A°, $ext(A)$ ***and*** $F_r(A)$ ***are mutually disjoint and*** $X = A^\circ \cup ext(A) \cup F_r(A)$. ***Also,*** $F_r(A)$ ***is a closed set.***

PROOF. Let (X, d) be a metric space. We know that $\text{ext}(A) = (A')^\circ$

Also, $A^\circ \subset A$, $(A')^\circ \subset A'$ and $A \cap A' = \phi$

$\Rightarrow$ $A^\circ \cap \text{ext}(A) = A^\circ \cap (A')^\circ = \phi$.

Now, $x \in F_r(A) \Leftrightarrow x \notin A^\circ$ and $x \notin \text{ext.}(A)$

$\Leftrightarrow x \notin A^\circ \cup \text{ext}(A)$

$\Leftrightarrow x \in [A^\circ \cup \text{ext}(A)]'$.

Therefore, $F_r(A) = [A^\circ \cup \text{ext}(A)]'$

$\Rightarrow$ $F_r(A) \cap A^\circ = \phi$ and $F_r(A) \cap \text{ext}(A) = \phi$

and $x = A^\circ \cup \text{ext}(A) \cup F_r(A)$.

Finally, since A° and ext (A) both are open.

Therefore, $(A^\circ)'$ and $(\text{ext}(A))'$ both are closed.

Hence, $F_r(A)$ is closed.

THEOREM 5. ***Let (X, d) be a metric space and A, B be any subsets of X. Then***

(i) $X^\circ = X$, $\phi^\circ = \phi$ ***(ii)*** $A^\circ \subset A$ ***(iii)*** $A \subset B \Rightarrow A^\circ \subset B^\circ$

(iv) $(A \cap B)^\circ = A^\circ \cap B^\circ$ ***(v)*** $A^\circ \cup B^\circ \subset (A \cup B)^\circ$ ***(vi)*** $A^{\circ\circ} = A^\circ$.

PROOF. (i) Since we know that X and ϕ both are open sets.

Therefore $X^\circ = X$ and $\phi^\circ = \phi$ ($\because A$ is open iff $A^\circ = A$)

(ii) Let $x \in A^\circ \Rightarrow x$ is an interior point of A

$\Rightarrow A$ is a nbd of x

$\Rightarrow x \in A \Rightarrow A^\circ \subset A$

(iii) Let $A \subset B$ and let $x \in A^\circ$

$\Rightarrow x$ is an interior point of A

$\Rightarrow x$ is an interior point of B ($\because A \subset B$)

$\Rightarrow x \subset B^\circ \Rightarrow A^\circ \subset B^\circ$. ($\because$ x is arbitrary)

(iv) Since we know that

$$\left.\begin{aligned} & A \cap B \subset A \Rightarrow (A \cap B)^\circ \subset A^\circ \\ \text{and} \quad & A \cap B \subset B \Rightarrow (A \cap B)^\circ \subset B^\circ \end{aligned}\right\} \quad \text{[using (iii)]}$$

This implies $(A \cap B)^\circ \subset A^\circ \cap B^\circ$. ...(1)

Now let $x \in A^\circ \cap B^\circ$

$\Rightarrow$ $x \in A^\circ$ and $x \in B^\circ$

$\Rightarrow$ x is an interior point of A and x is an interior point of B

$\Rightarrow$ x is an interior point of $A \cap B$

Therefore $(A^\circ \cap B^\circ) \subset (A \cap B)^\circ$. ...(2)

From (1) and (2), we conclude that

$$A^\circ \cap B^\circ = (A \cap B)^\circ.$$

(v) We know that

$A \subset A \cup B \Rightarrow A^\circ \subset (A \cup B)^\circ$

and $B \subset A \cup B \Rightarrow B^\circ \subset (A \cup B)^\circ$.

This implies that

$$A^\circ \cup B^\circ \subset (A \cup B)^\circ.$$

(vi) Since A° is always open and we know that A is open if and only if $A^\circ = A$. Apply the above result for A°, we get

$$(A^\circ)^\circ = A^\circ$$

$$\Rightarrow \quad A^{\circ\circ} = A^\circ.$$

REMARK

- In result (v) $A^\circ \cup B^\circ \neq (A \cup B)^\circ$.

For Example : If $A = [0, 1[$ and $B = [1, 2[$

$\Rightarrow \quad A^\circ =]0, 1[$ and $B^\circ =]1, 2[$

$\Rightarrow \quad A^\circ \cup B^\circ =]0, 1[\cup]1, 2[=]0, 2[- [1]$

also $\quad A \cup B = [0, 2[\;\Rightarrow\; (A \cup B)^\circ =]0, 2[.$

Therefore, $\quad (A \cup B)^\circ \neq A^\circ \cup B^\circ$

THEOREM 6. ***Let (X, d) be a metric space and let A, B be subsets of X. Then***

(i) $\text{ext}(X) = \phi$, $\text{ext}(\phi) = X$ ***(ii)*** $\text{ext}(A) \subset A'$

(iii) $\text{ext}(A) = \text{ext}[(\text{ext}(A))']$ ***(iv)*** $A \subset B \Rightarrow \text{ext}[B] \subset \text{ext}[A]$

(v) $A^\circ \subset \text{ext}[\text{ext}(A)]$ ***(vi)*** $\text{ext}(A \cap B) = \text{ext}(A) \cap \text{ext}(B)$.

PROOF. (i) $\text{ext}(X) = (X')^\circ = \phi^\circ = \phi \qquad (\because X' = \phi)$

$\text{ext}(\phi) = (\phi')^\circ = X^\circ = X \qquad (\because \phi' = X)$

(ii) $\text{ext}(A) = (A')^\circ \subset A'$.

(iii) Here, we have

$$\text{ext}[(\text{ext}(A))'] = \text{ext}(A'^{\circ})'] = \text{ext}(A'^{\circ\prime})$$

$$= [(A'^{\circ\prime})']^\circ = (A'^{\circ\prime\prime})^\circ = (A'^{\circ})^\circ \qquad (\because A'' = A)$$

$$= A'^{\circ\circ} = A'^{\circ} = (A')^\circ = \text{ext}(A) \qquad (\because A^{\circ\circ} = A^\circ)$$

(iv) Now $\quad A \subset B \quad \Rightarrow B' \subset A'$

$$\Rightarrow (B')^\circ \subset (A')^\circ$$

$$\Rightarrow \text{ext}(B) \subset \text{ext}(A).$$

(v) We have $\text{ext}(A) \subset A' \;\Rightarrow \text{ext}(A') \subset \text{ext}(\text{ext}(A))$

$$\Rightarrow A^\circ \subset \text{ext}(\text{ext}(A)) \qquad (\because A^\circ = \text{ext}(A'))$$

(vi) We have

$$\text{ext}(A \cup B) = [(A \cup B)']^\circ = [A' \cap B']^\circ$$

$$= A'^{\circ} \cap B'^{\circ} = \text{ext}(A) \cap \text{ext}(B).$$

THEOREM 7. ***Let (X, d) be a metric space and $A \subset X$. Then***

(i) closure of the complement of A is the complement of the interior of A.

(ii) the interior of A is the complement of the closure of the complement of A.

(iii) the closure of A is the complement of the interior of the complement of A.

Proof. (i) Since we know that

$$A^{o} = A'^{-}{}'.$$

Therefore $(A^{o})' = (A'^{-}{}') = A'^{-}{}'' = A^{o}{}' = A'^{-}.$

(ii) Using (i) $A^{o}{}' = A'^{-}$

$\Rightarrow$ $A^{o}{}'' = A'^{-}{}'.$ (By taking complement)

$\Rightarrow$ $A^{o} = A'^{-}{}'.$

(iii) Using (i) $A''^{-} = A^{o}{}'.$

Replacing A by A', we get

$$(A')'^{-} = (A')^{o}{}'$$

$$A'^{-} = A'^{o}{}' \Rightarrow \bar{A} = A^{o}{}'.$$

Theorem 8. ***Let (X, d) be a metric space and AX. Then*** $A = A^{o} \cup F_r(A)$

Proof. Since we know that

$$\bar{A} = \cap\,[F : F \text{ is closed and } A \subset F\,]$$

$\Rightarrow$ $(-\,\bar{A})' = \cup\,[F' : F' \text{ is open and } F' \subset A'\,] = \text{ext}\,(A)$

$\Rightarrow$ $(\bar{A})'' = [\text{ext}\,(A)]'$ (By taking complement)

$\Rightarrow$ $\bar{A} = A^{o} \cup F_r(A).$

Hence, we have $\bar{A} = A^{o} \cup F_r(A).$

Theorem 9. ***Let (X, d) be a metric space and let A, B be subsets of X. Then***

(i) $F_r(A) = \bar{A} \cap A'^{-} = \bar{A} - A^{o}$ ***(ii)*** $A^{o} = A - F_r(A)$

(iii) $[F_r(A)]' = A^{o} \cup A'^{o}$ ***(iv)*** $F_r(A^{o}) \subset F_r(A)$

(v) $F_r(\bar{A}) \subset F_r(A)$ ***(vi)*** $F_r(A \cup B) \subset F_r(A) \cup F_r(B)$

Proof. Let (X, d) be a metric space and A and B be any two subset of X.

(i) We know that

$$F_r(A) = [A^{o} \cup \text{ext}\,(A)]' = A'^{\,o} \cap [\text{ext}\,(A)]'$$

(By De-Morgan's law)

$$= A'^{-}{}''' \cap A^{-}{}'' = A'^{-} \cap \bar{A} \qquad (\because A'' = A)$$

Now $A \cap A'^{-} = \bar{A} - A'^{-}{}'$ $[\because A \cap B' = A - B]$

$$= \bar{A} - A^{o}$$

which implies $F_r(A) = \bar{A} \cap A'^{-} = A' - A^{o}.$

(ii) Since $F_r(A) = A' - A^{o}$

$\Rightarrow$ $A - F_r(A) = A' - (A' - A^{o}) = A^{o}.$

(iii) $[F_r(A)]' = (\bar{A} \cap A'^{-})' = A^{-}{}' \cup A'^{-}{}'$

Now using $A'^{o} = (A')^{o} = (A')^{-}{}' = A''^{-}{}' = A^{-}{}'$

(iv) Now $Fr(A^{o}) = A^{o-} \cap A^{o}{}'^{-} = A^{o-} \cap A'^{-}{}''^{-}$ $[\because A'^{-}{}' = A^{o}]$

$$= A^{o-} \cap A' \subseteq F_r(A).\ \phi$$

(v) $F_r(\bar{A}) = \bar{\bar{A}} \cap A^{-}{}'^{-} = \bar{A} \equiv A^{-}{}'^{-}$ $\left[\because \bar{\bar{A}} = \bar{A}\right]$

Now, $A \subset \bar{A} \Rightarrow \left(\bar{A}\right)' \subset A'$

$$\Rightarrow A^{-'-} \subset A'^{-}$$

Hence $F_r\left(\bar{A}\right) \subset \bar{A} \cap A'^{-} = F_r(A)$

$$\Rightarrow F_r\left(\bar{A}\right) \subset F_r(A).$$

(vi) Consider

$$F_r(A \cup B) = \left(\overline{A \cup B}\right) \cap \left(A \cup B\right)'^{-}$$

$$= \left(\bar{A} \cup \bar{B}\right) \cap \left(A' \cup B'\right)^{-} \subset \left(\bar{A} \cup \bar{B}\right) \cap \left(A'^{-} \cup B'^{-}\right)$$

$$= \left[\bar{A} \cap \left(A'^{-} \cap B'\right)\right] \cup \left[\bar{B} \cap \left(A'^{-} \cap B'^{-}\right)\right]$$

$$= \left[\left(\bar{A} \cup A'^{-}\right) \cap B'\right] \cup \left[\left(\bar{B} \cap B'^{-}\right) \cap A'\right] \subset F_r(A) \cup F_r(B).$$

THEOREM 10. ***Let (X, d) be a metric space and A be any subset of X. Then***

(i) If A is open, $F_r(A) = \bar{A} - A$.

(ii) $F_r(A) = \phi$ ***if and only if A is open as well as closed.***

(iii) A is open if and only if $A \cap F_r(A) = \phi$.

(iv) A is closed if and only if $F_r(A) \subset A$.

PROOF. (i) Since we know that

$$F_r(A) = \bar{A} - A.$$

If A is a open then $A = A^\circ$.

$$\therefore \quad F_r(A) = \bar{A} - A^\circ.$$

(ii) Let us first suppose $F_r(A) = \phi$. Then, we have

$$F_r(A) = \phi \Rightarrow \bar{A} - A^\circ = \phi \Rightarrow \bar{A} \subset A^\circ \Rightarrow \bar{A} \subset A \qquad (\because A^\circ \subset A)$$

$$\Rightarrow D(A) \subset A \qquad (\because \bar{A} = A \cup D(A))$$

$\Rightarrow A$ is closed.

Also, $F_r(A) \Rightarrow \phi = \bar{A} - A^\circ = \phi \Rightarrow \bar{A} \subset A^\circ$

$$\Rightarrow A \cup D(A) \subset A^\circ \Rightarrow A \subset A^\circ.$$

But $A^\circ \subset A$. Therefore $A^\circ = A$

$\Rightarrow$ A is open ($\because A$ is open if and only if $A^\circ = A$)

Hence $Fr(A) = \phi$ then A is closed as well as open.

Conversely, let A is open as well as closed. To show $F_r(A) = \phi$.

Since we know that $F_r(A) = \bar{A} - A^\circ$. ...(1)

Also if A is closed, then $\bar{A} = A$ and if A is open, then $A^\circ = A$.

Put the above values in (1), we get $F_r(A) = \phi$

(iii) We know that $F_r(A) = \bar{A} \cap A'^{-}$

If A is open, then A' is closed,

$\Rightarrow$ $A'^{-} = A'$.

Now $A \cap F_r(A) = A \cap [\bar{A} \cap A'^{-}] = A \cap [\bar{A} \cap A']$

$$= [A \cap \bar{A}] \cap A' = A \cap A' = \phi$$

Conversely, let $A \cap F_r(A) = \phi$.

Then, $A \cap F_r(A) = \phi \Rightarrow A \cap (\bar{A} \cap A'^{-}) = \phi$

$\Rightarrow (A \cap A') \cap A'^{-} = \phi.$

$\Rightarrow A \cap A'^{-} = \phi \Rightarrow A \subset A'^{-'} \Rightarrow A \subset A^0.$

Now $A^o \subset A$ gives $A^\circ = A$

$\Rightarrow$ A is open.

(iv) Let A is closed. Then $\bar{A} = A$ ($\because A$ is closed iff $\bar{A} = A$)

$\Rightarrow$ $F_r(A) = \bar{A} \cap A'^{-} = A \cap A'^{-} \subset A.$

Conversely, let $F_r(A) \subset A$ Then $A \cup F_r(A) = A$

But $A \cup F_r(A) = \bar{A} \Rightarrow A = \bar{A} \Rightarrow$ A is closed. ($\because A$ is closed iff $\bar{A} = A$)

RECAPITULATIONS

➠ The smallest closed set containing A is called closure of A. It is denoted by $\bar{A}$.	➠ A is closed $\Leftrightarrow \bar{A} = A$.
➠ $\bar{\phi} = \phi; A \subset \bar{A}; A \subset B \Rightarrow \bar{A} \subset \bar{B}; \overline{A \cup B} = \bar{A} \cup \bar{B}$	➠ $\overline{A \cap B} \subseteq \bar{A} \cap \bar{B}; \bar{\bar{A}} = \bar{A}.$
➠ The-interior of A is the largest open set contained in A. It is denoted by A°.	➠ A is open $\Leftrightarrow A = A^\circ$.

SOLVED EXAMPLES

EXAMPLE 1. *Consider the usual metric space* (R, *d*). *Find the closure of the following sets*

(*i*) $A = \left[\frac{1}{n} : n \in N\right]$ (*ii*) Z (*iii*) Q (*iv*)]0, 1[.

SOLUTION. (i) Since 0 is the only limit point of A

$\therefore$ $D(A) = [0]$ and so

$$\bar{A} = \left\{\frac{1}{n} : n \in N\right\} \cup [0].$$

(ii) Since $D(Z) = \phi$, therefore $\bar{Z} = Z \cup \phi = Z$;

(iii) D(Q)=R.

$\therefore$ $\bar{Q} = Q \cup D(Q) = Q \cup R = R.$

(iv) $\overline{]0,1[} = [0,1].$

EXAMPLE 2. *Given an example to show that in a metric space it is not necessary that* $\overline{(A \cap B)} = \bar{A} \cap \bar{B}$.

SOLUTION. Let us consider the usual metric space (R, d)

and $A = [0, 1[$ and $B =]1, 2]$

Clearly, $A \cap B = \phi \Rightarrow \overline{(A \cap B)} = \bar{\phi} = \phi$

and $\bar{A} = [0,1]$ and $\bar{B} = [1,2]$

$\Rightarrow$ $\bar{A} \cap \bar{B} = \{1\}.$

Hence $\overline{(A \cap B)} \neq \bar{A} \cap \bar{B}.$

EXAMPLE 3. *Define a dense set and give an example.*

SOLUTION. By definition of dense set we have

A subset A of metric space (X, d) is said to be dense in X iff $\bar{A} = X$.

Example of dense set: let (R, d) be a usual metric space on R and Q be set of rational number then $Q \subseteq R$

$$\bar{Q} = Q \cup D(Q) = Q \cup R = R$$

$\Rightarrow$ $$\bar{Q} = R$$

$\Rightarrow$ Q is dense in R.

EXAMPLE 4. *Define a nowhere dense set by giving a suitable example.*

SOLUTION. Let A be non-empty subset A of X, then A is said to be nowhere dense in X iff $(\bar{A})^\circ = \phi$.

EXAMPLE. Let (R, d) be usual metric on R and Z is the set of integers, $Z \subseteq R$.

$$\bar{Z} = Z \cup D(Z) = Z \cup \phi = Z$$

$\Rightarrow$ $(\bar{Z})^\circ = Z^\circ = \phi \Rightarrow$ Z is nowhere dense in R.

EXAMPLE 5. *Show that Cantor's set E is a non-dense set.*

SOLUTION. Since we know that Cantor set is always a closed set.

$\therefore$ $\bar{E} = E$ $\qquad (\because A \text{ is closed iff } \bar{A} = A)$

$\Rightarrow$ $(\bar{E})^\circ = (E)^\circ = \cup [G \subset [0, 1] : G \subset E] = \phi$

$\Rightarrow$ $(\bar{E})^\circ = \phi$

$\Rightarrow$ E is non-dense.

EXAMPLE 6. *In any metric space, show that* $(\bar{A})' = (A')^\circ$.

SOLUTION. Let (X, d) be a metric space and A be any subset of X.

Now $(\bar{A})' = X - \bar{A} = X -$ intersection of all closed set F_i

$= X - \cap F_i$ where F_i is closed and $A \subset F_i$

$= \cup [X - F_i]$ where $X - F_i$ is open.

Since $X - F_i \subset X - A$

Therefore $(\bar{A})'$ = Union of open subsets of $X - A = A' = (A')^\circ$.

EXAMPLE 7. *Consider the usual metric $d(x, y) = |x - y|$ on R and find (i) interior, (ii) exterior, (iii) frontier, (iv) and boundary of each of the following subsets of R*

(a) $A =]0, 1[$ (b) $B = [0, 1[$ (c) $C = \left\{\frac{1}{n} : n \in N\right\}$ (d) N

SOLUTION. (a) (i) Since A is an open set

$\Rightarrow$ A is a nbd of each of its points

$\Rightarrow$ every point of A is an interior point of A.

$\Rightarrow$ $A^\circ = A =]0, 1[$.

(ii) $A' =]-\infty, a[\cup [1, \infty[$

Here A' is a nbd of each of its points except 0 and 1

$\Rightarrow$ ext $(A) = (A')^\circ =]-\infty, 0[\cup]1, \infty[$.

(iii) $F_r(A) = [A^\circ \cup \text{ext}(A)]' = \{0, 1\}$

$\Rightarrow$ $F_r(A)$ contains two points 0 and 1.

(iv) $b(A) = \phi$. $\qquad (\because$ no frontier point is a point of $A)$

(b) (i) $B^\circ =]0, 1[$ (same as in (a))

(ii) $B' =]-\infty, 0[\cup [1, \infty[$

$\Rightarrow$ ext(B) = $(b')^\circ$ = $]-\infty, 0[\cup]1, \infty[$

(iii) $F_r(B) = \{0, 1\}$

(iv) $b(B) = \{0\}$ ($\because$ 0 is the only frontier point which belongs to B)

(c) (i) Since C can not be a nbd of any of its point $\frac{1}{n}$, $n = 1, 2, \ldots$, since $\exists \varepsilon > 0$ such that

$$\left]\frac{1}{n} - \varepsilon, \frac{1}{n} + \varepsilon\right[\subset D$$

$\Rightarrow$ no point of D can be its interior point so that $D^\circ = \phi$.

(ii) Clearly $D' = R - D$ is a nbd of each of its point except 0. Hence ext$(D) = (D')^\circ = R - \{D \cup (0)\}$.

(iii) $F_r(D) = [D^\circ \cup \text{ext}(D)]' = D \cup \{0\}$.

(iv) $b(D) = D$. Since all the points of D are frontier point of D.

(d) (i) Since N is not a nbd of any of its point $\Rightarrow N^\circ = \phi$.

(ii) Since N has no limit points, therefore we have $\overline{N} = N$.

$\Rightarrow$ N is a closed set

$\Rightarrow$ N' is open set. $\Rightarrow$ ext$(N) = (N')^\circ = N' = R - N$.

(iii) $F_r(N) = [N^\circ \cup \text{ext}(N)]' = [\phi \cup \text{ext}(N)]' = [\text{ext}(N)]' = (N')' = N$.

(iv) $b(N) = N$ (because every point of N is a frontier point of N).

EXAMPLE 8. *Show that a subset F of a metric space X is closed iff* $\{x \in X : d(x, F) = 0\} \subseteq F$.

SOLUTION. Let (X, d) be a metric space and $F \subseteq X$.

We know that $\overline{F} = \{x \in X : d(x, F) = 0\}$.

Let F be closed. Now we have to show that $\{x \in X : d(x, f) = 0\} \subseteq F$.

Since F is closed $\Rightarrow \overline{F} = F \Rightarrow \overline{F} \subseteq F$ (by the definition of closure)

$\Rightarrow \{x \in X : d(x, F) = 0\} \subseteq F$.

Conversely let $\{x \in X: d(x, F)\} = 0 \subseteq F$ then we have to prove that F is closed.

Since we have $\{x \in X: d(x, F)\} = 0 \subseteq F \Rightarrow \overline{F} \subseteq F$

But $F \subseteq \overline{F} \Rightarrow F = \overline{F}$

$\Rightarrow F$ is closed.

2.14 SOME MORE DEFINITIONS

(1) Separable spaces. A metric space (X, d) is said to be separable if X contains a countable dense subset *i.e.* there exists a countable subset A of X such that $\overline{A} = X$.

For Example: The usual metric space (R, d) is separable, because the set Q of rational numbers is a countable dense subset of R.

(2) Bases for the neighbourhood system of a point. Let (X, d) be a metric space and N(x) denotes the family of all neighbourhoods of a point $x \in X$. A sub family $\beta(x)$ of N(x) is said to be a base for N(x) if to each $N \in$ N(x) there exist $B \in \beta(x)$ such that $B \subset N$.

REMARKS

- Here, β(x) is said to be local base at x or a fundamental system of neighbourhoods of x.
- The set of all open interval in R form a base for the family of open subsets of R.

(3) Bases for the open sets of a metric space.

Let (X, d) be a metric space and let G be the family of all open subsets of a metric space (X, d). Then a subfamily β of G is said to be base if for each point $x \in X$ and each nbd N of $x \exists B \in \beta$ such that $x \in B \subset N$.

(4) First and second countable spaces.

(a) A metric space (X, d) is said to satisfy the first axiom of countability if each point $x \in X$ possesses a countable local base. Such a space is said to be *first countable*.

(b) A metric space (X, d) is said to satisfy the second axiom of countability if there exist a countable local base for G, where G denotes the family of all open subsets of X. Such a space is said to be *completely separable space*.

2.14.1 SOME MORE RESULTS

- A second countable space is also called completely separable space.
- The usual metric space (R, d) is a first as well as a second countable spaces.
- Every metric space (X, d) is first countable.
- A metric space is separable if and only if it is second countable.
- Let (X, d) be a metric space and Y be a proper subset of X. Let d^* denote the restriction of d on $Y \times Y$ *i.e.*, $d^*(x, y) = d(x, y)$ whenever x and y are points of Y. Then d^* is a metric for Y called the induced metric. Then $(Y, d^*)$is said to be subspace of (X, d).
- A property of a metric space is said to be heridIitary if and only if every sub-space of that space has that property.
- Every subspace of a separable metric space is separable *i.e.* separability is a heriditary properly in a metric space.

EXERCISE 2.3

1. Let (X, d) be a metric space and A be any subset of X. Then show that the following statements are equivalent:

(a) A is closed.

(b) A contains all its limit points.

(c) $\overline{A} = A$.

2. Let (X, d) be a metric space and G be any open set in X, Show that G is disjoint from A iff G is disjoint from $\overline{A}$.

3. Let (X, d) be a metric space and $A \subset X$. Then show that A° equals the set of all those points of A, which are not limit point of A'.

4. Show that the frontier of a subset of a metric space is closed.

5. Let (X, d) be a metric space and $A \subset X$. Find (i) A°, (ii) ext (A), (iii) $F_r(A)$, (iv) $b(A)$.

6. Are the following subsets of R, d-nbds of 3 when d denotes the usual metric defined by $d(x, y) = |x-y|$ for R?

(i)]2, 4[(ii) [1, 3[(iii) [3, 4[(iv) N.

7. Show that every subspace of a discrete metric space is discrete.

8. Consider the following subset of R. Find their ds relative to usual metric $d(x, y) = |x-y|$

(i) $A = \{1, 2, 3, 4\}$ (ii) $B = [1]$

(iii) $C =]2, \infty[$ (iv) $D =]1, 2[\cup]3, 4[$

(v) $E = \left\{\frac{n+1}{n} : n \in N\right\}$

9. Show that the diameter of a subset of a metric space is equal to the diameter of its closure.

10. Show that a closed set is nowhere dense if and only if its complement is everywhere open.

11. Let A be subset of a metric space (X, d). Prove that A is non-dense in X if and only if $X - \overline{A}$ is dense in X.

12. Let (X, d) be any metric space and let A be any subset of X. Prove that

(i) $\overline{X - A} = X - A^\circ$

(ii) $X - \overline{A} = (X-A)^\circ$.

13. In the metric space (R, d) where d is the usual metric on R, find the boundary of the set of integers Z.

14. Give an example of two subsets A and B of R such that

$$D(A \cap B) \neq D(A) \cap D(B)$$

the metric on R is the usual metric.

15. In a metric space prove that $(\overline{A})' = (A')^{\circ}$

HINT TO SELECTED PROBLEMS

9. Define $f(A) = \sup\{d(x, y): x, y \in A\}$.

By definition of closure, we have $\overline{A}$ is the smallest closed set containing A.

Also A is closed $\Rightarrow \overline{A} = A \Rightarrow d(\overline{A}) = d(A)$.

10. Since A is dense in X if $\overline{A}$ =X and A is nowhere dense in X if int $(\overline{A}) = \phi$.

$\therefore$ A is non-dense in X

$\Leftrightarrow$ int $(\overline{A}) = \phi$

$\Leftrightarrow$ $\exists$ no open neigbourhood of any point of $\overline{A}$ such that $N \subset \overline{A}$

$\Leftrightarrow$ $\overline{A}$ contains no nbd

$\Leftrightarrow$ $S(r, x) \subset \overline{A}$, $r > 0$ $\forall x \in X$

$\Leftrightarrow$ $S(r, x) \cap (\overline{A})' \neq \phi$

$\Leftrightarrow$ x is abherent point of $(\overline{A})'$ $\forall x \in X$.

$\Leftrightarrow$ $\left(\overline{(A')}\right) = X \Leftrightarrow \overline{(A)}'$ is dense in X.

ANSWERS

5.	(i) A	(ii) A'	(iii) ϕ	(iv) ϕ	
6.	(i) Yes	(ii) No	(iii) No	(iv) No	
8.	(i) A	(ii) B	(iii) $[2, \infty[$	(iv) $[1, 2] \cup [3, 4]$	(v) $E \cup [1]$.

2.15 COMPLETE METRIC SPACE

Definition 1. *A metric space (X, d) is said to be complete if every Cauchy sequence in X converges.*

A metric space, which is not complete is said to be incomplete.

SOLVED EXAMPLES

EXAMPLE 1. *Show that the real line is a complete metric space.*

SOLUTION. Let $\langle s_n \rangle$ be a Cauchy sequence in R. Then by definition given any $\varepsilon > 0$ there exist a positive integer k such that

$$|x_n - x_k| < \varepsilon, \ n \geq k.$$

Let $\varepsilon = 1$.

Then $|x_n - x_{k_1}| < 1, \forall n \geq k_1$ for some k_1

$\Rightarrow$ $x_{k_1} - 1 < x_n < x_{k_1} + 1, \ \forall n \geq k_1$.

Let us define

$$p = \min\{x_1, x_2, \ldots, x_{k_1} - 1, x_{k_1} + 1\}$$

and $q = \min\{x_1, x_2, \ldots, x_{k_1} - 1, x_{k_1} + 1\}$

Then $p \leq x_n \leq q, \forall \ n \in N$

$\Rightarrow$ $\langle x_n \rangle$ is bounded sequence of real numbers.

Let x= lim inf $\langle x_n \rangle$ and y = lim sup $\langle x_n \rangle$.

Let if possible $x \neq y$ and let $y - x = s$ then $s \geq 0$.

Now since $\langle x_n \rangle$ is a Cauchy sequence, therefore, there exists a positive integer n such that

$$|x_l - x_m| < \varepsilon/2, \ \forall \ l, m \geq n. \quad ...(1)$$

Since x is the limit inferior of $\langle x_n \rangle$, therefore $x \leq x_n < x+s/4$ for infinitely many values of n. In particular, we can find a positive integer $u>n$ such that

$$x \leq x_u < x+s/4. \quad ...(2)$$

Again since y is the limit superior of $\langle x_n \rangle$, therefore

$$y - s/4 < x_n \leq y \text{ for infinitely many values of } n.$$

In particular, we can find a positive integer $v>n$ such that

$$y - s/4 < x_v \leq y. \quad ...(3)$$

From (2) and (3), we conclude that there exist positive integers $u, v > n$ such that

$$|x_u - x_v| > s/2. \quad ...(4)$$

This contradicts (1) which says that for all positive integers $l, m \geq n$

$$|x_l - x_m| < s/2. \quad ...(5)$$

In view of the above contradiction, we must have $s \not> 0$. Thus, $s=0$ and therefore $x=y$ which gives

$$\liminf \langle x_n \rangle = \limsup \langle x_n \rangle.$$

Therefore $\lim\limits_{n\to\infty} \langle x_n \rangle$ exists.

Hence, the real line is complete.

EXAMPLE 2. *The set Z of integers with the usual metric ($d(x, y)=|x-y|$ $\forall$ $x, y \in$ Z) is a complete metric space.*

SOLUTION. Let $\langle x_n \rangle$ be any Cauchy sequence in Z. Take $\varepsilon = \frac{1}{2}$.

Then we can find a positive integer p such that $|x_m - x_n| < \frac{1}{2}$ $\forall$ $m, n \geq p$.

Since $x_n \in$ Z $\forall$ $n \in$ N, therefore $|x_m - x_n|$ must be a non-negative integer $\forall$ $m, n \in$ N. Therefore, we have

(i) $|x_m - x_n|$ $\forall$ $m, n \geq p$.

(ii) $|x_m - x_n|$ is a non-negative integer $\forall$ $m, n \in$ N.

Hence, we find that

$$|x_m - x_n| = 0 \ \forall \ m, n \geq p \ \text{ i.e. } \ x_m = x_n \ \forall \ m, n \geq p.$$

Consequently $\quad x_n = x_p \ \forall \ n \geq p$

$\Rightarrow$ $\quad \langle x_n \rangle$ is eventually a constant sequence

$\Rightarrow$ $\quad \langle x_n \rangle$ must converges to x_p.

Since, an arbitrary Cauchy sequence in Z converges.

Therefore, the set Z of integers equipped with the usual metric is a complete metric space.

EXAMPLE 3. *Let $X =]0, 1[$ and let $d(x, y)=|x-y|$ for all $x, y \in X$. Then show that (X, d) is not complete.*

SOLUTION. Let $\langle x_n \rangle$ be the sequence in X defined by setting $x_n = \frac{1}{n}$ for all $n \in$ N. Let ε be any arbitrary positive number.

Now $\qquad d(x_m, x_n) = |x_m - x_n| = \left|\frac{1}{m} - \frac{1}{n}\right| \leq \frac{1}{m} + \frac{1}{n}.$ ...(1)

Let p be any positive integer greater than $\frac{2}{\varepsilon}$, then

$$\frac{1}{m} + \frac{1}{n} < \frac{\varepsilon}{2} + \frac{\varepsilon}{2} \ \forall \ m, n \geq p.$$

Therefore, from (1) we can find that

$$d(x_m, x_n) < \varepsilon \quad \forall \quad m, n \geq p$$

$\Rightarrow$ $\langle x_n \rangle$ is a Cauchy sequence in X.

Now let $x \in X$. Then by the Archimedean property of real numbers, we can find a positive integer n such that

$$n \leq \frac{1}{n} < n+1 \Rightarrow \frac{1}{n} \geq x > \frac{1}{n+1}$$

Now there are two different cases :

Case (i) If $x = \frac{1}{n}$, then $\left]\frac{1}{(n-1)}, \frac{1}{(n+1)}\right[$ is a nbd of x in $]0, 1[$ which contains only one element of the sequence $\left\langle \frac{1}{n} \right\rangle$.

Case(ii) If $x \neq \frac{1}{n}$ then $\left]\frac{1}{n+1}, \frac{1}{n}\right[$ is a nbd of x in $]0, 1[$ which does not contain any element of the sequence.

In either case the sequence can not converge.

Since x is any point of X, therefore the sequence $\langle x_n \rangle$ can not converge to any point of X. Hence (X, d) is an incomplete metric space.

THEOREM 1. ***Let (X, d) be a complete metric space and Y be a subspace of X. Then Y is complete if and only if Y is closed.***

PROOF. Let (X, d) be a complete metric space. Let us first suppose Y be a complete subspace of X. To show Y is closed in X.

Let $p \in X$ be any limit point of Y. Then, by definition of limit point, for every positive integer n, the open sphere $s\left(p, \frac{1}{n}\right)$ must contains a point q_n of Y such that the sequence $\langle q_n \rangle$ converges to p

$\Rightarrow$ $\langle q_n \rangle$ is Cauchy in Y $\qquad$ ($\because$ Every convergent sequence is Cauchy)

Since Y is complete therefore $p \in Y$.

Now since $p \in Y$ is arbitrary, therefore we can say Y contains all its limit points and hence Y is closed.

Conversely, let Y is closed. To show Y is complete.

Let $\langle s_n \rangle$ be any Cauchy sequence in Y.

$\Rightarrow$ $\langle s_n \rangle$ is a Cauchy sequence in X. $\qquad$ ($\because$ Y is a subspace of X)

Also, since X is complete, therefore $\langle s_n \rangle$ must converge to a point $s_0 \in X$. We want to show $s_0 \in Y$. If the range set of $\langle s_n \rangle$ consist of finite number of distinct points then $\langle s_n \rangle$ must be of the form $\langle s_1, s_2, \ldots, s_n, s_0, s_0, s_0, \ldots \rangle$ where n is finite and hence $s_0 \in Y$. If the range set of $\langle s_n \rangle$ contains infinite many points, then s_0 is a limit of the range set of $\langle s_n \rangle$.

($\because$ If the range set of a convergent sequence in a metric space consists of infinitely many distinct points, then the limit of the sequence is a limit point of the range set of the sequence)

$\Rightarrow$ s_0 is also a limit point of Y

$\Rightarrow$ $s_0 \in Y$ ($\because$ Y is closed)

$\Rightarrow$ every Cauchy sequence in Y converges in Y. Hence, Y is complete.

THEOREM 2. **(Cantor's Intersection Theorem).** ***Let (X, d) be a metric space and let $\langle F_n \rangle$ be a nested sequence of non-empty closed subset of X such that $\delta(F_n) \to 0$ as $n \to \infty$, Then X is complete if and only if $\bigcap_{n=1}^{\infty} F_n$ consist of exactly one point.***

PROOF. (i) **Necessary condition.** Let (X, d) be a complete metric space and let $\langle F_n \rangle$ be the sequence of non-empty closed subsets of X such that $F_{n+1} \subset F_n$ $\forall n \in \mathbb{N}$ (by definition of nested sequence) and $\delta(F_n) \to 0$ as $n \to \infty$. To show $\bigcap_{n \in \mathbb{N}} F_n$ consist of exactly one point.

Step 1. For each $n \in \mathbb{N}$, choose a point $x_n \in F_n$. It can be easily seen that $\langle x_n \rangle$ is a Cauchy sequence of points of X. In fact, given $\varepsilon > 0$, we can find a positive integer m such that

$$\delta(F_n) < \varepsilon \;\forall\; n \geq m. \qquad \ldots(1)$$

Since the sequence $\langle F_n \rangle$ is nested, therefore $F_n \subseteq F_m$ $\forall$ $n \geq m$ and consequently

$$x_n \in F_m \;\forall\; n \geq m. \qquad \ldots(2)$$

Since we know that for any non-empty bounded set S in a metric space (X,d)

$$\delta(S) = \sup\{d(x, y) : x, y \in S\}$$

so that
$$d(x, y) \leq \delta(S) \;\forall\; x, y \in S \qquad \ldots(3)$$

From (2) and (3), we conclude that

$$d(x_n, x_m) \leq \delta(F_m) \;\forall\; n \geq m \qquad \ldots(4)$$

Now (1) and (4) gives

$$d(x_n, x_m) < \varepsilon \;\forall\; x \geq m$$

$\Rightarrow$ $\langle x_n \rangle$ is a Cauchy sequence in X.

Step 2. Since the metric space (X, d) is complete, therefore the Cauchy sequence $\langle x_n \rangle$ converges *i.e.* there exist $x_0 \in X$ such that $x_n \to x_0$

We shall show that $x_0 \in F_n$ $\forall$ $x \in \mathbb{N}$ and consequently $\bigcap_{n \in \mathbb{N}} F_n$ is non-empty. Let $k \in \mathbb{Z}$ be fixed. Now from (2) $x_n \in F_k$ $\forall$ $n \geq k$

i.e. $\langle x_k, x_{k+1}, x_{k+2}, \ldots \rangle$ is a sequence in F_k.

The sequence $\langle x_k, x_{k+1}, \ldots \rangle$ converges to x_0 and therefore $x_0 \in F_k$.
Since F_k is closed $\Rightarrow X \in F$.
Now $x_0 \in F_k$ and k is arbitrary, therefore $x_0 \in F_n \quad \forall\, n \in \mathrm{N}$

$$\Rightarrow \qquad x_0 \in \bigcap_{n \in \mathrm{N}} F_n.$$

Step 3. (Uniqueness). Let if possible x_0, y_0 be two points in the intersection of F_n's. To show $x_0 = y_0$
Suppose if possible $x_0 \neq y_0$ and let $d(x_0, y_0) = \varepsilon > 0$.
Since $\delta(F_n) \to 0$, therefore we can find a positive integer m such that

$$\delta(F_n) < \varepsilon/2 \quad \forall\, n \geq m.$$

Since x_0 , $y_0 \in F_n \; \forall\, n \in \mathrm{N}$, therefore in particular $x_0, y_0 \in F_m$

$$\Rightarrow \qquad \delta(F_n) < \varepsilon/2 \;\forall\, n \geq m.$$

Since $d(x_0, y_0) = \varepsilon$, we have $\varepsilon = d(x_0, y_0) < \varepsilon/2$
which is a contradiction and therefore $x_0 = y_0$ and so $\bigcap_{n \in \mathrm{N}} F_n$ consist of exactly one point.

(ii) Condition is sufficient. Let us suppose that every nested sequence of closed sets with diameter tending to zero has non-empty intersection.

To show X is complete.

Let $\langle x_n \rangle$ be a Cauchy sequence in X.

Correspond to $\varepsilon = 1/2$, we can find $n_1 \in \mathrm{N}$ such that

$$d(x_{n_1}, x_n) < \frac{1}{2} \;\forall\, n \geq n_1 \text{ and for } n_2 > n_1$$

$$d(x_{n_2}, x_n) < \frac{1}{2^2} \;\forall\, n \geq n_2\,.$$

Proceeding in the same manner, we can construct a strictly increasing sequence $\langle n_1, n_2, n_3, \ldots \rangle$ of positive integers such that

$$d\left(x_{n_1}, x_n\right) < \frac{1}{2}, \forall\, n \geq n_1$$
$$d\left(x_{n_2}, x_n\right) < \frac{1}{2^2}, \forall\, n \geq n_2$$
$$\ldots \quad \ldots \quad \ldots \quad \ldots \quad \ldots$$
$$d\left(x_{n_k}, x_n\right) < \frac{1}{2^k}, \forall\, n \geq n_k$$
$$\ldots \quad \ldots \quad \ldots \quad \ldots \quad \ldots$$

Let us write $\qquad F_k = S^*\left(x_{n_k}, 2^{-k+1}\right), k = 1, 2, 3, \ldots$

Then $\langle F_n \rangle$ is a sequence of closed sets with diameters tending to zero.

To see that $\langle x_k \rangle$ is nested *i.e.* $F_{k+1} \subset F_k \;\forall\, k \geq 1$, take a point $\mathrm{y} \in F_{k+1}$.

Then $\qquad d\left(x_{n_k}, y\right) \leq d\left(x_{n_k}, x_{n_{k+1}}\right) + d\left(x_{n_{k+1}}, y\right) \qquad$ (By triangle inequality)

$$\leq 2^{-k} + 2^{-k} = 2^{-k+1}$$

$$\Rightarrow \qquad y \in S\left(x_{n_k}, 2^{-k+1}\right) \subset F_k$$

$$\Rightarrow \qquad F_{k+1} \subset F_k.$$

Thus, we find that $\langle F_n \rangle$ is a nested sequence of non-empty closed sets with diameters tending to zero therefore there exist $x_0 \in X$ such that $x_0 \in F_k \; \forall$ N.

Consider the sequence $\langle x_{n_1}, x_{n_2}, x_{n_3}, \ldots \rangle$.

Since $n_1 < n_2 < n_3$ therefore the above sequence is a subsequence of $\langle x_n \rangle$.

Also, $\left(x_{n_k}, x_0\right) < 2^{-k} \quad \forall\, k \in \mathrm{N}$

Since $\langle x_n \rangle$ is a Cauchy sequence in x and $\langle x_{n_k} \rangle$ is a subsequence of $\langle x_n \rangle$ converging to x_0, therefore $\langle x_n \rangle$ converges to x_0.

Hence (X, d) is complete.

2.16 METRIC SPACE OF FIRST AND SECOND CATEGORY

Definition 1. *Let (X, d) be a metric space. A subset of a metric space is said to be of first category if and only if it can be written as the union of a countable family of nowhere dense sets: otherwise it is said to be the second category.*

For example. Set of rational numbers Q is of second category.

Definition 2. (Contracting Mapping). *Let (X, d) be a complete metric space.*

A mapping $f: X \to X$ is called a contracting, mapping (or contraction on X) if there exist a real number α with $0 \le \alpha < 1$ such that

$$d[f(x), f(y)] \le \alpha\, d(x, y) < d(x, y) \;\forall\, x, y \in X.$$

Definition 3. *Let X be a non-empty set and let $f{:}X \to X$ be a mapping. A point $x \in X$ is said to be a fixed point of f if $f(x) = x$.*

For example. Let $f: X \to X$ be the identity mapping on X such $f(x) = x$

Therefore every point of X is a fixed point.

THEOREM 1. ***Let A be a subset of a metric space X, then following statements are equivalent:***

(i) A is non-dense in X.

(ii) A contains no neighbourhoods

(iii) $(\overline{A})'$ is dense in X.

PROOF. Let (X, d) be a metric space and A be any subset of X.

Here, we first prove (i) $\Leftrightarrow$ (ii).

A is non-dense in $X \Leftrightarrow (\overline{A})^\circ = \phi$

$\Leftrightarrow$ No point of X is an interior point of $\overline{A}$

$\Leftrightarrow \overline{A}$ is not a nbd of any of its points

$\Leftrightarrow \overline{A}$ contains no nbd.

We now prove (ii) $\Leftrightarrow$(iii)

$\overline{A}$ contains no nbd $\Leftrightarrow$ For every $x \in X$, $S(x_0, r) \not\subset \overline{A}$, $r > 0$

$\Leftrightarrow S(x, r) \cap (\overline{A})' \ne \phi$ for every $x \in X$ and every $r > 0$

$\Leftrightarrow$ every nbd of x contain a point of $(\overline{A})'$ for every $x \in X$

$\Leftrightarrow x$ is an adherent point of $(\overline{A})'$ for every $x \in X$

$\Leftrightarrow \left[\overline{(\overline{A'})}\right] = X$ [By definition of adherent point]

$\Leftrightarrow (\overline{A})'$ is dense in X.

REMARKS

- Since $(\overline{A})'=\text{ext}(A)$, it follows, from that A is non-dense if and only if ext (A) is everywhere dense.
- If A is non-dense in X, then A' is dense in X.
- If A is nowhere dense, then $\overline{A}$ is not the entire space X.

THEOREM 2. ***The union of a finite number of nowhere dense sets is no where dense.***

PROOF. Let A and B be two nowhere dense subsets of a metric space (X, d).

Let us write $G = \left(\overline{A \cup B}\right)^{\circ}$ so that $G \subset \overline{A \cup B} = \overline{A} \cup \overline{B}$.

It follows that

$$G \cap (\overline{B})' \subset (\overline{A} \cup \overline{B}) \cap (\overline{B})'$$
$$= [\overline{A} \cap (\overline{B})'] \cup [\overline{B} \cap (\overline{B})'] \quad \text{(By distributivity)}$$
$$= \overline{A} \cap (\overline{B})' \quad [\because \overline{B} \cap (\overline{B})' = \phi]$$
$$\subset \overline{A}$$

$$\Rightarrow \quad [G \cap (\overline{B})']^{\circ} \subset (\overline{A})^{\circ} = \phi \quad \text{...(1)}$$

[$\because$ A is non-dense *i.e.* $(A)^{\circ}=\phi$]

But $[G \cap (\overline{B})']^{\circ} = G \cap (\overline{B})'$ ...(2)

[$\because G \cap (\overline{B})'$ is open so use $A^{\circ}=A$]

From (1) and (2), we conclude that

$$G \cap (\overline{B})' = \phi$$

$\Rightarrow$ $G \subset \overline{B} \Rightarrow G^{\circ} = (\overline{B})^{\circ} = \phi$ [$\because B$ is non-dense in X so $(\overline{B})^{\circ} = \phi$]

But $G^{\circ} = (A \cup B)^{\circ\,\circ} = \left(\overline{A \cup B}\right)^{\circ}$

so that $(A \cup B)^{\circ} = \phi$.

Hence, $A \cup B$ is a non-dense.

In general, the union of a finite number of no where dense set is no where dense.

THEOREM 3. **(Bair's Category Theorem).** ***Every complete metric space is of second category.***

PROOF. Let (X, d) be a complete metric space. To show X is of second category. Let if possible, X is not of second category *i.e.* X is of first category.

$\Rightarrow$ X can be expressed as a countable union of nowhere dense sets arranged in a sequence $\langle A_n \rangle$. Now, since A_1 is non-dense and so there exists a closed sphere K_1 of radius $r_1 < \frac{1}{2}$ *s.t.* $K_1 \cap A_1 = \phi$.

Let the open sphere, with same centre and redius as r_1, be denoted by S_1.

In S_1, we can find a closed sphere K_2 of radius $r_2 < \left(\frac{1}{2}\right)^2$ such that

$$K_2 \cap A_2 = \phi \text{ and so } K_2 \cap A_1 = \phi.$$

Continuing this process, we constitute a nested sequence $\langle K_n \rangle$ of closed spheres which have the following properties :

(i) For each positive integer, n, K_n does not intersect $A_1, A_2, \ldots, A_n\ldots$

(ii) The radius of K_n tends to zero as $n \to \infty$. $\left(\text{For } \frac{1}{2^n} \to 0 \text{ as } n \to \infty\right)$

Since (X, d) is complete, therefore by Cantor's intersection theorem $\bigcap_n K_n$ contains a single point x_0, so that

$$x_0 \in K_n, \forall\, n.$$

By (i), $x_0 \notin A_n \ \forall\ n$. By assumption, it is not possible, because X is the union of A_n's.Hence, X is not of first category, which is a contradiction.

Hence, X is of second category.

REMARKS

- The Bair's category theorem can also be stated as follows :
"If $\langle A_n \rangle$ is a sequence of nowhere dense sets in a complete metric space (X, d) $\exists$ a point in X, which is not in A_n's. or "If a complete metric space is the union of a sequence of its subsets, then the closure of at least one set in the sequence must have non-empty interior.

THEOREM 4. ***Every contracting mapping is continuous.***

PROOF. Let f be a contracting mapping on a metric space (X, d), therefore there exists a positive real number $\alpha < 1$ such that

$$f:(X, d) \to (X, d),\ \forall\ x, y \in X$$

and
$$d\,[f(x), f(y)] \le \alpha\, d(x, y) < d(x, y). \qquad ...(1)$$

Taking $d(x, y) < \varepsilon$, we get $d[f(x), f(y)] < \varepsilon$.

Given $\varepsilon > 0\ \exists\ \delta > 0$ such that $d[f(x), f(y)] < \varepsilon$, whenever $d(x, y) < \delta$.

Here $\varepsilon = \delta$.

Hence, f is a continuous mapping.

THEOREM 5. **(Banach Fixed Point Theorem).** ***If f is a contracting mapping on a complete metric space (X, d), then there exists a unique point p in X such that f(p)=p.***

PROOF. Let (X, d) be a complete metric space. Define a contracting mapping $f:(X,d) \to (X,d)$. Then by definition of contracting mapping there exists a positive real number $\alpha < 1$ such that

$$d\,[f(x), f(y)] \le \alpha\, d(x, y)\ \forall\ x, y \in X. \qquad ...(1)$$

Then
$$d[f^2(x), f^2(y)] = d[f(f(x)), f(f(y))]$$
$$\le \alpha\, d[f(x), f(y)] \qquad \text{[using (1)]}$$
$$\le \alpha.\alpha.d(x, y) \qquad \text{[using (1)]}$$
$$\Rightarrow \quad d[f^2(x), f^2(y)] \le \alpha^2\, d(x, y).$$

In general, we get

$$d[f^n(x), f^n(y)] \le \alpha^n.d(x, y),\ n \in \mathrm{N}. \qquad ...(2)$$

Next, we suppose that $x_0 \in X$ and let

$$x_1 = f(x_0),\ x_2 = f(x_1) = ff(x_0) = f^2(x_0)$$
$$x_n = f(x_{n-1}) = f^n(x_0).$$

Thus
$$x_n = f^n(x_0),\ x_{n+1} = f^{n+1}(x_0) = f^n[f(x_0)] = f^n(x_1)$$
$$\Rightarrow \quad x_n = f^n(x_0),\ x_{n+1} = f^n(x_1). \qquad ...(3)$$

We claim that $\langle x_n \rangle$ is a Cauchy sequence.

Let $m, n \in \mathrm{N}$ be arbitrary such that $m > n$ and $m = n + p$, p being a positive integer ≥ 1.

Now
$$d(x_n, x_m) = d(x_n, x_{n+p})$$
$$\le d(x_n, x_{n+1}) + d(x_{n+1}, x_{n+2}) + \ldots + d(x_{n+p-1}, x_{n+p})$$
(By triangle inequality)

Now using (3)

$$d(x_n, x_n) \le d[f^n(x_0), f^n(x_1)] + d[f^{n+1}(x_0), f^{n+1}(x_1)] + d[f^{n+p-1}(x_0), f^{n+p-1}(x_1)]$$

On using (2)

$$d(x_n, x_m) \le \alpha^n d(x_0, x_1) + \alpha^{n+1} d(x_0, x_1) + \ldots + \alpha^{n+p-1} d(x_0, x_1)$$

$$= \alpha^n d(x_0, x_1)[1 + \alpha + \alpha^2 + \ldots + \alpha^{p-1}]$$

$$= d(x_0, x_1).\alpha^n \frac{(1-\alpha^p)}{(1-\alpha)} \quad \text{[being the sum of G.P.]}$$

$$\le d(x_0, x_1)\frac{\alpha^n}{1-\alpha} \quad \text{for } 0<\alpha<1.$$

Thus $\quad d(x_n, x_m) \le \dfrac{\alpha^n}{1-\alpha} d(x_0, x_1).$

Therefore $\quad 0<\alpha<1 \Rightarrow \lim\limits_{n\to\infty} \alpha^n = 0$

Now (4) gives $\quad d(x_n, x_m) < \varepsilon.$

$\Rightarrow \langle x_n \rangle$ is a Cauchy sequence.

Now since (X, d) is complete, therefore by definition, every Cauchy sequence in X converges to some point in X

$\Rightarrow \exists\, p \in X$ s.t. $x_n \to p$ as $n \to \infty$.

$\Rightarrow \quad \lim x_n = p.$

Also, f is contracting mapping $\Rightarrow f$ is continuous.

($\because$ Every contracting mapping is continuous)

Therefore $\quad x_n \to p \Rightarrow f(x_n) \to f(p) \Rightarrow \lim f(x_n) = f(p).$

But $\quad x_{n+1} = f(x_n).$

$\therefore \quad f(p) = \lim f(x_n) = \lim x_{n+1} = \lim x_n = p$

or $\quad f(p) = p$

$\Rightarrow p$ is a fixed point.

Now to show the uniqueness of the point p.

Let if possible $\exists$ another fixed point $q \in X$ such that $p \ne q$ and $f(q) = q$.

Consider $\quad d[f(p), f(q)] \le \alpha\, d(p, q)$

$\Rightarrow d(p, q) \le \alpha\, d(p, q)$. But $d(p, q) \ne 0$

$\Rightarrow \alpha \ge 1$, which is a contradiction $\quad (0 < \alpha < 1)$

Hence, p is unique.

2.17 COMPLETION OF A METRIC SPACE

Definition 1. *A metric space (X, d) is said to be isometric to a space (Y, d_1) if there exist one-one mapping $f : X \to Y$ which preserves the distance i.e.*

$$d(x, y) = d_1[f(x), f(y)] \;\forall\; x, y \in X.$$

Definition 2. *If (X, d) is not complete metric space, then we extend the metric space (X, d) to (X_1, d_1) which is complete and (X, d) is a dense subspace.*

A metric space (X_1, d_1) is called completion of a metric space (X, d) if X_1 is complete and X is isometric to a dense subset of X.

THEOREM 1. ***The metric space (R, d) is complete where d denotes usual metric for R.***

PROOF. Let $\langle s_n \rangle$ be a Cauchy sequence of real numbers. Now, we define a sequence $\langle n_k \rangle$ of positive integers as n_{k+1} is the smallest integer greater than n_k such that

$$n, m \geq n_k \Rightarrow |s_n - s_m| < \frac{1}{2^{k+1}} \text{ (Since } \langle s_n \rangle \text{ in Cauchy sequence)}$$

Now let I_k be the closed interval $\left[S_{n_k} - 2^{-k}, S_{n_k} + 2^{-k}\right]$.

Here, it is easy to see that $I_{k+1} \subset I_k$, we have

$$\left|s_{n_k} - s_{n_{k+1}}\right| < \frac{1}{2^{k+1}}$$

Also, the length of $I_k \to 0$ as $k \to \infty$. So by nested interval theorem, $\bigcap_{k=1}^{\infty} I_k$ consists of exactly one point, say $a \in$ R.

Thus $a \in I_k$, $\forall\, k \in$ N so that $\left|a - s_{n_k}\right| < \dfrac{1}{2^k}$ for all $k \in$ N. ...(2)

Now for $n \geq n_k$, we have from (1)

$$\left|s_{n_k} - s_n\right| < \frac{1}{2^{k+1}} < \frac{1}{2^k}. \quad ...(3)$$

Hence, for all $n \geq n_k$.

$$\begin{aligned} |a - s_n| &= \left|a - s_{n_k} + s_{n_k} - s_n\right| \\ &\leq \left|a - s_{n_k}\right| + \left|s_{n_k} - s_n\right| \\ &< \frac{1}{2^k} + \frac{1}{2^k} = \frac{1}{2^{k-1}} \end{aligned} \quad \text{[by (2) and (3)]}$$

It follows that $\lim_{n \to \infty} s_n = a$.

Thus every Cauchy sequence in R converges to a point in R.

Hence, R is complete.

THEOREM 2. ***The set R^n of all n-tuples x = (x_1, x_2,...,x_m) of real numbers is a complete metric space with respect to the usual metric***

$$d(x, y) = \left[\sum_{i=1}^{n} (x_i - y_i)^2\right]^{1/2}.$$

PROOF. Here, we shall denote the elements of R^n by a functional notation. Thus, an element of R^n will be a real function defined on the set {1, 2, 3,..., n}. Let $f(m)$ stand for the element $[f_m(1), f_m(2) \ldots f_m(n)]$ of R^n. Now we shall show that R^n is complete.

Let $\langle f_m \rangle$ be a Cauchy sequence in R^n, then for a given $\varepsilon > 0$, $\exists$ a positive integer $n(\varepsilon)$ such that

$$\begin{aligned} p, q \geq n(\varepsilon) &\Rightarrow d(f_p, f_q) < \varepsilon \\ &\Rightarrow d^2(f_p, f_q) < \varepsilon^2 \\ &\Rightarrow \sum_{i=1}^{n} \left[f_p(i) - f_q(i)\right]^2 < \varepsilon^2 \\ &\Rightarrow [f_p(i) - f_q(i)]^2 < \varepsilon^2 \quad (i = 1, 2, 3, ..., n) \\ &\Rightarrow [f_p(i) - f_q(i)] < \varepsilon \quad (i = 1, 2, 3, ..., n) \end{aligned}$$

So $\langle f_m(i)\rangle$ is a Cauchy sequence of real numbers. Since R is complete, so every Cauchy sequence in R converge to a point in R.
i.e. The sequence $\langle f_m\rangle$ converges pointwise to a limit function *f* defined by

$$\lim f_m\ (i) = f(i).$$

Since the set {1, 2,..., *n*} is finite, this convergence is uniform. Hence there exists a positive integer n_0 such that

$$\left|f_m(i) - f(i)\right| < \frac{\varepsilon}{\sqrt{n}}, \text{ for all } m \geq n_0 \text{ and } \forall\ i.$$

Now, squaring and adding the above result for $i = 1, 2, \ldots, n$, we get

$$\sum_{i=1}^{n}\left|f_m(i) - f(i)\right|^2 < \frac{\varepsilon^2}{n}.n = \varepsilon^2$$

or, $$d^2(f_m, f) < \varepsilon^2\ \forall\ m \geq n_0 \text{ thus } d(f_m, f) < \varepsilon,\ \forall\ m \geq n_0$$

which shows that the Cauchy sequence $\langle f_m\rangle$ converges to the limit *f*.

Hence, R^n is complete.

THEOREM 3. ***The Hilbert space (l_2, d) is complete.***

PROOF. Let $\langle s_n\rangle$ be a Cauchy sequence in l_2. Since each $<s_n>$ is a sequence. So we have

$$s_n = \left\langle s_1{}^n, s_2{}^n, s_3{}^n \ldots\right\rangle.$$

So that $s_k{}^n$ denotes the kth term of the sequence $<s_n>$. Now

$$d(s_m, s_n) = \left[\sum_{i=1}^{n}\left(s_i{}^m - s_i{}^n\right)^2\right]^{1/2}.$$

Now, Since $\langle s_n\rangle$ in a Cauchy sequence, so for a given $\varepsilon > 0\ \exists$ a positive integer *m* such that

$$m, n \geq m_0 \Rightarrow d(s_m - s_n) < \varepsilon \Rightarrow d^2(s_m - s_n) < \varepsilon^2$$

$$\Rightarrow \sum_{i=1}^{n}\left(s_i{}^m - s_i{}^n\right)^2 < \varepsilon^2 \Rightarrow \left(s_i{}^m - s_i{}^n\right) < \varepsilon^2, \forall\, i \in N$$

$$\Rightarrow |s_i{}^m - s_i{}^n| < \varepsilon\ \forall\ i \in N.$$

So, $\{\left\langle s_i{}^n\right\rangle : i = 1, \ldots, n\}$ is a Cauchy sequence of real numbers for all $i \in N$.
and it must converge to some real number x_i, let $X = \langle x_1, x_2, \ldots, \rangle$.
Now, we shall show that the sequence $\langle s_n\rangle$ converges to *X*, and that $X \in l_2$.
Now for a fixed integer *M*, we have

$$\sum_{i=1}^{m}\left(s_i{}^m - s_i{}^n\right) < \varepsilon^2,\ \forall\ m, n \geq m_0.$$

If we fix *n* and let $m \to \infty$, we get

$$\sum_{i=1}^{m}\left(s_i - s_i{}^n\right)^2 < \varepsilon^2. \qquad \ldots(1)$$

Since, (1) hold for all *m*, we have for $n \geq m_0$.

$$\sum_{i=1}^{\infty}\left(s_i - s_i{}^n\right)^2 < \varepsilon^2. \qquad \ldots(2)$$

i.e., $d^2(x, s_n) < \varepsilon^2 \Rightarrow d(x, s_n) < \varepsilon$.

Hence, $\lim_{n\to\infty} s_n = x$.

Now we shall show that $x \in l_2$, we have

$$x_i^2 = \left(x_i - s_i^{m_0} + s_i^{m_0}\right)^2$$

$$= \left(x_i - s_i^{m_0}\right)^2 + \left(s_i^{m_0}\right)^2 + 2\left(x_i - s_i^{m_0}\right)\left(s_i^{m_0}\right)$$

$$\geq 2\left(x_i - s_i^{m_0}\right)^2 + 2\left(s_i^{m_0}\right)^2. \qquad (\because 2ab \leq a^2 + b^2)$$

So
$$\sum_{i=1}^{\infty} x_i^2 \leq 2\sum_{i=1}^{\infty}\left(x_i - s_i^{m_0}\right)^2 + 2\sum_{i=1}^{\infty}\left(s_i^{m_0}\right)^2$$

$$< 2\varepsilon^2 + 2\sum_{i=1}^{\infty}\left(s_i^{m_0}\right)^2 \qquad \text{[by (2)]} \qquad \text{...(3)}$$

Since $s_{m_0} \in l_2$, the series $\sum_{i=1}^{\infty}\left(s_i^{m_0}\right)^2$ converges so that for some $B>0$, we have

$$\sum_{i=1}^{\infty}\left(s_i^{m_0}\right)^2 < B \qquad \text{...(4)}$$

Now from (2) and (4), we have

$$\sum_{i=1}^{\infty} x_i^2 < 2\varepsilon^2 + 2B.$$

So the series $\sum_{i=1}^{\infty} x_i^2$ is convergent and consequently $x \in l_2$.

Thus, every Cauchy sequence, $\langle s_n \rangle$ in l_2 converges to a point x in l_2.

Hence, (l_2, d) is complete.

Completeness of $C[a, b]$. Let $C[a, b]$ is the set of all real valued continuous funcions defined on I $(I=[a, b])$. For $f \,.\, g \in C[a, b]$. We define $d(f, g) = \sup\{[f(x) - g(x) : x \in I\}$. We have already seen that d is matric for $C[a, b]$. So this space is known as the space of continuous functions on I

THEOREM 4. ***The space C[a, b] is complete.***

PROOF. Let $I=[a, b]$ and let $\langle f_m \rangle$ be a Cauchy sequence in $C[a, b]$.

Let for a given $\varepsilon>0$, there exist a positive integers m_0 such that

$$m, n \geq m_0 \Rightarrow d(f_m, f_n) < \varepsilon$$

$$\Rightarrow \sup\{|f_m(x) - f_n(x)| : x \in I\} < \varepsilon$$

So $|f_m(x) - f_n(x)| < \varepsilon$, $\forall\, m, n \geq m_0$ and for all $x \in I$.

This is the condition for uniform convergence. Thus the sequence $\langle f_m \rangle$ is a uniform convergent sequence of continuous functions and it must converge to a continuous function f on I.

Thus every Cauchy sequence $\langle f_m \rangle$ in $C[a, b]$ converges to a point f in $C[a, b]$. Hence, $C[a, b]$ is complete.

SOLVED EXAMPLE

EXAMPLE. *Show that the set C of complex numbers with usual metric is a complete metric space.*

SOLUTION. Let C is the set of complex number and $z_1, z_2 \in C$ such that

$$z_1 = x_1 + iy_1 \text{ and } z_2 = x_2 + iy_2 \text{ where } x_1, y_1, x_2, y_2 \in R$$

Let us definedan metric on C as

$$d(z_1, z_2) = |z_1 - z_2|.$$

Let $\langle z_n \rangle$ be a Cauchy sequence in C.

To show that C is a complete metric space, we have to show that $\langle z_n \rangle$ converges to a point $z \in C$. Let for given $\varepsilon > 0$, there exist $m \in N$ such that

$$|z_n - z_m| < \varepsilon, \ \forall\, n \geq m$$

$$\Rightarrow \quad |(x_n + iy_n) - (x_m + iy_m)| < \varepsilon$$

$$\Rightarrow \quad |(x_n - x_m) + i(y_n - y_m)| < \varepsilon$$

$$\Rightarrow \quad |(x_n - x_m) + i(y_n - y_m)|^2 < \varepsilon^2$$

$$\Rightarrow \quad |x_n - x_m|^2 + |y_n - y_m|^2 < \varepsilon^2$$

$$\Rightarrow \quad |x_n - x_m| < \varepsilon \text{ and } |y_n - y_m| < \varepsilon, \ \forall\, n \geq m$$

$\Rightarrow$ $\langle x_n \rangle$ and $\langle y_n \rangle$ are Cauchy sequence in R

and we know that every Cauchy sequence converges to a point.

$\Rightarrow$ $x_n \to x$ and $y_n \to y$ in R

$\Rightarrow$ $z_n \to x + iy = z$ in C

$\Rightarrow$ $\langle z_n \rangle$ converges to a point z in C

$\Rightarrow$ C is convergent

$\Rightarrow$ (C, d) is a complete metric space.

THEOREM 5. ***Let C[a, b] be the set of all continuous functions on [a, b]. For f, g ∈ C[a, b], define***

$$d(f, g) = \left\{ \int_a^b |f(x) - g(x)|^2 \, dx \right\}^{1/2}$$

then ρ is a metric for C[a, b], which is not complete.

PROOF. It is easy to see that ρ is a metric for C[a, b].

Now we shall show that the space is not complete. For this, consider the sequence of continuous functions $\langle f_n \rangle$ defined on [–1, 1] by

$$f_n(x) = \begin{cases} 0 & \text{if } -1 \leq x \leq 0 \\ nx & \text{if } 0 \leq x \leq \dfrac{1}{n} \\ 1 & \text{if } \dfrac{1}{n} \leq x \leq 1 \end{cases}$$

Then for $m > n$, we have

$$d^2(f_m, f_n) = (m-n)^2 \int_0^{1/m} x^2 dx + \int_{1/m}^{1/n} (1-nx)^2 dx$$

$$= \frac{(m-n)^2}{3m^3} + \left[\frac{1}{3n}\left(1-\frac{n}{m}\right)^3\right]$$

$$= \frac{(m-n)^2}{3m^3} + \frac{(m-n)^3}{3nm^3} = \frac{(m-n)^2}{3nm^2}$$

$$< \frac{1}{3n} < \varepsilon \text{ if } n > \frac{1}{3\varepsilon}$$

Therefore, the sequence is a Cauchy sequence.

Suppose, if possible, this Cauchy sequence converges to a continuous function f so that

$$d(f_n, f) = \int_{-1}^{1} |f_n(x) - f(x)|^2 dx \to 0.$$

This implies that the integral with any limits between ± 1 also tends to 0. Thus

$$\int_{-1}^{0} |f_n(x) - f(x)|^2 dx \to 0.$$

But $f_n(x) = 0$ when $x \le 0$ and hence this interval is independent of n. So the continuous function f is such that $\int_{-1}^{0} |f(x)|^2 dx = 0$.

It follows that $f(x) = 0$ when $x \le 0$. Again, if $c > 0$, then

$$\int_c^1 |f_n(x) - f(x)|^2 dx \to 0 \text{ as } n \to \infty.$$

If we choose $n \to \frac{1}{c}$, we have

$$\int_c^1 |1 - f(x)|^2 dx \to 0 \text{ as } n \to \infty.$$

As the interval is independent of n, is vanishes, and since $f(x)$ is continuous we have $f(x) = 1$ for $x \ge c$. But we can choose as near to zero as we want, thus there exists a continuous function which vanishes when $x \le 0$ which is equal to 1 when $x > 0$. So the Cauchy sequence does not converge to a point of C[a, b]. Hence the space is not complete.

THEOREM 6. ***Let (X, d) and (Y, e) be two complete metric space then the product space Z=X×Y with metric***

$$\rho(z_1, z_2) = \sqrt{\left[d^2(x_1, x_2) + e^2(y_1, y_2)\right]}$$

is complete where $z_1 = (x_1, y_1)$ and $z_2 = (x_2, y_2)$.

PROOF. We know that ρ is a metric for Z. Now we show that the product space (Z, ρ) is complete. Let $\langle z_n \rangle$ be a Cauchy sequence in Z. Then for a given $\varepsilon > 0$, there exists a positive integer m_0 such that

$$m, n \ge m_0 \Rightarrow \rho(z_m, z_n) < \varepsilon$$

$$\Rightarrow \rho^2(z_m, z_n) < \varepsilon^2$$

$$\Rightarrow d^2(x_m, x_n) + e^2(y_m, y_n) < \varepsilon^2$$

$$\Rightarrow d^2(x_m, x_n) < \varepsilon^2 \text{ and } e^2(y_m, y_n) < \varepsilon^2$$

$$\Rightarrow d(x_m, x_n) < \varepsilon \text{ and } e(y_m, y_n) < \varepsilon$$

It follows that $\langle x_n \rangle$ and $\langle y_n \rangle$ are Cauchy sequences in the space X and Y respectively. Since these space are complete, the sequences $\langle x_n \rangle$ and $\langle y_n \rangle$ converges respectively to point $x \in X$ and $y \in Y$. It follows that the sequence $\langle z_n \rangle$ converges to $z=(x, y) \in Z$ and consequently the product space $Z=X \times Y$ is complete.

REMARK

- The space defined in the above theorem is known as product metric space.

EXERCISE 2.4

1. Show that set of real numbers is complete.

2. Show that the Hilbert space is complete.

3. Define a complete metric space. Give an example of a complete metric space.

4. Define an incomplete metric space. Give an example of an incomplete metric space.

5. Show that set C^n of all ordered n-tuples $z = z_1, z_2,..., z_n)$ of complex number is a complete metric space with respect to the usual metric d defined by

$$d(z, u)=\left[\sum_{i=1}^{n}|z_i - u_i|^2\right]^{1/2}$$

where $(z=z_1, z_2,..., z_n)$ and $u=(u_1, u_2,..., u_n)$

6. Show that the metric space of rational numbers with the usual metric is incomplete.

7. Let D (a, b) denote the set of all functions f on $[a, b]$ which have continuous derivatives at all points of $I=[a, b]$. For $f, g \in D[a, b]$, define $d(f,g)=|f(a)-g(b)|+\sup\{|f'(x)-g'(x)|: x\in Z\}$. Show that d is a metric for $D(a, b)$ and that the space $(D[a, b], d)$ is complete.

8. Show that the completeness is preserved under isometrics.

9. Let X consist of all ordered pairs $X=(x_1, x_2)$ be real numbers with metric

$$d(x, y)=\max\{|x_1-y_1|, |x_2-y_2|\}.$$

Prove that X is complete.

10. Let X consist of all bounded sequences $X= \langle x_n \rangle$ in R. Prove that $d(x, y)=\sup\{|x_i-y_i| : i \in N\}$ is a metric on X and X is complete.

11. Which of the following are Cauchy sequence in R with respect to the usual metric d

(i) $\left\langle \frac{1}{n} \right\rangle$ (ii) $\left\langle \frac{n-1}{n+1} \right\rangle$

(iii) $\left\langle \frac{2^{n+1}}{2^n}+1 \right\rangle$ (iv) $\left\langle 1+\frac{(-1)^n}{n} \right\rangle$.

12. Let X consist of all sequence $X= \langle x_n \rangle$ in R. Show that $d(x,y)=\sum_{1}^{\infty}\frac{|x_n - y_n|}{[1+|x_n - y_n|]}$ is a metric on X and X is complete.

HINT TO SELECTED PROBLEMS

10. Let $\langle x_n \rangle$ be a cauchy sequence in the given space X so that given $\varepsilon>0$, $\exists\, n_0 \in N$ such that

$m, n \geq n_0 \Rightarrow d(x_m-x_n)<\varepsilon$

$\Rightarrow \quad \sup\{|x_m-x_n| : n \in N\}<\varepsilon$

$\Rightarrow \langle x_n \rangle$ is uniformly convergent

$\Rightarrow \langle x_n \rangle$ is convergent.

$\Rightarrow X$ is complete.

9. Do same as above.

11. Let $s_n=\frac{1}{n}$

$$d(s_n,s_m)=|s_n - s_m|=\left|\frac{1}{n}-\frac{1}{m}\right|\leq\frac{1}{n}+\frac{1}{m}$$

Now there may exists a positive integer p greater then $\frac{2}{\varepsilon}$ then

$$\frac{1}{n}+\frac{1}{m}<\frac{\varepsilon}{2}+\frac{\varepsilon}{2} \quad \forall\, m, n \geq p$$

$\therefore \quad d(s_n,s_m)\leq\frac{1}{n}+\frac{1}{m}<\frac{\varepsilon}{2}+\frac{\varepsilon}{2} \quad \forall\, m, n \geq p$

$\Rightarrow \quad d(s_n, s_m)<\varepsilon$

$\Rightarrow \quad \langle s_n \rangle$ is a Cauchy sequence.

Chapter Review: A Competitive Approach

SELECTED TERMS AND RESULTS

▶ TERMS

- **Metric :** Let X be a non-empty set and d be a function defined on $X \times X$ into the set R (*i.e.* set of reals) such that the image of any ordered pair (x, y) of $X \times X$ is denoted by $d(x, y)$, then d is called a metric or distance function iff d satisfies the following axioms :

  $M_1 : d(x, y) > 0 \ \forall \ x, y \in X$ (*Non-Negativity*)

  i.e. distance between any two points is a non-negative real number.

  $M_2 : d(x, y) = 0 \Leftrightarrow x = y \ \forall \ x, y \in X.$

  i.e. distance between two points is zero iff the points coincide.

  $M_3 : d(x, y) = d(y, x) \ \forall \ x, y \in X$ (*Symmetry*)

  i.e. distance between x and y is the same as the distance between y and x.

  $M_4 : d(x, y) \le d(x, z) + d(z, y) \ \forall \ x, y, z \in Y.$ (*Triangle Inequality*)

  i.e. any side of a triangle is always less than or equal to the sum of the remaining two sides. The equality occurring when all the three points forming a triangle are collinear.

- **Metric Space:** The set X along with metric d defined as above is called metric space and is written as (X, d). By defining the function d in various manners we can have various metric spaces on a given set X.

- **Metrizable:** A non-empty set X is said to be metrizable iff a metric d can be defined on it satisfying the four axioms M_1, M_2, M_3, M_4.

- **Pseudo Metric:** The mapping $d : X \times X \to$ R is called a pseudo metric for X iff d satisfies M_1, M_3, M_4 axioms written above and in place of axiom M_2 it satisfies $M_2{}'$ *i.e.* $d(x, x) = 0$

- **(Usual Metric on R).** If R be the set of real members then the mapping d : R × R→R defined as $d(x, y) = |x - y| \ \forall \ x, y \in$ R is a metric on R (known as usual metric on R)

- **(Discrete or Trivial Metric).** Let X be a non-empty set and $d : X \times X \to$ R defined as

  $$d(x, y) = \begin{cases} 0 \text{ if } x = y \\ 1 \text{ if } x \ne y \end{cases} \quad \forall x, y \in \mathrm{R}$$

  then d is a metric on X, (called the discrete or trivial metric.)

- **Distance between a point $p \in X$ and a subset A of X.**

  $$d(p, A) = \inf.\{d(p, x) : x \in A\}$$

- **Diameter of subset A.**

  $$\delta(A) = \sup.\ \{d(x, y) : x, y \in A\}$$

- **Open sphere $S(x_0, r)$:** Let (X, d) be a metric space and $x_0 \in X$ and r is any +ive real number then the set of all those points of X which are within a distance of r from x_0 is called an open sphere (ball) centred at x_0 and radius r. It is denoted by

  $$S(x_0, r) \text{ or } S_r(x_0) \text{ or } B(x_0, r) \text{ or } B_r(x_0).$$

  $$\therefore \quad S(x_0, r) = \{x \in X : d(x, x_0) < r\}$$

- **Closed sphere $S[x_0, r]$:**

  $S[x_0, r] = \{x \in X : d(x, x_0) \le \mathrm{r}\}$ (Note the sign of equality) is called a closed sphere or ball centred at x_0 and of radius r. It may be noted that a ball is non-empty as it always contains its centre x_0 as $d(x_0, x_0) = 0 \le r$, as r is a +ive real number.

- **Subspace.** Let (X, d) be a metric space, so that $d : X \times X \to$ R. And if Y be a non-empty subset of X then the same mapping d when restricted to point of Y *i.e.* $d : Y \times Y \to$ R will be a metric on Y and space (Y, d).

- **Range set of a sequence :** The range set X of a sequence may be finite or infinite.

- **Sub-sequence:** Let $<s>$ and $<t>$ be two sequences in a set then the sequence $<t>$ is said to be sub-sequence of $<s>$ if there exists

a mapping $\phi : N \to N$ such that
(1) $t = s \circ \phi$
(2) For each n in N there exists an m in N such that $\phi(i) \geq n$ for every $i \geq m$ in N.

- **Cluster point of a sequence.** Let (X, d) be a metric space. A point $x \in X$ is said to be cluster point of sequence s in X if s is frequently in every neighbourhood of X.
- **Convergence of a sequence :** A sequence $\langle x_n \rangle = \langle x_1, x_2, x_n, \ldots \rangle$ in a metric space (X, d) is said to converge to a point x of X if and only if for each $\varepsilon > 0$, $\exists$ a +ive integer $n \in N$ such that
$$n \geq n(\varepsilon) \Rightarrow d(x_n, x) < \varepsilon.$$
- **First and Second Category :** A subset of a metric space is said to be of first category if and only if it can be written as a union of a countable number of nowhere dense sets, otherwise it is said to be of second category.
- **Nowhere dense set :** A subset A of a metric space (X, d) is said to be nowhere dense if its closure has empty interior *i.e.,* $(\bar{A})^\circ = \phi$.
- In a metric space (X, d), a mapping $f : X \to X$ is called a contracting mapping or simply a contraction on X if and only if there exists a real number $0 \leq \alpha < 1$ such that
$d(f(x), f(y)) \leq \alpha d(x, y) < d(x, y)$ for every $x, y \in X$ ...(1)

▸ RESULTS

- In a metric space $d(x, y) = 0$ iff $x = y$ but in pseudo metric space $d(x, y)$ may be zero even though x may not be equal to y. In other words $x = y \Rightarrow d(x, y) = 0$ but $d(x, y) = 0$ does not necessarily imply that $x = y$. Then every pseudo metric space is not a metric space but every metric space is a pseudo metric space For example, if we consider the set R and $d : R \times R \to R$ such that $d(x, y) = |x^2 - y^2|$ $\forall\, x, y \in R$ then d is a pseudo metric because $d(x, y) = 0 \Rightarrow |x^2 - y^2| = 0 \Rightarrow x^2 - y^2 = 0 \Rightarrow x = \pm y$. In other words $d(x, y) = 0$ does not necessarily imply that $x = y$.
- A metric space (X, d) is said to be bounded metric space if there exists a positive real number k such that $d(x, y) < k$ $\forall\, x, y \in X$.
- A metric space which is not bounded is called unbounded.
- (*An alternative set of axioms for a metric*). Let X be a non-empty set and d be a real valued function of $X \times Y$ into R. then d is a metric if and only if the following two conditions hold
(1) $d(x, y) = 0 \Leftrightarrow x = y$
(2) $d(x, y) \leq d(x, z) + d(y, z)$.
- If $1 < p < \infty$ and $1 < q < \infty$ such that $\frac{1}{p} + \frac{1}{q} = 1$ and a, b be two non-negative real numbers then $a^{1/p} b^{1/q} \leq \frac{a}{p} + \frac{b}{q}$.
- **Holder's Inequality :** If x_i, y_i $(i = 1, 2, \ldots, n)$ are non-negative real numbers, then
$$\sum_{i=1}^{n} x_i y_i \leq \left(\sum_{i=1}^{n} x_i^p\right)^{1/p} \left(\sum_{i=1}^{n} y_i^q\right)^{1/q}$$
where $\frac{1}{p} + \frac{1}{q} = 1$.
- **Cauchy Inequality.** If we choose $p = q = 2$ so that $\frac{1}{p} + \frac{1}{q} = \frac{1}{2} + \frac{1}{2} = 1$ in Holder's inequality then it becomes
$$\sum_{i=1}^{n} (x_i y_i) \leq \left(\sum_{i=1}^{n} x_i^2\right)^{1/2} \left(\sum_{i=1}^{n} y_i^2\right)^{1/2}$$
- **Cauchy-Schwarz Inequality:** If x_i, y_i $(i = 1, 2, \ldots, n\}$ are complex numbers, then
$$\left|\sum x_i y_i\right| \leq \left[\sum |x_i|^2\right]^{1/2} \left[\sum |y_i|^2\right]^{1/2}, \text{ where } \sum = \sum_{i=1}^{n}.$$
- **Minkowski Inequality:** If x_i, y_i $(i = 1, 2, \ldots, n)$ are non-negative real numbers, then
$$\left[\sum (x_i + y_i)^p\right]^{1/p} \leq \left(\sum x_i^p\right)^{1/p} + \left(\sum y_i^p\right)^{1/p}$$
where $p > 1$ and $\sum = \sum_{i=1}^{n}$.
$$\sum (x_i + y_i)^p = \sum (x_i + y_i)(x_i + y_i)^{p-1}$$
$$= \sum x_i (x_i + y_i)^{p-1} + \sum y_i (x_i + y_i)^{p-1} \quad \ldots(1)$$
- Let (X, d) be a given metric space and A, B be subsets of X. Let $x \in A$ and $y \in B$ then we know that :
$$d(x, y) \geq 0.$$
Hence $\{d(x, y) : x \in A \text{ and } y \in B\}$ is a set of real numbers which is bounded below by zero and consequently this set must have a greatest lower bound *i.e.* infimum.
Thus if $A \cap B \neq \phi$ then $d(A, B) = 0$.
- Let A, B be two subsets of a metric space (X, d), then $\delta(A \cup B) \leq \delta(A) + \delta(B) + \delta(A, B)$ and if $A \cap B \neq \phi$, then
$$\delta(A \cup B) \leq \delta(A) + \delta(B)$$

- A is a non-empty subset of a metric space (X, d) and x, y are any two points of X, then
  $|d(x, A)-d(y, A)| \leq d(x, y)$.
- Let (X, d) be a metric space and B be the collection of all open spheres $S(x, r)$ where r is any +ive real number and x any point in X. Then B is a base for some topology on X.
- In a metric space (X, d) a non-empty subset G of X is open relative to d-metric topology for X if and only if for each $x \in G$ there exists an open sphere $S(x, r)$ centred at x and contained in G *i.e.* $S(x, r) \subset G$.
- In a metric space (X, d), $S=S(x_0, r)$ is an open sphere and p is any point of $S(x_0, r)$. Then there exists an open sphere $T=S(p, \in)$ centred at p and contained in $S(x_0, r)$ *i.e.* $T \subset S$.
- Let (X, d) be a metric space and let $p \in X$. If S and T are open spheres such that $p \in S \cap T$ then there exists an open sphere $S(p)$ centred at p such that $S(p) \subset S \cap T$.
- In a metric space (X, d) each open sphere is an open set.
- In a metric space (X, d), ϕ and X are open sets.
- Let X be a metric space then a subset G of X is open iff G is union of open spheres.
- In a metric space (X, d), Then
  (a) union of an arbitrary collection of open sets is open.
  (b) any finite intersection of open sets in X is open.
- Let (X, d) be a metric space, then every closed sphere in X is a closed set relative to the metric topology for X.
- In a metric space (X, d) all finite sets are closed.
- Let (X, d) be a metric space and f be a bijection from X to a set X^*. Then there exists a metric d^* for X^* such that the metric spaces (X, d) and (X^*, d^*) are isometric.
- Every metric is first countable.
- A metric space is separable if and only if it is second countable
- The function f is said to be continuous iff it is continuous at every point of x_1.
- If G_1 is an open set in (X_1, d_1) and G_2 is an open set in (X_2, d_2) then $G_1 \times G_2$ is open in the product space $(X_1 \times X_2, d)$.
- Let (X, d) be a metric space, then a sequence s in X is a function from set N of all +ive integers into X.
- Let s be a sequence in a non empty set X and Y be a sub set of X, then the sequence s is said to be eventually in Y if there exists a natural number m such that
  $s(n) \in Y$ for every $n \geq m$
- The sequence s is said to be frequently in Y if for each natural number m, there exists a natural number $n \geq m$ such that $s(n) \in Y$.
- The limit of a sequence in a metric space if it exists is unique.
- Let x_0, y_0 be any two points of a metric space (X, d) and $<y_n>$ be a sequence converging to y_0 then $<d(x_n, y_n)>$ converges to $d(x_0, y_0)$.
- Let (X, d) be a metric space and $<x_n>$ and $<y_n>$ are sequences in X such that $x_n \to x$ and $y_n \to y$ then $d(x_n, y_n) \to d(x, y)$.
- Let x be a limit point of a subset A of a metric space (X, d) then there exists a sequence $<x_n>$ of points of A all distinct from x which converges to x.
- A sequence $<x_n>$ in a metric space (X, d) is said to be a Cauchy sequence if and only if for each $\varepsilon>0$ there exists a +ive integer n_ε, such that $d(x_m, x_n)<\varepsilon \ \forall \ m, n \geq n_\varepsilon$.
- A Cauchy sequence is also called a fundamental sequence.
- In a metric space (X, d) every convergent sequence is a Cauchy sequence.
- If $<x_n>$ is a Cauchy sequence in metric space then any cluster point of $<x_n>$ is the limit of $<x_n>$.
- Let (X, d) be a metric space and $<x_n>$ be a Cauchy sequence in X. If $<x_n>$ be a sequence in X such that $d(x_n, y_n)<(1/n)$ for every positive integer n, then
  (a) $<y_n>$ is also a Cauchy sequence.
  (b) $<y_n>$ converges to a point $x \in X$ iff $<x_n>$ converges to x.
- A metric space (X, d) is said to be complete if every Cauchy sequence in X converges to a point in X.
- Let Y be a subspace of a complete metric space (X, d), then Y is complete $\Leftrightarrow$ Y is closed.
- Let Y be a subspace of a complete metric space (X, d), then Y is complete $\Leftrightarrow$ Y is closed.
- Let $<A_n>$ be a sequence of subsets of X in a metric space (X, d), then this sequence is called a decreasing sequence if $A_1 \supset A_2 \supset A_3 \supset ......$
  This sequence is sometimes called Nested Sequence.
- (*Cantor's Intersection Theorem*). Let $<A_n>$ be a nested sequence, *i.e.* monotonic decreasing sequence of non-empty closed subsets of a metric space (X, d) such that $\delta(A_n) \to 0$ as $n \to \infty$ (*i.e.*, whose diameters tend to zero) then X is complete if and only if $\cap \{A_n : n \in N\} \neq \phi$ and contains exactly one point.
- (*Baire Category Theorem*). Every complete metric space is of second category.

- The metric space (R, d) is complete where d denotes the usual metric given by
$$d(x, y) = |x-y| \ \forall\ x, y \in R.$$
- The set R^n of all n-tuples $(x_1, x_2, ..., x_n)$ of real numbers is a complete metric space with respect to the usual metric d defined by
$$d(x, y) = \left[\sum_{i=1}^{n} (x_i - y_i)^2\right]^{\frac{1}{2}}$$
- The set C of all complex numbers $z=x+iy$ is a complete metric space with respect to the usual metric d defined by $d(z_1, z_2) = |z_1-z_2|$.
- The set C^n of all n-tuples $z=(z_1, z_2, ..., z_n)$ of complex numbers is a complete metric space with respect to the usual metric space defined by
$$d(z, u) = \left[\sum_{i=1}^{n} |z_i - u_i|^2\right]^{\frac{1}{2}}$$
where $z=(z_1, z_2, ..., z_n)$ and $u=(u_1, u_2, ..., u_n)$.
- The Hilbert space (l_2, d) is complete.
- The space $C[a, b]$ is complete with the metric defined as
$$d(f, g) = \sup\{|f(x)-g(x)| : x \in [a, b]\}$$
- Every contraction mapping is continuous.
- (*Banach Fixed Point Theorem*): Every contractive mapping in a complete metric space has one and only one fixed point such that $f(x)=x$.

REVIEW QUESTIONS

1. Let (X, d) be a discrete metric space and let $x \in X$. Show that every subset of X containing x is a nbd of x.

2. Show that every subset of a discrete metric space is closed.

3. Show that in a metric space the compliment of a finite set is open.

4. For any two subsets A and B of a metric space $[X, d]$ show that $\delta(A \cup B) = \delta(A) \cup \delta(B)$.

5. Show that the metric space (R^n, d) is complete where d is usual metric on R^n

OBJECTIVE TYPE QUESTIONS

▶ FILL IN THE BLANKS

1. In a metric space (X, d), $d(x, y)=0$ iff __________.

2. In a metric space (X, d), $x, y, z \in X$ then we have $d(x, y) \leq d(x, z) +$ __________ .

3. Let (X, d) be a metric space and A be non-empty subset of X then diameter of A denoted as $\delta(A)$ = __________ .

4. Let $X=R^n$, the set of all ordered n-tuples of real numbers let $x=(x_1, x_2,..., x_n)$ and $y=(y_1, y_2,..., y_n)$ then $d(x, y)$ = __________ .

5. In a metric space the union of an arbitrary collection of open set is __________ .

6. In a metric space intersection of finite number of open set is __________ .

7. In a metric space intersection of infinite number of open set is __________ .

8. In a metric space, a subset in it is open iff it is of each of its points.

9. A subset A of metric space is closed iff $D(A)$ __________ .

10. Let A be a subset of metric space then $\overline{A} = A \cup$ __________ .

11. Let A and B are two closed sets then :
(i) $A \cap B$ is __________ .
(ii) $A \cup B$ is __________ .

12. Let A and B both are open set then :
(i) $A \cap B$ is __________ .
(ii) $A \cup B$ is __________ .

13. A subset A of X is said to dense set if $\overline{A}$ = __________ .

14. A subset A of X is said to be nowhere dense set if $(\overline{A})^\circ$ __________ .

15. The open interval $]a, b[$ is a nbd of __________ of its point.

16. The set of real numbers R is a nbd of __________ of its point.

17. The set of integers Z is __________ a nbd of any of its points.

18. The empty set of ϕ is a nbd of __________ of its points.

19. Every open interval is an __________ set.

20. Every closed interval is a __________ set.

21. If A and B are two open sets in a metric space (X, d), then $A \cap B$ is __________ .

22. The set $[1, 2] \cup [3, 4]$ is a __________ set.

23. In a metric space, every finite set is __________ set.

24. On the real line, the set Z of integers has __________ limit points.

25. On the real line, the set $S=\left\{\frac{1}{n}: n \in \mathrm{N}\right\}$ has __________ limit points.
26. A subset of a metric space is __________ iff it contains all its limit point.
27. In a metric space, the derived set of a metric space is __________.
28. If $S \subset T$ then $S' \subset$ __________.
29. In a metric space, the set of all adherent points is called __________.
30. On the real line, the closure of each of the sets]0, 1[,]0, 1[and [0, 1] is __________.
31. A set A is closed iff $\overline{A}$ __________.
32. A set A is open iff A^o __________.
33. In a metric space (X, d) a point $x \in X$ is an __________ point of a set $S \subset X$ iff every open set containing x contain a point of S.
34. The __________ of a set A is the intersection of all closed supersets of A.
35. A sequence $\langle s_n \rangle$ is said to be __________ if $\exists$ a number $M>0$ such that $|s_n|<M \ \forall \ n \in \mathrm{N}$.
36. A sequence which is not convergent is said to be __________.
37. A sequence $\langle s_n \rangle$ in X is said to converge to a point $s_0 \in X$ if for each $\varepsilon>0$ of a positive integer m such that $d\langle s_n, s_0 \rangle < \varepsilon$ for all __________.
38. In a metric space the convergent sequence has a __________ limit.
39. Every convergent sequence in a metric space is __________ sequence.
40. A metric space (X, d) is said to be __________ iff every Cauchy sequence in X converges to a point in X.
41. The open interval is __________.
42. A subspace of a metric space is complete iff it is __________.

▶ TRUE/FALSE

Write 'T' for True and 'F' for False statement.

1. In case of real line, the open sphere $S(x, r)$ is the open interval $]x-r, x+r[$. **(T/F)**
2. For the metric space (X, d), the open sphere centerd at z_0 and having radius r is the set $\{z: |z-z_0|<r\}$. **(T/F)**
3. The closed interval $[a, b]$ is a nbd of each of its points. **(T/F)**
4. In a metric space (X, d), the intersection of the family of all nbds of a point $p \in X$ is $\{p\}$. **(T/F)**
5. A finite set in a metric space can be open. **(T/F)**
6. A finite non-empty set in a metric space can be open. **(T/F)**
7. The intersection of any arbitrary family of open sets in a metric space is always open. **(T/F)**
8. Union of an arbitrary family of open sets in a metric space is always open. **(T/F)**
9. Every infinite set in a metric space is open. **(T/F)**
10. In a discrete metric space, every subset is open. **(T/F)**
11. Every closed interval is a closed set. **(T/F)**
12. In a metric space, every singleton is a closed set. **(T/F)**
13. The arbitrary union of closed set in a metric space is necessarily closed. **(T/F)**
14. The arbitrary union of open set in a metric space is necessarily open. **(T/F)**
15. If S is open and T is closed then $S-T$ is open. **(T/F)**
16. If S is closed and T is open then $S-T$ is closed. **(T/F)**
17. A subset of a metric space is closed if it containing all its limit points. **(T/F)**
18. On the real line, the set of integers has no limit point. **(T/F)**
19. The closure of a set A is the smallest closed set. **(T/F)**
20. The interior of a set A is the largest closed set. **(T/F)**
21. In a metric space every convergent sequence is cauchy. **(T/F)**
22. In a metric space, every Cauchy sequence is convergent. **(T/F)**
23. In a complete metric space, every Cauchy sequence is convergent. **(T/F)**
24. A sequence in a metric space can converge to at most one point of that space. **(T/F)**
25. If a Cauchy sequence in a metric space has a convergent subsequences, then the sequence is convergent. **(T/F)**
26. Completeness is not preserved under isometrics. **(T/F)**
27. The real line is complete. **(T/F)**
28. The image of a Cauchy sequence under a uniformly continuous mapping is a Cauchy sequence. **(T/F)**

▶ MULTIPLE CHOICE QUESTIONS

Choose the most appropriate one:

Problem Set-1

1. The interior of the set of rational numbers Q is:
(a) ϕ (b) Q
(c) R (d) None of these

2. The interior of the set of integers Z is:
(a) Z (b) ϕ
(c) R (d) None of these

3. Which of the following is not true:
(a) int.$\phi=\phi$
(b) int X=X
(c) $S \subset T \Rightarrow \text{int } S \subset \text{int } T$
(d) None of these

4. If p is an interior point of S then S is:
(a) empty (b) nbd of p
(c) exterior of p (d) None of these

5. Which is not true:
(a) X–int $(A)=$ cl $(X–A)$
(b) cl$(A)=X\sim$int$(X–A)$
(c) X–cl$(A)=$int $(X–A)$
(d) None of these

6. A complete subspace of a metric space is:
(a) Closed
(b) Open
(c) Neither open nor closed
(d) None of these

7. The set of integers with the usual metric is:
(a) Incomplete (b) Complete
(c) Both are true (d) None of these

8. Set of rational numbers with the usual metric is:
(a) Always complete
(b) May or may not be complete
(c) Not complete
(d) None of these

9. The set of irrational with the usual metric is:
(a) Always complete
(b) May or may not be complete
(c) Not complete
(d) None of these

10. Which space is complete, with the usual metric:
(a) $X=]0, 1[$ (b) $X=[0, 1[$
(c) $X=]0, \infty[$ (d) None of these

11. The triangle inequality in a metric space (X, d) holds equality sign when three points (x, y), (y, z) and (x, z) are :
(a) Collinear (b) non-collinear
(c) on the triangle (d) None of these

12. If X=R and $d(x, y)=|x–y|$ $\forall\, x, y \in X$ then d is a metric on X. This metric is known as:
(a) Pseudo metric (b) usual metric
(c) trival metric (d) None of these

13. If A is a closed subset of a complete metric space, then:
(a) A is complete (b) A is incomplete
(c) undefined (d) None of these

14. Let A be the subset of a metric space X. A point $x \in A$ is an interior point of A if A is a nbd of X. What is the correct option for the interior point of A *i.e.*, A^o:
(a) A^o is the union of all open set contained in A
(b) A^o is the largest open set contained in A
(c) both (a) and (b)
(d) None of these

15. A closed sphere is defined by:
(a) $S(x_0, r)=\{x \in X{:}d(x_0, x)=r\}$
(b) $S(x_0, r)=\{x \in X{:}d(x_0, x) \leq r\}$
(c) $S(x_0, r)=\{x \in X{:}d(x_0, x)<r\}$
(d) None of these

16. Definition of closure of a set states that:
(a) $\bar{A}=\cup\, [F : F \text{ is closed}, F \supset A]$
(b) $\bar{A}=\cap\, [F : F \text{ is closed}, F \subset A]$
(c) $\bar{A}=\cap\, [F : F \text{ is closed}, F \supset A]$
(d) None of these

17. If every cauchy sequence $<s_n>$ is also:
(a) Convergent (b) Divergent
(c) Oscillatory (d) None of these

18. Every subspace of X containing $x \in X$ in a discrete space (X, d) is:
(a) not a nbd of x (b) is a nbd of x
(c) is a closed set (d) None of these

19. A sequence $<s_n>$ in a metric space (X, d) is a Cauchy sequence in X such that $\forall\, \varepsilon>0$, $\exists$ a positive integer n such that:
(a) $m, n \geq n(\varepsilon) \Rightarrow d(s_m, s_n)<\varepsilon$
(b) $m, n \geq n(\varepsilon) \Rightarrow |(s_m, s_n)|<\varepsilon$
(c) $m, n \geq n(\varepsilon) \Rightarrow d(s_m, s_n)>\varepsilon$
(d) None of these

20. In the usual metric space R, the derived set Q of all rational numbers is:

(a) Q (b) Z
(c) I (d) R

21. Let $[X, d]$ be a metric space. If $\exists$ a positive integer M such that $d(x, y) \le M \;\forall\; x, y \in X$. Then X is called:
(a) Bounded metric space
(b) unbounded
(c) complete
(d) None of these

22. If $\overline{A}$ denote the closure of A in a metric space (X, d), which one of the following is false:
(a) $A \subset \overline{A}$ (b) $A \subset B \Rightarrow \overline{A} \subset \overline{B}$
(c) $\overline{A \cup B} = \overline{A} \cup \overline{B}$. (d) $\overline{A \cap B} = \overline{A} \cap \overline{B}$

23. If (X, d) be a metric space and A be a non-empty subset of X, the diameter of A *i.e.*, $\delta(A)$ is:
(a) $\delta(A) = \sup\{d(x, y) : x, y \in A\}$
(b) $\delta(A) = \inf\{d(x, y) : x, y \in A\}$
(c) $\delta(A) = \{d(x, y) : x, y \in A\}$
(d) None of these

24. The sequence $< \frac{1}{n} : n \in N >$ in metric space with usual metric R is:
(a) Convergent (b) divergent
(c) not cauchy (d) None of these

25. A metric space X is bounded iff its diameter is:
(a) not-defined (b) finite
(c) infinite (d) 0

26. A metric space X satisfy Bolzano-Weirstrass property, then:
(a) every finite sequence $\langle x_n \rangle$ in Y has no cluster point
(b) every infinite sequence has a cluster point
(c) X is not sequentially compact
(d) None of these

27. A is said to be no where dense if :
(a) $(\overline{A})^\circ \ne \phi$ (b) $(\overline{A})^\circ = R$
(c) $(\overline{A})^\circ = \phi$ (d) None of these

28. By BWP, every bounded sequence in a metric space has a:
(a) limit (b) limit point
(c) interior point (d) None of these

29. If A is a non-empty subset of a metric space X is complete, then:
(a) A is closed (b) A is open
(c) $A = \phi$ (d) None of these

30. The open sphere of radius 2 about (1, 1) in R^2 with metric

$$d(z_1, z_2) = \sqrt{(x_1 - x_2)^2 + (y_1 - y_2)^2} :$$

$z_1 = (x_1, y_1),\; z_2 = (x_2, y_2)$ is :
(a) $(x-1)^2 + (y-1)^2 = 4$
(b) $(x-1)^2 + (y-1)^2 = 2$
(c) $(x-1)^2 + (y-1)^2 = 1$
(d) None of these

Problem Set-2

1. A set $E \subset R$ is compact, then it is :
(a) closed
(b) bounded
(c) closed and bounded
(d) none of these

2. A compact subset A of a metric space (X, d) implies :
(a) A is closed
(b) A is bounded
(c) Closure of B is compact for every $B \subseteq A$
(d) all are true

3. If S and T are subsets of R, then which one of the following is not true :
(a) $S^\circ \cap T^\circ = (S \cap T)^\circ$
(b) $S^\circ \cup T^\circ \subseteq (S \cup T)^\circ$
(c) $\overline{S}$ is closed in R
(d) none of these

4. If every sequence $<x_n>$ in a metric space (X, d) is convergent then every cauchy sequence in $<x_n>$ is :
(a) convergent (b) divergent
(c) oscillatory (d) none of these

5. A collection F of sets have finite intersection property of :
(a) any finite sub collection of F has empty, intersection
(b) any finite sub collection of F has non empty intersection
(c) any finite sub collection of F has empty intersection
(d) none of these

6. If A is an open subset of complete metric space X then :
(a) A is closed (b) A is open
(c) A' is closed (d) none of these

7. The union of any finite collection of non empty closed sets is :
(a) closed (b) open
(c) null (d) none of these

8. If A is a closed subset of complete metric space, then :
(a) A is a complete (b) A is a not complete
(c) Can't say (d) none of these

9. If X is a complete metric space, E is a non empty open subset of X, then E is of :
(a) first category (b) second category
(c) Can't say (d) none of these

10. A set E is said to be of first category if :
(a) E is the union of collection of nowhere dense sets
(b) E is the union of countable collection of nowhere dense sets
(c) Can't say
(d) none of these

11. A set is said to be of second category if it is :
(a) of first category
(b) not of first category
(c) empty
(d) none of these

12. The empty set of a metric space is :
(a) open
(b) closed
(c) both open and closed
(d) none of these

13. The function $d : X \to R$ is said to be pseudo metric if :
(a) $d(x, y)=0 \Rightarrow x=y$
(b) $d(x, y)=0$ for some $x \neq y$
(c) $d(x, y) \neq 0$ for $x=y$
(d) none of these

14. A set of metric space (X, d) is said to be closed if it :
(a) contains all its limit points
(b) it is not open
(c) its compliment is open
(d) all are true

15. A function of mapping $f : (X, d_1) \to (Y, d_2)$ is said to be homomorphism if :
(a) $f(x)= y$ (b) $f(x) \subset Y$
(c) $f(x) \neq y$ (d) none of these

16. If X and Y are metric spaces and for every open set O of Y, $O \subset Y, f^{-1}(O) \subset X$ is open then :
(a) f is continuous (b) f is constant
(c) f is discontinuous (d) none of these

17. Which of the following is/are true ?
(a) Q is of first category in R w.r.t. usual metric
(b) If X is of second category and if $X=A \cup B$, then either A or B must be of second category.
(c) every countable subspace of R is of first category in R
(d) all are true

18. Let (X, d) be a metric space and x, y, z be points of X then $|d(x, y)-d(y, z)|$:
(a) $\leq d(x, z)$ (b) $\geq d(x, z)$
(c) $= d(x, z)$ (d) none of these

19. Let (X, d) be a metric space and $d^*(x, y)= \min\{1, d(x, y)\}$ then d^* is :
(a) a metric on x (b) pseudo metric
(c) not a metric (d) none of these

20. If d is metric an X and $d^* : X \times X \to R$ defined as $d^*(x, y)=\dfrac{d(x,y)}{1+d(x,y)}$ then d^* is :
(a) a metric on X (b) pseudo metric
(c) not a metric (d) none of these

21. Let X be the set of all real valued continuous function defined on the closed unit interval (0, 1) and d be a mapping of $X \times X \to R$ defined by $d(f, g)=$
(a) a metric on X (b) pseudo metric
(c) not a metric (d) none of these

22. Let $C[0, 1]$ denote the collection of all real valued bounded continuous function defined on [0, 1]. Let d be defined by
$$d[f, g]=\{|f(x)-g(x):x \in [0, e]|\}$$
then d is :
(a) a metric on x (b) pseudo metric
(c) not a metric (d) none of these

23. If A and B be two subsets of a metric space (X, d) then :
(a) $\delta(A \cup B) \leq \delta(A)+\delta(B)+\delta(A, B)$
(b) $\delta(A \cup B) \leq \delta(A)+\delta(B)$ if $A \cap B \neq \phi$
(c) both (a) and (b) are true
(d) none of these

24. If A is a non empty subset of a metric space (X, d) and x, y are any two points of X then $|d(x, A)-d(y, A)|$:
(a) $\leq d(x, y)$ (b) $\geq d(x, y)$
(c) $= d(x, y)$ (d) none of these

25. The closure of A can be defined by :
(a) $\overline{A}=\{x \in X : d(x, A)=0\}$
(b) $\overline{A}=\{x \in X : d(x, A)=1\}$
(c) $\overline{A}=\{x \in X : d(x, A) \neq 0\}$
(d) none of these

26. A subset A of metric space (X, d) is closed if and only if $\{x \in X : d(x, A)=0\}$:
(a) $< A$ (b) $> A$
(c) $= A$ (d) none of these

27. The metric space (R, d) is :
(a) complete
(b) not complete
(c) may or may not be complete
(d) none of these

28. The set R^n of all n tuples $(x_1, x_2, ..., x_n)$ of real numbers is :
(a) complete
(b) not complete
(c) may or may not be complete
(d) none of these

29. Let C be the set of complex numbers and $d(z_1, z_2) = |z_1-z_2|$ then d is :
(a) complete
(b) not complete
(c) may or may not be complete
(d) none of these

30. If $d[f, g]=\{|f(x)-g(x): x\in[a,b]|\}$ then d is :
(a) complete
(b) not complete
(c) may or may not be complete
(d) none of these

31. There is a countable family $\{O_i\}$ of open sets such that for any open set $O \subset X$, $O=\cup O_i$ then metric space is :
(a) separable (b) not separable
(c) can't say (d) none of these

32. Let (X, d) be a complete metric space. Then "no non empty open subset of X is of first category *i.e.* the union of a countable collection of no where dense subsets" is called :
(a) Bair's category theorem
(b) Bolzano-Weirstrass theorem
(c) both (a) and (b) are true
(d) none of these

33. Which one of the following is not correct :
(a) $S \subset R^n$ is closed, if it contains all its limit points
(b) $S \subset R^n$ is closed if it does not contains all its limit points
(c) $S \subset R^n$ is closed if it contains all isolated points
(d) none of these

34. Canter ternary set is :
(a) closed
(b) uncountable
(c) closed and uncountable
(d) none of these

35. If x is an accumulation point of $S \subset R^n$, then :
(a) every open sphere contains infinitely many points
(b) every open sphere contains finitely many points
(c) every open sphere contains no points
(d) none of these

36. Which of the following is not true :
(a) The union of any collection of open sets is closed
(b) The union of any collection of open sets is not open
(c) The union of any collection of open sets is open
(d) none of these

37. Which one of the following is true :
(a) The Intersection of an arbitrary collection of closed sets is closed.
(b) The intersection of an arbitrary collection of closed sets is open.
(c) The intersection of an arbitrary collection of closed sets is empty.
(d) none of these

38. Which of the following is not true :
(a) $S \subset R^n$ is closed $\Rightarrow S=\overline{S}$
(b) $S=\overline{S} \Rightarrow S \subset R^n$ is closed
(c) $S=\overline{S} \Rightarrow S \subset R^n$ is open
(d) none of these

39. Let A and B be two sets of positive real numbers bounded above. Let $a=\sup(A)$ and $b=\sup(B)$ and $C=[xy : x \in A, y \in B]$ then :
(a) $x>0, y>0 \Rightarrow xy<ab$
(b) $ab=\sup(C)$
(c) both (a) and (b) are true
(d) none of these

40. If d is an extended metric on the set X, then for some $x, y \in X$:
(a) $d(x, y)>\infty$ (b) $d(x, y)<\infty$
(c) $d(x, y)=\infty$ (d) none of these

ANSWERS

▶ FILL IN THE BLANKS

1. $x=y$	**2.** $d(z, y)$	**3.** $\sup\{d(x, y) : x \in A, y \in A\}$		**4.** $\left[\sum (x_i - y_i)^2\right]^{1/2}$	
5. open	**6.** open	**7.** not necessarily open		**8.** neighbourhood	
9. $\subseteq A$	**10.** $D(A)$	**11.** (i) closed	(ii) closed	**12.** (i) open	(ii) open
13. X	**14.** ϕ	**15.** each	**16.** each	**17.** not	**18.** each
19. open	**20.** closed	**21.** open	**22.** closed	**23.** closed	**24.** no
25. one	**26.** closed	**27.** closed	**28.** T′	**29.** closure	**30.** [0, 1]
31. A	**32.** A	**33.** adherent	**34.** closure	**35.** Bounded	**36.** Divergent
37. $n \geq m$	**38.** Unique	**39.** Cauchy	**40.** Complete	**41.** Complete	**42.** Closed

▶ TRUE/FALSE

1. T	**2.** T	**3.** F	**4.** F	**5.** T	**6.** T	**7.** F	**8.** F	**9.** F
10. T	**11.** T	**12.** T	**13.** F	**14.** T	**15.** T	**16.** T	**17.** T	**18.** T
19. T	**20.** T	**21.** T	**22.** F	**23.** T	**24.** T	**25.** T	**26.** F	**27.** T
28. T								

▶ MULTIPLE CHOICE QUESTIONS

Problem Set-1

1. (a)	**2.** (b)	**3.** (d)	**4.** (b)	**5.** (d)	**6.** (a)	**7.** (b)	**8.** (c)	**9.** (c)
10. (d)	**11.** (a)	**12.** (b)	**13.** (a)	**14.** (c)	**15.** (b)	**16.** (b)	**17.** (a)	**18.** (c)
19. (a)	**20.** (d)	**21.** (a)	**22.** (d)	**23.** (a)	**24.** (a)	**25.** (b)	**26.** (b)	**27.** (c)
28. (b)	**29.** (a)	**30.** (a)						

Problem Set-2

1. (c)	**2.** (d)	**3.** (d)	**4.** (a)	**5.** (a)	**6.** (c)	**7.** (a)	**8.** (a)	**9.** (a)
10. (b)	**11.** (b)	**12.** (c)	**13.** (b)	**14.** (d)	**15.** (a)	**16.** (a)	**17.** (d)	**18.** (a)
19. (a)	**20.** (a)	**21.** (a)	**22.** (a)	**23.** (a)	**24.** (a)	**25.** (a)	**26.** (a)	**27.** (a)
28. (a)	**29.** (a)	**30.** (a)	**31.** (a)	**32.** (a)	**33.** (a)	**34.** (c)	**35.** (a)	**36.** (c)
37. (a)	**38.** (c)	**39.** (c)	**40.** (a)					

Verify each of the following :

1. The function d defined by $d(z_1, z_2) = |z_1 - z_2|$: $z_1, z_2 \in C$ is a metric on the set C of complex numbers.
2. Let $X = \left\{1, \frac{1}{2}, \frac{1}{3}, \ldots, \frac{1}{n}, \ldots\right\}$ and d is the usual metric defined on X.

   Let $A = \left\{1, \frac{1}{3}, \frac{1}{5}, \ldots, \frac{1}{2n-1}, \ldots\right\}$ and

   $B = \left\{\frac{1}{2}, \frac{1}{4}, \frac{1}{6}, \ldots, \frac{1}{2n}, \ldots\right\}$ then $d(A, B) = 0$.
3. The function $d : C \times C \to R$ defined by $d(x, y) = \dfrac{2|x-y|}{\sqrt{1+|x|^2}.\sqrt{1+|y|^2}}$ is a metric on the set of complex numbers.
4. Let $R[0, 1]$ be the set of all R-integrable function defined on [0, 1] and $d(f, g) = \int_0^1 |f(x) - g(x)| dx \ \forall f, g \in R[0, 1]$ then d is not a metric on $R[0, 1]$.
5. In the metric space (R, d) of real numbers with the usual metric d, the open sphere $S(r, a)$ is the open interval $]a-r, a+r[$ and the closed sphere $S[r, a]$ is the closed interval $[a-r, a+r]$, $a \in R, r > 0$.
6. The subset $A = [0, 1]$ of the metric space $[X, d]$ where $X = [0, 2[$ and d is the usual metric, is an open set.
7. The subset $A = [(x, y) : y^2 < x, x, y \in R]$ of R^2 with the Euclidean metric is an open set.
8. The derived set of every subset of a discrete space is empty.
9. Every real number is a limit point of the set of rationals.
10. A finite set and the set of integers has no limit point.
11. The set $\left[\frac{1+i}{n} : n \in N\right]$ is neither open nor closed with respect to the usual metric in the complex plane.
12. Let $X = R$ and d is the usual metric, $A = Q$ then $int(A) = \phi$ and $ext(A) = \phi$, $F_r(A) = R$ and $bd(A) = \phi$.
13. Let $[X, d]$ be a discrete metric space and $A \subseteq X$, then int $(A) = A$, $ext(A) = A'$, $F_r(A) = bd(A) = \phi$.
14. The Euclidean space R^n is separable.
15. The Cantor set is nowhere dense.
16. The discrete space $[X, d]$ and the space (R, d) are complete metric space.
17. Any closed interval with the usual metric space is compact.
18. The open interval $]0, 1[$ with the usual metric is not compact.
19. For any non-empty subset A of a metric space $[X, d]$ the infimum $f : X \to R$ given by $f(x) = d(x, A) \ \forall x \in X$ is uniformly continuous. Also $f(x) = 0 \Leftrightarrow x \in \overline{A}$.
20. If $n \geq 2$ then there are open subset of R^n which cannot be expressed as the union of a countable family of pairwise disjoint open spheres in R^n.

●●●●●●

Chapter 3

Linear Spaces

3.1 INTRODUCTION

We are familiar with the concept of semigroup, a group and a ring. A group was obtained from a semigroup by imposing certain restrictions on the composition of a semigroup. A ring was obtained by defining a certain composition on a group structure and by giving rules that connected the group composition with the new composition. We shall now discuss another algebraic structure called a linear space or linear space which is going to involve a group structure, a ring structure and an operation connecting the elements of these two structures.

In order to discuss a linear space we need two basic things. One of them is the set of vectors and the other is the set of scalars. Therefore, to define a linear space we need a field F. The elements of field F are called the scalars. In addition, we need two binary operations. One of them is internal composition and the other is external composition. Now we distinguish the internal and external compositions as follows:

Internal Composition. Let R be any set. If $a * b \in$ R for all $a, b \in$ R and $a*b$ is a unique, then '*' is known as the internal composition. That is, a binary operation defined over the vectors is called internal composition (vector addition).

External Composition. Let V be the set of vectors and F be a field. Then a binary operation defined between the vectors and scalars is called external composition. That is, if $a \circ \alpha \in V$ for all $\alpha \in V$ and $a \in$ F and $a \circ \alpha$ is unique, then o is called an external composition or (scalar multiplication).

3.2 LINEAR SPACES

Definition. *Let V be a non-empty set of vectors and F be a field. Then an algebraic structure (V, +, .) together with two binary operations vectors addition and scalar multiplication is said to be linear space over F if this structure satisfies the following conditions :*

(i) $(V, +)$ *is an abelian group*

(ii) $a(\alpha + \beta) = a\alpha + a\beta,\ \forall\ \alpha, \beta \in V$ *and* $\forall\ a \in F$

(iii) $(a + b)\alpha = a\alpha + b\alpha,\ \ \forall\ \alpha \in V$ *and* $\forall\ a,b \in F$

(iv) $(ab)\alpha = a(b\alpha),\ \forall\ a \in V$ *and* $\forall\ a,b \in F$

(v) $I\alpha = \alpha,\ \forall\ a \in V$ *and* $1 \in F$

This linear space V over F is denoted by $V(F)$ or $L(F)$.

For example, if $F = R$, the field of real numbers, then $V(R)$ is a linear space and it is called a real linear space.

☛ ILLUSTRATIONS

(1) Let $R^2 = \{(a_1, a_2): a_1 \in R, a_2 \in R)$. The set R^2 is a linear space over R with addition and scalar multiplication defined as follows :

$(a_1, a_2) + (b_1, b_2) = (a_1+b_1, a_2+b_2)$

$c(a_1, a_2) = (ca_1, ca_2), \forall\, a_1, a_2, b_1, b_2, c \in R$

(2) Vector in 3-dimensional space form a linear space over R with respect to addition and scalar multiplication of vectors.

(3) Let R^n be the set of n-tuples of real numbers, *i.e.*,

$$R^n = \{(a_1, a_2,...,a_n : a_i \in R\}$$

Then R^n is a linear space over R with pointwise addition and scalar multiplication as defined in (1).

(4) Let C^n be the set of all ordered n-tuples of complex numbers. Then $\mathbf{C}^n$ is a linear space over C with addition and scalar multiplication.

3.3 ELEMENTARY PROPERTIES OF LINEAR SPACES

THEOREM 1. ***Let V(F) be a linear space over a field F and 0 be the zero (null) vector of V. Then prove that***

(i) $a0 = 0, \forall\ a \in F$ **(ii)** $0\alpha = 0, \forall\ \alpha \in V$

(iii) $a(-\alpha) = -(a\alpha), \forall\ a \in F, \alpha \in V$

(iv) $(-a)\alpha = -(a\alpha)\ \forall\ a \in F, \alpha \in V$

(v) $a(\alpha - \beta) = a\alpha - a\beta, \forall\ a \in F, \alpha, \beta \in V$

(vi) $a\alpha = 0 \Rightarrow a = 0$ *or* $\alpha = 0$

(vii) $a\alpha = a\beta \Rightarrow \alpha = \beta, \forall\ a \in F, \alpha, \beta \in V, a \neq 0$

(viii) $a\alpha = b\alpha \Rightarrow a = b\ \forall\ a, b \in F, \alpha \in V$ **and** $\alpha \neq 0$.

PROOF. (i) $a0 = 0$

we have $a0 = a\,(0 + 0)$ $[\because 0+0=0]$

$= a0 + a0$ [By property of linear space]

or $0 + a0 = a0 + a0$ $[\because\ 0 + a0 = a0]$

or $0 = a0$

(ii) $0\alpha = 0$

We have $0\alpha = (0+0)\alpha$ $[\because 0 + 0 = 0, 0 \in F]$

$= 0\alpha + 0\alpha$ [By the definition of $V(F)$]

or $0 + 0\alpha = 0\alpha + 0\alpha$ $[\because 0\alpha \in V \therefore 0 + 0\alpha = 0\alpha]$

or $0 = 0\alpha$

(iii) $a(-\alpha) = -(a\alpha)$

We have,

$a0 = 0$ [From(i)]

or $a(-\alpha + \alpha) = 0$ $[\because\ -\alpha + \alpha = 0]$

or $a(-\alpha) + a\alpha = 0$ [By the definition of $V(F)$]

(iv) $(-a)\,\alpha = -(a\alpha)$

Since, we have

$0\alpha = 0$ [From (ii)]

or $(-a + a)\alpha = 0$ [$\because -a + a = 0, a \in F$]

or $(-a)\alpha + a\alpha = 0$ [By the definition of $V(F)$]

or $(-a)\,\alpha = -(a\alpha)$

(v) $a(\alpha - \beta) = a\alpha - a\beta$

We have,

$$a(\alpha - \beta) = a\,[\alpha + (-\beta)\,]$$

$$= a\alpha + a(-\beta)$$ [By definition of $V(F)$]

$\therefore$ $a(\alpha - \beta) = a\alpha - a\beta$ [$\because a(-\beta) = -(\alpha\beta)$]

(vi) $a\alpha = 0 \Rightarrow a = 0$ or $\alpha = 0$

Suppose $0 \neq a \in F$, then a^{-1} exists in F.

Now $a\alpha = 0$ [given]

$\Rightarrow$ $a^{-1}a\alpha = a^{-1}0$

$\Rightarrow$ $(a^{-1}a)\alpha = 0$ [$\because a^{-1}0 = 0$]

$\Rightarrow$ $1\alpha = 0$ [$aa^{-1} = 1$]

$\Rightarrow$ $\alpha = 0$ [$\because 1\alpha = \alpha$]

Suppose $\alpha \neq 0$, then to prove $a = 0$, let us assume that $a \neq 0$, then a^{-1} exists. Since we have

$$a\alpha = 0$$

$\Rightarrow$ $a^{-1}\,(a\alpha) = a^{-1}\,0$

$\Rightarrow$ $(a^{-1}a)\alpha = 0$ [$\because a^{-1}\,a = 1$]

$\Rightarrow$ $1\alpha = 0$

$\Rightarrow$ $\alpha = 0$ [$\because 1\alpha = \alpha$]

$\Rightarrow$ This gives a contradiction because we have taken $\alpha \neq 0$.

Hence, $a = 0$.

(vii) We have,

$$a\alpha = a\beta \Rightarrow a\alpha - a\beta = 0$$

$\Rightarrow$ $a(\alpha - \beta) = 0$

$\Rightarrow$ $\alpha - \beta = 0$ [$\because a \neq 0$ and from (vi)]

$\Rightarrow$ $\alpha = \beta$.

Hence, $a\alpha = a\beta \Rightarrow$ for all $a \neq 0 \in F$ and $\alpha, \beta \in V$

(viii) We have,

$$a\alpha = b\alpha \Rightarrow a\alpha - b\alpha = 0$$

$\Rightarrow$ $(a-b)\alpha = 0$

$\Rightarrow$ $a - b = 0$ [$\because \alpha \neq 0$ and from (vi)]

$\Rightarrow$ $a = b$

Hence, $a\alpha = b\alpha \;\Rightarrow\; a = b$.

3.4 LINEAR SUBSPACES: LINEAR SPACES WITHIN LINEAR SPACE

Just like a subgroup and a subring, we do have the concept of a linear subspace which is generally addresses as a subspace.

Definitions. *Let W be a non-empty subset of V, where V is a linear space over a field F. Then W is said to be a linear subspace of V(F) if W is itself a linear space aver F with respect to the same operations as defined on V.*

For example: The set $W = \{ (a, 0, b) : a, b \in \mathrm{R}\}$ is a subspace of $R^3(\mathrm{R})$.

3.5 ELEMENTARY PROPERTIES OF LINEAR SUBSPACES

THEOREM 1. ***The necessary and sufficient conditions for a non-empty subset W of V(F) to be a subspace are that:***

(i) $\alpha \in W, b \in W \Rightarrow \alpha - \beta \in W$

(ii) $a \in F, \alpha \in W \Rightarrow a\alpha \in W.$

PROOF. Suppose W is a subspace of a linear space $V(F)$. Then if

$$\beta \in W \Rightarrow -\beta \in W$$

$\therefore$ $\alpha \in W, -\beta \in W \Rightarrow a + (-\beta) \in W$ [$\because W$ is closed under vector addition.]

$\Rightarrow$ $\alpha - \beta \in W$

and $a \in F, \alpha \in W \Rightarrow a\alpha \in W$ [$\because$ W is closed under scalar multiplication.]

Conversely, Suppose W is a subset of V and

(i) $\alpha \in W, \quad \beta \in W \Rightarrow \alpha - \beta \in W.$

(ii) $a \in F, \quad \alpha \in W \Rightarrow a\alpha \in W.$

Now we have to show that W is a subspace. For this purpose we proceed as follows:

$\alpha \in W, \alpha \in W \Rightarrow \alpha - \alpha \in W.$ [From (i)]

$\Rightarrow 0 \in W \Rightarrow$ identity exists.

and $0 \in W, \alpha \in W \Rightarrow 0 - \alpha \in W.$ [From (i)]

$\Rightarrow -\alpha \in W \Rightarrow$ inverse exists.

Now, $\alpha \in W, -\beta \in W \Rightarrow \alpha - (-\beta) \in W.$

$\Rightarrow \alpha + \beta \in W$ [From (i)]

$\therefore$ W is closed under vector addition.

Also vectror addition is always associative and commutative. Thereore $(W, +)$ is an abelian group.

From (ii) it is obvious that W is closed under multiplication. Space V is a linear space over F, therefore remaining properties will also hold in W. Hence W is a linear space and Hence W is itself a subspace.

THEOREM 2. ***The necessary and sufficient condition for a non-empty subset of W of a linear space V(F) to be a subspace of V is***

$$a, b \in F, \alpha, \beta \in W \Rightarrow a\alpha + b\beta \in W$$

PROOF. Suppose W is a subspace of a linear space $V(F)$. Then W is closed under vector addition and multiplication, therefore we have

$a \in F, \alpha \in W \Rightarrow a\alpha \in W$

and $b \in F, \beta \in W \Rightarrow b\beta \in W$

$\therefore$ $a\alpha \in W, b\beta \in W \Rightarrow a\alpha + b\beta \in W.$

Conversely, Suppose W is a subset of $V(F)$ and

$a, b \in F, \alpha, \beta \in W \Rightarrow a\alpha + b\beta \in W$ is given.

Then we have to show that W is a subset of $V(F)$.
Now taking $a=1$, $b=1$, then

$$1 \in F;\ \alpha,\beta \in W \Rightarrow 1\alpha + 1\beta \in W$$

$$\Rightarrow \quad \alpha + \beta \in W \qquad [1\alpha = \alpha,\ 1\beta = \beta]$$

$\therefore$ W is closed under vector addition.
and taking $a=0$, $b=-1$, we have

$$0\alpha + (-1)\beta \in W$$

$$\Rightarrow \quad 0 + (\beta) \in W \qquad [\because (-1)\beta = -\beta]$$

$$\Rightarrow \quad -\beta \in W$$

$\therefore$ Additive inverse exists in W.
Again, taking $a=0, b=0$, we have

$$0\alpha + 0\beta \in W$$

$$\Rightarrow \quad 0 \in W \qquad [\because 0\alpha = 0.\text{Similarly},\ 0\beta = 0]$$

Since $W \subseteq V$, theerfore vector addition is associative and commutative.
Thus, W is an abelian group under vector addition.
Further, taking $\beta=0$, we have

$$a\alpha + b0 \in W$$

$$\Rightarrow \quad a\alpha \in W \qquad [\because b0 = 0 \text{ and } a\alpha + 0 = a\alpha]$$

$\therefore$ W is closed under scalar multiplication.
The rest properties will hold in W because $W \subseteq V$ and these properties hold in V.
Hence W is a linear space and consequently W is a subspace of $V(F)$.

3.6 ALGEBRA OF SUBSPACES

THEOREM 1. ***The intersection of any two subspaces of a linear space is a subspace .***

PROOF. Let $V(F)$ be a linear space over F and W_1, W_2 be two subspaces of $V(F)$. Then, we have to show that $W_1 \cap W_2$ is a subspace of $V(F)$

Let $\quad \alpha,\beta \in W_1 \cap W_2 \Rightarrow \alpha,\beta \in W_1$ and $\alpha,\beta \in W_2$

Since, W_1 and W_2 are subspaces of V, so we have

$$a, b \in W \text{ and } \alpha, \beta \in W_1 \Rightarrow a\alpha + b\beta \in W_1 \qquad ...(1)$$

and, $\quad a, b \in F$ and $\alpha, \beta \in W_2 \Rightarrow a\alpha + b\beta \in W_2 \qquad ...(2)$

From (1) and (2), we get

if $\quad a,b \in F$ and $\alpha,\beta \in W_1 \cap W_2 \Rightarrow a\alpha + b\beta \in W_1 \cap W_2$

Hence, $W_1 \cap W_2$ is a subspace of V.

THEOREM 2. ***The intersection of an arbitrary collection of a subspaces of a linear space is also a subspace.***

PROOF. Let $\{W_\lambda : \lambda \in \Lambda\}$ be an arbitrary collection of subspaces of a vcector space V(say). Then we have to show that $\cap \{W_\lambda : \lambda \in \Lambda\}$ is a subspace of V.

Let $\quad \alpha,\beta \in \cap \{W_\lambda : \lambda \in \Lambda\}$

$\Rightarrow \quad \alpha,\beta \in W_\lambda$ for each $\lambda \in \Lambda$.

Since, each W_λ is a subspace of V, then for any two scalars $a,b \in F$, we have

$$a\alpha + b\beta \in W_\lambda \text{ for each } \lambda \in \Lambda$$

$$\Rightarrow \quad a\alpha + b\beta \in \cap \{W_\lambda : \lambda \in \Lambda\}$$

Hence, $\cap \{W_\lambda : \lambda \in \Lambda\}$ is a subspace of V.

THEOREM 3. ***The union of two subspaces of a linear space is not necessarily a subspace.***

PROOF. Let W_1, W_2 be two subspaces of a linear space V and suppose that

$$W_1 = \{(a_1, a_2, 0) : a_1, a_2 \in F\}$$

and $$W_2 = \{(a_1, 0, a_3) : a_1, a_3 \in F\}$$

Obviously, W_1 and W_2 are subspaces of $R^3(R)$. By definition of W_1 and W_2, we have $W_1 \cup W_2$ containing all triads i.e., 3-tuples of the form $(a_1, a_2, 0)$ and those of the form $(a_1, 0, a_3)$.

Now, if we consider the elements $\alpha = (1,2,0)$ and $\beta = (3,0,5)$ of $W_1 \cup W_2$, then for scalars $a = 1$ and $b = 2$.

$$a\alpha + b\beta = 1(1,2,0) + 2(3,0,5) = (1,2,0) + (6,0,10)$$

$$= (7,2,10) \notin W_1 \cup W_2$$

Thus, if $\alpha \in W_1 \cup W_2$ and $\beta \in W_1 \cup W_2$, then it is not necessarily implied that $a\alpha + b\beta \in W_1 \cup W_2$ for some $a, b \in F$

Hence, $W_1 \cup W_2$ is a subspace of $R^3(R)$.

THEOREM 4. ***The union of two subspaces of a linear space is a subspace iff one is contained in the other.***

PROOF. Let V(F) be a linear space and W_1, W_2 be two subspaces of V.

Suppose $W_1 \subseteq W_2$ or $W_2 \subseteq W_1$. Then we have to show that $W_1 \cup W_2$ is a subspace of V.

Now $W_1 \cup W_2 = W_2$ if $W_1 \subseteq W_2$ and W_2 is a subspace, therefore $W_1 \cup W_2$ is subspace.

Also, $W_1 \cup W_2 = W_1$ if $W_1 \subseteq W_2$ and since W_1 is a subspace, therefore $W_1 \subseteq W_2$ is a subspace V.

Conversely, Suppose $W_1 \cup W_2$ is a subspace of V. Then we have to show that

$$W_1 \subseteq W_2 \text{ or } W_2 \subseteq W_1.$$

Let us assume that W_1 is not a subset of W_2 and W_2 is not a subset of W_1.

Now, W_1 is not a subset of W_2, this implies that there exists an element α in W_1 which is not in W_2.

Also, W_2 is not a subset of W_1, therefore there exists an element β in W_2 which is not in W_1. But we have $\alpha \in W_1 \cup W_2$ and $\beta \in W_1 \cup W_2$ and since $W_1 \cup W_2$ is a subspace of V, we have.

$$a, b \in F, \alpha, \beta \in W_1 \cup W_2 \Rightarrow a\alpha + b\beta \in W_1 \cup W_2$$

Now taking $a = 1$, $b = 1$, we have

$$1\alpha + 1\beta \in W_1 \cup W_2 \qquad [\because 1\alpha \in W_1 \cup W_2 \subseteq V \therefore 1\alpha = \alpha]$$

$$\Rightarrow \quad \alpha + \beta \in W_1 \cup W_2$$

$$\Rightarrow \quad \alpha + \beta \in W_1 \text{ or } \alpha + \beta \in W_2$$

Suppose $\alpha + \beta \in W_1$ and $\alpha \in W_1$, then $(\alpha + \beta) - \alpha \in W_1$, because W_1 is a subspace of V. Therefore $\beta \in W_1$ and this gives a contradiction.

Now, suppose $\alpha + \beta \in W_2$ and $\beta \in W_2$, then

$$(\alpha + \beta) - \beta \in W_2$$

$$\Rightarrow \quad \alpha \in W_2 \qquad [\because W_2 \text{ is a subspace.}]$$

Hence, either $\quad W_1 \subseteq W_2$ or $W_2 \subseteq W_1$

3.7 LINEAR SUM OF TWO SUBSPACES

Let W_1 and W_2 be two subspaces of a vector $V(F)$. Then the linear sum of W_1 and W_2 is the set of all those elements each one of which is expessible as the sum of an element of W_1 and an element of W_2. The linear sum of W_1 and W_2 can be written as W_1+W_2. That is

$$W_1+W_2=\{\alpha+\beta:\alpha\in W_1,\beta\in W_2\}$$

REMARK

- If $a\in W_1$, then $a=a+0$ with $0\in W_1$ and $0\in W_2$ so $W_1\subseteq W_1+W_2$, Similarly, $W_2\subseteq W_1+W_2$.

THEOREM 1. ***The linear sum of two subspaces of a linear space is also a subspace.***

PROOF. Let W_1 and W_2 be two subspaces of a linear space $V(F)$. Then we have to show that W_1+W_2 is a subspace of $V(F)$.

Let α,β be any two arbitrary elements of W_1+W_2,

Then, $\alpha,\beta\in W_1+W_2$

$\Rightarrow$ $\alpha=\alpha_1+\alpha_2$ and $\beta=\beta_1+\beta_2$, where $\alpha_1,\beta_1\in W_1$ and $\alpha_2,\beta_2\in W_2$.

Since, $\alpha_1,\alpha_2,\beta_1,\beta_2\in V$

$\therefore$ $W_1+W_2\subseteq V$.

Since, W_1 and W_2 are subspaces of V. Then

$\alpha_1,\beta_1\in W_1 \quad\Rightarrow a\alpha_1+b\beta_1\in W_1$ for some $a,b\in F$

and $\alpha_2,\beta_2\in W_2 \quad\Rightarrow a\alpha_2+b\beta_2\in W_2$

Now, $a\alpha_1+b\beta_1\in W_1$ and $a\alpha_2+b\beta_2\in W_2$

$\Rightarrow (a\alpha_1+b\beta_1)+(a\alpha_2+b\beta_2)\in W_1+W_2$

$\Rightarrow a(\alpha_1+\alpha_2)+b(\beta_1+\beta_2)\in W_1+W_2$

$\Rightarrow a\alpha+b\beta\in W_1+W_2$

$\therefore$ $\alpha,\beta\in W_1+W_2,\ a,b\in F\Rightarrow a\alpha+b\beta\in W_1+W_2$

Hence, W_1+W_2 is a subspace.

3.8 DIRECT SUM OF VECTOR SUBSPACES

Definition. *Let W_1 and W_2 be two subspaces of a linear space V. Then V is said to be the direct sum of W_1 and W_2 if each element of V can be uniquely expressed as the sum of an element of W_1 and an element of W_2. If V is direct sum of W_1 and W_2, then it can be written as $V=W_1\oplus W_2$.*

In general, if V is the direct sum of $W_1,W_2,\ldots.W_n$, then

$$V=W_1\oplus W_2\oplus\ldots..\oplus W_n$$

Here $W_1,W_2,\ldots,W_n$ *are called complementary spaces.*

THEOREM 1. ***The necessary and sufficient condition for a linear space V to be the direct sum of two of its subspaces W_1 and W_2 are:***

(i) $V=W_1+W_2$ ***(ii)*** $W_1\cap W_2=\{0\}$

PROOF. **Condition is necessary:**

Suppose V is the direct sum of W_1 and W_2, then each element of V can be uniquely experessed as sum of an element of W_1 and an element of W_2, so in particular each element of V is expressible as the sum of an element of W_1 and element of W_2, this concludes that

$$V=W_1+W_2$$

Next, we shall show that $W_1 \cap W_2 = \{0\}$, for this let, if possible, there be a non-zero vector in $W_1 \cap W_2$ and let it be $\alpha \in W_1 \cap W_2$. Then we may write

$$\alpha = \alpha + 0 \text{ with } \alpha \in W_1 \text{ and } 0 \in W_2.$$

and $\quad \alpha = 0 + \alpha$ with $0 \in W_1$ and $\alpha \in W_2$.

Since $\quad W_1 + W_2 \in V$, so $\alpha \in V$ and $V = W_1 \oplus W_2$ therefore, α can be uniquely expessed as the sum of an element of W_1 and an element of W_2. Thus contains only zero vector. This implies $W_1 \cap W_2 = \{0\}$.

Condition is Sufficient:

Suppose the conditions:

(i) $V = W_1 + W_2$

(ii) $W_1 \cap W_2 = \{0\}$

hold then we shall show that $V = W_1 \oplus W_2$.

From (i) we conclude that each element of V can be expressed as the sum of an element of W_1 and an element of W_2. Therefore, we shall only show that this representation is unique.

Let, if possible, an element $\alpha \in V$ has two representations, that is,

$$\alpha = \alpha_1 + \alpha_2 \text{ with } \alpha_1 \in W_1 \text{ and } \alpha_2 \in W_2.$$

and $\quad \alpha' = \alpha'_1 + \alpha'_2$ with $\alpha'_1 \in W_1$ and $\alpha'_2 \in W_2$.

$\Rightarrow \quad \alpha_1 + \alpha_2 = \alpha'_1 + \alpha'_2$

$\Rightarrow \quad \alpha_1 - \alpha'_1 = \alpha'_2 - \alpha_2$

$\quad \alpha_1, \alpha'_1 \in W_1 \Rightarrow \alpha_1 - \alpha'_1 \in W_1$

and $\quad \alpha_2, \alpha'_2 \in W_2 \Rightarrow \alpha'_2 - \alpha_2 \in W_2 \quad$ [$\because$ W_1 and W_2 are subspaces]

$\therefore \quad \alpha_1 - \alpha'_1 = \alpha'_2 - \alpha_2 \in W_1 \cap W_2$.

But $\quad W_1 \cap W_2 = \{0\}$, this implies

$\quad \alpha_1 - \alpha'_1 = 0 = \alpha'_2 - \alpha_2$

$\Rightarrow \quad \alpha_1 = \alpha'_1$ and $\alpha'_2 = \alpha_2$.

This shows that each element of V can be uniquely expressed as the sum of an element of W_1 and an element of W_2. Hence, $V = W_1 \oplus W_2$ *i.e.*, V is the direct sum of W_1 and W_2.

RECAPITULATIONS

- An algebraic structure $(V, +, .)$ is said to be linear space over a field F if
  - (i) $(V, +)$ is an abelian group.
  - (ii) $a(\alpha + \beta) = a\alpha + a\beta \ \forall \ \alpha, \beta \in V, a \in F$
  - (iii) $(a+b)\alpha = a\alpha + b\alpha \ \forall \ \alpha \in V, \ a, b \in F$
  - (iv) $(ab)\alpha = a(b\alpha) \ \forall \ \alpha \in V, \ a, b \in F$
  - (v) $1.\alpha = \alpha \ \forall \ \alpha \in V, \ 1 \in F$
- The necessary and sufficient conditions for a non-empty subset W of $V(F)$ to be a subspace is that

$$a, b \in F, \alpha, \beta \in W \Rightarrow a\alpha + b\beta \in W$$

- $W_1 + W_2 = \{\alpha + \beta : \alpha \in W_1, \beta \in W_2\}$
- The necessary and sufficient condition for a linear space V to be the direct sum of two of its subspaces W_1 and W_2 are
  - (i) $V = W_1 + W_2$
  - (ii) $W_1 \cap W_2 = \{0\}$

SOLVED EXAMPLES

EXAMPLE 1. *In $V=R^3$. Let W_1 be the xy-plane and let W_2 be the z-plane given by*

$$W_1=\{(x, y, 0): x, y\in R\}$$

and $$W_2=\{(0,y,z): y, z\in R\}$$

Show that $$V=\{W_1 \oplus W_2\}$$

SOLUTION. Let $(x, y, z) \in V$, then this element can be writen as the sum of an element of W_1 and an element of W_2 on one and only one way *i.e.,*

$$(x,y,z)= (x, y, 0)+ (0,0, z)$$

Accordingly, V is the direct sum of W_1 and W_2, that is $V=\{W_1 \oplus W_2\}$

EXAMPLE 2. *In $V=R^3$ and W_1 be the xy-plane and let W_2 be the yz-plane:*

$$W_1=\{(x,y,0): x, y\in R\}$$

and $$W_2=\{(0,y,z): y, z\in R\}$$

then V is not the direct sum of W_1 and W_2.

SOLUTION. Let (a,b,c) be any element of R^3, then

$$(a,b,c)= (x ,y, 0)+ (0,y, z)= (x,2y,z)$$

$$\Rightarrow \quad a=x, b=2y, c=z$$

$$\Rightarrow \quad a,b,c \in R \qquad [\because x, y, z \in R]$$

$\Rightarrow$ every element of V can be written as the sum of an element of W_1 and an element of W_2.

But such sums are not unique, for example:

Let $(3,5,7)\in R^3$, then

$$(3,5,7)=(3,2,0)+ (0,3,7),\ (3,2,0) \in W_1 \text{ and } (0,3,7)\in W_2.$$

Also $$(3,5,7)=(3,1,0)+ (0,4,7),\ (3,1,0) \in W_1 \text{ and } (0,4,7)\in W_2.$$

Hence, V is not the direct sum of W_1 and W_2.

EXAMPLE 3. *If $V_3(R)$ is a linear space and $W_1=\{(a,0,c):a, c \in R\}$ and $W_2=\{(0,b,c):b,c \in R\}$ are two subspaces of $V_3(R)$, then show that $V = W_1 + W_2$ and $V \neq W_1 \oplus W_2$.*

SOLUTION. Let (x,y,z) be an arbitrary element of $V_3(R)$, then

$$(x,y,z)= (a,0,c)+ (0,b,c)= (a,b,2c)$$

$$\Rightarrow \quad x =a, y=b, z =2c$$

$$\Rightarrow \quad x,y,z \in R \qquad [\because a, b, c \in R]$$

$\Rightarrow$ every element of $V_3(R)$ can be written as the sum of an element of W_1 and an element of W_2.

$$\Rightarrow \quad V = W_1 + W_2$$

But such representations is not unique, for example:

Let $(3,5,6)\in V_3(R)$, then

$$(3,5,6)=(3,0,2)+ (0,5,4);\ (3,0,2) \in W_1 \text{ and } (0,5,4)\in W_2.$$

Also $$(3,5,6)=(3,0,5)+ (0,5,1);\ (3,0,5) \in W_1 \text{ and } (0,5,1)\in W_2.$$

So here $(3,5,6)$ can be written as the sum of elements of W_1 and W_2 in two ways.

Hence, $$V \neq W_1 \oplus W_2.$$

EXAMPLE 4. *Let V be a linear space of all functions from R into R; let V_e be the subset of even functions, such that $f(-x)= f(x)$; let V_0 be the subset of odd functions $f(-x)= -f(x)$. Prove that*

(a) V_e and V_0 are subspaces of V.

(b) $V_e + V_0 = V$.

(c) $V_e \cap V_0 = \{0\}$.

SOLUTION. (a) Since V is a linear space of all functions, therefore

$$V_e \subseteq V, V_0 \subseteq V.$$

Let $f(x)g(x) \in V_e$, so that $f(-x)= f(x)$ *and* $g(-x)= g(x)$ and let $a,b \in R$, then consider,

$$h(x) = af(x)+bg(x)$$

Now $h(-x) = af(-x)+bg(-x)$

or $h(-x) = af(x)+bg(x)$ $[f(-x)= f(x)\ ,\ g(-x)= g(x)]$

$= h(x)$

$\Rightarrow$ $h(x) \in V_e$

$\Rightarrow af(x)+bg(x) \in V_e$

Consequently, if $f(x),g(x) \in V_e$, then $af(x)+bg(x) \in V_e$. Hence, V_e is a subspace of V.

Similarly, we can prove that V_0 is a subspace of V.

(b) Let $f(x)$ be any element of V.

Consider $f(x)= \frac{1}{2}\ [f(x)+f(-x)] + \frac{1}{2}\ [f(x) - f(-x)]$.

Let $\alpha(x)= \frac{1}{2}\ [f(x)+f(-x)]$ and $\beta(x)= \frac{1}{2}\ [f(x)+f(-x)]$

$\therefore$ $f(x)= \alpha(x)+ \beta(x)$.

Also $\alpha(-x)= \frac{1}{2}\ [f(-x)+f(x)]= \alpha(x)$

$\therefore$ $\alpha(x)= V_e$.

and $\beta(-x)= \frac{1}{2}\ [f(-x)+f(x)]= \frac{1}{2}\ [f(x)+f(-x)] = \beta(x)$

$\therefore$ $\beta(x)= V_0$.

Consequently, $f(x)= \alpha(x)+ \beta(x)$ where, $\alpha(x)= V_e$ and $\beta(x)= V_0$. Hence every element of V can be expressed as the sum of an element of V_e and an element of V_0. That is,

$$V = V_e + V_0.$$

(c) $V_e \cap V_0 = \{0\}$.

Let if possible, there exist a non-zero function $f(x)$, which belongs to $V_e \cap V_0$ Therefore, if $f(x) \in V_e$, then $f(-x)= f(x)$ and $f(x) \in V_0$, then $f(-x)= - f(x)$. So that,

$$f(x)= - f(x)$$

$\Rightarrow$ $2\,f(x)= 0 \Rightarrow f(x)= 0$

This gives a contradiction, because $f(x)$ is assumed to be non-zero function. Hence every function of $V_e \cap V_0$ is a zero function. Consequently, $V_e \cap V_0=\{0\}$.

EXAMPLE 5. *If W_1 and W_2 are subspace of a linear space $V(F)$, then show that W_1+W_2 is also a subspace of $V(F)$.*

SOLUTION. Let us define

$$W_1+W_2=\{\alpha_1+\alpha_2;\alpha_1 \in W_1, \alpha_2 \in W_2\}.$$

We have to show that W_1+W_2 is a subspace of V.

Let $\alpha \in W_1+W_2 \Rightarrow \alpha=\alpha_1+\alpha_2$ for some $\alpha_1 \in W_1$ and $\alpha_2 \in W_2$

$$\Rightarrow \alpha=\alpha_1+\alpha_2 \text{ for some } \alpha_1, \alpha_2 \in V \Rightarrow \alpha \in V$$

$\therefore \quad W_1+ W_2 \subseteq V.$

Now, let $\alpha_1+\alpha_2 \in W_1+W_2$ and $\beta_1+\beta_2 \in W_1+W_2$, where, $\alpha_1+\alpha_2 \in W_1$ and $\beta_1+\beta_2 \in W_2$.

Since, W_1 and W_2 are subspaces of V, then

$$a, b \in F, \alpha_1,\beta_1 \in W_1 \Rightarrow a\alpha_1+b\beta_1 \in W_1$$

$$a, b \in F , \alpha_2,\beta_2 \in W_0 \Rightarrow a\alpha_2+b\beta_2 \in W_2$$

$$a\alpha_1+b\beta_1 \in W_1 , a\alpha_2+b\beta_2 \in W_2$$

$$\Rightarrow (a\alpha_1+b\beta_1)+(a\alpha_2+b\beta_2) \in W_1 + W_2$$

$$\Rightarrow a(\alpha_1+\alpha_2)+b(\beta_1+\beta_2) \in W_1+W_2$$

Since, $\alpha_1+\alpha_2 \in W_1+W_2$ and $\beta_1+\beta_2 \in W_1+W_2$. Thus we have

$$a,b \in F , \alpha_1+ \alpha_1,\beta_1+ \beta_1 \in W_1+ W_2$$

$$\Rightarrow a(\alpha_1+\alpha_2)+b(\beta_1+\beta_2) \in W_1+W_2$$

Hence, W_1+W_2 is subspace of $V(F)$.

EXAMPLE 6. *Let R be the filed of real numbers, show that the set $W=\{(x,2y,3z):x,y,z \in R\}$ is a subspace of $V_3(R)$.*

SOLUTION. Since $W=\{(x,2y,3z):x,y,z\in R\}$.

Let $\alpha,\beta \in W$, where $\alpha=\{x_1,2y_1,3z_1\}$ and $\beta=\{x_2,2y_2,3z_2\}$ and $x_1,y_1,z_1,x_2,y_2,z_2 \in$ R,

If a, $b \in R$, then

$$a\alpha+b\beta=a(x_1,2y_1,3z_1)+ b(x_2,2y_2,3z_2)$$

$$=(ax_1,2ay_1,3az_1)+ (bx_2,2by_2,3bz_2)$$

$$=(ax_1+bx_2),2(ay_1+by_2),3(az_1+bz_2)$$

$\therefore \quad a\alpha+b\beta \in W,$

Because $ax_1+bx_2,ay_1+by_2,az_1+bz_2 \in R$.

Hence, W is a subspace of $V_3(R)$.

EXAMPLE 7. *Show that the set of all real valued continuous functions defined on* [0,1] *is a linear space over field of reals.*

SOLUTION. Let V be the set of all real valued continuous functions defined on [0,1]. Now we have to show that V is a vector spsce over R (field of real numbers) under vector addition and scalar multiplication which is defined as follows:

$$(f+g)(x)= f(x)+g(x), \forall f, g \in V$$

and $(af)(x) = af(x), \forall f \in V$ and $a \in R$.

First we shall show that $(V,+)$ is an abelian group.

Let $f, g \in V$, then

$$(f+g)(x) = f(x)+g(x), \forall f, g \in V$$

$\therefore$ $f+g \in V$. Thus V is closed under vector addition.

Now let $0(x) \in V$, we have

$$f(x)+0(x) = (f+0)(x) = f(x), \forall f \in V$$

$\therefore$ $0(x)$ is the additive identity in V.

Let $-f \in V$, then we have

$$-f(x)+f(x) = (-f+f)(x) = 0(x), \forall f \in V$$

$\therefore$ $-f$ is the additive inverse of V.

Since vector addition is always associative as well as commutative, consequently $(V,+)$ is an abelian group.

Further, since V is closed under scalar multiplication therefore af is a real valued continuous function defined on $[0,1]$.

(i) If $a \in R$ and $f, g \in V$, then we have

$$a[(f+g)x] = a[f(x)+g(x)]$$
$$= af(x)+ag(x) = (af+ag)(x)$$

$\therefore$ $a(f+g) = af+ag$.

(ii) If $a,b \in R$ and $f \in V$, then we have

$$[(a+b)f](x) = (a+b)f(x) = af(x)+bf(x) = (af+bf)(x)$$

$\therefore$ $(a+b)f = af+bf$.

(iii) If $a,b \in R$ and $f \in V$, then we have

$$[(ab)f](x) = (ab)f(x) = a[bf(x)] = [a(bf)](x).$$

$\therefore$ $(ab)f = a(bf)$.

(iv) If $1 \in R$ and $f \in V$, then we have

$$(1f)(x) = 1f(x) = f(x)$$

$\therefore$ $1f = f, \ \forall f \in V$

Hence, V is a linear space over R.

EXAMPLE 8. *Show that R^2(R) is not a linear space when addition and scalar multiplication composition are defined by*

$$(a_1,a_2)+(b_1,b_2) = (a_1+b_1, a_2+b_2)$$

and $$a(a_1,a_2) = (aa_1,a_2), \forall a,a_1,a_2,b_1,b_2 \in \text{R}.$$

SOLUTION. Suppose $a=1$ and $b=2$ and $(a_1,a_2)=(3,4)$, then using the given compositions, we have

$$(a+b)(a_1,a_2) = (1+2).(3+4)$$
$$= 3.(3,4) = (3.3,4) = (9,4)$$

and $a.(a_1,a_2)+b.(a_1,a_2) = 1.(3,4)+2.(3,4)$

$$= (3.1,4)+(3.2,4)$$

$$= (3,4)+(6,4)$$
$$= (3+6,4+4) = (9,8)$$

$\therefore \qquad (a+b)\cdot(a_1, a_2) \neq a\cdot(a_1,a_2) + b\cdot(a_1, a_2).$

Hence, R^2(R) is not a linear space.

EXAMPLE 9. *Show that* the set

$$W=\{(a,b,c): a–3b+4c=0\}$$

is a subspace of 3-*tuple space* R^3(R).

SOLUTION. Let $\alpha=(a_1,b_1,c_1)$ and $\beta=(a_2,b_2,c_2)$ be any two elements of W, such that

$$a_1–3b_1+4c_1=0 \text{ and } a_2–3b_2+4c_2=0$$

For $a,b \in$ R, we have

$$a\alpha+b\beta = a(a_1,b_1,c_1)+b(a_2,b_2,c_2)$$
$$= (aa_1,ab_1,ac_1)+(ba_2,bb_2,bc_2)$$
$$= (aa_1+ ba_2,ab_1+ bb_2,ac_1+bc_2)$$

Now, $\qquad (aa_1+ba_2) - 3(ab_1+ bb_2)+4(ac_1+bc_2)$

$$=(aa_1–3ab_1+ 4ac_1)+ (ba_2 - 3bb_2+4bc_2)$$
$$=a(a_1–3\, b_1+ 4c_1)+ b(a_2– 3b_2+4c_2)$$
$$=a.0+ b.0=0$$

So, $\qquad a\alpha+b\beta \in W.$

Thus $\qquad \alpha \in W, \beta \in W \Rightarrow a\alpha+b\beta \in W \ \forall \ a,b \in$ R.

Hence, W is a subspace of R^3(R).

EXAMPLE 10. *Prove the solution set W of the differential equation*

$$2\frac{d^2y}{dx^2} - 9\frac{dy}{dx} + 2y = 0$$

is a subspace of linear space of all real valued functions of R.

SOLUTION. Let $W = \left\{y : 2\frac{d^2y}{dx^2} - 9\frac{dy}{dx} + 2y = 0\right\}$, be the set of all solutions of the given differential equation where, $y=f(x)$.

Now if we define a real valued function denoted by 0 on R by $0(x)=0, \forall \, x \in$ R, then $0(x)$ satisfies the given differential equation, so that $0(x) \in W$.

Let $y_1=f(x)$ and $y_2=g(x)$ be any two elements of W, then we have

$$2\frac{d^2 f(x)}{dx^2} - \frac{df(x)}{dx} + 2f(x) = 0 \qquad .... (1)$$

and

$$2\frac{d^2 g(x)}{dx^2} - \frac{dg(x)}{dx} + 2g(x) = 0 \qquad .... (2)$$

Let a, b be any two scalars.

Now, multiplying (1) by a and (2) by b and then adding, we get

$$2\frac{d^2}{dx^2}[af(x)+bg(x)] - 9\frac{d}{dx}[af(x)+bg(x)] + 2[af(x)+bg(x)] = 0$$

which shows that $af(x)+bg(x)$ is also the solution of the given differential equation.

So that $[af(x) + bg(x)] \in W$.

$$\therefore \quad f(x) \in W, g(x) \in W \Rightarrow af(x) + bg(x) \in W \ \forall \ a, b \in \mathrm{R}.$$

Hence W is a subspace of a linear space of all real valued functions of R.

EXAMPLE 11. *Let V be the linear space of all functions from the real field R into R. Show that the set $W = \{f : f(7) = 2 + f(1)\}$ is not a subspace of V.*

SOLUTION. Let f and g be any two elements of W, *i.e.*,

$$f(7) = 2 + f(1) \text{ and } g(7) = 2 + g(1).$$

Then
$$\begin{aligned}(f+g)(7) &= f(7) + g(7)\\ &= 2 + f(1) + 2 + g(1)\\ &= 4 + f(1) + g(1)\\ &= 4 + (f+g)(1) \neq 2 + (f+g)(1).\end{aligned}$$

Hence $f + g \neq W$, and so W is not a subspace of V.

EXAMPLE 12. *Let V be the linear space of all square $n \times n$ matrices over a field of reals R. Show that W is a subspace of V, where*

(i) W consists of the symmetric matrices.

(ii) W consists of all matrices which commute with a given matrix M, i.e.,

$$W = \{A \in V : AM = MA\}.$$

SOLUTION. (i) The null matrix $0 \in W$, as all its entries being zero and it is symmetric.

Let $A = [a_{ij}]$ and $B = [b_{ij}]$ be any two elements of W.

Then $a_{ij} = a_{ji}$ and $b_{ij} = b_{ji}$.

For any scalars $a, b \in \mathrm{R}$, we have

$$aA + bB = a[a_{ij}] + b[b_{ij}] = [aa_{ij} + bb_{ij}] = [c_{ij}] = C$$

where, $\quad c_{ij} = aa_{ij} + bb_{ij}$

Now, $\quad c_{ij} = aa_{ij} + bb_{ij} = aa_{ji} + bb_{ji} = c_{ji}$

Thus $aA + bB$ is symmetric so it belong to W. Hence W is subspace of V.

(ii) The null matrix $0 \in W$ as $OM = MO$.

Now suppose A and B be any two elements of W then, $AM = MA$ and $BM = MB$.

For $a, b \in \mathrm{R}$, we have

$$\begin{aligned}(aA + bB)M &= (\alpha A)M + (bB)M = a(AM) + b(BM)\\ &= a(MA) + b(MB) = M(aA + bB)\end{aligned}$$

$$\therefore \quad aA + bB \in W, A\,B \in W \text{ and } \forall \ a, b \in R.$$

Hence W is a subspace of V.

EXAMPLE 13. *Let $V = R^3$. Show the set $W = \{(a, b, c) : a^2 + b^2 + c^2 \leq 1\}$ is not a subspace of* V.

SOLUTION. Let $\alpha = (1, 0, 0) \in W$, $\beta = (0, 1, 0) \in W$. But we have

$$\begin{aligned}\alpha + \beta &= (1, 0, 0) + (0, 1, 0)\\ &= (1, 1, 0) \notin W \quad \text{as} \quad 1^2 + 1^2 + 0^0 = 2 > 1.\end{aligned}$$

Hence W is not a subspace of V.

EXAMPLE 14. *Let V be the linear space of all 2×2 matrices over the real field R. Show that W is not a subspace of V, where*

(i) W consists of all matrices with zero determinant,

(ii) W consists of all matrices A from which $A^2 = A$.

SOLUTION. (i) Let $A = \begin{pmatrix} 1 & 0 \\ 0 & 0 \end{pmatrix}$ and $B = \begin{pmatrix} 0 & 0 \\ 0 & 1 \end{pmatrix}$ be two elements of W. But,

$$A + B = \begin{pmatrix} 1 & 0 \\ 0 & 0 \end{pmatrix} + \begin{pmatrix} 0 & 0 \\ 0 & 1 \end{pmatrix} = \begin{pmatrix} 1 & 0 \\ 0 & 0 \end{pmatrix} \notin W \text{ as } |A + B| = 1 \neq 0$$

$\therefore$ W is not a subspace of V.

(ii) The unit matrix $I = \begin{pmatrix} 1 & 0 \\ 0 & 0 \end{pmatrix} \notin W$ as $I^2 = I$.

But $2I = \begin{pmatrix} 2 & 0 \\ 0 & 2 \end{pmatrix} \notin W$ as

$$(2I)^2 = \begin{pmatrix} 2 & 0 \\ 0 & 2 \end{pmatrix}\begin{pmatrix} 2 & 0 \\ 0 & 2 \end{pmatrix} = \begin{pmatrix} 4 & 0 \\ 0 & 4 \end{pmatrix} \neq 2I.$$

Hence W is not a subspace of V.

EXERCISE 3.1

1. Show that a field K can be regarded as a linear space over any subfield F of K.
2. Show that the complex field **C** is a linear space over the field R of reals.
3. Let V be the set of all ordered pairs (x, y) of reals and let F be the field of real numbers, then show that V is not a linear space over F with respect to addition and multiplication defined as
$(x, y) + (x_1, y_1) = (x + x_1, y + y_1)$ and $c(x, y) = (cx, cy)$
4. Let V be the set of all pairs of real numbers and let F be the field of real numbers and define
$(x, y) + (x_1, y_1) = (3y + 3y_1, -x - x_1)$ and $c(x, y) = (3cy, -cx)$.
5. Show that the set $W = \{(a_1, a_2, 0) : a_1, a_2 \in F\}$ is a subspace of $V_3(F)$.
6. Show that the set W of the elements of the linear space $V_3(R)$ of the form $(x + 2y, y, -x+3y)$ where $x, y \in R$ is a subspace of $V_3(R)$.
7. Prove that the set of all solutions (a, b, c) of the equation $a+b+2c = 0$ is a subspace of linear space $V_3(R)$.
8. Prove that the arbitrary intersection of subspaces of a linear space is a subspace.
9. Let $V = R^3$. Show that the set $W = \{(a,b,c) : a,b,c \in Q]$ is not a subspace of R^3.

HINT TO SELECTED PROBLEMS

2. Let C be the set of vectors and R the set of scalars, Since C is a field so that $a\alpha \in C$, this show that C is closed under scalar multiplication.

$1 \in R$ so $1 \in C$.

Also for $a, b \in C$ and $a,b \in R$,

$$a, b \in R \Rightarrow a,b \in C \qquad (\because R \subseteq C)$$

$\therefore \quad a(\alpha + \beta\} = a\alpha + a\beta$

and $(a + b)\alpha = a\alpha + b\alpha$.

also, $(ab)\alpha = a(b\alpha)$, $1 . \alpha = \alpha, \forall \alpha \in C$.

Hence C is a linear space over R.

3. Let $\alpha = (x, y)$ and $a,b \in F$, then

$$(a + b)\alpha = (a + b)(x, y) = ((a+b)\, x, y) = \{ax + bx, y\}$$

and $a\alpha + b\alpha = a(x, y) + b(x, y) = (ax, y) + (bx, y) = (ax + bx, 2y)$.

$\therefore (a + b)\alpha \neq a\alpha + b\alpha$. Hence V is not a linear space.

6. $W = t(x+2y, y, -x+3y) : x, y \in R)$, then show that $a\alpha + b\beta \in W \; \forall \; \alpha, \beta \in W$ and $a, b \in R$.

$$a\alpha+b\beta = a(x_1+2y_1, y_1, -x_1+3y_1] + b(x_2+2y_2, y_2, -x_2+3y_2]$$
$$= (ax_1+2ay_1, ay_1, -ax_1+3ay_1) + (bx_2+2by_2, by_2, -bx_2+by_2)$$
$$= (ax_1+bx_2+2(ay_1+by_2), ay_1+, by_2, -(ax_1+bx_2)+3(ay_1+by_2)) \in W$$

$\therefore a\alpha+b\beta \in W \Rightarrow W$ is a subspace.

9. $(a, b, c) \in W$, $(a, b, -c) \in W$ as $-c \in Q$.
Now $(a, b, c) + (a, b, -c) = (2a, 2b, 0) \in W$.
Hence W is not a subspace.

3.9 LINEAR COMBINATION OF VECTORS

Definition. *Let V be a linear space over a field F and $\alpha_1, \alpha_2, \ldots, \alpha_n \in V$, then any vector $\alpha \in V$ can be expressed as below:*

$$\alpha = a_1\alpha_1 + a_2\alpha_2 + \ldots. + a_n\alpha_n$$

where $a_1, a_2, \ldots, a_n \in F$, is said to be the linear combination of vectors $\alpha_1, \alpha_2, \ldots, \alpha_n$.

Definition. *Let V(F) be a linear space aver F and let S be any non-empty subset of V, then the set of all linear combination of finite elements of S, is called the linear span of S. It is denoted by L(S). Therefore, we have*

$$L(S) = \{a_1\alpha_1 + a_2\alpha_2 + \ldots. + a_n\alpha_n : a_1, a_2, \ldots, a_n \in F$$

and $\alpha_1, \alpha_2, \ldots., \alpha_n$ are a finite elements of $S\}$

THEOREM 1. ***The linear span L(S) of a non-empty subset S of a linear space V(F) is the smallest subspace of V containing S.***

PROOF. By the definition of $L(S)$, we have

$$L(S) = \{a_1\alpha_1 + a_2\alpha_2 + \ldots + a_n\alpha_n : a_i \in V\}$$

[Let $\alpha \in S$, then $\alpha = 1.\alpha$, $1 \in F$, so $\alpha \in L(S)$

$$\therefore \quad S \subseteq L(S)$$

Now, we shall show that $L(S)$ is a subspace.

Let α, β be any two arbitrary elements of $L(S)$, then

$\alpha = a_1\alpha_1 + a_2\alpha_2 + \ldots + a_n\alpha_n$; for $\alpha_1, \alpha_2, \ldots., \alpha_n \in S$ and for $a_1, a_2, \ldots., a_n \in F$

Also, $\beta = b_1\beta_1 + b_2\beta_2 + \ldots + b_m\beta_m$ for all $a, b \in F$, we have

$$a\alpha + b\beta = a(a_1\alpha_1 + a_2\alpha_2 + \ldots + a_n\alpha_n) + b(b_1\beta_1 + b_2\beta_2 + \ldots + b_m\beta_m)$$
$$= (aa_1)\alpha_1 + (aa_2)\alpha_2 + \ldots + (aa_n)\alpha_n + (bb_1)\beta_1 + (bb_2)\beta_2 + \ldots + (bb_m)\beta_m$$

This implies that $a\alpha + b\beta$ is a linear combination of finite numbers of elements of S, so $a\alpha + b\beta \in L(S)$. Hence $L(S)$ is a subspace of V.

Next, we shall show that $L(S)$ is the smallest subspace containing S.

For this there is a subspace W of V containing S. Let $\alpha_1, \alpha_2, \ldots, \alpha_t \in S \subset W$ and W being a subspace, then

$$a_1\alpha_1 + a_2\alpha_2 + \ldots + a_t\alpha_t \in W; \text{ for all } a_i \in F$$

This implies that W contains all linear combinations of finite elements of S, therefore $L(S) \in W$.

Hence $L(S)$ is the smallest sub space of V containing S.

THEOREM 2. ***If S, T are two subsets of a linear space V, then***

(i) $S \subseteq T \Rightarrow L(S) \subseteq L(T)$ ***(ii)*** $L(S \cup T) = L(S) + L(T)$

(iii) $L[L(S)] = L(S)$

PROOF. Let α be an arbitrary element of $L(S)$, then

$$\alpha \in L(S) \Rightarrow \alpha = a_1\alpha_1 + a_2\alpha_2 + \ldots. + a_n\alpha_n$$

where, $\alpha_1, \alpha_2, \ldots, \alpha_n \in S$ and $a_1, a_2, \ldots, a_n \in F$.

(i) Since $S \subseteq T$, so that $\alpha_1, \alpha_2, \ldots, \alpha_n \in T$, therefore α is also the linear combination of finite elements of T. This implies $a \in L(T)$.

Thus $\alpha \in L(S) \Rightarrow \alpha \in L(T)$

Hence $L(S) \subseteq L(T)$ if $S \subseteq T$.

(ii) Since $S \subseteq S \cup T$ and $T \subseteq S \cup T$, then from (i), we have

$$L(S) \subseteq L(S \cup T)$$

and $$L(T) \subseteq L(S \cup T)$$

$$\Rightarrow L(S) + L(T) \subseteq L(S \cup T) \quad \ldots(1)$$

Next, let α be an arbitrary element of $L(S \cup T)$, then α is a linear combination of finite elements of $S \cup T$. This implies that some of $a_i \in S$ or some of $\alpha_j \in T$. This shows that α is a linear combination of finite elements of S and finite elements of T, therefore $\alpha \in L(S) + L(T)$. Thus,

$$L(S \cup T) \subseteq L(S) + L(T) \quad \ldots(2)$$

From (1) and (2), we get

$$L(S \cup T) \subseteq L(S) + L(T) \quad \ldots(3)$$

(iii) Since $S \subseteq T(S)$, then from (i), we have

$$L(S) \subseteq L[(S)],$$

Next, let α be any arbitrary element of $L[L(S)]$, then α is a linear combination of finite elements of $L(S)$. Suppose, we have

$$\alpha = b_1\beta_1 + b_2\beta_2 + \ldots + b_n\beta_n = \sum_{i=1}^{n} b_i\beta_i \quad \ldots(2)$$

where, each $\beta_i \in L(S)$ for all $b_1, b_2, \ldots, b_n \in F$. Also, each β_i is a linear combination of finite elements of S, so that

$$\beta_1 = a_{11}\alpha_1 + a_{12}\alpha_2 + \ldots + a_{1m}\alpha_m$$
$$\beta_1 = a_{21}\alpha_1 + a_{22}\alpha_2 + \ldots + a_{2t}\alpha_t$$
$$\ldots \quad \ldots \quad \ldots \quad \ldots \quad \ldots \quad \ldots \quad \ldots$$

On putting the values of $\beta_1, \beta_2, \ldots$, etc. in (2), we see that α is a linear combination of finite elements of S. Thus $\alpha \in L(S)$

$$\therefore \quad L[L(S)] \subseteq L(S) \quad \ldots(3)$$

From (1) and (3), we get $L(S) = L[L(S)]$.

THEOREM 3. ***The linear sum of two subspaces W_1 and W_2 of a linear space $V(F)$ is generated by their union. That is, $W_1+W_2 = L(W_1 \cup W_2)$.***

PROOF. We have already proved that the linear sum of two subspaces is aiso a subspace and linear span of a subset of a linear space is also a subspace.

Therefore, W_1+W_2 and $L(W_1 \cup W_2)$ are subspaces of $V(F)$.

Let α be any arbitrary element of W_1+W_2, then

$$\alpha \in W_1+W_2$$

$$\Rightarrow \quad \alpha = \alpha_1 + \alpha_2 \text{ for some } \alpha_1 \in W_1 \text{ and } \alpha_2 \in W_2$$

Since $\alpha_1 \in W_1$ and $\alpha_2 \in W_2$, so, $\alpha_1 + \alpha_2 \in W_1 + W_2$

Also, we may write $\alpha = \alpha_1 + \alpha_2 = 1.\ \alpha_1 + 1.\alpha_2$. This implies that α is a linear combination of finite elements namely α_1 and α_2 of $W_1 \cup W_2$, so that $\alpha \in L(W_1+W_2)$

$$\therefore \quad \alpha \in W_1+W_2 \Rightarrow a \in L(W_1 \cup W_2)$$

Thus, $W_1 + W_2 \subseteq L(W_1 \cup W_2)$. ... (1)

But $L(W_1 \cup W_2)$ being the smallest subspace containing $W_1 \cup W_2$, and since $W_1 + W_2$ is a subspace containing $W_1 \cup W_2$, therefore

$$L(W_1 \cup W_2) \subseteq W_1 \cup W_2 \quad \text{... (2)}$$

From (1) and (2), we get

$$W_1 + W_2 = L(W_1 \cup W_2)$$

3.10 LINEAR DEPENDENCE AND INDEPENDENCE OF VECTORS

In this section, we shall discuss the concept of linear dependence or linear independence which lays the foundations for the key notions (viz., dimensions) of the theory of linear spaces.

Definition 1. *Let $V(F)$ be a linear space over a field F. Then a finite set $\{\alpha_1, \alpha_2, ..., a_n\}$ of vectors of V is said to be linearly dependent if there exists scalars $a_1, a_2, ..., a_n$ not all of them equal to zero such that*

$$a_1 \alpha_1 + a_2 \alpha_2, ..., a_n \alpha_n = 0$$

Definition 2. *Let V(F) be a linear space over F. Then a finite set of vectors $\{\alpha_1, \alpha_2, ..., \alpha_n\}$ of V is said to linearly independent if for every expressions of the type.*

$$a_1 \alpha_1 + a_2 \alpha_2 + ... + a_n \alpha_n = 0$$

where $a_1, a_2, ..., a_n \in F$ implies $a_1 = 0, a_2 = ... = a_n$.

REMARKS

- *S* is linearly independent ⇔ S is not linearly dependent
  - ⇔ no finite subset of *S* is linearly dependent.
  - ⇔ each finite subset of S is linearly independent.
  - ⇔ whenever $\sum_{i=1}^{n} a_i \alpha_i = 0, a_i \in F, \alpha_i \in S, i = 1, 2, ..., n$ and $n \in N$, then each $a_i = 0$.
- Any infinite set is linearly independent if it's every finite subset is linearly independent otherwise it is linearly dependent.

THEOREM 1. ***If $\alpha_1, \alpha_2, ..., \alpha_n \in V$ are linearly independent, then every element in their linear span has a unique representation in the form $a_1 \alpha_1 + a_2 \alpha_2, ..., + a_2 \alpha_n$ with $a_i \in F$.***

PROOF. By definition of linear span, we know that every element in the linear span is of the form $a_1 \alpha_1 + a_2 \alpha_2 + ... + a_n \alpha_n$. Therefore, we only show the uniqueness of the representation.

Let if possible,

$$a_1 \alpha_1 + a_2 \alpha_2 + ... + a_n \alpha_n \text{ and } b_1 \alpha_1 + b_2 \alpha_2 + ... + b_n \alpha_n$$

be two forms of an element in linear span of $\alpha_1, \alpha_2, ..., \alpha_n$, then

$$a_1 \alpha_1 + a_2 \alpha_2 + ... + a_n \alpha_n = b_1 \alpha_1 + b_2 \alpha_2 + ... + b_n \alpha_n$$

$$\Rightarrow \quad (a_1 - b_2)\alpha_1 + (a_2 - b_2)\alpha_2 + ... + (a_n - b_n)\alpha_n = 0$$

Since, $\alpha_1, \alpha_2, ..., \alpha_n$ are linearly independent so, we have

$$a_1 - b_1 = 0, \; a_2 - b_2 = 0, ..., a_n - b_n = 0$$

$$\Rightarrow \quad a_1 = b_1, \quad a_2 = b_2, ..., \quad a_n = b_n.$$

Hence, every element in the linear span of linearly independent vectors $\alpha_1, \alpha_2, ...,\alpha_n$ has a unique form of $a_1 \alpha_1 + a_2 \alpha_2 + ... + a_n \alpha_n$.

THEOREM 2. ***If $\alpha_1, \alpha_2, ..., \alpha_n \in V$ (are in V, then either they are linearly independent or some α_k is a linear combination of preceding ones $\alpha_1, \alpha_2, ..., \alpha_{k-1}$.***

PROOF. $\alpha_1, \alpha_2, ..., \alpha_n \in V$ are linearly independent, then nothing is to prove. So assume that $\alpha_1, \alpha_2, ..., \alpha_n$ are not linearly independent. Then, there are some $a_i \in F$ which are non-zero such that

$$a_1 \alpha_1 + a_2 \alpha_2 + ... + a_n \alpha_n = 0$$

Let k be the largest integer for which $a_k \neq 0$. Since $a_i = 0$ for $i > k$, and

$$a_1 \alpha_1 + a_2 \alpha_2 + ... + a_k \alpha_k = 0$$

$$\Rightarrow \quad a_k = a_k^{-1}(-a_1\alpha_1 - a_2\alpha_2 - .... - a_{k-1}\alpha_{k-1}) \qquad [\because a_k \neq 0]$$

$$\Rightarrow \quad a_k = (-a_k^{-1}a_1)\alpha_1 + (-a_k^{-1}a_2)\alpha_2 + ... + (-a_k^{-1}a_{k-1})\alpha_{k-1}$$

$\Rightarrow$ a_k is a linear combination of preceding ones $\alpha_1, \alpha_2, ..., \alpha_{k-1}$.

3.11 BASIS OF A LINEAR SPACE

Definition. *Let V be a linear space over a field F and let S be any non-empty subset of V. Then S is said to be a basis of V if*

(i) S is linearly independent.

(ii) $L(S) = V$, *i.e.*, every element of V is a linear combination of finite elements of S.

For example: The set $S = \{(1, 0, 0), (0, 1, 0), (0, 0, 1)\}$ forms a basis of $V_3(\text{R})$, and is called usual basis.

REMARKS

- The zero space has no basis.
- Every finitely generated linear space has a basis.
- Every non-zero linear space has a basis.
- A linear space many have more than one basis.

3.12 FINITE DIMENSIONAL LINEAR SPACE

Definition. *Let V(F) be a linear space over a field F and let S be any non-empty subset of V, then V(F) is said to be finite dimensional if S is finite subset of V such that L(S) = V. If this set contains n elements, then the dimension of V is n.*

THEOREM 1. ***If $S = \{\alpha_1, \alpha_2, ..., \alpha_n\}$ is the basis of a linear space V(F), then each element of V is uniquely expressible as a linear combination of elements of S.***

PROOF. Since S is the basis of a linear space $V(F)$, then by the definition of basis, each element of V is a linear combination of elements of S. Thus, we only show the uniqueness. Let there be two different sets $\{a_1, a_2, ..., a_n\}$ and $\{b_1, b_2, ..., b_n\}$ of scalars corresponding to an element $\alpha \in V$ such that and

$$\alpha = a_1\alpha_1 + a_2\alpha_2 + ... + a_n\alpha_n$$

and $$\alpha = b_1\alpha_1 + b_2\alpha_2 + ... + b_n\alpha_n \qquad ....(1)$$

$$\Rightarrow a_1 \alpha_1 + a_2 \alpha_2 + ... + a_n\alpha_n = b_1 \alpha_1 + b_2 \alpha_2 + ... + b_n\alpha_n$$

$$\Rightarrow a_1 \alpha_1 - b_1 \alpha_1 + a_2 \alpha_2 - b_2 \alpha_2 + ... + a_n\alpha_n - b_n\alpha_n = 0$$

$$\Rightarrow \quad (a_1 - b_1) \alpha_1 + (a_2 - b_2)\alpha_2 + ... + (a_n - b_n)\alpha_n = 0$$

Since the set $S = \{\alpha_1, \alpha_2, ..., \alpha_n\}$ is linearly independent so that

$$a_1 - b_1 = 0, \; a_2 - b_2 = 0, ..., a_n - b_n = 0$$

$\Rightarrow \quad a_1 = b_1, \quad a_2 = b_2, ..., \quad a_n = b_n.$

Hence, the expression (1) is unique.

THEOREM 2. **(Existence Theorem).** ***Every finitely generated linear space has a finite basis.***

Step Outlines : To make the proof easier, use the following steps :

Step 1. *If S is linearly independent, that theorem is obvious.*

Step 2. *Assume S is linearly dependent and obtained the set* $S_1 = (\alpha_1, \alpha_2, ...\alpha_{k-1}, \alpha_{k+1}, ...\alpha_n)$ *by eliminating* α_k *from S.*

Step 3. *If* S_1 *is linearly independent, result is obvious. And if* S_1 *is linearly dependent, proceed same as in step-2.*

PROOF. Let V be a linear space over F which is generated by a finite set (say) $S = \{\alpha_1, \alpha_2, .., \alpha_n\}$ of vectors of V. Without loss of any generality we may assume that all the elements in S are non-zero, because zero vector in the linear combination of elements of S is zero.

If S is linearly independent, then S forms a finite basis for V and in this case theorem is proved. So, we assume that S is linearly dependent, then there exists some α_k $(2 < k \le n)$ in S such that α_k is a linear combination of preceding vectors $\alpha_1, \alpha_2, ... \alpha_{k-1}$. Therefore,

$$\alpha_k = a_1\alpha_1 a_2\alpha_2 + ... + a_{k-1}\alpha_{k-1} = \sum_{i=1}^{k-1} a_i\alpha_i \qquad ... (1)$$

for some scalars $a_i's \in F$.

But S generates V so that an arbitrary element $\alpha \in V$ is expressible as a linear combination of elements of S.

$$\therefore \quad \alpha = b_1\alpha_1 + b_2\alpha_2 + ... + b_k\alpha_k + ... + b_n\alpha_n$$

$$= \sum_{i \ne k} b_i\alpha_i + b_k\alpha_k, \text{for some } b_i's \in F$$

$$= \sum_{i \ne k} b_i\alpha_i + b_k \sum_{i=1}^{k-1} a_i\alpha_i$$

$$= (b_1 + b_k a_1)\alpha_1 + (b_2 + b_k a_2)\alpha_2 + ...$$

$$+ (b_{k-1} + b_k a_{k-1})\alpha_{k-1} + (b_{k+1})\,\alpha_{k+1} + ... + b_n\alpha_n$$

$\Rightarrow$ α is a linear combination of $a_1\alpha_1 + a_2\alpha_2 + ... + a_{k-1}\alpha_{k-1}$

Thus the set,

$$S_1 = \{\alpha_1, \alpha_2, ..., \alpha_{k+1} ..., \alpha_k\}.$$

Obtained by eliminating some α_k from S also generates V.

If S_1 in linearly independent, then S_1 will form a basis of V and the theorem is proved in this case. If S_1 is linearly dependent, then by above process, we obtain a new set

$$S_2 = \{\alpha_1, \alpha_2, ..., \alpha_{k-1}, \alpha_{k+1}, ..., \alpha_{1-1}, \alpha_{1+1}, ..., \alpha_n\}$$

by eliminating some α_1 $(1 > k)$ from S_1, which generates V. If S_2 is linearly independent, then S_2 will form a basis. If S_2 is linearly dependent, then we continue the above process, till after a finite number of steps we obtain a linearly independent set which generates V. At the most by repeating the above process we may obtain a singleton set which is always linearly independent and it generating V and will form a basis of V. Hence, in every finitely generated linear space there exists a finite basis.

THEOREM 3. ***If V is a finite-dimensional linear space and if $a_1, a_2, ..., a_m$ span V, then some subset of $\alpha_1, \alpha_2, ..., \alpha_m$ forms a basis of V.***

PROOF. Since a finite-dimensional linear space has a basis of V containing a finite number of elements. Let these vectors be $\alpha_1, \alpha_2, ..., \alpha_n$. Thus every element in V has a unique representation of the form

$$a_1\alpha_1 + a_2\alpha_2 + a_n\alpha_n; \text{ for } a_1, a_2, ..., a_n \in F.$$

If $a \in V$, then

$$\alpha = a_1\alpha_1 + a_2\alpha_2 + ... + a_n\alpha_n$$

Now define a map ϕ from V into $F^{(n)}$ by

$$\phi\,(a_1\alpha_1 + a_2\alpha_2 + ... + a_n\alpha_n) = (a_1, a_2, ..., a_n)$$

Since $a_1\alpha_1 + a_2\alpha_2 + ... + a_n\alpha_n$ is a unique representation so that ϕ is well defined, one-to-one and onto and also preserves the composition.

Thus V is isomorphic to $F^{(n)}$ for some n, where n is the number of elements in some basis of V over F. If some other basis of V has m elements, then V would be isomorphic to $F^{(m)}$'. Since both $F^{(n)}$ and $F^{(m)}$ are isomorphic to V, therefore, $F^{(n)}$ and $F^{(m)}$ are isomorphic to each other. This implies $n = m$. Hence the theorem.

THEOREM 4. ***If $\{\alpha_1, \alpha_2, ..., \alpha_n\}$ is a basis of V(F) and if $\beta_1, \beta_2, ..., \beta_m \in V$, are linearly independent over F, then $m \leq n$.***

PROOF. Since the set $\{\alpha_1, \alpha_2, ..., \alpha_n\}$ is a basis of $V(F)$, then every element in V is a linear combination of the elements of basis, so in particular, β_m is a linear combination of $\alpha_1, \alpha_2, ..., \alpha_n$. Therefore, the set $\{\beta_m, \alpha_1, \alpha_2, ..., \alpha_n\}$ is linearly dependent. Also, this set spans V because $\alpha_1, \alpha_2, ..., \alpha_n$ span V. Thus proper subset of the set $\{\beta_m, \alpha_1, \alpha_2, ..., a_n\}$ forms a basis of V. Let this proper set be $\{\beta_m, \alpha_1, \alpha_2, ..., \alpha_n\}$ with $k \leq n-1$. In forming this new basis at least one α_i is replaced by some one β_j. Repeat this procedure with the set $\{\beta_{m-1}, \alpha_{i_1}, \alpha_{i_1}, ..., \alpha_{i_k}\}$ which is obviously linearly dependent, so we can extract a basis of the form $\{\beta_{m-1}, \beta_m, \alpha_{j_1}, \alpha_{j_2}, ..., \alpha_{j_s}\}$ with $s \leq n-2$. Continuing this procedure with the set $\{\beta_2, \beta_3, ..., \beta_{m-1}, \beta_m, \alpha_x, \alpha_y\}$. Since β_1 is not a linear combination of $\beta_2, \beta_3, ..., \beta_{m-1}, \beta_m, \alpha_x, \alpha_y\}$, therefore the basis $\{\beta_2, \beta_3, ..., \beta_{m-1}, \beta_m, \alpha_x, \alpha_y\}$ must contain some α's. To get this basis. We have introduced $m-1$ β's and each such introduction costs at least one α's and yet there is an 'α' left. Thus $m - 1 \leq n - 1$ implying $m \leq n$.

THEOREM 5. ***If V is a finite-dimensional linear space over F, then any two bases of V have the same number of elements.***

PROOF. Let S_1 and S_2 be any two bases of $V(F)$ and let

$$S_1 = \{\alpha_1, \alpha_2, ..., \alpha_m\} \text{ and } S_2 = \{\beta_1, \beta_2, ..., \beta_n\}.$$

Then we shall have to show that $m = n$.

Since S_1 forms a basis of V, so that every element of V is uniquely expressible as a linear combination of the elements of S_1. In particular β_1 is uniquely expressible as a linear combination of S_1. Thus the set $S_3 = \{\beta_1, \alpha_1, \alpha_2, ..., \alpha_{k-1}\}$ is now linearly dependent.

Therefore, there exists an element α_k in S_s which is linear combination of proceeding ones, $\beta_1, \alpha_1, \alpha_2, ..., \alpha_{k-1}$. But every element of V can be expressed as a linear combination of $\alpha_1, \alpha_2, ..., \alpha_k, ..., \alpha_m$. Also, α_k is a linear combination of $\beta_1, \alpha_1, \alpha_2, ..., \alpha_{k-1}$. This implies that each element of V is expressible as a linear combination of

$\{\beta_1, \alpha_1, \alpha_2, \ldots, \alpha_{k-1} \ldots \alpha_m\}$. Thus the set

$$S_4 = \{\beta_1, \alpha_1, \alpha_2, \ldots, \alpha_{k-1}, \alpha_{k+1}, \ldots, \alpha_m\}$$

generates V. This set is obtained by adjoining β_1 to S_3 and eliminating α_k and S_3. Since $\beta_2 \in V$ so it is the linear combination of elements of S_4, therefore, the set is obtained by adjoining β_2 to S_4 and eliminating α_1 as before. Thus, the set $S_5 = \{\beta_2, \beta_1, \alpha_1, \alpha_2, \ldots, \alpha_{k-1}, \alpha_{k+1}, \ldots, \alpha_{l-1}, \alpha_{l+1}, \ldots, \alpha_m\}$ generates V.

Continuing the above manner, we observe that each step consists of an inclusion of one β' s and the exclusion of an α's and the resulting set generates V. Since all the α's can not be exhausted before the β's. If it is so then a proper subset of S_2 generates V which is a contradiction of linear independence of S_2. Hence $m < n$. Similarly, if we change the role of S_1 and S_2, we obtain $n < m$. Hence $m = n$.

THEOREM 6 **(Extension Theorem).** ***If V(F) is a finite dimensional linear space, then every linearly independent subset of V is either a basis of V or can be extended to form a basis of V.***

PROOF. Suppose the dimension of $V = n$. Let $S = \{a_1, a_2, \ldots, a_n\}$ be a basis of V and let $S_1 = \{\beta_1, \beta_2, \ldots, \beta_m\}$ be a linearly independent subset of V.

Since S is a basis of V so that every element of V is expressible as a linear combination of elements of S, in particular, β_m is a linear combination of elements of S, therefore, the set obtained by adjoining β_m to S is linearly dependent. Since, the superset of a linearly dependent set of linearly dependent, it follows that the set,

$$S_2 = \{\beta_1, \beta_2, \ldots, \alpha_1, \alpha_2, \ldots, \alpha_n\}$$

is linearly dependent.

Since the set S_1 is linearly independent, then there exists an element α_k in S_2 which can be expressed as a linear combination of preceding ones $\beta_1, \beta_2, \ldots, \beta_m, \alpha_1, \alpha_2, \ldots, \alpha_{k-1}$, therefore each element in V can be expressed as a linear combination of $\beta_1, \beta_2, \ldots, \beta_m, \alpha_1, \alpha_2, \ldots, \alpha_{k-1}, \alpha_{k+1}, \ldots, \alpha_n$. Thus the set

$$S_3 = \beta_1, \beta_2, \ldots, \beta_m, \alpha_1, \alpha_2, \ldots, \alpha_{k-1}, \alpha_{k+1}, \ldots, \alpha_n.$$

is obtained be eliminating α_k from S_2 which generates V. If S_3 is linearly independent, then S_3 forms a basis of V containing S_1. Thus, in this case the theorem is proved.

If S_3 is linearly dependent, then repeat the above process of adjoining and eliminating, till after a finite number of steps we obtain a linearly independent set which generates V and contains S_1. At the most by repeating the above process we may obtain the set S_1 itself which generates V and being linearly independent will form a basis of V.

Hence, either S_1 is a basis of V or can be extended to form a basis of V.

THEOREM 7. ***Let V be a finite-dimensional linear space and let dim. V= n. Then***

(i) any subset of V which contains more than n vectors is linearly dependent.

(ii) no subset of V which contains less than n vectors can span V.

PROOF. (i) Since $V(F)$ is n-dimensional, every basis of V will contain n vectors. Let S be any subset of V which contains more than n vectors. Let, if possible, S be linearly independent, then by previous theorem either S is a basis of V or can be extended to form a basis of V. But in both cases the basis of V contains more than n elements which is contradictory to the fact that V is n-dimensional. Hence S is linearly dependent.

(ii) Let S be a subset of V which contains less than n elements and which can span V. Then every element of V can be expressed as a linear combination of elements of S. If S is linearly independent, then S forms a basis of V which show that dim. $V < n$ which is contradictory to the fact that dim. $V = n$. On the other hand it S is linearly dependent, then S cannot span V which is again a contradiction because we have assumed that S spans V. Hence in both cases such subset S cannot exist. Consequently no subset of V containing less than n vectors can span V.

3.13 DIMENSION OF A SUBSPACE OF LINEAR SPACE

THEOREM 1. ***Let S be a linearly independent subset of a linear space V. Suppose β is a vector in V which is not in the subspace spanned by S. Then the set obtained by adjoining β to S is linearly independent.***

PROOF. Let $S = \{\alpha_1, \alpha_2, \ldots \alpha_n\}$ be a linearly independent subset of V. Then we shall show that the set

$$S_1 = \{\beta, \alpha_1, \alpha_2, \ldots \alpha_n\}$$

obtained by adjoining β to S is also linearly independent where $\beta \in V$. but not in the subspace of V which is spanned by S.

Since $\alpha_1, \alpha_2, \ldots, \alpha_n$ are distinct vectors in S such that

$$a_1 \alpha_1 + a_2 \alpha_2 + \ldots + a_n \alpha_n + b\beta = 0 \qquad \ldots(1)$$

where, all a' s. are zero.

We actually show that $b = 0$. Let, if possible, $b \neq 0$. Then from (1), we have

$$\beta = \left(-\frac{a_1}{b}\right)\alpha_1 + \left(-\frac{a_2}{b}\right)\alpha_2 + \ldots + \left(-\frac{a_n}{b}\right)\alpha_n$$

$\Rightarrow$ β is a linear combination of $\alpha_1, \alpha_2, \ldots, \alpha_n$.

$\Rightarrow$ β is in the subspace of V spanned by $\alpha_1, \alpha_2, \ldots, \alpha_n$.

But it is contradictory to the hypothesis that β is not in the subspace spanned by S. Hence $b = 0$. Consequently, the set S_1 is linearly independent.

THEOREM 2. ***If W is a subspace of a finite-dimensional linear space V, every linearly independent subset of W is finite and is a part of a (finite) basis for W.***

PROOF. Let S be a linear independent subset of W and let S_1 be a linearly independent subset of W containing S. Then S_1 contains not more than dim. V elements.

If S spans W, then S is a basis for W and in this case, theorem is proved. If S does not span W, then we find a vector β_1 in W such that the set

$$S_2 = S \cup \{\beta_1\}$$

is linearly independent (By theorem 1). If S_2 spans W, it will form a basis for W, If S_2 does not span W, then again we find a vector β_2 in W such that the set

$$S_3 = S \cup \{\beta_1, \beta_2\}$$

is linearly independent.

Continuing above process to finite number of steps less than dim V, we get a set ,

$$S_m = S \cup \{\beta_1, \beta_2, \ldots, \beta_{m-1}\}$$

which is linearly independent and is a basis for W or a part of basis for W.

THEOREM 3. ***If V(F) be a finite-dimensional linear space and W be a subspace of V, then W is finite dimensional and dim. W≤ dim V. In particulars, if W is a proper subspace of V, then dim W< dim V. Also V = W if and only if dim V = dim W.***

PROOF. Let dim $V = n$. Then every basis of V will contain n vectors of V, therefore, every subset having vectors more than n will be linearly dependent. Thus, a linearly independent set of vectors in W contains at most n-elements. Let $S = \{\alpha_1, \alpha_2, ..., \alpha_m\}$, with $m \geq n$ being a maximal linearly independent set in W. If α is an arbitrary element in W, then the set

$$S_1 = \{\alpha_1, \alpha_2, ,...,\alpha_m\}$$

will be linearly dependent because S being a maximal linearly independent set. Therefore, α is a linear combination of $\alpha_1, \alpha_2, ,...,\alpha_m$ which shows that S spans V. Hence W is finite-dimensional. Also, dim. $W = m \leq n =$ dim V.

$\therefore$ $$\dim W < \dim V.$$

Next, if $V = W$, then every basis of V is also the basis of W which shows that dim. $V =$ dim W. On the other hand, if dim. $V =$ dim W, then every basis of W will contain the vectors equal to dim V so it will also generate V. Thus each one of V and W is generated by some basis. Hence $V = W$.

THEOREM 4. **(Existence of Complementary Subspace).** ***Every subspace of a finite dimensional linear space has a complement.***

PROOF. Let $V(F)$ be a finite dimensional linear space and let W_1 be its subspace. Then our aim is to find out a subspace W_2 of V such that $V = W_1 \oplus W_2$. Since $V(F)$ is a finite dimensional linear space so that W_1 will be finite dimensional. Let $S_1 = [\alpha_1, \alpha_2 ,..., \alpha_n\}$ be the basis of W_1, then by extension theorem, we have that, S_1 can be extended to form a basis of V. Let this extended set be

$$S_2 = \{\alpha_1, \alpha_2, ,...,\alpha_n, \beta_1, \beta_2, ,...,\beta_m\} \text{ be the basis of } V.$$

Let us suppose that the set $\{\beta_1, \beta_2, ,...,\beta_m\}$ generates a subspace and let this subspace be W_2.

Now we shall show that $V = W_1 \oplus W_2$ or equivalently $V = W_1 + W_2$ and $W_1 \cap W_2 = \{0\}$.

Let γ be an arbitrary element of V and S_2 being the basis for V, then there exist scalars $a_1, a_2,...,a_n, b_1, b_2,...,b_m$ such that

$\gamma = a_1\alpha_1 + a_2\alpha_2 + ... + a_n\alpha_n + b_1\beta_1 + b_2\beta_2 + ... + b_m\beta_m = \alpha + \beta$

Where, $\alpha = \sum_{i=1}^{n} a_i\alpha_i$ and $\beta = \sum_{j=1}^{m} b_j \beta_j$

Since $\{\alpha_1, \alpha_2, ,...,\alpha_n\}$ generates W_1 and $\{\beta_1, \beta_2, ,...,\beta_m\}$ generates W_2, therefore $a \in W_1$ and $\beta \in W_2$

Thus every element of V is expressible as the sum of an element of W_1 and an element of W_2 .

$\therefore$ $V = W_1 + W_2$

Again, $\alpha = \sum_{i=1}^{n} a_i\alpha_i \in W_1$ and $\beta = \sum_{j=1}^{m} b_j \beta_j \in W_2$

Let if possible $W_1 \cap W_2 \neq \{0\}$. Then three exists a non-zero element which belong to both W_1 and W_2 . Let it be x. Then $x \in W_1$ and $x \in W_2$.

$$\therefore \quad x = \sum_{i=1}^{n} a_i'.\alpha_i \text{ and } x = \sum_{j=1}^{m} b_j'.\beta_j$$

$$\Rightarrow \quad \sum_{i=1}^{n} a_i'.\alpha_i = \sum_{j=1}^{m} b_j'.\beta_j$$

$$\Rightarrow \quad \sum_{i=1}^{n} a_i'.\alpha_i + \sum_{j=1}^{m} (-b_j').\beta_j = 0$$

Since the set S_2 is linearly independent, so that

$$a_i' = 0, \text{ for each } i = 1, 2 ,..., n$$

and $$b_j' = 0, \text{ for each } i = 1, 2 ,..., m$$

$$\Rightarrow \quad x = 0$$

which is a contradiction, because we have taken $x \neq 0$.

Thus the contradiction arises by assuming that $W_1 \cap W_2 \neq \{0\}$

Hence, $W_1 \cap W_2 = \{0\}$

Consequently

$$V = W_1 \oplus W_2.$$

REMARK

- Here W_2 is the subspace complementary to the subspace W_2 of finite dimensional linear space V.

THEOREM 5. ***If W_1 and W_2 are two finite-dimensional subspaces of a linear space V(F), then W_1+W_2 is finite dimensional and***

$$\textit{dim.}W_1 + \textit{dim.}W_2 = \textit{dim}(W_1 \cap W_2) + \textit{dim.}(W_1 + W_2)$$

PROOF. Since W_1 and W_2 are subspaces of V so that $W_1 \cap W_2$ will be a subspace of V and its dimension is finite. Let dim. $W_1 = m$, dim. $W_2 = n$ and dim.$(W_1 \cap W_2) = r$. Let $\{\alpha_1, \alpha_2, ..., \alpha_r\}$ be a basis of $W_1 \cap W_2$. Therefore, we can extend this basis to a basis of W_1 and also to a basis of W_2.

Let, $S_1 = \{\alpha_1, \alpha_2, \ldots \alpha_r, \beta_1, \beta_2, \ldots \beta_{m-r}\}$

and $S_2 = \{\alpha_1, \alpha_2, \ldots \alpha_r, \gamma_1, \gamma_2, \ldots \gamma_{n-r}\}$

be the basis of W_1 and W_2 respectively. Consider the set

$S = \{\alpha_1, \alpha_2, \ldots \alpha_r, \beta_1, \beta_2, \ldots \beta_{m-r}, \gamma_1, \gamma_2, \ldots \gamma_{n-r}\}$

Now we have to show that S will form a basis for W_1+W_2. For this, we shall show that S is linearly independent and spans $W_1 + W_2$. For this, suppose

$$\sum a_i\alpha_i + \sum b_j\beta_j + \sum c_k\gamma_k = 0 \text{ for } a_i' \text{ s}, b_j' \text{ s}, c_k' \text{ s} \in F.$$

Then $\sum c_k\gamma_k = \sum a_i\alpha_i + \sum b_j\beta_j \Rightarrow \sum c_k\gamma_k \in W_1$

Also, $\sum c_k\gamma_k \in W_2$. It follows that $\sum c_k\gamma_k \in W_1 \cap W_2$ and we have $\sum c_k\gamma_k \in d_i\alpha_i$ for some scalars $d_1, d_2, ..., d_r$. Since the set $\{\alpha_1, \alpha_2, \ldots \alpha_r, \gamma_1, \gamma_2, \ldots \gamma_{n-r}\}$ is linearly independent hence all the scalars $c_1 = 0 = c_2 = \ldots = c_{n-r}$.

Thus $\sum a_i\alpha_i + \sum b_j\beta_j = 0$

and since the set $\{\alpha_1, \alpha_2, \ldots \alpha_r, \beta_1, \beta_2, \ldots \beta_{m-r}\}$ is also linearly independent, then

$$a_1 = 0 = a_2 = \dots = a_r$$

and $$b_1 = 0 = b_2 = \dots = b_{m-r}$$

Thus the set $S = \{\alpha_1, \alpha_2, \dots \alpha_r, \beta_1, \beta_2, \dots \beta_{m-r}, \gamma_1, \gamma_2, \dots \gamma_{n-r}\}$ is linearly independent. Now we shall show that S spans $W_1 + W_2$.

Let α be an arbitrary element of $W_1 + W_2$ then it can be written as $\alpha = \beta + \gamma$ with $\beta \in W_1$ and $\gamma \in W_2$. Now S_1 and S_2 being the basis of W_1 and W_2 respectively, β and γ can be expressed uniquely in the form

$$\beta = \sum_{i=1}^{r} a_i\alpha_i + \sum_{j=1}^{n-r} b_j\beta_j \text{ , for some } a_i\text{'s and } b_j\text{'s.}$$

and $$\gamma = \sum_{i=1}^{r} e_i\alpha_i + \sum_{j=1}^{n-r} c_j\gamma_j \text{ , for some } e_i\text{'s and } c_j\text{'s.}$$

$$\therefore \quad \alpha = \beta + \gamma = \sum_{i=1}^{r}(a_i + e_i)\alpha_i + \sum_{j=1}^{m-r} b_j\beta_j + \sum_{j=1}^{m-r} c_j\gamma_j .$$

$\Rightarrow$ α is a linear combination of elements of S.

$\Rightarrow$ S spans $W_1 + W_2$.

Hence, S is basis of $W_1 + W_2$ so that $W_1 + W_2$ is finite-dimensional, with dimensional $(m + n - r)$.

Finally,

$$\dim W_1 + \dim W_2 = m + n = r + (m + n - r)$$
$$= \dim .(W_1 \cap W_2) + \dim .(W_1 + W_2).$$

THEOREM 6. ***If a finite dimensional linear space V(F) be the direct sum of its two subspaces W_1 and W_2, then $\dim.V = \dim.W_1 + \dim.W_2$.***

PROOF. Since V is finite-dimensional, therefore W_1 and W_2 are also finite-dimensional. Let

$$\dim.W_1 = m, \ \dim.W_2 = n$$

Also $V = W_1 \oplus W_2$, implying that

(i) $V = W_1 + W_2$

(ii) $W_1 \cap W_2 = \{0\}$.

Let $S_1 = \{a_1, a_2, \dots, a_m\}$ be a basis of W_1 and the set $S_2 = \{b_1, b_2, \dots, b_n\}$ is a basis of W_2.

Now consider a set

$$S_3 = \{a_1, a_2, \dots a_m, b_1, b_2, \dots, b_n\}$$

We claim that S_3 forms a basis of V.

For some scalars $a_1, a_2, \dots, a_m, b_1, b_2, \dots, b_n \in F$, we have

$$a_1\alpha_1 + a_2\alpha_2 + \dots + a_m\alpha_m + b_1\beta_1 + b_2\beta_2 + \dots + b_n\beta_n = 0$$

$\Rightarrow$ $a_1\alpha_1 + a_2\alpha_2 + \dots + a_m\alpha_m = -(b_1\beta_1 + b_2\beta_2 + \dots + b_n\beta_n) = 0$

$\Rightarrow$ $a_1\alpha_1 + a_2\alpha_2 + \dots + a_m\alpha_m \in W_2$ as $b_1\beta_1 + b_2\beta_2 + \dots + b_n\beta_n \in W_2$

and $b_1\beta_1 + b_2\beta_2 + \dots + b_n\beta_n \in W_1$ as $a_1\alpha_1 + a_2\alpha_2 + \dots + a_m\alpha_m \in W_1$

$\Rightarrow$ $a_1\alpha_1 + a_2\alpha_2 + \dots + a_m\alpha_m \in W_1 \cap W_2$

$\Rightarrow \quad b_1\beta_1+b_2\beta_2+...+b_n\beta_n \in W_1 \cap W_2$

But from(i) $W_1 \cap W_2=\{0\}$.

$\Rightarrow \quad a_1\alpha_1+a_2\alpha_2+...+a_m\alpha_m=0$

and $\quad b_1\beta_1+b_2\beta_2+...+b_n\beta_n=0$

Since S_1 and S_2 both are linearly independent, therefore

$$a_1=0=a_2= ....=a_m, \; b_1=0=b_2= ....=b_n.$$

$\Rightarrow$ S_1 is linearly independent.

Next, let γ be an arbitrary element of V, then

$$\gamma=\alpha+\beta,\ \alpha \in W_1,\ \beta \in W_2 \qquad [\because V=W_1+W_2]$$

Since $\quad \alpha \in W_1 \Rightarrow \alpha \in a_1\alpha_1+a_2\alpha_2+...+a_m\alpha_m$ for some a_is $\in F$.

and $\quad \beta \in W_2 \Rightarrow \beta \in b_1\beta_1+b_2\beta_2+...+b_m\beta_m$ for some b_is $\in F$.

$\therefore \quad \gamma= \alpha+\beta = a_1\alpha_1+a_2\alpha_2+...+a_m\alpha_m+b_1\beta_1+b_2\beta_2+.....+b_n\beta_n$

$\Rightarrow$ S_3 generates V, thus S_3 forms a basis of V.

Accordingly, $dim.V = m+n = dim.W_1+ dim.W_2$.

REMARK

- It should be noted that in case of finite dimensional spaces, since a basis is a maximal linearly independent subset of the linear space, so the dimension of a finite dimensional linear spaces may be regarded as the maximum of numbers of elements in all linearly independent subsets. If this is adopted as definition of the dimension of a linear space, is n iff it has a basis consisting of n elements.

3.14 COSETS

Let V be a linear space and W be its any subspace,then the sets,

$$\alpha+W =\{\alpha+\beta: \forall\beta \in W\}$$

and $\quad W+\alpha=\{\beta+\alpha: \forall\beta \in W\}$,

where a is any arbitrary element of V, These are called left and right cosets respectively.

Since addition is commutative in V, so

$$\alpha+W=W+\alpha \ \forall\alpha \in V$$

Therefore, there is no matter we take right coset or left coset.

REMARKS

- If $W+\alpha=W+\beta$,then $\alpha-\beta \in W$ and so conversely.
- Any two cosets are either disjoint or identical.
- Since W is an additive subgroup of V, so we take additive cosets.

3.15 ADDITION AND MULTIPLICATION OF TWO COSETS

Let V be a linear space and W be its any subspace and let α,β be any two arbitrary elements of V, and let $\alpha+W$ and $\beta+W$ be two cosets of W in V, then

$$(\alpha+W) + (\beta+W)=(\alpha+\beta)+W.$$

and $$(\alpha+W)(\beta+W)=(\alpha\beta)+W.$$

REMARKS

- Zero elements of the set $\alpha+W=\{\alpha+\beta : \beta \in W\}$ is W.
- Unit elements of this set $1+W$.

3.16 QUOTIENT SPACES

Let V be a linear space over a field F and let W be any subspace of V. Let $\alpha \in V$.

Then, $W+\alpha = \{w+\alpha : w \in W\}$ is called the right coset of W in V.

Similarly $\alpha+W = \{\alpha+w : w \in W\}$ is called the left coset of W in V.

Clearly, the sets of $W+\alpha$ and $\alpha+W$ are both subsets of the linear space $V(F)$. Now since $(V,+)$ is an abelian group, therefore $W+\alpha$ and $\alpha+W$ are equal.

Thus, we can say that $W+\alpha$ is a coset of W in V generated by α. Let V/W represents the set of all cosets of W in V, *i.e.*,

$$V/W=\{W+\alpha : \alpha \in V\}$$

Further, we define vector addition and scalar multiplication of V/W as follows

$$(W+\alpha)+(W+\beta)= W+(\alpha+\beta)$$

and $$a(W+\alpha)= W+a\alpha \ \forall\ \alpha,\beta \in V,\ a \in F$$

Then, V/W is a linear space over F for these compositions. This linear space is known as *quotient space* or *factor space.*

THEOREM 1. ***If V is a linear space over a field F and if W is a subspace of V, then the set***

$$V/W=\{\alpha+W : \forall\ \alpha \in V\}$$

is a linear space over the field F with respect to the linear compositions.

(i) $(\alpha+W)+(\beta+W)=(\alpha+\beta)+W,\ \forall\ \alpha,\beta \in V$

(ii) $a(\alpha+W)=a\alpha+W,\ \forall\ \alpha \in V$ ***and*** $a \in F$

PROOF. First of all, we shall show that the above compositions are well defined. For this, let $\alpha+W=\alpha'+W$ and $\beta+W=\beta'+W$, then we have

$$\alpha+W=\alpha'+W \text{ and } \beta+W=\beta'+W$$

$$\Rightarrow \quad \alpha-\alpha' \in W \text{ and } \beta-\beta' \in W$$

$$\Rightarrow \quad (\alpha-\alpha')+(\beta-\beta') \in W$$

$$\Rightarrow \quad (\alpha+\beta)-(\alpha'+\beta') \in W$$

$$\Rightarrow \quad (\alpha+\beta)+W=(\alpha'+\beta')+W$$

$$\Rightarrow \quad (\alpha+W)+(\beta+W)=(\alpha'+W)+(\beta'+W)$$

$\Rightarrow$ The composition given in (i) is well defined.

Also, $$\alpha+W=\alpha'+W$$

$$\Rightarrow \quad \alpha-\alpha' \in W$$

$$\Rightarrow \quad a(\alpha-\alpha') \in W \text{ for some } a \in F.$$

$$\Rightarrow \quad (a\alpha-a\alpha') \in W$$

$$\Rightarrow \quad (a\alpha+W)=(a\alpha'+W).$$

$\Rightarrow$ The composition given in (ii) is well defined.

(i) Next, $\alpha+W$, $\beta+W$ and $\gamma+W$ be any three elements of V/W, then

$$\begin{aligned}[(\alpha+W)+(\beta+W)]+(\gamma+W) &= [(\alpha+\beta)+W]+(\gamma+W).\\ &= [(\alpha+\beta)+\gamma]+(W).\\ &= [\alpha+(\beta+\gamma)]+(W).\\ &= (\alpha+W)+[(\beta+\gamma)+(W)].\\ &= (\alpha+W)+[(\beta+W)+(\gamma+W)].\end{aligned}$$

$\therefore$ The first composition given in (i) is associative in V/W.

(ii) Let $\alpha+W$ be an element of V/W, then

$$(\alpha+W)+(0+W) = (\alpha+0)+W = \alpha + W.$$

and $$(0+W)+(\alpha+W) = (0+\alpha)+W = \alpha + W.$$

Thus, $0+W$ is an additive identity in V/W.

(iii) Since, if $\alpha \in V$ and $-\alpha \in V$, so $\alpha + W \in V$ and $(-\alpha) + W \in V/W$. Then

$$(\alpha+W)+[(-\alpha)+W] = [\alpha+(-\alpha)]+W = 0 + W=W.$$

and $$(0+W)+(\alpha+W) = (0+\alpha)+W = \alpha + W.$$

Thus, $(-\alpha)+W$ is an additive inverse of $\alpha + W$ in V/W.

(iv) Let $\alpha+W$, $\beta+W$ be any two elements of V/W, then

$$(\alpha+W)+(\beta+W) = (\alpha+\beta)+W = (\beta+\alpha)+W$$

[$\because$ addition is commutative in V.]

$$= (\beta+W)+(\alpha+W).$$

Thus addition is commutative in V/W.

(v) Let $\alpha+W$, $\beta+W$ be any elements of V/W and $a \in F$, then

$$\begin{aligned} a.[(\alpha+W)+(\beta+W)] &= a.[(\alpha+\beta)+W] \\ &= a.(\alpha+\beta)+W \\ &= (a.\alpha+a.\beta)+W \\ &= (a\alpha + W) + (a\beta + W) \\ &= a(\alpha + W) + a(\beta + W). \end{aligned}$$

(vi) Let $\alpha + W$ be an arbitrary element of V/W and $a, b \in F$, then

$$\begin{aligned} (a + b)(\alpha + W) &= (a + b)\alpha + W = (a\alpha + b\alpha) + W \\ &= (a\alpha + W) + (b\alpha + W) \\ &= a(\alpha + W) + b(\alpha + W) \end{aligned}$$

(vii) Let $\alpha + W$ be an arbitrary element of V/W and $a, b \in F$, then

$$\begin{aligned} (ab)(\alpha + W) &= (ab)\alpha + W \\ &= a(b\alpha) + W \quad \text{[By elementary property of (v)]} \\ &= a\,(b\alpha + W) = a(b(\alpha+W)). \end{aligned}$$

(viii) Let $\alpha + W$ be an arbitrary element of V/W and $1 \in F$, then

$$1\,.\,(\alpha + W) = (1.\,\alpha) + W = \alpha + W. \qquad [\because 1.\,\alpha = 1]$$

Hence V/W is a linear space and this linear space is known as Quotient space.

THEOREM 2. ***If V is finite-dimensional and if W is a subspace of V, then W is finite-dimensional and***

$$\textbf{dim. (V/W) = dim.V - dim. W.}$$

PROOF. Let dim. $V = n$, then any $n+1$ elements in V are linearly dependent, in particular any $n+1$ elements in W are also linearly dependent. Thus we can find a maximal linearly independent subset of W.

Let, $S = \{\alpha_1, \alpha_2,..., \alpha_m\}$ with $m \leq n$.

If $a \in W$, then the set $\{\alpha_1, \alpha_2,..., \alpha_m\}$ is linearly dependent such that

$$a\alpha + a_1\alpha_1 + a_2\alpha_2 + ... + a_m\alpha_m = 0$$

where, not all of the $a_i's$ are zero, If $a = 0$, then we get each $a_i = 0$ as S being linearly independent, which contradicts that $a = 0$. Thus $a \neq 0$ so that

$$\alpha = -a^{-1}(a_1\alpha_1 + a_2\alpha_2 + \ldots + a_m\alpha_m)$$

$\Rightarrow$ $\alpha_1, \alpha_2, \ldots, \alpha_m$ span W

$\Rightarrow$ W is finite dimensional with dim . $W = m$.

Since S forms a basis of W so it can be extended to form a basis of V. Let the extended set $S_1 = (\alpha_1, \alpha_2, \ldots, \alpha_m, \beta_1, \beta_2, \ldots, \beta_{n-m}\}$ be the basis of V, as n being the dim. V. Consider a set $S_2 = \{W + \beta_1, W + \beta_2, \ldots, W + \beta_{n-m}\}$ of cosets in V/W. Now, we have to prove that S_2 forms a basis of V/W.

For some scalars $b_1, b_2, \ldots, b_{n-m}$, we have

$$b_1(W + \beta_1) + b_2(W + \beta_2) + \ldots + b_{n-m}(W + \beta_{n-m}\} = W + 0$$

where, $W + 0$ being the zero element of V/W.

$$\Rightarrow \quad (W + b_1\beta_1) + (W + b_2\beta_2) + \ldots + (W + b_{n-m}\beta_{n-m}\} = W + 0$$

$$\Rightarrow \quad W + (b_1\beta_1 + b_2\beta_2 + \ldots + b_{n-m}\beta_{n-m}) = W + 0$$

$$\Rightarrow \quad b_1\beta_1 + b_2\beta_2 + \ldots + b_{n-m}\beta_{n-m} \in W$$

Since S generates W so that for some scalars $c_1, c_2, \ldots, c_m$ we have

$$b_1\beta_1 + b_2\beta_2 + \ldots + b_{n-m}\beta_{n-m} = c_1\alpha_1 + c_2\alpha_2 + \ldots + c_m\alpha_m$$

$$\Rightarrow \quad c_1\alpha_1 + c_2\alpha_2 + \ldots + c_m\alpha_m + (-b_1)\beta_1 + (-b_2)\beta_2 + \ldots + (-b_{n-m})\beta_{n-m} = 0$$

$$\Rightarrow \quad c_1 = 0 = c_2 = \ldots = c_m,\ b_1 = 0 = b_2 = \ldots = b_{n-m},$$

[$\because$ S_1 is linearly independent.]

$\Rightarrow$ each $b_i = 0$ for $i = 1, 2, \ldots, n - m$

$\Rightarrow$ S_2 is linearly independent.

Also, if $W + \beta$ is an arbitrary element of V/W, then $\beta \in V$ and S_1 being the basis of V, therefore, β can be expressible as a linear combination of elements of S_1. That is, for some scalars $d_1, d_2, \ldots, d_m, e_1, e_2, \ldots, e_{n-m}$

$$\beta = d_1\alpha_1 + d_2\alpha_2 + \ldots + d_m\alpha_m + e_1\beta_1 + e_2\beta_2 + \ldots + e_{n-m}\beta_{n-m}$$

$$= \sum_{i=1}^{m} d_i\alpha_i + \sum_{j=1}^{n-m} e_j\beta_j$$

$$\therefore \quad W + \beta = W + \left(\sum_{i=1}^{m} d_i\alpha_i + \sum_{j=1}^{n-m} e_j\beta_j\right)$$

$$W + \beta = \left(W + \sum_{i=1}^{m} d_i\alpha_i\right) + \left(W + \sum_{j=1}^{n-m} e_j\beta_j\right) \quad \left[\because \sum_{i=1}^{m} d_i\alpha_i \in W\right]$$

$$= W + \sum_{i=1}^{n-m} e_i\beta_i = W + (e_1\beta_1 + e_2\beta_2 + \ldots + e_{n-m}\beta_{n-m})$$

$$= (W + e_1\beta_1) + (W + e_2\beta_2) + \ldots + (W + e_{n-m}\beta_{n-m})$$

$$= e_1(W + \beta_1) + e_2(W + \beta_2) + \ldots + e_{n-m}(W + \beta_{n-m}).$$

$\Rightarrow$ S_2 generates V/W.

Thus, S_2 forms a basis of V/W and having $(n - m)$ elements so that

$$\dim. V/W = n - m$$

Accordingly,

$$\dim .V/W = n - m = \dim. V - \dim. W.$$

3.17 ISOMORPHISM

Let U and V be too linear spaces over the same field F. Then a mapping $T : U \to V$ which associates to each element $\alpha \in U$ to a unique element $T(\alpha) \in V$ such that

$$T(a\alpha + b\beta) = aT(\alpha) + bT(\beta), \forall\ \alpha, \beta \in U, a, b \in F$$

is called a linear transformation of U into V.

A linear transformation T of U onto V is called an isomorphism if it is one-one. Therefore, we can say that U is isomorphic to V and write this as $U \cong V$.

THEOREM 1. ***Every n-dimensional linear space V(F) is isomorphic to $F^n(F)$.***

PROOF. Since the linear space V is n-dimensional. Let the set

$$S = \{\alpha_1, \alpha_2, ..., \alpha_n\} \text{ form the basis for } V.$$

Then every vector V can be expressed as the linear combination of the elements of S. Let a be any arbitrary vector in V, then there exist a unique order set $\{a_1, a_2, ..., a_n\}$ of scalars such that

$$\alpha = \sum_{i=1}^{n} a_i \alpha_i .$$

Consider a mapping

$$T : V \to F^n \text{ by } T(\alpha) = (a_1, a_2, ..., a_n)\ \forall\ \alpha \in V.$$

(i) T is linear.

Let $\alpha = \sum_{i=1}^{n} a_i \alpha_i$ and $\beta = \sum_{i=1}^{n} b_i \alpha_i$ be any two elements of V, then for all $a, b \in F$, we have

$$T(a\alpha + b\beta) = T\left(a\sum_{i=1}^{n} a_i\alpha_i + b\sum_{i=1}^{n} b_i\alpha_i\right) = T\left(\sum_{i=1}^{n}(a_i\alpha_i + b_i\alpha_i)\alpha_i\right)$$

$$= (aa_1 + bb_1, aa_2 + bb_2, ..., aa_n + bb_n)$$

$$= (aa_1, aa_2, ..., aa_n) + (bb_1, bb_2, ..., bb_n)$$

$$= a(a_1, a_2, ..., a_n) + b(b_1, b_2, ..., b_n) = a\,T(\alpha) + bT(\beta).$$

$\therefore$ T is linear.

(ii) T is one-one.

Suppose that

$$T(\alpha) = T(\beta)$$

$$\Rightarrow \quad T\left(\sum_{i=1}^{n} a_i\alpha_i\right) = T\left(\sum_{i=1}^{n} b_i\alpha_i\right)$$

$$\Rightarrow \quad (a_1, a_2, ..., a_n) = (b_1, b_2, ..., b_n)$$

$$\Rightarrow \quad a_i = b_i, \text{ for each } i = 1, 2, 3, ..., n$$

$$\Rightarrow \qquad \sum_{i=1}^{n} a_i\alpha_i = \sum_{i=1}^{n} b_i\alpha_i \quad \Rightarrow \quad \alpha=\beta$$

$\therefore$ T is one-one.

(iii) T is onto

Since, corresponding to each element $(aa_1, aa_2, ..., aa_n) \in F^n$, there exists a vector

$$\sum_{i=1}^{n} a_i\alpha_i \in V \text{ such that } T\left(\sum_{i=1}^{n} a_i\alpha_i\right) = (a_1, a_2, ..., a_n)$$

$\Rightarrow$ T is onto.

Hence, V is isomorphic to F^n, *i.e.*, $V \cong F^n$.

THEOREM 2. ***If W_1 and W_2 are the complementary subspaces of a linear space V(F), then the correspondance that assigns to each other $a \in W_2$, the coset $W_1 + \alpha$ is an isomorphism between W_2 and V/W_1.***

PROOF. Since W_1 and W_2 are the complementary subspaces of V, then $V = W_1 \oplus W_2$, or we have

$$V = W_1 + W_2 \text{ and } W_1 \cap W_2 = \{0\}$$

Consider a mapping

$$T : W_2 \to V/W_1$$

defined by $\quad T(\alpha) = W_1 + \alpha, \ \forall \ \alpha \in W_2$

First to show that T is linear.

Let α, β be any two elements in W_2, then for all $a, b \in F$, we have

$$T(a\alpha + b\beta) = W_1 + (a\alpha + b\beta) = (W_1 + a\alpha) + (W_1 + b\beta)$$
$$= a(W_1 + \alpha) + b(W_1 + \beta) = aT(\alpha) + bT(\beta)$$

$\therefore$ T is linear.

Second, to show that T is one-one.

Suppose that

$$T(\alpha) = T(\beta)$$
$$\Rightarrow \quad W_1 + \alpha = W_2 + \beta \Rightarrow \alpha - \beta \in W_1$$
$$\Rightarrow \quad \alpha - \beta \in W_1 \cap W_2 \text{ as } \alpha - \beta \in W_2$$
$$\Rightarrow \quad \alpha - \beta = 0 \qquad [\because W_1 \cap W_2 = \{0\}]$$
$$\Rightarrow \quad \alpha = \beta.$$

$\therefore$ T is one-one.

Finally, to show that T is onto.

Let $W_1 + \gamma$ be an arbitrary element of V/W_1 so that $\gamma \in V$. But V is the direct sum of W_1 and W_2, therefore, γ can be uniquely expressed as the sum of an element of W_1 and an element of W_2. So there exists $\alpha \in W_1$ and $\beta \in W_2$ such that

$$\gamma = \alpha + \beta$$
$$\Rightarrow \quad W_1 + \gamma = W_1 + (\alpha + \beta) = (W_1 + \alpha) + (W_1 + \beta)$$
$$= (W_1 + 0) + (W_1 + \beta) \qquad [\because \alpha \in W_1 \Rightarrow W_1 + \alpha = W_1 = W_1 + 0]$$
$$= (W_1 + \beta) = T(\beta)$$

which shows that to each element $W_1+\gamma$ in V/W_1, there exists a unique element $\beta \in W$ such that

$$T(\beta) = W_1+\gamma$$

$\therefore$ T is onto.

Hence, W_2 is isomorphic to V/W_1.

THEOREM 3. ***Any two finite dimensional linear spaces over the same field are isomorphic if and only if they are of same dimension.***

PROOF. Let $U(F)$ and $V(F)$ be two spaces over same field F. Suppose that both are of same dimension n (say).

Let $S_1=\{\alpha_1,\alpha_2,\ldots\ldots,\alpha_n\}$ and $S_2=\{\beta_1,\beta_2,\ldots\ldots,\beta_n\}$ be the basis for U and V respectively. So, corresponding to each element $\alpha \in U$, there exists unique scalars $a_1,a_2,\ldots\ldots,a_n \in F$ such that

$$a=a_1\alpha_1+a_2\alpha_2+\ldots+a_n\alpha_n=\sum_{i=1}^{n} a_i\alpha_i$$

Consider a mapping $T : U \to V$.

by $$\mathrm{T}(\alpha)=\sum_{i=1}^{n} a_i\beta_i \;, \forall\, a_i \in F \text{ and for all } \alpha \in U$$

First we show that T is linear.

Let α, β be any two elements of U, then $\alpha=\sum_{i=1}^{n} a_i\alpha_i$ and $\beta=\sum_{i=1}^{n} b_i\alpha_i$.

For $a,b \in F$, we have

$$T(a\alpha+b\beta) = T\left(a\sum_{i=1}^{n} a_i\alpha_i + b\sum_{i=1}^{n} b_i\alpha_i\right)$$

$$= T\left(\sum_{i=1}^{n}(aa_i+bb_i)\alpha_i\right) = \sum_{i=1}^{n}(aa_i+bb_i)\beta_i$$

$$= (\sum_{i=1}^{n} aa_i\beta_i + \sum_{i=1}^{n} bb_i)\beta_i$$

$$= a\sum_{i=1}^{n} a_i\beta_i + \mathrm{b}\sum_{i=1}^{n} b_i\beta_i$$

$$= aT(\alpha)+bT(\beta)$$

$\therefore$ T is a linear.

Now, we show that T is one-one.

Suppose for $\alpha,\beta \in U$

$$\Rightarrow \quad \sum_{i=1}^{n} a_i\beta_i = \sum_{i=1}^{n} b_i\beta_i$$

$$\Rightarrow \quad \sum_{i=1}^{n}(a_i-b_i)\beta_i = 0 \text{ for each } \mathrm{i}=1,2,3,\ldots.,n$$

[$\because$ S_2 is linearly indepenmdent.]

$$\Rightarrow \quad \sum_{i=1}^{n} a_i\beta_i = \sum_{i=1}^{n} b_i\beta_i \Rightarrow \alpha=\beta.$$

$\therefore$ T is a one-one.

Finally, we show that T is onto.

Let γ be any element of V, then there exist scalars $c_1,c_2,\ldots,c_n$ of F such that

$$g=\sum_{i=1}^{n} c_i\beta_i\ ,$$

Now, $$T\left(\sum_{i=1}^{n} c_i\alpha_i\right)=\sum_{i=1}^{n} c_i\beta_i=\gamma$$

$\Rightarrow$ T-image of $\sum_{i=1}^{n} c_i\alpha_i \in$ U is $\gamma=\sum_{i=1}^{n} c_i\beta_i$

$\therefore$ T is a onto.

Hence T is an isomorphism, *i.e.*, $U\cong V$

Conversely, suppose that $U\cong V$ and let T be the corresponding isomorphism.

Let $S_1=\{\alpha_1,\alpha_2,\ldots,\alpha_n\}$ be the basis of U.

We claim that the set

$$S_2=\{T(\alpha_1),T(\alpha_2),\ldots,T(\alpha_n)\}$$

is the basis for V.

First we show that S_2 is linearly independent.

For some scalars $a_1,a_2,\ldots,a_n \in F$,the relation

$$a_1T(\alpha_1)+a_2T(\alpha_2)+\ldots\, a_nT(\alpha_n)=0$$

$\Rightarrow$ $T(a_1\alpha_1+a_2\alpha_2+\ldots\, a_n\alpha_n)=0$ [$\because$ T is a linear]

$T\left(\sum_{i=1}^{n} a_i\alpha_i\right)=T(0)$ [$\because T(0)=(0)$]

$\sum_{i=1}^{n} a_i\alpha_i=0$ [$\because T$ is one-one]

$\Rightarrow$ $a_i=0$ for each $i=1,2,3,\ldots,n$. [$\because S_1$ is linearly independent.]

$\therefore$ S_2 is linearly independent.

Secondly, We show that S_2 spans V.

Let γ be any element of V. Since T is one-one onto mapping, then there exists a unique vector $\alpha=\sum_{i=1}^{n} a_i\alpha_i \in U$ such that

$$\gamma=T(\alpha)=T\left(\sum_{i=1}^{n} a_i\alpha_i\right)=\sum_{i=1}^{n} a_iT(\alpha_i)\qquad [\because T \text{ is a linear.}]$$

Thus, $\gamma\in V$ is expressible as a linear combinations of elements of S_2, therefore S_2 spans V.

Consequently $\dim V=n=\dim U$.

Hence, both the vectors $U(F)$ and $V(F)$ are finite dimensional of dimension n.

REMARK

- The dimension of the solution space W of the homogeneous system of linear equations $AX=O$ is $(n-r)$ where, n is the number of unknowns and r is the rank of the coefficient matrix A.

RECAPITULATIONS

- A finite set of vectors $\{\alpha_1, \alpha_2,...,\alpha_n\}$ of vectors of V is said to be linearly dependant if $\exists$ scalars $a_1,a_2,....,a_n$ not all them equal to zero such that $a_1\alpha_1+ a_2\alpha_2+...+ a_n\alpha_n=0$ and it is said to be linearly independent if every expression of the form $a_1\alpha_1+ a_2\alpha_2,+...+ a_n\alpha_n=0$ implies $a_1=0,a_2=0,....,a_n=0$
- **Left Cosets** $\alpha+W= \{\alpha+\beta, \forall\ b\in W\}$
- A non-empty subset S of $V(F)$ is said to be a basis if S is linearly independent and $L(S)=V$.
- $\dim(W_1)+ \dim(W_2) = \dim(W_1 \cap W_2)+ \dim(W_1 + W_2)$
- If $V(F)$ is the direct sum of its two subspaces W_1 and W_2. Then $\dim V = \dim W_1+\dim W_2$
- **Right Cosets** $W+\alpha= \{\beta+\alpha, \forall\ b\in W\}$

SOLVED EXAMPLES

EXAMPLE 1. *Prove that if two vectrors are linearly dependent, one of them is a scalar multiple of the other.*

SOLUTION. Let V be a linear space over F and α, β be two vectors of V. These vectors are linearly dependent, then there exists scalars $a,b \in F$ not both equal to zero such that

$$a\alpha+b\beta =0.$$

If $a \neq 0$, then a^{-1} exists in F, we have

$$a^{-1}(a\alpha)+ a^{-1}(b\beta) = a^{-1}0$$

$$\Rightarrow \quad (a^{-1}a)\alpha+ (a^{-1}b)\beta = 0 \qquad [\because a^{-1}0=0]$$

$$\Rightarrow \quad 1\alpha+ a^{-1}b\beta = 0 \qquad [\because aa^{-1}=1]$$

$$\Rightarrow \quad \alpha+ (a^{-1}b)\beta = 0 \qquad [\because 1\alpha=\alpha]$$

$$\Rightarrow \quad \alpha = -(a^{-1}b)\beta \qquad [\because 1\alpha=\alpha]$$

$\Rightarrow$ α is a scalar multiple of β.

Similarly, if $b \neq 0$, then we obtain β as a scalar multiple of α. Hence, one of α or β is scalar multiple of the other.

EXAMPLE 2. *Show that the set $S=\{(1,2,1),(3,1,5),(3,-4,7)\}$ is linearly dependent where $S\subseteq V_3(\mathrm{R})$*

SOLUTION. Let a, b, $c \in \mathrm{R}$ such that

$$a(1,2,1)+b(3,1,5)+c(3,-4,7)= (0,0,0)$$

or

$$(a+3b+3c,2a+b-4c,a+5b+7c)= (0,0,0)$$

or

$$\left.\begin{aligned} a+3b+3c&=0\\ 2a+b-4c&=0\\ a+5b+7c&=0 \end{aligned}\right\} \qquad ...(1)$$

Equation (1) is a system of linear homogeneous equations.

$\therefore$ Coefficient matrix is given by

$$A=\begin{bmatrix} 1 & 3 & 3\\ 2 & 1 & -4\\ 1 & 5 & 7 \end{bmatrix}$$

Now,

$$\det(A)= 1(7+20)-3(14+4)+3(10-1)$$
$$= 27-54+27=0$$

$\therefore$ Rank of A is less than 3, i.e., rank of A is less than the number of variables a,b and c so there exist a non-zero solutions. Thus, a,b,c are all not equal to zero. Hence, the given set S is linearly dependent.

EXAMPLE 3. *Show that the vectors* (1,1,2,4), (2,–1,–5,2), (1,–1,–4,0) *and* (2,1,1,6) *are linearly dependent in* R^4.

SOLUTION. Let $a, b, c, d \in$ R such that

$$a(1,1,2,4)+b(2,-1,-5,2)+c(1,-1,-4,0)+d(2,1,1,6)=(0,0,0,0)$$

or
$$(a+2b+c+2d, a-b-c+d, 2a-5b-4c+d, 4a+2b+6d)=(0,0,0,0)$$

or
$$\left.\begin{aligned} a+2b+c+2d&=0\\ a-b-c-d&=0\\ 2a-5b-4c+d&=0\\ 4a+2b+6d&=0\end{aligned}\right\} \qquad ...(1)$$

Therefore, equation (1) represents a system of linear homogeneous equation. Now the coefficient matrix of these equation is

$$A=\begin{bmatrix}1 & 2 & 1 & 2\\ 1 & -1 & -1 & 1\\ 2 & -5 & -4 & 1\\ 4 & 2 & 0 & 6\end{bmatrix}$$

performing $R_2 \to R_2-R_1$, $R_3 \to R_3-2R_1$, $R_4 \to R_4-4R_1$, we get

$$\sim\begin{bmatrix}1 & 2 & 1 & 2\\ 0 & -3 & -2 & -1\\ 0 & -9 & -6 & -3\\ 0 & -6 & -4 & -2\end{bmatrix}$$

performing $R_3 \to R_3-3R_2$, $R_4 \to R_4-2R_2$

$$\sim\begin{bmatrix}1 & 2 & 1 & 2\\ 0 & -3 & -2 & -1\\ 0 & 0 & 0 & 0\\ 0 & 0 & 0 & 0\end{bmatrix}$$

Thus, this matrix is in Echeleon form and having two non-zero rows so the rank of A is 2 which is less than the number of unknown. Therefore, the system of equation has non-zero solution. That is a, b, c, d are not all zero, Hence the given vectors are linearly dependent.

EXAMPLE 4. *Show that the vectors* (1,1,2), (1,2,5), (5,3,4) *do not form a basis of* R^3.

SOLUTION. Since we know that the set (1,0,0), (0,1,0), (0,0,1) forms a basis of R^3 so dimension of $R^3=3$.

Let $a, b, c \in$ R such that

$$a(1,1,2)+b(1,2,5)+c(5,3,4)=(0,0,0)$$

or
$$(a+b+5c, a+2b+3c, 2a+5b+4c)=(0,0,0)$$

or
$$\left.\begin{aligned} a+b+5c&=0\\ a+2b+3c&=0\\ 2a+5b+4c&=0\end{aligned}\right\} \qquad ...(1)$$

The equation (1) represents the system of linear homogeneous equations. Then the coefficient matrix of these equation is given by

$$A=\begin{bmatrix}1 & 1 & 5\\ 1 & 2 & 3\\ 2 & 5 & 4\end{bmatrix}$$

$$\therefore \quad |(A)| = 1(8-15)-1(4-6)+5(5-4)=-7+2+5=0$$

Thus A is a singular matrix and therefore the rank of A is less that of 3, *i.e.*, less than the number of unknown a, b, c. Then the scalars a, b, c are all not equal to zero.

Hence, the given get of vectors is not linearly independent, and hence this given set does not form a basis of R^3.

EXAMPLE 5. *Show that the vectors (2,1,4), (1,–1,2), (3,1,–2) form a basis of R^3.*

SOLUTION. Since we know that the set {(1,0,0), (0,1,0), (0,0,1)} forms a basis of R^3. Then dim of R^3 =3. Therefore, if the given set {(2,1,4), (1,–1,2), (3,1,–2)} is linearly independent, then it will form the basis of R^3.

Let a,b,c ∈ R such that

$$a(2,1,4)+b(1,-1,2)+c(3,1,-2)=(0,0,0)$$

or
$$(2a+b+3c,a-b+c,4a+2b-2c)= (0,0,0)$$

or
$$\left.\begin{aligned}2a+b+3c&=0\\a-b+c&=0\\4a+2b-2c&=0\end{aligned}\right\} \quad ...(1)$$

The equation (1) represents a system of three linear homogeneous equation. Then the coefficient matrix of these equation is given by

$$A=\begin{bmatrix}2 & 1 & 3\\1 & -1 & 1\\4 & 2 & -2\end{bmatrix}$$

Now
$$|(A)| = 2(2-2)-1(-2-4)+3(2+4)$$
$$= 0+6+18=24 \neq 0$$

∴ The matrix A is a non-singular matrix and thus the rank of A is 3 which is equal to the number of unknowns *a*, *b*, *c*. Hence, the system of equations has only zero solution, *i.e.*, $a=0$, $b=0$, $c=0$. Consequently the given set is linearly independent and hence forms a basis of R^3.

EXAMPLE 6. *Find the dimension of the solution W of the system of linear equations.*

$$x+2y-4z+3r-s=0;\; x+2y-2z+2r+s=0;\; 2x+4y-2z+3r+4s=0$$

SOLUTION. Above equation can be written as AX=O

where, $A=\begin{bmatrix}1 & 2 & -4 & 3 & -1\\1 & 2 & -2 & 4 & 1\\2 & 4 & -2 & 3 & 4\end{bmatrix}$ and $X=\begin{bmatrix}x\\y\\z\\r\\s\end{bmatrix}$, $O=\begin{bmatrix}0\\0\\0\end{bmatrix}$.

Now reduce A to Echelon form as follows:

$$A\sim\begin{bmatrix}1 & 2 & -4 & 3 & -1\\0 & 0 & 2 & 1 & 2\\0 & 0 & 6 & 3 & 6\end{bmatrix} \qquad \begin{bmatrix}R_2 \to R_2 - R_1\\R_3 \to R_3 - 2R_1\end{bmatrix}$$

$$\sim\begin{bmatrix}1 & 2 & -4 & 3 & -1\\0 & 0 & 2 & 1 & 2\\0 & 0 & 0 & 0 & 0\end{bmatrix} \qquad [R_3 \to R_3 - 3R_2]$$

Echelon form of A has two non-zero rows, so that rank of A=2 and number of unknowns = 5. Hence dim.W =5 – 2 = 3.

EXAMPLE 7. *Prove that a set of non-zero vectors {x1,x2,.....,xn} is linearly dependent if some of these vectors say xi is a linear combination of the preceding vectors x1,x2,.....,xi-1 and conversely.*

SOLUTION. Let us first assume that x_i can be expressed as a linear combination of $x_1, x_2, \ldots, x_{i-1}$. Then we have

$$x_i = a_1x_1 + a_2x_2 + \ldots + a_{i-1}x_{i-1} \quad \ldots(1)$$

For some scalars $a_1, a_2, \ldots, a_{i-1} \in F$.

Now equation (1) can be written as:

$$a_1x_1 + a_2x_2 + \ldots + a_{i-1}x_{i-1} + (-1)\, x_i = 0$$

or $$a_1x_1 + a_2x_2 + \ldots + a_{i-1}x_{i-1} + (-1)x_i + 0\, x_{i+1} + 0\, x_{i+2} + \ldots + 0\, x_n = 0$$

This equation shows that this relation has at least one non-zero coefficient (scalar) of x_i which is –1. It shows that the set $\{x_1, x_2, \ldots, x_n\}$ is linearly dependent. Then we have a relation.

$$a_1x_1 + a_2x_2 + \ldots + a_nx_n = 0$$

in which not all a_i's are zero.

If i is the greatest positive integer less than or equal to n such that $a_i \neq 0$, so that $a_{i+1} = 0 = a_{i+2} = \ldots = a_n$, then we have

$$a_1x_1 + a_2x_2 + \ldots + a_{i-1}x_{i-1} + a_ix_i + 0.\, x_{i+1} + 0.\, x_{i+2} + \ldots + 0.\, x_n = 0$$

$$\Rightarrow \quad x_i = -a_i^{-1}(a_1x_1 + a_2x_2 + \ldots + a_{i-1}x_{i-1}).$$

$\Rightarrow x_i$ is the linear combination of preceding vectors $x_1, x_2, \ldots, x_{i-1}$.

EXAMPLE 8. *Prove that the vector* (α_1, α_2), $(\beta_1, \beta_2) \in R \times R$ *are linearly dependent iff* $\alpha_1\beta_1 - \alpha_2\beta_2 = 0$

SOLUTION. The vectors (α_1, α_2) and (β_1, β_2) are linearly dependent if and only if there exist scalar $a, b \in R$ such that

$$a(\alpha_1, \alpha_2) + b(\beta_1, \beta_2) = (0,0)$$

where, (0,0) is the vector in $R \times R$

$$\Leftrightarrow \quad (a\alpha_1, a\alpha_2) + (b\beta_1, b\beta_2) = (0,0)$$

$$\Leftrightarrow \quad (a\alpha_1 + b\beta_1, a\alpha_2 + b\beta_2) = (0,0)$$

$$\Leftrightarrow \quad a\alpha_1 + b\beta_1 = 0,\ a\alpha_2 + b\beta_2 = 0$$

The equations $a\alpha_1 + b\beta_1 = 0$ and $a\alpha_2 + b\beta_2 = 0$ have non-trival (non-zero) solutions iff the determinant of coefficients.

$$\begin{vmatrix} \alpha_1 & \beta_1 \\ \alpha_2 & \beta_2 \end{vmatrix} = 0, \text{ i.e., } \alpha_1\beta_2 - \beta_1\alpha_2 = 0.$$

Hence, the vectors (α_1, α_2) and (β_1, β_2) are linearly dependent if $\alpha_1\beta_2 - \beta_1\alpha_2 = 0$.

EXAMPLE 9. *Prove that the vectors* (2, *i*, –*i*), (2*i*, –1, 1), (1, 2, 3) *are linearly independent in* $V_3(R)$ *but linearly dependent in* $V_3(C)$.

SOLUTION. Let $\alpha_1 = (2, i, -i)$, $\alpha_2 = (2i, -1, 1)$ and $\alpha_3 = (1, 2, 3)$ and let $\alpha_1, \alpha_2, \alpha_3$ be three scalars, Then

$$\Rightarrow \quad a_1\alpha_1 + a_2\alpha_2 + a_3\alpha_3 = 0$$

$$\Rightarrow a_1\,(2, i, -i) + \alpha_2(2i, -1, 1) + \alpha_3(1, 2, 3) = (0, 0, 0)$$

$$\Rightarrow (2a_1 + 2ia_2 + a_3,\ ia_1 - a_2 + 2a_3,\ -ia_1 + a_2 + 3a_3) = (0,0,0)$$

$$\therefore \quad 2a_1 + 2ia_2 + a_3 = 0 \quad \ldots(1)$$

$$ia_1 - a_2 + 2a_3 = 0 \quad \ldots(2)$$

$$-ia_1 - a_2 + 3a_3 = 0 \quad ...(3)$$

Now adding (2) and (3), we get

$$5a_3 = 0 \Rightarrow a_3 = 0.$$

Putting $a_3 = 0$. in any one of above equations, we get

$$ia_2 - a_2 = 0 \text{ or } a_2 = ia_1. \quad ...(4)$$

Case I: In $V_3(R)$, equation (4) is satisfied when $a_1 = 0, a_2 = 0$.

Thus in case of $V_3(R)$, equation (4) is satisfied when $a_1 = 1, a_2 = i$.

$$\therefore \quad 1(2, i, -i) + i(2i, -1, 1) + 0(1, 2, 3) = (0,0,0).$$

Thus in case of $V_3(C)$, the vectors α_1, α_2 and α_3 are linearly dependent.

EXAMPLE 10. *Prove that the vectors* (1, 1, 0) (3, 1, 3) *and* (5, 3, 3) *are linearly dependent.*

SOLUTION. Let $\alpha_1 = (1,1,0), \alpha_2 = (3,1,3)$ and $a_3 = (5,3,3)$ and let α_1, α_2 and α_3 be some scalars such that

$$a_1\alpha_1 + a_2\alpha_2 + a_3\alpha_3 = 0$$

$$\Rightarrow \quad a_1(1,1,0) + a_2(3,1,3) + a_3(5,3,3) = (0,0,0)$$

$$\Rightarrow \quad (a_1 + 3a_2 + 5a_3, a_1 + a_2 + 3a_3, 3a_2 + 3a_3) = (0,0,0)$$

$$\Rightarrow \quad a_1 + 3a_2 + 5a_3 = 0 \quad ...(1)$$

$$a_1 + a_2 + 3a_3 = 0 \quad ...(2)$$

$$3a_2 + 3a_3 = 0 \quad ...(3)$$

Coefficient matrix of above equation is $A = \begin{bmatrix} 1 & 3 & 5 \\ 1 & 1 & 3 \\ 0 & 3 & 3 \end{bmatrix}$

Reduce this matrix into Echelon from as follows:

$$A \sim \begin{bmatrix} 1 & 3 & 5 \\ 0 & -2 & -2 \\ 0 & 3 & 3 \end{bmatrix} \quad [R_2 \to R_2 - R_1]$$

$$\sim \begin{bmatrix} 1 & 3 & 5 \\ 0 & -2 & -2 \\ 0 & 0 & 0 \end{bmatrix} \quad \left[R_3 \to R_3 + \frac{3}{2}R_2\right]$$

This is the Echelon from and having two non-zero rows so that rank of A=2. Thus the equation (1), (2) and (3) have non-zero solutions. Hence the vectors (1, 1, 0), (3, 1, 3) and (5, 3, 3) are linearly dependent.

EXAMPLE 11. *Under what conditions on the scalar 'a' are the vectors* $(a, 1, 0)$, $(1, a, 1)$ *and* $(0, 1, a)$ *in* R^3 *linearly dependent?*

SOLUTION. Let $\alpha_1 = (\alpha, 1, 0)$, $\alpha_2 = (1, a, 1)$ and $\alpha_3 = (0, 1, a)$ and let a_1, a_2 and a_3 be three scalars such that

$$a_1\alpha_1 + a_2\alpha_2 + a_3\alpha_3 = 0$$

$$\Rightarrow \quad a_1(1,1,0) + a_2(1,a,1) + a_3(0,1,a) = (0,0,0)$$

$$\Rightarrow \quad (a_1 a + a_2, a_1 + a_2 a + a_3, a_2 + a_3 a) = (0,0,0)$$

$$\Rightarrow \quad \begin{aligned} a_1 a + a_2 &= 0 \quad ...(1) \\ a_1 + a_2 a + a_3 &= 0 \quad ...(2) \\ a_2 + a_3 a &= 0 \quad ...(3) \end{aligned}$$

Let A be the coefficient matrix of the equations (1), (2), (3)

$$\therefore \quad A = \begin{bmatrix} a & 1 & 0 \\ 1 & a & 1 \\ 0 & 1 & a \end{bmatrix}$$

For non-trivial solution of (1), (2) and (3), we must have

$$\begin{vmatrix} a & 1 & 0 \\ 1 & a & 1 \\ 0 & 1 & a \end{vmatrix} = 0$$

$$\Rightarrow \quad a[a^2-1]-1[a-0] = 0 \Rightarrow a(a^2-2) = 0$$

$$\Rightarrow \quad a = 0, a = \pm\sqrt{2}$$

Hence the vectors $(a, 1, 0)$,$(1, a, 1)$ and $(0, 1, a)$ are linearly dependent when $a = 0; \pm\sqrt{2}$.

EXAMPLE 12. *For what value of m, the vector (m, 3, 1) a linear combination of* $e_1 = (3, 2, 1)$ *and* $e_2 = (2, 1, 0)$.

SOLUTION. For some scalars a_1 and a_2 such that

$$(m, 3, 1) = a_1e_1 + a_2e_2 = a_1(3,2,1) + a_2(2,1,0)$$
$$= (3a_1+2a_2, 2a_1+a_2, a_1)$$

$$\Rightarrow \quad \begin{aligned} 3a_1 + 2a_2 &= m \quad &...(1)\\ 2a_1 + a_2 &= 3 \quad &...(2)\\ a_1 &= 1 \quad &...(3) \end{aligned}$$

Putting the value of $a_1 = 1$ in (2), we get $2(1) + a_2 = 3$, $a_2 = 3 - 2 = 1$.
Now putting the values of a_1 and a_2 in (l),we get m = 3(1) +2(1) = 3 + 2 = 5

EXAMPLE 13. *Show that the set* {(1,2,1), (2,1,0), (I, –1,2)} *forms a basis for* V_3(R).

SOLUTION. Since we know that the set {(1, 0, 0), (0, 1, 0),(1, –1,2)} is linearly independent, then it will form the basis of V_3(R).

Let $a, b, c \in R$ such that $a(1,2,1) + b(2,1,0) + c(1,-1,2) = (0,0,0)$

$$\Rightarrow \quad (a + 2b + c, 2a + b - c, a + 2c) = (0, 0, 0)\ 0$$

$$\Rightarrow \quad \begin{aligned} a + 2b + c &= 0 \quad &...(1)\\ 2a + b - c &= 0 \quad &...(2)\\ a + 2c &= 0 \quad &...(3) \end{aligned}$$

The equation (1), (2), (3) represent a system of three linear homogeneous equations. Then the coefficient matrix of these equations is given by :

$$A = \begin{bmatrix} 1 & 2 & 1 \\ 2 & 1 & -1 \\ 1 & 0 & 2 \end{bmatrix}$$

Now, $$|A| = 1(2-0) - 2(4+1) + 1(0-1) = 2 - 10 - 1 = -9 \neq 0.$$

Therefore, matrix A is non-singular and thus the rank of A is 3 which is equal to the number of unknowns a, b, c. Hence the system of equations has only zero solution, *i.e.*, $a = 0$, $b = 0$, $c = 0$. Consequently, the given set is linearly independent and hence forms a basis of V_3(R).

EXAMPLE 14. *Under what conditions on the scalar 'a', do the vectors* (1, 1, 1) *and* (1,*a, a*2) *form a basis of* C^3(C).

SOLUTION. Since dim. $C^3(\mathbf{C}) = 3$ so that every basis of $C^3(\text{C})$ must have three vectors. Therefore, a must have two values. Let a_1 and a_2 be the two values of a such that $a_1^2 = a_2^2$. In other words, we are to find the condition on the scalar 'a' such that the vectors (1,1,1), $(1, a, a^2)$, $(1, -a, a^2)$ form a basis of $C^3(\text{C})$. Since dim. $C^3(\text{C}) = 3$. therefore, we find the condition such that (1,1,1), $(1, a, a^2)$ and $(1, -a, a^2)$ are linearly independent.

For some scalar $a_1, a_2, a_3 \in \text{C}$, we have

$$a_1 (1, 1, 1) + a_2(1,a,a^2) + a_3(1,-a, a^2) = (0, 0, 0)$$

$$\Rightarrow \quad (a_1 + a_2 + a_3 , a_1 + aa_2 - aa_3 , a_1 + a^2a_2 + a^2a_3) = (0, 0, 0)$$

$$\Rightarrow \quad a_1 + a_2 + a_3 = 0 \quad ...(1)$$

$$a_1 + aa_2 + aa_3 = 0 \quad ...(2)$$

$$a_1 + a^2a_2 + a^2a_3 = 0 \quad ...(3)$$

This system of homogeneous equations has a trivial solution if and only if

$$\begin{vmatrix} 1 & 1 & 1 \\ 1 & a & -a \\ 1 & a^2 & a^2 \end{vmatrix} \neq 0$$

$$\Rightarrow \quad 1(a^3+a^3) - 1(a^2 + a) + 1(a^2 - a) \neq 0$$

$$\Rightarrow \quad 2a^3 - 2a \neq 0 \Rightarrow 2a(a^2 - 1) \neq 0 \Rightarrow a \neq 0, a \neq \pm 1.$$

Hence the vectors (1,1,1) and $(1, a, a^2)$ form a basis of $C^3(\mathbf{C})$ if any only if $a \neq 0, a \neq \pm 1$.

EXERCISE 3.2

1. Show that the set $S = \{(1, 0, 0), (0, 1, 0), (0, 0, 1)\}$ spans the linear space $V_3(\text{R})$.
2. Prove that every superset of a linearly dependent set is linearly dependent.
3. Show that the set $S = \{(1, 2, 4), (1, 0, 0), (0, 1, 0), (0, 0, 1)\}$ is linearly dependent subset of the linear space $V_3(\text{R})$.
4. Show that the set $\{1, x, 1 + x + x^2\}$ is linearly independent set of vectors in the linear space of all polynomials over real number field.
5. Show that the vectors $(1, 1, -1)$, $(2, -3, 5)$, $(-2, 1, 4)$ of R^3 are lineally independent.
6. Show that the vectors $(1, 3, 2)$, $(1, -7, -8)$, $(2, 1, -1)$ of $V_3(\text{R})$ are linearly dependent.
7. Is the vector $(3, -1, 0, 1)$ in the subspace of R^4 spanned by the vectors $(2, -1, 3, 2)$, $(-1, 1, 2, -3)$, $(1, 1, 9, -5)$?
8. Is the vector $(2, -5, 3)$ in the subspace of R^3 spanned by the vectors $(1, -3, 2)$, $(2, -4, -1)$, $(1, -5, 7)$?
9. If F is the field of complex numbers, prove that the vectors (a_1, a_2) and (b_1, b_2) in $V(F)$ are linearly dependent iff $a_1b_2 - a_2b_1 = 0$.
10. Show that every subset of a linearly independent set of vectors is linearly independent.
11. Examine each of the following sets of vectors for linear dependent in the linear space $V_3(\text{R})$
    (i) $\{(1, 2, 0), (0, 3, 1), (-1, 0, 1)\}$
    (ii) $\{(-1, 2, 1), \{3, 0, -1), (-5, 4, 3)\}$

3.18 LINEAR TRANSFORMATION

The concept of homomorphism is easily carried to linear spaces. To begin with linear transformation (linear space homomorphism) is a function from one linear space to another. Like a ring homomorphism, it is supposed to preserve both the linear space operations. The process of taking functional values and performing the linear space operation should be commutative. This requires that the scalar field in case of either space should be the same.

Definition. *Let U and V be two linear spaces over the same field F. A mapping $T : U \to V$ is said* *to be a linear transformation from U into V which associates to each element α of U to a unique element T(α) of V such that*

$$T(a\alpha + b\beta) = aT(\alpha) + bT(\beta)$$

for all α and β in *U* and all scalars *a, b* in *F*.

REMARKS

- Linear transformation is also known as linear space homomorphism.
- If a linear transformation is onto, then it is known as isomorphism.

ILLUSTRATIONS

(1) If *V* is any linear space over *F*, then the identity transformation *I*, defined by $I(\alpha)=\alpha$, $\forall\ \alpha\in V$ is a linear transformation from *V* into *V*. Also the zero transformation 0 denoted by $0(\alpha) = 0$, is a linear transformation.

(2) Let *F* be a field of real numbers and let *V* be the linear space of all polynomials, then a mapping

$$D : V \to V$$

given by $D[f(x)] = \frac{d}{dx}[f(x)]$, $\forall\ f(x) \in V$ is a linear transformation.

Since for any *f*(*x*) and $g(x) \in V$ and $a, b \in F$

$$D[af(x)+bg(x)] = \frac{d}{dx}[af(x)+bg(x)] = \frac{d}{dx}[af(x)] + \frac{d}{dx}[bf(x)]$$

$$= a\frac{d}{dx}[f(x)] + b\frac{d}{dx}[g(x)] = aD[f(x)] + bD[g(x)]$$

(3) Let R be the field of real numbers and let *V* be the linear space of all functions from R into R which are continuous. Then a mapping $T : V \to V$ given by

$$T[f(x)] = \int_0^x f(t)\,dt$$

is a linear transformation.

For any $f(x), g(x)\in V$ and $a,b \in$ R

$$\therefore \quad T[af(x)+bg(x)] = \int_0^x [af(t) + \int_0^x bg(t)]dt$$

$$= \int_0^x af(t)\,dt + \int_0^x bg(t)\,dt$$

$$= a\int_0^x f(t)\,dt + b\int_0^x g(t)\,dt = aT[f(x)] + bT[g(x)]$$

(4) Let *V* be the linear space of all $m\times n$ matrices over a field *F* and let *P* be a fixed $m\times n$ matrix and *Q* be a fixed matrix of order $n\times n$.

Then a mapping $T : V \to V$ given by $T(A) = PAQ$, $\forall A\in V$ is a linear transformation.

For any two matrices, $A,B\in V$ and $a,b\in F$

$$T(aA + bB) = P(aA + bB)Q = (aPA + bPB)Q$$

$$= aPAQ + bPBQ = aT(A) + bT(B).$$

THEOREM 1. ***Let U and V be two finite-dimensional linear spaces over the same field F and let*** $\{\alpha_1, \alpha_2, \ldots, \alpha_n\}$ ***be an ordered basis for U and let*** $\{\beta_1, \beta_2, \ldots, \beta_n\}$ ***be an ordered set in V. Then there is precisely one linear transformation T from U into V such that*** $T(\alpha_j) = \beta_j$, $j = 1, 2, 3, \ldots n$.

PROOF. Since the set $\{\alpha_1, \alpha_2, ..., \alpha_n\}$ is a basis of $U(F)$, then for each $\alpha \in U$,there are some scalars $a_1, a_2, ..., a_n$ such that

$$\alpha = a_1\alpha_1 + a_2\alpha_2 + ... + a_n\alpha_n = \sum_{i=1}^{n} a_i\alpha_i$$

For this vector α we define $T : U \to V$ given by

$$T(\alpha) = a_1\beta_1 + a_2\beta_2 + ... + a_n\beta_n = \sum_{i=1}^{n} a_i\beta_i$$

Then T is well defined for each vector α in U and a vector $T(\alpha) \in V$. From the definition it is clear that $T(\alpha_j) = \beta_j$ for each j.

Now we shall show that T is linear. For this if $\alpha = \sum_{i=1}^{n} a_i\alpha_i$ and $\beta = \sum_{i=1}^{n} b_i\alpha_i$ are any two vectors in U, then for all $a, b \in F$, we have

$$T(a\alpha + b\beta) = \left[a\sum_{i=1}^{n} a_i\alpha_i + b\sum_{i=1}^{n} b_i\alpha_i\right] = T\left(\sum_{i=1}^{n} aa_i\alpha_i + \sum_{i=1}^{n} bb_i\alpha_i\right)$$

$$= T\left(\sum_{i=1}^{n} aa_i + \sum_{i=1}^{n} bb_i\right)\alpha_i = \sum_{i=1}^{n} (a_ia_i + b_ib_i)\beta_i$$

$$= a\sum_{i=1}^{n} a_i\beta_i + b\sum_{i=1}^{n} b_i\beta_i$$

$$= aT\left(\sum_{i=1}^{n} a_i\alpha_i\right) + bT\left(\sum_{i=1}^{n} b_i\alpha_i\right) = aT(\alpha) + bT(\beta)$$

Now, we shall show the uniqueness of T.

Let if possible, T_1 be another linear transformation from U into V such that $T_1(\alpha_j) = B_j$, $j = 1, 2, ..., n$.

Then for any vector $\alpha = \sum_{i=1}^{n} a_i\alpha_i$, we have

$$T_1(\alpha) = T_1\left(\sum_{i=1}^{n} a_i\alpha_i\right) = \sum_{i=1}^{n} a_iT_1(\alpha_i) \qquad [\because T_1 \text{ is linear}]$$

$$= \sum_{i=1}^{n} a_i\beta_i \qquad [\because T_1(\alpha_i) = \beta_i]$$

$$= T\left(\sum_{i=1}^{n} a_i\alpha_i\right) = T(\alpha)$$

$\Rightarrow \quad T_1 = T \qquad [\because \alpha$ is an arbitrary vector.]

Hence, T is unique.

3.19 ALGEBRA OF LINEAR TRANSFORMATIONS

THEOREM 1. ***Let U and V be two linear spaces over the field F. Let T_1 and T_2 be two linear transformations from U into V. then the function (T_1+T_2) defined by***

$$(T_1+T_2)(\alpha) = T_1(\alpha) + T_2(\alpha), \forall\ \alpha \in U$$

is a linear transformation from U into V. If c is any element of F, then the function (cT) defined by

$$(cT)(\alpha) = cT(\alpha)$$

is a linear transformation from U into V.

The set of all transformations L(U, V) from U into V, together with the addition and scalar multiplication defined above, is a linear space over the field F.

PROOF. For $\alpha, \beta \in U$ and $a, b \in F$, we have

$$(T_1+T_2)(a\alpha+b\beta) = T_1(a\alpha+b\beta) + T_2(a\alpha+b\beta) \quad \text{[By definition]}$$

$$= [aT_1(\alpha) + bT_1(\beta)] + [aT_2(\alpha) + bT_2(\beta)]$$

$$[\because T_1 \text{ and } T_2 \text{ are linear transformations.}]$$

$$= [aT_1(\alpha) + aT_2(\alpha)] + [bT_1(\beta) + bT_2(\beta)]$$

$$= a(T_1+T_2)(\alpha) + b(T_1+T_2)(\beta)$$

$\therefore$ T_1+T_2 is a linear transformation.

Again, T is linear transformation and c is any scalar, then for $\alpha, \beta \in U$ and $a, b \in F$, We have

$$(cT)(a\alpha+b\beta) = c[T(a\alpha+b\beta)] \quad \text{[By definition]}$$

$$= c[aT(\alpha) + bT(\beta)] \quad [\because T \text{ is linear transformation.}]$$

$$= c[aT(\alpha)] + c[bT(\beta)] = (ca)T(\alpha) + (cb)T(\beta)$$

$$= (ac)T(\alpha) + (bc)T(\beta)]$$

$$[\because \text{Multiplication is commutative in } F.]$$

$$= a(cT)(a) + b(cT)(\beta)$$

$\therefore$ cT is a linear transformation.

Now we shall show that the set of all linear transformations $L(U, V)$ from U into V forms a linear space with respect to the above defined compositions. First we show that $(L(U, V), +)$ is an abelian group :

(i) Closure Property.

If $T_1, T_2 \in L(U, V)$, then we have already proved that T_1+T_2 is linear transformation, so that $T_1, T_2 \in L(U,V)$.

(ii) Associative Property.

For all $T_1, T_2, T_3 \in L(U,V)$ and for all $\alpha \in U$, we have

$$[(T_1+T_2)+T_3](\alpha) = (T_1+T_2)(\alpha) + T_3(\alpha)$$

$$= [T_1(\alpha)+T_2(\alpha)]+T_3(\alpha)$$

$$= T_1(\alpha)+[T_2(\alpha)+T_3(\alpha)]$$

$$[\because \text{Addition is associative in } V.]$$

$$= T_1(\alpha)+(T_2+T_3)(\alpha) = [T_1+(T_2+T_3)](\alpha)$$

$$\therefore \quad (T_1+T_2)+T_3 = T_1+(T_2+T_3).$$

(iii) Commutative Property.

For all $T_1, T_2, \in L(U,V)$ and $\alpha \in U$, we have

$$(T_1+T_2)(\alpha) = T_1(\alpha) + T_2(\alpha)$$
$$= T_2(\alpha) + T_1(\alpha) \quad [\because \text{Addition is commutative in } V.]$$
$$= (T_2+T_1)(\alpha)$$
$$\therefore \quad T_1 + T_2 = T_2 + T_1$$

(iv) Existence of Identity .

The zero transformation, denoted by 0 and defined by $0(\alpha) = 0, \forall \alpha \in U$ is a linear transformation.

Also if $T \in L(U, V)$, then

$$(T + 0) = 0 + T = T, \text{ for all } T.$$

$\therefore$ $0 \in L(U, V)$ and is identity transformation.

(v) Existence of Inverse.

For each $T \in L(U, V)$, there exists $(-T) \in L(U,V)$, defined by $(-T)(\alpha) = -T(\alpha), \forall \alpha \in U$, $(-T)$ is linear and $T + (-T) = (-T) + T = 0$.

$\therefore$ $(-T)$ is the additive inverse of T.

(vi) Distributive Property.

For all $T_1, T_2 \in L(U, V)$, $\alpha \in U$ and $a \in F$,

$$a[(T_1 + T_2)](\alpha) = a(T_1 + T_2)(\alpha) = a[(T_1(\alpha) + T_2)(\alpha)]$$
$$= aT_1(\alpha) + aT_2(\alpha) = (aT_1 + aT_2)(\alpha)$$
$$\therefore \quad a(T_1 + T_2) = aT_1 + aT_2$$

Also, for all $T \in L(U, V)$ and $\alpha \in U$, $a, b \in F$,

$$[(a+b)T](\alpha) = (a+b)T(\alpha) = aT(\alpha) + bT(\alpha) = (aT + bT)(\alpha)$$
$$\Rightarrow \quad (a+b)T = aT + bT$$

(vii) For all $T \in L(U,V)$, $\alpha \in U$ and $a, b \in F$,

$$[(ab)T](\alpha) = (ab)T(\alpha) = a[bT(\alpha)] - a(bT)(\alpha)$$
$$(ab).T = a.(b.T.)$$

(viii) If 1 is the unity in F, then for all $T \in L(U, V)$ and $\alpha \in U$

$$(1.T)(\alpha) = 1.T(\alpha) = T(\alpha)$$

$\therefore$ Hence, $L(U, V)$ is a linear space.

REMARK

- The linear space $L(U, V)$ is also denoted by Hom. (U, V), *i.e.*, (the set of all homomorphism from U into V).

THEOREM 2. ***Let U be an m-dimensional linear space over the field F, and let V be an n-dimensional linear space over F. Then the linear space L(U, V) is finite-dimensional and has dimension mn.***

PROOF. Since U and V both are finite-dimensional linear spaces of dimensions m and n respectively, therefore, let

$$\beta = \{\alpha_1, \alpha_2, \ldots \alpha_m\} \text{ and } \beta' = \{\beta_1, \beta_2, \ldots \beta_n\}$$

be the ordered basis of U and V respectively.

For each pair of integers (i, j) with $1 \le i \le m$ and $1 \le j \le n$, we define a linear transformation T_{ij} from U into V by

$$T_{ij}(\alpha_k) = \begin{cases} 0; & \text{if } k \neq j \\ \beta_i; & \text{if } k = j \end{cases}$$

The existence and uniqueness of above linear transformations follows from preceding theorem. It is obvious that there are *mn* linear transformations of the type T_{ij}, so we claim that these *mn* transformations form a basis of $L(U, V)$.

(i) For *mn* scalars a_{ij}, we have

$$\sum_{i=1}^{n} \sum_{j=1}^{m} a_{ij} T_{ij} = 0 \qquad \text{[zero transformation]}$$

$$\Rightarrow \quad \sum_{i=1}^{n} \sum_{j=1}^{m} a_{ij} T_{ij}(\alpha_k) = 0(\alpha_k), \forall \alpha_k \in U, 1 \leq k \leq n$$

$$\Rightarrow \quad \sum_{i=1}^{n} \sum_{j=1}^{m} a_{ij} T_{ij}(\alpha_k) = 0$$

$$\Rightarrow \quad \sum_{i=1}^{n} \left(\sum_{j=1}^{m} a_{ij} T_{ij}(\alpha_k) \right) = 0$$

$$\Rightarrow \quad \sum_{i=1}^{n} [a_{i1} T_{i1}(\alpha_k) + a_{i2} T_{i2}(\alpha_k) + ... + a_{im} T_{im}(\alpha_k)] = 0$$

$$\Rightarrow \quad \sum_{i=1}^{n} [a_{i1} T_{i1}(\alpha_k) + \sum_{i=1}^{n} a_{i2} T_{i2}(\alpha_k) + ... + \sum_{i=1}^{n} + a_{im} T_{im}(\alpha_k)] = 0$$

$$\begin{aligned} \Rightarrow \quad & a_{11} T_{11}(\alpha_k) + a_{21} T_{21}(\alpha_k) + ... + a_{n1} T_{n1}(\alpha_k) \\ & + a_{12} T_{12}(\alpha_k) + a_{22} T_{22}(\alpha_k) + ... + a_{n2} T_{n2}(\alpha_k) \\ & + ... + a_{1m} T_{1m}(\alpha_k) + a_{2m} T_{2m}(\alpha_k) + ... + a_{nm} T_{nm}(\alpha_k) = 0 \end{aligned}$$

$$\begin{aligned} = a_{11}\beta_1 + a_{21}\beta_2 + ... + a_{n1}\beta_n + a_{12}\beta_1 + a_{22}\beta_2 + ... + a_{n2}\beta_n \\ + \alpha_{1m}\beta_1 + \alpha_{2m}\beta_2 + ... + a_{nm}\beta_n = 0 \end{aligned}$$

....

$$\left(\begin{aligned} \because T_{ij}(\alpha_k) &= 0, (j \neq k) \\ T_{ij}(\alpha_k) &= \beta_i, (j = k) \end{aligned} \right)$$

Since $\beta' = \{\beta_1, \beta_2, ..., \beta_n)$ is a basis of V, therefore it is linearly independent so that

$$a_{11} = 0 = a_{21} = ... = a_{n1}$$
$$a_{12} = 0 = a_{22} = ... = a_{n2}$$

....

....

$$a_{1m} = 0 = a_{2m} = ... = a_{nm}$$

Thus, $\{T_{ij} : 1 \leq i \leq m, i \leq j \leq n\}$ is linearly independent.

(ii) Now we show that $\{T_{ij} : 1 \leq i \leq m, i \leq j \leq n\}$ spans $L(U, V)$. For this, let T be an arbitrary linear transformation from U into V, *i.e.*, $T \in L(U, V)$.

For $\alpha_j \in U$, $T(\alpha_j) \in V$ and $\beta' = \{\beta_1, \beta_2, ..., \beta_n\}$ is a basis of V so that

$$T(\alpha_j) = a_{1j}\beta_1 + a_{2j}\beta_2 + ... + a_{nj}\beta_n \,;\, 1 < j < n$$

where $a_{1j}, a_{2j}, ..., a_{nj}$ are the coordinates of vector $T(\alpha_j)$ in β'.

$$T(\alpha_j) = \sum_{i=1}^{n} a_{ij}\beta_i = \sum_{i=1}^{n}\sum_{j=1}^{m} a_{ij}T_{ij}(\alpha_j)$$

$$\Rightarrow \qquad T = \sum_{i=1}^{n}\sum_{j=1}^{m} a_{ij}T_{ij}$$

$\Rightarrow \quad \{T_{ij} : 1 \le i \le m, 1 \le j \le n\}$ generates $L(U, V)$.

$\Rightarrow \quad \{T_{ij} : 1 \le i \le m, 1 \le j \le n\}$ is a basis of $L(U, V)$.

Hence, $L(U, V)$ is finite-dimensional and dim. $L(U, V) = mn$.

THEOREM 3. ***Let U, V and W be linear spaces aver the field F. Let T_1 be a linear transformation from U into V and T_2 be a linear transformations from V into W, then the composed function T_2T_1 is defined by***

$$(T_2T_1)(\alpha) = T_2[T_1(\alpha)], \text{ for all } \alpha \in U$$

is a linear transformation from U into W.

PROOF. For $\alpha, \beta \in U$ and $a, b \in F$, we have

$$(T_2T_1(a\alpha + b\beta)) = T_2[T_1(a\alpha + b\beta)] = T_2[aT_1(\alpha) + bT_1(\beta)) \quad [\because T_1 \text{ is linear.}]$$

$$= a(T_2T_1)(\alpha) + b(T_2T_1)(\beta) \quad [\because T_2 \text{ is linear.}]$$

$\therefore$ T_2T_1 is a linear transformation from U into W.

3.20 LINEAR OPERATOR

Definition. *If V is a linear space over the field F, then a linear transformation from V into V is called a linear operator.*

In case of above theorem if U, V and W are replaced by V, then T_1 and T_2 are linear operators on the space V and $T_2\,T_1$ is also a linear operator on V. Thus the linear space $L(V, V)$ has a 'multiplication' defined on it by composition. In this case the operator $T_1\,T_2$ is also defined but in general $T_2T_1 \neq T_1T_2$. Therefore, if T is a linear operator on V, then we can compose T with T as follows :

$$T^2 = T\,T$$

$$T^3 = T\,T\,T$$

in general, $\quad T^n = T\,T...T$ (n times) for $n = 1, 2, 3, ...$

REMARK

- If $T \neq 0$, then we define $T^0 = 1$ (Identity transformation).

3.21 ALGEBRA OF LINEAR OPERATORS

THEOREM 1. ***Let V be a linear space over the field F and let T, T_1, T_2 and T_3 be linear operators on V and let c be an element in F, then***

(i) ***IT = TI = T being an identity operator.***

(ii) $T_1(T_2 + T_3) = T_1T_2 + T_1T_3$; $(T_2 + T_3)T_1 = T_2T_1 + T_3T_1$

(iii) $T_1(T_2T_3) = (T_1T_2)T_3$ ***(iv)*** $c(T_1T_2) = (cT_1)T_2 = T_1(cT_2)$

(v) $T0 = 0T = 0$, ***0 being zero linear operator.***

PROOF. (i) For $\alpha \in V$

$$(IT)(\alpha) = I[T(\alpha)] = T(\alpha) \qquad [\because I(\alpha) = \alpha]$$

$$IT = T$$

Also, $(TI)(\alpha) = T[I(\alpha)] = T(\alpha)$

$\Rightarrow$ $TI = T$

Thus, $IT = TI = T$

(ii) For any $\alpha \in V$

$$[T_1(T_2 + T_3)](\alpha) = T_1[(T_2+T_3)(\alpha)] = T_1[T_2(\alpha) + T_3(\alpha)$$
$$= (T_1T_2)(\alpha) + (T_1T_3)(\alpha) = (T_1T_2+T_1T_3)(\alpha)$$

$\therefore$ $T_1(T_2+T_3) = T_1T_2+T_1T_3$

Similarly,

$$(T_2+T_3)T_1 = T_2T_1+T_3T_1$$

(iii) For any $\alpha \in V$

$$[T_1(T_2T_3)](\alpha) = T_1[(T_2T_3)(\alpha)] = T_1[T_2(T_3(\alpha))]$$
$$= (T_1T_2)[T_3(\alpha)] = [(T_1T_2)T_3](\alpha)$$

$\therefore$ $T_1(T_2T_3) = (T_1T_2)T_3$

(iv) For any $\alpha \in V$, $c \in F$

$$[c(T_1T_2)](\alpha) = c[(T_1T_2)(\alpha)] = c\,[T_1(T_2(\alpha))]$$
$$= (cT_1)[T_2(\alpha)] = [(cT_1)T_2](\alpha)$$

$\therefore$ $c(T_1T_2) = (cT_1)T_2$

Also, $[c(T_1T_2)](\alpha) = (cT_1)[T_2(\alpha)] = T_1(cT_2(\alpha)] = T_1[(cT_2)](\alpha)$

$\therefore$ $c(T_1T_2) = T_1(cT_2)$

Thus, $c(T_1T_2) = (cT_1)T_2 = T_1(cT_2)$

(v) For any $\alpha \in V$,

$$(T0)(\alpha) = T[0(\alpha)] = T(0) \qquad [\because 0(\alpha) = 0]$$
$$= 0$$

Similarly, $0T = 0$.

3.22 RANGE AND NULL SPACE OF A LINEAR TRANSFORMATION

(i) Range space of a linear transformation. If T is a linear transformation from U into V, then the range of T is a subspace of V. Let R_T be the range of T, that is, the set of all vectors β in V such that $T(\alpha) = \beta$ for some $\alpha \in U$,

i.e., $R_T = \{\beta \in V : T(\alpha) = \beta, \text{ for some } \alpha \in U\}$.

If U is finite-dimensional, then the dimension of range of T is called rank of T and is denoted by $\rho(T)$.

(ii) Null space of a linear transformation. If T is a linear transformation from a linear space U into a linear space V, then the null space of T denoted by $N(T)$ is the set of all vectors α in U such that $T(\alpha) = 0$, where 0 is the zero vector in V, *i.e.,*

$N(T) = \{\alpha \in U : T(\alpha) = 0\}$.

If U is finite-dimensional, then the dimension of null space $N(T)$ is called nullity of T and is denoted by $n(T)$.

REMARK

- Kernel of T is also known as null space of T.

THEOREM 1. ***Let U and V be linear spaces over the field F and let T be a linear transformation from U into V. Suppose U is finite-dimensional. Then***

$$\textbf{rank } (T) + \textbf{nullity } (T) = \textbf{dim. } U$$

i.e., $$\rho(T) + n(T) = \text{dim. } U$$

PROOF. Let $\{\alpha_1, \alpha_2, ..., \alpha_k)$ be the basis of N_T, the null space of T. Let the dimension of U be n, so that $\alpha_{k+1}, \alpha_{k+2}, ..., \alpha_n \in U$ such that $\{\alpha_1, \alpha_2, ..., \alpha_n)$ forms a basis of U. Therefore, dim. N_T, $= k$ and dim. $U = n$.

We claim that $[T(\alpha_{k+1}), T(\alpha_{k+2}),..., T(\alpha_n)]$ is a basis of range of T.

For scalars $a_i \in F$ we have

$$a_{k+1}T(\alpha_{k+1})+a_{k+2}\,T(\alpha_{k+2})+ \,...\, +a_n\,T(\alpha_n)=0$$

$$\Rightarrow \quad \sum_{i=k+1}^{n} a_i T(\alpha_i) = 0 \quad \Rightarrow \quad T\left(\sum_{i=k+1}^{n} a_i\alpha_i\right) = 0$$

$$\Rightarrow \quad \sum_{i=k+1}^{n} a_i\alpha_i \in N_T$$

Since $\{\alpha_1, \alpha_1, ..., \alpha_k\}$ is the basis of N_T, so that for some scalars $b_1, b_2,,..., b_k$, we have

$$\sum_{i=k+1}^{n} a_i\alpha_i = b_1\alpha_1 + b_2\alpha_2 + ... + b_k\alpha_k$$

$$\Rightarrow \quad b_1\alpha_1 + b_2\alpha_2 + ... + b_k\alpha_k - \sum_{i=k+1}^{n} a_i\alpha_i = 0$$

$$\Rightarrow \quad b_1\alpha_1 + b_2\alpha_2 + ... + b_k\alpha_k + (-a_{k+1})\alpha_{k+1} + ... + (-a_n)\alpha_n = 0$$

Since $\alpha_1, \alpha_2, ...,\alpha_n$ are linearly independent, we must have

$$b_1 = 0 = b_2 = ... = b_k = a_{k+1} = a_n.$$

$\therefore$ $[T(\alpha_{k+1}), T(\alpha_{k+2}), ... , T(\alpha_n)]$ is linearly independent.

Now, we shall show that $T(\alpha_{k+1}), T(\alpha_{k+2}), ... , T(\alpha_n)]$ spans range of T.

For this, let $T(\alpha) \in R_T$ (range of T) for some $\alpha \in U$.

Since $\{\alpha_1, \alpha_2,..., \alpha_n\}$ spans U so that

$$\alpha = a_1\alpha_1 + a_2\alpha_2 + ... + a_n\alpha_n$$

For $a_i'\,s \in F$, we have

$$\begin{aligned} T(\alpha) &= T(a_1\alpha_1 + a_2\alpha_2 + ... + a_n\alpha_n) \\ &= T(a_1\alpha_1) + T(a_2\alpha_2) + ... + T(a_n\alpha_n) \qquad [\because T \text{ is linear.}] \\ &= a_1T(\alpha_1) + a_2T(\alpha_2) + ... + a_kT(\alpha_k) + a_{k+1}T(\alpha_{k+1}) + ... + a_nT(\alpha_n) \\ &= a_{k+1}T(\alpha_{k+1}) + a_{k+2}T(\alpha_{k+2}) + ... + a_nT(\alpha_n) \\ &\qquad [\because T(\alpha_i) = 0,\ 1 \le i \le k] \end{aligned}$$

Thus, $\quad T(\alpha_{k+1})...T(\alpha_n)$ spans R_T.

Hence, $\quad [T(\alpha_{k+1}),..., T(\alpha_n)]$ is a basis of R_T.

Accordingly, $\dim. R_T = n - k = \dim. U - \dim. N_T$

$\therefore$ $\dim. R_T + \dim. N_T = \dim. U$

Hence, $\text{rank}(T) + \text{nullity}(T) = \dim. U$.

3.23 INVERTIBLE LINEAR TRANSFORMATION

A linear transformation T from a linear space $U(F)$ into $V(F)$ is called invertible or regular if : there exists a unique linear transformation T^{-1} (called the inverse of T) from $V(F)$ into $U(F)$ such that (T^{-1}) is the identity linear transformation on U and (TT^{-1}) is the identity transformation on V.

Furthermore, T is invertible iff

(i) T is one-to-one

(ii) T is onto, *i.e.*, $RT = V$

THEOREM 1. ***Let U and V be linear spaces over the same field F and left T be a linear transformation from U into V. If T is invertible, then T^{-1} is a linear transformation from V into U.***

PROOF. Since T is invertible, so for each $\beta \in V$, there is a unique $\alpha \in U$ such that

$$T(\alpha) = \beta \Leftrightarrow T^{-1}(\beta) = \alpha$$

Now, we shall show that T^{-1} is linear.

For $\alpha_1, \alpha_2 \in U$ and $a, b \in F$

$$T(a\alpha_1 + b\alpha_2) = aT(\alpha_1) + bT(\alpha_2) \qquad [\because T \text{ is linear.}]$$

But for β_1 and β_2 in V, there are unique $\alpha_1, \alpha_2 \in U$ respectively, such that

$$T(\alpha_1) = \beta_1 \Leftrightarrow T^{-1}(\beta_1) = \alpha_1$$

and $$T(\alpha_2) = \beta_2 \Leftrightarrow T^{-1}(\beta_2) = \alpha_2$$

Thus, we have

$$T(a\alpha_1 + b\alpha_2) = a\beta_1 + b\beta_2$$

$$\Rightarrow \quad a\alpha_1 + b\alpha_2 = T^{-1}(a\beta_1 + b\beta_2) \qquad [\because a\alpha_1 + b\alpha_2 \text{ is unique in } V.]$$

$$\Rightarrow \quad aT^{-1}(\beta_1) + b\,T^{-1}(\beta_2) = T^{-1}(a\beta_1 + b\beta_2)$$

Hence, T^{-1} is a linear transformation.

THEOREM 2. ***Let T_1 be an invertible linear transformation from U(F) into V(F) and T_2 an invertible linear transformation from V(F) into W(F). Then T_1T_2 is invertible and $(T_2T_1)^{-1} = T_1^{-1}T_2^{-1}$***

PROOF. To show T_2T_1 is invertible, we shall show that it is one-one and onto.

If $\alpha_1, \alpha_2 \in U$ such that $(T_2T_1)(\alpha_1) = (T_2T_1)(\alpha_2)$, then

$$T_2T_1)(\alpha_1) = (T_2T_1)(\alpha_2) \Rightarrow T_2[T_1(\alpha_1)] = T_2[T_1(\alpha_2)]$$

$$\Rightarrow T_1(\alpha_1) = T_1(\alpha_2) \qquad [\because T_2 \text{ is one-one.}]$$

$$\Rightarrow \alpha_1 = \alpha_2 \qquad [\because T_1 \text{ is one-one.}]$$

Thus, T_2T_1 is one-one.

Also, T_1 and T_2 being onto, then for each $\beta \in V$, there exists a unique $\alpha \in U$ such that

$$T_1(\alpha) = \beta$$

and for each $\gamma \in W$, there exists a unique $\beta \in V$ such that $T_2(\beta) = \gamma$.

Thus, $\gamma \in W \Rightarrow$ there exists $\beta \in V : \gamma = T_2(\beta)$.

$\Rightarrow$ there exists $\alpha \in U : \gamma = T_2(T_1(\alpha))$ $[\because T_1(\alpha) = \beta]$

$\Rightarrow$ there exists $\alpha \in U : \gamma = (T_2 T_1)(\alpha)$.

Therefore $(T_2 T_1)$ is onto. Hence, $(T_2 T_1)$ is invertible.

Also, $(T_2T_1)(T_1^{-1}T_2^{-1}) = T_2(T_2T_1^{-1})T_2^{-1} = (T_2I)T_2^{-1} = T_2T_2^{-1} = I$

Similarly,

$$(T_1^{-1}T_2^{-1})(T_2T_1) = T_1^{-1}(T_2^{-1}T_2)T_1 = T_2^{-1}(IT_1) = T_1^{-1}T_1 = I$$

Hence, $(T_2T_1^{-1}) = T_1^{-1}T_2^{-1}$

3.24 NON-SINGULAR LINEAR TRANSFORMATIONS

Let U and V be linear spaces over the field F. Then a linear transformation T from U into V is called non-singular if the null space of T is (0).

Thus, if T is non-singular

$$T(\alpha) = 0 \Rightarrow \alpha = 0$$

Also, when T is non-singular and $\alpha, \beta \in U$

$$T(\alpha) = T(\beta) \qquad \Rightarrow T(\alpha) - T(\beta) = 0$$

$\Rightarrow$ $T(\alpha-\beta) = 0$ $[\because T$ is linear.$]$

$\Rightarrow$ $\alpha-\beta = 0$ $[\because T$ is non-singular.$]$

$\Rightarrow$ $\alpha = \beta$

Hence, when T is non-singular, implies that T is one-one.

THEOREM 1. ***Let T be a linear transformation from U(F) into V(F). Then T is non-singular if and only if T carries each linearly independent subset of U onto a linearly independent subset of V.***

PROOF. Let us first suppose that T is non-singular. Now, let

$$S = \{\alpha_1, \alpha_2,, ..., \alpha_k)$$

be an arbitrary linearly independent subset of U. Then we have to show that the set

$$S_1 = \{T(\alpha_1), T(\alpha_2), ..., T(\alpha_k)\}$$

is linearly independent subset of V.

For scalars $a_1, a_2, ... a_k \in F$ we have

$$a_1 T(\alpha_1) + a_2T(\alpha_2) + ... + a_kT(\alpha_k)\} = 0$$

$\Rightarrow$ $T(a_1\alpha_1) + T(a_2\alpha_2) + ... + T(a_k\alpha_k)\} = 0$ $[\because T$ is linear.$]$

$\Rightarrow$ $T(a_1\alpha_1 + a_2\alpha_2 + ... + a_k\alpha_k) = 0$ $[\because T$ is linear.$]$

$\Rightarrow$ $a_1\alpha_1 + a_2\alpha_2 + ... + a_k\alpha_k = 0$ $[\because T$ is non-singular.$]$

$\Rightarrow$ $a_1 = a_2 = ... = a_k = 0$ $[\because S$ is linear independent.$]$

Hence, S_1 is linearly independent.

Conversely, suppose that T carries each linearly independent subset of U into a linearly independent subset of V. Let α be a non-zero vector in U, then $\{\alpha\}$ is linearly independent so is $\{T(\alpha)\}$. Consequently, $T(\alpha)$ # 0 because the set consisting of the zero vector alone is dependent. Therefore, the null space of T is the zero space and hence T is non-singular.

THEOREM 2. ***Let U and V be finite-dimensional linear spaces over the field F such that dim. U= dim. V. If T is a linear transformation from U into V, then the following are equivalent:***

(i) T is invertible.

(ii) T is non-singular.

(iii) T is onto, that is, the range of T is V.

(iv) If $\{\alpha_1, \alpha_2, ..., \alpha_n\}$ ***is a basis of U, then*** $\{T(\alpha_1), T(\alpha_2), ..., T(\alpha_2)\}$ ***is a basis of V.***

PROOF. **(i) ⇒ (ii) :** Since T is invertible, so it is one-one and onto, therefore, T is non-singular.

(ii) ⇒ (iii): Let T be non-singular and let $\{\alpha_1, \alpha_2, ..., \alpha_n\}$ be the basis of U, then the set $\{T(\alpha_1), T(\alpha_2), ..., T(\alpha_n)\}$ is linearly independent subset of V, but dim. U = dim. V, therefore $\{T(\alpha_1), ..., T(\alpha_n)\}$ is a basis for V.

For any $\beta \in V$, $a_1, a_2, ..., a_n \in F$, we have

$$\beta = a_1 T(\alpha_1) + a_2 T(\alpha_2) + ... + a_n T(\alpha_n)$$

$$\beta = T(a_1 \alpha_1 + a_2 \alpha_2 + ... + a_n \alpha_n) \quad [\because T \text{ is linear.}]$$

$$\Rightarrow \quad \beta \in R_T$$

Thus, $V \subseteq R_T$, but $R_T \subseteq V$

$\therefore \quad R_T = V$

i.e., the range of $T = V$

(iii) ⇒ (iv) : Suppose range of $T = V$. Let the set $\{\alpha_1, \alpha_2, ..., \alpha_n\}$ be a basis of U so that an arbitrary element $\alpha \in U$ is expressible as linear combination of $\alpha_1, \alpha_2, ..., \alpha_n$.

$\therefore \quad \alpha = b_1 \alpha_1 + b_2 \alpha_2 + ... + b_n \alpha_n$ for some scalars, $b_1, b_2, ..., b_n \in F$

$$\Rightarrow T(\alpha) = T(b_1 \alpha_1 + b_2 \alpha_2 + ... + b_n \alpha_n)$$

$$= b_1 T(\alpha_1) + b_2 T(\alpha_2) +, ...,+ b_n T(\alpha_n) \quad [\because T \text{ is linear.}]$$

This shows that each element of range of T is expressible as a linear combination of $\{T(\alpha_1) + T(\alpha_2), ..., T(\alpha_n)\}$.Thus, the set

$$\{T(\alpha_1) + T(\alpha_2), ..., T(\alpha_n)\}.$$

spans R_T. Since $R_T = V$. Also, dim. U = dim. $V = n$.

Hence, $\{T(\alpha_1), T(\alpha_2), ..., T(\alpha_n)\}$ forms a basis of V.

(iv) ⇒ (i): Let $\{\alpha_1, \alpha_2, ..., \alpha_n\}$ be a basis of U such that $\{T(\alpha_1), T(\alpha_2), ..., T(\alpha_n)\}$ is a basis of V.

Let α be an arbitrary element of U, then for $b_1, b_2, ..., b_n \in F$, we have

$$\alpha = b_1 \alpha_1 + b_2 \alpha_2 + ... + b_n \alpha_n$$

Now, $\quad T(\alpha) = 0$

$$\Rightarrow \quad T(b_1 \alpha_1 + b_2 \alpha_2 + ... + b_n \alpha_n) = 0$$

$$\Rightarrow \quad b_1 T(\alpha_1) + b_2 T(\alpha_2) + ... + b_n T(\alpha_n) = 0 \quad [\because T \text{ is linear.}]$$

$$\Rightarrow \quad b_1 = b_2 = ... = b_n = 0$$

$[\because \{T(\alpha_1), T(\alpha_2), ..., T(\alpha_n)\}$ is linearly independent.]

$$\Rightarrow b_1 \alpha_1 + b_2 \alpha_2 + ... + b_n \alpha_n = 0 \Rightarrow \alpha = 0$$

Hence, T is non-singular and therefore T is one-one.

Also, $\{T(\alpha_1), T(\alpha_2), ..., T(\alpha_n)\}$ spans V and range of T is V. Consequently, T is one-one and hence T is invertible.

3.25 COORDINATE VECTOR

Let V be a finite-dimensional linear space over a field F and let dim. $V = n$, then $B = \{\alpha_1, \alpha_2, ..., \alpha_n\}$ is a basis of V and for $\alpha \in V$, suppose that

$$\alpha = a_1 \alpha_1 + a_2\alpha_2 + ... + a_n\alpha_n$$

for $a_i's \in F$. Then the coordinate vector of α relative to β, which we write as a column vector unless otherwise specified or implied, is

$$[\alpha]_B = \begin{bmatrix} a_1 \\ a_2 \\ \vdots \\ a_n \end{bmatrix}$$

3.26 MATRIX REPRESENTATION OF A LINEAR TRANSFORMATION

Let U be an m-dimensional linear space over a field F and let V be an n-dimensional linear space over the field F. Let $B = \{\alpha_1, \alpha_2, \alpha_m\}$ and $B' = \{\beta_1, \beta_2, ..., \beta_n)$ be the basis of U and V respectively. If T is a linear transformation from U into V, then $T(\alpha_1)$, $T(\alpha_2)$, ...,$T(\alpha_m)$ are vectors in V. Since $B' = \beta_1, \beta_2, ..., \beta_n\}$ is a basis of V so that each $T(\alpha_i)$ is a linear combination of the elements of B'. For $a_{ij} \in F$, $1 \le i \le m$, $1 \le j \le n$, we have

$$T(\alpha_1) = \{a_{11}\beta_1 + a_{12}\beta_2 + ,..., + a_{1n}\beta_n)$$

$$T(\alpha_2) = \{a_{21}\beta_1 + a_{22}\beta_2 + ,..., + a_{2n}\beta_n)$$

....

$$T(\alpha_m) = \{a_{m1}\beta_1 + a_{m2}\beta_2 + ,..., + a_{mn}\beta_n)$$

Definition. *The transpose of the above matrix of coefficients, denoted by $[T]_B$ is called the matrix representation of T relative to the ordered basis B.*

Thus,
$$[T]_B = \begin{bmatrix} a_{11} & a_{21} & \cdots & a_{m1} \\ a_{12} & a_{22} & \cdots & a_{m1} \\ \cdots & \cdots & \cdots & \cdots \\ a_{1n} & a_{2n} & \cdots & a_{mn} \end{bmatrix}_{n \times m}$$

For example : Let V be the linear space of polynomials in 't' over the field of reals R, of degree ≤ 3, and let

$$D : V \to V$$

be the differential operator defined by $D\,[p(t)] = \alpha = \sum_{i=1}^{n} a_i\alpha_i$

We compute the matrix of D in the basis $B = [1, t, t^2, t^3]$ as follows

$$D(1) = 0 = 0 + 0t + 0t^2 + 0t^3$$

$$D(t) = 1 = 1+0t+0t^2+0t^3$$
$$D(t^2) = 2t = 0+2t+0t^2+0t^3$$
$$D(t^3) = 3t^2 = 0+0t+3t^2+0t^3$$

Thus, the matrix of D relative to B is given by

$$[D]_B = \begin{bmatrix} 0 & 1 & 0 & 0 \\ 0 & 0 & 1 & 0 \\ 0 & 0 & 0 & 3 \\ 0 & 0 & 0 & 0 \end{bmatrix}$$

THEOREM 1. ***Let U be an m-dimensional linear space aver the field F and V an n-dimensional linear space over the field F. Let B be an ordered basis for U and B′ an ordered basis for V. Let T be any linear transformation from U into V. Then for any vector α ∈ U.***

$$[T]_B\,[\alpha]_B = [T(\alpha)]_{B'}$$

PROOF. Let $B = \{\alpha_1, \alpha_2,..., \alpha_m\}$ be an ordered basis for U and $B' = \{\beta_1, \beta_2,..., \beta_n\}$ an ordered basis for V. T is linear transformation from U into V, then T is determined by its action on the vectors α_j, $1 \le i \le m$. Each of m vectors $T(\alpha_i)$ is uniquely expressible as a linear combination of elements of B' :

$$T(a_i) = \sum_{j=1}^{n} a_{ij}\beta_j \qquad ...(1)$$

where $a_{i1}, a_{i2},..., a_{in}$ are the coordinates of $T(\alpha_i)$ in the ordered basis B'.
If α be any vector in U, then

$$\alpha = a_1\alpha_1 + a_2\alpha_2 + \ ...\ + a_m\alpha_n$$

$$\therefore \qquad [\alpha]_B = \begin{bmatrix} a_1 \\ a_2 \\ \vdots \\ a_m \end{bmatrix}$$

Now,
$$\begin{aligned} T(\alpha) &= T(a_1\alpha_1 + a_2\alpha_2 + \ ...\ + a_m\alpha_m) \\ &= a_1T(\alpha_1) + a_2T(\alpha_2) + \ ...\ + a_mT(\alpha_m) \\ &= a_1 \sum_{j=1}^{n} a_{1j}\beta_j + a_2 \sum_{j=1}^{n} a_2a_{2j}\beta_j + ... + a_m \sum_{j=1}^{n} a_ma_{mj}\beta_j \qquad \text{[using (1)]} \\ &= \sum_{j=1}^{n} a_1a_{1j}\beta_j + \sum_{j=1}^{n} a_2a_{2j}\beta_j + ... + \sum_{j=1}^{n} a_ma_{mj}\beta_j \\ &= a_1a_{11}\beta_1 + a_1a_{12}\beta_2 + ... + a_1a_{1n}\beta_n + a_2a_{21}\beta_1 + a_2a_{22}\beta_2 + \ ...\ + a_2a_{2n}\beta_n \\ &\quad .... \\ &\quad + a_ma_{m1}\beta_1 + a_ma_{m2}\beta_2 + \ ...\ + a_ma_{mn}\beta_n \end{aligned}$$

$$\therefore \quad [T(\alpha)]_{B'} = \begin{bmatrix} a_1a_{11} + a_2a_{21} + ... + a_ma_{m1} \\ a_1a_{12} + a_2a_{22} + ... + a_ma_{m2} \\ .... \\ a_1a_{1n} + a_2a_{2n} + ... + a_ma_{mn} \end{bmatrix}_{m\times n}$$

$$= \begin{bmatrix} a_{11} & a_{21} & \cdots & a_{m1} \\ a_{12} & a_{22} & \cdots & a_{m2} \\ \cdots & \cdots & \cdots & \cdots \\ a_{1n} & a_{2n} & \cdots & a_{mn} \end{bmatrix}_{n\times m} \begin{bmatrix} a_1 \\ a_2 \\ \vdots \\ a_m \end{bmatrix}_{m\times 1}$$

$[T(\alpha)]_{B'} = [T]_B[\alpha]_B$.

THEOREM 2. ***Let U, V and W be linear spaces over the field F of respective dimensions n, m and p. Let T_1 be a linear transformation from U into V and T_2 a linear transformation from V into W. If B, B′ and B″ are the ordered bases for the spaces U, V and W respectively, if A is the matrix of T_1, relative to the pair B, B′ and B is the matrix of T_2 relative to the pair B′ and B″, then the matrix of (T_2T_1) relative to the pair B, B″ is the product matrix C = BA.***

PROOF. Let $B = \{\alpha_1, \alpha_2,..., \alpha_n\}$, $B' = \{\beta_1, \beta_2, \ldots, \beta_m\}$ and $B'' = \{\gamma_1, \gamma_2, ..., \gamma_p]$ be the bases of U, V and W respectively. If α is any vector in U, then

$$[T_1\{\alpha)]_{B'} = [T_1]_B[\alpha]_B \qquad \text{[By above theorem]}$$
$$= A[\alpha]_B \qquad [\because A = [T_1]_B]$$

and $\quad [T_2(T_1(\alpha))]_{B''} = [T_2]_{B'}[T_1(\alpha)]_{B'} = B[T_1(\alpha)]_{B'} \qquad [\because B = [T_2]B']$

$\therefore \quad [(T_2T_1](\alpha)]_{B''} = BA[\alpha]_B.$

Hence, by the definition and uniqueness of the representing matrix, we must have $C = BA$ as the matrix of (T_2T_1) relativeto B, B''.

THEOREM 3. ***Let V be an n-dimensional linear space over the field F and B be an ordered basis of V. If T_1 and T_2 are linear operators from V into V, then***

(i) $[T_1+T_2]_B = [T_1]_B+[T_2]_B$

(ii) $[cT_1]_B = c[T_1]_B$, ***for*** $c \in F$

(iii) $[T_2T_1]_B = [T_1]_B[T_2]_B$.

PROOF. Let $\{a_1, a_2,..., a_n\}$ be the basis of V, then for $a_{ij} \in F$ and $b_{ij} \in F$, $1 \le i \le n$, $1 \le j \le n$, we have

$$T_1(\alpha_1) = a_{11}\alpha_1 + a_{12}\alpha_2 + ... + a_{1n}\alpha_n$$
$$T_1(\alpha_2) = a_{21}\alpha_1 + a_{22}\alpha_2 + ... + a_{2n}\alpha_n$$
$$\cdots\cdots\cdots\cdots\cdots$$
$$T_1(\alpha_n) = a_{n1}\alpha_1 + a_{n2}\alpha_2 + ... + a_{nn}\alpha_n$$

$$\therefore \quad [T_1]_B = \begin{bmatrix} a_{11} & a_{21} & \cdots & a_{n1} \\ a_{12} & a_{22} & \cdots & a_{n2} \\ \cdots & \cdots & \cdots & \cdots \\ a_{1n} & a_{2n} & \cdots & a_{nn} \end{bmatrix}$$

Also,

$$T_2(\alpha_2) = b_{11}\alpha_1 + b_{12}\alpha_2 + ... + b_{1n}\alpha_n$$
$$T_2(\alpha_2) = b_{21}\alpha_1 + b_{22}\alpha_2 + ... + b_{2n}\alpha_n$$
$$\cdots\cdots\cdots\cdots\cdots$$
$$T_2(\alpha_n) = b_{n1}\alpha_1 + b_{n2}\alpha_2 + ... + b_{nn}\alpha_n$$

$$[T_2]_B = \begin{bmatrix} b_{11} & b_{21} & \cdots & b_{n1} \\ b_{12} & b_{22} & \cdots & b_{n2} \\ \cdots & \cdots & \cdots & \cdots \\ b_{1n} & b_{2n} & \cdots & b_{nn} \end{bmatrix}$$

$\therefore$

(i) $(T_1 + T_2)(\alpha_1) = (T_1)(\alpha_1) + (T_2)(\alpha_1)$

$= (a_{11} + b_{11})\alpha_1 + (a_{12} + b_{12})\alpha_2 + \ldots + (a_{1n} + b_{1n})\alpha_n$

$(T_1 + T_2)(\alpha_2) = (T_1)(\alpha_2) + (T_2)(\alpha_2)$

$= (a_{21} + b_{21})\alpha_1 + (a_{22} + b_{22})\alpha_2 + \ldots + (a_{2n} + b_{2n})\alpha_n$

$(T_1 + T_2)(\alpha_n) = (T_1)(\alpha_n) + (T_2)(\alpha_n)$

$= (a_{n1} + b_{n1})\alpha_1 + (a_{n2} + b_{n2})\alpha_2 + \ldots + (a_{nn} + b_{nn})\alpha_n$

$$[T_1 + T_2]_B = \begin{bmatrix} (a_{11}+b_{11}) & (a_{21}+b_{21}) & \cdots & (a_{n1}+b_{n1}) \\ (a_{12}+b_{12}) & (a_{22}+b_{22}) & \cdots & (a_{2n}+b_{2n}) \\ \ldots & \ldots & & \ldots \\ (a_{1n}+b_{1n}) & (a_{2n}+b_{2n}) & \cdots & (a_{nn}+b_{nn}) \end{bmatrix}$$

$$= \begin{bmatrix} a_{11} & a_{21} & \cdots & a_{n1} \\ a_{12} & a_{22} & \cdots & a_{2n} \\ \ldots & \ldots & \ldots & \ldots \\ a_{1n} & a_{2n} & \cdots & a_{nn} \end{bmatrix} + \begin{bmatrix} b_{11} & b_{21} & \cdots & b_{n1} \\ b_{12} & b_{22} & \cdots & b_{2n} \\ \ldots & \ldots & \ldots & \ldots \\ b_{1n} & b_{2n} & \cdots & b_{nn} \end{bmatrix}$$

$= [T_1]_B + [T_2]_B$

(ii) $(cT_1)(\alpha_1) = cT_1(\alpha_1) = ca_{11}\alpha_1 + ca_{12}\alpha_2 + \ldots + ca_{1n}\alpha_n$

$(cT_1)(\alpha_2) = cT_1(\alpha_2) = ca_{21}\alpha_1 + ca_{22}\alpha_2 + \ldots + ca_{2n}\alpha_n$

....

$(cT_1)(\alpha_n) = cT_1(\alpha_n) = ca_{n1}\alpha_1 + ca_{n2}\alpha_2 + \ldots + ca_{nn}\alpha_n$

$$\therefore \quad (cT_1)_B = \begin{bmatrix} ca_{11} & ca_{21} & \cdots & ca_{n1} \\ ca_{12} & ca_{22} & \cdots & ca_{2n} \\ \ldots & \ldots & \ldots & \ldots \\ ca_{1n} & ca_{2n} & \cdots & ca_{nn} \end{bmatrix} = \begin{bmatrix} a_{11} & a_{21} & \cdots & a_{n1} \\ a_{12} & a_{22} & \cdots & a_{2n} \\ \ldots & \ldots & \ldots & \ldots \\ a_{1n} & a_{2n} & \cdots & a_{nn} \end{bmatrix} = c[T_1]_B$$

(iii) $(T_2T_1)(\alpha_1) = T_2(T_1(\alpha_1))$

$= T_2(a_{11}\alpha_1 + a_{12}\alpha_2 + \ldots + a_{1n}\alpha_n)$

$= a_{11}T_2(\alpha_1) + a_{12}T_2(\alpha_2) + \ldots + a_{1n}T_2(\alpha_n)$ [$\because T_2$ is linear.]

$= a_{11}(b_{11}\alpha_1 + b_{12}\alpha_2 + \ldots + b_{1n}\alpha_n)$

$+ a_{12}(b_{21}\alpha_1 + b_{22}\alpha_2 + \ldots + b_{2n}\alpha_n)$

....

$+ a_{1n}(b_{n1}\alpha_1 + b_{n2}\alpha_2 + \ldots + b_{nn}\alpha_n)$

$= (a_{11}b_{11} + a_{12}b_{21} + \ldots + a_{1n}b_{n1})\alpha_1$

$+ (a_{11}b_{12} + a_{12}b_{22} + \ldots + a_{1n}b_{n2})\alpha_2$

....

$+ (a_{11}b_{1n} + a_{12}b_{2n} + \ldots + a_{1n}b_{nn})\alpha_n$

$(T_2T_1)(\alpha_2) = T_2(T_1(\alpha_2)) = T_2(a_{21}\alpha_1 + a_{22}\alpha_2 + \ldots + a_{2n}\alpha_n)$

$= a_{21}T_2(\alpha_1) + a_{22}T_2(\alpha_2) + \ldots + a_{2n}T_2(\alpha_n)$

$= a_{21}(b_{11}\alpha_1) + b_{12}\alpha_2 + \ldots + b_{1n}\alpha_n)$

$$
\begin{aligned}
&+a_{22}(b_{21}\alpha_1+b_{22}\alpha_2+\ldots+b_{2n}\alpha_n)\\
&\ldots\ldots\ldots\ldots\ldots\ldots\\
&+a_{2n}(b_{n1}\alpha_1+b_{n2}\alpha_2+\ldots+b_{nn}\alpha_n)\\
&= a_{21}b_{11}+a_{22}b_{21}+\ldots+a_{2n}b_{n1})\alpha_1\\
&\quad+(a_{21}b_{12}+a_{22}b_{22}+\ldots+a_{2n}b_{n2})\alpha_2\\
&\ldots\ldots\ldots\ldots\ldots\ldots\\
&\quad+(a_{21}b_{1n}+a_{22}b_{2n}+\ldots+a_{2n}b_{nn})\alpha_n
\end{aligned}
$$

Similarly,

$$
\begin{aligned}
(T_2T_1)(\alpha_n) &= T_2(T_1(\alpha_n)) = T_2(a_{n1}\alpha_1+a_{n2}\alpha_2+\ldots+a_{nn}\alpha_n)\\
&= a_{n1}T_2(\alpha_1)+a_{n2}T_2(\alpha_2)+\ldots+a_{nn}T_2(\alpha_n)\\
&= a_{n1}(b_{11}\alpha_1+b_{12}\alpha_2+\ldots+b_{1n}\alpha_n)\\
&\quad+a_{n2}(b_{21}\alpha_1+b_{22}\alpha_2+\ldots+b_{2n}\alpha_n)\\
&\ldots\ldots\ldots\ldots\ldots\ldots\\
&\quad+a_{nn}(b_{n1}\alpha_1+b_{n2}\alpha_2+\ldots+b_{nn}\alpha_n)\\
&= (a_{n1}b_{11}+a_{n2}b_{21}+\ldots+a_{nn}b_{n1})\alpha_1\\
&\quad+(a_{n1}b_{12}+a_{n2}b_{22}+\ldots+a_{nn}b_{n2})\alpha_2\\
&\ldots\ldots\ldots\ldots\ldots\ldots\\
&\quad+(a_{n1}b_{1n}+a_{n2}b_{2n}+\ldots+a_{nn}b_{nn})\alpha_n
\end{aligned}
$$

$\therefore$

$$
[T_2T_1]_B = \begin{bmatrix}
a_{11}b_{11}+a_{12}b_{21}+\ldots+a_{1n}b_{n1} & a_{21}b_{11}+\ldots+a_{2n}b_{n1} & \ldots & a_{n1}b_{11}+\ldots+a_{nn}b_{n1}\\
a_{11}b_{12}+a_{12}b_{22}+\ldots+a_{1n}b_{n2} & a_{21}b_{12}+\ldots+a_{2n}b_{n2} & \ldots & a_{n1}b_{12}+\ldots+a_{nn}b_{n2}\\
\ldots\ldots & \ldots\ldots & \ldots & \ldots\ldots\\
a_{11}b_{1n}+a_{12}b_{2n}+\ldots+a_{1n}b_{nn} & a_{21}b_{1n}+\ldots+a_{2n}b_{nn} & \ldots & a_{n1}b_{1n}+\ldots+a_{nn}b_{nn}
\end{bmatrix}
$$

$$
= \begin{bmatrix}
b_{11} & b_{21} & \cdots & b_{n1}\\
b_{12} & b_{22} & \cdots & b_{n2}\\
\cdots & \cdots & \cdots & \cdots\\
b_{1n} & b_{2n} & \cdots & b_{nn}
\end{bmatrix}
\begin{bmatrix}
a_{11} & a_{21} & \cdots & a_{n1}\\
a_{12} & a_{22} & \cdots & a_{n2}\\
\cdots & \cdots & \cdots & \cdots\\
a_{1n} & a_{2n} & \cdots & a_{nn}
\end{bmatrix} = [T_2]_B[T_1]_B
$$

3.27 CHANGE OF BASIS

It has been shown that we can represent vectors by tuples (column vectors) and linear operators by matrix once we have selected a basis.

In this section we will see how the representation of matrix of linear transformation changes if we take another basis?

Let $\{\alpha_1, \alpha_2, \ldots, \alpha_n\}$ be a basis of V and let $\{\beta_1, \beta_2, \ldots, \beta_n\}$ be another basis of V and suppose

$$
\begin{aligned}
\beta_1 &= a_{11}\alpha_1+a_{12}\alpha_2+\ldots+a_{1n}\alpha_n\\
\beta_2 &= a_{21}\alpha_1+a_{22}\alpha_2+\ldots+a_{2n}\alpha_n\\
&\ldots\ldots\ldots\ldots\\
\beta_n &= a_{n1}\alpha_1+a_{n2}\alpha_2+\ldots+a_{nn}\alpha_n
\end{aligned}
$$

Then the transpose of the coefficient matrix of above equation is called the transition matrix from the basis $\{\alpha_1, \alpha_2,..., \alpha_n\}$ to the basis $\{\beta_1, \beta_2,..., \beta_n\}$

$$P = \begin{bmatrix} a_{11} & a_{21} & \cdots & a_{n1} \\ a_{12} & a_{22} & \cdots & a_{n2} \\ \cdots & \cdots & \cdots & \cdots \\ a_{1n} & a_{2n} & \cdots & a_{nn} \end{bmatrix}$$

REMARK

- P is invertible and its P–1 is the transition matrix from matrix new basis to old basis.

For example: Let $\{(1,0),(0,1)\}$and $\{(1,1),(-1,0)\}$ be two bases of R^2, then $(1,1)= 1.(0,1)+1(0,1)$ and $(-1,0)=-1(1,0)+0.(0,1)$

$$\therefore \quad P=\begin{bmatrix} 1 & -1 \\ 1 & 0 \end{bmatrix}$$

THEOREM 1. ***Let P be the transition matrix from a basis B to a basis B′ in a linear space V. Then for any vector $a \in V$, $P[\alpha]B'=[\alpha]B$ and $[\alpha]B'=P^{-1}[a]B$.***

PROOF. Let V be an n-dimensional linear space and let,

$$B= \{\alpha_1,\alpha_2,...,\alpha_n\} \text{ and } B'=\{\beta_1,\beta_2,...,\beta_n\}$$

be two bases of V and let P be the transition matrix from B to B'. Then we have,

$$\beta_1= a_{11}\alpha_1+a_{12}\alpha_2+...+a_{1n}\alpha_n$$
$$\beta_2= a_{21}\alpha_1+a_{22}\alpha_2+...+a_{2n}\alpha_n$$
$$\cdots\cdots\cdots\cdots\cdots\cdots$$
$$\beta_n=a_{n1}\alpha_1+a_{n2}\alpha_2+...+a_{nn}\alpha_n; \text{ for } \alpha_{ij} \in F$$

$$\therefore \quad P=\begin{bmatrix} a_{11} & a_{21} & \cdots & a_{n1} \\ a_{12} & a_{22} & \cdots & a_{n2} \\ \cdots & \cdots & \cdots & \cdots \\ a_{1n} & a_{2n} & \cdots & a_{nn} \end{bmatrix}.$$

Now suppose $\alpha \in V$ such that

$$\alpha=b_1\beta_1+b_2\beta_2+...+b_n\beta_n.$$

Substituting β's from above, we obtain,

$$\alpha=b_1(a_{11}\alpha_1+a_{12}\alpha_2+...+a_{1n}\alpha_n)+ b_2(a_{21}\alpha_1+a_{22}\alpha_2+...+a_{2n}\alpha_n)+$$
$$\cdots\cdots\cdots\cdots$$
$$+b_n(a_{n1}\alpha_1+a_{n2}\alpha_2+...+a_{2n}\alpha_n)$$
$$= (b_1a_{11}+b_2a_{12}+...+b_na_{1n})\alpha_1+ (b_1a_{12}+b_2a_{22}+...+b_na_{2n})\alpha_2$$
$$\cdots\cdots\cdots\cdots$$
$$+(b_1a_{1n}+b_2a_{2n}+...+b_na_{2n})\alpha_n$$

$$\text{Thus, } [\alpha]_{B'}= \begin{bmatrix} b_1 \\ b_2 \\ \cdots \\ b_n \end{bmatrix} \text{ and } [\alpha]_B= \begin{bmatrix} b_1a_{11}+b_2a_{21}+......+b_na_{n1} \\ b_1a_{12}+b_2a_{22}+......+b_na_{2n} \\ \cdots\cdots\cdots\cdots \\ b_1a_{1n}+b_2a_{2n}+......+b_na_{nn} \end{bmatrix}$$

Accordingly,

$$P[\alpha]_{B'}=\begin{bmatrix} a_{11} & a_{21} & \cdots & a_{n1} \\ a_{12} & a_{22} & \cdots & a_{n2} \\ \cdots & \cdots & \cdots & \cdots \\ a_{1n} & a_{2n} & \cdots & a_{nn} \end{bmatrix} \cdot \begin{bmatrix} b_1 \\ b_2 \\ \cdots \\ b_n \end{bmatrix} = \begin{bmatrix} b_1a_{11}+b_2a_{21}+\ldots\ldots+b_na_{n1} \\ b_1a_{12}+b_2a_{22}+\ldots\ldots+b_na_{2n} \\ \ldots\ldots\ldots\ldots\ldots\ldots\ldots\ldots \\ b_1a_{1n}+b_2a_{2n}+\ldots\ldots+b_na_{nn} \end{bmatrix} = [\alpha]_B$$

Furthermore, since P is invertible , hence

$$P[\alpha]_{B'}=[\alpha]_B \quad \Rightarrow \quad P^{-1}P[\alpha]_{B'}=P^{-1}[\alpha]_B$$

$$\Rightarrow \quad I[\alpha]_{B'}=P^{-1}[\alpha]_B \quad \Rightarrow \quad [\alpha]_{B'}=P^{-1}[\alpha]_B$$

THEOREM 2. ***Let P be the transition matrix from a basis to a basis B′ in a linear space V. Then for any linear operator T on V,***

$$[T]_{B'}=P^{-1}[T]_BP.$$

PROOF. Let α be any vector in V. then we have

$$[T]_B[\alpha]_B=[T(\alpha)]_B \qquad \ldots(1)$$

and $$P[\alpha]_{B'}=(\alpha)_B \qquad \ldots(2)$$

$\Rightarrow$ $[T]_BP[\alpha]_{B'}=[T]_B[\alpha]_B=[T(\alpha)]_B$ [using (1)]

$\Rightarrow$ $P^{-1}[T]_BP[\alpha]_{B'}=P^{-1}[T(\alpha)]_B=[T(\alpha)]_{B'}$ [By theorem (1)]

$=[T]_{B'}[\alpha]_{B'}$ [using (1)]

$\Rightarrow$ $P^{-1}[T]_BP=[T]_{B'}$ [$\because [\alpha]_{B'}\in F$ are arbitrary]

RECAPITULATIONS

- A mapping $T:U \to V$ is said to be linear transformation from U into V if $T(a\alpha+b\beta)=aT(\alpha)+bT(\beta)\ \forall\ \alpha,\beta \in U,\ a,b \in F$.
- Rank (T) + nullity (T) = dim(U).
- T is invertible iff it is bijective.
- T is non-singular if $T(\alpha)=0 \Rightarrow \alpha=0$.

SOLVED EXAMPLES

EXAMPLE 1. *Show that the mapping T defined by*

$T(a,b)=(\alpha+\beta,\alpha-\beta,\beta),\ \forall(\alpha,\beta)\in V_2(\mathrm{R})$ is a linear transformation.

SOLUTION. Obviously, T is a mapping from $V_2(\mathrm{R})$ into $V_3(\mathrm{R})$, because

$$(\alpha+\beta,\alpha-\beta,\beta)\in V_3(\mathrm{R})\ \forall(\alpha,\beta)$$

For each $a,b \in F$ and (α_1,β_1), $(\alpha_2,\beta_2) \in V_2(\mathrm{R})$

$$T[a(\alpha_1,\beta_1)+(\alpha_2,\beta_2)]=T(a\alpha_1+b\alpha_2,\ a\beta_1+a\beta_2)$$
$$=[(a\alpha_1+b\alpha_2)+(a\beta_1+a\beta_2),(a\alpha_2+b\alpha_2)-(a\beta_1+a\beta_2),\ (a\beta_1+a\beta_2)]$$
$$=a(\alpha_1+\beta_1,\alpha_1-\beta_1,\beta_1)+b(\alpha_2+\beta_2,\alpha_2-\beta_2,\beta_2)$$
$$=aT(\alpha_1,\beta_1)+bT(\alpha_2,\beta_2)$$

Hence, T is linear.

EXAMPLE 2. *Which of the following function T from R^2 into R^2 are linear transformation?*

(i) $T(x_1,x_2)=(x_1^2,x_2)$ (ii) $T(x_1,x_2)=(\sin x_1,x_2)$

(iii) $T(x_1,x_2)=(x_1^2-x_2,0)$.

SOLUTION. (i) Let (x_1,x_2) and (y_1,y_2) be any two vectors in R^2 and $a,b \in \mathrm{R}$, then

$$T[a(x_1,x_2)+(y_1,y_2)]=T[(ax_1+by_1),(ax_2+by_2)]$$

$=[(ax_1+by_1)^2,(ax_2+by_2)]$ $\quad [\because T(x_1, x_2)=(x_1{}^2, x_2)]$

$= [a^2x_1{}^2 + b_2y_1{}^2+2abx_1y_1,\ ax_2+by_2]$

$\neq aT(x_1, x_2)+bT(y_1 +y_2)$

$\therefore$ T is not a linear transformation.

(ii) Let $(x_1 , x_2), (y_1 , y_2) \in R^2$ be an arbitrary vector and let $a, b \in R$, then

$T[a(x_1, x_2)+(y_1, y_2)]=T(ax_1+by_1),(ax_2+by_2)]$

$=[\sin(ax_1+by_1),\ ax_2+by_2]\ \{\because T\{x_1, x_2)=(sin\ x_1 , x_2)\}$

$\therefore$ T is not a linear transformation.

(iii) Let $(x_1,x_2),(y_1, y_2) \in R^2$ be an arbitrary vector and let $a,b \in R$, then

$T[a(x_1,x_2)+(y_1, y_2)] = T[(ax_1+by_1),(ax_2+by_2)]$

$= (ax_1+by_1)-(ax_2+by_2),0]$ $\quad [\because T(x_1,x_2) = (x_1-x_2,0)]$

$=[a(x_1-x_2)+ b(y_1-y_2),0]$

$= aT(x_1,x_2)+bT(y_1, y_2)$

Hence, T is a linear transformation.

EXAMPLE 3. *Show that the mapping $T : R^3\to R^2$, defined by*

$$T(\alpha,\beta,\gamma) = (\alpha,\beta),\ \forall(\alpha,\beta,\gamma)\in R^3$$

is a homomorphism (linear transformation) of the linear space R^3(R) onto R^2(R).

SOLUTION. Let $(\alpha_1,\beta_1,\gamma_1),(\alpha_2,\beta_2,\gamma_2)\in R^3$(R) be an arbitrary vector and let $a,b\in$R, then

$T[a(\alpha_1,\beta_1,\gamma_1)+b(\alpha_2,\beta_2,\gamma_2)]$

$= T[(a\alpha_1+b\alpha_2,a\beta_1+b\beta_2, a(\gamma_1+\gamma_2)]$

$= (a\alpha_1 +b\alpha_2, a\beta_1+b\beta_2)$ $\quad [\because T(\alpha,\beta,\gamma)=(\alpha,\beta)]$

$= a(\alpha_1,\beta_1)+b(\alpha_2,\beta_2)=aT(\alpha_1,\beta_1,\gamma_1)+bT(\alpha_2,\beta_2,\gamma_2)$

Thus, T is a homomorphism.

Further, if $(\alpha,\beta) \in R^2$,then $(\alpha,\beta,\gamma) \in R^3$ and we have $T(\alpha,\beta,\gamma)=(\alpha,\beta)$.

Hence, T is onto.

EXAMPLE 4. *Let F be the field of complex numbers and let T be the function from R^3 onto R^3 defined by $T(a_1,a_2,a_3) = (a_1-a_2+2a_3,2a_1+a_2-a_3,-a_1-2a_2)$. Verify that T is a linear transformation. Describe the null space of T.*

SOLUTION. Let $\alpha = (a_1,a_2,a_3)$and $\beta=(b_1,b_2,b_3)$ be any two vectors in R^3 and $a,b\in$R, then

$a\alpha+b\beta=a(a_1,a_2,a_3)+b(b_1,b_2,b_3)$

$=(aa_1, aa_2,aa_3)+(bb_1,bb_2,bb_3)$

$=(aa_1+bb_1,aa_2+bb_2,aa_3+bb_3)$

Now, $T(a\alpha+b\beta)=T(aa_1+bb_1,aa_2+bb_2,aa_3+bb_3)$

$=[(aa_1+bb_1)-(aa_2+bb_2)+2(aa_3+bb_3),2(aa_1+bb_1)+(aa_2+bb_2)$
$-(aa_3+bb_3),-(aa_1+bb_1)-2(aa_2+bb_2)]$

$=[a(a_1-a_2+2a_3)+b(b_1-b_2+2b_3),a(2a_1+a_2-a_3)$
$+b(2b_1+b_2-b_3),a(-a_1-2a_2)+b(-b_1-2b_2)]$

$=[a(a_1-a_2+2a_3,\ 2a_1+a_2-a_3,-a_1-2a_2)$
$+b(b_1-b_2+2b_3,2b_1+b_2-b_3,-b_1-2b_2)]$

Thus, T is a linear transformation.

Next, By the definition of null space of T, we have

$$N_T = \{\alpha \in R^3 : T(\alpha) = 0 = (0,0,0)\}$$

Let $\quad \alpha = (a_1, a_2, a_3) \in R^3$

$\therefore \quad T(\alpha) = T(a_1, a_2, a_3) = (0,0,0)$

$\Rightarrow \quad (a_1 - a_2 + 2a_3, 2a_1 + a_2 - a_3, -a_1 - 2a_2) = (0,0,0)$

$$\Rightarrow \quad \left.\begin{aligned} a_1 - a_2 + 2a_3 &= 0 \\ 2a_1 + a_2 - a_3 &= 0 \\ -a_1 - 2a_2 &= 0 \end{aligned}\right\} \qquad \ldots(1)$$

Now, we find the solution of system of equation (1).

Let A be the coefficient matrx of (1), then we have,

$$A = \begin{bmatrix} 1 & -1 & 2 \\ 2 & 1 & -1 \\ -1 & -2 & 0 \end{bmatrix}$$

Performing $R_2 \to R_2 - 2R_1$, $R_3 \to R_3 + R_1$, we get

$$A = \begin{bmatrix} 1 & -1 & 2 \\ 0 & 3 & -5 \\ 0 & -3 & 2 \end{bmatrix}$$

Performing $R_3 \to R_3 + R_2$, we get

$$\sim \begin{bmatrix} 1 & -1 & 2 \\ 0 & 3 & -5 \\ 0 & 0 & -3 \end{bmatrix}$$

This matrix is in Echelon form and having three non-zero rows, thus its rank=3, which is equal to the number of unknowns. Hence, the system of equations (1) has only trival solution, *i.e.*, $a_1 = 0$, $a_2 = 0$, $a_3 = 0$. Consequently, $N_T = \{(0, 0, 0)\}$.

EXAMPLE 5. *Describe explicitly the linear transformation $T : R^2 \to R^3$ such that $T(2, 3) = (4, 5)$ and $T(1, 0) = (0, 0)$.*

SOLUTION. Let $\alpha = (2, 3)$, $\beta = (1,0)$ and let $a, b \in R$, then

$$a\alpha + b\beta = 0(0,0) \qquad \Rightarrow a(2,3) + b(1,0) = (0,0)$$

$\Rightarrow \quad (2a + b, 3a) = (0,0)$

$\Rightarrow \quad 2a + b = 0$

$\qquad 3a = 0$

$\Rightarrow \quad a = 0, b = 0$

$\Rightarrow$ the set $\{\alpha, \beta\} = \{(2,3), (1,0)\}$ is linearly independent.

Also, dim. $R^2 = 2$, thus $\{(2, 3), (1, 0)\}$ forms a basis of R^2.

Let (x, y) be any element of R^2 and for some scalars p and q in R, we have

$$(x,y) = p\alpha + q\beta = p(2,3) + q(1,0) = (2p + q, 3p)$$

$$\Rightarrow \quad x = 2p + q,\ y = 3p \Rightarrow p = \frac{y}{3},\ q = \frac{3x - 2y}{3}$$

$$\therefore \quad (x, y) = \frac{y}{3}(2,3) + \frac{3x - 2y}{3}(1,0)$$

Now, $$T(x,y) = T\left[\frac{y}{3}(2,3) + \frac{3x-2y}{3}(1,0)\right]$$

$$= \frac{y}{3}T(2,3) + \frac{3x-2y}{3}T(1,0) \qquad [\because T \text{ is linear}]$$

$$= \frac{y}{3}(4,5) + \frac{3x-2y}{3}(0,0) = \left(\frac{4y}{3}, \frac{5y}{3}\right).$$

EXAMPLE 6. *If a map $T: V_2(R) \to V_3(R)$ defined by $T(a, b) = (a+b, a-b, b)$ is a linear transformation. Find the range, rank, null-space and nullity of T.*

SOLUTION. **Determination of range of T, *i.e.*, R_T and rank :**

Since the ordered set $\{(1, 0), (0, 1)\}$ forms a basis of $V_2(R)$. Then by definition of T, we have,

$$T(1,0) = (1+0, 1-0, 0) = (1,1,0)$$

and $$T(0,1) = (0+1, 0-1, 1) = (1,-1,1).$$

Since (1,0), (0,1) generates $V_2(R)$. Therefore

$T(1,0)$, $T(0,1)$ will generate $T(V_2(R)) = R_T$

$\Rightarrow$ (1,1,0), (1,–1,1) generates R_T

Also, for some scalars $a, b \in R$, such that

$$a(1,1,0) + b(1,-1,1) = (0,0,0)$$

$$\Rightarrow \quad (a+b, a-b, b) = (0,0,0)$$

$$\Rightarrow \quad a+b = 0,\ a-b = 0,\ b = 0$$

$$\Rightarrow \quad a = 0,\ b = 0.$$

$\therefore$ $\{(1,1,0), (1,-1,1)\}$ is linearly independent and spans R_T, so it forms a basis of R_T. Hence, dim. $R_T = 2$.

Determination of null space and nullity of T.

Since T is a linear transformation from $V_2(R)$ into $V_3(R)$. Therefore,

$$\text{dim. } R_T + \text{dim. } N_T = \text{dim. } V_2(R)$$

$$2 + \text{dim. } N_T = 2 \Rightarrow \text{dim. } N_T = 0$$

Thus, nullity of $T = 0$.

Since dim. $N_T = 0$

$\Rightarrow$ Null space of T, *i.e.*, N_T is a zero space.

$$\Rightarrow \quad N_T = \{(0, 0)\}.$$

EXAMPLE 7. *Let T be a linear operator on a linear space V(F). If $T_2 = 0$, what can you say about the relation of the range of T to the null space of T? Give an example of a linear operator on $V_2(R)$ such that $T^2 = 0$ but $T \neq 0$.*

SOLUTION. Since $T^2 = 0$, then for $\alpha \in V$

$$T^2(\alpha) = 0(\alpha)$$

$$\Rightarrow \quad T[T(\alpha)] = 0$$

$$\Rightarrow \quad T(\alpha) \in N_T \qquad \text{[By the definition of null space]}$$

But $$T(\alpha) \in R_T \ \forall\ \alpha \in V$$

$\therefore \qquad R_T \subset N_T$

Hence, when $T^2 = 0$, the range of T is contained in null space of T.

Next, let T be a linear map from $V_2(R)$ into $V_2(R)$ such that

$$T(a,b) = (0,a), \forall (a,b) \in V_2(R)$$

Obviously, $T \neq 0$.

Also, $T^2(a, b) - T[T(a, b)] = T[(0, a)] = \{0,0) = 0\ (a, b)$

$\Rightarrow \qquad T^2 = 0.$

EXAMPLE 8. *Find a linear transformation $T : R^2 \to R^2$ such that $T(1, 0) = (1,1)$ and $T(0, 1) = (-1, 2)$. Prove that T maps the square with vertices $(0, 0)$, $(1, 0)$, $(1, 1)$ and $(0,1)$ into a parallelogram.*

SOLUTION. Since the ordered set $\{(1, 0), (0, 1)\}$ forms a basis of R^2, so that for some scalars p and q in R and $(x, y) \in R^2$ such that

$$(x,y) = p(1,0) + q(0,1) \qquad \text{....(1)}$$

$$\Rightarrow \qquad T(x,y) = T[p(1,0) + q(0,1)]$$

$$= pT(1,0) + qT(0,1) = p(1, 1) + q(-1,2) = (p-q, p + 2q) \qquad \text{...(2)}$$

From (1), we have

$$(x. y) = (p,q) \quad \Rightarrow \quad p = x, q = y.$$

From (2), we get

$$T(x, y) = (x-y, x + 2y). \qquad \text{...(3)}$$

This is the required linear transformation.

Next, let A, B, C and D be the vertices of a square with A (0, 0), B (1, 0), C (1, 1)and D (0, 1) and let P, Q, R and S be the T-images of A, B, C and D respectively.Then we have,

$$P = T(A) = T(0, 0) = (0, 0) \qquad \text{[using (3)]}$$

$$Q = T(B) = T(1,0) = (1,1) \qquad \text{[using (3)]}$$

$$R = T(C) = T(1,1) = (0,3) \qquad \text{[using (3)]}$$

and $\qquad S = T(D) = T(0,1) = (-1, 2) \qquad$ [using (3)]

Now, PQ = Distance between (0, 0) and (1, 1) $= \sqrt{(1-0)^2 + (1-0)^2} = \sqrt{2}$

PS = Distance between (0,0) and (–1,2) $= \sqrt{(-1-0)^2 + (2-0)^2} = \sqrt{1+4} = \sqrt{5}$

RS = Distance between (0,3) and (–1,2) $= \sqrt{(-1-0)^2 + (2-3)^2} = \sqrt{1+1} = \sqrt{2}$

and

QR = Distance between (1, 1) and (0, 3) $= \sqrt{(0-1)^2 + (3-1)^2} = \sqrt{1+4} = \sqrt{5}$

Hence, $PQRS$ is a parallelogram.

EXAMPLE 9. *Let $T: V_3(R) \to V_3(R)$ be a linear transformation defined by*

$$T(a,b,c) = (3a, a-b, 2a + b + c), \forall a,b,c \in R.$$

Prove that T is invertible and find T^{-1}. Also prove that $(T^2 - I)(T - 3I) = 0$.

SOLUTION. Let $\alpha = (a_1,b_1,c_1)$ and $\beta = (a_2,b_2,c_2)$ be two vectors in $V_3(R)$. Suppose that

$$T(\alpha) = T(\beta) \Rightarrow T(a_1,b_1,c_1) = T(a_2,b_2,c_2)$$

$$\Rightarrow \quad (3a_1, a_1 - b_1, 2a_1 + b_1 + c_1) = (3a_2, a_2 - b_2, 2a_2 + b_2 + c_2)$$

$$\Rightarrow \left.\begin{aligned} 3a_1 &= 3a_2 \\ a_1 - b_1 &= a_2 - b_2 \\ 2a_1 + b_1 + c_1 &= 2a_2 + b_2 + c_2 \end{aligned}\right\}$$

$$\Rightarrow \qquad a_1 = a_2,\ b_1 = b_2 \text{ and } c_1 = c_2 \Rightarrow \alpha = \beta.$$

Thus, T is one-one.

Since, V_3 (R) is a 3-dimensional linear space, then T is onto also. Thus, T is one-one onto mapping. Hence T is invertible.

Determination of T^{-1}.

Let $\quad T(a, b, c) = (p, q, r)$, then $T^{-1}(p, q, r) = (a, b, c)$

$$T(a, b, c) = (p, q, r) \Rightarrow (3a, a - b, 2a + b + c) = (p, q, r)$$

$$\Rightarrow \begin{cases} p = 3a \\ q = a - b \\ r = 2a + b + c \end{cases}$$

$$\Rightarrow \qquad a = \frac{p}{3},\ b = \frac{p}{3} - q, c = r - p + q$$

$$\therefore \qquad T^{-1}(p,q,r) = \left(\frac{p}{3}, \frac{p}{3} - q, r - p + q\right)$$

To prove that $(T^2 - 1)(T - 3I) = 0$

$$\begin{aligned}
(T - 3I)(a, b, c) &= T(a, b, c) - 3I(a, b, c) \\
&= T(a,b,c) - 3(a, b, c) \quad [\because I \text{ is identity transformation.}] \\
&= (3a, a - b, 2a + b + c) - 3(a, b, c) \\
&= (3a, a - b, 2a + b + c) - (3a, 3b, 3c) \\
&= (0, a - 4b, 2a + b - 2c)
\end{aligned}$$

$$\begin{aligned}
[(T^2 - I)(T - 3I)](a, b, c) &= (T^2 - I)[(T - 3I)(a, b, c)] \\
&= (T^2 - I)(0, a - 4b, 2a + b - 2c) \\
&= T^2(0, a - 4b, 2a + b - 2c) - I(0, a - 4b, 2a + b - 2c) \\
&= T[T(0, a - 4b, 2a + b - 2c)] - (0, a - 4b, 2a + b - 2c) \\
&= T[T(0, -a + 4b, 0 + a - 4b + 2a + b - 2c)] \\
&\qquad - (0, a - 4b, 2a + b - 2c) \\
&= T[0, -a + 4b, 3a - 3b - 2c)] - (0, a - 4b, 2a + b - 2c) \\
&= (0, 0 - (-a + 4b), 0 + (-a + 4b) + (3a - 3b - 2c) \\
&\qquad - (0, a - 4b, 2a + b - 2c) \\
&= (0, a - 4b, 2a + b - 2c) - (0, a - 4b, 2a + b - 2c) \\
&= (0, 0, 0) = 0 = 0(a, b, c)
\end{aligned}$$

$$\Rightarrow \qquad (T^2 - I)(T - 3I) = 0.$$

EXAMPLE 10. *If T is a linear transformation on a linear space V such that $T^2 - T + I = 0$, then show that T is invertible.*

SOLUTION. Since $T^2 - T + I = 0$, then

$$T^2 = T - I$$

For every $\alpha_i \in V$, we have

$$T^2(\alpha_i) = (T - I)(\alpha_i) \Rightarrow T[T(\alpha_i)] = T(\alpha_i) - I(\alpha_i)$$

$$\Rightarrow \quad T[T(\alpha_i)] = T(\alpha_i) - \alpha_i.$$

Now, for some $\beta_i \in V$, such that $T(\alpha_i) = \beta_i$

$$\Rightarrow \quad T(\beta_i) = \beta_i - \alpha_1 \quad ...(1)$$

To show that T is one-one :

For $\beta_1, \beta_2 \in V$, suppose that

$$T(\beta_1) = T(\beta_2) \Rightarrow \beta_1 - \alpha_1 = \beta_2 - \alpha_1 \quad \text{[using (1)]}$$

$$\Rightarrow \quad \beta_1 = \beta_2$$

$\therefore$ T is one-one.

To show T is onto :

For every $\beta_i \in V$, there exists $\beta_i - \alpha_i \in V$ such that $T(\beta_i) = \beta_i - \alpha_i$.

Thus T is also onto. Hence, T is one-one and onto. Hence, T is invertible.

EXAMPLE 11. *Lei $T : R^2(R) \to R^2(R)$, where for any $(x,y) \in R^2, T(x, y) = \left(2x, \frac{1}{2}y\right)$. Find the matrix associated with T w.r.t. the ordered basis $\{(1,0), (0,1)\}$.*

SOLUTION. Let $B = \{(1,0), (0,1)\}$ be an ordered basis of $R^2(R)$ and $T(x,y) = \left(2x, \frac{1}{2}y\right)$, then

$$T(1, 0) = (2, 0) \text{ and } T(0, 1) = \left(0, \frac{1}{2}\right)$$

Now, $$T(1,0) = (2,0) = 2(1,0) + 0(0,1) \text{ and } T(0, 1) = \left(0, \frac{1}{2}\right) = 0(1,0) + \frac{1}{2}(0,1)$$

Hence, the matrix associated with T w.r.t., B is $[T]_B = \begin{bmatrix} 2 & 0 \\ 0 & \frac{1}{2} \end{bmatrix}$

EXAMPLE 12. *Find the matrix representation of a linear map $T : R^3 \to R^3$ defined by $T(x, y, z) = (z, y + z, x + y + z)$ relative to the basis $\{(1, 0, 1), (-1, 2, 1), (2, 1, 1)\}$.*

SOLUTION. Suppose $B = \{(1,0,1), (-1,2,1), (2,1,1\}$ is the basis of R^3 and $T(x, y, z) = (z, y+z, x+y+z)$ then

$$\left.\begin{aligned} T(1,0,1) &= (1,1,2) \\ T(-1,2,1) &= (1,3,0) \\ T(2,1,1) &= (1,2,4) \end{aligned}\right\} \quad ...(1)$$

For some $a, b, c \in$ R and $(x, y, z) \in R^3$ we have

$$(x, y, z) = a(1,0,1) + b(-1,2,1) + c(2,1,1) = (a - b + 2c, 2b + c, a + b + c)$$

$$\Rightarrow \quad x = a - b + 2c, \; y = 2b + c, \; z = a + b + c$$

$$\Rightarrow \quad a = \frac{1}{4}(-x - 3y + 5z), \; b = \frac{1}{4}(-x + y + z), c = \frac{1}{4}(2x + 2y - 2z)$$

So $$(x, y, z) = \frac{1}{4}(-x - 3y + 5z)(1,0,1) + \frac{1}{4}(-x + y + z)(-1,2,1) + \frac{1}{4}(2x + 2y - 2z)(2,1,1) \quad ...(2)$$

Putting $x = 1, y = 1, z = 1$ in (2) and using (1), we get

$$T(1,0,1) = (1,1,2) = \frac{3}{2}(1,0,1) + \frac{1}{2}(-1,2,1) + 0(2,1,1) \quad ...(3)$$

Putting $x = 1, y = 3, z = 0$ in (2)and using (1), we get

$$T(-1,2,1) = (1,3,0) = 0.(1,0,1)+1.(-1,2,1)+1.(2,1,1) \quad ...(4)$$

Putting $x = 1, y = 2, z = 4$ in (2) and using (1), we get

$$T(2,1,1) = (1,2,4) = \frac{13}{4}(1,0,1)+\frac{5}{4}(-1,2,1) - \frac{1}{2}(2,1,1) \quad ...(5)$$

Now, the matrix of coefficients of equations (3), (4) and (5) is,

$$\begin{bmatrix} \frac{3}{2} & \frac{1}{2} & 0 \\ 0 & 1 & 1 \\ \frac{13}{4} & \frac{5}{4} & -\frac{1}{2} \end{bmatrix}$$

Thus, the matrix representation of T relative to B is the transpose of above matrix:

$$[T]_B = \begin{bmatrix} \frac{3}{2} & 0 & \frac{13}{4} \\ \frac{1}{2} & 1 & \frac{5}{4} \\ 0 & 1 & -\frac{1}{2} \end{bmatrix}$$

EXAMPLE 13. *Let T be linear operator in R^3 defined by*

$$T(x_1, x_2, x_3) = (3x_1 + x_3, -2x_1 + x_2, -x_1 + 2x_2 + 4x_3)$$

What is the matrix of T in the ordered basis $\{\alpha_1, \alpha_2, \alpha_3\}$, where

$$\alpha_1 = (1,0,1),\ \alpha_2 = (-1,2,1) \text{ and } \alpha_3 = (2,1,1)$$

SOLUTION. Suppose $B = \{\alpha_1, \alpha_2, \alpha_3\}$ is the basis of R^3 where, $\alpha_1 = (1,0,1)$, $\alpha_2 = (-1,2,1)$ and $\alpha_3 = (2,1,1)$ also $T : R^3 \to R^3$ defined by

$$T(x_1, x_2, x_3) = (3x_1 + x_3, -2x_1 + x_2, -x_1 + 2x_2 + 4x_3)$$

Then, we get

$$\left.\begin{aligned} T(\alpha_1) &= T(1,0,1) = (4,-2,3) \\ T(\alpha_2) &= T(-1,2,1) = (-2,4,9) \\ T(\alpha_3) &= T(2,1,1) = (7,-3,4) \end{aligned}\right\} \quad ...(1)$$

Let $(x, y, z) = a\alpha_1 + b\alpha_2 + c\alpha_3$ for some $a, b, c \in$ R. Then

$$(x, y, z) = a(1,0,1) + b(-1,2,1) + c(2,1,1) = (a - b + 2c, 2b + c, a + b + c)$$

$$\Rightarrow \quad x = a - b + 2c,\ y = 2b + c,\ z = a + b + c$$

$$\Rightarrow \quad a = \frac{1}{4}(-x - 3y + 5z),\ b = \frac{1}{4}(-x + y + z),\ c = \frac{1}{4}(2x + 2y - 2z)$$

$$\therefore \quad (x, y, z) = \frac{1}{4}(-x - 3y + 5z)\alpha_1 + \frac{1}{4}(-x + y + z)\alpha_2 + \frac{1}{4}(2x + 2y - 2z)\alpha_3 \quad ...(2)$$

Putting $x = 4, y = -2, z = 3$ in (2) and using (1), we get

$$T(\alpha_1) = (4,-2,3) = \frac{17}{4}\alpha_1 - \frac{3}{4}\alpha_2 - \frac{1}{2}\alpha_3 \qquad ...(3)$$

Putting $x = -2, y = 4, z = 9$ in using (1), we get

$$T(\alpha_2) = (-2,4,9) = \frac{35}{4}\alpha_1 + \frac{15}{4}\alpha_2 - \frac{7}{2}\alpha_3 \qquad ...(4)$$

Putting $x = 7, y = -3, z = 4$ in (2) and using (1), we get

$$T(\alpha_3) = (7,-3,4) = \frac{11}{2}\alpha_1 - \frac{3}{2}\alpha_2 + 0.\alpha_3 \qquad ...(5)$$

Now, the coefficient matrix of the system of equation (3),(4) and (5) is

$$\begin{bmatrix} 17/4 & -3/4 & 11/2 \\ 35/4 & 15/4 & -7/2 \\ 11/2 & -3/2 & 0 \end{bmatrix}$$

Thus, the matrix of T relative to B is obtained by taking the transpose of above coefficeint matrix :

$$[T]_B = \begin{bmatrix} 17/4 & 35/4 & 11/2 \\ -3/4 & 15/4 & -3/2 \\ -1/2 & -7/2 & 0 \end{bmatrix}.$$

EXAMPLE 14. *If the matrix of a linear transformation T on a linear space $V_2(\mathbf{C})$ w.r.t. the ordered basis $B = \{(1, 0, (0, -1)\}$ is $\begin{bmatrix} 1 & 1 \\ 1 & 1 \end{bmatrix}$ what is the matrix w.r.t. the ordered basis $B' = \{(1, 1), (1, -1)\}$.*

SOLUTION. Since $B = \{(1, 0, (0, -1)\}$ and $B' = \{(1,1), (1, -1)\}$ and

$$[T]_B = \begin{bmatrix} 1 & 1 \\ 1 & 1 \end{bmatrix}$$

Determination of $[T]_{B'}$.

Let $\alpha_1 = (1,0), \alpha_2 = (0,-1)$

Since, $[T]_B = \begin{bmatrix} 1 & 1 \\ 1 & 1 \end{bmatrix}$ is the matrix of T w.r.t. B, then

$$T(\alpha_1) = T(1,0) = 1.\alpha_1 + 1.\alpha_2$$
$$= 1.(1, 0) + 1.(0, -1) = (1, 0) + (0, -1) = (1, -1)$$
$$T(\alpha_2) = T(0,-1) = 1.(\alpha_1) + 1.\alpha_2$$
$$= 1.(1,0) + 1.(0,-1) = (1,-1)$$

If $(a, b) \in V_2(\mathbf{C})$, then we can write, for some $p, q \in \mathbf{C}$

$\Rightarrow$ $(a,b) = p(1, 0) + q(0,-1) = (p,-q)$

$\Rightarrow$ $p = a, q = -b$

Now, $T(a,b) = T(p\alpha_1 + q\alpha_2\} = pT(\alpha_1) + qT(\alpha_2)$

$\therefore$ $T(a,b) = (a-b, a+b)$. ...(1)

Further, since $B' = \{(1,1), (1, -1)\}$ is another basis of $V_2(\mathbf{C})$, let

$\beta_1 = (1,1), \beta_2 = (1,-1)$

$$\left.\begin{aligned} T(\beta_1) &= T(1,1) = (0,0) \quad &[\text{using } (1)] \\ T(\beta_2) &= T(1,-1) = (0,0) \quad &[\text{using } (2)] \end{aligned}\right\} \quad \text{...(2)}$$

Let $(x,y) \in V_2(C)$ such that

$$(x, y) = p_1\beta_1 + q_1\beta_2, \text{ for some } p_1, q_1 \in C$$
$$= p_1(1,1) + q_1(1,-1) = (p_1+q_1, p_1-q_1)$$

$$\Rightarrow \quad x = p_1+q_1, y = p_1-q_1 \Rightarrow \quad p_1 = \frac{x+y}{2}, q_1 = \frac{x-y}{2}$$

$$\therefore \quad (x,y) = \left(\frac{x+y}{2}\right)\beta_1 + \begin{bmatrix} 0 & 0 \\ 0 & 2 \end{bmatrix}\beta_2 = \frac{x+y}{2}(1,1) + \frac{x-y}{2}(1,-1) \quad \text{...(3)}$$

Putting $x = 0, y = 0$ in (3) and using (2), we get

$$T(1,1) = (0, 0) = 0.(1,1) + 0.(1,-1) \quad \text{...(4)}$$

Putting $x = 2, y = -2$ in (3) and using (2), we get

$$T(1, -1) = (2, -2) = 0.(1, 1) + 2(1, -1) \quad \text{...(5)}$$

Now, coefficient matrix of the system of equations (4) and (5) is

$$\begin{bmatrix} 0 & 0 \\ 0 & 2 \end{bmatrix}$$

Thus, the matrix of T relative to B' is the transpose of above coefficient matrix :

$$[T]_B = \begin{bmatrix} 0 & 0 \\ 0 & 2 \end{bmatrix}$$

EXAMPLE 15. *Let $B = \{(1,0), (0,1)\}$ and $B' = \{(1,2), (2,3)\}$ be any two bases of R^2*

(i) Find the transition matrices P from B to B'

(ii) Verify that $[\alpha]_B = P^{-1}[\alpha]_{B'}$ $\forall\, \alpha \in R^2$

(iii) Verify that $P^{-1}[T]_B P = [T]_{B'}$ where $T(x, y) = (2x - 3y, x + y)$.

SOLUTION. Let $\alpha_1 = (1, 0)$, $\alpha_2 = (0, 1)$, $\alpha'_1 = (1, 2)$ *and* $\alpha'_2 = (2, 3)$.

Since $\quad T(x, y) - (2x - 3y, x + y)$

(i) $\quad \alpha'_1 = (1,2) = 1.(1,0) + 2(0,1) = 1.\alpha_1 + 2\alpha_2 \quad$ ...(1)

and $\quad \alpha'_2 = (2,3) = 2.(1,0) + 3(0,1) = 2.\alpha_1 + 3\alpha_2 \quad$ ...(2)

The coefficient matrix of the system of equation (1) and (2) is given by

$$\begin{bmatrix} 1 & 2 \\ 2 & 3 \end{bmatrix}.$$

$\therefore$ The transition matrix P is the transpose of the coefficient matrix is given by

$$P = \begin{bmatrix} 1 & 2 \\ 2 & 3 \end{bmatrix}.$$

Let α be an arbitrary element of R^2, then

$$\alpha = a_1\alpha_1 + a_2\alpha_2, \text{ for some } a_1, a_2 \in \text{R}.$$
$$= a_1(1, 0) + a_2(0, 1) = (a, a)$$

If $\alpha = \{x,y\}$; $x,y \in$ R, then

$$(x,y) = (a_1, a_2) \Rightarrow \quad x = a_1, y = a_2$$

$$\therefore \quad (x,y) = x\alpha_1 + y\alpha_2$$

Thus, $[\alpha]_\text{B} = \begin{bmatrix} x \\ y \end{bmatrix}$, which is the co-ordinate vector of *a w.r.t. B.*

Again, $(x, y) = a'_1\alpha'_1 + a'_2\alpha'2$

$$= a'_1(1, 2)+a'_2(2, 3)=(a'_1+2a'_2, 2a'_2+3a'_2)$$

$\Rightarrow$ $x = a'_1+2a'_2, y=2a'_1+3a'_2$

$\Rightarrow$ $a'_1 = 2y - 3x, a'_2= 2x - y.$

$\therefore$ $(x, y) = (2y-3x)\alpha'_1 + (2x-y)\alpha'_2$ ...(3)

Thus, $[\alpha]_{B'} = \begin{bmatrix} 2y-3x \\ 2x-y \end{bmatrix}$

$\therefore$ $P[\alpha]_{B'} = \begin{bmatrix} 1 & 2 \\ 2 & 3 \end{bmatrix}\begin{bmatrix} 2y-3x \\ 2x-y \end{bmatrix}$

$$= \begin{bmatrix} 1(2y-3x)+2(2x-y) \\ 2(2y-3x)+3(2x-y) \end{bmatrix} = \begin{bmatrix} x \\ y \end{bmatrix} = [\alpha]_B$$

(ii) Since $P= \begin{bmatrix} 1 & 2 \\ 2 & 3 \end{bmatrix}$

$$|P|=(1\times3-2\times2) = -1.$$

Matrix of co-factors of element of P is

$\begin{bmatrix} 3 & -2 \\ -2 & 1 \end{bmatrix}$ so, $adj\ P=\begin{bmatrix} 3 & -2 \\ -2 & 1 \end{bmatrix}$

$\therefore$ $P^{-1}=\frac{1}{P}\begin{bmatrix} 3 & -2 \\ -2 & 1 \end{bmatrix}$

$$= \frac{1}{P}\begin{bmatrix} 3 & -2 \\ -2 & 1 \end{bmatrix} = -1\begin{bmatrix} 3 & -2 \\ -2 & 1 \end{bmatrix} = \begin{bmatrix} -3 & 2 \\ 2 & -1 \end{bmatrix}$$

Since $T(x, y) = (2x-3y, x+y)$

$\therefore$ $T(\alpha_1)=T(1,0)=(2,1)=2(1,0)+1.(0, 1)=2\alpha_1+\alpha_2$ ...(4)

$T(\alpha_2)=T(0,1)=(-3,1)=-3(1,0)+1.(0, 1)=-3\alpha_1+\alpha_2$ ...(5)

$\therefore$ $[T]_B$ = Transpose of the coefficient matrix of the system of equations (4) and (5)

$$=\begin{bmatrix} 2 & -3 \\ 1 & 1 \end{bmatrix}$$

Also, $\left.\begin{array}{l} T(\alpha'_1) = T(1,2) = (-4,3) \\ T(\alpha'_2) = T(2,3) = (-5,5) \end{array}\right\}$ ...(6)

From (3), we have

$(x, y) = (2x-3x)\ \alpha'_1+(2x-y)\ \alpha'_2$ ...(7)

Putting $x = -4, y = 3$ in (7) and using (6), we get

$T(\alpha'_1)=(-4, 3) = 18\alpha'_1-11\alpha'_2$ ...(8)

Putting $x = -5, y = 5$ in (7) and using (6), we get

$T(\alpha'_2\}=(-5, 5) = 25\alpha'_1-15\alpha'_2$ ...(9)

$\therefore$ $[T]_{B'}$ = Transpose of coefficient matrix of the system of equations (8) and (9)

$$= \begin{bmatrix} 18 & -11 \\ 25 & -15 \end{bmatrix} = \begin{bmatrix} 18 & 25 \\ -11 & -15 \end{bmatrix}$$

So, $P^{-1}[T]_B P= \begin{bmatrix} -3 & 2 \\ 2 & -1 \end{bmatrix}\begin{bmatrix} 2 & -3 \\ 1 & 1 \end{bmatrix}\begin{bmatrix} 1 & 2 \\ 2 & 3 \end{bmatrix}$

$$= \begin{bmatrix} -4 & 11 \\ 3 & -7 \end{bmatrix}\begin{bmatrix} 1 & 2 \\ 2 & 3 \end{bmatrix} = \begin{bmatrix} 18 & 25 \\ -11 & -15 \end{bmatrix} = [T]_{B'}$$

EXAMPLE 16. *If T_1, T_2 and T_3 are linear transformations on a linear space V(F) such that $T_1T_2 = T_3T_1 = I$, then T_1 is invertible and $T_1^{-1} = T_2 = T_3$.*

SOLUTION. First we shall show that T_1 is one-one :

For $\alpha, \beta \in V$, suppose that

$$T_1(\alpha) = T_1(\beta) \quad \Rightarrow T_3(T_1(\alpha)) = T_3(T_1(\beta))$$

$$\Rightarrow \quad (T_3T_1)(\alpha) = (T_3T_1)(\beta) \Rightarrow \quad I(\alpha) = I(\beta) \qquad [\because T_3T_1 = I]$$

$$\Rightarrow \quad \alpha = \beta.$$

Thus, T_1 is one-one.

Secondly, we shall show that T_1 is onto : Let $\beta \in V$ be an arbitrary vector.

Since $T_2 : V \to V$, then $T_2(\beta) \in V$.

Let us take $\quad T_2(\beta) = \alpha \in V$

Now, $\quad T_2(\beta) = \alpha \quad \Rightarrow T_1(T_2(\beta)) = T_1(\alpha)$

$$\Rightarrow \quad (T_1T_2)(\beta)) = T_1(\alpha) \Rightarrow \quad I(\beta) = T_1(\alpha) \qquad [\because T_1T_2 = 1]$$

$\therefore$ For any β, there exists $\alpha \in V$ such that $T_1(\alpha) = \beta$. Hence T_1 is onto.

Thus, T_1 is one-one and onto and hence T_1 is invertible.

Next, since $\quad T_1T_2 = T_3T_1 = I$

$$\therefore \quad T_1T_2 = I \quad \Rightarrow T_1^{-1}(T_1)T_2 = T_1^{-1}I$$

$$\Rightarrow \quad (T_1^{-1}T_1)T_2 = T_1^{-1} \quad \Rightarrow \quad IT_2 = T_1^{-1}I$$

$$\Rightarrow \quad T_2 = T_1^{-1}$$

Also, $\quad T_3T_1 = I \quad \Rightarrow (T_3T_1)T_1^{-1} = I\,T_1^{-1}$

$$\Rightarrow \quad T_3(T_1T_1^{-1}) = T_1^{-1} \quad \Rightarrow \quad T_3I = T_1^{-1} \Rightarrow T_3 = T_1^{-1}.$$

Hence, $\quad T_1^{-1} = T_2 = T_3.$

EXAMPLE 17. *Let T be an invertible linear operator on a linear space V(F). Then show that*

(i) aT is also an invertible linear operator, when $a \neq 0$ and $a \in F$

(ii) $(aT)^{-1} = \left(\frac{1}{a}\right)T^{-1}$, where $a \neq 0$ and $a \in F$.

(iii) T^{-1} is invertible and $(T^{-1})^{-1} = T$.

SOLUTION. Since T is invertible so that it is one-one and onto.

(i) Let $\alpha, \beta \in V$ and $0 \neq a \in F$ and suppose that

$$(aT)(\alpha) = (aT)(\beta) \quad \Rightarrow \quad a(T(\alpha)) = a(T(\beta))$$

$$\Rightarrow \quad T(\alpha) = T(\beta) \quad \Rightarrow \quad \alpha = \beta \qquad [\because T \text{ is one-one.}]$$

$\therefore$ *T is one-one.*

Let β be any vector in V, then there exists a vector α in V such that

$$T(\alpha) = \beta \qquad [\because T \text{ is onto.}]$$

$$\Rightarrow \quad a(T)(\alpha) = a(\beta) \Rightarrow (aT)(\alpha) = a\beta$$

$$\Rightarrow \quad (aT)(\alpha) = \gamma \text{ for some } \gamma = a\beta \in V.$$

$\therefore$ aT is onto.

Thus, aT is one-one and onto and hence aT is invertible.

(ii) Consider,

$$(aT)\left(\frac{1}{a}T^{-1}\right) = a.\frac{1}{a}\left(TT^{-1}\right)$$

$$= 1.(TT^{-1})=I \qquad [\because T \text{ is invertible.}]$$

Also, $\left(\frac{1}{a}T^{-1}\right)(aT)=I$

Hence, $(aT)^{-1}=\left(\frac{1}{a}\right)T^{-1}$

(iii) Since $T^{-1}T=TT^{-1}=I$, this implies that inverse of T^{-1} is T,

i.e., $(T^{-1})^{-1}=T.$

$\therefore$ $(T^{-1})^{-1}T^{-1}=T^{-1}(T^{-1})^{-1}=I$

$\Rightarrow$ T^{-1} exists.

Hence, T^{-1} exists and $(T^{-1})^{-1} =T.$

EXERCISE 3.3

1. Show that the following mappings are linear:
(i) $T : R^2 \to R^2$ defined by $T(x, y) =(2x-y, x)$.
(ii) $T:R^3 \to R^2$ defined by $T(x, y, z)=(z, x+y)$.
(iii) $T : R \to R^2$ defined by $T(x) = (2x, 3x)$.
(iv) $T : R^2 \to R^2$ defined by $T(x,y) = (ax+by, cx+dy)$ where $a, b, c, d \in R$.

2. Show that the following mappings are linear:
(i) $T : R^2 \to R^2$ defined by $T(x, y) = (x^2, y^2)$.
(ii) $T : R^3 \to R^2$ defined by $T(x, y, z) = (x+1, y+z)$.
(iii) $T : R^2 \to R^2$ defined by $T(x, y) = |x-y|$.

3. Show that the map $T : R^2 \to R^3$ defined by $T(a, b) = (a-b, b-a, -a)$ is linear transformation. Find the range, rank, null space and nullity.

4. Let $F : R^3 \to R^2$ be a map given by $F(a, b, c) = (a, b), \forall (a, b, c) \in R^3$. Prove that F is a (homomorphism) linear tranformation. Also, find the kernel of F (null space of F).

5. Give an exmple of a linear tranformation T on $V_3(R)$ such that $T \neq 0$, $T^2 \neq 0$ but $T^3 = 0$ where 0 is the zero tranformation.

6. Let there be a linear operator on R^3 given by $T(x, y, z)= (2x, 2x-5y, 2y + z)$. Find T^{-1}.

7. Let $T : R^3 \to R^2$ be a linear map defind by $T(x, y, z)= (x+ 2y-z, y+z, x+y-2z)$. Find basis and dimension of (i) range of T (ii) null space of T.

8. Prove that a linear transformation T on a finite-dimensional linear space $V(F)$ is invertible if and only if T is nonsingular.

9. Describe explicitly a linear transformation from $V_3(R)$ into $V_3(R)$ which has its range subsace spanned by $(1, 0, -1)$ and $(1, 2, 2)$.

10. Let T be the linear oeprator on R^2 defined by $T(x, y)= (4x-2y, 2x + y)$. Compute the matrix of T relative to the basis $\{\alpha_1, \alpha_2\}$, where $\alpha_1=(1, 1), \alpha_2=(-1, 0)$.

11. Let $V = R^3$ and $T : V \to V$ be a linear map defined by $T(x, y, z)= (x+z, -2x + y, -x+2y+z)$. What is matrix of T *w.r.t.* basis $B =\{(1, 0, 1), (-1, 1, 1), (0, 1, 1)\}$?

12. If the matrix of a linear transformation on $V_3(C)$ with respect to the basis $B= \{(1, 0\ 0),$ $(0, 1, 0), (0, 0, 1)\}$ is $\begin{bmatrix} 0 & 1 & 1 \\ 1 & 0 & -1 \\ -1 & -1 & 0 \end{bmatrix}$. What is the matrix of T w.r.t. basis $B'=\{(0,1,-1), (1, -1, 1), (-1, 0, 1)$ and $B' = (1, 1, -1), (-1, 0, 1)\ (1, 2, 1)\}$

13. Find the matirx representation of the linear mappings relative to the usual basis $B = \{(1, 0, ..., 0), (0, 1, ...,1), (0, 0,..., 1)\}$. For R^n
(i) $T : R^3 \to R^2$ defined by $T(x, y, z) = (2x -4y + 9z, 5x + 3y-2z)$
(ii) $T : R \to R^2$ defined by $T(x) = (3x ,5x)$.

14. Let T be a linear operator on R^3 defined by $T(x, y) = (2y, 3x -y)$. Find the matrix representation of T relative to the basis $B = \{(1, 3), (2, 5)\}$

15. Let F be a linear operator on R^3 defined by $F(x, y, z)= (2y+z, x-4y, 3x)$
(i) Find the matrix of F in the basis $B' = \{(1, 1, 1), (1, 1, 0), (1, 0, 0)\}$
(ii) Verify that $[F]_B =[F(\alpha)]_{B'} = [F(\alpha)]_{B'}$, $\forall \alpha \in R^3$.

16. Let $B = \{(1, 0, 0), (0, 1, 0), (0, 0, 1)\}$ and $B' = \{(1, 1, 1), (1, 1, 0), (1, 0, 0)\}$ be two basis of R^3

(i) Show that $P[\alpha]_{B'} = [\alpha]_{B'}\ \forall \alpha \in R^3$.

(ii) Show that $[T]_{B'} = P^{-1}[T]_B P$, where $T(x, y, z) = (2y+z, x-4y, 3x)$.

17. Find two linear tranformation T and S on a linear space $R^2(R)$ such that $TS=0$ but $ST \neq 0$.

HINT TO SELECTED PROBLEMS

1. (i) Let $\alpha = (x_1, y_1, z_1)$, $\beta = (x_2, y_2, z_2) \in R^3$ and $a, b \in R$.

Then $a\alpha + b\beta = a(x_1, y_1, z_1) + b(x_2, y_2, z_2)$
$= (ax_1 + bx_2, ay_1 + by_2, az_1 + bz_2)$

$\therefore\ T(a\alpha + b\beta)$
$= T(ax_1 + bx_2, ay_1 + by_2, az_1 + bz_2)$
$= (az_1 + bz_2, ax_1 + bx_2 + ay_1 + by_2)$
$= \{az_1 + bz_2, a(x_1 + y_1) + b(x_2 + y_2)]$
$= \{az_1, a(x_1 + y_1)\} + \{bz_2, b(x_2 + y_2)\}$
$= a(z_1, x_1 + y_1) + b(z_2, x_2 + y_2)$
$= aT(\alpha) + bT(\beta)$

Hence T is linear.

2. (iii) Let $\alpha = (x_1, y_1)$, $\beta = (x_2, y_2) \in R^2$ and $a, b \in R$.

Then $a\alpha + b\beta = a(x_1, y_1) + b(x_2, y_2)$
$= (ax_1 + bx_2, ay_1 + by_2)$

$\therefore\ T(a\alpha + b\beta) = T(ax_1 + bx_2, ay_1 + by_2)$
$= |(ax_1 + bx_2) - ay_1 + by_2)|$
$= |(a(x_1 - y_1) + b(x_2 - y_2)|$
$\leq |a(x_1 - y_1)| + |b(x_2 - y_2)|$ (by triangular inequality)
$\leq aT(\alpha + bT(\beta)$.

$\therefore\ T(a\alpha + b\beta) \neq aT(\alpha) + bT(\beta)$.

Hence T is not linear.

3. $B = \{(1, 0)\ (0, 1)\}$ be the basis of R^2. Then $T(1, 0) = (1, -1, -1)$ and $T(0, 1) = (-1, 1, 0)$.

Range space is generated by the set $B_1 = \{(1, -1, -1), (-1, 1, 0)\}$.

Clerly B_1 is linearly independent.

Thus rank $(T) = 2$.

For Null space.

$N(T) = \{\alpha \in R^2 : T(\alpha) = 0\}$
$= \{(x, y) \in R^2 : T(x, y) = 0\}$

Now, $T(x, y) = (x - y, y - x, -x) = (0, 0, 0)$
$\Rightarrow x - y = 0, y - x = 0, -x = 0$
$\Rightarrow x = 0, y = 0, -x = 0$

$\therefore$ Null space of T $= \{0\}$ and nullity $(T) = 1$.

4. Let $\alpha = (a_1, b_1, c_1)$, $\beta = (a_2, b_2, c_2)$ and $a, b, c \in R$. Then

$a\alpha + b\beta = (aa_1 + ba_2, ab_1 + bb_2, ac_1 + bc_2)$

$\therefore\ F(a\alpha + \beta) = (aa_1 + ba_2, ab_1 + bb_2)$
$= (aa_1 + ab_1) + (ba_1 + bb_2)$
$= a(a_1 + b_1) + b(a_2 + b_2)$
$= aF(a_1, b_1, c_1) + bF(a_2, b_2, c_2)$
$= aF(\alpha) + bF(\beta)$

$\therefore$ F is linear.

Now Ker $F = \{\alpha \in R^3 : F(\alpha) = 0\}$
$\{(a, b, c) \in R^3 : F(a, b, c) = 0\}$
$= \{(a, b, c)\} \in R^3 : (a, b) = (0, 0)\}$
$= \{(a, b, c)\} \in R^3 : a = 0, b = 0\}$
$= \{(0, 0, c) : c \in R\}$.

6. Since $T(x, y, z,) = (2x, 2x - 5y, 2y + z)$
$(x, y, z) = T^{-1}(2x, 2x - 5y, 2y + z)$

Let $p = 2x$, $q = 2x - 5y$, $r = 2y + z$

so that $x = \dfrac{p}{2}, y = \dfrac{p-q}{5}$ and $z = \dfrac{5r - 2p + 2q}{5}$

Then $T^{-1}(p, q, r) = \left(\dfrac{p}{2}, \dfrac{p-q}{5}, \dfrac{5r - 2p + 2q}{5}\right)$

9. Since range space is generated by $(1, 0, -1)$ and $(1, 2, 2)$.

So, we take $T(1, 0, 0) = (1, 0, -1)$, $T(0, 1, 0) = (1, 2, 2)$ and $T(0, 0, 1) = (0, 0, 0)$.

Let $(x, y, z) \in V_3(R)$, then

$(x, y, z) = x(1, 0, 0) + y(0, 1, 0) + z(0,0,1)$
$T(x, y, z) = xT(1,0,0) + yT(0,1,0) + zT(0,0,1)$
$= x(1, 0, -1) + y(1, 2, 2) + z(0, 0, 0)$
$= (x+y, 2y, -x+2y)$.

10. $T : R^2 \to R^2$ given by $T(x, y) = (4x - 2y, 2x + y)$

$T(\alpha_1) = T(1,1) = (2,3) = a(1,1) + b(-1, 0)$;
$T(\alpha_2) = T(-1,0) = (-4,-2) = c(1, 1) + d(-1, 0)$.

$\therefore a - b = 2, a = 3$ and $c - d = -4$. $c = -2$,

i.e., $a = 3, b = 1$ anc $c = -2, d = 2$

Hence, $[T]_{(\alpha_1, \alpha_2)} = \begin{bmatrix} 3 & -2 \\ 1 & 2 \end{bmatrix}$

12. $B = (1, 0, 0), (0, 1, 0), (0, 0, 1)$ and

$[T]_{B'} = \begin{bmatrix} 1 & 0 & 0 \\ 0 & 0 & 0 \\ 0 & 0 & -1 \end{bmatrix}$

$\therefore\ T(1, 0, 0) = (0, 1, -1)$; $T(0,1,0) = (1,0,-1)$; $T(0, 0, 1) = (1, -1, 0)$.

Let $(x, y, z) \in V_3(\mathbf{C})$ Then

$(x, y, z) = x(1, 0, 0) + y(0, 1, 0) + z(0, 0, 1)$.

$\therefore T(x, y, z) = xT(1, 0, 0) + yT(0, 1, 0) + zT(0, 0, 1)$

$= x(0, 1, -1) + yT(1, 0, -1) + z(1, -1, 0)$

$= (y + z, x - z, -x - y)$.

Now for $B_1 = \{(0, 1, -1, (1, -1, 1), (-1, 0, 1)\}$

$T(0, 1, -1) = (0. 1, -1)$

$= a_1(0, 1, -1) + b_1(1, -1, 1) + c_1(-1, 0, 1)$

$T(1, -1, 1) = (0, 0, 0)$

$= a_2(0, 1, -1) + b_1(1, -1, 1) + c_2(-1, 0, 1)$

$\therefore \quad b_1 - c_1 = 0,\ a_1 - b_1 = 1,\ -a_1 + b_1 + c_1 = -1$

$b_2 - c_2 = 0,\ a_2 - b_2 = 0,\ -a_2 + b_2 + c_2 = 0$

and $\quad b_3 - c_3 = 1,\ a_3 - b_3 = -2,\ -a_3 + b_3 + c_3 = 1$.

Solving these equations, we get $a_1 = 1, b_1 = 0, c_1 = 0; a_2 = 1, b_2 = 0, c_2 = 0; a_3 = 0, b_3 = 0, c_3 = -1$.

$$\therefore \quad [T]_{B'} = \begin{bmatrix} 1 & 0 & 0 \\ 0 & 0 & 0 \\ 0 & 0 & -1 \end{bmatrix}$$

15 (ii) From (i)

$$[F]_{B'} = \begin{bmatrix} 3 & 3 & 3 \\ -6 & -6 & -2 \\ 6 & 5 & -1 \end{bmatrix}$$

Let $a \in R^3$ so taking $\alpha = (a, b, c)$

$\therefore \quad (a, b, c) = p_1(1, 1, 1) + q_1(1, 1, 0) + r_1(1, 0, 0)$

$\Rightarrow \quad p_1 = c, q_1 = b - c, r_1 = a - b$.

$$\therefore \quad [\alpha]_{B'} = \begin{bmatrix} p_1 \\ q_1 \\ r_1 \end{bmatrix} = \begin{bmatrix} c \\ b-c \\ a-b \end{bmatrix}$$

$F(\alpha) = F(a, b, c) = (2b + c, a - 4b, 3a)$

$= x(1, 1, 1) + y(1, 1, 0) + z(1, 0, 0)$

$\therefore \quad x = 3a, y = -2a - 4b, z = -a + 6b + c$

$$\text{So that} \quad F[\alpha]_{B'} = \begin{bmatrix} x \\ y \\ z \end{bmatrix} = \begin{bmatrix} 3a \\ -2a-4b \\ -a+6b+c \end{bmatrix}$$

Now

$$[F]_{B'}[\alpha]_{B'} = \begin{bmatrix} 3 & 3 & 3 \\ -6 & -6 & -2 \\ 6 & 5 & -1 \end{bmatrix} \begin{bmatrix} c \\ b-c \\ a-b \end{bmatrix} = \begin{bmatrix} 3a \\ -2a-4b \\ -a+6b+c \end{bmatrix} = [F(\alpha)]_{B'}$$

ANSWERS

5. Kernel of $F = \{(0, 0, 0) \in R^3 : T(0, 0, c) = (0, 0)\}$

5. $T : V_3(\mathrm{R}) \to V_3(\mathrm{R})$ defined by $T(a, b, c) = (0, a, b)$ such that $T \neq 0$, $T^2 \neq 0$ and $T^3 \neq 0$.

6. $= (-1,0,7) - \frac{25}{25}(3,0,5) = (-1,0,7) - (3,0,4)$

7. (i) $\{(1, 0, 1), (2, 1, 1)\}$ is a basis of R_T and dim. $R_T = 2$

(ii) $\{(3, -1, 1\}$ is a basis of N_T and dim. $N_T = 1$.

9. $T(a,b,c) = (a+b, 2b, 2b-a)$

10. Matrix of T relative to $\{\alpha_1, \alpha_2\}$ is $\begin{bmatrix} 3 & -2 \\ 1 & 2 \end{bmatrix}$ **11.** $[T]_B = \begin{bmatrix} 2 & 1 & 2 \\ 0 & 1 & 1 \\ -2 & 2 & 0 \end{bmatrix}$

12. $[T]_B = \begin{bmatrix} 1 & 0 & 0 \\ 0 & 0 & 0 \\ 0 & 0 & -1 \end{bmatrix}$ **13. (i)** $[T]_B = \begin{bmatrix} 2 & -4 & 9 \\ 5 & 3 & -2 \end{bmatrix}$ (ii) $[T]_B = \begin{bmatrix} 3 \\ 5 \end{bmatrix}$

14. $[T]_B = \begin{bmatrix} -30 & -48 \\ 18 & 29 \end{bmatrix}$ **15.** $[T]_{B'} = \begin{bmatrix} 3 & 3 & 3 \\ -6 & -6 & -2 \\ 6 & 5 & -1 \end{bmatrix}$ **16.** (i) $P = \begin{bmatrix} 1 & 1 & 1 \\ 1 & 1 & 0 \\ 1 & 0 & 0 \end{bmatrix}$

17. $T(a, b) = (2a, 0)$ and $S(a, b) = (0, 2a)$

3.28 LINEAR FUNCTIONALS

We shall study the linear mappings from a linear space V into its field F of scalars. Naturally all the theorems and results for arbitrary linear mappings on V hold for this special case. However, such type of mappings are treated separately because of their fundamental importance and because of the relationship between V and F, But all the theorems and results do not apply in the general case.

Definition. *Let V be a linear space over a field F. Then a linear transformation $\phi : V \to F$ is called a linear functional (or linear form) if for every $\alpha, \beta \in V$ and every $a, b \in F$.*

$$\phi(a\alpha + b\beta) = a\phi(\alpha) + b\phi(\beta).$$

ILLUSTRATIONS

(1) Let V be the linear space of polynomials in t over R. Then, the integral operator $\phi : V \to \mathbb{R}$ defined by

$$\phi[p(t)] = \int_0^1 p(t)\,dt \text{ is a linear functional.}$$

(2) Let V be the linear space of n-square matrices over the field F. Let $T : V \to F$ be the trace mapping defined by $T(A) = a_{11} + a_{12} + \ldots + a_{1n}$ where $A = [a_{ij}]$. This mapping is a linear functional.

(3) If V is a vector over the field F, then a mapping $T : V \to F$ defined by

$$T(\alpha) = 0, \forall a \in V$$

is a linear functional. This functional is also known as zero functional.

3.29 DUAL SPACES

Definition. *The set of linear functionals on a linear space V over the field F is also a linear space over F with addition and scalar multiplication defined by,*

(i) $(\phi_1 + \phi_2)(\alpha) = \phi_1(\alpha) + \phi_2(\alpha), \forall \alpha \in V$

(ii) $(c\phi)(\alpha) = c\phi(\alpha), c \in F$

where ϕ_1, ϕ_2, ϕ are linear functionals on V.

This space is called the dual space (conjugate space) of V and is denoted by V^*. We also write $V^* = L(V, F)$.

THEOREM 1. ***Let V be a finite-dimensional linear space over the field F. Then, dim. V* = dim. V.***

PROOF. Let V be n-dimensional linear space. Let V^* be its dual space, so that

$$\text{dim. } V = n, L(V, F) = V^*, \text{dim. } F = 1.$$

Since we know that

$$\text{dim. } L(U, V) = (\text{dim. } U)(\text{dim. } V)$$

$$\Rightarrow \quad \text{dim. } L(V, F) = (\text{dim.} V)(\text{dim. } F)$$

$$\Rightarrow \quad \text{dim. } V^* = (\text{dim.} V).1 \quad \Rightarrow \quad \text{dim. } V^* = \text{dim. } V$$

THEOREM 2. ***Let V be a finite dimensional linear space and a ≠ 0 in V, then there is an element f ∈ V* such that f(a) ≠ 0.***

PROOF. Let V be n-dimensional linear space over the field F and let $\alpha \neq 0$ be arbitrary non-zero vector of V and let $\{\alpha_1, \alpha_2, \ldots, \alpha_n\}$ be the basis of V. Then there exists unique scalars $a_i \in F$ such that

$$\alpha = a_1\alpha_1 + a_2\alpha_2 + \ldots + a_n\alpha_n = \sum_{j=1}^{n} a_j\alpha_j$$

Suppose $\{\phi_1, \phi_2, ..., \phi_n\}$ is dual basis of V^*, then

$$\phi_i(\alpha_j) = \begin{cases} 1, & i = j \\ 0, & i \neq j \end{cases} = \delta_{ij}.$$

Now, $$\phi_i(\alpha) = \phi_i\left(\sum_{j=1}^{n} a_j\alpha_j\right)$$

$$= \sum_{j=1}^{n} a_j\phi_i(\alpha_j) = \sum_{j=1}^{n} a_j\delta_{ij} = a_i$$

$\Rightarrow$ $\phi_1(\alpha) = a_1, \phi_2(\alpha) = a_2, .., \phi_n(\alpha) = a_n$

Since all the scalars a_i are not zero, this implies that there exists a linear functional $\phi \in V^*$ such that $\phi(\alpha) \neq 0$.

3.30 DUAL BASIS

Let $B = \{\alpha_1, \alpha_2, ..., \alpha_n\}$ be a basis for V, then there exists a unique linear functional f on V for each i such that

$$\phi_i(\alpha_j) = \begin{Bmatrix} 1, & \text{if } i = j \\ 0, & \text{if } i \neq j \end{Bmatrix} = \delta_{ij} \text{(Kronecker delta)}$$

Thus we obtain from B, a set of n distinct linear functional $\phi_1, \phi_2, .., \phi_n$ on V. These functionals are also linearly independent and generate V^*.

Therefor, the set $B^* = \{\phi_1, \phi_2, .., \phi_n\}$ foms a basis for V^*. This basis is called **dual basis** of B.

THEOREM 1. ***If $\{\alpha_1, \alpha_2, ..., \alpha_n\}$ is a basis of a linear space V over the field F. Let $\phi_1, \phi_2, ..., \phi_n \in V^*$ be the linear functionals defined by***

$$\phi_i(\alpha_j) = \delta_{ij} = \begin{cases} 1, & \text{if } i = j \\ 0, & \text{if } i \neq j \end{cases}$$

Then $\{\phi_1, \phi_2, ..., \phi_n\}$ is a basis of V^*

PROOF. First we show that $\{\phi_1, \phi_2, ..., \phi_n\}$ is linarly independent.

For $a_i \in F$ such that $a_1\phi_1 + a_2\phi_2, ..., a_n\phi_n = 0$.

Applying both sides to α_1, we get

$$(a_1\phi_1 + a_2\phi_2, ..., a_n\phi_n)(\alpha_1) = 0(\alpha_1) = 0$$

$\Rightarrow$ $(a_1\phi_1(\alpha_1) + a_2\phi_2(\alpha_1), ..., a_n\phi_n(\alpha_1) = 0$

$\Rightarrow$ $a_1.1 + a_2.0 + ... + a_n.0 \quad \Rightarrow \quad a_1 = 0$

Similarly, for $i = 2, 3, ..., n$ we have

$a_1\phi_1(\alpha_i) + a_2\phi_2(\alpha_i) + ... + a_n\phi_n(\alpha_i) + ... + a_n\phi_n(\alpha_i) = 0 \Rightarrow a_i = 0$. Thus

$a_1 = 0, a_1 = 0, ... a_n = 0$. Hence $\{\phi_1, \phi_2, ..., \phi_n\}$ is linearly independent.

Now we show that $\{\phi_i\}$ spans V^*.

For this, let ϕ be an arbitrary element of V^* and suppose that

$$\phi(\alpha_1) = c_1, \phi(\alpha_2) = c_2, ..., \phi(\alpha_n) = c_n$$

and set $$\psi = c_1\phi_1 + c_2\phi_2 + ... + c_n\phi_n \text{, then}$$

$$\psi(\alpha_1) = c_1\phi_1(\alpha_1) + c_2\phi_2(\alpha_1) + ... + c_n\phi_n(\alpha_1) = c_1.$$

Similarly, for $i = 2, 3, ..., n$. We have

$$\psi(\alpha_i) = c_i.$$

Thus, $\phi(\alpha_i) = \psi(\alpha_i)$ for $i = 1, 2, 3, ..., n$. Since ϕ and ψ both agree on the basis vectors of V.

$\therefore$ $$\phi = \psi = c_1\phi_1 + c_2\phi_2 + ... + c_n\phi_n$$

$\Rightarrow$ $\{\phi_1, \phi_2, ..., \phi_n\}$ spans V^*.

Hence, $\{\phi_1, \phi_2, ..., \phi_n\}$ forms a basis of V^*.

THEOREM 2. ***Let $B = \{\alpha_1, \alpha_2, ..., \alpha_n\}$ be a basis of V and let $B^* = \{\phi_1, \phi_2, ..., \phi_n\}$ be the dual basis of V^*. Then, for any vector $\alpha \in V$, such that***

$$\alpha = \phi(\alpha)\alpha_1 + \phi_2(\alpha)\alpha_2 + ... + \phi_n(\alpha)\alpha_n$$

and for any linear functional $\phi \in V^*$, $\phi = \phi(\alpha)\phi_1 + \phi(\alpha)\phi_2 + ... + \phi_n(\alpha)\phi_n$.

PROOF. Since $B = \{\alpha_1, \alpha_2, ..., \alpha_n\}$ is a basis for V and $B^* = \{\phi_1, \phi_2, ..., \phi_n\}$ is a basis for V^*, then there is a unique ϕ_j for each j such that

$$\phi_j(\alpha_i) = \delta_{ij} \quad ... (1)$$

Since $\{\phi_1, \phi_2, ..., \phi_n\}$ generates V^*, then for some scalar $a_i \in F$ and for $\phi \in V^*$ such that

$$\phi = a_1\phi_1 + a_2\phi_2 + ... + a_n\phi_n = \sum_{j=1}^{n} a_j\phi_j \quad ...(2)$$

Then $$\phi(\alpha_i) = \sum_{j=1}^{n} a_j\phi_j(\alpha_i) = \sum_{j=1}^{n} a_j\delta_{ij} \quad \text{[using (1)]}$$

$$\phi(\alpha_i) = a_i, i = 1, 2, ..., n.$$

Thus (2) becomes

$$\phi = \phi(\alpha_1)\phi_1 + \phi(\alpha_2)\phi_2 + ... + \phi(\alpha_n)\phi_n.$$

Similarly, $B = (\alpha_1, \alpha_2, ..., \alpha_n)$ generates V, then for $\alpha \in V$ some scalars $b_i \in F$ such that

$$\alpha = b_1\alpha_1 + b_2\alpha_2 + ... + b_n\alpha_n = \sum_{i=1}^{n} b_i\alpha_j \quad ... (3)$$

Then, $$\phi_j(\alpha) = \sum_{i=1}^{n} b_i\alpha_j(\alpha_i) = \sum_{i=1}^{n} b_i\delta_{ij} \quad \text{[using (1)]}$$

$$= b_j \quad \text{for } j = 1, 2, ..., n.$$

Hence, (3) becomes

$$\alpha = \phi_1(\alpha)\alpha_1 + \phi_2(\alpha)\alpha_2 + ... + \phi_n(\alpha)\alpha_n.$$

3.31 SECOND DUAL SPACE : BIDUAL SPACE

It has been shown that every linear space V has a dual space V^* which consists of all the linear functionals on V. Therefore, V^* itself has a dual space V^{**}, this dual space of V^* is called the **second dual** of V.

Furthermore, $$\dim. V = \dim. V^* = \dim. V^{**}.$$

3.32 NATURAL MAPPING

Let V be an n-dimensional linear space, V^* its dual and V^{**} the dual of V^*. Then a mapping $v \to \bar{v}, v \in V, \bar{v} \in V^{**}$, where, v is a linear functional on V^* defined by $\bar{v}(\phi) = \phi(v), \phi \in V^*$ is an isomorphism. This mapping is called a natural mapping.

REMARK

- If V is not finite dimensional, then the natural mapping can never be onto V^{**}. However, it is always linear and one-one.

3.33 ANNIHILATOR

Definition. *Let V be a linear space over the field F and V* its dual. Let W be a subset of V which is not necessarily a subspace. Then a linear functional $\phi \in V^*$ is called an annihilator of W if* $\phi(\alpha) = 0$ for every $\alpha \in W$, which is denoted by W^0.

That is, the set of all linear functional ϕ on V such that $\phi(\alpha)=0, \forall \alpha \in V$, *i.e.*, $\phi(W)=\{0\}$ is called annihilator of W.

Also, $$W^0 = \{\phi \in V^* : \phi(\alpha) = 0, \forall \alpha \in V\}$$

REMARKS

- Annihilator of V is the zero functional on V.
- $\{0\}^0 = V^*$.

THEOREM 1. ***W_0 is a subspace of V^*.***

PROOF. By definition of W^0, it is clear that $\mathbf{0} \in W^\circ$ and $W^\circ \subseteq V^*$.

Now suppose $\phi_1, \phi_2, \in W^0$ and for any scalars $a, b \in F$ and for any $a \in W$.

$$(a\phi_1 + a\phi_2)(\alpha) = a\phi_1(\alpha) + b\phi_2(\alpha) = a.0 + b.0 \qquad [\because \phi_1, \phi_2 \in W^0]$$
$$= 0$$

$$\Rightarrow \qquad a\phi_1 + b\phi_2 \in W^0$$

Hence, W^0 is a subspace of V^*.

THEOREM 2. ***Let V be a finite-dimensional linear space over the field F and let W be a subspace of V. Then dim.W + dim.W_0=dim.V.***

PROOF. Let V be n-dimensional linear space over the field F. Let Dim. $W = m$.

Let W be a subspace of V. Then W^0 is a subspace of V^*(dual of V).

Since W is a subspace of V so that

$$\dim . W \le \dim . V, \text{ i.e., } m \le n$$

Let $\{a_1, a_2, ..., a_n\}$ be a basis of W so it can be extended to form a basis of V, therefore choose vectors $\{\alpha_{m+1}, \alpha_{m+2}, \ldots, \alpha_n\}$ in V such that

$$B=\{\alpha_1, \alpha_2, ..., \alpha_m, \alpha_{m+1}, \ldots, \alpha_n)$$

is a basis of V. Let $\{\phi_1, \phi_2, .., \phi_n)$ be the basis of V^* which is the dual of B.

Now we claim that $\{\phi_{m+1}, \phi_{m+2}, .., \phi_n)$ is a basis of W^0. Obviously, $\phi_i \in W^0$ for $i \ge m+1$, because

$$\phi_i(\alpha_j) = \delta_{ij} = \begin{cases} 0, & i \ne j \\ 1, & i = j \end{cases}$$

and $\qquad \delta_{ij}=0$ if $i \ge m+1$ and $j \le m$.

Since $\{\phi_{m+1}, \phi_{m+2}, ..., \phi_n\}$ is a subset of linearly independent set $\{\phi_1, \phi_2, ..., \phi_n\}$ hence $\{\phi_{m+1}, \phi_{m+2}, ..., \phi_n\}$ is linearly independent. Now we shall show that $\{\phi_{m+1}, \phi_{m+2}, ..., \phi_n\}$ spans W^0.

Let $\phi \in W^0$ be an arbitrary linear functional, so that

$$\phi(\alpha_i) = 0, \text{ for } 1 \le i \le m \qquad ...(1)$$

Since $W^0 \subseteq V^*$, then $\phi \in V^*$

But $\{\phi_1, \phi_2, ..., \phi_n\}$ generates V*, therefore, we have

$$\phi = \sum_{i=1}^{n} \phi(\alpha_i)\phi_i \qquad \text{[By theorem]}$$

$$= \phi(\alpha_1)\phi_1 + \phi(\alpha_2)\phi_2 + \ldots + \phi(\alpha_m)\phi_m + \phi(\alpha_{m+1})\phi_{m+1} + \phi(\alpha_{m+2})\phi_{m+2} + \ldots + \phi(\alpha_n)\phi_n$$

$$= \phi(\alpha_{m+1})\phi_{m+1} + \phi(\alpha_{m+2})\phi_{m+2} + \ldots + \phi(\alpha_n)\phi_n = \sum_{i=1}^{n} \phi(\alpha_i)\phi_i$$

This shows that $\{\phi_{m+1}, \phi_{m+2}, \ldots, \phi_n\}$ spans W^0.

Thus $\{\phi_{m+1}, \phi_{m+2}, \ldots, \phi_n\}$ forms a basis of W^0.

Accordingly, $\dim. W^0 = n - m = \dim. V - \dim. W$

Hence, $\dim. W + \dim. W^0 = \dim. V$.

Corollary. *If W and W_1 are two subspaces of a linear space V which are annihilated by the subspace W^0, then $\dim. W = \dim. W_1$.*

PROOF. Since W and W_1 are both annihilated by the W^0, and both are subspaces of V, then by above theorem we have

$$\dim. W + \dim. W^0 = \dim. V \quad \ldots(1)$$

and

$$\dim. W_1 + \dim. W^0 = \dim. V \quad \ldots(2)$$

On using (1) and (2), we get $\dim. W = \dim. W_1$

THEOREM 3. ***Let W_1 and W_2 be subspaces of a finite dimensional linear space over the field F, then $W_1 = W_2$ if and only if $W_1^0 = W_2^0$***

PROOF. If $W_1 = W_2$, then obviously, $W_1^0 = W_2^0$. Conversely, if $W_1^0 = W_2^0$, then we can show that $W_1 = W_2$.

Let if possible, $W_1 \neq W_2$, then there is at least one vector in W_1 which is not in W_2. Suppose $\alpha \in W_2$ and $\alpha \neq W_1$. then there is a linear functional ϕ such that $\phi(\beta) = 0 \ \forall \ \beta \in W$ but $\phi(\alpha) \neq 0$. This implies that $\phi \in W_1^0$ but $\phi \notin W_2^0$ and thus $W_1^0 \neq W_2^0$. Hence if $W_1^0 = W_2^0$, then $W_1 = W_2$.

3.34 ANNIHILATOR OF AN ANNIHILATOR

Definition. *Let V be a linear space over the field F. V^* its dual space and let V^{**} be the dual space of V^*.*

Let W be a subset of a linear space $V(F)$. Then W^0 is a subspace of V^*. Therefore,

$$(W^0)^0 = W^{00} = \{\psi \in V^{**} : \psi(\phi) = \mathbf{0}, \forall \ \phi \in W^0\}$$

Thus W^{00} is called an annihilator of W^0.

Since V^* is natural isomorphism to V^{**}, then we can write W^{00} as follows

$$W^{00} = \{\alpha \in V : \phi(\alpha) = 0 \ \forall \ \phi \in W^0\}$$

THEOREM 1. ***If W is a subspace of a finite-dimensional linear space V over the field F, then $W = W^{00}$.***

PROOF. In order to prove $W = W^{00}$, we prove that

(i) $W \subset W^{00}$ (ii) $W^{00} \subset W$

By definition, we have

$$W^0 = \{\phi \in V^* : \phi(\alpha) = 0, \forall \phi \in W^0\} \quad \ldots(1)$$

and

$$W^{00} = \{\psi \in V^{**} : \psi(\phi) = 0, \forall \phi \in W^0\}$$

or

$$W^{00} = \{\alpha \in V : \phi(\alpha) = 0, \text{ for all } \phi \in W^0\} \quad \ldots(2)$$

Let $\alpha \in W$ and $\phi \in W^0$ be arbitrary. Then

$$\alpha \in W \Rightarrow \phi(\alpha)=0 \ \forall \ \phi \in W^0 \Rightarrow \alpha \in W^{00} \quad \text{[from (2)]}$$

$\therefore$ $W \subset W^{00}$

Further, we know that

$$\dim. W + \dim. W^0 = \dim. V \quad ...(3)$$

and $$\dim. W^0 + \dim. W^{00} = \dim. V^*. \quad ... (4)$$

From (3) and (4), we get $\dim. W = \dim. W^{00}$ $[\because \dim. V = \dim. V^*]$

As $W \subset W^{00} \Rightarrow W$ is a subspace of W^{00} with the property that $\dim W = \dim W^{00}$.

Hence $W = W^{00}$

RECAPITULATIONS

- A linear transformation $\phi : V \rightarrow F$ is called linear functional if for every $\alpha, \beta \in V$ and $a, b \in F$, $\phi(a\alpha + b\beta) = a\phi(\alpha) + b\phi(\beta)$
- $\dim.(W) + \dim (W^0) = \dim .V$
- The set of linear functionals on a linear space V over the field F is also a linear space such that $(\phi_1 + \phi_2) = \phi_1(\alpha) + \phi_2(\alpha) \forall \alpha \in V$ and $\phi(c\alpha) = c\phi(\alpha)$, $c \in F$, where ϕ_1, ϕ_2 are linear functional on V. This space is called the dual space or conjugate space.

SOLVED EXAMPLES

EXAMPLE 1. *If $B = \{(-1, 1, 1), (1, -1, 1), (1, 1, -1)\}$ is a basis of $V3(\mathrm{R})$, then find the dual basis of B.*

SOLUTION. Let $\alpha_1 = (-1, 1, 1), \alpha_2 = (1, -1, 1), \alpha_3 = (1, 1, -1)$.

$\therefore$ $B = \{\alpha_1, \alpha_2, \alpha_3\}$ is a basis of $V_3(\mathrm{R})$.

Let $B^* = (\phi_1, \phi_2, \phi_3)$ be the dual of B such that

$$\phi_i(\alpha_j) = \begin{cases} 1 \text{ if } & i = j \\ 0 \text{ if } & i \neq j \end{cases} \quad ...(1)$$

Now write,

$$\phi_1(x, y, z) = a_1x + b_1y + c_1z$$
$$\phi_2(x, y, z) = a_2x + b_2y + c_2z$$
$$\phi_3(x, y, z) = a_3x + b_3y + c_3z$$

Determination of ϕ_1.

$$\phi_1(\alpha_1) = \phi_1(-1, 1, 1) = -a_1 + b_1 + c_1$$
$$\phi_1(\alpha_2) = \phi_1(1, -1, 1) = a_1 - b_1 + c_1$$
$$\phi_1(\alpha_3) = \phi_1(-1, 1, -1) = a_1 + b_1 - c_1$$

But using (1), we have $\phi_1(\alpha_1) = 1, \phi_1(\alpha_2) = 0, \phi_1(\alpha_3) = 0,$

$$\therefore \quad \left.\begin{aligned} -a_1 + b_1 + c_1 &= 1 \\ a_1 - b_1 + c_1 &= 1 \\ a_1 + b_1 - c_1 &= 1 \end{aligned}\right\} \quad ... (2)$$

From last two equations of (2), we have

$$\frac{a_1}{(-1)(-1)-(1)(1)} = \frac{b_1}{(1)(1)-(1)(-1)} = \frac{c_1}{(1)(1)-(-1)(1)} = k$$

$$\Rightarrow \quad \frac{a_1}{1-1} = \frac{b_1}{1+1} = \frac{c_1}{1+1} = k$$

$$a_1 = 0, b_1 = 2k, c_1 = 2k.$$

Putting the values of a_1, b_1 and c_1 in the first equation of (2), we get

$$0 + 2k + 2k = 1 \Rightarrow 4k = 1 \Rightarrow k = \frac{1}{4}$$

$$\therefore \quad a_1 = 0,\ b_1 = \frac{1}{2}, c_1 = \frac{1}{2}$$

Thus $\quad \phi_1(x, y, z) = \frac{1}{2}(y + z)$

Determination of ϕ_2.

$$\phi_2(\alpha_1) = \phi_2(-1, 1, 1) = -a_2 + b_2 + c_2$$
$$\phi_2(\alpha_2) = \phi_2(1, -1, 1) = a_2 - b_2 + c_2$$
$$\phi_2(\alpha_3) = \phi_2(-1, 1, -1) = a_2 + b_2 - c_2$$

Again, $\quad \phi_2(\alpha_1) = 0,\ \phi_2(\alpha_2) = 1,\ \phi_2(\alpha_3) = 0.$

So that above equations become:

$$-a_2 + b_2 + c_2 = 0$$
$$a_2 - b_2 + c_2 = 1$$
$$a_2 + b_2 - c_2 = 0$$

Solving these equations, we get

$$a_2 = \frac{1}{2}, b_2 = 0,\ c_2 = \frac{1}{2}$$

$$\therefore \quad \phi_2(x, y, z) = \frac{1}{2}(x + z)$$

Determination of ϕ_3.

$$\phi_3(\alpha_1) = \phi_3(-1, 1, 1) = -a_3 + b_3 + c_3$$
$$\phi_3(\alpha_2) = \phi_3(1, -1, 1) = a_3 - b_3 + c_3$$
$$\phi_3(\alpha_3) = \phi_3(-1, 1, -1) = a_3 + b_3 - c_3$$

Again, $\quad \phi_3(\alpha_1) = 0,\ \phi_3(\alpha_2) = 1,\ \phi_3(\alpha_3) = 1.$

So that above equations become:

$$-a_3 + b_3 + c_3 = 0$$
$$a_3 - b_3 + c_3 = 0$$
$$a_3 + b_3 - c_3 = 1$$

Solving these equations, we get $a_3 = \frac{1}{2}, b_3 = \frac{1}{2}, c_3 = 0$

$$\therefore \quad \phi_3(x, y, z) = \frac{1}{2}(x + y)$$

Hence $\{\phi_1, \phi_2, \phi_3\}$ is the dual basis,

where $\quad \phi_1(x, y, z) = \frac{1}{2}(y + z), \phi_2(x, y, z) = \frac{1}{2}(x + z), \phi_3(x, y, z) = \frac{1}{2}(x + y).$

EXAMPLE 2. *If $\{\alpha_1 = (1, -2, 3), \alpha_2 = (1, -1, 1), \alpha_3 = (2, -4, 7)\}$ is a basis of R^3, then find the dual basis $\{\phi_i\}$.*

SOLUTION. Let $\{\phi_1, \phi_2, \phi_3\}$ be the dual to the basis $\{\alpha_1, \alpha_2, \alpha_3\}$ such that

Let $\quad \phi_i(\alpha_j) = \begin{cases} 1 \text{ if } & i = 1 \\ 0 \text{ if } & i \neq j \end{cases} \quad ...(1)$

$$\phi_1(x, y, z) = a_1x+b_1y+c_1z$$
$$\phi_2(x, y, z) = a_2x+b_2y+c_2z$$
$$\phi_3(x, y, z) = a_3x+b_3y+c_3z$$

Determination of ϕ_1.

$$\phi_1(\alpha_1) = \phi_1(1, -2, 3) = -a_1+2b_1+3c_1$$
$$\phi_2(\alpha_2) = \phi_1(1, -1, 1) = a_1-b_1+c_1$$
$$\phi_3(\alpha_3) = \phi_1(2, -4, 7) = 2a_1-4b_1-7c_1 \qquad \text{[Using (1)]}$$

But $\phi_1(\alpha_1) = 1, \phi_1(\alpha_2) = 0, \phi_1(\alpha_3) = 0$,

So that above equations become:

$$a_1-2b_1+3c_1 = 1$$
$$a_1-b_1+c_1 = 0$$
$$2a_1-4b_1-7c_1 = 0$$

Solving these equations, we get

$$a_1 = -3, b_1 = -5, c_1 = 2.$$

$$\therefore \qquad \phi_1(x, y, z) = -3x-5y-2z.$$

Determination of ϕ_2.

$$\phi_2(\alpha_1) = \phi_2(1, -2, 3) = a_2-2b_2+3c_2$$
$$\phi_2(\alpha_2) = \phi_2(1, -1, 1) = a_2-b_2+c_2$$
$$\phi_1(\alpha_3) = \phi_3(2, -4, 7) = 2a_2-4b_2+7c_2$$

Since, $\phi_2(\alpha_1) = 0, \phi_2(\alpha_2) = 1, \phi_2(\alpha_3) = 0$, the above equation become

$$a_2-2b_2+3c_2 = 0$$
$$a_2-b_2+c_2 = 1$$
$$2a_2-4b_2+7c_2 = 0$$

Solving these equations, we get

$$a_2 = 2, b_2 = 1, c_2 = 0$$

$$\therefore \qquad \phi_2(x, y, z) = 2x+y$$

Determination of ϕ_3.

$$\phi_3(\alpha_1) = \phi_3(1, -2, 3) = -a_3-2b_3+3c_3$$
$$\phi_3(\alpha_2) = \phi_3(1, -1, 1) = a_3-b_3+c_3$$
$$\phi_3(\alpha_3) = \phi_3(2, -4, 7) = 2a_3-4b_3+7c_3$$

Again, $\phi_3(\alpha_1) = 0, \phi_3(\alpha_2) = 0, \phi_3(\alpha_3) = 1$, the above equations become

$$-a_3-2b_3+3c_3 = 0$$
$$a_3-b_3+c_3 = 0$$
$$2a_3-4b_3+7c_3 = 0$$

Solving the equations, we get

$$a_3 = 1, b_3 = 2, c_3 = 1$$

$$\therefore \qquad \phi_3(x, y, z) = x+2y+z.$$

Hence, $\{\phi_1, \phi_2, \phi_3\}$ is the dual basis, where

$$\phi_1(x, y, z) = -(3x+5y+2z), \phi_2(x, y, z) = 2x+y, \phi_3(x, y, z) = 2x+2y+2z$$

EXAMPLE 3. *Find the dual basis of the set*

$B = \{(1, -1, 3), (0, 1, -1), (0, 3, -2)\}$ for V_3 (R)

SOLUTION. Let us suppose $\alpha_1 = (1, -1, 3)$, $\alpha_2 = (0, 1, -1)$ and $\alpha_3 = (0, 3, -2)$ and let $B^* = \{(\phi_1, \phi_2, \phi_3)\}$ be the dual to the basis B such that

$$\phi_i(\alpha_j) = \begin{cases} 1 \text{ if } & i = j \\ 0 \text{ if } & i \neq j \end{cases} \quad \text{...(1)}$$

Suppose

$$\phi_1 (x, y, z) = a_1 x + b_1 y + c_1 z$$
$$\phi_2 (x, y, z) = a_2 x + b_2 y + c_2 z$$
$$\phi_3 (x, y, z) = a_3 x + b_3 y + c_3 z$$

Determination of ϕ_1.

$$\phi_1 (\alpha_1) = \phi_1 (1, -1, 3) = a_1 - b_1 + 3c_1$$
$$\phi_1 (\alpha_2) = \phi_1 (0, 1, -1) = b_1 - c_1$$
$$\phi_1 (\alpha_3) = \phi_1 (0, 3, -2) = 3b_1 - 2c_1$$

Since $\phi_1 (\alpha_1) = 1, \phi_1 (\alpha_2) = 0, \phi_1 (\alpha_3) = 0,$ [Using (1)]

then above equations become:

$$a_1 - b_1 + 3c_1 = 1$$
$$b_1 - c_1 = 0$$
$$3b_1 - 2c_1 = 0$$

Solving these equations, we get

$$a_1 = 1, b_1 = 0, c_1 = 0.$$

$\therefore$ $\phi_1 (x, y, z) = x$

Determination of ϕ_2.

$$\phi_2 (\alpha_1) = \phi_2 (1, -1, 3) = a_2 - b_2 + 3c_2$$
$$\phi_2 (\alpha_2) = \phi_2 (0, 1, -1) = b_2 - c_2$$
$$\phi_2 (\alpha_3) = \phi_3 (0, 3, -2) = 3b_2 - 2c_2$$

Since $\phi_2 (\alpha_1) = 0, \phi_2 (\alpha_2) = 1, \phi_2 (\alpha_3) = 0,$

the above equation become

$$a_2 - b_2 + 3c_2 = 0$$
$$b_2 - c_2 = 1$$
$$3b_2 - 2c_2 = 0$$

Solving these equations, we get

$$a_2 = 7, b_2 = -2, c_2 = -3$$

$\therefore$ $\phi_2(x, y, z) = 7x - 2y - 3z$

Determination of ϕ_3.

$$\phi_3 (\alpha_1) = \phi_3 (1, -1, 3) = -a_3 - 2b_3 + 3c_3$$
$$\phi_3 (\alpha_2) = \phi_3 (0, 1, -1) = b_3 - c_3$$
$$\phi_3 (\alpha_3) = \phi_3 (0, 3, -2) = 3b_3 - 2c_3$$

Since, $\phi_3 (\alpha_1) = 0, \phi_3 (\alpha_2) = 0, \phi_3 (\alpha_3) = 1,$

the above equations become

$$a_3 - b_3 + 3c_3 = 0$$
$$b_3 - c_3 = 0$$

$$3b_3 - 2c_3 = 0$$

Solving the equations, we get

$$a_3 = -2, b_3 = 1, c_3 = 1$$

$\therefore$ $$\phi_3 (x, y, z) = -2x + y + z.$$

Hence, $\{\phi_1, \phi_2, \phi_3\}$ is the dual basis, where

$$\phi_1 (x, y, z) = x, \phi_2 (x, y, z) = 7x - 2y - 3z, \phi_3 (x, y, z) = -2x + y + z$$

EXAMPLE 4. *A basis of the linear space $R^3(R)$ is*

$$B = \{\alpha_1 = (1,1,0), \alpha_2 = (1,0,1), \alpha_3 = (0,1,1)\}$$

and f is a linear functional on R^3 such that $f(\alpha_1) = 1, f(\alpha_2) = -1, f(\alpha_3) = 3$, then find when $\alpha = (1, -1, 3)$.

SOLUTION. Since $B = \{\alpha_1 = (1,1,0), \alpha_2 = (1,0,1), \alpha_3 = (0,1,1)\}$ is a basis of R^3(R).

Let $B^* = \{\phi_1, \phi_2, \phi_3\}$ be the basis. Then

$$\phi_i(\alpha_j) = \begin{cases} 1 \text{ if } i = j \\ 0 \text{ if } i \neq j \end{cases} \quad ...(1)$$

Now write

$$\phi_1(x, y, z) = a_1x + b_1y + c_1z$$
$$\phi_2(x, y, z) = a_2x + b_2y + c_2z$$
$$\phi_3(x, y, z) = a_3x + b_3y + c_3z$$

Determination of ϕ_1.

$$\phi_1(\alpha_1) = \phi_1(1,1,0) = a_1 + b_1$$
$$\phi_1(\alpha_2) = \phi_1(1,0,1) = a_1 + c_1$$
$$\phi_1(\alpha_3) = \phi_1(0,1,1) = b_1 + c_1$$

Since $$\phi_1(\alpha_1) = 1, \phi_1(\alpha_2) = 0, \phi_1(\alpha_3) = 0 \quad \text{[using (1)]}$$

Then, the above equations become

$$a_1 + b_1 = 1$$
$$a_1 + c_1 = 0$$
$$b_1 + c_1 = 0.$$

Solving these equations, we get

$$a = \frac{1}{2}, b_1 = \frac{1}{2}, c_1 = -\frac{1}{2}$$

$\therefore$ $$\phi_1(x, y, z) = \frac{1}{2}(x + y - z).$$

Determination of ϕ_2.

$$\phi_2(\alpha_1) = \phi_2(1,1,0) = a_2 + b_2$$
$$\phi_2(\alpha_2) = \phi_2(1,0,1) = a_2 + c_2$$
$$\phi_2(\alpha_3) = \phi_3(0,1,1) = b_2 + c_2.$$

Using (1) we have $\phi_2(\alpha_1) = 1, \phi_2(\alpha_2) = 1, \phi_2(\alpha_3) = 0$,

the above equations become

$$a_2 + b_2 = 0$$
$$a_2 + c_2 = 1$$

$$b_2 + c_2 = 0.$$

Solving these equations, we get

$$a_2 = \frac{1}{2}, b_2 = -\frac{1}{2}, c_2 = \frac{1}{2}.$$

$$\therefore \quad \phi_2(x, y, z) = \frac{1}{2}(x - y + z).$$

Determination of ϕ_3.

$$\phi_3(\alpha_1) = \phi_3(1,1,0) = a_3 + b_3$$
$$\phi_3(\alpha_2) = \phi_3(1,0,1) = a_3 + c_3$$
$$\phi_3(\alpha_3) = \phi_3(0,1,1) = b_3 + c_3$$

Since $\phi_3(\alpha_1) = 0, \phi_3(\alpha_2) = 0, \phi_3(\alpha_3) = 1.$

Then, the above equations become

$$a_3 + b_3 = 0$$
$$a_3 + c_3 = 0$$
$$b_3 + c_3 = 1$$

Solving these equations, we get

$$a_3 = -\frac{1}{2}, b_3 = \frac{1}{2}, c_3 = \frac{1}{2}$$

$$\phi_3(x,y,z) = \frac{1}{2}(-x+y+z)$$

Since f is a linear functional R^3 such that $f(\alpha_1) = 1, f(\alpha_2) = -1, f(\alpha_3) = 3.$

Then by theorem, we have

$$f = f(\alpha_1)\phi_1 + f(\alpha_2)\phi_2 + f(\alpha_3)\phi_3$$
$$f = \phi_1 - \phi_2 + 3\phi_3.$$

$$\therefore \quad \phi(x, y, z) = \phi_1(x, y, z) - \phi_2(x, y, z) + 3\phi_3(x, y, z).$$
$$= \frac{1}{2}(x + y - z) - \frac{1}{2}(x - y + z) + \frac{3}{2}(-x + y + z)$$
$$= -\frac{3}{2}x + \frac{5}{2}y + \frac{1}{2}z.$$

Determination of when $\alpha = (1, -1, 3)$

$$f(\alpha) = f(1,-1,3) = -\frac{3}{2}(1) + \frac{5}{2}(-1) + \frac{1}{2}(3).$$
$$= -\frac{3}{2} - \frac{5}{2} + \frac{3}{2} = -\frac{5}{2}.$$

EXAMPLE 5. *If W_1 and W_2 are subspace of a vector V over a field F and $W_1 \subset W_2$, find $W_2^0 \subset W_1^0$.*

SOLUTION. Since W_1 and W_2 are subspace of V with the condition that $W_1 \subset W_2$. Then we shall show that $W_2^0 \subset W_1^0$.

Let $\phi \in W_2^0$ be an arbitrary linear functional.

Then, we have $\phi \in W_2^0 \Rightarrow \phi(\alpha) = 0, \forall\, \alpha \in W_1$

$$\Rightarrow \phi(\alpha) = 0, \forall \alpha \in W_2 \qquad [\because W_1 \subset W_2]$$

$$\Rightarrow \phi \in W_1^0 .$$

Hence, $W_2^0 \subset W_1^0$.

EXAMPLE 6. *If W is a subset of a linear space V(F), then*

(i) $W^0 = [L(W)]^0$ (ii) $W^{00} = L(W)$.

SOLUTION. (i) Let $W = \{\alpha_1, \alpha_2, \ldots \alpha_n\}$ be a subset of a linear space $V(F)$. We shall show that $W^0 = [L(W)]^0$

By the definition of W^0, we have

$$W^0 = \{\phi \in V^* : \phi(\alpha) = 0, \alpha \in W\}.$$

Since $W \subseteq L(W)$, therefore

$$[L(W)]^0 \subset W^0 \qquad \ldots(1)$$

Now, $\phi \in W^0 \Rightarrow \phi(\alpha) = 0, \forall \alpha \in W$

$$\Rightarrow \phi(\alpha) = 0, \forall \alpha \in L(W) \qquad [\because W \subset L(W)]$$

$$\Rightarrow \phi \in [L(W)]^0.$$

$\therefore$ $W^{00} \subset [L(W)]^0 \qquad \ldots(2)$

From (1) and (2), we get

$$W^0 = [L(W)]^0$$

(ii) Since we know that

$$W^{00} = W. \qquad \ldots(3)$$

$$\therefore \quad W^{00} = (W^0)^0 = [[L(W)]^0]^0 \qquad [\because W^0 = [L(W)]^0]$$

$$= L(W) \qquad [\text{using } (3)]$$

Hence, $W^{00} = L(W)$.

EXAMPLE 7. *Let W_1 and W_2 be subspaces of a finite-dimensional linear space V aver the field F. Prove that*

(i) $(W_1 + W_2)^0 = W_1^0 \cap W_2^0$ (ii) $W_1^0 \cap W_2^0 = (W_1 \cap W_2)^0$

SOLUTION. (i) To prove that : $(W_1 + W_2)^0 = W_1^0 \cap W_2^0$.

Since we know that : $W_1 \subset W_2 \Rightarrow W_2^0 \subset W_1^0$.

Obviously, $W_1 \subset W_1 + W_2$ and $W_2 \subset W_1 + W_2$

$\therefore$ $(W_1 + W_2)^0 \subset W_1^0$ and $(W_1 + W_2)^0 \subset W_2^0$

$\Rightarrow$ $(W_1 + W_2)^0 \subset W_1^0 \cap W_2^0 \qquad \ldots(1)$

Now, let $\phi \in W_1^0 \cap W_2^0$ be an arbitrary linear functional

$$\phi \in W_1^0 \cap W_2^0 \quad \Rightarrow \phi \in W_1^0 \text{ and } \phi \in W_2^0$$

$$\Rightarrow \phi(\alpha) = 0, \forall \alpha \in W_1 \text{ and } \phi(\beta) = 0, \forall \beta \in W_2$$

Let $\gamma \in W_1 + W_2$ be an arbitrary vector so that

$$y = \alpha + \beta, \text{ for some } \alpha \in W_1, \beta \in W_2$$

$$\therefore \quad \phi(\gamma) = \phi(\alpha + \beta) = \phi(\alpha) + \phi(\beta) \qquad [\because \phi \text{ is linear.}]$$

$$= 0 + 0 = 0$$

$\Rightarrow \qquad \phi(\gamma) = 0$, for all $\gamma \in W_1 + W_2$

$\Rightarrow \qquad \phi \in (W_1+W_2)^0$

Thus, $\qquad W_1^0 \cap W_2^0 \subset (W_1 + W_2)^0 \qquad$ ... (2)

From (1) and (2), we get

$$(W_1+W_2)^0 = W_1^0 \cap W_2^0$$

(ii) Above result (i) is taken for the linear space $V^*(F)$ in place of $V(F)$, then we have

$$(W_1^0 + W_2^0)^0 = W_1^{00} \cap W_2^{00}$$

$\Rightarrow \qquad (W_1^0 + W_2^0)^0 = W_1 \cap W_2 \qquad [\because W^{00} = W]$

$\Rightarrow \qquad (W_1^0 + W_2^0)^{00} = (W_1 \cap W_2)^0$

$\Rightarrow \qquad W_1^0 + W_2^0 = (W_1 \cap W_2)^0 \qquad [\because W^{00} = W]$

EXERCISE 3.4

1. Let $\phi = R^2 \to R$ and $\psi = R^2 \to R$ be the linear functionals defined by $\phi(x, y) = x + 2y$ and $\psi(x,y) = 3x - y$. Find
   (i) $\phi + \psi$ (ii) 4ϕ
   (iii) $2\phi - 5\psi$
2. Let $\phi = R^3 \to R$ and $\psi = R^3 \to R$ be the linear functionals defined by $\phi(x, y, z) = 2x - 2y + z$ and $\psi(x, y, z) = 4x - 2y + 3z$. Find
   (i) $\phi + \psi$ (ii) 3ϕ (iii) $2\phi - 5\psi$
3. Let ϕ be the linear functional on R^2 defined by $\phi(2,1) = 15$ and $\phi(1, -2) = -10$. Find $\phi(x, y)$ and, in particular, find $\phi(-2, 7)$.

3.35 EIGENVALUES AND EIGENVECTOR OF A LINEAR TRANSFORMATION

Let V be a finite-dimensional linear space over a field F and let T be a linear operator on V, then a scalar $\lambda \in F$ is called an eigenvalue of T if there exists a non-zero vector $\alpha \in V$ such that

$$T(\alpha) = \lambda\alpha.$$

Also, each such non-zero vector $\alpha \in V$ is called an eigenvector of T corresponding to λ.

The set E_λ of all such vectors is a subspace of V called the eigenspace of λ.

THEOREM 1. ***Non-zero eigenvectors belonging to distinct eigenvalues are linearly independent.***

PROOF. Let $T : V \to V$ be a linear operator and let $\alpha_1, \alpha_2, \ldots, \alpha_n$ be non-zero eigenvectors of T corresponding to distinct eigenvalues $\lambda_1, \lambda_2, \ldots, \lambda_n$. Then we have to show that $\alpha_1, \alpha_2, \ldots, \alpha_n$ are linearly independent.

We shall prove by induction on n. If $n = 1$ then α_1 is linearly independent as $\alpha_1 \neq 0$, so assume that for $n > 1$, we have $\alpha_1, \alpha_2, \ldots, \alpha_{n-1}$ are linearly independent.

Suppose $\qquad a_1\alpha_1 - a_2\alpha_2 + \ldots + a_n\alpha_n = 0 \qquad$ ...(1)

for $a_1, a_2, \ldots, \in F$.

Applying T to (1), we get

$$T(a_1\alpha_1 + a_2\alpha_2 + + \ldots + a_n\alpha_n) = T(0) = 0$$

$\Rightarrow \qquad a_1T(\alpha_1) + a_2T(\alpha_2) + \ldots + a_nT(\alpha_n) = 0 \qquad (\because T$ is linear.) $\qquad$ ...(2)

But by given hypothesis

$$T(\alpha_i) = \lambda_i\alpha_i, \forall\, i.$$

Therefore, (2) becomes

$$a_1\lambda_1\alpha_1 + a_2\lambda_2\alpha_2 + ... + a_n\lambda_n\alpha_n = 0 \qquad ...(3)$$

Now multiplying (1) by λ_n, we get

$$a_1\lambda_n\alpha_1 + a_2\lambda_n\alpha_2 + ... + a_n\lambda_n\alpha_n = 0 \qquad ...(4)$$

Subtracting (4) from (3), we get

$$a_1(\lambda_1 - \lambda_n)\,\alpha_1 + a_2\{\lambda_2 - \lambda_n)\,\alpha_2 + ... + a_{n-1}\{\lambda_{n-1} - \lambda_n)\,\alpha_{n-1} = 0 \qquad ...(5)$$

Since $\alpha_1, \alpha_2, ..., \alpha_{n-1}$ are linearly independent, also λ_i are distinct, therefore from (5) we obtain

$$a_1 = a_2 = ... = a_{n-1} = 0.$$

Substituting these values of $a_1, a_2 ... a_{n-1}$ into (1), we get

$$a_n\alpha_n = 0 \Rightarrow a_n = 0 \qquad [\because \alpha_n \neq 0]$$

So, $a_1 = a_2 = ... = a_n = 0$, hence $\alpha_1, \alpha_2, .., \alpha_n$ are linearly independent.

THEOREM 2. ***Let $T : V \to V$ be a linear operator on a finite dimensional linear space over a field F. Then $\lambda \in F$ is an eignvalue of T if and only if the operator $(\lambda I - T)$ is singular. The eigenspace of X is then the kernel of $(\lambda I - T)$.***

PROOF. Suppose $\lambda \in F$ is an eigenvalue of T, then we shall show that $(\lambda I - T)$ is singular.

Let α be any non-zero vector of V, then by definition of eigenvalue of T, we have

$$T(\alpha) = \lambda\alpha \quad \text{or} \quad \lambda\alpha - T(\alpha) = 0$$

$$\Rightarrow \quad (\lambda I)(\alpha) - T(\alpha) = 0 \Rightarrow (\lambda I - T)(\alpha) = 0$$

$$\Rightarrow \quad |(\lambda I - T)| = 0$$

$$\Rightarrow \quad \lambda I - T = 0 \Rightarrow \lambda I - T \text{ is singular.}$$

Conversely, suppose $(\lambda I - T)$ is singular, then we shall show that λ is an eigenvalue of T.

Let $\alpha \neq 0$ be an element of V so it is linearly independent.

Since $(\lambda I - T)$ is singular so that

$$|(\lambda I - T)| = 0 \Rightarrow (\lambda I - T)(\alpha) = 0 \quad [\because \alpha \text{ is linearly independent.}]$$

$$\Rightarrow \quad (\lambda I)(\alpha) - T(\alpha) = 0 \Rightarrow \lambda I - T(\alpha) = 0 \qquad [\because I(\alpha) = \alpha]$$

$$\Rightarrow \quad T(\alpha) = \lambda\alpha$$

$\Rightarrow$ λ is an eigenvalue of T.

Also, $(\lambda I - T)(\alpha) = 0 \Rightarrow \alpha \in$ kernel of $(\lambda I - T)$.

Definition. *Let T be a linear operator on a n-dimensional linear space V over a field F. If B is an ordered basis of V, then $|(xI - T) = (xI - [T]_B)|$ is catted a monic polynomial of degree n in F[x]. The monic polynomial $|(xI - T)|$ in the variable x is also called characteristic polynomial of T, it is denoted by $\Delta(x)$.*

By above theorem, λ is an eigenvalue of T if and only if λ is a root of this polynomial in F.

REMARK

- The degree of characteristic polynomial of T is exactly equal to n, then n-dimension of V, T cannot have more than n eigenvalues, counted with multiplicity.

THEOREM 3. **(Cayley-Hamilton Theorem).** ***If T be a linear transformation on n-dimensional linear space and $\Delta(\lambda)$ be its characteristic polynomial then $\Delta(T) = 0$.***

PROOF. Let $T : V(F) \to V(F)$ be a linear operator and let B be an ordered basis of V and the matrix of T corresponding to B be V.

$$[T]_B = A = [a_{ij}]_{n\times n} \text{ for } a_{ij}\text{'s} \in F$$

where, $$T(\alpha_j) = \sum_{i=1}^{n} a_{ij}\alpha_i .$$

If I is an identity matrix of order $n\times n$ and λ an indeterminate scalar, then the characteristic polynomial of T is given by

$$\Delta(\lambda) = |T - \lambda I| = 0 = |[T]_B - \lambda I| = 0 = |A - \lambda I| = 0$$

$$\therefore \quad \Delta(\lambda) = \begin{vmatrix} a_{11}-\lambda & a_{12} & \dots & a_{1n} \\ a_{21} & a_{21}-\lambda & \dots & a_{2n} \\ \dots & \dots & \dots & \dots \\ a_{n1} & a_{n2} & \dots & a_{nm}-\lambda \end{vmatrix}$$

$$= b_n\lambda^n + b_{n-1}\lambda^{n-1} + \dots b_1\lambda + b_0 \text{ (say).}$$

Thus the characteristic polynomial of A is

$$\Delta(\lambda) = 0, \text{ i.e., } b_n\lambda^n + b_{n-1}\lambda^{n-1} + \dots b_1\lambda + b_0 = 0. \qquad \dots(1)$$

Since the elements of $A - \lambda I$ are polynomials of degree at most one in λ so that each element of adj$(A-\lambda I)$is the polynomial at most of degree $n-1$, in λ. Therefore,

$$\text{adj } (A - \lambda I) = B_0 + B_1\lambda + \dots + B_{n-1}\lambda^{n-1} \qquad \dots(2)$$

where B_i's are square matrices of degree $n \times n$.

Since $(A - \lambda I_n)$ adj $(A - \lambda I_n) = |A - \lambda I_n| I_n$

or $$(A - \lambda I)(B_0 + B_1\lambda + B_2\lambda^2 + \dots + B_{n-1}\lambda^{n-1}) = (b_n\lambda^n + b_{n-1}\lambda^{n-1} + \dots + b_0) I.$$

Comparing the coefficients of like powers in λ, on both sides, we have

$$AB_0 = b_0 I$$
$$AB_1 - B_0 = b_1 I$$
$$AB_2 - B_1 = b_2 I$$
$$\dots \quad \dots \quad \dots \quad \dots$$
$$AB_{n-1} - B_{n-1} = b_{n-1}I - B_{n-1} = b_n I.$$

Multiplying above equation by $I, A, A^2, \dots, A^n$ respectively and then adding, we get

$$0 = b_0 I + b_1 A + b_2 A^2 + \dots + b^n A^n$$

or $$\Delta(A) = 0 \qquad \text{[using (1)]}$$

or $$\Delta(T) = 0.$$

THEOREM 4. ***If $\lambda \in F$ is a characteristic root (eigenvalue) of T, then for any polynomial $p(x) \in F[x]$, $p(\lambda)$ is a characteristic root of $p(T)$.***

PROOF. Since $\lambda \in F$ is a characteristic root of T, there exists a non-zero vector a in V such that

$$T(\alpha) = \lambda\alpha. \qquad \dots(1)$$

Now, $$T^2(\alpha) = T[T(\alpha)] = T(\lambda\alpha) \text{ [using (1)]}$$
$$= \lambda T(\alpha) \qquad [\because T \text{ is linear.}]$$
$$= \lambda^2\alpha. \qquad \text{[using (1)]}$$

Continuing in this way, we get

$$T^k(\alpha) = \lambda^k\alpha \qquad \dots(2)$$

for all positive integer k.

Let $$p(x) = a_0x^n + a_1x^{n-1} + \ldots + a_{n-1}x + a_n,\ a_i \in F$$

Then $$p(T) = a_0T^n + a_1T^{n-1} + \ldots + a_{n-1}T + a_nI.$$

Now
$$\begin{aligned}[p(T)](\alpha) &= (a_0T^n + a_1T^{n-1} \ldots + a_{n-1}T + a_nI)(\alpha)\\ &= a_0T^n(\alpha) + a_1T^{n-1}(\alpha) + \ldots + a_{n-1}T(\alpha) + a_nI(\alpha)\\ &= a_0\lambda^n\alpha + a_1\lambda^{n-1}\alpha + \ldots + a_{n-1}\lambda\,\alpha + a_n\,\alpha\\ &= (a_0\lambda^n + a_1\lambda^{n-1} + \ldots + a_{n-1}\lambda + a_n)(\alpha)\\ &= [p(\lambda)](\alpha)\end{aligned}$$

$\Rightarrow$ $$[p(\lambda)I - p(T)](\alpha) = 0$$

$\Rightarrow$ $$[p(\lambda)I - p(T)] = 0$$

$\Rightarrow$ $p(\lambda)$ is a characteristic root of $p(T)$.

3.36 MINIMAL POLYNOMIAL

Definition. *A monic polynomial $m_T(x) \in F[x]$ of least degree, is said to be a minimal polynomial of T if $m_T(T) = 0$.*

THEOREM 1. ***If $\lambda \in F$ is a characteristic root of T then λ is a root of the minimal polynomial of T. In particular, T only has a finite number of characteristic roots in F.***

PROOF. Let $m(x) \in F[x]$ be a minimal polynomial of T, then we shall show that λ is a root of $m(x) = 0$.

If $\lambda \in F$ is a characteristic root of T, then there exists a non-zero vector α in V such that
$$T(\alpha) = \lambda\alpha.$$

By above theorem,
$$[m(T)](\alpha) = [m(\lambda)](\alpha) = m(\lambda)\,\alpha.$$

But $m(T) = 0$ as $m(x)$ is a minimal polynomial of T.

$\therefore$ $$0\,\alpha = m(\lambda)\alpha$$

$\Rightarrow$ $$m(\lambda)\,\alpha = 0 \qquad \text{[By elementary property of linear space]}$$

Since $\alpha \neq 0$ so that $m(\lambda) = 0$.

Thus λ is a root of $m(x)$.

Also, V is a finite dimensional linear space; if dim $V = n$ then degree of $m(x) \le n^2$, therefore $m(x)$ has atmost n^2 roots implying that $m(x)$ has only a finite number of roots in F, thus there can only be finite number of characteristic roots of T in F.

THEOREM 2. ***If T and S are two linear operators with S invertible, then T and STS^{-1} have the same minimal polynomial.***

SOLUTION. Now,
$$\begin{aligned}(STS^{-1})^2 &= (STS^{-1})((STS^{-1}) = ST(S^{-1}S)TS^{-1}\\ &= ST^2S^{-1} \qquad [S^{-1}S = I]\end{aligned}$$

Continuing in this way, we get
$$(STS^{-1})^k = ST^kS^{-1} \qquad \ldots(1)$$

Let $m_T(x)$ be a minimal polynomial of T, thus $m_T(T) = 0$ and let
$$m_T(x) = a_0x^n + a_1x^{n-1} + \ldots + a_{n-1}x + a_n.$$

Then, $\quad a_0T^n + a_1T^{n-1} + ... + a_{n-1}T + a_nI = 0.$

Now, $\quad m_T(STS^{-1}) = a_0(STS^{-1})^n + a_1(STS^{-1})^{n-1} + ... + a_n(STS^{-1})$

$= a_0ST^n S^{-1} + a_1ST^{n-1}S^{-1} + ... + a_nSTS^{-1}$ [using (1)]

$= S(a_0T^n + a_1ST^{n-1} + ... + a_nI)S^{-1}$

$m_T(STS^{-1}) = S \,.\, 0 \,.\, S^{-1}.$ [using (2)]

$= 0.$

Hence, $m_T(x)$ is also the minimal polynomial of STS^{-1}.

THEOREM 3. ***Let V be a finite dimensional linear space over a field F and let T be a linear operator on V. Then there exists a vector αj in V such that $m_T^{\alpha_j}(x) = mT(x)$.***

PROOF. For any vector $\beta \in V$, the polynomial $m_T^{\beta}(x)$ divides $m_T(x)$. Thus each $m_T^{\beta}(x)$ is a factor of $m_T(x) \in F[X]$. Since V is finite-dimensional so for all $\beta \in V$, we get only finite number of polynomials. Let these polynomials be $p_1(x), p_2(x), ..., p_k(x)$ corresponding to the vectors $\alpha_1, \alpha_2, ..., \alpha_k$, that is

$$m_T^{\alpha_i}(x) = p_i(x) \qquad \text{for } i = 1, 2, ..., k.$$

Let $\quad V_i = \ker p_i(T)$.

Then $V \subseteq \bigcup_{i=1}^{k} V_i$ and so $V = V_j$ for some j, therefore,

$$(p_j(T))_\beta = 0, \forall \beta \in V$$

$$\Rightarrow \qquad p_j(T) = m_T^{\alpha_i}(x) = m_T(x).$$

3.37 INVARIANCE OF LINEAR OPERATOR

Definition. *Let $T : V \to V$ be a linear operator. A subspace W of V is said to be T-invariant or invariant under T if T maps W into itself, that is, if $a \in W$ implies $T(\alpha) \in W$. In this case T restricted to W defines a linear operator on W; that is T induces a linear operator $\hat{T}(\alpha) = T(\alpha)$, $\forall \alpha \in W$.*

THEOREM 1. ***Let $T : V \to V$ be a linear operator and let p(x) be any polynomial. Then the kernel of p(T) is T-invariant.***

PROOF. Let $\alpha \in \ker p(T)$, then $(p(T))(\alpha) = 0$.

Now we shall show that $T(\alpha) \in \ker p(T)$.

Since $\quad p(x)x = x\,p(x)$ so that $p(T)T = Tp(T)$.

$\therefore \quad (p(T)T)(\alpha) = T(p(T)(\alpha)) = T(0) = 0$

$\Rightarrow \quad p(T)(T(\alpha)) = 0.$

Thus, $\quad T(\alpha) \in \ker p(T)$.

THEOREM 2. ***Let $T : V \to V$ be a linear operator, and suppose that p(x) = f(x) g(x) are polynomials such that p(T) = 0 and f(x) and h(x) are relatively prime. Then V is the direct sum of the T-invariant subspace W_1 and W_2 where, W_1 = ker f(T) and W_2 = ker g(T).***

PROOF. Since $f(x)$ and $g(x)$ are relatively prime, then there exist two polynomials $r(x)$ and $s(x)$ such that

$$r(x)f(x) + s(x)g(x) = I.$$

So, for the operator T, we get

$$r(T)f(T) + s(T)g(T) = I. \quad ...(1)$$

Let $a \in V$, then form (1), we have

$$\alpha = [r(T)f(T)](\alpha) + [s(T)g(T)](\alpha) \quad ...(2)$$

Since $\quad [g(T)(r(T)f(T))](\alpha) = r(T)[f(T)g(T)(\alpha)]$

$$= r(T)[p(T)(\alpha)]f = r(T)0.\ \alpha \qquad [\because p(T) = 0]$$

$$= 0.$$

$\therefore \quad [r(T)f(T)](\alpha) \in W_2 = \ker g(T).$

Similarly, $[s(T)g(T)](\alpha) \in W_1 = \ker f(T)$.

Thus, α is the sum of an element of W_1 and an element of W_2. Hence

$$V = W_1 + W_2.$$

Now, to show that $V = W_1 \oplus W_2$, we must show that a sum $\alpha = \beta + \gamma$ with $\beta \in W_1$ and $\gamma \in W_2$, is uniquely determined by α.

Applying the operator $r(T)f(T)$ to $\alpha = \beta + \gamma$ and using $f(T)(\beta) = 0$, we get

$$[r(T)f(T)](\alpha) = [r(T)f(T)](\beta) + [r(T)f(T)](\gamma) = [r(T)f(T)](\gamma) \quad ...(3)$$

Also, from (1), we get

$$[r(T)f(T)](\gamma) + [s(T)g(T)]\gamma = \gamma$$

$$\Rightarrow \quad \gamma = [r(T)f(T)](\gamma) \qquad [\because (g(T))(\gamma) = 0] \quad ...(4)$$

From (3) and (4), we get

$$\gamma = [r(T)f(T)](\alpha)$$

which shows that γ is uniquely determined by α.

Similarly, β is uniquely determined by α. Hence $V = W_1 \oplus W_2$.

THEOREM 3. ***If T_1 is the restriction of T to W_1 and T_2 is the restriction of T to W_2, where, $T : V \to V$ is a linear operator and $p(x) = f(x)g(x)$ such that $p(x)$ is the minimal polynomial of T and $f(x)$ and $g(x)$ are monic. Then $f(x)$ and $g(x)$ are minimal polynomials of T_1 and T_2 respectively.***

PROOF. Let $m_1(x)$ and $m_2(x)$ be the minimal polynomials of T_1 and T_2 respectively, then we shall show that $m_1(x) = f(x)$ and $m_2(x) = g(x)$.

Since $\quad W_1 = \ker f(x)$ and $W_2 = \ker g(x)$.

$\therefore \quad f(T_1) = 0$ and $g(T_2) = 0$.

Thus, $m_1(x)$ divides $f(x)$ and $m_2(x)$ divides $g(x)$.

Let $p(x)$ be the least common multiple of $m_1(x)$ and $m_2(x)$. But $m_1(x)$and $m_2(x)$ are relatively prime. Accordingly, we have

$$p(x) = m_1(x)\, m_2(x).$$

But we have,

$$p(x) = f(x)g(x).$$

Since $m_1(x)$ is monic and divides $f(x)$ and $m_2(x)$is also monic and divides $g(x)$.

Thus $f(x) = m_1(x)$ and $g(x) = m_2(x)$.

THEOREM 4. **(Primary Decomposition Theorem).** ***Let T : V→ V be a linear operator with minimal polynomial***

$$m(x) = p_1(x)^{n_1} p_2(x)^{n_2} \dots p_r(x)^{n_r}$$

where, the $p_i(x)$ are distinct monic irreducible polynomials. Then V is the direct sum of T-invariant subspaces $W_1, W_2, \dots, W_r$, where W_i = ker $p_i(T)^{n_i}$. Moreover, $p_i(x)^{n_i}$ is the minimal polynomial of the restriction of T to W_i.

PROOF. We shall prove the theorem by induction on r. If $r = 1$, then the theorem is trivially true.

Suppose, that the theorem is true for $r - 1$.

By theorem 1, we may write V as the direct sum of T- invariant subspaces W_1 and V_1 where W_1 = ker $p_1(T)^{n_1}$ and $V_1 = \ker(p_2(T)^{n_2} \dots p_r(T)^{n_r})$.

Now by theorem 1, the minimal polymials of the restrictions of T to W_1 and V_1 are respectively $p_1(x)^{n_1}$ and $p_2(x)^{n_2} \dots p_r(x)^{n_r}$.

Let T_1 be the restriction of T to V_1 . Then by induction hypothesis, V_1 is the direct sum of subspaces $W_2, W_3, \dots, W_r$ such that

$$W_i = \ker p_i(T)^{n_i}$$

where $p_i(x)^{n_i}$ is the minimal polynomial for the restriction of T_1 to W_i.

But $\ker p_i(T)^{n_i} \subseteq V_1$ for $i = 2, 3, \dots, r$.

Since $p_i(x)^{n_1} | p_2(x)^{n_2} \dots p_r(x)^{n_r}$. Thus the $\ker p_i(T)^{n_i}$ is the same as the $\ker p_i(T_1)^{n_i} = W_1$. Also the restriction of T to W_i is the same as the restriction of T_1 to $W_i (i = 2,3,\dots,r)$. Hence $p_i(x)^{n_i}$ is also the minimal polynomial for the restriction of T to W_i.

Hence, $V = W_1 \oplus W_2 \oplus \dots \oplus W_r$.

RECAPITULATIONS

- A scalar $\lambda \in F$ is called an eigen value of a linear operator T if their exists a non-zero vector $\alpha \in V$ such that $T(\alpha) = \lambda \cdot \alpha$.
- λ is an eigen value of T if and only if λ is the root of characteristic polynomial.
- Let T be a linear transformation on n-dimensional linear space and $\Delta(\lambda)$ be its characteristic equation then $\Delta(T) = 0$ [Cayley-Hamilton theorem]

SOLVED EXAMPLES

EXAMPLE 1. *Let $I : V \to V$ be the identity mapping on any non-zero linear space V. Show that $\lambda=1$ is an eigenvalue of I. What is the eigenspace E1 of $\lambda = 1$?*

SOLUTION. Let $\alpha \neq 0$ be any vector of V, then

$$I(\alpha) = \alpha = 1\alpha.$$

$\Rightarrow \quad \lambda = 1$ is an eigenvalue of I.

Also, every vector in V is an eigenvector corresponding to 1, therefore $E_1 = V$.

EXAMPLE 2. *Show that 0 is an eigenvalue of T if and only if T is singular.*

SOLUTION. Let α be a non-zero vector of V.

0 is an eigenvalue of $T \Leftrightarrow$ there exists a non-zero vector such that

$$T(\alpha) = 0 . \alpha$$

$$\Leftrightarrow \quad T(\alpha) = 0$$

$\Leftrightarrow \quad T$ is singular.

EXAMPLE 3. *Let λ be an eigenvalue of a linear operator $T : V \to V$. Let $E\lambda$ be the eigenspace of A. Show that $E\lambda$ is a subspace of V.*

SOLUTION. In order to show E_λ to be subspace of V, we shall show that

(i) if $\alpha \in E_\lambda$ then $k\alpha \in E_\lambda$ for any scalar $k \in F$,

(ii) if $\alpha, \beta \in E_\lambda$, then $\alpha + \beta \in E_\lambda$

(i) Since $\alpha \in E_\lambda$, so we have

$$T(\alpha) = \lambda\alpha$$

then $$T(k\alpha) = kT(\alpha) = k(\lambda x) = \lambda(k\alpha).$$

$$k\alpha \in E_\lambda.$$

(ii) Since $\alpha, \beta \in E_\lambda$, we have

$$T(\alpha) = \lambda\alpha, \quad T(\beta) = \lambda\beta.$$

Then, $$T(\alpha+\beta) = T(\alpha) + T(\beta) = \lambda\alpha + \lambda\beta = \lambda(\alpha+\beta).$$

$\therefore$ $$\alpha + \beta \in E_\lambda.$$

Hence, E_λ is a subspace of V.

EXAMPLE 4. *Let V be the vecfor space of differentiable functions on R and $D:V \to V$ be the differential operator. Show that functions $e^{a_1 t}, e^{a_2 t}, \ldots, e^{a_n t}$, where a1,a2,...,an are distinct non-zero scalars are eigenvectors of D, to which eigenvalue λi does $e^{a_i t}$ belong ?*

SOLUTION. Since $e^{a_i t} \neq 0$ for all $a_i \neq 0$.

Then, we have

$$D(e^{a_i t}) = a_i e^{a_i t}, \qquad \forall\, i = 1, 2, 3, \ldots, n.$$

This equation shows that $e^{a_i t}$ is the eigenvector of D corresponding to $\lambda_i = a_i$.

EXAMPLE 5. *Suppose λ is an eigenvalue of an invertible operator T. Show that λ^{-1} is an eigenvalue of T^{-1}.*

SOLUTION. Since T is invertible so that it is non-singular, therefore, $\lambda \neq 0$. Also λ is an eigenvalue of T, then there exists a non-zero vector $\alpha \in V$ such that

$$T(\alpha) = \lambda\alpha.$$

Multiplying both sides by T^{-1}, we get

$$T^{-1}[T(\alpha)] = T^{-1}(\lambda\alpha)$$

$$\Rightarrow \quad (T^{-1}T)(\alpha) = T^{-1}(\lambda\alpha) \quad \Rightarrow \quad I(\alpha) = \lambda[T^{-1}(\alpha)]$$

$$\Rightarrow \quad \alpha = \lambda T^{-1}(\alpha) \Rightarrow T^{-1}(\alpha) = \frac{1}{\lambda}\alpha = \lambda^{-1}\alpha.$$

Hence, λ^{-1} is an eigenvalue of T^{-1}.

EXAMPLE 6. *Suppose α is a none-zero eigenvector of linear maps S and T. Show that α is an eigenvector of $S + T$.*

SOLUTION. Since α is an eigenvector of S and T. Let λ_1 and λ_2 be eigenvalues of S and T corresponding to α respectively.

Then, $$S(\alpha) = \lambda_1\alpha \text{ and } T(\alpha) = \lambda_2\alpha.$$

Now, $$(S+T)(\alpha) = S(\alpha) + T(\alpha) = \lambda_1\alpha + \lambda_2\alpha$$

$$(S+T)(\alpha) = (\lambda_1 + \lambda_2)\alpha.$$

Thus, α is an eigenvector $S + T$ corresponding to the eigenvalue $\lambda_1 + \lambda_2$.

EXAMPLE 7. *Let A be a square matrix of order $n\times n$ with its elements belong to F. The left multiplication by A defines a linear operator*

$$T_A : F^{n\times m} \to F^{n\times m}$$

such that $$T_A(B) = AB.$$

A scalar $\lambda \in F$ is an eigenvale of T_A if and only if λ is an eigenvalue of A.

SOLUTION. Suppose λ is an eigenvalue of A, then there exists a non-zero vector $\alpha \in F^n$ such that

$$A(\alpha) = \lambda\alpha.$$

Let B be an $n \times m$ matrix whose first column is α and all other columns zero, then B is a non-zero matrix and

$$T_A(B) = AB = \lambda B \qquad [\because A(\alpha) = \lambda\alpha]$$

Thus, λ is an eigenvalue of T_A.

Conversely, if λ, is eigenvalue of T_A, then there exists a non-zero matrix $B \in F^{n\times m}$ such that

$$T_A(B) = \lambda B = AB.$$

Let α be one of the non-zero columns of B, then

$$AB = \lambda B.$$

$$\therefore \qquad A(\alpha) = \lambda\alpha.$$

Thus, λ is an eigenvalue of A.

3.38 DIAGONALIZATION

Let V be a finite dimensional linear space over a field F. Then a linear operator $T : V \to V$ is said to be diagonalizable if V has a basis consisting of eigenvectors of T only. Equivalently, T has n linearly independent eigenvectors, if V is n-dimensional linear space.

For example: A scalar multiple of identity operator is diagonalizable.

THEOREM 1. ***A linear operator $T : V \to V$ can be represented by a diagonal matrix A iff V has a basis consisting of eigenvectors of T. In this case the diagonal elements of A are the corresponding eigenvalues.***

PROOF. Let V be n-dimensional linear space. Let $B = \{\alpha_1, \alpha_2, ..., \alpha_n\}$ be an ordered basis of V such that each α_j is the eigenvector of T.

If $\lambda_1, \lambda_2,..., \lambda_n$ are the corresponding eigenvalue of T, then

$$\left.\begin{aligned} T(\alpha_1) &= \lambda_1\alpha_1 \\ T(\alpha_2) &= \lambda_2\alpha_2 \\ \cdots \; &\cdots \; \cdots \; \cdots \\ T(\alpha_n) &= \lambda_n\alpha_n \end{aligned}\right\} \qquad ...(1)$$

System (1) can also be rewritten as

$$\begin{aligned} T(\alpha_1) &= \lambda_1\alpha_1 + 0\alpha_2 + 0\alpha_3 + ... + 0\alpha_n \\ T(\alpha_2) &= 0\alpha_1 + \lambda_2\alpha_2 + 0\alpha_3 + ... + 0\alpha_n \\ \cdots & \quad \cdots \quad \cdots \quad \cdots \quad \cdots \\ T(\alpha_n) &= 0\alpha_1 + 0\alpha_2 + 0\alpha_3 + ... + \lambda_n\alpha_n \end{aligned}$$

Thus, the matrix representation of T, with respect to B is given by

$$A = [T]_B = \begin{vmatrix} \lambda_1 & 0 & 0 \\ 0 & \lambda_2 & 0 \\ \vdots & \vdots & \vdots \\ 0 & 0 & \lambda_n \end{vmatrix}.$$

Hence, T can be represented by a diagonal matrix A whose elements are the eigenvalues of T.

Conversely, suppose T can be represented by a diagonal matrix A given by

$$A = \begin{vmatrix} \lambda_1 & 0 & \cdots & 0 \\ 0 & \lambda_2 & \cdots & 0 \\ \vdots & & & \\ 0 & 0 & \cdots & \lambda_n \end{vmatrix}.$$

Since V is n-dimensional, there exists a basis $\{\alpha_1, \alpha_2,...,\alpha_n\}$ of V for which

$$T(\alpha_i) = \lambda_i\alpha_i, \qquad \forall\, i = 1, 2,..., n.$$

This equation shows that each α_i in the basis is an eigenvector of T corresponding to each λ_i.

REMARK

- Let T be a linear operator on an n-dmiensional linear space V over a field F. If T has n distinct eigenvalues, then T is diagonalizable.

Definition. *Let T be a linear operator on V and let λ be an eigenvalue of T. Then $E_\lambda = \ker(T - \lambda I)$ is catted the eigenspace of T corresponding to λ.*

The dimension of E_λ is called the geometric multiplicity of λ and the multiplicity of λ as a root of the characteristic polynomial is called the algebraic multiplicity of λ.

THEOREM 2. ***Let λ be an eigenvalue of a linear operator $T : V \to V$. Then the geometric multiplicity of λ does not exceed its algebraic multiplicity.***

PROOF. Suppose the geometric multiplicity of λ be r. Then there are r linearly independent eigenvector of T corresponding to λ. Let these be $\alpha_1, \alpha_2,...,\alpha_r$.

If r = dim. V, then $\quad B = \{\alpha_1, \alpha_2,...,\alpha_r\}$ is an ordered basis of V and

$$[T]_B = \lambda I_r.$$

Thus $\quad |xI_r - [T]_B| = (x - \lambda)^r$ so that the algebraic multiplicity of λ is r.

If r <dim.$V = n$, then the set $B = \{\alpha_1, \alpha_2, ..., \alpha_r\}$ can be extended to form a basis of V. Let the extension of B is given by

$$B_1 = \{\alpha_1, \alpha_2, ..., \alpha_r, \beta_{r+1}, \beta_{r+2}, ..., \beta_n\}$$

We have

$$\begin{aligned}
T(\alpha_1) &= \lambda\alpha_1 \\
T(\alpha_2) &= \lambda\alpha_2 \\
\dots \quad & \dots \quad \dots \\
T(\alpha_r) &= \lambda\alpha_r \\
T(\beta_{r+1}) &= a_{11}\alpha_1 + ... + a_{1r}\alpha_r + a_{1(r+1)}\beta_{r+1} + ... + a_{1n}\beta_n \\
T(\beta_{r+2}) &= a_{21}\alpha_1 + ... + a_{2r}\alpha_r + a_{2(r+1)}\beta_{r+1} + ... + a_{2n}\beta_n \\
\dots \quad & \dots \quad \dots \quad \dots \quad \dots \quad \dots \quad \dots \\
T(\beta_n) &= a_{n1}\alpha_1 + ... + a_{nr}\alpha_r + a_{n(r+1)}\beta_{r+1} + ... + a_{nn}\beta_n.
\end{aligned}$$

Thus the matrix of T in the basis B_1, is given by

$$[T]_{\beta_1} = \begin{vmatrix}
\lambda & 0 & \dots & 0 & a_{11} & a_{21} & \dots & a_{n1} \\
0 & \lambda & \dots & 0 & a_{12} & a_{22} & \dots & a_{n2} \\
\dots & \dots & \dots & \dots & \dots & \dots & \dots & \dots \\
0 & 0 & \dots & \lambda & a_{1r} & a_{2r} & \dots & a_{nr} \\
0 & 0 & \dots & 0 & a_{1(r+1)} & a_{2(r+1)} & \dots & a_{n(r+1)} \\
0 & 0 & \dots & 0 & a_{1(r+2)} & a_{2(r+2)} & \dots & a_{n(r+2)} \\
\dots & \dots & \dots & \dots & \dots & \dots & \dots & \dots \\
0 & 0 & \dots & 0 & a_{1n} & a_{2n} & \dots & a_{nn}
\end{vmatrix}$$

$$= \begin{bmatrix} \lambda I_r & \vdots & A \\ \dots & \dots & \dots \\ O & \vdots & B \end{bmatrix}$$

Since $[T]_{\beta_1}$ is a block triangular matrix, the characteristic polynomial of $\lambda\mathbf{I}_r$ is $(x-\lambda)^r$, which must divide the characteristic polynomial of $[T]_{\beta_1}$ and hence T. Hence the algebraic multiplicity of λ for the operator T is at least r.

THEOREM 3. ***Let T be a linear operator on an n- dimensional linear space V over a field F and let $\lambda_1, \lambda_2, ..., \lambda_k$ be all the distinct eigenvalues of T. Then the following statements are equivalent :***

(i) T is diagonalizable

(ii) $V = E_{\lambda_1} \oplus E_{\lambda_2} \oplus ... \oplus E_{\lambda_k}$, $E_{\lambda_i} = \ker(T - \lambda_i I)$

(iii) the characteristic polynomial of T splits over F, and the algebraic multiplicity of each eigenvalue equals its geometric multiplicity

(iv) the minimal polynomial of T is $(x - \lambda_1)(x - \lambda_2)...(x - \lambda_k)$.

PROOF. (i) ⇒ (ii). If T is diagonalizable, the V has an ordered basis

$$B = \{\alpha_1, \alpha_{12}, ..., \alpha_{1n}, \alpha_{21}, \alpha_{22}, ..., \alpha_{2n_2}, ..., \alpha_{k1}, ..., \alpha_{kn_k}\}$$

Such that each α_{i_m} is an eigenvector of T corresponding to eigevalue λ_{i_m}. Obviously $n_1 + n_2 + ... + n_k = n$. Thus it is enough to show that for each m, the set $\{\alpha_{m1}, \alpha_{m2}, ..., \alpha_{mn_m}\}$ is a basis of E_{λ_m}.

Let if possible $\{\alpha_{m1}, \alpha_{m2}, ..., \alpha_{mn_m}\}$ is not a basis of $E_{\lambda m}$ for some $m = 1, 2, ..., k$. Then there exists vector $\beta_m \in E_{\lambda m}$ such that $\{\alpha_{m1}, \alpha_{m2}, ..., \alpha_{mn_m} \beta_m\}$ is a linealry independent subset of $E_{\lambda m}$. But $B \cup \{\beta_m\}$ is a linearly independent subset of V, which gives a contradiction as B spans V, therefore, $\{\alpha_{m1}, \alpha_{m2},...., \alpha_{mn_m}\}$ is a basis of $E_{\lambda m}$.

(ii) ⇒ (i) It is obvious.

(i) ⇒ (iii) If T is diagonalizable then there is an order basis B on V such that $[T]_B$ is a diagonal matrix with its diagonal elements as eigenvalues of T, so that characteristic ploynomial split over F. Since.

$$V = E_{\lambda_i} \oplus E_{\lambda_2} .... \oplus E_{\lambda_k}$$

so, if B_i *is a basis of* E_{λ_i} then

$$B = \bigcup_{i=1}^{k} B_i$$

is a basis of V and

$$[T]_B = \text{diag. } ([T_1]B_1 \cdots [T_k]_{B_k})$$

where T_i is the linear operator of E_{λ_i} induced by T.

Let dim. $E_{\lambda_i} = n_i$, since $T_i(\alpha) = T(\alpha) = \lambda_i$ for all $\alpha \in E_{\lambda_i}$) so that

$$[T_i]B_i = (\lambda_i I_{n_i}) \; .$$

Hence the characteristic polynomial is

$$(x-\lambda_1)^{n_1}(x-\lambda_2)^{n_2},...,(x-\lambda_k)^{n_k}$$

which shows that the algebraic multiplicity of each λ_i equals geometric multiplicity.

(iii) ⇒ (iv) Since characteristic polynomial of T splits over F and it is given by

$$(x-\lambda_1)^{n_1}(x-\lambda_2)^{n_2},...,(x-\lambda_k)^{n_k}$$

Also the geometic multiplicity and the algebraic multiplicity of each eigevalues are equal, so for each λ_i, there are n_i linearly independent eignvalues of T which are $\alpha_{i1}, \alpha_{i2},.... \alpha_{in_i}$,.

Therefore, $B = \{\alpha_{11},...\alpha_{1n_1}, \alpha_{21},...\alpha_{2n_2},...\alpha_{11},...\alpha_{kn_k}\}$ is a basis of V and the linear operator $(T-\lambda_1 I)(T-\lambda_2 I)....(T-\lambda_k I)$ maps all the basis elements of B to the zero vector.

Thus $(T-\lambda_1 I)(T-\lambda_2 I)....(T-\lambda_k I) = \mathbf{0}$, hence $(x-\lambda_1)(x-\lambda_2)...(x-\lambda_k)$ is minimal polynomial of T.

THEOREM 4. ***A linear operator $T : V \to V$ has a diagonal matrix respresentation if and only if its minimal polynomial m(x) is product of distinct linear polynomials.***

PROOF. Suppose $m(x)$ is a product of distinct linear polynomials, let

$$m(x) = (x-\lambda_1)\;(x-\lambda_2)...\;(x-\lambda_n) \qquad ...(1)$$

where $\lambda_1, \lambda_2, ..., \lambda_n$, are distinct scalars.

By primary decomposition theorem V is direct sum of subspaces $W_1, W_2,..., W_n$, where,

$$W_1 = \ker(T-\lambda_1 I),\; W_2 = \ker(T-\lambda_2 I),\; W_n = \ker(T-\lambda_n I).$$

If $\alpha \in W_i$, then

$$(T-\lambda_1 I)\ (\alpha)=0 \Rightarrow T(\alpha) = \lambda_i \alpha$$

$\Rightarrow$ every vector in W_i is an eignevector corresponding to the eigenvalue λ_i.

Thus, the union of bases for $W_1, W_2 \ldots W_n$, form a basis of V. This basis consists of eigen vectors and so T is a diagonlizable

Conversely, suppose, T has a diagonal matrix representation, implying that V has a basis consisting of eigevectors of T.

Let $\lambda_1, \lambda_2, \ldots \lambda_r$ be the distinct eigenvalus of T, then

$$P(T) = (T-\lambda_1 I)\ (T-\lambda_2 I) \ldots.\ (T-\lambda_r I).$$

Since $(T-\lambda_1 I)$ maps each bsais vector to 0 so that $p(T) = 0$.

Thus, the minimal polynomial $m(x)$ of T divides the polynomial $P(x) = (x-\lambda_1)(x-\lambda_2) \ldots (x-\lambda_r)$. According $m(x)$ is the product of distinct linear polynomials.

SOLVED EXAMPLES

EXAMPLE 1. *If T is a linear operator on R3 which is represented in the standard ordered basis by*

$$A = \begin{bmatrix} 5 & -6 & -6 \\ -1 & 4 & 2 \\ 3 & -6 & -4 \end{bmatrix}$$

then show that T is diagonalizable.

SOLUTION. The characteristic polynomial of A is

$$\Delta(x) = |xI - A| = \begin{vmatrix} x-5 & 6 & 6 \\ 1 & x-4 & -2 \\ -3 & 6 & x+4 \end{vmatrix}$$

$$= (x-5)\{(x-4)(x+4) + 12\} - 6\{x+4-6\} + 6\{6+3(x-4)\}$$

$$= (x-5)(x^2-4) - 6x + 12 + 18x - 36$$

$$= x^3 - 5x^2 - 4x + 20 - 6x + 12 + 18 - 36$$

$$= x^3 - 5x^2 + 8x - 4$$

$$\Delta(x) = (x-1)(x-2)^2$$

Now we find the minimal polynomial of T.

The possible form of the minimal polynomial $m(x)$ of T are :

(i) $m_1(x) = (x-1)(x-2)$

(ii) $m_2(x) = (x-1)(x-2)^2$

Since $m_2(A) = 0$ may be the minimal polyonmial of T of $m_1(A) \neq 0$, but if $m_1(A) = 0$, then the minimal polynomial will be $m_1(x)$ but not the $m_2(x)$. So we test whether $m_1(A) = 0$, or $m_1(A) \neq 0$.

Since, $m_1(A) = (A-I)(A-2I)$.

Now
$$A - I = \begin{bmatrix} 5-1 & -6 & -6 \\ -1 & 4-1 & 2 \\ 3 & -6 & -4-1 \end{bmatrix} = \begin{bmatrix} 4 & -6 & -6 \\ -2 & 3 & 2 \\ 3 & -6 & -5 \end{bmatrix}$$

And
$$A - 2I = \begin{bmatrix} 5-2 & -6 & -6 \\ -1 & 4-2 & 2 \\ 3 & -6 & -4-2 \end{bmatrix} = \begin{bmatrix} 3 & -6 & -6 \\ -1 & 2 & 2 \\ 3 & -6 & -6 \end{bmatrix}$$

$$\text{So} \quad (A - I)(A - 2I) = \begin{bmatrix} 4 & -6 & -6 \\ -2 & 3 & 2 \\ 3 & -6 & -5 \end{bmatrix} = \begin{bmatrix} 3 & -6 & -6 \\ -1 & 2 & 2 \\ 3 & -6 & -6 \end{bmatrix} = \begin{bmatrix} 0 & 0 & 0 \\ 0 & 0 & 0 \\ 0 & 0 & 0 \end{bmatrix} = 0$$

$$m_1(A) = 0$$

Thus the minimal polynomial of T is $m_1(x) = (x-1)(x-2)$ which is the product of monic linear polynomial over F. Hence T is diagonalizable.

EXAMPLE 2. *Let $T : R_3 \to R_3$ be definded by $T(x, y, z) = (2x + 3y - 2z, 5y + 4z, x - z)$. Find the characteristic polynomial D (t) of T.*

SOLUTION. Let $B = \{(1, 0, 0), (0, 1, 0), (0, 0, 1)\}$ be the standard basis of R^3. So that

$$T(1, 0, 0) = (2, 0, 1)$$
$$T(0, 1, 0) = (3, 5, 0)$$
$$T(0, 0, 1) = (-2, 4, -1)$$

$$\text{Thus,} \quad [T]_B = \begin{bmatrix} 2 & 3 & -2 \\ 0 & 5 & 4 \\ 1 & 0 & -1 \end{bmatrix}$$

The characteristic polynomial of T is given by

$$\Delta(t) = \begin{vmatrix} t-2 & -3 & 2 \\ 0 & t-5 & -4 \\ -1 & 0 & t+1 \end{vmatrix}$$

$$= (t-2)\{(t-5)(t+1) - 0\} + 3(0-4)\} + 2\{0 + t - 5\}$$
$$= (t-2)(t^2 - 4t - 5) - 12 + 2t - 10$$
$$= t^3 - 2t^2 + 4t^2 + 8t - 5t + 10 - 12 + 2t - 10$$
$$= t^3 - 6t^2 + 5t - 12$$

EXAMPLE 3. *If A be a square matrix given by* $A = \begin{bmatrix} 3 & 0 & 0 \\ 0 & 2 & -5 \\ 0 & 1 & -2 \end{bmatrix}$

then find all the eigenvalues of A viewed as matrics over (i) Real field R (ii) Complex field C. Also, find in which case the matrix A is diagonalizable.

SOLUTION. The characteristic polynomial of A is given by

$$\Delta(t) = |tI - A| = \begin{vmatrix} t-3 & 0 & 0 \\ 0 & t-2 & 5 \\ 0 & -1 & t+2 \end{vmatrix}$$

$$= (t-3)\{(t-2)(t+2) + 5\}$$

$$\Delta(t) = (t-3)(t^2+1) = (t-3)(t-i)(t+i)$$

The roots of $\Delta t = 0$ are $t = 3, -i, i$.

(i) If A is a matrix over the field of reals, then A has only one eigenvalue which is 3. Thus, A can not be diagonalized.

(ii) If A is a matrix over the field of complex numbers, then A has three distinct eigenvalues, so that the minimal polynomial of A is equal to $\Delta(t)$ which is the product of linear polynomials. Hence A is digonalizable.

EXAMPLE 4. *If T be a linear tranformation on V (F). Then the following are equivalent :*

(i) λ is the characteristic value of T

(ii) *the transformation* $T - \lambda I$ *is singular*

(iii) $|(T - \lambda I)| = 0$.

SOLUTION. **(i) $\Rightarrow$ (ii).** If λ is a characteristic value of T, then there exists a non-zero vector $\alpha \in V$ such that

$$T(\alpha) = \lambda\alpha \Rightarrow T(\alpha) = \lambda I(\alpha) \qquad [\because I(\alpha) = \alpha]$$

$$\Rightarrow \quad T(\alpha) - \lambda I(\alpha) = 0 \Rightarrow (T - \lambda I)(\alpha) = 0 \text{ with } \alpha \neq 0.$$

$\therefore$ $T - \lambda I$ is singular.

(ii) $\Rightarrow$ (iii). Since $(T - \lambda I)$ is singular, *i.e.*, it is not invertible so that

$$|(T - \lambda I)| = 0$$

(iii) $\Rightarrow$ (i). Since $|(T - \lambda I)| = 0$, therefore $(T - \lambda I)$ is not invertible, *i.e.*,

$$(T - \lambda I)\alpha = 0 \Rightarrow T(\alpha) - \lambda I(\alpha) = 0$$

$$\Rightarrow \quad T(\alpha) - \lambda\alpha = 0 \Rightarrow T(\alpha) = \lambda\alpha$$

$\therefore$ λ is an eigenvalue of T.

EXAMPLE 5. *Show that the matrix* $A = \begin{bmatrix} 1 & 2 \\ 0 & 1 \end{bmatrix}$ *is diagonalizable over the field of complex numbers.*

SOLUTION. The characteristic polynomial of A is

$$\Delta(t) = |tI - A| = \begin{vmatrix} t-1 & -2 \\ 0 & t-1 \end{vmatrix} = (t-1)^2$$

Thus A has only one eignvalue which is I.

The eigenvectors of A of $\lambda = 1$ are given by the solution of the homogeneous system

$$(I - A)\begin{bmatrix} x_1 \\ x_2 \end{bmatrix} = \begin{bmatrix} 0 \\ 0 \end{bmatrix} \text{ or } \begin{bmatrix} 0 & -2 \\ 0 & 0 \end{bmatrix}\begin{bmatrix} x_1 \\ x_2 \end{bmatrix} = \begin{bmatrix} 0 \\ 0 \end{bmatrix} \text{ or } -2x_2 = 0$$

Above equation gives $x_2 = 0$ and x_1 is assigned let $x_2 = 1$, thus A has only one linearly independent eigenvector $\begin{bmatrix} 1 \\ 0 \end{bmatrix}$. Hence A is not diagonlizable.

EXAMPLE 6. *Let V be the linear space of functions which have B = {sinq , cosq} as a basis and let D be the differential operator on V. Find the characterstic polynomial D(t) of D.*

SOLUTION. First, we fine the matrix A which represents D in the basis B :

$$D(\sin\theta) = \cos\theta = 0(\sin\theta) + 1.(\cos\theta)$$

$$D(\cos\theta) = -\sin\theta = (-1)\sin\theta + 0(\cos\theta)$$

Thus, $\quad [D]_B = A \begin{bmatrix} 0 & 1 \\ -1 & 0 \end{bmatrix}$

Now, $\quad \Delta(t) = \begin{bmatrix} t & -1 \\ 1 & t \end{bmatrix} = t^2 + 1$

This is the required characteristic polynomial of A.

EXAMPLE 7. *What is the algebraic and geometric multiplicity of l = –2, where l = –2 is one of the eigenvalue of the matrix*

$$A = \begin{bmatrix} -3 & 1 & -1 \\ -7 & 5 & -1 \\ -6 & 6 & -2 \end{bmatrix}$$

SOLUTION. The characteristic polynomial of A is given by.

$$\Delta(t) = |tI - A| = \begin{vmatrix} t+3 & -1 & 1 \\ 7 & t-5 & 1 \\ 6 & -6 & t+2 \end{vmatrix}$$

$$= (t+3)\{(t-5)(t+2)+6\}+1\{7(t+2)-6\}+1\{-42-6(t-5)\}$$
$$= (t+3)\{t^2-3t-10+6\}+7t+8-42-6t+30$$
$$=(t+3)(t^2-3t-4)+t-4\}$$
$$= t^3+3t^2-3t^2-9t-4-12+t-4$$
$$\Delta(t) = t^3-12t-16=(t+2)^2-(t-4).$$

Since the factor $(t + 2)$ occurs twice in $\Delta(t)$, so that the algebaric multiplicity of $\lambda = -2$ is two.

Now we find a basis of the eigenspace of $\lambda = -2$.

Let $X = \begin{bmatrix} x_1 \\ x_2 \\ x_3 \end{bmatrix}$ be non- zero eigenvector corresponding to $\lambda = -2$, then

$$(\lambda I - A)X = 0 \text{ or } (-2I - A)X = 0$$

or
$$\begin{bmatrix} 1 & -1 & 1 \\ 7 & -7 & 1 \\ 6 & -6 & 0 \end{bmatrix}\begin{bmatrix} x_1 \\ x_2 \\ x_3 \end{bmatrix} = \begin{bmatrix} 0 \\ 0 \\ 0 \end{bmatrix}$$

or
$$x_1 - x_2 + x_3 = 0$$
$$7x_1 - 7x_2 + x_3 = 0$$
$$6x_1 - 6x_2 = 0$$

or
$$x_1 - x_2 + x_3 = 0$$
$$7x_1 - 7x_2 + x_3 = 0$$
$$x_1 - x_2 = 0.$$

This system has only one independent solution, *i.e.*, $x_1 = 1$, $x_2 = 1$ and $x_3 = 0$. Thus $\alpha = (1, 1, 0)$ forms, a basis of the eigenspace E_2. Hence the geometirc multiplicity of $\lambda = -2$ is one as dim $E_2 = 1$.

EXERCISE 3.5

1. Let α and β be eigenvectors of T corresponding to two eigenvalues λ and μ respectively. Show that for non-zero scalars a and b, $a\alpha+b\beta$ is not an eigenvector. of T.
2. Suppose $T : V \to V$ is a linear operator on a linear space V with dimension n. Define the characteristic polynomial $\Delta(t)$ of T.
3. Suppose a linear map $T : V \to V$ may have matrix representations, is it possible for T to have many characteristic polynomials?
4. Let $T: R^2 \to R^2$ be the linear operator which rotates each vector $\alpha \in R^2$ by an angle $\theta=\pi/2$. Show geometrically that T has no eigenvalues and hence no eigenvectors.
5. Let λ be an eigenvalue of a linear operator $T:V\to V$. Let E_λ be the set of all eigenvectors of T belonging to λ. Show that E_λ is a subspace of V.
6. Let λ be an eigenvalue of a linear operator T: $V \to V$. Define the algebraic multiplicity and the geometric multiplicity of λ.

7. Let A and B be n–square matrices. Show that AB and BA have the same eigenvalues.

8. Show that if λ is an eigenvalue of T and $p(x) \in F[x]$, then $p(\lambda)$ is an eigenvalue of $p(T)$.

9. Find eigenvalues and eigenvectors for the following matrices over **C**, the field of complex numbers:

(i) $A=\begin{bmatrix} 6 & -1 & 2 \\ 4 & 1 & 2 \\ -10 & 0 & 3 \end{bmatrix}$ (ii) $A=\begin{bmatrix} 4 & 2 & 2 \\ 3 & 3 & 2 \\ -3 & -1 & 0 \end{bmatrix}$

(iii) $A=\begin{bmatrix} 0 & 0 & 1 \\ 1 & 0 & -1 \\ 0 & 1 & 1 \end{bmatrix}$.

10. Prove that if every non-zero is an eigenvector of T, then T is a scalar multiple of the identity operator.

11. Prove the if char $F \neq 2$, then T can be expressed as the sum of two invertible linear operators.

12. If $V_2(R)$ is a linear space and T is a linear transformation on $V_2(R)$ whose matrix relative to the basis of $V_2(R)$ is $A=\begin{bmatrix} 1 & 2 \\ 3 & 2 \end{bmatrix}$.

Find the dimension of the eigenspace of T.

13. Find the minimal polynomial for real matrix

$$A=\begin{bmatrix} 7 & 4 & -1 \\ 4 & 7 & -1 \\ -4 & -4 & 4 \end{bmatrix}.$$

14. Find the minimal polynomial for real matrix

$$A=\begin{bmatrix} 1 & 1 & 1 \\ 1 & 1 & 1 \\ 1 & 1 & 1 \end{bmatrix}.$$

15. Find the invertible matrix P such that $P^{-1}AP$ is a diagonal matrix, where $A=\begin{bmatrix} 1 & 4 \\ 2 & 3 \end{bmatrix}$.

16. If T be the linear transformation on $V_3(R)$ which is represented in the standard ordered basis by the matrix $\begin{bmatrix} -9 & 4 & 4 \\ -8 & 3 & 4 \\ -16 & 8 & 7 \end{bmatrix}$. Prove that T is diagonalizable.

17. Show that the matrix $\begin{bmatrix} 1 & 1 \\ 0 & 1 \end{bmatrix}$ is not diagonalizable.

18. If T is linear operator on $V_3(R)$ which is represented in the standard ordered basis by the matrix: $A=\begin{bmatrix} 5 & -6 & -6 \\ -1 & 4 & 2 \\ -3 & -6 & -4 \end{bmatrix}$. Find the eigenvalues of A and prove that T is diagonalizable.

19. Let λ be an eigenvalue of $T \in L(V)$ with algebraic multiplicity m. Prove that T is not diagonalizable if rank $(T-\lambda I)>n-m$, where $n=\dim. V$.

20. A linear operator $T:R^3 \to R^3$ defined by $T(x,y,z) = \{2x + y, y-z, 2y+4z)$.

(i) Find the characteristic polynomial $\Delta(t)$ of T

(ii) Find the eigenvalues of T.

(iii) Is T diagonalizable ?

21. Suppose α is a non-zero eigenvector of T. Show that, for any $k \in F$, is an eigenvector of kT.

22. Suppose λ is an eigenvalue of a linear operator T :

(i) Show that λ^2 is an eigenvalue of T^2

(ii) Show that λ^n is an eigenvalue of T^n for $n \geq 1$.

HINT TO SELECTED PROBLEMS

1. Since α,β are the eigenvectors of T corresponding to two distinct eigenvalues λ and μ. Then we have $T(\alpha) = \lambda\alpha$ and $T(\beta) = \mu\beta$. Now for a, b (non-zero) scalars,

$$T(a\alpha + b\beta) = aT(\alpha) + bT(\beta) = a\lambda\alpha + b\mu\beta$$

$$\neq v(a\alpha + b\beta), \text{ for some scalar } v.$$

$\therefore$ $a\alpha + b\beta$ is not an eigenvector of T.

5. $E_\lambda = \{\alpha \in V: T(\alpha) = \lambda\alpha\}$. Let $\alpha,\beta \in E_\lambda$ and $a,b \in F$, then

$$T(a\alpha + b\beta) = aT(\alpha) + bT(\beta) = a\lambda a + b\mu\beta \quad [\because T(\alpha) = \lambda\alpha, T(\beta) = \lambda\alpha]$$

$$= \lambda(a\alpha + b\beta).$$

$\therefore$ $a\alpha + b\beta$ is also an eigenvector of T corresponding to λ.

$\therefore$ $a\alpha + b\beta \in E_\lambda$. Hence E_λ is subspace of V.

8. Since λ is an eigenvalue of T so that for a non-zero vector $\alpha \in V, T(\alpha) = \lambda\alpha$. For $p(x) = F[x]$.

We have $(p(T))(\alpha) = p(T)(\alpha)) = p(\lambda\alpha)$ $[\because T(\alpha)=\lambda\alpha]$

$$= p(\lambda)\alpha.$$

This show that $p(\lambda)$ is an eigenvalue of $p(T)$.

10. Let α be any non-zero vector and if α is an eigenvector of T, then $T(\alpha) = \lambda\alpha$, for some scalar λ

$$(T(\alpha)) = \lambda\alpha = \lambda I(\alpha) \qquad [\because I(\alpha)=\alpha]$$

$(T - \lambda I)(\alpha) = 0, \forall \alpha \in V$

$\Rightarrow \quad T = \lambda I$

12. Since matrix representation of T is given by $A = \begin{bmatrix} 1 & 2 \\ 3 & 2 \end{bmatrix}$.

Eigenvalues of A are the root of the equation

$\Delta(t) = |A - tI| = 0$ or $\begin{vmatrix} 1-t & 2 \\ 3 & 2-t \end{vmatrix} = 0$

or $(1-t)(2-t)-6 = 0$ or $t^2-3t-4 = 0$

or $(t-4)(t+1) = 0$ or $t = -1, 4$.

Let $X_1 = \begin{bmatrix} x_1 \\ x_2 \end{bmatrix}$ be an eigenvector corresponding to $\lambda = -1$

$\therefore (A - \lambda I)X_1 = 0$ or $\begin{bmatrix} 2 & 2 \\ 3 & 3 \end{bmatrix}\begin{bmatrix} x_1 \\ x_2 \end{bmatrix} = \begin{bmatrix} 0 \\ 0 \end{bmatrix}$.

or $2x_1 + 2x_2 = 0, 3x_1 + 3x_2 = 0$, *i.e.*, $x + x_2 = 0$.

So we have one equation in two variables, then set $x_2 = -1, x_1 = 1$.

$\therefore \quad X_1 = \begin{bmatrix} 1 \\ -1 \end{bmatrix}$. Again $X_2 = \begin{bmatrix} x_1 \\ x_2 \end{bmatrix}$ be an eigenvector corresponding to $\lambda = 4$.

$\therefore \quad (A - \lambda I)X_1 = 0$

or $\begin{bmatrix} -3 & 2 \\ 3 & -2 \end{bmatrix}\begin{bmatrix} x_1 \\ x_2 \end{bmatrix} = \begin{bmatrix} 0 \\ 0 \end{bmatrix}$.

or $\quad -3x_1 + 2x_2 = 0, 3x_1 - 2x_2 = 0$

$\therefore$ We have one equation in two variables so set $x_2 = 3$, and $x_1 = 2$.

$\therefore X_2 = \begin{bmatrix} 2 \\ 3 \end{bmatrix}$.

Hence eigenspace of T is generated by two vectors so the dimension of eigenspace of T is 2.

16. The characteristic polynomial of the matrix of T is given by $\Delta(t) = \begin{vmatrix} -9-t & 4 & 4 \\ -8 & 3-t & 4 \\ -16 & 8 & 7-t \end{vmatrix}$

$= (-9-t)\{(3-t)(7-t) - 32\}$
$\quad + 4\{-64 + (-8(7-t)\}$
$\quad + 4\{-64 + 16(3-t)\}$

$= -(9+t)\{21 - 10t + t^2 - 32\}$
$\quad + 4\{-8 - 8t\} + 4\{-16 - 16t\}$

$= -(9+t)(t_2 - 10t - 11) - 32 - 32t - 64 - 64t$

$= -(t^3 - 10t^2 - 11t + 9t^2 - 90t - 99) - 96 - 96t$

$= -t^3 + t^2 + 101t + 99 - 96 - 96t$

$\therefore \Delta(t) = -t^3 + t^2 + 5t + 3 = (t+1)^2(3-t)$.

There are two polynomials $m_1(t) = (t+1)^2$ (3–1) and $m_2(t) = (t+1)(3-t)$ one of them may be minimal polynomial.

Now $(A + I)(3I - A)$

$$= \begin{bmatrix} -8 & 4 & 4 \\ -8 & 4 & 4 \\ -16 & 8 & 8 \end{bmatrix}\begin{bmatrix} 12 & -4 & -4 \\ 8 & 0 & -4 \\ 16 & -8 & -4 \end{bmatrix}$$

$$= \begin{bmatrix} 0 & 0 & 0 \\ 0 & 0 & 0 \\ 0 & 0 & 0 \end{bmatrix} = \mathbf{O} . \therefore m_2(A) = O$$

Thus $m_2(t) = (t+1)(3-t)$ is a minimal polynomial of A which is a product of linear polynomials.

Hence the matrix of T is diagonalizable,

21. Let λ be an eigenvalue of T corresponding to which α is its eigenvector. Thus $T(\alpha) = \lambda\alpha$.

For $k \in F$. $(kT)(\alpha) = k(T(\alpha)) = k\lambda\alpha = (k\lambda)\alpha$.

Hence α is also an eigenvector of T corresponding to eigenvalue $k\lambda$.

22. Let $\alpha \neq 0$ be an eigenvector of T corresponding to an eigenvalue k. Then $T(\alpha) = \lambda\alpha$.

(i) $(T^2)(\alpha) = T(T(\alpha)) = T(\lambda\alpha)$
$\qquad = \lambda(T(\alpha)) = \lambda(\lambda\alpha) = \lambda^2\alpha$.

$\therefore \quad \lambda^2$ is an eigenvalue of T^2. Similarly we may prove that λ^n is an eigenvalue of T^n.

ANSWERS

3. No. **9.** (*ii*) 1,2,4;[0, 1, –1]′,[1, 1, –2]′[1, 1, –1]′. (*iii*) [1,0,1]′;[1–(1+*i*)*i*]′,[1–(1–*t*)–1]′.

12. Two. **13.** $m(x) = x^2 - 15 + 36$. **14.** 0,3;[10–1]′,[*aaa*]′, $a \in F$. **15.** $p = \begin{bmatrix} 1 & 2 \\ 1 & -1 \end{bmatrix}$.

18. 1,2,2. **20.** (*i*) $\Delta(t) = t^3 - 7t^2 + 16t - 12$. (*ii*) 2,2,3. (*iii*) No

Chapter Review: A Competitive Approach

SELECTED TERMS AND RESULTS

▶ TERMS

- **Linear space:** An algebraic structure $(V, +, .)$ is said to be linear space over a field F if
  - (i) $(V, +)$ is an abelian group
  - (ii) $a(\alpha+\beta) = a\alpha+a\beta \; \forall\alpha, \beta\in V, a\in F$
  - (iii) $(a+b)\alpha = a\alpha + b\alpha \; \forall\alpha\in V, a, b\in F$
  - (iv) $(ab)\alpha = a(b\alpha) \; \forall\alpha\in V, a, b\in F$
  - (v) $1. \alpha = \alpha\forall\in V$
- **Linear subspace:** A non-empty subset W of a linear space $V(F)$ which itself is a linear space is called linear subspace of $V(F)$
- **Linear sum:** The linear sum of two subspaces W_1 and W_2 is the set of all those elements each one of which is expressible as the sum of an element of W_1 and an element of W_2.
- **Direct sum:** V is said to be direct sum of W_1 and W_2 if each element of V can be uniquely expressed as the sum of an element of W_1 and an element of W_2.
- **Linear combination of vectors:** Let $V(F)$ be a linear space and $\alpha_1, \alpha_2, ...\alpha_n\in V$. Then any vector $\alpha \in V$ can be expressed as $\alpha = a_1\alpha_1+a_2\alpha_2+...+a_n\alpha_n$, where $a'_is \in F$ is said to be linear combination of vectos $\alpha_1, \alpha_2, ..., \alpha_n$.
- **Linear span:** Let $V(F)$ be a linear space and S be any non-empty subset of V, then set of all linear combination of finite elements of S is called the linear span of S.
- **Linearly dependent vectors:** A finite set $\{\alpha_1, \alpha_1, ..., \alpha_n\}$ of vectors of V is said to be linearly dependent if there exists scalars $\alpha_1, \alpha_2, ..., \alpha_n$ not all of them equal to zero such that
$$a_1\alpha_1+a_2\alpha_2+ , ..., a_n\alpha_n = 0$$
- **Linearly Independant vectors:** A finite set of vectors $\{\alpha_1, \alpha_2,..., \alpha_n\}$ of the linear space $V(F)$ is said to be linearly independent if for every expression of the type $a_1\alpha_1+a_2\alpha_2+ ...+ a_n\alpha_n = 0$, $a_i's \in F$ implies $a_1= a_2= ...= a_n = 0$.
- **Basis of a linear space:** A non-empty subset S of a linear space $V(F)$ is said to be basis if
  - (i) S is linearly independent.
  - (ii) $L(S) = V$
- **Finite-dimensional linear space:** Let S be a non-empty subset of a linear space $V(F)$, then $V(F)$ is said to be finite dimensional if S is finite subset of V such that $L(S) =V$.
- **Cosets:** Let W be a subspace of a linear space $V(F)$ then the set $\alpha+W = \{\alpha+\beta \; \forall\beta\in W\}$ and $W+\alpha = \{\beta + \alpha\forall\beta\in W\}$ are called left and right cosets respectively.
- **Quotient space:** Let $V/W = \{ W+\alpha : \alpha\in V\}$ be the set of all cosets of W in V such that $(W+\alpha)+(W+\beta)= W+(\alpha+\beta)$ and $a(W+\alpha) = W+a\alpha$. Then linear space V/W is called quotient space.
- **Linear transformation:** Let U and V be two linear spaces over the same field F, then a mapping $T : U \to V$ which associates to each element $\alpha\in U$ to a unique element $T(\alpha)\in V$ such that $T(a\alpha+b\beta)=aT(\alpha) + bT(\beta)$ is called a linear transformation of U into V.
- **Isomorphism:** A linear transformation $T : U\to V$ which is one-one and onto.
- **Linear operator:** Let $V(F)$ be a linear space, then a linear transformation T from V to V is called linear operator.
- **Range of a linear Transformation**
$$R_T = \{\beta\in V : T(\alpha)\beta \text{ for some } \alpha\in U\}$$
- **Null space of a linear transformation:**
$$N_T = \{\alpha\in U : T(\alpha) = 0\}$$
- **Non-singular linear transformation:**
$$T(\alpha) = 0 \Rightarrow \alpha = 0$$
- **Linear functionals:** Let $V(F)$ be a linear space. Then a linear transformation $f : V\to F$ is called linear functional if ;
$$f(a\alpha+b\beta) = af(\alpha)+bf(\beta) \; \forall\alpha, \beta\in V, a, b\in F.$$

- **Dual space:** The set of linear functionals on a linear space $V(F)$ is also a linear space, called dual space.
- **Annhilators:** The set of all linear functional ϕ on V such that $\phi(\alpha) = 0 \forall \in V$ is called an annihilator of W.
- **Eigen values of linear transformation:** Let $V(F)$ be finite dimensional linear space and T be a linear operator on V then a scalar $\lambda \in F$ is called an eigen value of T if there exists a non-zero vector $\alpha \in V$ such that $T(\alpha) = \lambda\alpha$.
- **Diagonalization** : A linear operator $T : V \to V$ is said to be diagonalizable if V has a basis consisting of eigen vectors of T only.

▶ RESULTS

- The intersection of any two subspaces of a linear space is a subspace.
- The union of two subspaces of a linear space is a subspace iff one is contained in the other.
- The linear span $L(S)$ of a non-empty subset S of a linear space $V(F)$ is the smallest subspace of V containing S.
- The zero space has no basis.
- Every finitely generated linear space has a finite basis **(Existence theorem)**.
- Every non-zero linear space has a basis. A linear space may have more than one basis.
- Every linearly independent subset of a linear space V is either a basis of V or can be extended to form a basis of V **(Extension theorem)**.
- Every subspace of a finite-dimensional linear space has a compliment.
- Any two cosets are either disjoint or identical.
- Every n-dimensional linear space $V(F)$ is isomorphic to $F^n(F)$.
- Any two finite-dimensional linear space over the same field are isomorphic if and only if they are of same dimension.
- If two vectors are linearly dependent, then one of them is a scalar multiple of other.
- A set of non-zero vector $[x_1, x_2, \ldots, x_n]$ is linearly dependent if some of these vectors, say x_i is a linear combination of the preceding vectors $x_1, x_2, \ldots, x_{i-1}$ and conversely.
- If W_1 and W_2 are subspaces of a linear space $V(F)$, then $W_1 + W_2$ is also a subspace of $V(F)$.
- The set of all real valued continuous function defined in [0, 1] is a linear space over field of reals.
- The complex field C is a linear space over the field of reals.
- Arbitrary intersection of subspaces of a linear space is a subspace.
- Linear transformation is also known as linear space homomorphisrn.
- Let V be an m-dimensional linear space over the field F and let V be an n-dimensional linear space over F. Then the linear space $L(U, V)$ is finite dimensional and has dimension mn.
- If V is a linear space over a field F, then a linear transformation from V into V is called a linear operator.
- Kernel of a linear transformation is also known as null space of T.
- A linear transformation T is said to be invertible if T is one-one and T is onto.
- A linear transformation T from U into V is called non-singular if the null space of T is $\{0\}$.
- If T is non-singular, then T is one-one.
- If V is not finite dimensional, then the natural mapping can never be onto V^{**}. However, it is always linear and one-one.
- Non-zero eigen vectors belonging to distinct eigen values are linearly independent.
- 0 is the eigen value of T if and only if T is singular.
- If λ is an eigen value of T, then λ^{-1} is an eigen value of T^{-1}.
- A linear space V together with an inner product is called an inner product space.
- A finite dimensional real inner product space is called an Euclidean space.
- A complete inner product space is called a unitary space.
- Any two vectors $\alpha, \beta \in V$ is said to be orthogonal to each other if $(\alpha, \beta) = 0$.
- An orthonormal set of non-zero vectors is linearly independent.
- A maximal orthonormal set of vectors in an inner product space V is called a complete orthonormal set.
- In a finite dimensional inner product space, a complete orthonormal set is basis.

REVIEW QUESTIONS

1. If V is a linear space over an infinite field F then show that it is not possible to write V as union of a finite number of proper subspaces.
2. Show that $L(S)$ is the smallest subspace of V containing S.
3. Show that following vectors are linearly

dependent :

(i) (1, 1, 2), (–3, 1, 0), (1, –1, 1) (1, 2, –3) in R^3(R)

(ii) (1, –1, 2, 0), (3, 0, 0, 1), (2, 1, –1, 0), (1, –1, 2, 0) in R^4(R)

4. Show that following vectors are linearly independent :

(i) (1, 1, 0), (1, 0, 1), (0, 1, 1) in R^3(R)

(ii) (1, 0, 0), (1, 1, 1), (1, 2, 3) in R^3(R)

5. Show that the vectors (v_1, v_2) and (w_1, w_2) in **C** are linearly dependent if $v_1 w_2 = v_2 w_1$

6. Let S be a finite subset of a linear space such that S is linearly independent and every proper superset of S in V is linearly dependent show that S is a basis of V.

7. Verify that following is an inner product on R^2.

$$(u, v) = x_1y_1 - 2x_1y_2 - 2x_2y_1 + 5x_2y_2$$

8. If W is a subspace of V and $v \in V$ satisfies

$$(v, w) + (w, v) \leq (w, w) \forall\, w \in W$$

prove that $(v, w) = 0\ \forall w \in W$, where V is an inner product space over F.

9. Let T be a linear operator on V and let Rank T^2 = Rank T then show that

$$\text{Range } (T) \subseteq \text{Ker } (T) = \{0\}$$

10. Show that a necessary and sufficient condition for the map $T : F^2 \to F^2$ such that $T(x_1, x_2) = (\alpha x_1 + \beta x_2)$, ($\alpha, \beta, \gamma, \delta$, are some fixed element of F) to be an isomorphism is that $\begin{vmatrix} \alpha & \beta \\ \gamma & \delta \end{vmatrix} \neq 0$

11. Let A be $n \times n$ matrix over F. Show that A is invertiable if and only if rows of A are linearly independent over F.

12. Let $u, v \in V$ and that $f(u) = 0 \Rightarrow f(v) = 0$ for all $f \in V^*$. Show that $v = \alpha u$ for some scalar α.

13. Let T be a linear operator on R^2 which is represented in the standard ordered basis by the matrix $A = \begin{bmatrix} 0 & -1 \\ 1 & 0 \end{bmatrix}$, show that T has no eigen values in R.

14. Let V be the linear space of all real valued continuous functions. Define $T : V \to V$ by $Tf(x) = \int_0^x f(t)dt$ show that T has no eigen values.

15. Let a, b, c be elements of the field F and

$$A = \begin{bmatrix} 0 & 0 & c \\ 1 & 0 & b \\ 0 & 1 & a \end{bmatrix}$$

Prove that the characteristic polynomial of A is same as that of its minimal polynomial.

OBJECTIVE TYPE QUESTIONS

▸ FILL IN THE BLANKS

1. In a linear space (V, '+', '.') the external composition is also known as __________ .

2. In a linear space $V(F)$, the vector addition is also known as __________ .

3. The elements of a field F for $V(F)$ are called __________ .

4. In a linear space (V, '+', '.'), (V, +) must be __________ .

5. The additive identity for the linear space $V(F)$ is __________ .

6. If a, $\alpha \in V$ for $a \in F$ and $\alpha \in V$, then V is closed under __________ .

7. If $a \in F$ and $\alpha \in V$, then $(ab)\alpha$ = __________ .

8. If $\alpha, \beta, \gamma \in V$ and $\alpha + \beta = \gamma$, then $\alpha + \beta - \gamma$ = __________ .

9. If F is any field, then F is a linear space over __________ .

10. If $V(F)$ is a linear space and $\mathbf{0} \in V$, then __________ $= 0\ \forall\ \alpha \in V$.

11. If $V(F)$ is a linear space and $a \in F$, $\alpha \in V$, then $a\alpha = 0 \Rightarrow a =$ __________ or $\alpha =$ __________ .

12. If W be a subset of a linear space $V(F)$ and $a\alpha + b\beta \in W$, for all $a, b \in F$ and $\alpha, \beta \in W$, then W is a __________ .

13. If $W = \{(a_1, a_2, 0) : a_1, a_2 \in F\}$, then W is a __________ of linear space.

14. R(**C**) is __________ .

15. For any non-empty subset W of $V(F)$, if $\alpha - \beta \in W$ and $a\alpha \in W$ for all $a \in F$ and $\alpha, \beta \in W$, then W is __________ .

16. If W_1 and W_2 are two subspaces of $V(F)$, then $W_1 \cap W_2$ is a __________ .

17. If W_1 and W_2 are two subspaces of $V(F)$ and either $W_1 \subseteq W_2$ or $W_2 \subseteq W_1$, then __________ is a subspace of $V(F)$.

18. $L(\phi)$ = __________ .

19. If $S = \{(1,0,0), (0,1,0), (0, 0, 1)\}$ is a subset of $V_3(F)$, then $L(S)$ = __________ .

20. If $a \neq 0$, $b \neq 0 \in F$ and $a\alpha + b\beta = \mathbf{0}$ for $\alpha, \beta \in V$, then α, β are __________ .

21. The vectors (1, 0, 0), (0, 1, 0), (0, 0, 1) are linearly ______.

22. For any subset S of $V(F)$, $L(S) = V$, then S is a basis of $V(F)$ if S is linearly ______.

23. Every superset of a linearly dependent set of vectors is linearly ______.

24. The vectors in a basis are linearly ______

25. Every linear space has a ______.

26. Let $S = \{\alpha\}$ and $a \neq 0 \in V(F)$, then S is always linearly______.

27. Any infinite set of vectors of V is linearly independent if its every finite subset is linearly ______.

28. If a basis of linear space has 4 elements, then dimension of the linear space is ______.

29. The set {(1, 0, 0), (0, 1, 0), (0, 0, 1)} forms a ______ of $V_3(F)$.

30. The set S = {(1, 0), (0, 1)} is a basis of $V_n(F)$ for n = ______.

31. Any subset containing $(n+1)$ vectors of an n-dimensional linear space is linearly ______.

32. M and N are two subspaces of a linear space V, then $V = M \oplus N$ if (i) ______ and (ii) ______

33. If V^* is a dual space of V, then dim. V = ______.

34. Let V be a linear space and V^* be its dual space. Let W be a non-empty subset of V, then W^* is a subspace of ______.

35. If W is a subspace of a finite-dimensional linear space $V(F)$, then W^{00} ______.

36. M is an m-dimensional subspace of a n-dimensional linear space V and M' is an annihilator of M°, then dim. M° = ______.

37. If W is a subset of a linear space $V(F)$, then $[L(W)]^\circ$ = ______.

38. If W_1 and W_2 are subsets of a linear space with, $W_1 \subset W_1$, then ______.

39. V is a finite-dimensional linear space, V^* is its dual space, $x, y \in V$, $x \neq y$ then there is an $f \in V^*$ such that ______.

40. Let W_1 and W_2 be subspace of a linear space over a field F, then $W_1^\circ + W_2^\circ$ = ______.

▸ TRUE/FALSE

Write 'T' for True and 'F' for False statement.

1. In a vector space $V(F)$, the vector addition is also called an internal composition. **(T/F)**
2. The elements of V are scalars. **(T/F)**
3. The elements of F are vectors. **(T/F)**
4. Let $V(F)$ be a linear space. Then the zero space $\{\mathbf{0}\}$ is called a trivial subspace. **(T/F)**
5. The set $W = \{(a, 0, b) : a, b \in R\}$ is not a subspace of $R^3(R)$. **(T/F)**
6. Let $V(F)$ be a linear space and $\alpha \in V$, $\alpha \in F$, then $a\mathbf{0} = 0$ and $\mathbf{0}\alpha = 0$. **(T/F)**
7. If K is a field and $F \subseteq K$, then $K(F)$ is a linear space. **(T/F)**
8. The field of complex numbers is not a linear space over a field of real number. **(T/F)**
9. The field of real number is a linear space over a field of complex numbers. **(T/F)**
10. If any subset W of V is closed under addition and scalar multiplication in V, then W is a subspace of V. **(T/F)**
11. The intersection of two subspaces of a linear space is also a subspace. **(T/F)**
12. If $\{(1, 0), (0, 1)\} \subseteq V_2(F)$, then L {1, 0), (0, 1)} = F^2. **(T/F)**
13. $L(\phi) = \{0\}$ **(T/F)**
14. The single non-zero vector is always linearly dependent. **(T/F)**
15. If W_1 and W_2 are two subspaces of $V(F)$, then $W_1 + W_2$ is also a subspace of $V(F)$. **(T/F)**
16. In a linear space, every subset of a linearly independent set is linearly dependent. **(T/F)**
17. The set containing the zero vector is linearly dependent. **(T/F)**
18. The vectors $(1 + i, 2i)$, $(1, 1 + i)$ in $C^2(\mathbf{C})$ are linearly dependent but in $C^2(\mathrm{R})$ are linearly independent. **(T/F)**
19. For a non-empty subset S of $V(F)$, $L(S) = V$ and S is linearly independent, then S is a basis of $V(F)$. **(T/F)**
20. Every linear space has a finite basis. **(T/F)**
21. The vectors in a basis are linearly independent. **(T/F)**
22. Every linear space has a basis. **(T/F)**
23. Let V^* be the dual space of V and V^{**} be the dual of V^*, then

    dim. V^* < dim. V^{**} **(T/F)**
24. If W is any subset of a finite-dimensional linear space V, then $W^{\circ\circ} = L(W)$. **(T/F)**
25. If W_1 and W_2 are subspace of $V(F)$ with $W_1 \subset W_2$, then $\overset{\circ}{W_1} \subset \overset{\circ}{W_2}$ **(T/F)**
26. Every linear functional is a linear transformation. **(T/F)**

27. If W is a subset of V with $W = V$, then $W^{\circ\circ} = \{0\}$. **(T/F)**

28. If W is a subset of V with $W = \{0\}$, then $W^{\circ} = V$. **(T/F)**

29. If W_1 and W_2 are two subspaces of $V(F)$, then $W_1 = W_2$ iff $W_1^{\circ} = W_2^{\circ}$ **(T/F)**

30. If W is a subspace of a finite dimensional linear space $V(F)$, then $W = W^{\circ\circ}$ **(T/F)**

▶ MULTIPLE CHOICE QUESTIONS

Choose the most appropriate option :

1. In a linear space $V(F)$, a 0 equals :
(a) 0 (b) 0
(c) a (d) 1

2. In a linear space $V(F)$, $\alpha \in V$ and $a, b \in \mathbf{F}$, then $(ab)\,\alpha$ equals :
(a) $a(b\alpha)$ (b) ab
(c) a (d) $\alpha(ab)$

3. If W_1 and W_2 are two subspaces of $V(F)$, then $W_1 \cup W_2$ is a subspace if:
(a) $W_1 - W_2$ (b) $W_1 \subseteq W_2$
(c) $W_1 \cap W_2$ (d) None of these

4. $L(\phi)$ equals :
(a) 0 (b) ϕ
(c) $\{0\}$ (d) None of these

5. If $S = \{(1, 0), (0, 1) \subseteq V_2(\mathrm{R})\}$, then $L(S)$ equals :
(a) R (b) S
(c) R^3 (d) R^2

6. If $\alpha = k\beta$ for $\alpha, \beta \in V$ and $k \in F$, then $\{\alpha, \beta\}$ is linearly :
(a) dependent (b) independent
(c) None of these

7. The set $S = \{(1, 0, 0), (0, 1, 0), (0, 0, 1)\}$ forms a basis for $V_n(\mathrm{R})$ if 'n' equals :
(a) 2 (b) 3
(c) 4 (d) 1

8. The dimension of a linear space $R^3(\mathrm{R})$ is :
(a) 2 (b) 4
(c) 1 (d) 3

9. Which of the sets is linearly dependent ?
(a) $\{0\}$ (b) $\{\phi\}$
(c) $\{1\}$ (d) $\{\alpha\}$

10. A subset S of $V(F)$ forms a basis of $V(F)$ if S is linearly independent and :
(a) $L(S) = S$ (b) $L(S) = V$
(c) $L(S) = F$ (d) None of these

11. Which of the following is not a linear space ?
(a) R(R) (b) **C**(**C**)
(c) R(**C**) (d) **C**(R)

12. Condition that vectors (a_1, a_2) and (b_1, b_2) are linearly dependent is :
(a) $a_1b_1 + a_2b_2 = 0$ (b) $a_1b_2 + a_2b_1 = 0$
(c) $a_1b_2 - a_2b_1 = 0$ (d) $a_1b_1 - a_2b_2 = 0$

13. If W_1 and W_2 are subsets of a linear space V such that $W_2 \subset W_1$ then:
(a) $W_1^{\circ} \subset W_2^{\circ}$ (b) $W_1^{\circ} = W_2^{\circ}$
(c) $W_2^{\circ} \subset W_1^{\circ}$ (d) None of the above

14. If V is a finite dimensional linear space and V^* is its dual and V^{**} is the dual of V^*, then:
(a) V is isomorphic to V^* and but not to V^{**}
(b) V is isomorphic to V^{**} but not to V^*
(c) V is not isomorphic to both V^* and V^{**}
(d) None of the above

15. If W is a subspace of a finite-dimensional linear space $V(F)$, then dim. V+dim. W° is equal to:
(a) dim. V (b) 0
(c) 1 (d) None of the above

16. If W_1 and W_2 are subspace of a linear space $V(F)$, then $(W_1 \cap W_2)^{\circ}$ is equal to:
(a) $W_1^{\circ} + W_2^{\circ}$ (b) $W_1^{\circ} \cup W_2^{\circ}$
(c) $W_1^{\circ} \cap W_2^{\circ}$ (d) None of these

17. If W_1 and W_2 are subspace of a linear space $V(F)$, then $(W_1 + W_2)^{\circ}$ is equal to:
(a) $W_1^{\circ} \cup W_2^{\circ}$ (b) $W_1^{\circ} \cap W_2^{\circ}$
(c) $W_1^{\circ} + W_2^{\circ}$ (d) None of the above

18. If W_1 and W_2 are subspace of a linear space $V(F)$, such that $V = W_1 \oplus W_2$, then $W_1^{\circ} \oplus W_2^{\circ}$ is equal to:
(a) V^* (b) V
(c) V^{**} (d) None of the above

19. If W_1 and W_2 are subspace of a linear space $V(F)$ which are annihilated by the subspace W°, then dim. W_1 + dim.W_2 is equal to:
(a) 2 (dim. V – dim. W°)
(b) dim. V – dim. W°
(c) $\frac{1}{2}$ (dim. V – dim. W°)
(d) 2 (dim. W° – dim. V)

ANSWERS

▶ FILL IN THE BLANKS

1. scalar multiplication **2.** internal composition **3.** scalars **4.** abelian group **5.** zero vectors **6.** scalar multiplication. **7.** $a(b\alpha)$ **8.** zero vector *i.e.*, 0 **9.** F **10.** 0α **11.** 0, **0** **12.** subspace **13.** subspace **14.** not a linear space **15.** subspace **16.** subspace **17.** $W_1 \cup W_2$ **18.** {0} **19.** V_3 **20.** linearly dependent **21.** independent **22.** independent **23.** dependent **24.** independent **25.** basis **26.** independent **27.** independent **28.** 4 **29.** basis **30.** 2 **31.** dependent **32.** $V = M+N, M \cup N = \{0\}$ **33.** div V^* **34.** V^* **35.** W **36.** $(n-m)$ **37.** W^* **38.** $\overset{\circ}{W_2} \subset \overset{\circ}{W_1}$ **39.** $f(x) \neq f(y)$ **40.** $(W_1 \cap W_2)^*$

▶ TRUE/ FALSE

1. T	**2.** F	**3.** F	**4.** T	**5.** F	**6.** F	**7.** T	**8.** F	**9.** F
10. T	**11.** T	**12.** T	**13.** T	**14.** T	**15.** T	**16.** F	**17.** F	**18.** T
19. T	**20.** F	**21.** T	**22.** T	**23.** F	**24.** T	**25.** F	**26.** T	**27.** T
28. F	**29.** T	**30.** T						

▶ MULTIPLE CHOICE QUESTIONS

1. (b)	**2.** (a)	**3.** (b)	**4.** (c)	**5.** (d)	**6.** (a)	**7.** (b)	**8.** (d)	**9.** (a)
10. (b)	**11.** (c)	**12.** (c)	**13.** (a)	**14.** (c)	**15.** (a)	**16.** (a)	**17.** (b)	**18.** (a)
19. (a)								

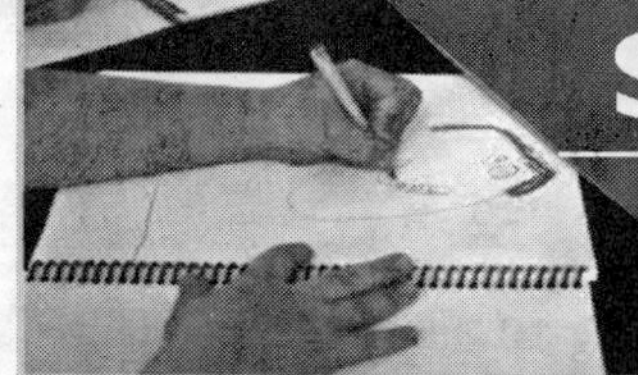

Self Assessment Test

1. In V_3(R), examine each of the following set of vectors for linear dependence :
   (i) {(1,2,0), (0,3,1), (–1,0,1)}
   (ii) {(–1,2,1), (3,0,–1), (–5,4,3)}
   (iii) {(1,3,2), (1,–7,–8), (2,1,–1)}
   (iv) {(1,1,–1), (2,–3,5), (–2,1,4)}
2. Prove that the four vectors (1,0,0), (0,1,0), (0,0,1), (1,1,1) in V_3(C) form a linearly dependent set but any of them are linearly independent.
3. If α,β and γ are vectors such that $\alpha+\beta+\gamma=0$ then show that α and β span the same subspaces as β and γ.
4. In the linear space R^3, let $\alpha=(1,2,1)$, $\beta=(3,1,5)$, $\gamma=(3,-4,7)$. Show that the subspace spanned by $S=\{\alpha,\beta\}$ and $T=\{\alpha,\beta,\gamma\}$ are the same.
5. Let *f* be a linear transformation from a linear space *U* into a linear space *V*. If *S* is a subspaces of *U*, show that *f(S)* will be a subspaces of *V*.
6. Show that the space of all real functions is the direct sum of the subspaces of odd functions and even functions.
7. Show that the vectors *S* and α of the linear space R over Q are linearly independent iff α is an irrational number but that the same is not true in the real linear space of R.
8. Show that $\{a+ib,c+id\}$ forms a basis of the linear space of complex numbers over the field of real numbers.
9. Let $W=(1,2,3)$ be a vector in Euclidean space R^3. Find an orthonormal basis of $W^{\perp}$.
10. Let *P* be orthogonal, prove that $\|P_u\|=\|u\|$ for -every $u\in V$.
11. If *A* is orthogonally equivalent to *B*. Show that *B* is orthogonally equivalent to *A*.
12. Find an orthonormal basis for the subspaces *U* of R^4 spanned by $V_1=(1,1,1,1)$, $V_2=(1,2,4,5)$, $V_3=(1,-3,-4,-2)$.
13. Let *V* be a linear space of polynomials *f(t)* with inner product $(f,g)=\int_{-1}^{1} f(t)g(t)d(t)$. Apply the Gram-schmidt algorithm to the set $\{1, t, t^2, t^3\}$ to obtain an orthonormal set $\{f_0, f_1, f_2, f_3\}$.
14. Find the matrix *A* which represents the given inner product on R^2 with respect to usual basis {(1,0), (0,1)} of R^2.
15. Find the matrix relative to the basis $\{1+i,1+2i\}$.
16. Let $F: R^2 \to R^2$ be defined by $F(1,0)=(2,4)$ and $F(0,1)=(5,8)$.
    Find the matrix *A* representing *F* with respect to the usual basis for R^2.
17. Let *V* be the finite dimensional and *T* be a linear operator on *V*. Show that *T* is invertible if and only if *T* is non-singular.
18. If a linear space has one basis that contains infinitely many elements, prove that every basis contains infinitely many element.
19. Let *T* be a linear transformation from *V* to *W*. Prove that the image of *V* under *T* is a subspacce of *W*.
20. If $\{u,v,w\}$ is a linearly independent subset of a linear space, show that $\{u,u+v,u+v+w\}$ is also linearly independent.

4.1 INTRODUCTION

In this chapter we first describe the basic framework in which we shall be working with structure by ′norm′ on a linear space over the scalar field K.

4.2 CONCEPT OF NORM

The concept of norm introduced in order to give a method for measuring the magnitude of a vector. For example, if $x = (-1, -2, -3, -5, -11)$ is in R^5, then $\|x\| = 11$ is the vector norm which is the length of the largest coordinate.

The Euclidean norm in C^n wil be defined by

$$\|z\|_2 = \left(\sum_{K=1}^{n} |z_K|^2 \right)^{\frac{1}{2}}, z = (z_1, z_2, \ldots, z_n) \in C^n.$$

which is the same as the Euclidean distance of the point $z \in C^n$ from the origin, for example if $n = 1$, we have

$$x \in R \quad \Rightarrow \quad \|x\| = |x|$$

which is the absolute value of the real number x and that

$$x + iy = z \in C \quad \Rightarrow \quad \|z\| = |z| = \sqrt{x^2 + y^2}$$

4.3 NORMED LINEAR SPACE

(DELHI–2011, KANPUR–2001, MEERUT–2005, 06, 06BP, 07, 07BP, 09BP, 2011)

Let N be a linear space and let $\|.\| : N \to R$ be a function such that for all $x, y \in N$ and for all α (may be real or complex) following conditions are satisfied.

(1) $\|x\| \geq 0$; (Positivity)

(2) $\|x\| = 0$, if and only if $x = 0$

(3) $\|x + y\| \leq \|x\| + \|y\|$ (Triangle inequality, or sub additivity)

(4) $\|\alpha x\| = |\alpha| . \|x\|$ (Homogeneity)

Then the function $\|.\|$ is called a norm on N and ordered pair $(N, \|.\|)$ is called a normed linear space.

Remarks

- If we dropped the second condtion, then the function $\|.\|$ is called a semi or pseudo norm on N and N is said to be a semi normed linear space.
- The following are the immediate consequences of the definition of norm :

  (i) If $\alpha = -1$, then by property (4), we have

$$\|-x\| = \|x\|$$

In particular

$$\|y - x\| = \|x - y\|$$

(ii) Clearly

$$|\,\|x\| - \|y\|\,| \le \|x - y\|$$

(iii) A p-norm is defined by replacing the property (4) in the definition of norm by the following one

$$\|\alpha x\| = |\alpha|^p \|x\|$$

- It must be noted that a p-norm with $p = 1$ is just a norm.

Definition. *Let $(N, \|\ \|)$ be a normed linear space then a sequence $<x_n>$ in N is said to converge to an element $x_0 \in N$ if for given $\in > 0$ $\exists$ a positive integer m_0 such that*

$$\|x_n - x_0\| < \in \text{ whenever } n \ge m_0.$$

Here, x_0 is said to be the limit of x_n, i.e., $\lim_{n\to\infty} x_n = x_0$.

It can also be written as

$$x_n \to x_0 \text{ iff } \|x_n - x_0\| \to 0.$$

Definition. *A sequence $<x_n>$ in N is said to be a Cauchy sequence if for given $\in > 0$, $\exists$ a positive integer m_0 such that*

$$\|x_n - x_m\| < \in \textit{ whenever } m, n \ge m_0. \qquad \text{(MEERUT–2006BP)}$$

Definition. *A normed linear space N is said to be complete if and only if every Cauchy sequence in N is convergent.*

THEOREM 1. ***Let N be a normed linear space and let d be a function from N × N into R defined by d(x, y) = ‖x – y‖, then d is a metric on N.***

(AGRA–2002, MEERUT–2006BP, 07, GARHWAL–2002, KANPUR–85, 2010)

PROOF. As per given $d : N \times N \to R$ be a function such that $d(x, y) = \|x - y\|$. To show d is a metric on N.

(i) Let $x, y \in N$ and N is a linear space, therefore $x - y \in N$

$\Rightarrow \quad \|x - y\| \ge 0$ [By the property of norm]

$\Rightarrow \quad d(x, y) \ge 0$

(ii) We have

$$d(x, y) = 0 \Leftrightarrow \|x - y\| = 0 \Leftrightarrow x - y = 0 \Leftrightarrow x = y$$

(iii) We have

$$d(x, y) = \|x - y\| = \|(-)(y - x)\| = |-1|\,\|y - x\| = 1.\|y - x\|$$
$$= \|y - x\| = d(y, x)$$

(iv) For $x, y, z \in N$, we have

$$\|x - y\| = \|x - z + z - y\| \le \|x - z\| + \|z - y\|$$

Thus $d(x, y) \le d(x, z) + d(z, y)$

Hence, d is a metric on N.

THEOREM 2. ***In a normed linear space, every convergent sequence is a Cauchy sequence.***

(MEERUT–2006BP)

PROOF. Let N be a normed linear space. Suppose that $<x_n>$ is a convergent sequence converge to $x_0 \in N$.

To show that $<x_n>$ is Cauchy.

Since, the given sequence $<x_n>$ is convergent therefore by definition for given $\in > 0$; $\exists$ a positive integer m_0 such that

$$\|x_n - x_0\| < \frac{\in}{2}; \text{ for } n \ge m_0 \qquad \ldots(1)$$

In particular for $n = m$, we have

$$\|x_m - x_0\| < \frac{\in}{2}; \text{ for } m \geq m_0 \qquad \ldots(2)$$

Now, consider

$$\begin{aligned}\|x_m - x_n\| &= \|x_m - x_0 + x_0 - x_n\| \\ &\leq \|x_m - x_0\| + \|x_0 - x_n\| = \|x_m - x_0\| + \|x_n - x_0\| \\ &< \frac{\in}{2} + \frac{\in}{2} \qquad \text{[Using (1) and (2)]} \\ &= \in\end{aligned}$$

$$\Rightarrow \qquad \|x_m - x_n\| < \in, \forall\, m, n \geq m_0$$

Hence, the convergent sequence $<x_n>$ in N is cauchy.

4.4 BANACH SPACES

(DELHI–2008, MEERUT–2006, 07BP, 08, 09BP, GARHWAL–2003, 06, KANPUR–1984, 90, 2010, 11, BUNDELKHAND–1985)

Definition 1. *A normed linear space which is complete as a metric space is called a Banach space.*

WORKING PROCEDURE

For showing any set to be a Banach space	
Step 1.	**The given set is a linear space.** We can easily verify the postulates of a linear space for the given definition of addition and scalar multiplication.
Step 2.	**The linear space is normed linear space.** For this we should be define norm of a vector x in the vector space and show that with this definition of norm all the four properties of the norm hold good.
Step 3.	**Completeness.** Lastly, we have to establish the completeness of the above normed linear space and for this, we will take up any Cauchy sequence and show that it is convergent.

Definition 2. (Summability in Normed linear space)

A series Σs_n in a normed linear space is said to be summable to a norm s if s is in the space and the sequence of partial sums of the series converges to s, i.e.,

$$\left\| s - \sum_{i=1}^{n} s_i \right\| \to 0 \text{ as } n \to \infty.$$

If we write $s = \sum_{i=1}^{\infty} s_i$. Then, the series Σs_n is said to be absolutely summable if $\sum_{i=1}^{\infty} \|s_i\| < \infty$.

REMARK

- For real numbers, every absolutely summable series is summable. This result is not true, in general.

Definition 3. *Let N and M be two normed linear spaces, a mapping $f : N \to M$ is said to be continuous at $x_0 \in N$ if and only if to each $\in > 0$, $\exists$ a $\delta > 0$ such that*

$$\|x - x_0\| < \delta \qquad \Rightarrow \qquad \|f(x) - f(x_0)\| < \in$$

Definition 4. *A function $f : N \to M$ is said to be continuous if and only if it is continuous at each point of N.*

Definition 5. *Let N and M be two normed linear spaces. A mapping $f : N \to M$ is said to be continuous at $x_0 \in N$ if and only if for every sequence $<x_n>$ in N converging to x_0, the sequence $<f(x_0)>$ in M converges to $f(x_0) \in M$*

i.e.,
$$x_n \to x_0 \to f(x_n) \to f(x_0)$$

Definition 6. *Let X, Y and Z be metric spaces, then a mapping $f : X \times Y \to Z$ is said to be jointly continuous if and only if $f(X_n, Y_n) \to f(x, y)$ whenever $x_n \to x$ and $y_n \to y$ as $n \to \infty$. $(x \in X, y \in Y)$.*

REMARK

- Every jointly continuous function is continuous in each of the variable x and y but the converse is not necessarily true.
  For example $f : R \times R \to R : f(x, y) = \dfrac{xy}{x^2 + y^2}, f(0,0) = 0.$

ILLUSTRATIONS

(1) The set R of real numbers with the norm $\|x\| = |x|$ is a Banach space.

(2) The n-dimensional Euclidean space R^n is a Banach space with the norm

$$\|x\| = \left(\sum_{k=1}^{n} |x_k|^2 \right)^{\frac{1}{2}}, \text{ where } (x_1, x_2, ..., x_n) \in R^n$$

(3) The n-dimensional unitary space C^n is a Banach space with the norm

$$\|x\| = \left(\sum_{k=1}^{n} |x_k|^2 \right)^{\frac{1}{2}}, \text{ where } (x_1, x_2, ..., x_n) \in C^n.$$

THEOREM 1. ***A normed linear space N is a Banach space iff every absolutely summable series in N is summable.***

(MEERUT–2001, 07, 08, 10, 11, KUMAON–2010, KANPUR–2002, 03, 07)

PROOF. Let us first suppose that N be a Banach space and Σf_n be an absolutely summable series of N. To show that Σf_n is summable.

By definition of absolute summability, there exists $k > 0$ such that

$$\Sigma \|f_n\| = k < \infty$$

Thus, for given $\in > 0 \ \exists$ a positive integer M such that

$$\sum_{n=M}^{\infty} \|f_n\| < \in$$

Let $S_n = \sum_{i=1}^{n} f_i$ be the partial sum of series Σf_n. Therefore, for $n \geq m \geq M$, we have

$$\|S_n - S_m\| = \left\| \sum_{i=m}^{n} f_i \right\| \leq \sum_{i=m}^{n} \|f_i\| \leq \sum_{i=M}^{\infty} \|f_i\| \leq \in$$

$\Rightarrow$ The sequence $<S_n>$ of partial sums is a Cauchy sequence in N. Also, since N is complete, therefore $<S_n>$ converges to an element s in N.

Hence, Σf_n is summable.

Conversely, let us suppose that every absolutely summable series in N be summable. To show that N is a Banach space. Let $<f_n>$ be a Cauchy sequence in N. By definition of Cauchy sequence we have that for every integer k, there exists an integer n_k such that

$$\|f_n - f_m\| < 2^{-k} \ \forall \ n, m \geq n_k \qquad ...(1)$$

Choose n_k's such that $n_{k+1} > n_k$, then $<f_{n_k}>$ is a subsequence of $<f_n>$

Hence, we get $g_1 = f_{n_1}$ and $g_k = f_{n_k} - f_{n_{k-1}}$ for $k > 1$

Then, we have

$$g_1 + g_2 + ... + g_k = f_{n_1} + (f_{n_2} - f_{n_1}) + (f_{n_3} - f_{n_2}) + ... + (f_{n_k} - f_{n_{k-1}}) = f_{n_k}$$

Therefore, we get a series Σg_k with k^{th} partial sum is f_{n_k}.

Also, $\|g_k\| = \left\|f_{n_k} - f_{n_{k-1}}\right\| < 2^{-k+1}$ [Using (1)]

$\Rightarrow$ $\Sigma\|g_k\| \leq \|g_1\| + \Sigma 2^{-k+1} = \|g_1\| + 1$

$$\left(\because \sum_{k=2}^{\infty} 2^{-k+1} = \frac{1}{2} + \frac{1}{2^2} + ... = \frac{1/2}{1-\frac{1}{2}} = 1\right)$$

$\Rightarrow$ The series Σg_k is absolutely summable and hence by our hypothesis it is summable, therefore there exists f in N to which the series Σg_k converges.

$\Rightarrow$ The subsequence $<f_{n_k}>$ of the partial sum of the series Σg_k converges to f.

It remains to prove that the sequence $<f_n>$ converges to f.

Since $<f_n>$ is a cauchy sequence, therefore by definition, for given $\in > 0 \ \exists$ a $M > 0$, such that

$$\|f_n - f_m\| < \frac{\in}{2}, \forall n, m \geq M \quad ...(2)$$

Since, $f_{n_k} \to f$, therefore

$$\left\|f_{n_k} - f\right\| < \frac{\in}{2}, \forall k \geq L \quad ...(3)$$

Choose k so large such that $k > L$ and $n_k > M$, then

$$\|f_n - f\| = \left\|f_n - f_{n_k} + f_{n_k} - f\right\| \leq \left\|f_n - f_{n_k}\right\| + \left\|f_{n_k} - f\right\| < \frac{\in}{2} + \frac{\in}{2}$$

[Using (2) and (3)]

$\Rightarrow$ for all $n > M$, we have $\|f_n - f\| < \in$

$\Rightarrow$ $<f_n>$ converges to f.

$\Rightarrow$ Every Cauchy sequence in N converges to an element of N.

$\Rightarrow$ N is complete.

Hence, N is a Banach space.

THEOREM 2. ***Let N be a normed linear space over a field F (C or R). Then the mapping***

$$f : N \times N \to f(x, y) = x + y$$

and $$g : F \times N \to g(a, x) = ax$$

are continuous. (DELHI–2004, 07, AGRA–2001, MEERUT–2000, 02, 04, 06)

PROOF. Let N be a normed linear space over a field F. Let $<x_n>$ and $<y_n>$ be two sequences in N and $<\alpha_n>$ be a sequence of scalars in F such that

$$x_n \to x, y_n \to y \text{ and } \alpha_n \to \alpha \text{ as } n \to \infty. \quad ...(1)$$

Consider $\|f(x_n, y_n) - f(x, y)\| = \|(x_n + y_n) - (x + y)\| = \|(x_n - x) + (y_n - y)\|$

$\leq \|x_n - x\| + \|y_n - y\| \to 0$ as $n \to \infty$ [Using (1)]

Thus $f(x_n, y_n) \to f(x, y)$ as $(\alpha_n - x_n)n \to \infty$

Further

$$\|g(\alpha_n, x_n) - g(\alpha, x)\| = \|\alpha_n x_n - \alpha x\| = \|\alpha_n x_n - \alpha_n x + \alpha_n x - \alpha x\|$$
$$\leq \|\alpha_n (x_n - x)\| + \|(\alpha_n - \alpha)x\|$$
$$= |\alpha_n| \|(x_n - x)\| + |\alpha_n - \alpha| \|x\|$$
$$\to 0 \text{ as } n \to \infty.$$

$\Rightarrow$ $g(\alpha_n, x_n) \to g(\alpha, x)$ as $n \to \infty$ [Using (1)]

Hence, f and g are continuous.

REMARK

- The above theorem can be restated as follows :
"Vector addition and scalar multiplication are jointly continuous"

THEOREM 3. ***Let N be a normed linear space and let x, y ∈ N, then*** $\left| \|x\| - \|y\| \right| \leq \|x - y\|$.

(MEERUT–2002, 03, 06, KANPUR–2006, GARHWAL–2004, DELHI–2011, 14, AGRA–2002, 03, ROHILKHAND–2008)

PROOF. We can write

$$\|x\| = \|(x - y) + y\| \leq \|x - y\| + \|y\|$$

$\Rightarrow$ $\|x\| - \|y\| \leq \|x - y\|$ ...(1)

Similarly, we can prove that

$\|y\| - \|x\| \leq \|y - x\|$ ...(2)

Also, $\|y - x\| = \|(-1)(x - y)\| = |-1| \|x - y\| = \|x - y\|$

Putting this value in (2), we get

$\|y\| - \|x\| \leq \|x - y\|$ ...(3)

Hence, from (1) and (3), we may get

$$\left| \|x\| - \|y\| \right| \leq \|x - y\|$$

THEOREM 4. ***Let N be a normed linear space. Then the mapping f : N → R such that f(x) = ‖x‖ is continuous, i.e., the norm ‖ ‖ on N is a continuous function.***

(MEERUT–2001, 06, 07)

PROOF. Let N be a normed linear space and $x \in N$ and let $<x_n>$ be a sequence in N such that

$x_n \to x$ ...(1)

Consider $|f(x_n) - f(x)| = \left| \|x_n\| - \|x\| \right|$ [By definition of f]

$\leq \|x_n - x\|$ (Using previous theorem)

$\to 0$ as $n \to \infty$ (Using (1))

$\Rightarrow$ $f(x_n) \to f(x)$ as $n \to \infty$. Hence, f is continuous.

THEOREM 5. ***Let*** $<x_n>$ ***and*** $<y_n>$ ***be two sequences such that*** $x_n \to x_0$ ***and*** $y_n \to y_0$ ***as*** $n \to \infty$***. Also, let α be any scalars, then***

(i) $\lim_{n\to\infty} (x_n + y_n) = x_0 + y_0$ ***(ii)*** $\lim_{n\to\infty} \alpha x_n = \alpha x_0$

(iii) $\lim_{n\to\infty} (x_n - y_n) = x_0 - y_0$ ***(iv) If*** $x_n < y_n$, $\forall$ ***n then*** $x_0 < y_0$

PROOF. (i) As per given, we have $x_n \to x_0$ and $y_n \to y_0$ as $n \to \infty$.

Therefore, by definition, for $\in > 0$, $\exists$ two positive integers m_1 and m_2 such that

$$\left|x_n - x_0\right| < \frac{\in}{2}, \forall n > m_1 \qquad \text{...(1)}$$

$$|y_n - y_0| < \frac{\in}{2}, \forall n > m_2 \qquad \ldots(2)$$

Let $\quad m_0 = \max\{m_1, m_2\}$

Then (1) and (2) reduce to

$$\left.\begin{aligned} |x_n - x_0| &< \frac{\in}{2}, \forall n \geq m_0 \\ \text{and} \quad |y_n - y_0| &< \frac{\in}{2}, \forall n \geq m_0 \end{aligned}\right\} \qquad \ldots(3)$$

Now consider

$$|(x_n + y_n) - (x_0 + y_0)| = |(x_n - x_0) + (y_n - y_0)|$$

$$\leq |(x_n - x_0)| + |(y_n - y_0)| < \frac{\in}{2} + \frac{\in}{2} = \in \qquad \text{[Using (3)]}$$

Thus, $\lim_{n\to\infty} (x_n + y_n) = x_0 + y_0$

(ii) For $\alpha = 0$, the result is obvious. Assume that $\alpha \neq 0$. Let $\in > 0$ be given. It is given that $x_n \to x_0$, by definition, there exist a positive integer m_0 such that

$$|x_n - x_0| < \frac{\in}{|\alpha|}, \forall n \geq m_0 \qquad \ldots(1)$$

Consider

$$|\alpha x_n - \alpha x_0| = |\alpha(x_n - x_0)| = |\alpha||x_n - x_0| < |\alpha|.\frac{\in}{|\alpha|} = \in \qquad \text{[Using (1)]}$$

$\Rightarrow \quad |\alpha x_n - \alpha x_0| < \in$

Thus, $\quad \lim_{n\to\infty} \alpha x_n = \alpha x_0$.

(iii) Consider

$$\lim_{n\to\infty} (x_n - y_n) = \lim_{n\to\infty} |x_n + (-y_n)| = \lim_{n\to\infty} x_n + \lim_{n\to\infty} (-y_n)$$

[Using result (1)]

$$= \lim_{n\to\infty} x_n + \lim_{n\to\infty} (-1)y_n = \lim_{n\to\infty} x_n - \lim_{n\to\infty} y_n = x_0 - y_0$$

(iv) We have

$$y_n \geq x_n, \ \forall\, n \in N$$

Therefore $\quad y_n - x_n \geq 0, \forall n \in N$

$\Rightarrow \quad \lim_{n\to\infty} (y_n - x_n) \geq 0 \quad \Rightarrow \quad \lim_{n\to\infty} y_n - \lim_{n\to\infty} x_n \geq 0$

$\Rightarrow \quad \lim_{n\to\infty} y_n \geq \lim_{n\to\infty} x_n$

4.5 SOME IMPORTANT INEQUALITIES

4.5.1 CAUCHY INEQUALITY

Statement. *If $x = (x_1, x_2, \ldots, x_n)$ and $y = (y_1, y_2, \ldots, y_n)$ be two n-tuples of real or complex numbers and if the norm of x be defined as* $\|x\| = \left(\sum_{i=1}^{n} |x_i|^2\right)^{1/2}$ *then*

$$\sum_{i=1}^{n} |x_i y_i| < \|x\|\|y\|,$$

i.e., $$\sum_{i=1}^{n} |x_i y_i| < (\Sigma|x_i|^2)^{1/2}(\Sigma|y_i|^2)^{1/2}$$

and hence deduce that $\|x + y\| \leq \|x\| + \|y\|$.

(MEERUT–2006, ROHILKHAND–2008, KANPUR–2008, AVADH–2009)

Proof. If $x = 0, y = 0$ then above result is obvious as both sides become zero. Hence we assume that $x \neq 0, y \neq 0$.

Now, define a_i, b_i as follow :

$$a_i = \left\{\frac{|x_i|}{\|x\|}\right\}^2, b_i = \left\{\frac{|y_i|}{\|y\|}\right\}^2$$

Clearly, a_i, b_i both are non-negative real numbers.

Since we know that G.M. ≤ A.M., therefore

$$\sqrt{(a_i b_i)} \leq \frac{a_i + b_i}{2} \text{ for each } i = 1, 2, ..., n.$$

or $$\frac{|x_i|}{\|x\|} \cdot \frac{|y_i|}{\|y\|} \leq \frac{1}{2}\left[\frac{|x_i|^2}{\|x\|^2} + \frac{|y_i|^2}{\|y\|^2}\right]$$

or $$\frac{|x_i y_i|}{\|x\|\|y\|} \leq \frac{1}{2}\left\{\frac{|x_i|^2}{\|x\|^2} + \frac{|y_i|^2}{\|y\|^2}\right\}$$

Summing the above inequality as i varies from 1 to n, we get

$$\sum_{i=1}^{n} \frac{|x_i y_i|}{\|x\|\|y\|} \leq \frac{1}{2}\left\{\frac{\Sigma|x_i|^2}{\|x\|^2} + \frac{\Sigma|y_i|^2}{\|y\|^2}\right\} = \frac{1}{2}\left[\frac{\|x\|^2}{\|x\|^2} + \frac{\|y\|^2}{\|y\|^2}\right] = 1 \text{ by definition of } \|x\|$$

Hence $\Sigma|x_i, yi| \leq \|x\| \|y\|$

Deduction. $\|x + y\| \leq \|x\| + \|y\|$.

Above is particular case of Minkowski inequality proved next.

Consider $$\|x+y\|^2 = \left(\sum_i |x_i + y_i|^2\right) = \sum_i (|x_i + y_i|)(|x_i + y_i|)$$

$$\leq \sum_i (|x_i + y_i|)(|x_i| + |y_i|) = \sum_i |x_i + y_i||x_i| + \sum_i |x_i + y_i||y_i|$$

$$\leq \sum_i \|x+y\|\|x\| + \|x+y\|\|y\|$$ [By Cauchy inequality]

$$\therefore \quad \|x+y\|^2 \leq (\|x+y\|)(\|x\| + \|y\|)$$

If $\|x + y\| = 0$ then above is clearly true.

In case $\|x + y\| \neq 0$ then we can divide both sides by $\|x + y\|$.

$$\therefore \quad \|x+y\| \leq \|x\| + \|y\|.$$

Lemma. *If* $1 < p < \infty$, $1 < q < \infty$ *such that* $\frac{1}{p} + \frac{1}{q} = 1$ *and* a, b *be two non negative real numbers then*

$$a^{1/p} b^{1/q} \leq \frac{a}{p} + \frac{b}{q}$$ (AGRA–2002)

Proof. If $a = 0$ and $b = 0$ then above is obviously true and hence we assume that $a \neq 0$, $b \neq 0$.

Consider a funciton $f(z)$ where $z \geq 0$ defined as under

$$f(z) = (1 - \lambda) + \lambda z - z^{\lambda} \text{ where } 0 < \lambda < 1 \quad ...(1)$$

$$f'(z) = \lambda - \lambda z^{\lambda - 1} = \lambda(1 - z^{\lambda - 1})$$

Since $\lambda - 1 < 0$, therefore, $z^{\lambda - 1} < 1$ if $z > 1$

or $\qquad z^{\lambda - 1} > 1$ if $z < 1$,

$\therefore \qquad 1 - z^{\lambda - 1} > 0$ if $z > 1$, $1 - z^{\lambda - 1} < 0$ if $z < 1$

$\therefore \qquad f'(z) = -$ve if $z < 1$

$\qquad = +$ve if $z > 1$.

Above relations show that $f(z)$ must be minimum at $z = 1$ as $f'(z)$ changes sign from –ve to +ve. Also at $z = 1$ we have from (1), $f(z) = 0$, which gives minimum value of $f(z)$.

$$\therefore \qquad f(z) = (1 - \lambda) + \lambda z - z^{\lambda} \geq 0.$$

$$\text{or} \qquad z^{\lambda} \leq (1 - \lambda) + \lambda z. \quad ...(2)$$

Put $z = \dfrac{a}{b}$ in (2) where a and b are non negative real numbers, we get

$$\frac{a^{\lambda}}{b^{\lambda}} \leq (1 - \lambda) + \lambda \frac{a}{b}.$$

$$\text{or} \qquad a^{\lambda} b^{1-\lambda} \leq (1 - \lambda) b + \lambda a.$$

Now choose $\lambda = \dfrac{1}{p}, 1 - \lambda = \dfrac{1}{q}$ so that $\dfrac{1}{p} + \dfrac{1}{q} = 1$

$$\therefore \qquad a^{1/p} b^{1/q} \leq \frac{a}{p} + \frac{b}{q} \text{ where } \frac{1}{p} + \frac{1}{q} = 1.$$

REMARK

- If we choose $p = q = 2$ then above result reduces to

$$\sqrt{ab} \leq \frac{a+b}{2} \quad \textit{i.e.}, \text{ GM} \leq \text{AM}$$

4.5.2 HOLDER'S INEQUALITY AND MINKOWSKI'S INEQUALITY

Statement. *If x, y be two elements of* l_p^n *space, i.e.,* $x = (x_1, x_2, ..., x_n), y = (y_1, y_2, ..., y_n)$. *and the norm of x is defined as*

$$\|x\|_p = \left[\sum_{i=1}^{n} |x_i|^p\right]^{1/p}$$

then Holder's Inequality states that

$$\sum_{i=1}^{n} |x_i y_i| \leq \left(\sum_{i=1}^{n} |x_i|^p\right)^{1/p} \left(\sum_{i=1}^{n} |y_i|^q\right)^{1/q}$$

or

$$\sum_{i=1}^{n} |x_i y_i| \leq \|x\|_p \|y\|_q, \text{ where } \frac{1}{p} + \frac{1}{q} = 1, 1 < p < \infty, 1 < q < \infty.$$

and hence deduce Minkowski Inequality $\|x + y\|_p \leq \|x\|_p + \|y\|_p$

(DELHI–2004, KANPUR–2000, 06, MADRAS–2006, ANNA–2004)

Proof. If $x = 0$ and $y = 0$ then above inequality is obviously true. Thus, we suppose that both x and y are not zero.

We know from Lemma proved above that

$$a^{1/p}b^{1/q} \le \frac{a}{p} + \frac{b}{q} \text{ where } \frac{1}{p} + \frac{1}{q} = 1 \qquad \ldots(1)$$

Let $$a_i = \left[\frac{|x_i|}{\|x\|_p}\right]^p, b_i = \left[\frac{|y_i|}{\|y\|_q}\right]^q$$

Using the result (1), for a_i and b_i we have

$$\frac{|x_i|}{\|x\|_p} \cdot \frac{|y_i|}{\|y\|_q} \le \frac{1}{p} \cdot \frac{|x_i|^p}{[\|x\|_p]^p} + \frac{1}{q} \cdot \frac{|y_i|^q}{[\|y\|_q]^q}$$

Above result is true for each i and hence summing w.r.t. i, we get

$$\frac{\Sigma|x_i||y_i|}{\|x\|_p \|y\|_q} \le \frac{1}{p} \cdot \frac{\Sigma|x_i|^p}{[\|x\|_p]^p} + \frac{1}{q} \cdot \frac{\Sigma|y_i|^q}{[\|y\|_q]^q} = \frac{1}{p} \cdot \frac{[\|x\|_p]^p}{[\|x\|_p]^p} + \frac{1}{q} \cdot \frac{[\|y\|_q]^q}{[\|x\|_q]^q}$$

[by def. of $\|x\|_p$]

$$= \frac{1}{p}.1 + \frac{1}{q}.1 = 1$$ [Using (1)]

$$\therefore \quad \sum_{i=1}^{n} |x_i y_i| \le \|x\|_p \|y\|_q$$

or $$\sum_{i=1}^{n} |x_i y_i| \le (\Sigma|x_i|^p)^{1/p} (\Sigma|y_i|^q)^{1/q} \text{ where } \frac{1}{p} + \frac{1}{q} = 1.$$

REMARK

- When $p = q = 2$ then the above inequality becomes

$$\sum_{i=1}^{n} |x_i y_i| \le (\Sigma|x_i|^2)^{1/2} (\Sigma|y_i|^2)^{1/2} \text{ where } \frac{1}{2} + \frac{1}{2} = 1$$

which is known as Cauchy Inequality.

4.5.3 MINKOWSKI'S INEQUALITY

(MEERUT–2010, KANPUR–2004, 07)

$\|\boldsymbol{x} + \boldsymbol{y}\|_p \le \|\boldsymbol{x}\|_p + \|\boldsymbol{y}\|_p$.

Proof. If $p = 1$, then $\|x\|_p = \|x\|_1 = \sum_{i=1}^{n} |x_i|$

$$\therefore \quad \|x + y\|_1 = \sum_{i=1}^{n} |x_i + y_i| \le \sum_{i=1}^{n} |x_i| + \sum_{i=1}^{n} |y_i| = \|x\|_1 + \|y\|_1$$

Thus the above inequality holds good for $p = 1$.

Let $p \ne 1$ then

$$(\|x + y\|_p)^p = \sum_{i=1}^{n} |x_i + y_i|^p$$ [By definition of norm]

$$= \sum_{i=1}^{n} |x_i + y_i||x_i + y_i|^{p-1} \le \sum_{i=1}^{n} (|x_i|) + (|y_i|)|x_i + y_i|^{p-1}$$

$$= \Sigma|x_i|(|x_i + y_i|^{p-1}) + \Sigma|y_i|(|x_i + y_i|^{p-1}) \qquad \ldots(1)$$

Now by Holder's Inequality

$$\Sigma|x_i y_i| \le \left(\Sigma|x_i|^p\right)^{1/p} + \left(\Sigma|y_i|^q\right)^{1/q} \text{ where } \frac{1}{p}+\frac{1}{q}=1.$$

or
$$\frac{1}{q} = 1 - \frac{1}{p} = \frac{p-1}{p}$$
$$(p-1)q = p$$

Applying the above inequaliuty on R.H.S. of (1) we get

$$(\|x+y\|_p)^p \le \left(\Sigma|x_i|^p\right)^{1/p}\left(\Sigma|x_i+y_i|^{(p-1)q}\right)^{1/q} + \left(\Sigma|y_i|^p\right)^{1/p}\left(\Sigma|x_i+y_i|^{(p-1)q}\right)^{1/q}$$

Put $(p-1)q = p$ and $\frac{1}{q} = \frac{p-1}{p}$ in R.H.S. we get

$$= \left(\Sigma|x_i|^p\right)^{1/p}\left(\Sigma|x_i+y_i|^p\right)^{p-1/p} + \left(\Sigma|y_i|^p\right)^{1/p}\left(\Sigma|x_i+y_i|^p\right)^{p-1/p}$$
$$= \|x\|_p (\|x+y\|_p)^{p-1} + \|y\|_p (\|x+y\|_p)^{p-1}$$
$$= (\|x\|_p + \|y\|_p)(\|x+y\|_p)^{p-1} \qquad \text{[By definition of norm]}$$

$$\therefore \qquad [\|x+y\|_p]^p \le (\|x\|_p + \|y\|_p)(\|x+y\|_p)^{p-1}$$

In case $\|x + y\|_p = 0$ then both sides vanish and above is obviously true. But if $\|x + y\|_p \ne 0$ then we can divide both sides by $\|x+y\|_p^{p-1}$ and we get

$$\|x+y\|_p \le \|x\|_p + \|y\|_p$$

REMARK

- If $p = q = 2$ and $\|x\| = \left(\sum_{i=1}^{n} |x_i|^2\right)^{1/2}$ then above reduces to $\|x + y\| \le \|x\| + \|y\|$.

4.5.4 HOLDER AND MINKOWKI'S INEQUALITIES FOR SEQUENCES

Let $x = <x_n>$, $y = <y_n>$ be sequence of scalars, such that

$$\sum_{n=1}^{\infty} |x_n|^p < \infty \text{ and } \sum_{n=1}^{\infty} |y_n|^p < \infty \ p \ge 1.$$

Define $\|x\|_p = \left(\sum_{i=1}^{\infty} |x_n|^p\right)^{1/p}$ then

(a) $\sum_{n=1}^{\infty} |x_n y_n| \le (\sum_{n=1}^{\infty} |x_n|^p)^{1/p} (\sum_{n=1}^{\infty} |y_n|^q)^{1/q} = \|x\|_p \|y\|_q$ where $\frac{1}{p}+\frac{1}{q}=1$

(Holder's Inequality)

(b) $\|x+y\|_p = \|x\|_p + \|y\|_q$ (Minkowski Inequality) (KANPUR–2004)

For any positive integer m we have from Holder's inequality proved before

$$\sum_{n=1}^{m} |x_n y_n| \le (\sum_{n=1}^{m} |x_n|^p)^{1/p} (\sum_{n=1}^{m} |y_n|^q)^{1/q} \le (\sum_{n=1}^{\infty} |x_n|^p)^{1/p} (\sum_{n=1}^{\infty} |y_n|^q)^{1/q} \qquad \ldots(1)$$

$$\le \infty \text{ by given definition.}$$

Above shows that partial sum g $\sum_{n=1}^{m} |x_n y_n|$ of $\sum_{n=1}^{\infty} |x_n y_n|$

are bounded and hence we conclude that $\sum_{n=1}^{\infty} |x_n y_n| < \infty$

Now we know that if $x_n < y_n$ therefore, $\lim_{n\to\infty} x_n \le \lim_{n\to\infty} y_n$, *i.e.*, $x_0 \le y_0$.

Hence, if in the result (1) we make $m \to \infty$ then by the above result we obtain

$$\sum_{n=1}^{\infty} |x_n y_n| \le (\sum_{n=1}^{\infty} |x_n|^p)^{1/p} (\sum_{n=1}^{\infty} |y_n|^q)^{1/q} = \|x\|_p \|y\|_q$$

Proof of Minkowski Inequality

In the proof of Minlowski ineqality, we had shown that

$$\sum_{i=1}^{n} |x_i + y_i|^p \le (\sum_{i=1}^{n} |x_i|^p)^{1/p} (\sum_{i=1}^{n} |x_i + y_i|^p)^{p-1/p} (\sum_{i=1}^{n} |y_i|^p)^{1/p} (\sum_{i=1}^{n} |x_i + y_i|^p)^{p-1/p}$$

$$\le (\sum_{i=1}^{\infty} |x_i|^p)^{1/p} (\sum_{i=1}^{\infty} |x_i + y_i|^p)^{p-1/p} (\sum_{i=1}^{\infty} |y_i|^p)^{1/p} (\sum_{i=1}^{\infty} |x_i + y_i|^p)^{p-1/p}$$

$$= \|x\|_p \|x+y\|_p^{p-1} + \|y\|_p \|x+y\|_p^{p-1} \quad \ldots(2)$$

Now if we allow n to tend to infinity then

L.H.S. $\sum_{i=1}^{n} |x_i + y_i|^p$ becomes $\sum_{i=1}^{\infty} |x_i + y_i|^p = \|x+y\|_p^p$

Hence from (2), we get

$$= \|x+y\|_p^p \le [\|x\|_p \|y\|_p][\|x+y\|_p]^{p-1}$$

In case $\|x + y\|_p = 0$ then both sides vanish and if $\|x + y\| \ne 0$ then we can divide, both sides by $\|x+y\|_p^{p-1}$ and we obtain

$$\|x+y\|_p \le \|x\|_p + \|y\|_p .$$

RECAPITULATIONS

- In a normed linear space every convergent sequence is Cauchy.
- A normed linear space which is Complete is called Banach space.
- A normed linear space is Banach space iff every absolutely summable series is summable.
- $\Sigma |x_i y_i| \le \|x\| \cdot \|y\|$
- $\|x + y\| \le \|x\| + \|y\|$
- $\Sigma |x_i y_i| \le \|x\|_q \cdot \|y\|_q$ where $\frac{1}{p} + \frac{1}{q} = 1$

SOLVED EXAMPLES

EXAMPLE 1. *Show that the real linear space* R *and the complex linear space* C *are Banach space under the norm.*

$$\|x\| = |x|, x \in \text{C } or \text{ R}$$

(MEERUT–2001, 02, 03, 06, 06BP, 07, 08, 09, RANCHI–2008, ROHILKHAND–2011 DELHI–2009, PATNA–2011)

SOLUTION. Firstly, we shall satisfy the axioms of norm

(i) $\|x\| = |x| \geq 0$ [By definition of modulus]

(ii) $\|x\| = 0 \Leftrightarrow |x| = 0 \Leftrightarrow x = 0$

(iii) $\|\alpha x\| = |\alpha x| = |\alpha|.|x| = |\alpha|.\|x\|$

(iv) $\|x + y\|^2 = |x + y|^2 = (x + y)\overline{(x + y)} = (x + y)(\bar{x} + \bar{y}) = x\bar{x} + y\bar{y} + x\bar{y} + y\bar{x}$

$\leq |x|^2 + |y|^2 + 2|x\bar{y}| = |x|^2 + |y|^2 + 2|x|.|y|$ $[\because |x\bar{y}| = |xy| = |x|.|y|]$

$= [|x| + |y|]^2$

Thus, $\|x + y\| \leq |x| + |y| = \|x\| + \|y\|$

Therefore, both C and R are normed linear spaces. Also, we know that every normed space is a metric space. Further, we know that these spaces are complete. Hence both C and R are Banach spaces.

EXAMPLE 2. *Show that the linear space* R^n *and* C^n *of all n-tuples* $x = (x_1, x_2, ..., x_n)$ *of real and complex numbers are Banach spaces under the norm*

$$\|x\| = \left(\Sigma |x_i|^2\right)^{1/2}$$

(The spaces defined on R^n *and* C^n*, has are known as Euclidean and unitary spaces respectively).* (MEERUT–1988, BUNDELKHAND–1987)

SOLUTION. Firstly, we shall prove the condition of normed linear spaces

(i) $\|x\|$ is clearly non-negative.

(ii) $\|x\| = 0 \Leftrightarrow \Sigma |x_i|^2 = 0 \Leftrightarrow x_i = 0$

$\Leftrightarrow x = (x_1, x_2, ..., x_n) = (0, 0, ..., 0) = 0$

(iii) $\|\alpha x\| = \left(\sum_{i=1}^{n} |\alpha x_i|^2\right)^{1/2} = \left(|\alpha|^2 \sum_{i=1}^{n} |x_i|^2\right)^{1/2} = |\alpha| \left(\sum_{i=1}^{n} |x_i|^2\right)^{1/2} = |\alpha|.\|x\|$

(iv) Let $x = (x_1, x_2, ..., x_n)$ and $y = (y_1, y_2, ..., y_n)$ be any two members of C^n or R^n.

Then $\|x + y\|^2 = \|(x_1, x_2, ..., x_n) + (y_1, y_2, ..., y_n)\|^2$

$= \|(x_1 + y_1, x_2 + y_2, ..., x_n + y_n)\|^2$

$= \sum_{i=1}^{n} |x_i + y_i|^2 = \sum_{i=1}^{n} |x_i + y_i||x_i + y_i| \leq \sum_{i=1}^{n} (|x_i + y_i|)(|x_i| + |y_i|)$

$= \sum_{i=1}^{n} |x_i + y_i|.|x_i| + \sum_{i=1}^{n} |x_i + y_i|.|y_i| = \|x + y\|.\|x\| + \|x + y\|.\|y\|$

$= \|x + y\|(\|x\| + \|y\|)$

$\Rightarrow \quad \|x + y\| \leq \|x\| + \|y\|$

Therefore, C^n or R^n are normed linear spaces.

Completeness

Let $<x_m> = <x_1, x_2, ..., x_m>$ be a Cauchy sequence in R^n and C^n where each x_m is a n-tuple and can be written as

$$x_m = (x_1^m, x_2^m, ..., x_n^m)$$

Also, x_i^m denotes the ith co-ordinate of the n-tuple x_m and belongs to C or R.

Now since, the given sequence is Cauchy, therefore by definition, for $\in > 0$, $\exists$ $n_0 > 0$ such that

$$\|x_m - x_p\| < \in, \forall\ m, p \geq n_0$$

$$\Rightarrow \quad \left(\sum_{i=1}^{n}\left|x_i^m - x_i^p\right|^2\right)^{1/2} < \epsilon \quad \Rightarrow \quad \left(\sum_{i=1}^{n}\left|x_i^m - x_i^p\right|^2\right) < \epsilon^2 \quad \ldots(1)$$

$\Rightarrow$ Each term in the above sigma is less than ϵ^2.

$$\Rightarrow \quad \left|x_i^m - x_i^p\right|^2 < \epsilon^2 \text{ for each } i = 1, 2, 3, \ldots, n.$$

$$\Rightarrow \quad \left|x_i^m - x_i^p\right| < \epsilon$$

Therefore, each of the following sequence

$$< x_1^j >, < x_2^j >, \ldots, < x_n^j >$$

is a Cauchy sequence of C or R. Also since C and R are complete therefore, each of these Cauchy sequence must converges to a point in C or R.

Thus, $\lim_{j\to\infty} x_1^j = z_1, \lim_{j\to\infty} x_2^j = z_2, \ldots, \lim_{j\to\infty} x_n^j = z_n$ ...(2)

where each $z_1, z_2, \ldots, z_n$ belongs to C and R, so that

$$z = (z_1, z_2, \ldots, z_n) \in C^n \text{ or } R^n$$

Now, it remains to prove that the Cauchy sequence $<x_n>$ in C^n or R^n converges to $z \in C^n$ or R^n.

Using (1), we have

$$\sum_{i=1}^{n}\left|x_i^m - x_i^p\right|^2 < \epsilon^2 \quad \ldots(3)$$

Taking limit $p \to \infty$ and using (2), we get

$$\lim_{p\to\infty} x_i^p = z_i$$

Also, from (3) we have

$$\sum_{i=1}^{n}\left|x_i^m - x_i^p\right|^2 < \epsilon^2 \quad \Rightarrow \quad \|x_m - z\|^2 < \epsilon^2$$

$$\Rightarrow \quad \|x_m - z\| < \epsilon$$

$\Rightarrow$ The Cauchy sequence $<x_m>$ in C^n or R^n congerges to $z \in C^n$ or R^n.

Thus, C^n or R^n are complete spaces and hence they are Banach spaces.

EXAMPLE 3. *(l_p^n space) Let p be a real number such that $1 \le p < \infty$ and l_p^n denote the linear space of all n-tuples of scalars with the norm of a vector space $x = (x_1, x_2, \ldots, x_n)$ defined by*

$$\|x\|_p = \left(\Sigma|x_i|^p\right)^{1/p}$$

Show that l_p^n is a Banach space. (KANPUR–2007, MEERUT–2002, 07BP)

SOLUTION. Firstly, we shall prove that l_p^n is a normed linear space.

(i) $\|x\|_p$ is clearly non-negative (By definition of norm)

(ii) $\|x\|_p = 0 \quad \Leftrightarrow \quad \left(\Sigma(x_i)^p\right)^{1/p} = 0 \quad \Leftrightarrow \quad \Sigma|x_i|^p = 0$

$\Leftrightarrow \quad |x_i| = 0$, for each i

$\Leftrightarrow \quad x = (x_1, x_2, \ldots, x_n) = (0, 0, \ldots, 0) = 0$

(iii) $\|\alpha x\|_p = \left(\Sigma|\alpha x_i|^p\right)^{1/p} = \left(\Sigma|\alpha|^p|x_i|^p\right)^{1/p} = \left(|\alpha|^p\,\Sigma|x_i|^p\right)^{1/p}$

$= |\alpha|^p\left(\Sigma|x_i|^p\right)^{1/p} = |\alpha|.\|x\|_p$

(iv) $\|x+y\|_p \leq \|x\|_p + \|y\|_p$ [By Minkowski´s inequlity]

Hence, the l_p^n space is normed linear space.

Also, we know that every normed linear space is a metric space such that

$$d(x, y) = \|x - y\|_p, \forall\, x, y \in l_p^n.$$

Completeness

Let $<x_m> = (x_1, x_2, ..., x_m)$ be a Cauchy sequence in l_p^n where each x_m is a n-tuple of scalars which belongs to the field of real or complex. Thus, we can write

$$x_m = (x_1^m, x_2^m, ..., x_m^m)$$

i.e., x_i^m denote the ith coordinate of n-tuple x_m and belongs to C or R.

Since $<x_m>$ is Cauchy, By definition, for $\in > 0 \exists n_0$ such that

$$\|x_m - x_i\|_p < \in, \forall\, m, l \geq n_0$$

$\Rightarrow$ $$\left[\|x_m - x_i\|_p\right]^{1/p} < \in^p$$

$\Rightarrow$ $$\sum_{i=1}^{n}\left|x_i^m - x_i^l\right|^p < \in^p$$ [By definition of norm]

$\Rightarrow$ Each term in the above sigma is less than $\in^p$.

$\Rightarrow$ $$\left|x_i^m - x_i^l\right| < \in^p \text{ for } i = 1, 2, 3, ..., n.$$

$\Rightarrow$ $$\left|x_i^m - x_i^l\right| < \in \text{ for } i = 1, 2, 3, ..., n. \quad ...(1)$$

Thus each of the following sequence $< x_1^j >, < x_2^j >, ..., < x_n^j >$ is a Cauchy sequence in C or R but C and R both are complete and therefore each Cauchy sequence must converges to a point in C or R.

Therefore, $\lim\limits_{j\to\infty} x_1^i = z_1, \lim\limits_{j\to\infty} x_2^i = z_2, ..., \lim\limits_{j\to\infty} x_n^i = z_n$ ...(2)

where $z = (z_1, z_2, ..., z_n) \in l_p^n$

Finally, we shall prove that Cauchy sequence $<x_m>$ in l_p^n converges to $z \in l_p^n$.

Using (1), we have

$$\sum_{i=1}^{n}\left|x_i^m - x_i^l\right|^p < \in^p$$

Now, taking limit $n \to \infty$ and using (2), we get

$$\sum_{i=1}^{n}\left|x_i^m - z_i\right|^p < \in^p \Rightarrow \left\|x_m - z\right\|_p^p < \in^p \Rightarrow \left\|x_m - z\right\|_p < \in$$

$\Rightarrow$ The Cauchy sequence $<x_m>$ in l_p^n converges to $z \in l_p^n$.

$\Rightarrow$ l_p^n is complete.

Hence, l_p^n is a Banach space.

EXAMPLE 4. *(Space l_2). Let l_2 denote the linear space of all sequences. Show that the set of all real sequences $s = \langle s_k \rangle$ such that $\sum_{k=1}^{\infty} s_k^2 < \infty$ with norm defined as*

$$\|s\|^2 = \left(\sum_{k=1}^{\infty} s_k^2\right)^{1/2}, \ \forall\, s \in l_2 \text{ is a Banach space.}$$

SOLUTION. The given set is closed for addition and scalar multiplication and is a linear space for these two compositions.

Now to show, it is a normed space.

(i) $\|s\| = \left(\sum_{k=1}^{\infty} s_k^2\right) \geq 0$

(ii) $\|s\| = 0 \Leftrightarrow \sum_{k=1}^{\infty} s_k^2 = 0 \Leftrightarrow s_k = 0$ for all k

$\Leftrightarrow s = \{0, 0, ..., 0\} = 0$

(iii) $\|\alpha s\| = \left(\sum_{k=1}^{\infty} (\alpha s_k)^2\right)^{1/2} = |\alpha|\left(\Sigma s_k^2\right)^{1/2} = |\alpha|.\|s\|$

(iv) $\|s+t\| = \left\{\sum_{k=1}^{\infty} (s_k + t_k)^2\right\}^{1/2} \leq \left(\Sigma s_k^2\right)^{1/2} + \left(\Sigma t_k^2\right)^{1/2}$

$= \|s\| + \|t\|$ [By Minkowski´s inequality]

Therefore, l_2-space is a normed linear space.

Completeness

Let $\langle s^{(n)} \rangle$ be any Cauchy sequence in l_2.

where $s^{(1)} = \left\{s_k^{(1)}\right\}$

$s^{(2)} = \left\{s_k^{(2)}\right\} ..., s^{(n)} = \left\{s_k^{(n)}\right\}$

Therefore, for each $\in > 0$, $\exists$ a positive integer n_0 such that

$\|(s^m - s^n)\| < \in, \forall\, m, n \geq n_0$

$\Rightarrow \left(\sum_{k=1}^{\infty} \left(s_k^m - s_k^n\right)^2\right)^{1/2} < \in$

$\Rightarrow$ Each of the following is less than $\in$, *i.e.*,

$\left|s_1^m - s_1^n\right| < \in, \left|s_2^m - s_2^n\right| < \in ...$

$\Rightarrow$ the sequence $\langle s_i^n \rangle$ for all values of i is a Cauchy sequence of real numbers each of which will converges to same real number because, the real line is always complete.

Thus, $\lim_{n\to\infty} s_1^n = x_1, \lim_{n\to\infty} s_2^n = x_2, ..., \lim_{n\to\infty} s_k^n = x_k$ ...(1)

Therefore, we get a sequence $x = \langle x_n \rangle$

Now, to show that $\langle s^{(n)} \rangle$ converges to $x = \langle x_n \rangle$

For a fixed integer p, we have

$$\left[\sum_{k=1}^{p}\left(s_k^m - s_k^n\right)^2\right]^{1/2} < \in \text{ for } m, n > n_0$$

Keeping n fixed and let $m \to \infty$, we have

$$\sum_{k=1}^{p}\left(x_k - s_k^n\right)^2 < \in \forall n \geq n_0$$

This relation hold for each p, thus making $p \to \infty$, we get

$$\left[\sum_{k=1}^{\infty}\left(x_k - s_k^n\right)^2\right]^{1/2} < \in \text{ for } n > n_0$$

$\Rightarrow$ $$d(x, s^n) < \in, \forall\, n \geq n_0$$

$\Rightarrow$ $$\|x - s^n\| < \in$$

$\Rightarrow$ $$s^n \to x$$

Now, it remains to prove that $x \in l_2$. We have

$$x_k^2 = \left[x_k - s_k^{n_0} + s_k^{n_0}\right]^2 = (x_k - s_k^{n_0})^2 + \left(s_k^{n_0}\right)^2 + 2\left(x_k - s_k^{n_0}\right)\left(s_k^{n_0}\right)$$

$$\leq 2\left(x_k - s_k^{n_0}\right) + 2\left(s_k^{n_0}\right)^2 \qquad [\because 2ab \leq a^2 + b^2]$$

Thus, $$\sum_{k=1}^{\infty} x_k^2 \leq 2\sum_{k=1}^{\infty}\left(x_k - s_k^{n_0}\right)^2 + 2\sum_{k=1}^{\infty}\left(s_k^{n_0}\right)^2 < 2\in^2 + 2\sum_{k=1}^{\infty}\left(s_k^{n_0}\right)^2$$

$\Rightarrow$ $$s_k^{n_0} = \left\{s_k^{n_0}\right\} \in l_2 \quad \Rightarrow \sum_{k=1}^{\infty}\left(s_k^{n_0}\right)^2 < \infty \quad \Rightarrow \quad x \in l_2$$

Hence, we conclude that every Cauchy sequence in l_2 converges to a point in l_2, this l_2 is complete which therefore is a Banach space.

l_p SPACE

Definition. *The set of sequence $s = \langle s_n \rangle$ such that $\sum_{n=1}^{\infty} |s_n|^p < \infty$ is called l_p space.*

REMARK

- We can easily verify that l_p is closed for addition and scalar multiplication.

EXAMPLE 5. *Show that the l_p space defined above, is a Banach space under the norm*

$$\|s\|_p = \left(\sum_{n=1}^{\infty} |s_n|^p\right)^{1/p}, \forall s = \langle s_n \rangle \in l_p$$

(GARHWAL–2003, 04, KANPUR–1996, 98, 2002, 07, 08, AVADH–2009)

SOLUTION. We can easily verify that l_p space is a normed linear space.

Now, we shall prove the completeness of l_p space.

Let $\langle s_n \rangle$ be any Cauchy sequence in l_p such that

$$s^{(1)} = \langle s_k^{(1)} \rangle, s^{(2)} = \langle s_k^{(2)} \rangle \ldots, s^{(n)} = \langle s_k^{(n)} \rangle.$$

Therefore for $\in > 0\ \exists$ a positive integer n_0 such that for all $m, n \geq n_0$

$$\|s^m - s^n\| < \in$$

$$\Rightarrow \left(\sum_{k=1}^{\infty}\left|s_k^m - s_k^n\right|^p\right)^{1/p} < \in$$

Above relation implies that each of the following is less than $\in$.

$$\left|s_1^m - s_1^n\right| < \in, \left|s_2^m - s_2^n\right| < \in \ldots$$

Therefore, we can say that sequence $< s_i^n >, \forall i$ is a Cauchy sequence on the real line each of which will converge to some real numbers as the real line is complete.

Thus, $$s_1^n \to x_1, s_2^n \to x_2, \ldots \qquad \ldots(1)$$

Therefore, we get a sequence $x = <x_n>$.

We want to show that $<s^n>$ converges to $x = <x_n>$ ...(2)

For a fixed integer i, we know that

$$\left[\sum_{k=1}^{i}\left|s_k^m - s_k^n\right|^p\right]^{1/p} < \in, \forall m, n \geq n_0$$

Keeping n fixed and let $m \to \infty$, we have,

$$\left(\sum_{k=1}^{i}\left|x_k - s_k^{(n)}\right|^p\right)^{1/p} < \in, n \geq n_0 \qquad \ldots(3)$$

Making $i \to \infty$, we get

$$\left(\sum_{k=1}^{\infty}\left|x_k - s_k^{n/p}\right|^p\right)^{1/p} < \in, n \geq n_0$$

$$\Rightarrow \qquad \|x - s^{(n)}\|_p < \in \text{ for } n \geq n_0 \Rightarrow s^{(n)} \to x$$

It remains to prove that $x \in l_p$

$$|x_k|^p = \left|x_k - s_k^{n_0} + s_k^{n_0}\right|^p \leq 2^p[\left|x_k - s_k^{n_0}\right|^p + \left|s_k^{n_0}\right|^p]$$

Therefore, $$\sum_{k=1}^{\infty}|x_k|^p \leq 2^p\left(\sum_{k=1}^{\infty}\left|x_k - s_k^{n_0}\right|^p + \sum_{k=1}^{\infty}\left|s_k^{n_0}\right|^p\right)$$

$$< 2^p \in^p + 2^p \sum_{k=1}^{\infty}\left|s_k^{n_0}\right|^p \quad \text{[Using (2)]} \qquad \ldots(3)$$

Also, $$s^{n_0} = < s_k^{n_0} > \in l_p \quad \Rightarrow \sum_{k=1}^{\infty}\left|s_k^{n_0}\right|^p < \infty$$

Thus from (3) $x = <x_k> \in l_p$. Hence, l_p is a Banach space.

EXAMPLE 6. *(Space l_∞^p) Consider the linear space of all n-tuples $x = (x_1, x_2, \ldots, x_n)$ of scalars and define the norm by*

$$\|x\|_\infty = max\{|x_1|, |x_2|, \ldots, |x_n|\}$$

This space is denoted by the symbol l_∞^n. Then, show that $(l_\infty^n, \|x\|_\infty)$ is a Banach space. (AGRA–2003, KANPUR–2011, CALCUTTA–2010, KUMAUN–2010)

SOLUTION. Firstly we shall prove that l_∞^p is a normed linear space.

(i) Since each $\|x_n\| \geq 0 \Rightarrow \quad \|x\|_\infty \geq 0$

(ii) $\|x\|_\infty = 0 \quad \Leftrightarrow \quad \max\{|x_1|, |x_2|, \ldots, |x_n|\}$

$\Leftrightarrow \quad |x_1| = 0, |x_2| = 0, \ldots, |x_n| = 0$

$\Leftrightarrow x_1 = 0, x_2, \ldots x_n = 0$

$\Leftrightarrow \quad (x_1, x_2, \ldots, x_n) = 0 \Leftrightarrow x = 0$

(iii) Let $x = (x_1, x_2, ..., x_n), y = (y_1, y_2, ..., y_n)$

Then $\|x + y\|_\infty = \max\{|x_1 + y_1|, |x_2 + y_2|, ..., |x_n + y_n|\}$

$\leq \max\{|x_1| + |y_1|, |x_2| + |y_2|, ..., |x_n| + |y_n|\}$

$\leq \max\{|x_1|, |x_2|, ..., |x_n|\} + \max\{|y_1|, |y_2|, ..., |y_n|\}$

$= \|x\|_\infty + \|y\|_\infty.$

Hence, $\|x + y\|_\infty \leq \|x\|_\infty + \|y\|_\infty$.

(iv) Let α be any scalar, then

$$\|\alpha x\|_\infty = \max\{|\alpha x_1|, |\alpha x_2|, ..., |\alpha x_n|\}$$
$$= \max\{|\alpha||x_1|, |\alpha||x_2|, ..., |\alpha||x_n|\}$$
$$= |\alpha| \max\{|x_1|, |x_2|, ..., |x_n|\} = |a|. \|x\|_\infty.$$

Thus, l_∞^n is normed linear space.

Completeness

Let $<x_n>$ be any Cauchy sequence in l_∞^n.

Let $x_m = \left(x_1^m, x_2^m, ..., x_n^m\right)$ be n-tuple of scalars.

Also, let $\in > 0$, then by definition of Cauchy sequence, there exists a positive integer m_0 such that

$$\|x_m - x_l\|_\infty < \in \forall\, l, m \geq m_0$$

$$\Rightarrow \quad \max\left\{\left|x_1^m - x_1^l\right|, \left|x_2^m - x_2^l\right|, ..., \left|x_n^m - x_n^l\right|\right\} < \in$$

$$\Rightarrow \quad \left|x_i^m - x_i^l\right| < \in, i \in \mathrm{N}$$

which implies that for fixed i, $< x_i^m >$ is a Cauchy sequence in C or R. Since, C and R both are complete, therefore, it must converge to some point $z_i \in$ C or R. Thus, we conclude that $<x_n>$ converges to $z = (z_1, z_2, ..., z_n)$.

Hence, l_∞^n is a Banach space.

EXAMPLE 7. *(Space l_∞). The space of all bounded sequences $x = <x_1, x_2, ..., x_n>$ of scalars is termed as l_∞ with norm of x defined as*

$$\|x\| = \sup. |x_n|$$

Show that l_∞ is a Banach space. (MEERUT–2002)

SOLUTION. Let $<x_m>$ be a Cauchy sequence in l_∞.

Then given $\in > 0\ \exists$ a positive integer $N = N(\in)$ such that

$$\|x_m - x_n\| < \in, \forall\, n, m \geq N$$

$$\Rightarrow \quad \left|x_{m_k} - x_{n_k}\right| < \in \forall n, m \geq N \text{ and for all } k \qquad ...(1)$$

Therefore, for each K, the sequence $< x_{n_k} >$ is a Cauchy sequence in C, but C is complete. Hence, there is a sequence $x = <x_k>$ such that $x_{n_k} \to x_k$ for all K as $n \to \infty$.

Now, we have to show that $x \in l_\infty$. Since, every Cauchy sequence is bounded therefore there exist $k > 0$ such that

$$\|x_n\| \leq k\ \forall n$$

$$\Rightarrow \quad \left\|x_{n_K}\right\| \leq k\ \forall n$$

$\Rightarrow \quad \lim_{n\to\infty} \left|x_{n_k}\right| = \left|x_k\right| \le k \ \forall n$

$\Rightarrow \quad x \in l_\infty$

Now, letting $m \to \infty$ in (1), we get

$$\left|x_k - x_{n_k}\right| \le \in \forall n \ge N \text{ and } \forall k$$

Hence,

$$\|x_n - x\| = \sup_k \left|x_k - x_{n_k}\right| \le \in \forall n \ge N$$

$\Rightarrow$ the arbitrary Cauchy sequence $<x_n>$ in l_∞ converges to an element $x \in l_\infty$.

$\Rightarrow$ l_∞ is complete.

EXAMPLE 8. *(The space C(X)). Let C(X) denote the linear space of all bounded continuous scalar valued function defined on a topological space X. Show that C(X) is a Banach space under the norm.*

$$\|f\| = \sup\{|f(x)| : x \in X\}, f \in C(X)$$

(MEERUT–2000, 01, 06, AMRITSAR–1982)

SOLUTION. We can easily verify that $C(X)$ is a linear space. Further, to show that $C(X)$ is a normed linear space.

(i) Since, $|f(x)| > 0, \forall x \in X$

Therefore, $\|f\| \ge 0$.

(ii) Now, $\|f\| = 0 \Leftrightarrow \sup\{|f(x)| : x \in X\} = 0$

$\Leftrightarrow |f(x)| = 0 \ \forall x \in X$

$\Leftrightarrow f(x) = 0 \ \forall x \in X \quad \Leftrightarrow f = 0$

(iii) $\|f + g\| = \sup\{|(f+g)(x)| : x \in X\} = \sup\{|f(x) + g(x)| : x \in X\}$

$\le \sup\{|f(x)| + |g(x)| : x \in X\}$

$\le \sup\{|f(x)| : x \in X\} + \sup\{|g(x)| : x \in X\} = \|f\| + \|g\|$

(iv) $\|\alpha f\| = \sup\{|(\alpha f)(x)| : x \in X\} = \sup\{|\alpha f(x)| : x \in X\}$

$= \sup\{|\alpha| . |f(x)| : x \in X\} = |\alpha| \sup\{|f(x)| : x \in X\} = |\alpha| . \|f\|$

$\Rightarrow$ $C(X)$ is a normed linear space.

Completeness

Let $<f_n>$ be any Cauchy sequence in $C(X)$, then for a given $\in > 0$, $\exists$ a positive integer m_0 such that

$$\|f_m - f_n\| < \in, \forall m, n \ge m_0$$

$\Rightarrow \sup\{|(f_m - f_n)(x)| : x \in X\} < \in$

$\Rightarrow \sup\{|(f_m(x) - f_n(x)| : x \in X\} < \in$

$\Rightarrow |f_m(x) - f_n(x)| < \in, \forall x \in X$

which is the required condition of uniform convergence of Cauchy sequence of bounded continuous scalar valued function. Thus, the sequence $<f_n>$ must converge to a bounded continuous function f on X.

$\Rightarrow$ $C(X)$ is complete. Hence, $C(X)$ is Banach space.

EXAMPLE 9. *The linear space $C(0, 1)$ of real valued continuous function on $[0, 1]$ is a normed space but not a Banach space with the norm of a function $f \in C[0, 1]$ defined as*

$$\|f\| = \int_0^1 |f(t)|\,dt$$

(GARHWAL–2002, KANPUR–1986)

SOLUTION. The first two properties of norm can be easily verify. Further

$$\|\alpha f\| = \int_0^1 |(\alpha f)(t)|\,dt = \int_0^1 |\alpha f(t)|\,dt = \int_0^1 |\alpha| . |f(t)|\,dt = |\alpha| \int_0^1 |f(t)| = |\alpha| \|f\|$$

Also, $\|f+g\| = \int_0^1 |(f+g)t|\,dt = \int_0^1 |f(t)+g(t)|\,dt$

$$\le \int_0^1 |f(t)|\,dt + \int_0^1 |g(t)|\,dt = \|f\| + \|g\|$$

Therefore, we conclude that $C(0, 1)$ with the given definition of norm is a normed linear space.

Completeness

Let $<f_n>$ be the sequence in $C(0, 1)$, defined as follows.

$$f_n(t) = \begin{cases} 1 & ; \quad 0 \le t \le \dfrac{1}{2} \\ -nt + \dfrac{n}{2} + 1 & ; \quad \dfrac{1}{2} \le t \le \dfrac{1}{2} + \dfrac{1}{n} \\ 0 & ; \quad \dfrac{1}{2} + \dfrac{1}{n} \le t \le 1 \end{cases}$$

where $n \ge 3$.

We can easily verify that each f_n is continuous in (0, 1). Thus, we can find a Cauchy sequence $<f_n>$ in $C(0, 1)$.

Now, if $m, n \ge 3$, then we have

$$\|f_m - f_n\| = \int_0^1 |(f_m - f_n)(t)|\,dt = \int_0^1 |f_m(t) - f_n(t)|\,dt$$

$$= \int_0^{1/2} |f_m(t) - f_n(t)|\,dt + \int_{1/2}^1 |f_m(t) - f_n(t)|\,dt$$

$$= \int_0^{1/2} |1-1|\,dt + \int_{1/2}^1 |f_m(t) - f_n(t)|\,dt$$

$$= \int_0^{1/2} |1-1|\,dt + \int_{1/2}^1 |f_m(t) - f_n(t)|\,dt$$

$$\le 0 + \int_{1/2}^1 |f_m(t)|\,dt + \int_{1/2}^1 |f_n(t)|\,dt \qquad \ldots(1)$$

Now, $\int_{1/2}^1 |f_n(t)|\,dt = \int_{1/2}^{1/2+1/n} \left|-nt + \dfrac{n}{2} + 1\right| dt + \int_{1/2+1/n}^1 0.dt$

$$= \int_{1/2}^{1/2+1/n} \left(-nt + \frac{n}{2} + 1\right) dt \qquad \left(\because -nt + \frac{n}{2} + 1 \ge 0\right)$$

Thus, $\int_{1/2}^1 |f_n(t)|\,dt = \left[-n\dfrac{t^2}{2} + \left(\dfrac{n}{2} + 1\right) dt\right]_{1/2}^{1/2+1/n}$

$$= -\frac{n}{2}\left[\left(\frac{1}{2} + \frac{1}{n}\right)^2 - \frac{1}{4}\right]^{1/2} + \left(\frac{n}{2} + 1\right)\left[\frac{1}{2} + \frac{1}{n} - \frac{1}{2}\right]$$

$$= -\frac{n}{2}\left(\frac{1}{n} + \frac{1}{n^2}\right) + \frac{1}{2} + \frac{1}{n} = -\frac{1}{2} - \frac{1}{2n} + \frac{1}{2} + \frac{1}{n} = \frac{1}{2n}$$

Similarly, we can find

$$\int_{1/2}^1 |f_m(t)|\,dt = \frac{1}{2m}$$

Therefore, from (1), we get

$$\|f_m - f_n\| \le \frac{1}{2m} + \frac{1}{2n} \to 0 \text{ as } m, n \to \infty.$$

$\Rightarrow$ $<f_n>$ is a Cauchy sequence in $C(0, 1)$, but this sequence is not convergent. For example, let $g \in C(0, 1)$ such that $\lim f_n = g$, where

$$g(t) = \begin{cases} 1 \ ; & 0 \le t \le 1/2 \\ 0 \ ; & 1/2 \le t \le 1 \end{cases}$$

Clearly, g is discontinuous at $t = \frac{1}{2}$. Thus, $g \notin C(0, 1)$.

$\Rightarrow$ $<f_n>$ is not a convergent sequence.

Hence, $C(0, 1)$ is not a Banach space.

EXAMPLE 10. *Let B be a Banach space which is the direct sum of two its sub-space M and N so that z in B is uniquely expressible in the form $z = x + y$ with $x \in M$, $y \in N$, then we define $\|z\|' = \|x\| + \|y\|$. Show that it is actually a norm. If B' denote the linear space equipped with this norm, show that $B'$is a Banach space if M and N are closed in B.*

(MEERUT–2005BP, KANPUR–2000, 06)

SOLUTION. **B' is a normed linear space**

(i) Consider $\|z\|' = \|x\| + \|y\| \ge 0$ [$\because$ both $\|x\|$ and $\|y\|$ are non-negative]

(ii) $\|z\|' = 0 \Leftrightarrow \|x\| + \|y\| = 0 \Leftrightarrow \|x\| = 0, \|y\| = 0$

$\Leftrightarrow x = 0, y = 0 \Leftrightarrow x + y = 0$

$\Leftrightarrow z = 0$

(iii) $\|\alpha z\|' = \|\alpha(x + y)\|' = \|\alpha x + \alpha y\|' = \|\alpha x\| + \|\alpha y\|$

$= |\alpha|\{\|x\| + \|y\|\} = |\alpha|.\|z\|'$

(iv) $\|z_1 + z_2\|' = \|x_1 + y_1 + x_2 + y_2\|' = \|(x_1 + x_2) + (y_1 + y_2)\|'$

$= \|x_1 + x_2\| + \|y_1 + y_2\| \le \{\|x_1\| + \|x_2\|\} + \{\|y_1\| + \|y_2\|\}$

$= \{\|x_1\| + \|y_1\|\} + \{\|x_2\| + \|y_2\|\} = \|z_1\|' + \|z_2\|'$

Thus, B' is a normed linear space.

Completeness

Let M and N are closed linear sub-spaces. To show B' is complete.

Let $<z_n>$ be a Cauchy sequence in B'. Then by definition of Cauchy sequence, for given $\in > 0$ there exists a positive integer n_0 such that for $m, n \ge n_0$ we have

$$\|z_m - z_n\|' < \in$$

$\Rightarrow$ $\|(x_m + y_m)\| - \|(x_n + y_n)\|' < \in$

$\Rightarrow$ $\|(x_m - x_n)\| + \|(y_m - y_n)\|' < \in$

$\Rightarrow$ $\|(x_m - x_n)\| + \|(y_m - y_n)\| < \in$

i.e., $\|(x_m - x_n) < \in$ and $\|(y_m - y_n)\| < \in$

$\Rightarrow$ $<x_n>$ and $<y_n>$ both are Cauchy sequence in M and N respectively. But since M and N are closed linear subspace of the Banach space B, therefore, both M and N are complete.

Thus, there exist points $x \in M$ and $y \in N$ such that $x_n \to x$ and $y_n \to y$.

Then by definition, there exist positive integers n_1 and n_2 such that

$\|x_n - x\| < \in/2$ for $n \ge n_1$

and $\|y_n - y\| < \in/2$ for $n \ge n_2$

If $z = x + y$ and $n_0 = \max\{n_1, n_2\}$. Then, for $n \ge n_0$

$$\|z_n - z\|' = \|(x_n + y_n) - (x + y)\|'$$
$$= \|(x_n - x) + (y_n - y)\|' = \|(x_n - x) + (y_n - y)\|$$
$$= \|(x_n - x) + (y_n - y)\| < \frac{\in}{2} + \frac{\in}{2} = \in$$

$\Rightarrow$ $z_n \to z$

$\Rightarrow$ B' is complete

Hence, B' is a Banach space.

EXAMPLE 11. *Let N be normed linear space and N^* be a conjugate space of N. Then show that N^* is a Banach space.* (KANPUR–2007)

SOLUTION. (i) Let N be a normed linear space over real numbers and N^* be a conjugate space of N. Then to show that $\lim_{n\to\infty} f_n \in B(N,R)$.

Consider the linear transformation

$$f : N \to R = B(N, R)$$

If $f, g \in N^*$ and α, β be any scalars of R.

Then $N^*R = B(N, R)$ is a linear space such that

$$(f + g)(x) = f(x) + g(x)$$
$$(\alpha f)(x) = \alpha f(x) \ \forall\ x \in N$$

(ii) Now to show that N^* is Banach space of a normed linear space N, in which norm is defined by

$$\|f\| = \sup\{|f(x)| : \|x\| \le 1\}$$
$$(N_1)\cdot \|f\| \ge 0 \text{ as } |f(x)| \ge 0$$
$$(N_2)\cdot \|f\| = 0 \text{ if and only if } f = 0$$

Then $\|f\| = 0$

$\Leftrightarrow$ $\sup\{|f(x)| : |x| \le 1\} = 0$

$\Leftrightarrow$ $|f(x)| = 0\ \forall\ x.$

$\Leftrightarrow$ $f(x) = 0\ \forall\ x$

$\Leftrightarrow$ $f = 0$

$$(N_3)\cdot \|\alpha \cdot f\| = |\alpha| \cdot \|f\|$$

Now $\|f\| = 0$

$\Leftrightarrow$ $\sup\{|(\alpha f)(x)| : \|x\| < 1\}$

$$= \sup\{|\alpha| \cdot |f(x)| : \|x\| \le 1\}$$
$$= |\alpha| \cdot \sup\{|f(x)| : \|x\| \le 1\} = |\alpha| \cdot \|f\|$$
$$N_4 \cdot \|f + g\| \le \|f\| + \|g\|$$

(iii) Now we have to prove that $B(N, R)$ is complete, where R is complete.

Now let $<f_n>$ be a Cauchy sequence in a Banach space $B(N, R)$.

Then $f_n : N \to R$ is continuous.

From definition of Cauchy sequence we know that

$$\|f_n - f_m\| \to 0 \text{ as } m, n \to \infty\ \forall\ x \in N, f_n(x) \in R\ \forall\ n$$

$\Rightarrow$ $\|f_n(x) - f_m(x)\| = \|f_n - f_m(x)\| \le \|f_n - f_m\|\,|x| \to 0$ as $m, n \to \infty$

Therefore $<f_n(x)>$ is a Cauchy sequence in R which is complete. Hence there exist $f(x) \in$ R such that

$$\lim_{n\to\infty} f_n(x) = f(x) \qquad \ldots(1)$$

Hence $\quad f : N \to R$

Now, show that f is linear and continuous.

Let x, y be any scalars of N and α, β be any scalars of R. Then

$$f(\alpha x + \beta y) = \lim_{n\to\infty} f_n(\alpha x + \beta y) = \lim_{n\to\infty} \alpha f_n(x) + \lim_{n\to\infty} \beta f_n(y)$$

(where f_n is linear)

$$= \alpha f(x) + \beta f(y)$$

$\Rightarrow \quad f$ is linear.

Since f_n is continuous and we know that every continuous function is bounded so that $\quad \exists\, k > 0$ such that

$$|f_n(x)| \le k\,\|x\|\ \forall\, x \in N \qquad \ldots(2)$$

$$\le \lim_{n\to\infty} k\|x\| = k\|x\|, \qquad \text{[From (2)]}$$

$\Rightarrow \quad f$ is bounded.

$\Rightarrow \quad f$ is continuous

$\Rightarrow \quad f \in B(N, R)$

$\Rightarrow \quad \lim_{n\to\infty} f_n \in B(N, R)$

Hence, we can say that N^* is complete.

$\Rightarrow \quad N^*$ is Banach space.

4.6 SOME GENERAL THEOREMS

THEOREM 1. ***A non-zero normed linear space N is a Banach space if and only if $S = \{x : \|x\| = 1\}$ is complete.***

(MEERUT–2005, 06, 09, MADRAS–2006, RAJASTHAN–2011, GARHWAL–2003, KANPUR–2005)

PROOF. Let us first suppose that N be a Banach space and $S = \{x : \|x\| = 1\}$ be the given set.

To show S is complete.

Since, N is Banach, therefore, N is complete.

Let $<x_n>$ be any Cauchy sequence in S, so that

$$\|x_n\| = 1,\ \forall n$$

To show $<x_n>$ converges to a point say $x \in S$.

Since $S \subset N$, therefore, the Cauchy sequence $<x_n>$ in S is also a Cauchy sequence in N, which is complete.

$\therefore \quad x_n \to x$ for some $x \in N$.

$\Rightarrow \quad \|x_n\| \to \|x\| \quad \Rightarrow \quad \|x\| = \lim_{n\to\infty} \|x_n\|$

$\Rightarrow \quad \|x\| = 1 \qquad [\because \|x\| = 1, \forall n]$

$\Rightarrow \quad x \in S \qquad$ [By definition of S]

Thus, S is complete.

Conversely, let us suppose that S is complete. To show N is a Banach space.

It is given that N be a normed space.

It remains to prove that N is complete.

Let $<y_n>$ be a Cauchy sequence in N, then for given $\in > 0$, there exists a positive integer n_0 such that $m, n \ge n_0$, we have

$$\|y_m - y_n\| < \in$$

Now, for each n, define $x_n = \dfrac{y_n}{\|y_n\|}$

$$\Rightarrow \qquad \|x_n\| = \left\|\frac{y_n}{\|y_n\|}\right\| = \frac{1}{\|y_n\|}.\|y_n\| = 1$$

Also, $\|\alpha x\| = |\alpha|.\|x\| = \alpha.\|x\|$, if α is positive.

Thus, $x_n \in S$.

Now, we want to show that $<x_n>$ is a Cauchy sequence.

Consider, $\|(x_m - x_n)\| = \left\|\dfrac{y_m}{\|y_m\|} - \dfrac{y_n}{\|y_n\|}\right\|$

$$= \left\|\left(\frac{y_m}{\|y_m\|} - \frac{y_n}{\|y_m\|}\right) + \left(\frac{y_n}{\|y_m\|} - \frac{y_n}{\|y_n\|}\right)\right\|$$

$$\Rightarrow \qquad \|x_m - x_n\| \le \left\|\frac{1}{\|y_m\|}(y_m - y_n)\right\| + \left\|\left(\frac{1}{\|y_m\|} - \frac{1}{\|y_n\|}\right)y_n\right\|$$

$$= \frac{1}{\|y_m\|}.\|y_m - y_n\| + \left|\frac{1}{\|y_m\|} - \frac{1}{\|y_n\|}\right|\|y_n\|$$

$$= \frac{1}{\|y_m\|}.\|y_m - y_n\| + \frac{\left|\|y_n\| - \|y_m\|\right|}{\|y_m\|}$$

$$\le \frac{1}{\|y_m\|}.\|y_m - y_n\| + \frac{1}{\|y_m\|}\|y_n - y_m\| = \frac{2\|y_m - y_n\|}{\|y_m\|} < \in$$

$\Rightarrow \quad <x_n>$ is a Cauchy sequence in S and S is complete.

$\Rightarrow \quad x_n \to x$ for some $x \in S$.

or $\qquad \dfrac{y_n}{\|y_n\|} \to x \qquad$ or $\qquad y_n \to \|y_n\|x$

Further $\qquad \left|\|y_m\| - \|y_n\|\right| < \|y_m - y_n\| < \in$

$\Rightarrow \quad <\|y_n\|>$ is a Cauchy sequence in R.

But R is complete $\quad \Rightarrow \quad \|y_n\| \to \alpha \in R$. Thus, $y_n \to \alpha x \in N$

Thus, Cauchy sequence $<y_n>$ in N converges to αx in N and hence, N is complete.

THEOREM 2. ***l_p space are normed linear space.*** (GARHWAL–2003)

PROOF. We know that the norm of a function f in l_p is given by

$$\|f\|_p = \left[\int_x |f(x)|^p \, d\mu\right]^{1/p} \quad \text{where } \int_x |f(x)|^p \, d\mu < \infty$$

(i) Since $|f(x)| \ge 0$

$\|f\|_p \ge 0, \forall f \in l_p$

(ii) $\|f\|_p = 0 \quad \Leftrightarrow \quad \left[\int_x |f(x)|^p \, d\mu\right]^{1/p} = 0 \quad \Leftrightarrow \int_x |f(x)|^p \, d\mu = 0$

$\Leftrightarrow \quad f = 0$ a.e.

(iii) $\|f + g\|_p \le \|f\|_p + \|g\|_p$ [By Minkowski inequality]

(iv) If α is any scalar, then

$$\|\alpha f\| = \left[\int_x |(\alpha f)(x)|^p \, d\mu\right]^{1/p} = \left[\int_x |\alpha f(x)|^p \, d\mu\right]^{1/p}$$

$$= |\alpha| \left[\int_x |f(x)|^p \, d\mu\right]^{1/p} = |\alpha| \|f\|_p$$

Hence, l_p-space is a normed linear space.

THEOREM 3. (Riesz-Fischer Theorem)

For $i \le p \le \infty$, l_p space is complete. (KANPUR–2001, 04, MEERUT–2003, 04)

PROOF. Let $<f_n>$ be a Cauchy space in l_p. Hence it is sufficient to prove that the sequence $<f_n>$ converges to a function $f \in l_p$. Since, $<f_n>$ is Cauchy, therefore for given $\in = \frac{1}{2} > 0$, there exist a positive integer n_1 such that

$$\|f_n - f_m\|_p < \frac{1}{2}, \forall n, m \ge n_1$$

Similarly for $\in = \left(\frac{1}{2}\right)^2$, we can choose a positive integer $n_2 > n_1$ such that

$$\|f_n - f_m\|_p < \left(\frac{1}{2}\right)^2, \forall n, m \ge n_2$$

Proceeding in the same manner choosing $n_1, \ldots, n_k$, let $n_{k+1} > n_k$ be such that

$$\|f_n - f_m\|_p \le \left(\frac{1}{2}\right)^{k+1}, \forall n, m \ge n_{k+1}$$

We want to show that subsequence $< f_{n_k} >$ converges a.e. to a limit function $f \in l_p$. Here, it is clear that

$$\sum_{k=1}^{\infty} \left\| f_{n_{k+1}} - f_{n_k} \right\| < \sum_{k=1}^{\infty} \left(\frac{1}{2}\right)^k = \frac{\frac{1}{2}}{1 - \frac{1}{2}} = 1 \qquad \ldots(1)$$

Define $\qquad g_k = \left|f_{n_1}\right| + \left|f_{n_2} - f_{n_1}\right| + \ldots + \left|f_{n_{k+1}} - f_{n_k}\right|$ for $k = 1, 2, 3, \ldots$

Then $<g_k>$ is an increasing sequence of non-negative measurable functions such that

$$\|g_k\|_1^p - \|g_k\|_p^p = \left[\left|f_{n_1}\right| + \left|f_{n_2} - f_{n_1}\right| + \ldots + \left|f_{n_{k+1}} - f_{n_k}\right|_p\right]^p$$

$$\le \left[\left\|f_{n_1}\right\|_p + \sum_{i=1}^{k} \left\|f_{n_{i+1}} - f_{n_i}\right\|_p\right]^p$$

[Using Minkowski's inequality]

$$\le \left[\left\|f_{n_1}\right\|_p + \sum_{i=1}^{\infty} \left\|f_{n_{i+1}} - f_{n_i}\right\|_p\right]^p = \left[\left\|f_{n_1}\right\|_p + 1\right]^p \qquad \text{[Using (1)]}$$

$$< \infty$$

Further let $g = \lim_{k \to \infty} g_k$

Then, using monotone convergence theorem, we have

$$\int g^p d\mu = \lim_{k \to \infty} \int g_k^p d\mu < \infty \quad \Rightarrow \quad g \in l_p$$

$$\Rightarrow \quad \int \left[\left| f_{n_i} \right| + \sum_{i=1}^{\infty} \left| f_{n_{i+1}} - f_{n_i} \right| \right]^p d\mu < \infty$$

$$\Rightarrow \quad \left[\left| f_{n_i} \right| + \sum_{i=1}^{\infty} \left| f_{n_{i+1}} - f_{n_i} \right| \right]^p < \infty \text{ a.e.}$$

Therefore, the series $\sum_{i=1}^{\infty} \left| f_{n_{i+1}}(x) - f_{n_i}(x) \right|$ converges a.e. and

Thus, the series $f_{n_1}(x) + \sum_{i=1}^{\infty} \left(f_{n_{i+1}}(x) - g_{n_i}(x) \right)$ converges a.e.

The kth partial sum of this series is $f_{n_{i+1}}(x)$ and so the sequence $< f_{n_k} >$ converges to a complex number $f(x)$ for all $x \in A$ where A is measurable and $\mu(A') = 0$. Further, define $f(x) = 0$ for all $x \in A'$. We can easily see that f is measurable. Now, to show that f is the limit in l_p of the sequence $<f_n>$. Let $\in > 0$ be given, choose l so large that

$$\|f_s - f_t\| < \in \ \forall\ s, t \geq n_i$$

$\Rightarrow$ for $k \geq l$ and $m > n_i$, we have $\left\| f_m - f_{n_k} \right\|_p < \in$

$$\Rightarrow \quad \left[\int \left| f_m - f_{n_k} \right|^p d\mu \right]^{1/p} < \in$$

$$\Rightarrow \quad \int \left| f_m - f_{n_k} \right|^p d\mu < \in^p$$

Then, using Fatou's lemma, we have

$$\int |f - f_m|^p \, d\mu = \int \lim_{k \to \infty} \left| f_{n_k} - f_m \right|^p dk < \in^p < \infty$$

Therefore, for each $m > n_i$, the function $f - f_m$ is in l_p and therefore, $f = f - f_m + f_m$ is also in l_p and $\lim_{n \to \infty} \|f - f_n\|_p = 0$.

Thus f is the limit in l_p of the sequence $<f_n>$. Hence, l_p is complete.

REMARK

- For $l \leq p < \infty$, l_p spaces are Banach spaces.

4.7 BANACH LIMIT

Let M be the linear space of all bounded sequences of real numbers, C is the subspace of all convergent sequences. If $x \in C$ then $L(x) = \lim_{n \to \infty} x_n$ is defined and L is a linear functional on C.

Definition 1. *A Banach limit is any linear functional L defined on linear space M such that*

(i) $L(x) \geq 0$ *if* $x_n \geq 0\ \forall\ n$

(ii) $L(x) = L(\sigma x)$ *where* $\sigma x = \sigma(x_1, x_2, ...) = (x_2, x_3, ...)$

(iii) $L(x) = 1$ *if* $x\ (1, 1, ...)$

Definition 2. *Let M be a linear space. Then $x \in M$ is called almost convergent and the number s is called F-limit of x if $L(x) = s$ for all Banach limits L.*

4.8 QUOTIENT AND SUBSPACE OF BANACH SPACES

Definition 1. *A non-empty subset M of a linear space L is a linear subspaces if and only if*

(i) $x \in M, y \in M \Rightarrow \quad x + y \in M$

(ii) $\alpha \in R, x \in M \Rightarrow \quad \alpha x \in M$

REMARKS

- The above condition can be equivalently written as : If $x, y \in M$, $\alpha, \beta \in R : \alpha x + \beta y \in M$
- A subspace of L is said to be proper if it is not all of L.
- The set consisting of null vector 0 alone is a subspace and is denoted by $\{0\}$.
- Every linear subspace is a linear space.

Definition 2. *Let L be a linear space and M be any subspace of L.*
Two elements $x_1, x_2 \in X$ are called equivalent modulo M if $x_1 - x_2 \in M$ and we can write

$$x_1 \equiv x_2 (\text{mod } M)$$

Here $\equiv$ is an equivalence relation on L and therefore L is divided into mutually disjoint equivalence classes. The set of all such equivalence classes be denoted by L/M.

In L/M vector addition and scalar multiplication are defined as follows

$$(x + M) + (y + M) = (x + y) + M \qquad \text{(Vector addition)}$$

$$\alpha(x + M) = \alpha x + M \qquad \text{(Scalar Multiplication)}$$

REMARK

- The zero element of L/M is M.

Definition 3. *The mapping $f : L \to L/M$ defined by $f(x) = x + M$ is called the canonical mapping of L onto L/M.*

THEOREM 1. ***Let M be a closed linear subspace of a normed linear space N. If the norm of the coset x + M in the quotient space N/M is defined by***

$$\|x + M\| = \inf \{\|x + M\| : m \in M\}$$

Then, N/M is a norm. Further, if N is a Banach space, then so is N/M.

(KANPUR–1990, 92, 95, 99, 2006, 11, BUNDELKHAND–1987, DELHI–1980, MEERUT–2006, AVADH–2008, DELHI–2006, 11, 12)

PROOF. **Axiom of Norm**

(i) $\|x\| \geq 0$: The elements of N/M are $x + M$, therefore,

$$\|x + m\| \geq 0$$

Now, $\|x + M\| = \inf \{\|x + m\| : m \in M\}$

Since norm is a non-negative real number, therefore $\|x + m\| \ \forall \ m \in M$ constitutes a set of non-negative real numbers and as such it is bounded below and therefore infimum exists and is greater than equal to zero.

$\Rightarrow \qquad \|x + M\| \geq 0$

(ii) We want to show that

$$\|x + M\| = 0 \qquad \Leftrightarrow x + M = \text{zero of } N/M, \text{ i.e., } M$$

Case-I : Let $x + M$ = zero of N/M, which is M.

i.e., $\qquad x + M = M$

Therefore, $x \in M$ because the cosets generated by an element x which belongs to M is the coset M itself. We shall show that $\|x + M\| = 0$

Now, $\qquad \|x + M\| = \inf \{\|x + M\| : m \in M \text{ and } x \in M\}$

$$= \inf \{\|y\| : y \in M\}$$

M is a subspace and as such it contains zero vector whose norm we know is zero and therefore.

$$\inf \{\|y\| : y \in M\} = 0$$

$\Rightarrow \qquad \|x + M\| = 0$, when $x + M = M$

Conversely, let $\|x + M\| = 0$. Then, to show $x + M = M$.

Now, $\qquad \|x + M\| = 0 \Rightarrow \quad \inf \{\|x + m\| : m \in M\} = 0$

$\Rightarrow \qquad \exists$ a $<m_k>$ in M such that $\|x + m_k\| \to 0$ as $k \to \infty$.

$\Rightarrow \qquad \lim m_k = -x$

But, $<m_k>$ is a sequence in M which is closed so that $\lim m_k \in M$ and therefore $-x \in M$ and hence $x \in M$.

$\therefore \qquad x + M = M$, because $x \in M$.

or $\qquad x + M =$ zero of N/M

(iii) $\|\alpha x\| = |\alpha| . \|x\|$

Consider

$$\begin{aligned}\|\alpha(x + M)\| &= \inf \{\|\alpha(x + m)\| : m \in M\}\\ &= \inf \{|\alpha| . \|x + m\| : m \in M\}\\ &= |\alpha| . \inf \{\|x + m\| : m \in M\} = |\alpha| . \|x + M\|\end{aligned}$$

(iv) $\|x + y\| \le \|x\| + \|y\|$

Consider

$$\begin{aligned}\|(x + M) + (y + M)\| &= \|x + y + M\|\\ &= \text{inf. } \{\|x + y + m\| : m \in M\}\\ &= \inf \{\|x + y + m_1 + m_2\| : m_1, m_2 \in M\}\\ &= \inf \{(\|x + m_1) + (y + m_2)\| : m_1, m_2 \in M\}\\ &\le \inf \{\|x + m_1\| + \|y + m_2\| : m_1, m_2 \in M\}\\ &= \inf \{\|x+m_1\| : m_1 \in M\} + \inf \{\|y+m_2\| : m_2 \in M\}\\ &= \|x + M\| + \|y + M\|\end{aligned}$$

Thus, $\|(x+M) + (y+M)\| \le \|x + M\| + \|y + M\|$

Hence, N/M satisfying all the axioms of norm.

Now, to show that N/M is complete when N is given to be complete. For this, we shall prove that any Cauchy sequence $<V_k + M>$ in N/M is convergent in N/M, where $V_k \in N$.

We know that a Cauchy sequence is convergent iff it has a convergent subsequence.

Since $<V_k + M>$ is a Cauchy sequence, therefore for $\in = \frac{1}{2}$, there exists an integer n_1, such that

$$\|(V_k + M) - (V_n + M)\| < \frac{1}{2} \forall k, n, \ge n_1$$

Let $V_{n_1} = x_1 \in N$.

Similarly, for $\in = \frac{1}{2^2}$, there exists an integer n_2 such that

$$\|(V_m + M) - (V_n + M)\| < \frac{1}{4} \forall m, n, \ge n_2$$

Clearly, $n_2 \ge n_1$

Let $V_{n_2} = x_2 \in N$ then $\|(x_1 + M) - (x_2 + M)\| < \frac{1}{2}$

Similarly, we can find $x_3 \in N$ such that

$$\|(x_2 + M) - (x_3 + M)\| < \frac{1}{2^2}$$

Continuing this process, we get

$$\|(x_{n-1} + M) - (x_n + M)\| < \frac{1}{2^{n-1}}$$

Therefore, we obtained a subsequence $<x_n + M>$ of Cauchy sequence $<V_k + M>$ such that

$$\|(x_n + M) - (x_{n+1} + M)\| < \frac{1}{2^n}, \forall n$$

Further, it remains to prove that this subsequence of Cauchy sequence is convergent in N/M.

Let y_1 is any vector in $x_1 + M$, *i.e.*, $y_1 = x_1 + m_1$, $m_1 \in M$.

Since, $\quad \| (x_1 + M) - (x_2 + M) \| < \frac{1}{2}$

$\Rightarrow \quad \inf \{\|x_1 - x_2 + m\| : m \in M\} < \frac{1}{2}$

$\Rightarrow \exists$ an $m_0 \in M$ such that $\|x_1 - x_2 + m_0\| < \frac{1}{2}$

$\Rightarrow \quad \| (x_1 + m_1) - (x_2 - m_0 + m_1) \| < \frac{1}{2}$

$\Rightarrow \quad \|y_1 - y_2\| < \frac{1}{2}$, where $\quad y_2 = x_2 - m_0 + m_1 \in x_2 + M$

Thus, for $y_1 \in x_1 + M$, we have obtained $y_2 \in x_2 + M$ such that $\|y_1 - y_2\| < \frac{1}{2}$

Similarly, we can find $y_3 \in x_3 + M$ such that $\|y_2 - y_3\| < \frac{1}{2^2}$

Continuing this process, we get $\|y_n - y_{n+1}\| < \frac{1}{2^n}$

Therefore, we have a sequence $<y_n>$ in N such that $\|y_n - y_{n+1}\| < \frac{1}{2^n}$

We will establish that sequence $<y_n>$ is a Cauchy sequence in N. For a given $\varepsilon > 0$, we may choose a positive integer n_0 so large that $\frac{1}{2^{n_0-1}} < \epsilon$.

Now for $m < n$ and $m, n \geq n_0$, we have

$$\|y_m - y_n\| = \|(y_m - y_{m+1}) + (y_{m+1} - y_{m+2}) + \ldots + (y_{n-1} - y_n)\|$$

$$\leq \|y_m - y_{m+1}\| + \|y_{m+1} - y_{m+2}\| + \ldots + \|y_{n-1} - y_n\|$$

$$< \frac{1}{2^m} + \frac{1}{2^{m+1}} + \ldots + \frac{1}{2^{n-1}} \quad \text{[Finite G.P.]}$$

$$< \text{sum of finite G.P.} < \frac{1}{2^{m-1}}$$

$\Rightarrow \quad <y_n>$ is a Cauchy sequence in N. Since N is complete, there exists a vector $y \in N$ such that $y_n \to y$. Now since $\| (x_n + M) + (y + M)\| \leq \|y_n - y\|$

$\Rightarrow \quad X_n + M$ converges to $y + M \quad \Rightarrow N/M$ is complete.

THEOREM 2. **(Riesz Lemma).** ***Let Y be a proper closed subspace of a normed linear space N over the field K. Let $0 < \alpha < 1$, then there exists some $x_\alpha \in N$ such that $\|x_\alpha\| = 1$ and $\inf_{y \in Y} \|x_\alpha - y\| \geq \alpha$.***

PROOF. Let N be a normed linear space and Y be a proper closed linear subspace of N.

Let $x \in X - Y$ and $d = \inf_{y \in Y} \| x - y \|$

Then $d \geq 0$, since Y is closed. Now, there exists $y_0 \in Y$ such that

$$0 \leq \|x - y_o\| < \frac{d}{\alpha}$$

Therefore, define $x_\alpha = \dfrac{(x - y_o)}{\|x - y_o\|}$

Clearly $\|x_\alpha\| = 1$ and $\inf_{y \in Y} \| x_\alpha - y \| \geq \alpha$.

THEOREM 3. ***Let N be a non-zero normed linear space and let $S = \{x \in N : \|x\| \leq 1\}$ be a linear subspace of N. Then N is Banach space if and only if S is complete.***

(MEERUT–1988, 2009, KANPUR–1991, 92, AGRA–1998, GORAKHPUR–2006)

PROOF. Let us first suppose that N is a Banach space and let $<x_n>$ be a Cauchy sequence in S such that $\|x_n\| \leq 1$.

To Show S is complete.

Since $S \subset N$

$\Rightarrow$ $< x_n >$ be a Cauchy sequence in N and N is complete.

$\Rightarrow$ $\exists\, x \in N$ such that $x_n \to x$

$\Rightarrow$ $\exists\, x \in N$ such that $\|x_n\| \to \|x\|$ [$\because \|.\|$ is a continuous function]

$\Rightarrow$ $\|x\| = \lim_{n \to \infty} \|x_n\| \leq 1$ [$\because \|x_n\| \leq 1, \ \forall\, n \in \mathbf{N}$]

$\Rightarrow$ $x \in S$.

Thus, S is complete.

Conversely, suppose that S is complete. To show that N is a Banach space. Clearly, N is given to be normed linear space.

Further let $< y_n >$ be any Cauchy sequence in N.

Then, by definition of Cauchy sequence, we have

$$\|y_m - y_n\| \to 0 \text{ as } m, n \to \infty$$

Let us define

$$x_n = \frac{y_n}{\|y_n\|}$$

Then
$$\|x_n\| = \left\| \frac{y_n}{\|y_n\|} \right\| = \frac{1}{\|y_n\|} . \|y_n\| = 1$$

$\Rightarrow$ $x_n \in S$

Now, we shall show that $<x_n>$ is a Cauchy sequence in S.

Consider
$$\|x_m - x_n\| = \left\| \frac{y_m}{\|y_m\|} - \frac{y_n}{\|y_n\|} \right\|$$

$$= \left\| \frac{y_m}{\|y_m\|} - \frac{y_n}{\|y_m\|} + \frac{y_n}{\|y_m\|} - \frac{y_n}{\|y_n\|} \right\|$$

$$\leq \left\| \frac{y_m}{\|y_m\|} - \frac{y_n}{\|y_m\|} \right\| + \left\| \frac{y_n}{\|y_m\|} - \frac{y_n}{\|y_n\|} \right\|$$

$$= \left\| \frac{1}{\|y_m\|}(y_m - y_n) \right\| + \left\| \left(\frac{1}{\|y_m\|} - \frac{1}{\|y_n\|} \right) y_n \right\|$$

$$= \frac{1}{\|y_m\|}\|y_m - y_n\| + \left| \frac{1}{\|y_m\|} - \frac{1}{\|y_n\|} \right| \|y_n\| \quad [\text{Using } \|\alpha x\| = |\alpha|.\|x\|]$$

$$= \frac{\|y_m - y_n\|}{\|y_m\|} + \frac{\big| \|y_n\| - \|y_m\| \big|}{\|y_m\|} \leq \frac{\|y_m - y_n\|}{\|y_m\|} + \frac{\|y_n - y_m\|}{\|y_m\|}$$

$$= \frac{2\|y_m - y_n\|}{\|y_m\|} \to 0 \text{ as } m, n \to \infty$$

Therefore, $<x_n>$ is a Cauchy sequence in S, and S is complete, therefore, there exists $x \in S$, such that

$$x_n \to x \Rightarrow \frac{y_n}{\|y_n\|} \to x$$

Further, we have $\big| \|y_n\| - \|y_m\| \big| \leq \|y_n - y_m\| \to 0$

$\Rightarrow$ $< \|y_n\| >$ is a Cauchy sequence of real numbers, but since R is complete thus, $\|y_n\| \to \alpha \in \mathrm{R}$.

$\Rightarrow$ $$y_n \to \alpha x \in N$$

Hence, N is complete.

THEOREM 4. ***Let N be a normed linear space and M be a subspace of N. Then, closure $\bar{M}$ of M is also a subspace of N.*** (MEERUT–2000, KANPUR–2008, 09, 11)

PROOF. Let N be a normed linear space and M be a subspace of N. Now since $\bar{M}$ is the closure of M, then by definition of closure of M, we have that every point of $\bar{M}$ is an adherent point of M.

$\Rightarrow$ for each $x \in M$ and $\in > 0$ $\exists$ $y \in \bar{M}$ such that

$$\|y - x\| < \in \qquad \ldots(1)$$

In order to show that $\bar{M}$ be a subspace of N, we shall show that any linear combination of elements in $\bar{M}$ is again in $\bar{M}$.

If $x, y \in \bar{M}$ and α, β be any scalars of N,

To show $\alpha x + \beta y \in \bar{M}$

Since, $x, y \in \bar{M}$, then there exist $x_1, y_1 \in M$ such that

$$\|x - x_1\| < \frac{\in}{2|\alpha|} \qquad \ldots(2)$$

and $$\|y - y_1\| < \frac{\in}{2|\beta|} \qquad \ldots(3)$$

Consider

$$\|(\alpha x + \beta y) - (\alpha x_1 + \beta y_1)\|$$

$$= \|\alpha(x - x_1) + \beta(y - y_1)\| \leq |\alpha|\|x - x_1\| + |\beta|\|y - y_1\|$$

$$< |\alpha|\frac{\in}{2|\alpha|} + |\beta|.\frac{\in}{2|\beta|}$$ [Using (2) and (3)]

$$= \in$$

Then, $\alpha x + \beta y \in M \Rightarrow \alpha x + \beta y \in \bar{M}$

Hence, $\bar{M}$ is a subspace.

THEOREM 5. ***In a normed linear space, the smallest closed subspace containing a given set of vectors is just the closure of the subspace spanned by that set.***

PROOF. Let N be a normed linear space and S be a subspace of N. Also, let M be the smallest closed subspace of N containing S. Let $[S]$ denote the subspaces spanned by S. To show $M = (\bar{S})$.

Since we know that $(\bar{S})$ is a closed subspace of N and it contain S (By Theorem-4). But we have assumed that M is the smallest closed subspace containing S, then

$$M \subset [\bar{S}]$$

But we know that

$$[S] \subset M \quad \Rightarrow \quad [\bar{S}] \subset \bar{M}$$

Hence, we conclude that $M = [\bar{S}]$.

THEOREM 6. ***Every complete subspace M of normed linear space N is closed.***

PROOF. Let N be a normed linear space and M be a complete subspace on N.

We have to show that M is closed.

Let $p \in N$ be an arbitrary limit point of M. Then by definition of limit point $\exists$ a sequence $<x_n>$ in M such that

$$\lim_{n \to \infty} x_n = p$$

Now, since, $<x_n>$ is a convergent sequence, therefore it is Cauchy.

($\because$ Every convergent sequence is Cauchy)

Further, $<x_n>$ is a Cauchy sequence in M which is complete and $x_n \to p$, therefore $p \in M$ (By property of completeness)

So p is a limit point of $M \Rightarrow p \in M$

or $D(M) \subset M$ ($\because$ p is arbitrary)

$\Rightarrow$ M contains all its limit point

Hence, M is closed.

THEOREM 7. ***A subspace M of a Banach space B is complete if and only if M is closed in B.***

PROOF. Let us first suppose M is a subspace of a Banach space B.

To prove, M is closed in B.

Since M is a subspace of a Banach space B.

Therefore B is complete as a metric space.

Now, since M is complete, so using previous theorem M is closed in B.

Conversely, suppose that M is closed in B. We have to prove that M is complete.

Let $<y_n>$ be a Cauchy sequence in $M \subset B$. Then $<y_n>$ is a Cauchy sequence in B, therefore $y_n \to y_0 \in B$ as B is complete.

Then $y_n \in M \subset \bar{M} \Rightarrow y_n \in \bar{M}$

$\Rightarrow$ $\lim y_n \in \bar{M} \Rightarrow y_0 \in \bar{M}$

Further, M is closed, therefore $M = \bar{M}$.

and hence $y_0 \in M$ ($\because M = \bar{M}$ and $y_0 \in \bar{M}$)

$\Rightarrow$ any Cauchy sequence $<y_n>$ in M converges to $y_0 \in M$.

Hence, M is complete.

4.9 CONTINUOUS LINEAR TRANSFORMATIONS

Definition 1. *Let N and N′ be two normed linear space with the same scalar and $T : N \to N'$ be the linear transformation. Then the linear transformation T is said to be continuous if*

$$x_n \to x \text{ in } N \Rightarrow T(x_n) \to T(x) \text{ in } N'$$

Definition 2. *The linear transformation $T : N \to N'$ is said to be bounded if there exists a real number $k \geq 0$ such that*

$$\|Tx\| \leq k\|x\| \quad \forall x \in N.$$ (MEERUT–2006, KANPUR–2010)

Definition 3. *If $T : N \to N'$ be a bounded linear transformation then, the norm of T is defined by*

$$\|T\| = \sup\left\{\frac{\|Tx\|}{\|x\|} : x \in N, x \neq 0\right\}$$

The norm of T can also be defined as

$$\|T\| = \sup\{\|Tx\| : x \in N; \|x\| = 1\}$$

REMARKS

- $\|Tx\| \leq \|T\|.\|x\| \quad \forall x \in N.$
- If T is a bounded linear transformation from a normed linear space N into itself, then T is called bounded linear operator.
- A bounded linear transformation from N into the field K is called a bounded linear functional. Also, it is called a real or complex bounded linear functional according as K is the real field R or complex field **C**.

☛ ILLUSTRATIONS

(1) The identity operator $I : N \to \text{N}$ on a normed linear space $N \neq \{0\}$ is a bounded linear operator with the norm $\| I \| = 1$.

(2) The zero transformation $0 : N \to N'$ on a normed linear space N is a bounded linear transformation and has the norm $\|0\| = 0$.

(3) The norm $\|.\| : N \to R$ on a linear space N, is not a linear functional, it is a sublinear functional.

(4) Let A be $m \times n$ matrix with entries a_{ij}. Then $A = \text{R}^n \to \text{R}^m$ is a linear transformation. Then A is a bounded linear transformation.

(5) The linear transformation $T : \text{C}[0, 1] \to \text{R}$ defined by

$Tf = \int_0^1 f(t)dt$ for each $f \in C[0,1]$ is a bounded linear transformation.

THEOREM 1. ***Let N and N′ be two normed linear spaces with the same scalars and $T : N \to N'$ be a linear transformation.***

Then the following conditions are equivalent to each other.

(i) T is continuous (MEERUT–2002, 03, 06)

(ii) T is continuous at the origin, in the sense that

$$x_n \to 0 \Rightarrow T(x_n) \to 0$$ (MEERUT–1995)

(iii) T is bounded, i.e., there exists a real number $k \geq 0$ with the property that

$$\|T(x)\| \leq k\|x\| \text{ for every } x \in N.$$

(AGRA–2005, KANPUR–1999,2001)

(iv) If $S = \{x : \|x\| \leq 1\}$ is the closed unit sphere in N then its image T(S) is a bounded set in N′. (MEERUT–2002, 03, 04, KANPUR–1994, AGRA–1988)

PROOF. **(i) $\Rightarrow$ (ii)**

Let T be continuous, then by definition

$$x_n \to x \Rightarrow T(x_n) \to T(x)$$

Let $z_n = x_n - x$ be a sequence of vectors in N and $z_n \to 0$. $[\because x_n \to x]$

Now, $T(z_n) = T(x_n - x) = T(x_n) - T(x) \to 0$

Therefore, $z_n \to 0 \Rightarrow T(z_n) \to 0$

Thus, T is continuous at the origin

(ii) $\Rightarrow$ (i) Let T is continuous at origin.

i.e. $x_n \to 0 \Rightarrow T(x_n) \to 0$

Let $x_n \to x$ so that $y = x_n - x \to 0$

$\therefore$ $T(y) \to 0 \Rightarrow T(x_n - x) \to 0$ $[\because T$ is continuous at the origin$]$

$\Rightarrow$ $T(x_n) - T(x) \to 0 \Rightarrow T(x_n) \to T(x)$

$\Rightarrow$ T is continuous.

(ii) $\Rightarrow$ (iii) It is given that $x_n \to 0 \Rightarrow T(x_n) \to 0$

i.e., T is continuous at the origin.

To show, T is bounded. For this we shall prove that there exists a real number $k \geq 0$ such that

$$\| Tx \| \leq k \| x \|, \ \forall x \in N$$

Let if possible, no such K exists. Therefore for each positive integer n, we can find a vector x_n such that

$$\| Tx_n \| \geq n \| x_n \| \qquad \Rightarrow \frac{\|Tx_n\|}{n\|x_n\|} \geq 1$$

Now, since $n\|x_n\|$ is a positive real scalar, thus we can write

$$\left\| T\left(\frac{x_n}{n\|x_n\|}\right) \right\| \geq 1$$

$$\Rightarrow \qquad \|Ty_n\| \geq 1, \text{ where } y_n = \frac{x_n}{n\|x_n\|}$$

Also $$\|y_n\| = \left\| \frac{x_n}{n\|x_n\|} \right\| = \frac{1}{n\|x_n\|} \cdot \|x_n\| = \frac{1}{n}$$

Therefore, $\|y_n\| \to 0$ as $n \to \infty$

But $\|Ty_n\| \geq 1$, *i.e.*, $\|Ty_n\|$ does not tend to zero where $y_n \to 0$ which is a contradiction, thus our assumption is wrong. Therefore, there exists $k \geq 0$ such that

$$\| Tx \| \leq k \| x \|, \ \forall x \in N$$

(iii) $\Rightarrow$ (ii)

Let T be bounded, *i.e.*, for $k \geq 0$

$$\|Tx\| \leq k\|x\| \quad \forall x \in N$$

To show that T is continuous at the origin

i.e., $x_n \to 0 \Rightarrow T(x_n) \to 0$

Now $x_n \to 0 \Rightarrow \|x_n\| \to 0$

$\therefore \quad \|T(x_n)\| \le k\|x_n\|$

Thus $\|T(x_n)\| \to 0$ as $\|x_n\| \to 0$

$\therefore \quad x_n \to 0 \Rightarrow T\|x_n\| \to 0$

Hence, T is continuous at the origin

(iii) $\Rightarrow$ (iv)

Let $\|Tx\| \le k\|x\|, \forall x \in N$.

To show that if $x \in S$, *i.e.*, $\|x\| = 1$

Then $T(S)$ is bounded, *i.e.*, $\|Tx\| \le k, \ \forall x \in S$

Since $\|Tx\| \le k\|x\| \le k.1 = k, \ \forall x \in S$ for which $\|x\| \le 1$.

(iv) $\Rightarrow$ (iii) Let $T(S)$ is bounded, *i.e.*,

$$\|Tx\| \le k \quad \forall x \in S \quad \text{or} \quad \forall \ x : \|x\| \le 1$$

To show T is bounded. For this, we shall prove that $\exists k \ge 0$ such that $\|Tx\| \le k\|x\|$ $\forall x \in N$

If $x = 0, \Rightarrow \|x\| = 0$ then result is obvious.

If $x \ne 0$, then $\|x\| > 0$

Choose $y = \dfrac{x}{\|x\|}$, where $x \in N$

$$\therefore \quad \|y\| = \left\|\frac{x}{\|x\|}\right\| = \frac{1}{\|x\|}.\|x\| = 1 \quad \Rightarrow \quad y \in S$$

Therefore by given condition

$$\|Ty\| \le k \quad \text{or} \quad \left\|T\left(\frac{x}{\|x\|}\right)\right\| \le k \quad \text{or} \quad \left\|\frac{1}{\|x\|}Tx\right\| \le k$$

$$\Rightarrow \quad \frac{1}{\|x\|}\|Tx\| \le k \quad \Rightarrow \quad \|Tx\| \le k\|x\| \quad \forall x \in N$$

Hence, T is bounded.

REMARK

- T is bounded if T is continuous.

THEOREM 2. ***Let N and N′ be normed linear space and T be a linear transformation of N into N′. Then, T^{-1} exists and is continuous on its domain if and only if there exists a constant k > 0 such that***

$$k\|x\| \le \|T(x)\| \quad \forall x \in N$$

(MEERUT–2001, 06BP, 07)

PROOF. Let us first suppose that

$$k\|x\| \le \|Tx\| \quad \forall x \in N \qquad \ldots(1)$$

To show that T^{-1} exists and is continuous.

***T* is one-one**

Let $Tx_1 = Tx_2 \Rightarrow Tx_1 - Tx_2 = 0$

$\Rightarrow T(x_1 - x_2) = 0 \Rightarrow \|T(x_1 - x_2)\| = 0$

$\Rightarrow k\|x_1 - x_2\| < 0$ [Using (1)]

$\Rightarrow x_1 - x_2 = 0$ [$\because \|y\| \geq 0$]

$\Rightarrow x_1 = x_2$

Hence T is one-one.

***T* is onto**

Now, since T^{-1} exists, therefore, for each $y \in N'$, there exists $x \in N$ such that

$$T(x) = y \Leftrightarrow T^{-1}(y) = x \Rightarrow T \text{ is onto} \quad \text{...(2)}$$

Therefore, from (1), we get

$$k\left\|T^{-1}y\right\| \leq \|y\| \Rightarrow \left\|T^{-1}y\right\| \leq \frac{1}{k}\|y\|$$

which implies that T^{-1} is bounded and hence it is continuous.

Conversely, let T^{-1} exists and be continuous. (Using above remark)

To show $\exists K > 0$ such that

$$k.\|x\| \leq \|Tx\|$$

Consider $\left\|T^{-1}y\right\| \leq m\|y\| \Rightarrow \|x\| \leq k\|T(x)\|$

$$\Rightarrow K\|x\| \leq \|T(x)\|, \text{ where } k = \frac{1}{m} > 0$$

THEOREM 3. ***Let N and N′ be normed linear spaces over the same scalar field and T be a linear transformation of N into N′. Then T is bounded if and only if it is continuous.***

PROOF. Let us first suppose T be bounded so that there exists $M > 0$ such that

$$\|Tx\| \leq m.\|x\| \in N \quad \text{...(1)}$$

To prove that T is continuous.

Let $x \in N$ be arbitrary. For any $\in > 0$, we choose $\delta = \in/m$. Then, for all $y \in N$ such that

$\|y - x\| < \delta$, we have

$$\|T(y) - T(x)\| = \|T(y - x)\| \leq m.\|y - x\| \quad \text{[Using (1)]}$$

$$= m.\frac{\in}{m} = \in$$

$\Rightarrow$ T is continuous at x. Since x is arbitrary. Thus, T is a continuous mapping.

Conversely, let T be continuous mapping. To show T is bounded. Let if possible, T is not bounded. Then $\exists$ no $k > 0$ such that

$$\|N(x)\| \leq k.\|x\|, \quad \forall x \in N$$

Then, for each positive integer $n, \exists\, x_n \in N$ such that

$$\|T(x_n)\| > n\|x_n\|$$

For each n, let $y_n = \dfrac{x_n}{n\|x_n\|}$

Then $\|y_n\| = \dfrac{1}{n} \to 0 \text{ as } n \to \infty$

$\Rightarrow \qquad y_n \to 0$ as $n \to \infty$

Now for every n, we have

$$\|T(y_n)\| = \left\|\left(\frac{T(x_n)}{n\|x_n\|}\right)\right\| = \left\|\frac{1}{n.\|x_n\|}.T(x_n)\right\| = \frac{1}{n.\|x_n\|}.\|T(x_n)\|$$

$\Rightarrow \qquad \|T(y_n)\| > 1$ because $\|T(x_n)\| > n\|x_n\|$

$\Rightarrow \quad T(y_n)$ does not tend to 0.

$\Rightarrow$ The sequence $< y_n >$ converges to 0 but the sequence $< T(y_n) >$ does not converge to $T(0)$. Thus, T is not continuous, which is a contradiction. Hence, T must be bounded.

THEOREM 4. ***Let T be a linear transformation of a normed linear space N into another normed linear space N′. Then, T is bounded if and only if T maps bounded sets in N into bounded sets in N′.***

PROOF. Let us first suppose that T be a bounded linear transformation and S be a bounded subset of N. Then $\exists$ k_1 and k_2 such that

$$\|Tx\| \le k_1 . \|x\|, \quad \forall\, x \in N$$

and $$\|x\| \le k_2, \quad \forall\, x \in S$$

$\Rightarrow$ $$\|T(x)\| \le k_1 . k_2, \ \forall\, x \in S$$

$\Rightarrow$ T is bounded.

Conversely, let T map bounded sets in N into bounded sets in N'. To show that T is a bounded linear transformation.

Let N be any bounded set and $S(1, 0)$ be the closed unit sphere centered at origin.

Then, by our assumption $T[S(1, 0)]$ must also be bounded so that $\exists$ a $k > 0$ such that

$$\|T(z)\| \le k, \ \forall\, z \in S(1, 0) \qquad \ldots(1)$$

Let $x \in N$ be any non-zero vector

Then, $\dfrac{1}{\|x\|} \in S(1, 0)$ and therefore by (1).

$$\left\|T\left(\frac{x}{\|x\|}\right)\right\| \le k \Rightarrow \left\|\frac{1}{\|x\|}.T(x)\right\| \le k = \frac{1}{\|x\|}.\|T(x)\| \le K$$

$\Rightarrow$ $$\|T(x)\| \le k\|x\|$$

The above inequality holds for $x = 0$ as well. Hence, we conclude that T is a bounded linear transformation.

THEOREM 5. ***Let M be a closed linear subspace of a normed linear space N and let ϕ be a natural mapping defined by $\phi(x) = x + M$. Then ϕ is a continuous linear transformation for which $\|\phi\| \le 1$.*** (MEERUT–2001,05BP, 08, 09BP, GARHWAL–2004, KANPUR–2002, 07, AVADH–2008, 09, AMRITSAR–1982)

PROOF. As per given, M is closed, N/M is a normed linear space with the norm of coset $x + M$ in N/M. Define $\phi : N \to N/M$ such that

$$\|x + M\| = \inf \{\|x + m\| : m \in M\}$$

(i) ϕ is linear

Let x, y be any two elements of N and α, β be any two scalars. Then

$$\phi(\alpha x+\beta y)=(\alpha x+\beta y)+M=(\alpha x+M)+(\beta y+M)$$
$$=\alpha(x+M)+\beta(y+M)=\alpha\phi(x)+\beta\phi(y)$$

(ii) ϕ is continuous

Consider $\|\phi(x)\|=\|x+M\|=\inf\{\|x+m\|: m\in M\}$

$\leq\|x+m\|,\ \forall\, m\in M$

In particular for $m = 0$, we have

$$\|\phi(x)\|\leq\|x\| = 1.\|x\|,\quad \forall\, x\in N$$

$\Rightarrow$ ϕ is bounded and hence continuous.

Finally, $\|\phi\|=\sup.\{\|\phi(x)\|: x\in N,\ \|x\|\leq 1\}$

$\leq\sup.\{\|x\|: x\in N,\ \|x\|\leq 1\}$

Hence, $\|\phi\|\leq 1$

THEOREM 6. ***Let N and N′ be normed linear operators and $T: N\to N'$ be a linear transformation. Then Ker(T) is a linear manifold and that Ker(T) is closed if T is continuous.*** (MEERUT–2001, 02, 04)

PROOF. By definition of kernel, we have

$$\text{Ker}(T)=\{x: x\in N,\ T(x)=0\}$$

Let $x, y\in \text{Ker}(T)$ and α, β be any scalars, then

$$T(x)=0,\ T(y)=0$$

Then, $T(\alpha x+\beta y)=\alpha T(x)+\beta T(y)=\alpha.0+\beta.0$

Thus, $\alpha x+\beta y\in \text{Ker}(T)$

$\Rightarrow$ Ker (T) is a linear manifold.

Now, let T be continuous. Then, since $\text{Ker}(T) = T^{-1}(0)$ and $\{0\}$ is a closed set in N'.

$\Rightarrow$ Ker(T) is closed.

THEOREM 7. ***Let N and N′ be normed linear space and T be a continuous linear transformation of N into N′. If M is the null space (Kernel) of T, then T induces a natural linear transformation T′ of N/M into N′ and***

$$\|T'\| = \|T\|$$ (MEERUT–1982, 2003)

PROOF. Using above theorem, M is a closed subspace of N and so N/M is a normed linear space with the norm of coset $x + M$ in N/M defined by

$$\|x+M\|=\inf\{\|x+M\|: m\in M\}$$

Define $T': N/M\to N'$

by $T'(x+M)=T(x)$

To show T' is a linear transformation such that $\|T'\|=\|T\|$.

Let $x + M, y + M$ be any two elements of N/M and α, β be any scalars, then

$$T'(\alpha(x+M)+\beta(y+M)) = T'((\alpha x+M)+(\beta y+M))=T'[(\alpha x+\beta y)+M)$$
$$=T(\alpha x+\beta y)=\alpha T(x)+\beta T(y)$$

[$\because$ T is linear]

$$=\alpha T'(x+M)+\beta T'(y+M)$$

$\Rightarrow$ T' is linear.

Also, $\|T'\|=\sup\{\|T'(x+M)\|: x\in N,\ \|x+m\|\leq 1\}$

$$= \sup \{\|T(x)\| : x \in N,\ \inf \{\|x+m\| : m \in M\} \le 1\}$$
$$= \sup \{\|T(x)\| : x \in N,\ m \in M,\ \|x+m\| \le 1\}$$
$$= \sup \{\|T(x)+T(m)\| : x \in N,\ m \in M,\ \|x+m\| \le 1\}$$
$$= \sup \{\|T(x+M)\| : x \in N,\ \|x\| \le 1\} = \|T\|$$

THEOREM 8. ***If $T : N(K) \to N'(K)$ is a linear mapping in a normed linear space and if T is continuous at the origin, then it is continuous everywhere and continuity is uniform.*** (KANPUR–2009, 11)

PROOF. Since, T is continuous at the origin, therefore there exists a sequence $<x_n>$ in N such that $x_n \to 0 \quad \Rightarrow \quad T(x_n) \to T(0) = 0$

We have to show that T is continuous everywhere. Let $y \in N$ be arbitrary such that $y_n \to y$ in N where $<y_n>$ is a sequence in N.

Now since $y_n \to y \quad \Rightarrow \quad y_n - y \to 0$

Also, T is continuous at the origin, therefore

$$y_n - y \to 0 \quad \Rightarrow T(y_n - y) \to T(0) = 0$$

$\Rightarrow \quad T(y_n) - T(y) \to 0 \qquad \Rightarrow \qquad T(y_n) \to T(y)$

$\Rightarrow \quad T$ is continuous at y

Since $y \in N$ is arbitrary, therefore, T is continuous everywhere in N and hence continuity is uniform.

THEOREM 9. ***If N be a normed linear space and X be a space of reals or complex, then functions***

$$f : N \times N \to N \text{ such that } f(x, y) = x + y \qquad \text{...(1)}$$

and $$g : X \times N \to N \text{ such that } g(\alpha, x) = \alpha . x \ \forall\ x, y \in N \text{ and } \alpha \in X \qquad \text{...(2)}$$

are continuous. (AGRA–2001, MEERUT–2004, 06)

PROOF. Let $x, y \in N$ and $\alpha, \beta \in X$. Then N is a metric space w.r.t. metric d_1 such that

$$d_1(x, y) = \|x - y\|$$

Similarly, X is also metric space w.r.t. metric d_2 such that

$$d_2(x, y) = \|x - y\|$$

Consider, convergent sequences $<x_n>$, $<y_n>$ in N and $<\alpha_n>$, $<\beta_n>$ in X such that

$$\lim_{n\to\infty} x_n = x,\ \lim_{n\to\infty} y_n = y,\ \lim_{n\to\infty} \alpha_n = \alpha,\ \lim_{n\to\infty} \beta_n = \beta$$

Then $\|x_n - x\| = 0, \|y_n - y\| = 0, \|\alpha_n - \alpha\| = 0, \|\beta_n - \beta\| = 0$ as $n \to \infty$

Now,
$$\|f(x_n, y_n) - f(x, y)\| = \|(x_n + y_n) - (x + y)\|$$
$$= \|(x_n - x) + (y_n - y)\|$$
$$\le \|x_n - x\| + \|y_n - y\|$$
$$= 0 + 0 \text{ as } n \to \infty$$

$\Rightarrow \qquad f(x_n, y_n) \to f(x, y)$ as $n \to \infty$

Now, by (2)
$$\|g(\alpha_n, x_n) - g(\alpha, x)\| = \|\alpha_n x_n - \alpha x\|$$
$$= \|\alpha_n x_n - \alpha_n x + \alpha_n x - \alpha x\|$$
$$= \|\alpha_n (x_n - x) + x(\alpha_n - \alpha)\|$$
$$\le |\alpha_n|\ \|x_n - x\| + \|x\|\ |\alpha_n - \alpha|$$
$$= \alpha.0 + \|x\|.0 \text{ as } n \to \infty$$

$\Rightarrow \qquad \lim_{n\to\infty} g(\alpha_n, x_n) = g(\alpha, x)$

Hence, f and g are jointly continuous.

RECAPITULATIONS

➠ A non-zero normed linear space is Banach space iff $S = \{x : \|x\| = 1\}$ is complete	➠ l_p space are normed linear space.
➠ l_p space is complete.	➠ Let N be a normed linear space and M be a subspace of N then $\bar{M}$ (closure of M) is also a subspace of N.
➠ In a normed linear space, the smallest closed subspace containing a given set of vectors is just the closure of the subspace spanned by that set.	➠ Every complete subspace of normed linear space is closed ➠ A subspace M of a Banach space is complete if and only if M is closed in B.

4.10 TOPOLOGICAL AND ISOMETRIC ISOMORPHISM

(1) Isometric Isomorphism

(MEERUT–2006BP, 07)

Let N and N' be normed linear spaces.

An isometric isomorphism of N into N' is a one-one linear transformation T of N into N' such that $\|T(x)\| = \|x\|$ for every $x \in N$.

REMARKS

- N is said to be isometrically isomorphic (or Congruent) to N' if there exists an isomorphism of N onto N'.
- If T is an isometric isomorphism of N into N' and x_1, x_2 be any two points of N, then $\|T(x_1) - T(x_2)\| = \|T(x_1 - x_2)\| = \|x_1 - x_2\|$ which shown that T preserve distances and therefore it is an isometry.

(2) Topological Isomorphism

Let N and N' be two normed linear spaces. A topological isomorphism of N into N' is a one-one linear transformation of N into N' such that T and T^{-1} are continuous.

REMARKS

- N is said to be topologically isomorphic to N' if there exists a topological isomorphism of N onto N'.
- Normed linear space N and N' are said to be topologically isomorphic provided there exists a homomorphism of N into N' which is also a linear transformation.
- The topological isomorphic spaces need not be isometrically isomorphic.

4.11 ALGEBRA

An algebra is a linear space whose vectors can be multiplied such that

(i) $x(yz) = (xy)z$

(ii) $x(y+z) = xy + xz$ *and* $(x+y)z = xz + yz$

(iii) $\alpha(xy) = (\alpha x)y = x(\alpha y)$ *for* $\alpha \in \mathbb{R}$

i.e., an algebra is a linear space that is also a ring in which (iii) holds.

4.11.1 NORMED ALGEBRA

If $\| xy \| \le \| x \| . \| y \|$ for all $x, y \in N$ then N is called a normed algebra.

4.11.2 BANACH ALGEBRA

A complex or real Banach space which is also a normed algebra is called a Banach algebra.

4.11.3 LINEAR OPERATOR ON NORMED LINEAR SPACES

If N be a normed linear space, then continuous or bounded linear transformation of N into itself is called an operator on N. The normed linear space of all operator on N is denoted by $B(N)$, which is also a Banach space.

4.12 NORM OF A CONTINUOUS LINEAR TRANSFORMATION

Let N and N' be two normed linear spaces and $T : N \to N'$ is a continuous linear transformation. Since T is continuous,

Therefore, $\|Tx\| \leq k\|x\| \quad \forall x \in N$ and $k \geq 0$

or $$\|Tx\| \leq k \quad \forall\, x : \|x\| \leq 1$$

Then, norm of T, *i.e.*, $\|T\|$ can be defined as follows:

$$\|T\| = \sup.\{\|Tx\| : \|x\| \leq 1\}$$

$$\|T\|_0 = \sup.\{\|Tx\| : \|x\| = 1\}$$

Clearly $$\|T\| = \|T\|_0$$

Other definition of $\|T\|$

(i) $\|T\| = \sup.\left\{\dfrac{\|Tx\|}{\|x\|} : x \in N, x \neq 0\right\}$

(ii) $\|T\| = \sup.\{k : k \geq 0 \text{ and } \|Tx\| \leq k\|x\| \forall x \in N\}$

THEOREM 1. ***Let N and N′ be two normed linear spaces. Then, N and N′ are topologically isomorphic if and only if there exists a linear transformation T of N onto N′ and positive constants m and k such that***

$$m\|x\| \;\leq\; \|Tx\| \;\leq\; k\|x\|, \quad \forall x \in N.$$

PROOF. Suppose that N is topologically isomorphic to N', therefore by definition T and T^{-1} are continuous.

To show, T is continuous $\Leftrightarrow$ T is bounded

$\Rightarrow$ $$\|Tx\| \;\leq\; k\|x\| \quad \forall x \in N,\ k > 0 \qquad \ldots(1)$$

Further T^{-1} is continuous $\Leftrightarrow$ there exists a positive constant m such that

$$m\|x\| \;\leq\; \|Tx\|, \quad \forall x \in N \text{ (Using Theorem-2)} \qquad \ldots(2)$$

Hence, from (1) and (2) we conclude that N and N' are topologically isomorphic if and only if $m\|x\| \leq \|Tx\| \leq k\|x\|$.

THEOREM 2. ***Let N and N′ be normed linear spaces and let B(N, N′) denote the set of all bounded (or continuous) linear transformation from N onto N′. Then B(N, N′) is itself a complete normed linear spaces with respect to point wise linear operations.***

$$(T+U)(x) = T(x) + U(x)$$

$$(\alpha T)(x) = \alpha T(x)$$

and the norm is defined by

$$\|T\| = \sup.\{\|T(x)\| : x \in N, \|x\| \leq 1\}$$

(MEERUT–2004,05, KANPUR–2007, AGRA–2001)

PROOF. Firstly, we shall prove that $B(N, N')$ is a linear space. Since we know that the set of all linear transformation from a linear space into another linear space is itself a linear space with respect to the point wise linear operations. Therefore, here it is sufficient to prove that $B(N, N')$ is a subspace of S.

Let $T_1, T_2 \in B(N, N')$

$\Rightarrow$ T_1, T_2 are bounded

$\Rightarrow$ $\exists\, k_1 \geq 0,\ k_2 \geq 0$ such that

$$\|T_1(x)\| \leq k_1 \|x\| \text{ and } \|T_2(x)\| \leq k_2 \|x\| \quad \forall x \in N$$

Now, let α, β be any scalars. Then consider

$$\|(\alpha T_1 + \beta T_2)(x)\| = \|(\alpha T_1)(x) + (\beta T_2)(x)\| = \|\alpha T_1(x) + \beta T_2(x)\|$$
$$\leq \|\alpha T_1(x)\| + \|\beta T_2(x)\| = |\alpha|\,\|T_1(x)\| + |\beta|\,\|T_2(x)\|$$
$$\leq |\alpha|.k_1\|x\| + |\beta|.k_2\|x\| = \{|\alpha|.k_1 + |\beta|.k_2\}\|x\|$$

$\Rightarrow$ $\alpha T_1 + \beta T_2$ is bounded.

$\Rightarrow$ $\alpha T_1 + \beta T_2 \in B(N, N')$.

$\Rightarrow$ $B(N, N')$ is a linear sub space of S.

Now, we shall prove that $B(N, N')$ is a normed linear space.

Let $T \in B(N, N')$

Then, by definition,

$$\|T\| = \sup.\{\|Tx\| : \|x\| \leq 1\}$$

(i) Since $\|Tx\| \geq 0 \quad \forall x \in N$

$\Rightarrow$ $\|T\| \geq 0$

(ii) We have $\|T\| = \sup.\{\|Tx\| : \|x\| \leq 1, x \in N\}$

$$= \sup.\left\{\frac{\|Tx\|}{\|x\|}, x \neq 0, x \in N\right\}$$

Therefore, $\|T\| = 0 \Leftrightarrow \dfrac{1}{\|x\|}\|Tx\| = 0, \quad \forall x \in N, x \neq 0$

$$\Leftrightarrow \|Tx\| = 0, \quad \forall x \in N, x \neq 0 \Leftrightarrow Tx = 0 \quad \Leftrightarrow T = 0$$

(iii) We have $\|(\alpha T)(x)\| = \|\alpha T(x)\| = |\alpha|.\|Tx\|$

Now, $\|\alpha T\| = \sup.\{\|(\alpha T)x\| : \|x\| \leq 1\}$

$$= \sup.\{|\alpha|.\|Tx\| : \|x\| \leq 1\} = |\alpha|.\sup\{\|Tx\| : \|x\| \leq 1\}$$

Therefore, $\|\alpha T\| = |\alpha|.\|T\|$

(iv) Consider

$$\|T_1 + T_2\| = \sup.\{\|(T_1 + T_2)(x)\| : \|x\| \leq 1\} = \sup.\{\|T_1 x + T_2 x\| : \|x\| \leq 1\}$$
$$\leq \sup\{\|T_1 x\| + \|T_2 x\| : \|x\| \leq 1\}$$
$$= \sup\{\|T_1 x\|; \|x\| \leq 1\} + \sup\{\|T_2 x\| : \|x\| \leq 1\} = \|T_1\| + \|T_2\|$$

$\Rightarrow$ $\|T_1 + T_2\| \leq \|T_1 + T_2\|$

Hence, $B(N, N')$ is a normed linear space.

Further, we shall prove that $B(N, N')$ is a Banach space, when N' is given to be Banach.

Let $< T_n >$ be any Cauchy sequence in $B(N, N')$.

To show $< T_n >$ is convergent to some point in B(N, N').

Since $< T_n >$ is Cauchy , therefore by definition

$$\|T_m - T_n\| \le \in \quad \forall m, n \ge n_0$$

Now, for fixed x in N, we have

$$\|T_m(x) - T_n(x)\| = \|(T_m - T_n)x\| \le \|T_m - T_n\|.\|x\| \qquad [\because \|Tx\| \le \|T\|.\|x\|]$$

$\Rightarrow \quad \|T_m(x) - T_n(x)\| \to 0$

$\Rightarrow < T_n(x) >$ is Cauchy in N', but N' is complete and hence it is convergent.

Thus, there exists a unique point $y = T(x)$ in N' such that

$$\lim_{n \to \infty} T_n(x) = y \text{ for a fixed } x$$

Therefore, we can define a mapping $T : N \to N'$ such that $Tx = y = \lim T_n(x)$

To show T is linear and bounded.

Let $x_1, x_2 \in N$, then for any scalar a, we have

$$T(ax_1 + x_2) = \lim\{T_n(ax_1 + x_2)\} = \lim\{aT_n(x_1) + T_n(x_2)\}$$
$$= a \lim T_n(x_1) + \lim T_n(x_2) = aT(x_1) + T(x_2)$$

Therefore, T is linear.

Further, we shall prove that T is bounded.

Consider $\|Tx\| = \|\lim T_n(x)\| = \lim \|T_n(x)\|$

$$\le \lim(\|T_n\|\|x\|) \le \sup.\{\|T_n\|.\|x\|\} = \sup\{\|T_n\|.\|x\|\}$$

But $< T_n >$ is a Cauchy sequence, therefore

$$\|T_m - T_n\| \le \in$$

$\Rightarrow \quad \big| \|T_m\| - \|T_n\| \big| \le \|T_m - T_n\| \le \in$

$\Rightarrow \quad < \|T_n\| >$ is a Cauchy sequence on real line and therefore it is convergent and bounded.

$\Rightarrow \quad \exists k \ge 0$ such that $\sup.\|T_n\| \le k$ for some $k \ge 0$

$\Rightarrow \quad \sup.\|T_n\| \le k$ for $k \ge 0$

Therefore, we have

$$\|Tx\| \le \sup.\{\|T_n\|\}.\|x\| \le k.\|x\| \forall x$$

Hence, T is bounded.

Since T is both linear and bounded. Thus, $T \in B(N, N')$.

Now, it remains to prove that $T_n \to T$

Since $< T_n >$ is a Cauchy sequence, By definition, we have

$$\|T_m - T_n\| < \in \quad \forall m, n \ge n_0$$

Choose $x \in N$ such that $\|x\| \le 1$. We are given that $T_n(x) \to T(x)$ and therefore by definition, we can find a positive integer $n_x > n_0$ such that $\left\|T(x) - T_{n_x}(x)\right\| < \in$.

Thus, for all $n \ge n_0$ and $\|x\| < 1$, we have

$$\|T_n(x) - T(x)\| = \left\|T_n(x) - T_{n_x}(x) + T_{n_x}(x) - T(x)\right\|$$
$$\le \left\|T_n(x) - T_{n_x}(x)\right\| + \left\|T_{n_x}(x) - T(x)\right\|$$
$$= \left\|(T_n - T_{n_x})x\right\| + \left\|T_{n_x}(x) - T(x)\right\|$$

$$\leq \left\|T_n - T_{n_x}\right\| . \|x\| + \left\|T_{n_x}(x) - T(x)\right\| \quad \text{[By using } \|Tx\| \leq \|x\| \text{]}$$

$$\leq \left\|T_n - T_{n_x}\right\| + \left\|T_{n_x}(x) - T(x)\right\| \quad [\because \|x\| \leq 1]$$

$$< \in + \in \; = 2\in$$

Therefore $\|T_n(x) - T(x)\| < 2\in, \; \forall n \geq n_0$ and $\|x\| \leq 1, x \in N$

$\Rightarrow \quad \sup.\{\|T_n(x) - T(x)\| : x \in N, \quad \|x\| \leq 1\} \quad \leq 2\in, \; \forall\, n \geq n_0$

$\Rightarrow \quad \sup.\{\|(T_n - T)x\| : x \in N, \quad \|x\| \leq 1\} < 2\in, \; \forall n > n_0$

Thus, $\qquad \|T_n - T\| < 2\in, \; \forall n > n_0$

$\Rightarrow \qquad T_n \to T$.

Hence, $B(N, N')$ is complete and therefore a Banach space.

THEOREM 3. ***Let N be a Banach space, then B(N) is a Banach algebra with the algebraic operators***

$$(T + U)(x) = T(x) + U(x)$$
$$(TU)(x) = T(U(x))$$
$$(\alpha T)(x) = \alpha T(x)$$

and the operator norm

$$\|T\| = \sup.\{\|T(x)\| : x \in N, \|x\| \leq 1\}$$

Also, multiplication in* B(*N*) *is jointly continuous.

(MEERUT–2000, 05, 14, KANPUR–2006)

PROOF. Using above theorem, we can easily say that $B(N)$ is a Banach space. Therefore, it remains to prove that $B(N)$ is a normed algebra.

Consider

$$[T(U + V)](x) = T(U + V)(x)$$
$$= T(U(x) + V(x)) = T(U(x) + T(V(x)) = (TU)(x) + (TV)(x)$$

$\Rightarrow \qquad T(U + V) = TU + TV$

Further $\qquad \|TU\| = \sup\{\|(TU)(x)\| : x \in N, \|x\| \leq 1\}$

$$= \sup\{\|T(Ux)\| : x \in N, \|x\| \leq 1\}$$
$$\leq \sup\{\|T\| . \|U(x)\| : x \in N, \|x\| \leq 1\}$$
$$= \|T\| \sup.\{\|U(x)\| : x \in N, \|x\| \leq 1\} = \|T\|\|U\|$$

Further, it remains to prove that multiplication in $B(N)$ is jointly continuous

i.e. $\quad T_n \to T, U_n \to U \Rightarrow T_n U_n \to TU$

Consider $\quad \|T_n U_n - TU\| = \|T_n(U_n - U) + (T_n - T)U\|$

$$\leq \|T_n\|\|U_n - U\| + \|T_n - T\|\|U\| \to 0$$

Hence $\qquad T_n U_n \to TU$

THEOREM 4. ***Let N be a normed linear space. Then each vector u in N includes a functional F_u on N^* defined by***

$$F_u(f) = f(u) \; \forall f \in N^* \text{ such that } \|F_u\| = \|u\|$$

Further the map $J : N \to N^*$ such that $J(u) = F_u \; \forall u \in N$ defines an isometric isomorphism of N into N^{**}.

(AGRA–2001, MEERUT–2003, 07, KUMAYON–2010)

PROOF. Let N and N^* be two normed linear spaces where $u \in N$ and $f \in N^*$. Then

$$f : N \to K \qquad \ldots(1)$$

is a linear continuous transformation. Now we have to prove that F_u is functional and also show that F_u is linear and continuous.

Let us assume that $f, g \in N^*$ and $\alpha, \beta \in K$ be arbitrary, where K is a field of scalars

$$F_u(\alpha f + \beta g) = (\alpha f + \beta g)(u)$$
$$= \alpha f(u) + \beta g(u) = \alpha F_u(f) + \beta F_u(g)$$

This implies that F_u is a linear transformation.

$$|F_u(f)| = |f(u)| \le \|f\| \cdot \|u\|$$

Now taking $\lambda = \|u\|$ = constant mapping of F_u,

$$|F_u(f)| \le \lambda \|f\|$$

Hence F_u is bounded and continuous. So F_u is linear and continuous such that F_u is a functional on N^*.

$$F_u \in (N^*)^* = N^{**}$$

Now $\|F_u\| = \sup\{|F_u(f)| : \|f\| \le 1\}$

$= \sup\{|f(u)| : \|f\| \le 1\}$ [From def of F_u]

$\le \sup\{\|f\| \cdot \|u\| : \|f\| \le 1\}$

$\Rightarrow$ $|f(x)| \le \|f\| \cdot \|u\| \le \|u\|$

$\Rightarrow$ $\|f_u\| \le \|u\| \ \forall\ u \in N$ ...(2)

Let u be any non-zero vector. There exist $g \in N^*$ such that $g(u) = \|u\|$ ...(3)

and $g(u) = 1$

Also, $g(u) = F_u(g)$ [From (i)]

and using eq^n (3), we get

$$F_u(g) = \|u\| = g(u)$$

$\Rightarrow$ $\|u\| = g(u) \le \sup\{|g(u)| : g \in N^*, \|g\| \le 1\}$

$= \sup\{|F_u(g)| : g \in N^*, \|g\| \le 1\} = \|F_u\|$

$\Rightarrow$ $\|u\| \le \|F_u\|$ ...(4)

$\Rightarrow$ $\|F_u\| = \|u\|$ [From eqn (2) and (4)] ...(5)

We have to show that

$$J : N \to N^*$$

such that $J(u) = F_u \ \forall\ u \in N.$ ...(6)

is an isomorphism. Consider

$$F(\alpha u + \beta v)(f) = f(\alpha u + \beta v) = \alpha f(u) + \beta f(v) = \alpha F_u(f) + \beta F_v(f)$$
$$= (\alpha F_u + \beta F_v)(f)$$

$\Rightarrow$ $F(\alpha u + \beta v) = \alpha F_u + \beta F_v$

$\Rightarrow$ $J(\alpha u + \beta v) = \alpha J(u) + \beta J(v)$

Hence, J is linear. Now,

$$\|J(u)\| = \|F(u)\| = \|u\|$$

$\Rightarrow$ $\|J(u)\| = \|u\|$ ...(7)

Hence, J preserves the norm.

$\Rightarrow$ $\|J(u - v)\| = \|u - v\|$ [From (7)]

$\Rightarrow$ $\|J(u) - J(v)\| = \|u - v\|$ ...(8)

$$J(u) = J(v)$$

$\Rightarrow \quad J(u) - J(v) = 0 \quad \Rightarrow \quad J(u - v) = 0$

$\Rightarrow \quad \|u - v\| = 0$ [From (8)]

$\Rightarrow \quad u = v$

Hence, J is one-one mapping.

4.13 EQUIVALENT NORMS

Let L be a normed linear space which is converted into a normed linear space in two ways *i.e.* $\|x\|_1$ and $\|x\|_2$ $\forall x \in L$. Then, these norms are said to be equivalent if and only if they generate the same topology on L. It is expressed as $\|x\|_1 \sim \|x\|_2$.

REMARK

- Since, the norm and metric are related to each other by $d(x, y) = \|x - y\|$, showing that norm on a linear space induces the metric which in turn a topology on L, called the metric topology. This is the topology by the norm.

THEOREM 1. ***Let N be a normed linear space and suppose two norms $\|\cdot\|_1$ and $\|\cdot\|_2$ are defined on N. Then these norms are equivalent if and only if there exists positive real numbers m and M such that***

$$m\|x\|_1 \leq \|x\|_2 \leq M\|x\|_1 \quad \forall x \in N$$

PROOF. Suppose that the normed linear space be denoted by N_1 and N_2 with $\|.\|_1$ and $\|.\|_2$.

Define $T : N_1 \to N_2$ by $T(x) = x \quad \forall x \in N$

i.e., T is identity transformation on N on which two norms are defined.

Thus $T^{-1} : N_2 \to N_1$ such that $T(x) = x, T^{-1}(x) = x$

Now, since T is continuous and therefore bounded

$\Leftrightarrow$ $\exists$ a positive number M such that

$$\|Tx\|_2 \leq M\|x\|_1$$

$\Leftrightarrow \quad \|x\|_2 \leq M\|x\|_1 \quad \forall x \in N \quad [\because Tx = x]$ ...(1)

Further, since T^{-1} is continuous and therefore, bounded

$\Leftrightarrow$ $\exists$ a positive number k such that

$$\left\|T^{-1}(x)\right\|_1 \leq k\|x\|_2 \quad \Leftrightarrow \|x\|_1 \leq k\|x\|_2 \quad [\because T(x) = x]$$

$\Leftrightarrow \quad \frac{1}{k}\|x\|_1 \leq \|x\|_2$

$\Leftrightarrow \quad m\|x\|_1 \leq \|x\|_2 \quad \left(\frac{1}{k} = m\right)$ ...(2)

Combining (1) and (2), we get

$$m\|x\|_1 \leq \|x\|_2 \leq M\|x\|_1 \qquad ...(3)$$

Now, it remains to prove that two norms induce the same topology on N.

Further, $\quad T : N_1 \to N_1$ and $T^{-1} : N_2 \to N_2$

$\Leftrightarrow$ Inverse images of open sets in N_2 and N_1 under T and T^{-1} are open sets in N_1 and N_2 respectively.

$\Leftrightarrow$ Open sets in N_1 and N_2 are the same because both T and T^{-1} are identity mappings.

$\Leftrightarrow$ $\|.\|_1$ and $\|.\|_2$ induce the same topology.

Hence, the two norms $\|.\|_1$ and $\|.\|_2$ are equivalent iff

$$m\|x\|_1 \leq \|x\|_2 \leq M\|x\|_1$$

THEOREM 2. ***All norms are equivalent on a finite dimensional space.*** (KANPUR–1991)

PROOF. Let N be a finite dimensional space with dimension $n\ (>0)$.

Let $\{e_1, e_2, ..., e_n\}$ be a basis for N.

Now, for $x \in N$, there exist unique scalars $\alpha_1, \alpha_2, ... \alpha_n$ such that

$$x = \sum_{i=1}^{n} \alpha_i e_i \qquad ...(1)$$

We can easily proved that

$$\|x\|_0 = max\|\alpha_i\| \qquad ...(2)$$

is a norm on N (zeroth norm). Now, it is sufficient to prove that any other norm on N is equivalent to this norm.

Let $\|.\|$ be any norm on N.

Then, using previous theorem we find two positive constants m and M such that

$$m\|x\|_0 \leq \|x\| \leq M\|x\|_0 \ \forall x \subset N \qquad ...(3)$$

For any $x \in N$, we have

$$\|x\| = \left\|\sum_{i=1}^{n} \alpha_i e_i\right\| \leq \sum_{i=1}^{n} |\alpha_i|\|e_i\| \leq \|x\|_0 \sum_{i=1}^{n} \|e_i\|$$

Take $M = \sum_{i=1}^{\infty} \|e_i\|$, therefore, for any $x \in N$, we have

$$\|x\| \leq M\|x\|_0 \qquad ...(4)$$

Now, consider the case of a one-dimensional space with basis $\{e_1\}$. Also, any vector x is written uniquely as $x = \alpha_1 e_1$ for some scalar α_1.

Thus $\|x\| = \|\alpha_1 e_1\| = |\alpha_1|\|e_1\| = \|x\|_0 \|e_1\|$

So in this case, the number m on the LHS of (3) can be taken to be just $\|e_1\|$.

Suppose that theorem is true for all spaces of dimension basis less than or equal to $n-1$, let $M = \{e_1, e_2, ..., e_{n-1}\}$ be the subspace spanned by the first $n-1$ basis vector. Then by our induction hypothesis

$$\|.\| \sim \|.\|_0 \text{ on } M$$

Let $< y_n >$ be a Cauchy sequence in M w.r.t. the norm $\|.\|$.

$\Rightarrow$ $< y_n >$ is a Cauchy sequence w.r.t. $\|.\|_0$. [$\because$ Two norms are equivalent]

Now, consider the jth term of this sequence

We have $$y_j = \alpha_1^{(i)} e_i + \alpha_2^{(i)} e_2 + ... + \alpha_{n-1}^{(i)} e_{n-1}$$

For some uniquely determined scalars $\alpha_1^{(i)} ... \alpha_{n-1}^{(i)}$

Since, $< y_n >$ is Cauchy, therefore by definition, we have

$$\|y_n - y_m\|_0 \to 0 \text{ as } m, n \to \infty \qquad ...(5)$$

But $$\|y_n - y_m\|_0 = max\left|\alpha_k^{(n)} - \alpha_k^{(m)}\right|$$

Therefore using (5) we can write

$$\left|\alpha_k^{(n)} - \alpha_k^{(m)}\right| \to 0 \text{ as } n, m \to \infty \qquad ...(6)$$

Now, since α_k 's $\in R$ (or C) each of which is complete, therefore, there must exist scalars $\alpha_1, \alpha_2, ..., \alpha_{n-1}$ such that

$$\alpha_k^{(m)} \to \alpha_k \text{ as } m \to \infty$$

Further, let $\quad y = \sum_{k=1}^{n-1} e_k \alpha_k$

Then clearly $y \in M$ and $y_m \to y$ with respect to zeroth norm. But $y_m \to y$ with respect to $\|.\|_0$ implies $y_m \to y$ with respect to $\|.\|$. Thus, we have proved that the subspace M is complete with respect to arbitrary norm and hence it is closed.

Now, consider the nth basis vector e_n and from the set

$$e_n + M = \{e_n + z : z \in M\}$$

Clearly, $e_n + M$ is isometric to M under the mapping $z \to e_n + z$

Since M is closed therefore, $e_n + M$ must also be closed

$\Rightarrow \quad (e_n + M)'$ is open.

To show $0 \notin (e_n + M)$ if $0 \in (e_n + M)$, then for some scalars $\beta_1, \beta_2, ... \beta_{n-1}$, we can write

$$0 = e_n + \beta_1 e_1 + \beta_2 e_2 + ... + \beta_{n-1} e_{n-1}$$

which is not possible because $e_n \neq 0$

$\therefore \quad 0 \notin (e_n + M) \quad \Rightarrow \quad 0 \in (e_n + M)'$

Thus, there exists a constant $C_n > 0$ such that the open sphere

$$S(0, C_n) = \{x : \|x\| < C_n\}$$

is contained in $(e_n + M)'$

$\Rightarrow \quad$ for any $x \in e_n + M$, we have $\|x\| \geq d$.

Therefore, for all scalars $\alpha_i (i = 1, 2, ... n-1)$, we have

$$\|\alpha_1 e_1 + \alpha_2 e_2 + ... + \alpha_{n-1} e_{n-1} + e_n\| \geq C_n \qquad ...(7)$$

If $\alpha_n \neq 0$, replacing α_i by $\alpha_i / \alpha_n \quad (i = 1, 2, ... n-1)$ in (7), we get

$$\left\| \frac{\alpha_1}{\alpha_n} e_1 + ... + \frac{\alpha_{n-1}}{\alpha_n} e_{n-1} + e_n \right\| \geq C_n$$

or

$$\|\alpha_1 e_1 + \alpha_2 e_2 + ... + \alpha_n e_n\| \geq |\alpha| C_n$$

Let us suppose we have taken $(e_1, e_2, ... e_{n-1}\}$ in place of $\{e_1, e_2, ... e_{n-1}\}$.

Since the only essential point about M in our discussion was that its dimension was $n - 1$, it is clear that in a similar manner, we could has arrived at some $C_i > 0$ such that

$$\|\alpha_1 e_1 + \alpha_2 e_2 + ... + \alpha_n e_n\| \geq C_i |\alpha_i|$$

Thus, for any $x = \sum_{i=1}^{n} \alpha_i e_i$, we have

$$\|\alpha_1 e_1 + \alpha_2 e_2 + ... + \alpha_n e_n\| \geq \min_i C_i \max_i |\alpha_i| = \min C_i \|x\|_o$$

Taking $m = \min\{C_i\}$, we have

$$m\|x\|_0 \leq \|\alpha_1 e_1 + ... + \alpha_n e_n\| = \|x\|$$

Hence, $\quad m\|x\|_0 \leq \|x\| \leq M\|x\|_0$

THEOREM 3. ***Every complete subspace of a normed linear space is closed.***

PROOF. Let N be a normed linear space and M be a complete subspace of N. To prove that M is closed. We know that a set is said to be closed if it contains all its limit points.

Let t be any limit point of M. Then by definition of limit point, for every positive integer n the open sphere

$$S\left(t,\frac{1}{n}\right)-\{x:\|x-t\|<\frac{1}{n}\}$$

must contain a point y_n of M.

Also, the sequence $< y_n >$ converges to t and therefore, it is a Cauchy sequence in M, also since M is complete .

Therefore $t \in M$.

Finally, since t is arbitrary, we conclude that every limit point of M belongs to M. Hence M is closed.

THEOREM 4. ***Every compact subset of a normed linear space is bounded.***

PROOF. Let A be a compact subset of a normed space $(N, \|.\|)$. Let $x \in A$ and $x \in A'$ simultaneously. We know that every normed space is a metric space and every metric space being normed is a Housdroff space, then N is a Housdroff space. Therefore, there exist distinct elements of N having disjoint nbds and hence for some real number r_x there exists an open sphere $S(x, r_x)$ and $S(y, r_x)$ such that

$$S(x, r_x) \cap S(y, r_x) = d \qquad \ldots(1)$$

Also, $\qquad A \subset \bigcup_{i=1}^{n} S(x,r_x)$

Assumming that A is a compact subset of a normed space N is not bounded, every open covering of A consists of unit-sphere $S(x_i, 1)$ with centres at each of its points $x_i : i = 1, 2, \ldots, n$ such that

$$A \subset \bigcup_{i=1}^{n} S(x_i,1)$$

Take $\qquad k = \max. \|x_i\| : 1 \le i \le n \qquad \ldots(2)$

Also, assume that $\exists$ an $x \in A$ such that $\|x\| > 1 + k \qquad \ldots(3)$

Since, A is not bounded, we must have an element x_i such that

$$x \in S(x_i, 1) \text{ for } x \in A \text{ and } A \subset \bigcup_{i=1}^{n} S(x_i,1)$$

$\Rightarrow \qquad \|x - x_i\| \subset I \qquad$ (By def. of open sphere)

Now,

$$\|x\| = \|x - x_i + x_i\|$$
$$\le \|x - x_i\| + \|x_i\|$$
$$\le 1 + \max. \|x_i\| = 1 + k \qquad \text{(Using (2))}$$

$\Rightarrow \qquad \|x\| \le 1 + k$, which is a contradiction. $\qquad$ (From (3))

Hence, A is bounded.

REMARK

- Converse of the above theorem is not true.

THEOREM 5. ***Every finite dimensional normed linear space is complete.***

(GARHWAL–2002)

PROOF. Let N be a finite dimensional normed linear space and $B = \{e_1, e_2, ..., e_n\}$ be a basis for N, therefore

$$x = \sum_{i=1}^{n} \alpha_i e_i \qquad ...(1)$$

Since, all norms on a finite dimensional linear space are equivalent.

Therefore it is sufficient to establish the complete zero of N with respect to zeroth norm defined as

$$\|x\|_0 = \max_i \|\alpha_i\| = \max_i |\alpha_i| \qquad ...(2)$$

Now, let $< y_n >$ be any Cauchy sequence in N, we have

$$y_k = \sum_{i=1}^{n} \alpha_i^k e_i = \alpha_1^k e_1 + \alpha_2^k e_2 + ... + \alpha_n^k e_n$$

such that all the n scalars α_i^k are unique.

Now $\quad y_n - y_m = \sum_{i=1}^{n} (\alpha_i^n - \alpha_i^m) e_i$

$$\therefore \quad \|y_n - y_m\|_0 = \max_i \left|\alpha_i^n - \alpha_i^m\right| \qquad ...(3)$$

Further, since $< y_n >$ is a Cauchy sequence in N, we have

$$\|y_n - y_m\|_0 \to 0 \text{ as } m, n \to \infty$$

Therefore, from (3), we have

$$\left|\alpha_i^n - \alpha_i^m\right| \to 0, \text{ as } m, n \to \infty$$

But α_i' belongs to either C or R each of which is complete so, there exist scalars $\alpha_1, \alpha_2 ... \alpha_n$ such that

$$\alpha_i^m \to \alpha_i \text{ as } n \to \infty$$

$$\Rightarrow \qquad y = \sum_{i=1}^{n} \alpha_i e_i \in N \text{ and } y_m \to y$$

Hence, N is complete.

THEOREM 6. ***Let N be a normed linear space and suppose that the surface of unit sphere S = {x∈N : ‖x‖ > 1} is compact, then N is finite dimensional.***

PROOF. Let N be a normed linear space and $S = \{x \in N : \|x\| = 1\}$ is a unit sphere which is compact.

Clearly, every sequence in S has a convergent subsequence.

We have to prove that N is finite dimensional.

Let if possible, N is not finite dimensional. Select $x_1 \in S$ and N_1 be subspace of N spanned by x_1. Therefore, N_1 is finite dimensional as dim $N_1 = 1$, and so it must be closed.

Since N_1 is a proper subspace of N which implies there exist $x_2 \in S$ such that

$$\|x_2 - x_1\| > 1 - \varepsilon \qquad \text{(By Riesz Lemma)}$$

Now taking $\varepsilon = 1/2$, $\quad \|x_2 - x_1\| > 1/2$

Further, let N_2 be a closed proper subspace of N generated by $\{x_1, x_2\}$ then again by Riesz Lemma $\exists\ x_3 \in S$ such that $\|x_3 - x_2\| > 1/2$ and $\|x_3 - x_1\| > 1/2$.

Proceeding in the same way, we get a sequence $<x_n>$ such that

$$\|x_m - x_n\| > 1/2$$

When m and n are positive integers such that $m \neq n$ and $\|x_n\| = 1 \ \forall \ n$. This sequence of points on surface of unit sphere has no convergent subsequence, which is a contradiction because S is compact.

Hence, N must be finite dimensional.

THEOREM 7. ***Every compact subset of a normed linear space is complete.***

(KANPUR–2005, BHOPAL–1998, 2007)

PROOF. Let $<x_n>$ be a Cauchy sequence belonging to a compact subset A of a normed linear space N. Since, A is compact, therefore, the sequence $<x_n>$ has a convergent subsequence $<x_m>$ and so $< x_{m_i} > \to x_0 \in A$.

Now for any i,

$$\left\|x_i - x_0\right\| = \left\|x_i - x_{n_i} + x_{n_i} - x_0\right\| \leq \left\|x_i - x_{n_i}\right\| + \left\|x_{n_i} - x_0\right\|$$

$$< \varepsilon/2 + \varepsilon/2 = \varepsilon \qquad [\because \left\|x_i - x_{n_i}\right\| < \varepsilon/2 \text{ as } <x_i> \text{ is a Cauchy}$$

$$\text{sequence and} \left\|x_{n_i} - x_0\right\| < \varepsilon / 2 \text{ as } \lim x_{n_i} = x_0]$$

$$\to 0 \text{ as } i \to \infty$$

$\Rightarrow \quad \lim\limits_{i \to \infty} x_i = x_0, \ x_0 \in A$

$\Rightarrow \quad <x_n>$ is a convergent sequence in A.

Hence, A is complete.

THEOREM 8. ***Let N and N′ be normed linear spaces and let $T : N \to N'$ be any linear transformation. If N is finite dimensional, then T is continuous.***

(MEERUT–2001, 02, 03, 05BP, GARHWAL–2005, KANPUR–2007, 08)

PROOF. As per given, we have N is finite dimensional, therefore it must here a basis

$$B = \{e_1, e_2, ..., e_n\}$$

Thus, $\quad x = \sum\limits_{i=1}^{n} \alpha_i e_i, \quad x \in N$

It is also given that $T : N \to N'$ is linear, therefore

$$T(x) = T\left(\sum_{i=1}^{n} \alpha_i e_i\right) = \sum_{i=1}^{n} \alpha_i T(e_i) \qquad ...(1)$$

Further, we know that all norms on a finite dimensional linear space are equivalent. Thus, we will prove the theorem with respect to zeroth norm on N, defined as

$$\| x \|_0 = \max_i |\alpha_i| \qquad ...(2)$$

Let the norm on N' be denote by $\|.\|$ and $T(x) \in N'$

Therefore,

$$\|T(x)\| = \left\|\sum_{i=1}^{n} \alpha_i T(e_i)\right\| \leq \sum |\alpha_i| \|T(e_i)\|$$

$$\leq \| x \|_0 \sum_{i=1}^{n} \|T(e_i)\| \qquad \text{[Using (2)]} \qquad \ldots(3)$$

But since basis B is fixed so that $\sum_{i=1}^{n} \|T(e_i)\|$ is a positive constant say M. Thus, from (3) we have

$$\|T(x)\| \leq M \| x \|_0$$

$\Rightarrow$ T is bounded.

Hence, T is continuous.

RECAPITULATIONS

➠ All norms are equivalent to a finite dimensional space.	➠ Every complete subspace of a normed linear space is closed.
➠ Every compact subset of a normed linear space is bounded.	➠ Every finite dimensional normed linear space is complete.
➠ Every compact subset of a normed linear space is complete.	

4.14 CONTINUOUS LINEAR FUNCTIONALS

(i) Functional. A linear bounded or continuous transformation from any arbitrary normed linear space N into R or C is called a functional.

4.14.1 NOTATIONS

(1) $B(N, N')$: The set of all linear transformation $T : N \to N'$, where N and N' both are normed linear spaces.

(2) $B(N)$: The set of all linear operators on N onto itself.

(3) The set $B(N, R)$ or $B(N, C)$ will be the set of all functional on N both there sets will be normed linear space. Also both are Banach spaces. These spaces are denoted by N^* and is called the conjugate of N.

(4) N^{**} : The second conjugate space of N.

We know that N^*, *i.e.*, $B(N, R)$ or $B(N, C)$ is again a normed linear space and the set of all linear continuous functionls on N^*, *i.e.*, $B(N^*, R)$ or $B(N^*, C)$ will be called the second conjugate of N and denoted by N^{**}.

☛ **ILLUSTRTIONS**

(1) The element of N^* are continuous linear functional on N.

(2) f is continuous iff it is bounded.

(3) If f is continuous at $x_0 \in N$, then it is continuous at every point of N.

4.14.2 NORM OF A FUNCTIONAL f ON N

We know that

$$\|T\| = \sup.\{\|Tx\| : \|x\| \leq 1\} = \inf.\{k : k \geq 0 \text{ and } \|Tx\| \leq k\|x\|, \ \forall x\}$$

Similarly, we can define

$$\|f\|_N = \sup.\{|f(x)| : x \in N \text{ and } \|x\| \leq 1\}$$
$$= \inf.\{k : k \geq 0 \text{ and } |f(x)| \leq k.\|x\| \forall x\}$$

4.14.3 EXTENSION OF A LINEAR FUNCTIONAL f

Let N be a normed linear space and M be a subspace of N and let f be a linear functional defined on M, then the functional f_0 on the whole space N will be called the extension of f if

$$f_0(x) = f(x) \quad \forall\ x \in M$$

It is also expressed as $f_0 \,|M| = f$, *i.e.*, the functional value of f_0 for all elements of M is the same as the corresponding functional value of f on M. It is also called the restriction of f_0 on M.

4.15 HAHN BANACH THEOREM

LEMMA. ***Let M be a linear subspace of a normed linear space N and let f be a functional defined on M. If x_0 is a vector not in M and if***

$$M_0 = M + \{x_0\} = (m + \alpha x_0 : m \in M, \alpha \in R)$$

is the linear subspace spanned by M and x_0, then f can be extended to a functional f_0 defined on M_0 such that

$$\| f_0 \| = \| f \|$$

(DELHI–2009, KANPUR–1984, 2009, 11, MEERUT–1985,88,87, 2004, 06, 09, 11, 14 SAGAR–2005, AGRA–2011, ROHILKHAND–2008, AVADH–2009)

PROOF. **Step-I**

When *N* is a real normed linear space :

Let N be a normed linear space over the field of scalars. It is given that f is a functional on M, which is a subspace of N.

Without loss of any generality, we may assume that $\| f \| = 1$

Since, $x_0 \notin M$, therefore each vector $y \in M_0$ can be expressed uniquely as

$$y = m + \alpha x_0 \text{ with } m \in M.$$

Define f_0 on M_0 as

$$\begin{aligned} &= f_0(y) = f_0(m + \alpha x_0) \\ &= f_0(m) + \alpha f_0(x_0) \\ &= f(m) + \alpha r_0 \end{aligned} \qquad \ldots(1)$$

When $r_0 = f_0(x_0)$ and by definition of extension $f_0(m) = f(m)$

To show f_0 is linear.

Let $y_1, y_2 \in M_0$ such that $y_1 = m_1 + \alpha_1 x_0$ and $y_2 = m_2 + \alpha_2 x_0$

$$\therefore \quad Cy_1 + y_2 = (Cm_1 + m_2) + C(\alpha_1 + \alpha_2)x_0$$

Also, from (1)

$$\begin{aligned} f_0(y_1) &= f_0(m_1 + \alpha_1 x_0) \\ &= f_0(m_1) + \alpha_1 f_0(x_0) \\ &= f(m_1) + \alpha_1 r_0 \end{aligned}$$

Similarly $f_0(y_2) = f(m_2) + \alpha_2 r_0$

$$\begin{aligned} \therefore\ f_0 C(y_1 + y_2) &= f_0(C(m_1 + m_2) + C(\alpha_1 + \alpha_2)x_0) \\ &= f(Cm_1 + m_2) + C(\alpha_1 + \alpha_2)r_0) \\ &= Cf(m_1) + f(m_2) + \alpha_1 C r_0 + \alpha_2 r_0 && [\because f \text{ is linear}] \\ &= C(f(m_1) + \alpha_1 r_0) + (f(m_2) + \alpha_2 r_0) \\ &= Cf_0(y_1) + f_0(y_2) && [\text{Using (1)}] \end{aligned}$$

$\Rightarrow$ f_0 is linear.

Further, f_0 is an extension of f because if $m \in M$.

Then $m = m + 0 \cdot x_0$ so that

$$f_0(m) = f_0(m + 0x_0) = f(m) + 0.r_0 = f(m)$$

$$\Rightarrow \quad f_0(m) = f(m) \quad \forall\, m \in M$$

$\Rightarrow$ f_0 is an extension of f over M.

Now, it remains to proved that $\| f_0 \| = \| f \|$ or $\| f_0 \| = 1$

we have assumed that $\| f \| = 1$

Now, let us assume that $\| f_0 \| = 1$

we know that $\|Tx\| \le \|T\|.\|x\|$

Therefore, $\quad |f_0 y| \le \|f_0\|.\|y\|, \quad y \in M_0$

$$\Rightarrow \quad |f_0(m + \alpha x_0| \le \|m + \alpha x_0\| \qquad [\because \|f_0\| = 1]$$

$$\Rightarrow \quad |f(m) + \alpha r_0| \le \|m + \alpha r_0\|$$

$$\Rightarrow \quad -\|m + \alpha r_0\| \le f(m) + \alpha r_0 \le \|m + \alpha x_0\|$$

$$\Rightarrow \quad -f(m) - \|m + \alpha x_0\| \le \alpha r_0 \le -f(m) + \|m + \alpha x_0\|$$

$$\Rightarrow -\alpha f\left(\frac{m}{\alpha}\right) - |\alpha|\left\|\frac{m}{\alpha} + x_0\right\| \le \alpha r_0 \le -\alpha f\left(\frac{m}{\alpha}\right) + |\alpha|.\left\|\left(\frac{m}{\alpha}\right) + x_0\right\| \qquad \ldots(2)$$

If α is positive, we can divide by α and using

$$kT(\alpha) = T(kx)$$

We get $-f\left(\frac{m}{\alpha}\right) - \left\|\frac{m}{\alpha} + x_0\right\| \le r_0 \le -f\left(\frac{m}{\alpha}\right) + \left\|\left(\frac{m}{\alpha}\right) + x_0\right\| \qquad \ldots(3)$

If α is negative, we say that $\alpha = -\beta$, where $\beta > 0$ then we can divide by α, *i.e.*, $-\beta$, we get

$$-f\left(\frac{m}{\alpha}\right) + \left\|\frac{m}{\alpha} + x_0\right\| \ge r_0 \ge -f\left(\frac{m}{\alpha}\right) - \left\|\frac{m}{\alpha} + x_0\right\|$$

$$-f\left(\frac{m}{\alpha}\right) - \left\|\frac{m}{\alpha} + x_0\right\| \le r_0 \le -f\left(\frac{m}{\alpha}\right) + \left\|\frac{m}{\alpha} + x_0\right\| \qquad \ldots(4)$$

Combining (3) and (4), we get

$$-f\left(\frac{m}{\alpha}\right) - \left\|\frac{m}{\alpha} + x_0\right\| \le r_0 \le -f\left(\frac{m}{\alpha}\right) + \left\|\frac{m}{\alpha} + x_0\right\| \qquad \ldots(5)$$

Now, we have to find r_0 such that inequality (5) holds.

Let x_1, x_2 be any two vectors in M, then

$$f(x_2) - f(x_1) = f(x_2) - f(x_1) \le |f(x_2 - x_1)|$$
$$\le \|f\|.\|x_2 - x_1\| = 1.\|x_2 - x_1\|$$

$$\Rightarrow \quad f(x_2) - f(x_1) \le \|(x_2 + x_0) - (x_1 + x_0)\| \le \|x_2 + x_0\| + \|-(x_1 + x_0)\|$$
$$= \|x_2 + x_0\| + |-1|\,\|x_1 + x_0\| = \|x_2 + x_0\| + \|x_1 + x_0\|$$

$$\Rightarrow \quad -f(x_1) - \|x_1 + x_0\| \le -f(x_2) + \|x_2 + x_0\|, \quad \forall\, x_1, x_2 \in M \qquad \ldots(6)$$

Define $$a = \sup.\{-f(x) - \|x + x_0\|,\ \ \forall x \in M\} \qquad \ldots(7)$$
$$b = \inf.\{-f(x) + \|x + x_0\|,\ \ \forall x \in M\} \qquad \ldots(8)$$

From (6) we have

$$a \le b$$

Now by denseness property of real numbers, we get

$$a \le r_0 \le b$$

Thus, for each $x \in M$, we have

$$-f(x) - \|x + x_0\| \le r_0 \le -f(x) + \|x + x_0\| \qquad \text{[Using (7), (8) and (9)]}$$

Choosing $x = \dfrac{m}{\alpha}, m \in M, \alpha \in F$

Therefore, $x = \dfrac{m}{\alpha} \in M$ and we get

$$-f\left(\frac{m}{\alpha}\right) - \left\|\frac{m}{\alpha} + x_0\right\| \le r_0 \le -f\left(\frac{m}{\alpha}\right) + \left\|\frac{m}{\alpha} + x_0\right\|$$

which shows that the required inequality (3) or (5) holds.

Hence, $\|f_0\| = 1$.

Step-II. Now, let N be any normed linear space and f is a complex valued functional on subspace M for which $\| f \| = 1$. Let g and h be real and imaginary parts of f, so that

$$f(x) = g(x) + ih(x),\ \text{for } x \in M \qquad \ldots(10)$$

To show g and h are linear.

Since $\quad f(cx + y) = cf(x) + f(y) \qquad [\because f \text{ is linear}]$

Therefore

$$g(cx + y) + ih(cx + y) = [c(g(x) + ih(x)] + [g(y) + ih(y)]$$

Equating real and imaginary parts, we get

$$g(cx + y) = cg(x) + g(y)$$
$$h(cx + y) = ch(x) + h(y)$$

Thus, g and h both are linear.

Also, both g and h are real valued linear functionals on real space M. Also, since $\| f \| = 1$, we have

$$|g(x)| \le |f(x)| \le \|f\|.\|x\| = \|x\|$$
$$|g(x)| \le \|g\|.\|x\| \qquad [\because \|g\| \le 1]$$

Now, $\quad f(x) = g(x) + ih(x)$

$\Rightarrow \quad f(ix) = g(ix) + ih(ix)$

$\therefore \quad f(x) = f(ix)$ implies $g(ix) + ih(ix) = i[(g(x) + ih(x)] = ig(x) - h(x)$

By equating real parts of both the sides, we get

$$h(x) = -g(ix)$$

$\therefore \quad f(x) = g(x) + ih(x)$

$\Rightarrow \quad f(x) = g(x) - ig(ix) \qquad \ldots(11)$

Since g is a functional on a real linear space M thus by step-I, it can extend to go defined on M_0 such that

$$\|g_0\| = \|g\|$$

Further, we define f_0 on M_0 as follows

$$f_0(x) = g_0(x) - ig_0(ix) \quad \forall x \in M_0 \qquad \ldots(12)$$

$\therefore$ for all $x \in M$, we have

$$f_0(x) = g_0(x) - ig_0(ix) = g(x) - ig(ix) = f(x) \qquad \text{[Using (11)]}$$

Thus, g_0 is an extension of g on M.

i.e., $g_0(x) = g(x) \quad \forall x \in M$ and $\| g_0 \| = \| g \|$

Since, $f_0(x) = f(x) \ \forall x \in M$, therefore f_0 is an extension of f from M to M_0.

Now, to show f_0 is linear.

Consider

$$f_0(x+y) = g_0(x+y) - ig_0(ix+iy) = g_0(x) + g_0(y) - ig_0(ix) - ig_0(iy)$$
$$= [g_0(x) - ig_0(ix)] + [g_0(y) - ig_0(iy)] = f_0(x) + f_0(y)$$

Further, if $a, b \in R$, then

$$f_0[(a+ib)x] = g_0(ax+ibx) - ig_0(iax - bx)$$
$$= ag_0(x) + bg_0(ix) - i\{ag_0(ix) - bg_0(x)\}$$
$$= a[g_0(x) - ig_0(ix)] + ib[g_0(x) - ig_0(ix)]$$
$$= af_0(x) + ibf_0(x)$$
$$= (a+ib)f_0(x)$$

$\Rightarrow$ f_0 is linear an complex linear space M_0.

Now, it remains to prove that $\| f_0 \| = 1$

where $\quad \| f_0 \| = \sup.\{|f_0(x)| : \| x \| = 1\} = 1$

For this, we shall prove that for all $x \in M_0$

$$\| x \| = 1, \ |f_0(x)| \le 1 \qquad \ldots(13)$$

If $f_0(x)$ is real, then $f_0(x) = g_0(x)$

$\therefore \quad |f_0(x)| = |g_0(x)| \le \|g_0\| \cdot \|x\| \le 1 \quad [\because \|g_0\| = \|g\| \le 1, \ \|x\| = 1]$

If $f_0(x)$ is complex, then let $f_0(x) = re^{i\theta}$

$\Rightarrow \quad |f_0(x)| = r = e^{-i\theta} f_0(x) = f_0(e^{-i\theta} \cdot x) = f_0(y) \quad$ (say)

$\Rightarrow \quad f_0(y) = r$, which is real.

Now, $\quad \| y \| = \left\| e^{-i\theta} \cdot x \right\| = \left| e^{-i\theta} \right| . \| x \| = 1 \cdot 1 = 1$

$$|f_0(y)| = \left| f_0(e^{-i\theta}.x) \right| \le 1 \qquad \text{[As proved for real]}$$

or $\quad \left| e^{-i\theta} f_0(x) \right| \le 1$

$\Rightarrow \quad \left| e^{-\theta} \right| . |f_0(x)| \le 1$, *i.e.*, $\quad 1 \cdot |f_0(x)| \le 1$

Finally, since $|f_0(x)| \le 1$, therefore

$$\|f_0\| = \{\sup.|f_0(x)| : \|x\| = 1\} = 1$$

Hence, we have extended f on M to a linear functional f_0 on M_0 such that $\|f_0\| = \|f\|$.

THEOREM 1. **(Main Hahn Banach Theorem).** ***Let M be a linear subspace of a normed linear space N and let f be a functional defined on M. Then, f can be extended to a functional f_0 on the whole space N such that $\| f \| = \| f_0 \|$.***

(MEERUT–1982, 2000, 04, 05, 05BP, 06, 09, 09BP, 11, KANPUR–1980, 82, 92, 2001, 09, 11, MADRAS–2006, DELHI–2007, GWALIOR–2009, CALCUTTA–2011, BUNDELKHAND–1983, 85, ROHILKHAND–2008, AGRA–2008, 11)

PROOF. Let M be a given linear subspace of N.

Using above lemma, we have, for any $x \in N$ and $x \notin M$ we can have an extension of M on $M + \{x\}$ such that $\| f \|$ is preserved for extension.

Now, consider the set G of all possible extension of f on all the subspace of N which contains M.

To show G is a partially ordered set,

Let $g_1, g_2 \in G$ and define the relation $\leq$ such that $g_1 \leq g_2$ means domain of g_1 is contained in the domain of g_2.

and $$g_1(x) = g_2(x) \quad \forall x \in \text{dom}(g_1)$$

(1) Reflexivity $g_1 \leq g_1 \ \ \forall \ g_1 \in G.$

(2) Antisymmetry Let $g_1 \leq g_2$ and $g_2 \leq g_1$

$\Rightarrow$ $\text{dom}(g_1) \subset \text{dom}(g_2)$ and $\text{dom}(g_2) \subset \text{dom}(g_1)$

$\Rightarrow$ $\text{dom}(g_1) = \text{dom}(g_2)$, and $g_1(x) = g_2(x) \ \ \forall \ x \in \text{dom}(g_1)$

$g_2(x) = g_1(x) \ \ \forall \ x \in \text{dom}(g_2)$

$\Rightarrow$ $g_1 = g_2$

(3) Transitivity

Let $g_1 \leq g_2$ and $g_2 \leq g_3$

$\Rightarrow$ $\text{dom}(g_1) \subset \text{dom}(g_2)$ with $g_1(x) = g_2(x) \ \forall x \in \text{dom}(g_1)$

and $\text{dom}(g_2) \subset \text{dom}(g_3)$ with $g_2(x) = g_3(x) \ \forall x \in \text{dom}(g_2)$

$\Rightarrow$ For all $x \in \text{dom}(g_1)$ and $x \in \text{dom}(g_1) \subset \text{dom}(g_2)$

$\Rightarrow$ $\text{dom}(g_1) \subset \text{dom}(g_3)$ with $g_1(x) = g_3(x) \ \ \forall x \in \text{dom}(g_1)$

$\Rightarrow$ $g_1 \leq g_3$

Thus, G is a partially ordered set.

So, by Zorn's lemma, there exists a maximal element in G say f_0. Now, it remains to show that f_0 is the required extension in N.

Let if possible there exists an $x \in N, x \notin M$ such that f_0 can be extended to domain of $f_0 + \{x\}$, *i.e.*, $M + \{x\}$ (By lemma), which is not possible, because this violates the maximality of f_0.

Hence, f_0 is the required extension of f to whole of N such that

$$\| f \| = \| f_0 \|$$

THEOREM 2. **(Generalised Hahn-Banach Theorem).** ***Let X be a real vector space, p is a real valued function on X such that***

$$\boldsymbol{p(u + v) \leq p(u) + p(v)}$$

$$\boldsymbol{p(\alpha u) = \alpha p(u) \text{ for } \alpha \geq 0}$$

and V is a subspace of V. If f is linear on V and f(v) ≤ p(v) ∀ v∈V then there exist a linear function F on U s.t. F(v) = f(v) on V and

F(u) ≤ p(u) on U. (ROHILKHAND–2009)

PROOF. Let U be a real vector space and V is a subspace of U. Then $V_1 = V \cup \{u_0\}$, where $u_0 \notin V$ but $u_0 \in U$

Then $$f(v) = g(v) \ \forall \ v \in V$$

We have to prove the existence of F.

Let F be defined on E where E is the family of extensions of g. Then

$$g_1 \le g_2, \text{ where } g_2 \text{ is an extension of } g_1.$$

Let $$g_1(u) = g_2(u) \ \forall \ u \in \text{Domain of } g_1 \subset \text{domain of } g_2$$

where g_1, g_2 are extensions of E.

Since $(E, \le)$ be a partially ordered set. Let C be the totally ordered subset of E. If $C \subset E$ then prove that T is a linear function such that

$$\text{Domain of } T = \text{Union of domains of all extensions } g \text{ for all } g \in C.$$

Now if u is any variable of domain of T. Then we have to prove that

$$T(u) = g(u) \qquad \dots(1)$$

where $g \in C$ and $x \in$ all extensions g

We know that T is a linear extension of f. Then

$$T(u) = g(u) = f(u) \le p(u)$$

$$T(u) \le p(u) \qquad \dots(2)$$

such that $T \in E$ and upper bound of C. From zorn's lemma if E has a maximal element F and F is an extension of f and domain U. Then

$$F(u) = f(u) \text{ on } V \qquad \text{[From (1)]}$$

and $$F(u) \le p(u) \text{ on } U \qquad \text{[From (2)]}$$

THEOREM 3. ***Let N be a normed linear space and $x_0 \in N$, $x_0 \neq 0$, then there exists a functional f_0 in N^* such that $f_0(x_0) = \|x_0\|$ and $\|f_0\| = 1$. In particular, if $x \neq y, (x, y \in N)$ then there exists an $f_0 \in N^*$ such that $f_0(x) \neq f_0(y)$.*** (MEERUT–2002, 04, 08, 09BP)

PROOF. Let $M = \{\alpha x_0\}$ be the space spanned by x_0, *i.e.*, every vector y in M can be uniquely written as $y = \alpha x_0$.

$$\therefore \quad \|y\| = |\alpha| \cdot \|x_0\|$$

Define f on M such that $f(y) = f(\alpha x_0) = \alpha\|x_0\|$

Clearly, $f : M \to f_0$, so that f is a functional.

Now, we shall prove that f is linear as well as continuous.

(i) *f* is linear

Let $y_1, y_2 \in M$ so that $y_1 = \alpha x_0, y_2 = \beta x_0$

Therefore

$$c_1 y_1 + c_2 y_2 = \{c_1(\alpha x_0) + c_2(\beta x_0)\} = (c_1\alpha + c_2\beta)\, x_0$$

$$\Rightarrow \quad f(c_1 y_1 + c_2 y_2) = f(c_1\alpha + c_2\beta)\, x_0 = (c_1\alpha + c_2\beta).\|x_0\|$$

$$= c_1\alpha\|x_0\| + c_2\beta\|x_0\| = c_1 f(y_1) + c_2 f(y_2)$$

$\Rightarrow \quad f$ is linear.

(ii) *f* is continuous (*i.e.*, bounded)

Consider $$|f(y)| = |f(\alpha x_0)| = |\alpha\|x_0\|| = |\alpha|\|x_0\| = \|y\|$$

$$\Rightarrow \quad |f(y)| = \|y\| \text{ or } |f(y)| \le \alpha\|y\|$$

$\Rightarrow$ f is bounded

$\Rightarrow$ f is continuous

Further

$$\|f\| = \sup.\{|f(y)| : \|y\| = 1\} = \sup\{\|y\| : \|y\| = 1\}$$

$$\Rightarrow \quad \|f\| = 1$$

Also $\quad f(x_0) = \|x_0\|$ [By choosing $\alpha = 1$]

Hence, by Hahn Banach theorem the functional f can be extended to a functional f_0 preserving norm, *i.e.*,

$$f_0(x_0) = f(x_0) = \|x_0\|$$

$$\Rightarrow \quad \|f_0\| = \|f\| = 1$$

Particularly, when $x \neq y$, *i.e.*, $x - y \neq 0 \Rightarrow \|x - y\| \neq 0$. Then, again by Hahn Banach theorem there exists a functional f_0 in N^* such that

$$f_0(x - y) = \|x - y\| \neq 0$$

$$\Rightarrow \quad f_0(x) - f_0(y) \neq 0$$

$$\Rightarrow \quad f_0(x) \neq f_0(y)$$

THEOREM 4. ***Let N be a normed linear space and x_0 is non-zero vector in N then there exists functional f_0 in N^* in such that***

$$\boldsymbol{f_0(x_0) = \|x_0\| \text{ and } \|f_0\| = 1}$$

(DELHI–2011, MEERUT–2008, 10, 11, AGRA–2001, 05, KANPUR–2005, 06, 08, GARHWAL–2006, AVADH–2008)

PROOF. Let N be the normed linear space over the field K of scalars and $x_0 \in N$ such that $x_0 \neq 0$.

Let $\quad M = \{\alpha x_0 : \alpha \in K\}$

Then M is a linear subspace of N and is generated by x_0. Now define a map

$$f : M \to K \text{ such that } f(\alpha x_0) = \alpha\|x_0\| \qquad \ldots(1)$$

Let $u, v \in M$, then $u = \alpha x_0$, $v = \beta x_0$, $\alpha, \beta \in K$.

Also, let $p, q \in K$ then

$$pu + qv = p\alpha x_0 + q\beta x_0 = (p\alpha + q\beta)x_0,\ p\alpha + q\beta \in K$$

$$\Rightarrow \quad f(pu + qv) = (p\alpha + q\beta)\|x_0\| = p\alpha\|x_0\| + q\beta\|x_0\|$$

$$= pf(\alpha x_0) + qf(\beta x_0) = pf(u) + qf(v)$$

$\Rightarrow$ f is linear.

Further,

$$|f(u)| = |\alpha|.\|x_0\| = \|\alpha x_0\| = \|u\| < 2\|u\| \qquad \ldots(2)$$

$$\Rightarrow \quad |f(u)| < 2\,\|u\|$$

$\Rightarrow$ f is bounded and hence continuous.

Now

$$\|f\| = \sup\{|f(u)| : u \in M, \|u\| \le 1\}$$

$$= \sup\{\|u\| : u \in M, \|u\| \le 1\} = 1$$

$$\Rightarrow \quad \|f\| = 1$$

Also, on taking $\alpha = 1$, $f(x_0) = \|x_0\|$

Therefore, we conclude that $f : M \to K$ such that $f(\alpha x_0) = \alpha\|x_0\|$ is continuous linear transformation. Hence, f is a functional on M such that

$$f(x_0) = \|x_0\| = \|f\| = 1. \qquad \ldots(3)$$

Now by Hahn-Banach theorem, the function f can be exended to a functional f_0 on N such that

$$\|f_0\| = \|f\| \qquad \ldots(4)$$

Using (3) and (4) we have

$$\|f_0\| = 1$$

Since, f_0 is an extension of f therefore, $f_0 = f$ on M

$$\Rightarrow \quad f_0(x_0) = f(x_0) \qquad (\because x_0 \in M)$$

$$\Rightarrow \quad f_0(x_0) = \|x_0\| \qquad \text{(Using (3))}$$

Finally, there exist $f_0 \in N^*$ such that $f_0(x) = \|x_0\|$ and $\|f_0\| = 1$.

REMARK

- Using the above theorem, we can also prove the following result
  "If $x \neq y$ and $x, y \in N$ then there exist $f_0 \in N^*$ such that $f_0(x) \neq f_0(y)$."

THEOREM 5. ***Let M be a closed linear subspace of a normed linear space N and let $x_0 \notin M$. If $d = d(x_0, M)$ then there exists a functional***

$$f_0 \in N^* \text{ such that } f_0(M) = 0,\ f_0(x_0) = 1,\ \|f_0\| = \frac{1}{d} \qquad \text{(MEERUT–2010)}$$

PROOF. Let M be a closed linear subspace of a normed linear space N and $x_0 \notin M$

Write $M_0 = (M \cup \{x_0\}) = \{x + \alpha x_0 : \alpha \in R\}$

$\Rightarrow$ M_0 is a linear subspace generated by M and $\{x_0\}$

Since, $x_0 \notin M$ therefore, $y \in M_0$ has a unique representation $y = x + \alpha x_0$ for some real number α. ...(1)

Further, let $\quad d =$ distance of x_0 from M, then

$$d = \text{inf. } (\|x - x_0\| : x \in M) \leq \|x - x_0\| \ \forall\, x \in M \qquad \ldots(2)$$

Since, $x_0 \notin M \Rightarrow x \neq x_0 \ \forall\, x \in M \Rightarrow \|x - x_0\| > 0 \ \forall\, x \in M$

$\Rightarrow d > 0$

Now, $d \leq \|x - x_0\| \quad \Rightarrow \quad \dfrac{1}{d} \geq \dfrac{1}{\|x - x_0\|}$

$$\Rightarrow \quad \frac{1}{\|x - x_0\|} \leq \frac{1}{d} \ \forall x \in M \qquad \ldots(3)$$

Now, define a map $f : M_0 \to K$ such that $f(y) = \alpha$ ...(4)

Since representation of y (given by (1)) is unique, therefore mapping f is well defined.

Now, $\quad f(x_0) = (0 + 1.x_0) = 1$ or $f(x_0) = 1$

and $\quad m \in M \quad \Rightarrow \quad f(m) = f(m + 0, 0.x_0) = 0$

$$\Rightarrow \quad f(M) = 0 \qquad \ldots(5)$$

So, we have proved that $f(M) = 0,\ f(x_0) = 1$

Now, $$|f(y)| = |\alpha| = \frac{|\alpha|\|y\|}{\|y\|} = \frac{|\alpha|.\|y\|}{\|x+\alpha x_0\|} = \frac{|\alpha|.\|y\|}{|\alpha|.\left\|\frac{x}{\alpha}+x_0\right\|}$$

$$= \frac{\|y\|}{\left\|x_0 - \left(\frac{-x}{\alpha}\right)\right\|}, -\frac{x}{\alpha} \in M$$

$$\leq \frac{\|y\|}{d} \qquad \text{(From (3))}$$

$$\Rightarrow \quad |f(y)| \leq \frac{\|y\|}{d} \qquad \ldots(6)$$

$$\Rightarrow \quad \|f\| = \sup\{|f(y)| : y \in M_0, \|y\| \leq 1\}$$

$$\leq \sup\left\{\frac{\|y\|}{d} : y \in M_0, \|y\| \leq 1\right\}$$

$$\Rightarrow \quad \|f\| = \frac{1}{d}$$

Therefore, we have shown that f is linear functional on M_0 such that

$$f(M) = 0, f(x_0) = 1, \|f\| = \frac{1}{d}$$

So, by Hahn Banach theorem, the function f can be extended to a functional f_0 on N such that $\|f_0\| = \|f\|$

i.e., $$\|f_0\| = \|f\| \text{ and } \|f\| = \frac{1}{d} \quad \Rightarrow \quad \|f_0\| = \frac{1}{d} \qquad \ldots(7)$$

Since, f_0 is extension of f, therefore, $f_0 = f$ on M_0

$$\Rightarrow \quad f_0 = f \text{ on } M_0 \supset M$$

$$\Rightarrow \quad f_0(M) = f(M) = 0 \qquad (\because f(M) = 0) \qquad \ldots(8)$$

Also, $f_0(x_0) = f(x_0)$

But, $f(x_0) = 1$

Hence, $f_0(x_0) = 1.$

SOLVED EXAMPLES

EXAMPLE 1. *Let N be a normed linear space and suppose that $f(x) = 0$ for all $f \in N^*$, then show that $x = 0$.* (MEERUT–1982)

SOLUTION. Let $x \neq 0$, then using previous theorem, there exists $f \in N^*$ such that $f(x) = \|x\| > 0$, which is a contradiction because, as per given $f(x) = 0$ for all $f \in N^*$. Thus, we must have $x = 0$.

EXAMPLE 2. *If M is a closed linear subspace of a normed linear space N and x_0 is a vector not in M, then show that there exists a functional f_0 in N^* such that $f_0(M) = 0$ and $f_0(x_0) \neq 0$.* (KANPUR–2006)

SOLUTION. We have already proved that $T : N \to N|M$ defined as $T(x) = x + M$ is a continuous linear transformation for which $\|T\| \le 1$.

If $m \in M$, then

$$T(m) = m + M = M = 0 \text{ of } N/M \quad ...(1)$$

If $\quad x_0 \notin M \Rightarrow T(x_0) = x_0 + M \ne M$, *i.e.*, 0 of N/M ...(2)

Thus, $T(x_0)$, *i.e.*, $x_0 + M \ne 0$, is a non-zero vector in N/M.

Hence, by theorem-4 (just before solved examples), there exists a functional *f* in $(N/M)^*$ such that

$$f(x_0 + M) = \| x_0 + M \| \ne 0 \quad ...(3)$$

Define f_0 on *N* as follows :

$$f_0(x) = f(T(x)) = f(x + M)$$

(i) **f_0 is linear**

Consider $\quad f_0(c_1x + c_2y) = f(T(c_1x + c_2y))$

$$= f((c_1x + c_2y) + M) = f[c_1(x + M) + c_2(y + M)]$$

$$= c_1 f(x + M) + c_2 f(y + M) \quad [\because f \text{ is linear}]$$

$$= c_1 f_0(x) + c_2 f_0(y)$$

$\Rightarrow \quad f_0$ is linear.

(ii) **f_0 is Bounded**

Consider

$$|f_0(x)| = |f(T(x)| \le \|f\| . \|Tx\|$$

$$\le \|f\|.\|T\|.\|x\| \le \|f\|.\|x\| \quad [\because \|T\| \le 1]$$

Now, as *f* is bounded, being a functional, therefore f_0, by above result f_0 is bounded.

$\Rightarrow \quad f_0$ is functional on *N*.

Also, $\quad f_0(m) = f(T(m)) = f(0) = 0 \ \forall m \in M$

$\Rightarrow \quad f_0(M) = 0$

and $\quad f_0(x_0) = f(T(x_0)) = f(x_0 + M) \ne 0$.

EXAMPLE 3. *Let M be a closed linear subspace of a normed linear space N and let x_0 be a vector not in M. If d is the distance from x_0 to M. Show that there exists a functional $F \in N^*$ such that*

$$F(M) = \{0\}, \ F(x_0) = d \text{ and } \|F\| = 1$$

(MEERUT–2003, 08)

SOLUTION. By definition, we have

$$d = \inf .\{\|x_0 - x\| : x \in M\}$$

Since, *M* is closed and $x_0 \notin M; d > 0$, consider the subspace

$$M_0 = \{x + \alpha x_0 : x \in M, \alpha \in R\}$$

spanned by *M* and x_0.

Since $x_0 \notin M$, thus for each vector *y* in M_0, we can write

$$y = x + \alpha x_0 \quad \text{[Unique representation]}$$

Now, define a map f_0 on M_0 by

$$f_0(y) = \alpha d \quad \text{where } y = x + \alpha x_0$$

It is clear that f_0 is linear on M_0.

Also $\quad f_0(x_0) = f_0(0+1.x_0) = 1.d = d$

If $m \in M$, then $\quad f_0(m) = f_0(m+0.x_0) = 0.d = 0$ so that $f_0(M) = \{0\}$

Now, to show that $\|f_0\| = 1$

Consider

$$\|f_0\| = \sup.\left\{\frac{|f_0(y)|}{\|y\|} : y \in M_0, y \neq 0\right\}$$

$$= \sup\left\{\frac{|f_0(x+\alpha x_0)|}{\|x+\alpha x_0\|} : x \in M, \alpha \in R, x \neq 0, \alpha \neq 0\right\}$$

$$= \sup\left\{\frac{|\alpha d|}{\|x+\alpha x_0\|} : x \in M, \alpha \in R,\ \alpha \neq 0\right\}$$

[By using $f(\alpha) = 0$ where $\alpha = 0$]

$$= \sup\left\{\frac{d}{\left\|x_0 + \dfrac{x}{\alpha}\right\|} : x \in M, \alpha \in R, \alpha \neq 0\right\}$$

[$\because d > 0, |\alpha d| = |\alpha| d$]

$$= d.\sup\left\{\frac{1}{\|x_0 - z\|} : z = -\frac{x}{\alpha} \in M\right\}$$

$$= d.[\inf.\{\|x_0 - z\| : z \in M\}^{-1} = d.\frac{1}{d} = 1$$

Therefore, f_0 is a linear functional on M_0, such that

$$f_0(m) = \{0\}, f_0(x_0) = d \text{ and } \|f_0\| = 1$$

Hence, by Hahn Banach theorem, there exists a functional F on N, such that

$$F(y) = f_0(y), \quad \forall y \in M$$

and $\quad \|F\| = \|f_0\|$.

EXAMPLE 4. *Let X and Y be normed linear mappings from a normed linear space N into N′. If S = {x ∈ N, ||x|| ≤ 1} is a closed unit sphere in N, then prove that its image T(S) is a bounded set in N′.*

(KANPUR–2008)

SOLUTION. **Step I.** If X and Y are normed linear transformation and N is a normed linear space into N'. Then

$$T : X \to Y$$

where T is bounded.

We have to prove that T is continuous. If there exist $\lambda > 0$ such that

$$\|T(x)\| \leq \lambda \|x\| \quad \forall\ x \in X \qquad \ldots(1)$$

Let us assume that $\langle s_n \rangle$ be a sequence in X.

Then $\quad \lim\limits_{n\to\infty} s_n = x$ where $x \in X$

Also, we know that a norm function is continuous function. Then

$$\lim_{n\to\infty} \|s_n\| = \|x\|$$

Therefore,

$$\lim_{n\to\infty} \|s_n - x\| = 0 \qquad \ldots(2)$$

$$\|T(s_n - x)\| \le \lambda \|s_n - x\| \qquad \text{[From eq}^n\text{ (1)]}$$

Taking $n \to \infty$ and then using eq^n (2), we have

$$\|T(s_n - x)\| \le 0 \text{ as } n \to \infty$$

But we know that $\|T(s_n - x)\| \ge 0 \ \forall\ n$

Then $\quad \lim_{n\to\infty} \|T(s_n - x)\| = 0$

$$\Rightarrow \quad \lim_{n\to\infty} T(s_n) = T(x)$$

$$\Rightarrow \quad s_n \to x \quad \Rightarrow \quad \lim_{n\to\infty} T(s_n) = T(x)$$

Hence, we can say that T is continuous map.

Step II From step (i), we can say that T is a continuous mapping. We have to prove that T is bounded. Now, let if possible T is not bounded for every positive integer n. Find $s_n \in X$ such that

$$\|T(s_n)\| > n\|s_n\|$$

$$\Rightarrow \quad \frac{1}{n\|s_n\|}\|T(s_n)\| > 1 \qquad \ldots(3)$$

Let us write

$$x_n = \frac{s_n}{n\|s_n\|} \qquad \ldots(4)$$

$$\Rightarrow \quad \|T(x_n)\| > 1 \ \forall\ n \qquad \text{[From (3)]}$$

Then $\quad \|x_n\| = \left\|\frac{s_n}{n\|s_n\|}\right\| = \frac{1}{n}$

$$\Rightarrow \quad \lim_{n\to\infty} \|x_n\| = \frac{1}{\infty} = 0$$

We know that a norm is continuous, then

$$\lim_{n\to\infty} x_n = 0$$

Then eq (4) gives

$$\lim_{n\to\infty} T(x_n) > 1$$

$$\Rightarrow \quad x_n \to 0 \quad \Rightarrow \quad T(x_n)$$

which is not tend to zero as $n \to \infty$ which is a contradiction.

Hence, T is bounded.

4.16 SEPARABLE SPACE

A normed linear space is said to be separable if there exists a denumerable sequence $f_1, f_2, ...,$ of its vectors which is dense in N.

THEOREM 1. ***Every subset of separable normed linear space is separable.***

PROOF. Let A be a subset of separable normed space N and $B = \{x_n\}$ is a countable dense set in N. We have to prove that A is separable.

For a numerical sequence $\varepsilon_n \to 0$ ($\varepsilon_n > 0$) we can find for each $i = 1, 2, ..., n$ an $a_{in} \in A$ such that

$$\|x_i - a_{im}\| < \inf \|x_i - y_{in}\| + \varepsilon_n \qquad ...(1)$$

If we take $x \in A$ and $\varepsilon > 0$ then there exists an $x_i \in B$ such that

$$\|x - x_i\| < \varepsilon/3. \qquad ...(2)$$

Taking n sufficiently large such that $\varepsilon_n < \varepsilon/3$ we have

$$\|x - a_m\| = \|x - x_i + x_i - a_{in}\| \le \|x - x_i\| + \|x_i - a_{in}\|$$
$$\le \varepsilon/3 + \text{inf. } \|x_i - a_{in}\| + \varepsilon_n \qquad \text{(Using (1) and (2))}$$
$$\le \varepsilon/3 + \|x_i - x\| + \varepsilon_n \text{ as } x, a_{in} \in A \text{ and } x \text{ is arbitrary}$$
$$\le \varepsilon/3 + \varepsilon/3 + \varepsilon/3 = \varepsilon$$

Hence, A is separable.

THEOREM 2. ***Every Compact subset of a normed linear space is separable.***

PROOF. We know that, every compact subset of a metric space is totally bounded and every normed linear space is a metric space. Also, given a metric space say N, a finite subset $A \subset N$ is an ε-net for N if and only if A is finite and $N = \cup[S(a, \varepsilon) : a \in A]$ where $S(a, \varepsilon)$ is an open sphere with centre at a and radius ε. As such a compact subset of a normed linear space N has a finite ε-net.

Therefore if $\varepsilon_r \to 0 \ \forall\ r = 1, 2, ...$ and A_r is a finite ε_r-net, then $A = \bigcup_{r=1}^{\infty} A_r$ is a countable dense subset of N, since it is the union of countable sets. Hence, N is separable.

THEOREM 3. ***A normed linear space is separable if its conjugate (or dual) space is separable.*** (MADRAS–2006, KANPUR–1994, 95, 96, 2004, DELHI–1980)

PROOF. Let N be a normed linear space whose conjugate space N^* is separable.

Now, consider the set

$$S = \{f : f \in N^*, \|f\| = 1\}$$

We know that every subspace of a metric space is separable, thus S must be separable.

By definition of separability, S contain countable dense subset, say

$$A = \{f_1, f_2, ..., f_n, ...\}$$

Since, each $f_n \in S$, we have $\|f_n\| = 1$ for all n.

Also, $\|f_n\| = \sup\{|f_n(x)| : \|x\| = 1\}$, therefore for each n, there must exist some vector x_n with $\|x_n\| = 1$ such that

$$|f_n(x)| > \frac{1}{2}$$

Now, let M be the closed linear subspace in N generated by the sequence $< x_n >$. We want to show that $M = N$.

Suppose $M \neq N$ and let $x_0 \in N - M$, *i.e.*, $x_0 \notin M$.

Then, using previous example (3) on page 317 there exists a $F \in N^*$ such that

$$\|F\| = 1,\ F(x_0) \neq 0 \text{ and } F(x) = 0 \text{ if } x \in M$$

Since $\|F\| = 1$, $F \in S$ and since each $x_n \in M$, we have

$$F(x_n) = 0\ ,\ n = 0,1,2,...$$

Thus $\frac{1}{2} < |f_n(x_n)| = |f_n(x_n) - F(x_n) + F(x_n)|$

$$\leq |f_n(x_n) - F(x_n)| + |F(x_n)|$$

$$= |(f_n - F)(x_n)| \qquad [\because F(x_n) = 0]$$

$$\leq \|f_n - F\| . \|x_n\| = \|f_n - F\| \qquad [\because \|x_n\| = 1]$$

$$\Rightarrow \qquad \|f_n - F\| \geq \frac{1}{2} \text{ for all } n$$

Further, since A is dense in S every point of S is an adherent point of A, so that each sphere centered at arbitrary $f \in S$ must contain a point of A. But the open sphere $\left\{f : \|f - F\| < \frac{1}{2}\right\}$ centered at $f \in S$ contains no point of A, which is a contradiction. Thus, we must have $M = N$. It then follows that the set of all linear combination of the x_n's, where coefficients are rational or if N is complex have rational real and imaginary parts, form a countable set everywhere dense in N and hence, N is separable.

4.17 THE OPEN MAPPING THEOREM

Notations

(1) $S(x,r) = x + S_r$

where $S(x, r)$ is an open sphere centered at x and of radius r and S_r is centered at origin with radius r.

(2) $S_r = rS_1$

where, $S_r = \{x : \| x \| < r\} = \left\{x : \frac{\| x \|}{r} < 1\right\} = \{ry : \| y \| < 1\} = r\, S_1$

Before consider the open mapping theorem, we shall prove the following lemma :

Lemma. ***Let B and B′ be two Banach spaces and T is a continuous linear transformation from B onto B′, then the image of every open sphere centered at origin in B contain an open sphere centered at origin in B′.***

(MEERUT–2000, KANPUR–1999)

Proof. Let $T : B \to B'$ be a continuous linear transformation and S_r and S'_r be open spheres of radii r and r' centered at origin of B and B' respectively. Also, let S_i be an open sphere of unit radius.

Since, we have $\qquad S_r = rS_1$

$$\Rightarrow \qquad T(S_r) = T(rS_1) = rT(S_1) \qquad ...(1)$$

we want to show that $T(S_1)$ contains some S'_r.

Step-I : First of all, we shall prove that $\overline{T(S_1)}$ contains an open sphere centered at the origin in B′.

Since T is onto

$$\therefore \quad B' = \bigcup_{n=1}^{\infty} T(S_n) \quad \ldots(2)$$

Let $y \in B'$, then since T is onto, there exists $x \in B$ such that $T(x) = y$.

Also, x is at a finite distance from the origin in B so that $x \in S_r$ for some positive integer r.

$$\therefore \quad T(x) \in T(S_n) \subset \bigcup_{n=1}^{\infty} T(S_n)$$

$$\Rightarrow \quad y \in \bigcup_{n=1}^{\infty} T(S_n)$$

Therefore,

$$B' \subset \bigcup_{n=1}^{\infty} T(S_n) \quad \ldots(3)$$

Conversely, let $y \in \bigcup_{n=1}^{\infty} T(S_n) \Rightarrow y \in T(S_n)$

But $S_n \subset B \Rightarrow T(S_n) \subset T(B) = B'$ [$\because$ T is onto]

So $y \in B'$.

Hence $\bigcup_{n=1}^{\infty} T(S_n) \subset B'$ ...(4)

From (3) and (4), we conclude that

$$B' = \bigcup_{n=1}^{\infty} T(S_n)$$

Since B' is complete, then by Bair's category theorem, it is of second category, thus it can not be a countable union of nowhere dense set. Then from (2), we conclude that there exists a positive integer n_0 such that $T(S_{n_0})$ is not nowhere dense set.

Now we know that A is nowhere dense if $(\overline{A})^0 = \phi$

$$\therefore \quad \left[\overline{T(S_{n_0})}\right]^0 \neq \phi \text{ for } n_0.$$

$\Rightarrow$ there exists an interior point y of $\left[\overline{T(S_{n_0})}\right]$ such that it belongs to $T(S_{n_0})$ as shown below.

Since, y is an interior point of $\left[\overline{T(S_{n_0})}\right]$, therefore, there exists an open set G such that

$$y \in G \subset \left[\overline{T(S_{n_0})}\right]$$

But $y \in \left[\overline{T(S_{n_0})}\right] \Rightarrow y$ is an adherent point of $T(S_{n_0})$.

$\Rightarrow$ There exists a nbd G of y must contain a point $y_0 \in T(S_{n_0})$

$\Rightarrow$ $y_0 \in T(S_{n_0})$ such that $y_0 \in G \subset \left[\overline{T(S_{n_0})}\right]$

$\Rightarrow$ y_0 is an interior point of $\left[\overline{T(S_{n_0})}\right]$ such that it belongs to $T(S_{n_0})$ ...(5)

Now, consider a map $f : B' \xrightarrow{\text{onto}} B'$ such that $f(y) = y - y_0$.

To show that f is a homomorphism of B' onto itself.

It is obvious that f is one-one and onto. It remains to prove that f and f^{-1} are continuous.

Let $< y_n >$ be any sequence in B' such that $y_n \to y$.

Then $$f(y_n) = y_n - y_0 \to y - y_0 = f(y)$$

which implies that
$$y_n \to y \Rightarrow f(y_n) \to f(y)$$

Therefore, f is continuous.

Further, consider
$$f(y_n + y_0) = y_n + y_0 - y_0 = y_n$$
$$\Rightarrow \quad f^{-1}(y_n) = y_n + y_0$$

Also, $\quad y_n \to y \Rightarrow f^{-1}(y_n) = y_n + y_0 \to y + y_0 = f^{-1}(y)$

$\Rightarrow$ f^{-1} is continuous.

Thus, f is a homeomorphism.

Step-II : Further, we want to prove that 0 is an interior point of $\overline{T(S_{n_0})} - y_0$.

Since, y_0 is an interior point of $\overline{T(S_{n_0})}$, there exists an open sphere centered at y_0 which is contained in $[\overline{T(S_{n_0})}]$.

Also, since f is an open map, therefore, $f\overline{T(S_{n_0})}$ has an open sphere centered at $f(y_0)$ and contained in it.

$\Rightarrow$ $\overline{T(S_{n_0})} - y_0$ has $f(y_0) = y_0 - y_0 = 0$ as its interior point ...(6)

Now, to show that $T(S_{n_0}) - y_0 \subset T(S_{2n_0})$

Let $y \in T(S_{n_0}) - y_0$

$\Rightarrow$ There exists $x \in S_{n_0}$ such that $y = T(x) - y_0$, where $\| x \| \leq n_0$.

Also, from (5), we get
$$y_0 \in T(S_{n_0}) \Rightarrow \exists\, x_0 \in S_{n_0} \text{ such that } y_0 = T(x_0) \text{ where } \| x_0 \| \leq n_0.$$

Thus $\quad y = T(x) - T(x_0) = T(x - x_0)$, where $\| x - x_0 \| \leq \| x \| + \| x_0 \| \leq 2n_0$.

$$\Rightarrow \quad (x - x_0) \in S_{2n_0} \Rightarrow y = T(x - x_0) \in T(S_{2n_0}) \qquad ...(7)$$

Therefore, $y \in T(S_{n_0}) - y_0 \Rightarrow y \in T(S_{2n_0})$

$$\Rightarrow \quad T(S_{n_0}) - y_0 \subseteq T(S_{2_{n_0}}) \qquad ...(8)$$

Now, since, f is a homeomorphism, so that
$$f \subset \overline{T(S_{n_0})} = \overline{f \subset T(S_{n_0})} \qquad [\because f(\bar{A}) = \overline{f(A)}]$$
$$\Rightarrow \quad \overline{T(S_{n_0})} - y_0 \subset \overline{T(S_{n_0}) - y_0} \qquad \text{[By definition of } f] \quad ...(9)$$

Using (8) and (9), we get
$$\overline{T(S_{n_0})} - y_0 \subset \overline{T(S_{2n_0})} \qquad ...(10)$$

Now, since 0 is an interior point of $\overline{T(S_{n_0})} - y_0$, therefore, 0 is an interior point of $\overline{T(S_{2n_0})}$ ...(By (10))

$\Rightarrow$ 0 is an interior point of $\overline{2n_0 T(S_1)}$

$\Rightarrow$ 0 is an interior point of $2n_0 \overline{T(S_1)}$

$\Rightarrow$ 0 is an interior point of $\overline{T(S_1)}$

$\Rightarrow$ There exists an open sphere of radius $\in > 0$ centered at origin in B', which is contained in $\overline{T(S_1)}$, i.e., $S'_{\in} \subset \overline{T(S_1)}$. ...(11)

$\Rightarrow$ $\overline{T(S_1)}$ contained an open sphere centered at origin in B'.

Step-III : Now, to show that $S'_\in \subset T(S_3)$

Let y be any arbitrary point of $S'_\in$ so that $\| y \| < \in$.

Using (11), we have $y \in S'_\in \subset \overline{T(S_1)} \Rightarrow y \in \overline{T(S_1)}$.

$\Rightarrow$ y is an adherent point of $T(S_1)$.

$\Rightarrow$ $\exists$ a y_1 in $T(S_1)$ such that $\|y - y_1\| < \in/_2$ ...(12)

But $y_1 \in T(S_1) \Rightarrow y_1 = T(x_1)$ for some $x_1 \in S_1$, *i.e.*, $\| x \| < 1$

Also, form (11), we have

$$\frac{1}{2} S'_\in \subset \frac{1}{2} \overline{T(S_1)}$$

$$\Rightarrow \qquad S'_{\in/_2} \subset \overline{T(S_{1/_2})}$$

and since by (12), $\|y - y_1\| < \in/_2$, we have

$$y - y_1 \in S'_{\in/_2} \subset \overline{T(S_{1/_2})}$$

Continuing this process, we can say that there exists a vector y_2 in $T(S_{1/2})$ such that $\| y - y_1 - y_2 \| < \in/_{2^2}$, where $y_2 = T(x_2)$ for some $x_2 \in S_{1/2}$ so that $\| x_2 \| < \frac{1}{2}$.

Continuing in the similar fashion, we get a sequence $<x_n>$ in B such that

$$\| x_n \| < \frac{1}{2^{n-1}}$$

and $\|y - (y_1 + y_2 + ... + y_n)\| < \in/_{2^n}$ where, $y_n = T(x_n)$

$$\Rightarrow \quad \lim(y_1 + y_2 + ... + y_n) = y$$

let $\qquad s_n = x_1 + x_2 + ... + x_n$ ...(13)

we claim that $< s_n >$ is a Cauchy sequence in B

Consider $\|s_n\| = \|x_1 + x_2 + ... + x_n\| \le \|x_1\| + \|x_2\| + ... + \|x_n\|$

$$< 1 + \frac{1}{2} + \frac{1}{2^2} + ... + \frac{1}{2^{n-1}} < 2$$

$\Rightarrow \quad \|s_n\| < 2$ ...(14)

Also, from $n > m$, we have

$$\|s_n - s_m\| = \|x_{m+1} + ... + x_n\| \le \|x_{m+1}\| + ... + \|x_n\|$$

$$< \frac{1}{2^m} + \frac{1}{2^{m+1}} + ... + \frac{1}{2^{n-1}} = \frac{\frac{1}{2^m}(1 - \frac{1}{2^{n-m}})}{1 - \frac{1}{2}} \quad \text{[Sum of G.P.]}$$

$$= \frac{1}{2^{m-1}} - \frac{1}{2^{n-m-1}} \to 0 \text{ as } n, m \to \infty$$

$\Rightarrow$ $< s_n >$ is a Cauchy sequence in B which is complete, therefore, there exists a vector x in B such that

$$\lim_{n \to \infty} s_n = x.$$

and $\|x\| = \|\lim s_n\| = \lim \|s_n\| \leq 2 < 3$

or $\|x\| < 3$

Since $\|x\| < 3 \Rightarrow x \in S_3$ ...(15)

Now, $y_1 + y_2 + ... + y_n = Tx_1 + Tx_2 + ... + Tx_n$

$= T(x_1 + x_2 + ... + x_n) = T(S_n)$ ...(16)

Also, since, T is continuous, therefore.

$$x = \lim s_n \Rightarrow T(x) = \lim T(s_n)$$

$\Rightarrow$ $T(x) = \lim(y_1 + y_2 + ... + y_n)$

$\Rightarrow$ $T(x) = y$ [Using (13)]

Thus, $y = T(x)$, where $\|x\| < 3$, *i.e.*, $x \in S_3$ by (15) so that $y \in T(S_3)$

Therefore, we have proved that $y \in S'_\in \Rightarrow y \in T(S_3)$.

$\therefore$ $S'_\in (TS_3)$ or $\frac{1}{2} S'_\in \subset \frac{1}{3} T(S_3)$

$\Rightarrow$ $S'_{\in/2}(TS_1)$ $[\because T(S_r) = rT(S_1)]$

There exists an open sphere $S'_{\in/2}$ centered at origin in B' and contained in $T(S_1)$.

MAIN OPEN MAPPING THEOREM

STATEMENT. ***Let B and B′ be two Banach spaces and if T is continuous linear transformation of B onto B′, then T is an open mapping.***

(MEERUT–1985, 2001, 02, 03, 06, 06BP, 07, 08, 09, 09BP, 10, 11, 13, 14 KANPUR–1995, 2005, 10, DELHI–1981, BUNDELKHAND–1985, AGRA–2011)

PROOF. Define $T : B \to B'$ be a continuous linear transformation. To show T is an open mapping. Let G be any open set in B. To show that its image $T(G)$ will be open in B'.

We know that $T(G)$ will be an open set in B' if for each point $y \in T(G)$ there exists an open sphere which is contained in $T(G)$.

Let $y \in T(G) \Rightarrow \exists\, x \in G : T(x) = y$

Now, $x \in G$, and G is open, therefore, by definition $\exists$ an open sphere $S(x, r)$ centered at x and with radius r, which is contained in G

$\therefore$ $S(x, r) \subset G$

We also have $S(x, r) = x + S_r$

where S_r is an open sphere of radius r centered at origin of B.

THEOREM 2. **(Banach Theorem).** ***Let B and B′ be Banach spaces and let T be a one-one continuous linear transformation of B onto B′. Then, T is a homomorphism. In particular, T^{-1} is automatically continuous.***

PROOF. Using the open mapping theorem, we can say that T is one-one onto and continuous. Hence, T is a homomorphism. (Give the same proof)

4.18 PROJECTIONS

4.18.1 PROJECTION ON A LINEAR SPACE

Let us consider the linear space R^2, whose elements are the points (x, y) in the Cartesian plane. Also, Consider two subspace $M = x$-axis whose elements are of the type $(x, 0)$ and $N = y$-axis whose element are of the type $(0, y)$.

Since, $(x,y) = (x,0) + (0,y)$

So that $z = \alpha + \beta \in R^2 \in M \cap N$

$\Rightarrow$ any element $z \in R^2$ is expressible as the sum of an element of M and an element of N.

Also, since $M \cap N = \{0,0\} = \{0\}$, as the axis have only the origin as the common point.

Thus, we conclude that vector space R^2 is the direct sum of the subspace M and N, *i.e.*, $R^2 = M \oplus N$.

Also, we know that the projection of the point (x, y) on the axis of x is $(x,0)$ which is an element of M and similarly projection of the point (x, y) on y-axis is $(0, y)$ which is an element of N.

So, we can define a linear operator T such that $T(x,y) = (x,0)$

Then, T is a projection on M or if $T(x,y) = (0,y)$, if, T is a projection on N.

Definition. *Let V be a linear space, which is the direct sum of two subspaces M and N, i.e.,* $V = M \oplus N, M \cap N = \{0\}$ *such that each element* $z \in V$ *can be expressed uniquely as* $x + y : x \in M, y \in N$, *then the projection of M along N is a linear operator E defined by* $E(z) = x$.

ILLUSTRATIONS

(1) A linear transformation E on V is a projection on some subspace if and only if $E^2 = E$.

(2) If V be the direct sum of its subspace M and N and E is a projection on M along N, then M and N are respectively the sets of all solutions of the equation

$$E(z) = Z, \quad E(z) = 0$$

(3) If $V = M \oplus N$ and E is a projection on M along N then

(i) Range of $E = M$

(ii) Null space of $E = N$.

(4) E is a projection on M along N if and only if $I - E$ is a projection on N along M.

4.18.2 PROJECTION ON A BANACH SPACE

Let B be a Banach space. A projection P on B is an idempotent linear operator on B, which is also continuous. Thus, a projection P on a Banach space is a projection on linear space B such that it is Continuous.

THEOREM 1. ***Let P be a projection on a Banach space and if M and N are its range and null spaces, then M and N are closed linear subspaces of B such that*** $\boldsymbol{B = M \oplus N}$. (KANPUR–2004, 06)

PROOF. As per given, P is a projection on B, which is a linear space. Therefore by definition $B = M \oplus N$. It remains to prove that M and N are closed

Since, P is a projection on B. Therefore, P is continuous

and $N = \{x : P(x) = 0\}$

Thus $N = P^{-1}\{0\}$

But $\{0\}$ is a closed subspace of every linear space $\Rightarrow P^{-1}\{0\} = N$ is also closed ($\because P$ is continuous)

Also, $M = \{x : P(x) = x\} = \{x : (I - P)x = 0\}$

$\Rightarrow$ M is a null space of $I - P$ but $I - P$ is continuous. Hence, M is also closed.

THEOREM 2. ***Let B be a Banach space and let M and N be two closed linear subspace of B such that $B = M \oplus N$.*** (KANPUR–1999, 2004, 06)
If $z = x + y$ is the unique represent of a vector in B as the sum of vectors in M and N, then the mapping P defined by $P(z) = x$ is a projection on B whose range and null space are M and N. (MEERUT–1980, KANPUR–2000)

PROOF. It is given that $B = M \oplus N$

Also, $P(z) = x$

$\Rightarrow$ $x \in B$ is a unique representation as $z = x + y$ with $x \in M$ and $y \in N$.

Also, the mapping $P(z) = x$ is an idempotent mapping, where range and null spaces are M and N respectively. (Using Previous Theorem)

Therefore, to prove that P is a projection on Banach spaces. It is sufficient to prove that P is continuous.

Step-I Define $\|z\|' = \|x\| + \|y\|$

We claim that $\|.\|'$ defined a norm and B' also a Banach space

(i) $\|z\|' = \|x\| + \|y\| > 0$ $\quad [\because \|x\| > 0 \text{ and } \|y\| > 0]$

(ii) $\|z\|' = 0 \Leftrightarrow \|x\| + \|y\| = 0 \Leftrightarrow \|x\| = 0, \|y\| = 0$
$\Leftrightarrow x = 0, y = 0 \Leftrightarrow x + y = z = 0$

(iii) $$\|\alpha z\|' = \|\alpha(x+y)\|' = \|\alpha x + \alpha y\|' = \|\alpha x\| + \|\alpha y\|$$
$$= |\alpha|.\{\|x\| + \|y\|\} = |\alpha|.\|z\|'$$

(iv) $$\|z_1 + z_2\|' = \|x_1 + y_1 + x_2 + y_2\|' = \|(x_1 + x_2) + (y_1 + y_2)\|'$$
$$= \|x_1 + x_2\| + \|y_1 + y_2\| \quad [\text{By definition of } \|\cdot\|']$$
$$\le \{\|x_1\| + \|x_2\|\} + \{\|y_1\| + \|y_2\|\}$$
$$= \{\|x_1\| + \|y_1\|\} + \{\|x_2\| + \|y_2\|\} = \|z_1\|' + \|z_2\|'$$

Thus $\|\cdot\|$ defines a norm on B which becomes B' with this new norm.

Step-II Now to show P is continuous transformation from B' onto B.

Consider
$$\|Pz\| = \|P(x+y)\| > \|x\| \le \|x\| + \|y\| = \|z\|'$$

$\Rightarrow$ $\|Pz\| \le \|z\|', \ \forall z \in B'$

$\Rightarrow$ P is bounded and hence, continuous from B' into B.

Step-III It remains to prove that P is continuous from B into itself, for this we shall prove that B and B' have the same topology.

Let T denotes the identity map of B' into B

Then $$\|Tz\| = \|z\| = \|x + y\| \le \|x\| + \|y\| = \|z\|'$$

$\Rightarrow$ $\|T_z\| \le \|z\|' \ \forall z \in B'$

$\Rightarrow$ T is continuous from B' into B.

Also, T is one to one.

$\Rightarrow$ B' and B have the same topology.

$\Rightarrow$ P is continuous from B' into B itself. Hence, P is a projection.

4.19 CLOSED GRAPH THEOREM

(i) Graph of a Linear Transformation. Let X and Y be two sets and D be a subset of X and $T: D \to Y$, then the set T_G of ordered pairs (x,Tx) such that $T_G = \{(x,Tx) : x \in D\}$ is called the graph of T.

REMARK

- If X and Y are normed linear space and D be a subspace of x and $T : D \to Y$ then $T_G = \{(x,Tx) : x \in D\}$.

(ii) Addition and Scalar Multiplication of Elements of T_G :

Let X and Y be two linear spaces, then by the set $X \times Y$ we mean the set of all ordered pairs (x,y) such that $x \in X$ and $y \in Y$. Therefore, we can define the addition and scalar multiplication as follows.

$$(x_1, y_1) + (x_2, y_2) = [(x_1 + x_2), (y_1 + y_2)]$$
$$\alpha(x,y) = (\alpha x, \alpha y)$$

(iii) Norm of an element of $X \times Y$

Let x and y be two normed linear spaces, then we define

$$\|(x,y)\| = \|x\| + \|y\|$$

we can also define the norm as follows

$$\|(x,y)\| = (\|x\|^p + \|y\|^p)^{1/P} \qquad (p \geq 1)$$

and $$\|(x,y)\| = \max.\{\|x\|, \|y\|\}$$

(iv) Closed Linear Transformation

Let X and Y be two normed linear spaces and let $D \subset X$. Also, let $T : D \to Y$.

Then, the linear Transformation T is known as closed if for every convergent sequence $< x_n >$ of points of D, where $x_n \to x \in X$ such that $< Tx_n >$ is a convergent sequence of points of Y where $Tx_n \to y \in Y$, the following two conditions hold good.

(i) $x \in D$ (ii) $y = Tx$

THEOREM 1. ***T is a closed linear transformation if and only if its graph T_G is a closed subspace.***

PROOF. Let $T : D \to Y$, where $D \subset X$ and T is closed.

To show $T_G = \{(x,Tx) : x \in D\}$ is closed.

For this we shall prove that every limit point of T_G belongs to T_G.

Let (x,y) be a limit point of T_G. Then, by definition of limit point there is a sequence of point in T_G, $< (x_n, Tx_n) >$, where $x_n \in D$ which converges to (x,y).

Thus,

$$(x_n, Tx_n) - (x,y) \to 0$$

or $$\|\{(x_n - x), (Tx_n - y)\}\| \to 0$$

$\Rightarrow$ $$\|x_n - x\| + \|Tx_n - y\| \to 0$$

$\Rightarrow$ $$\|x_n - x\| \to 0 \text{ and } \|Tx_n - y\| \to 0$$

$\Rightarrow$ $$x_n \to x \text{ and } Tx_n \to y$$

Since, T is closed, therefore $x \in T$ and $y = Tx$.

Thus, the limit point (x,y) can be written as (x,Tx) which belongs to T_G. Hence T_G is closed.

Case II Let T_G is closed and $x_n \to x$, $x_n \in D\ \forall n$ and $Tx_n \to y$. We have to show that T is closed.

For this we shall prove that $x \in D$ and $y = Tx$

The given condition implies that $< (x_n, Tx_n) > \to (x, y) \in \overline{T}_G$.

But since T_G is closed, therefore we have $\overline{T}_G = T_G$

$$\Rightarrow \quad (x, y) \in \overline{T}_G = T_G$$

$$\Rightarrow \quad (x, y) \in T_G$$

Thus, by definition of T_G , it follows that $x \in D$ and $y = Tx$. Hence, T is closed.

THEOREM 2. ***(Closed Graph Theorem) Let B and B′ be two Banach spaces and if T is a linear transformation of B into B′, then T is continuous if and only if its graph is closed.***

(MEERUT–1982, 85, 88, 2003, 06, 06BP, 07, 08, 09, 09 BP, 10, 11, DELHI–2012, MADRAS–2006 KANPUR–1996, 2000, 01, 09, 11, AGRA–2005, 10, 11, CALCUTTA–2004, RAJASTHAN–2009)

PROOF. **Necessary Condition**

Let $T : B \to B'$ given to be continuous, *i.e.*,

$$x_n \to x_0 \Rightarrow T(x_n) \to Tx_0$$

To show that graph of T, *i.e.*, $T_G = \{(x, T(x)) : x \in B\}$ is closed.

For this we shall prove that $\overline{T}_G = T_G$.

Let $(x_0, y_0) \in \overline{T}_G$

$\Rightarrow$ $\exists$ a sequence $< (x_n, Tx_n) >$ in T such that

$$(x_n, T(x)) \to (x_0, y_0)$$

Therefore, $\quad x_n \to x_0, T(x_n) \to y_0$

or $\quad T(x_n) \to T(x_0) = y_0 \qquad [\because T(x_n) \to y_0]$

Therefore

$$T(x_0) = y_0$$

Now, $\quad x_0 \in B \Rightarrow (x_0, T(x_0)) \in T_G$

$$\Rightarrow \quad (x_0, y_0) \in T_G$$

Thus, $\quad (x_0, y_0) \in \overline{T}_G \Rightarrow (x_0, y_0) \in T_G$

$$\Rightarrow \quad \overline{T}_G \subset T_G$$

But by definition $T_G \subset \overline{T}_G$. Hence, $\overline{T}_G = T_G$.

$\Rightarrow$ T_G is closed.

Sufficient Condition

Let T_G be closed. To show T is continuous.

Define a new norm on the space B to be denoted by $\|\cdot\|_1$ as follows:

For $\quad x \in B : \|x\|_1 = \|x\| + \|Tx\|$

Firstly, we shall prove that $\|.\|_1$ is a norm.

(1) Since $\|x\| \geq 0, \ \|Tx\| \geq 0$

$\Rightarrow$ $\|x\| + \|Tx\| \geq 0$, *i.e.*, $\|x\|_1 \geq 0$

(2) $\|x\|_1 = 0 \Leftrightarrow \|x\| + \|Tx\| = 0$

$$\Leftrightarrow \|x\| = 0 \text{ and } \|Tx\| = 0$$

$$\Leftrightarrow \quad x = 0 \text{ and } x = 0$$

$$\Leftrightarrow \quad x = 0$$

(3) $\|\alpha x\|_1 = \|\alpha x\| + \|T(\alpha x)\| = |\alpha| . \|x\| + |\alpha| . \|Tx\|$

$= |\alpha| \{\|x\| + \|Tx\|\} = |\alpha| . \|x\|_1$

(4) $\|x+y\|_1 = \|x+y\| + \|T(x+y)\|$

$= \|x+y\| + \|Tx+Ty\| \leq \|x\| + \|y\| + \|Tx\| + \|Ty\|$

$= (\|x\| + \|Tx\|) + (\|y\| + \|Ty\|) = \|x\|_1 + \|y\|_1$

Hence, $\|.\|_1$ is a norm

Now, let the linear space B equipped with this new norm denoted by B_1.

Therefore, B_1 is normed linear space.

Now, $\|Tx\| \leq \|x\|_1 = \|Tx\| \leq 1.\|x\|_1$

Therefore, $\|Tx\| \leq \|x\|_1$ or $\|Tx\| \leq 1.\|x\|_1$

Thus, T considered as a mapping from B_1 to B' is bounded and hence continuous.

To show $T : B_1 \to B'$ is continuous. For this we shall prove that B_1 and B' are homomorphism.

Firstly, we shall prove that B_1 is a Banach space.

Let $< x_n >$ be a Cauchy sequence in B_1

Therefore, by definition

$$\|x_n - x_m\|_1 \to 0 \text{ as } m,n \to \infty$$

$$\Rightarrow \quad \|x_n - x_m\| + \|Tx_n - Tx_m\| \to 0 \text{ as } m,n \to \infty$$

$$\Rightarrow \quad \|x_n - x_m\| \to 0 \text{ and } \|Tx_n - Tx_m\| \to 0 \text{ as } m,n \to \infty$$

Thus, $< x_n >$ is a Cauchy sequence in B_1 and $< Tx_n >$ is a Cauchy sequence in B', but both B and B' are Banach spaces and hence complete.

Thus $x_n \to x_0$ and $Tx_n \to y \in B_1$

Now $< (x_n, Tx_n) >$ is a Cauchy sequence in the graph of G, which is given to be closed. Thus,

$$(x_n, T(x_n)) \to (x, y) \in T_G$$

But $\quad (x, y) \in T_G \Rightarrow y = Tx$.

Now $\quad \|x_n - x\|_1 = \|x_n - x\| + \|T(x_n - x)\|$

$$= \|x_n - x\| + \|Tx_n - Tx\| \to 0$$

$\Rightarrow \quad x_n \to x$ in B_1

$\Rightarrow \quad B_1$ is complete.

Hence, B_1 is a Banach space.

Now, it remains to prove that there is a homeomorphism between B and B'

Let $I : B_1 \to B$ such that $I(x) = x \ \forall x \in B_1$.

$\Rightarrow \quad I$ is the identity mapping, which is one-one and onto.

Thus, I is a homeomorphism from one Banach space to another.

Also, T is continuous from B_1 to B'.

$\Rightarrow \quad T$ is continuous on its homeomorphic image B. Hence, $T : B \to B'$ is continuous.

4.20 THE NATURAL EMBEDDING OF N IN N**

(KANPUR–1980, 84)

4.20.1 ISOMETRIC ISOMORPHISM

(MEERUT– 2006BP, 07)

Let N and N' be two normed linear spaces and T is a linear transformation of N into N' such that

$$\|(Tx)\| = \|x\|, \quad \forall x \in N$$

or
$$\|Tx - Ty\| = \|x - y\|$$

i.e., T preserves norms.

Then, the mapping $T : N \to N'$ is called isometric isomorphism of N into N'.

Also, N and N' are said to be isometrically isomorphic to N'.

4.20.2 FIRST CONJUGATE AND SECOND CONJUGATE OF N

Let N be a normed linear space. Then first conjugate N^* of N, defined by $N^* = \{f : N \to R$ or C and f is linear), is a normed linear space consisting all functional on N and denoted by $B(N,R)$ or $B(N,C)$.

In a similar way N^{**}, will be the conjugate of N^* and its elements will be functional on N^*.

(KANPUR–1982, 84, 90, 92)

THEOREM 1. ***Let N be a normed linear space. Then each vector x in N induces a functional F_x on N^* defined by***

$$F_x(f) = f(x), \ \forall \ f \in N^*$$

such that $\|Fx\| = \|x\|$

further, the mapping

$$T : N \to N^{**} : T(x) = F_x, \ \forall x \in N$$

define an isometric isomorphism of N^{**}. (MEERUT–1980, 2006BP, 07)

PROOF. Let N be a normed linear space. To prove F_x is a functional on N^*. For this we shall prove that F_x is linear and bounded.

(1) F_x is linear

Let $f, g \in N^*$ and α, β be any scalars,

Then, $F_x(\alpha f + \beta g) = (\alpha f + \beta g)(x)$

$$= \alpha f(x) + \beta g(x) = \alpha F_x(f) + \beta F_x(g)$$

(2) F_x is bounded (Continuous)

For any $f \in N^*$, we have

$$|F_x(f)| = |f(x)| = \|f\|.\|x\| \qquad \ldots(1)$$
$$\leq \|x\|$$

Thus, F_x is bounded.

Therefore, F_x is a functional on N^*.

Next, we will prove that $\|F_x\| = \|x\|$

Consider

$$\|F\| = \sup\{|F_x(f)| : \|f\| \leq 1\}$$
$$= \sup\{\|f\|.\|x\| : \|f\| \leq 1\} \leq \|x\| \qquad \ldots(2)$$

To prove the reversed inequality, we first consider the case when $x = 0$, then (2) gives

$$\| F_0 \| = \| 0 \| = 0$$

But $\| F_x \| = 0$ always. Therefore, $\| F_x \| = \| 0 \|$, *i.e.*, $\| F_x \| = \| x \|$ when $x = 0$.

Now, let x be any non-zero vector. Then there exists a functional $F \in N^*$ such that

$$F(x) = \| x \| \quad \text{and} \; \| F \| = 1$$

But $\quad \| F_x \| = \sup\{| F_x(f) | \; : \; \| f \| = 1\}$

$$= \sup\{| f(x) | : \| f \| = 1\}$$

Also, since $\quad \| x \| = | F(x) | \leq \sup\{| f(x) | : \| f \| = 1\}$

$$\Rightarrow \quad \| F_x \| \geq \| x \| \quad \ldots(3)$$

Using (2) and (3), we conclude that

$$\| F_x \| = \| x \| \quad \ldots(4)$$

Now, it remains to prove that T is an isometric isomorphism. For this we shall prove that T is a linear transformation as well as an isometric.

(i) *T* is linear

Let $x, y \in N$ and for any scalar α

$$F_{x+y}(f) = f(x+y) = f(x) + f(y)$$

$$= F_x(f) + F_y(f) = (F_x + F_y)(f)$$

and $\quad F_{\alpha x}(f) = f(\alpha x) = \alpha f(x)$

$$= \alpha F_x(f) = (\alpha F_x)(f) \;\; \forall F \in N^*$$

$\Rightarrow \quad F_{x+y} = F_x + F_y \quad$ and $\quad F_{\alpha x} = \alpha F_x$

Therefore,

$T(x+y) = T_{x+y}$, by definition of $T = F_x + F_y = T(x) + T(y)$

and $\quad T(\alpha x) = F_{\alpha x}$, by definition of $T = \alpha F_x = \alpha T(x)$

$\Rightarrow \quad T$ is linear

(ii) *T* is isometry

Equation (4) shows that T is a norm preserving and hence isometry

If $x, y \in N$, then

$$\| T(x) - T(y) \| + \| F_x - F_y \| = \| F_{x-y} \| = \| x - y \| \quad \ldots(5)$$

$\Rightarrow \quad T$ preserves distances

$\Rightarrow \quad T$ is an isometry

Also, from (5), we have

$$T(x) - T(y) = 0$$

$$\Rightarrow \quad T(x-y) = 0 \Rightarrow x - y = 0$$

i.e., $\quad T(x) = T(y) = x = y$

$\Rightarrow \quad T$ is one-one.

Hence, T defines an isometric isomorphism of N into N^*.

REMARKS

- Every isometry is essentially a one-one mapping.
- The mapping T defined above is known as natural imbedding or Canonical mapping of N into N^{**}.
- The space N into N^{**} would be isomorphic if T is onto.

SOLVED EXAMPLE

EXAMPLE 1. *If N is a finite dimensional normed linear space over the field K then prove that*

$$\dim. N = \dim N^*$$

(KANPUR–2006, 11)

SOLUTION. Let us suppose

$$\dim. N = n \text{ (finite)}$$

and N^* be the dual space of N

Then $\dim K(K) = 1$

$$N^* = L(N, K) = \{T : T : N \rightarrow K \text{ is a linear map}\}$$

We know that

$$\dim. L(N, K) = (\dim N)(\dim K) = \dim(N) \,.\, 1 = \dim N$$

$\Rightarrow$ $\dim N^* = \dim N$

Because $L(V, U)$ denote the linear space of mappings T from $V(K)$ to $U(K)$ and $\dim L(U, V) = \dim(U).\dim(V)$.

4.21 UNIFORM BOUNDED PRINCIPLE

THEOREM 1. **(The Uniform Bounded Theorem)** ***Let B be a Banach space and N a normed linear space. If $\{T_i\}$ is a non empty set of continuous linear transformations of B into N with the property that $\{T_i(x)\}$ is a bounded subset of N for each vector x in B, then $\{\| T_i \|\}$ is a bounded set of numbers, i.e., $\{T_i\}$ is bounded as a subset of $B(B, N)$.***

(MEERUT–2003, 07BP, 09, KANPUR–1999, 2009, 10, 11, GARHWAL–2003, 05, AGRA–2005)

PROOF. Here, we are given the following

$T_i : B \rightarrow N$ and T_i is a continuous linear transformation such that $\{T_i(x)\}$ is a bounded subset of N for each $x \in B$. ...(1)

To show, that $\{\| T_i \|\}$ is bounded set of numbers, *i.e.*, $\{\| T_i \|\}$ is bounded as a subset of $B(B, N)$, which is the set of all linear transformations from B to N.

Define

$$F_n = [\{x : x \in B \text{ and } \|T_i(x)\| \leq n \forall i\} \; \forall n \in \mathrm{N}] \quad ...(2)$$

Step I F_n is a closed subset of B

Let $x \in F_n \Rightarrow \|T_i(x)\| \leq n \;\; \forall i$

$\Rightarrow$ $T_i(x) \in S_n^* \;\; \forall i$

where S_n^* denote the closed sphere centered at origin of N and of radius n

$\Rightarrow$ $x \in T_i^{-1}(S_n^*) \;\; \forall i$

$\Rightarrow$ $F_n = \bigcap_i T_i^{-1}(S_n^*)$ ...(3)

Now, each T_i is continuous.

$\Rightarrow$ $T_i(S_n^*)$ is closed for all i.

$\Rightarrow \quad \cap T_n^{-1}(S_n^*)$ is closed for all i.

$\Rightarrow \quad$ each F_n is closed subset of B. ...(4)

($\because$ Arbitrary intersection of closed sets is closed)

Step-II $\boldsymbol{B = \bigcup_{n=1}^{\infty} F_n}$

If $B \neq \bigcup_{n=1}^{\infty} F_n \Rightarrow$ there exists $x \in B$ such that $x \notin F_n \ \forall n$

$\Rightarrow \quad \|T_i(x)\| > n \ \ \forall i$, which is not possible, because $\{T_i(x)\}$ is bounded for each i. (Using (1))

Thus
$$B = \bigcup_{n=1}^{\infty} F_n \qquad \ldots(5)$$

Further, since, B is complete and is union of its subsets hence by Bair Category theorem (*i.e.*, if a complete metric space is the union of a sequence of its subsets, then the closure of at least one set in the sequence must have non-empty interior) there exists an integer n_0 such that $\bar{F}_{n_0}$ has non-empty interior.

Using (4) F_{n_0} is closed subset of B and hence

$$F_{n_0} = \bar{F}_{n_0} \Rightarrow F_{n_0} \text{ has non-empty interior.}$$

let x_0 be an interior point of F_{n_0}.

$\Rightarrow$ there exists a closed sphere S_0 centered at x_0 and radius $r > 0$ which is contained in F_{n_0}. [$\because$ F_{n_0} is closed]

i.e.,
$$S_0 \subset F_{n_0} \qquad \ldots(6)$$

Now, $S_0 = \{x \in B : \| x - x_0 \| \leq r_0) \subset F_{n_0}$ [Using (6)]

$\therefore x \in S_0 \subset F_{n_0} \ \Rightarrow \ \| x - x_0 \| \leq r_0$

and
$$\| T_i(x) \| \leq n_0 \quad \forall i \qquad \ldots(7)$$

Choose $z = r_0 y$, where $\| y \| < 1$.

Therefore
$$\| z \| = \| r_0 y \| = r_0 \| y \| \leq r_0$$

But
$$z = z + x_0 - x_0$$

or
$$\| z + x_0 - x_0 \| \leq r_0 \Rightarrow z + x_0 \in S_0 \subset F_{n_0}$$

$\Rightarrow$
$$\|T_i(z + x_0\| \leq n_0 \qquad \text{[By (7)]} \qquad \ldots(8)$$

Also,
$$x_0 \in S_0 \subset F_{n_0} \Rightarrow \| T_i(x_0) \| \leq n_0 \qquad \text{[By (7)]} \qquad \ldots(9)$$

$\therefore$
$$\|T_i(y)\| = \left\| T_i\left(\frac{z}{r_0}\right) \right\| \qquad [\because z = r_0 y]$$

$$= \left\| \frac{1}{r_0} T_i(z) \right\| = \frac{1}{r_0} \| T_i(z) \|$$

$$= \frac{1}{r_0} \| T_i(z + x_0 - x_0) \|$$

$$= \frac{1}{r_0} \| T_i(z + x_0) - T_i(x_0) \|$$

$$\leq \frac{1}{r_0} [\| T_i(z + x_0) \| + \| T_i(x_0) \|$$

$$\leq \frac{1}{r_0}(n_0 + n_0) = \frac{2n_0}{r_0} \qquad \text{[By (8) and (9)]}$$

$$\Rightarrow \qquad \| T_i(y) \| \leq \frac{2n_0}{r_0}, \text{ where } \| y \| \leq 1$$

$$\therefore \qquad \| T_i \| = \sup\{\| T_i(y) \| : \| y \| < 1\} \leq \frac{2n_0}{r_0}$$

$$\Rightarrow \qquad \| T_i \| \leq \frac{2n_0}{r_0} \quad \forall i$$

$\Rightarrow$ $\{\| T_i \|\}$ is a bounded set of numbers. Hence, $\{T_i\}$ is bounded as a subset of $\mathbf{B}(B,N)$.

THEOREM 2. ***A non-empty subset X of a normed linear space N is bounded if and only if f(x) is bounded set of numbers for each $f \in N^*$.***

PROOF. Let us first suppose that X is bounded subset of N

i.e., $\| x \| \leq k \ \forall x \in X$

To show, that $f(X)$ is bounded for each $f \in N^*$.

Let, $f \in N^* \Rightarrow f$ is continuous.

$\Rightarrow$ f is bounded

$\Rightarrow$ $\exists\, M > 0$ such that $|f(x)| \leq M.\|x\| \ \ \forall x \in N$ ...(2)

From (1) and (2), we conclude that

$$|f(x)| \leq Mk \ \ \forall x \in X.$$

Therefore, $f(x)$ is a bounded subset of numbers for each $f \in N^*$.

Conversely, let $f(x)$ be a bounded set of numbers for each $f \in N^*$.

Choose $X = \{x_\lambda : \lambda \in \Delta\}$

Then, by natural imbedding $\{x_\lambda : \lambda \in \Delta\}$ in N to pass from X to $\{F_{x_\lambda} : \lambda \in \Delta\}$ in N^{**}, we have

$$F_{x_\lambda}(f) = f(x_\lambda), \ \forall f \in N^*$$

But $\{f(x_\lambda) : \lambda \in \Delta\}$ is a bounded set of numbers for each $f \in N^*$. Hence, $\{F_{x_\lambda} : \lambda \in \Delta\}$ is bounded for each $f \in N^*$.

Also, N^* is complete. Hence, by uniform bounded principle, $\{\| F_{x_\lambda} \|\}$ is a bounded subset of numbers.

Also, we know that natural imbedding preserves norm, therefore we conclude that $\{\|x_\lambda\|\}$ is a bounded set of numbers. Hence, X is bounded in N.

THEOREM 3. ***Let B be a Banach space and $<f_n>$ is a sequence of continuous linear functionals on B such that $\{|f_n(x)|\}$ is bounded set for every $x \in B$, then sequence $<\|f_n\|>$ is bounded.***

PROOF. It is given that $f_n : B \to K$ is a continuous linear mapping, where K is the field of scalars.

Define $F_n = \{x \in B : |f_i(x)| < n \ \forall \ i\}$ ...(1)

Then F_n is a closed set for every positive integer n, therefore $F_n = \overline{F}_n$. We want to prove that

$$B = \bigcup_{n=1}^{\infty} F_n \quad \text{...(2)}$$

Let if possible (2) is not true, *i.e.*, $B \neq \bigcup_{n=1}^{\infty} F_n$

Therefore, there exist $x \in B$ such that $x \notin F_n$ for any n.

By definition of F_n, we have $|f_i(x)| > n \ \forall \ n$

$\Rightarrow$ $\{|f_i(x)|\}$ is not bounded

which is a contradiction.

Hence, $B = \bigcup_{n=1}^{\infty} F_n$

Further, since, B is a Banach space.

$\Rightarrow$ B is complete.

$\Rightarrow$ B is of second category (By Bair's Category theorem)

$\Rightarrow$ B is not of first category

$\Rightarrow$ One of F_n's say F_m is dense set

$\Rightarrow$ $\left(\overline{F}_m\right)^\circ \neq \phi$

$\Rightarrow$ $\left(F_m\right)^\circ \neq \phi$ ($\because f_m$ is closed, so $\overline{F}_m = F_m$)

$\Rightarrow$ $\exists \ x_0 \in \left(F_m\right)^\circ \Rightarrow x_0$ is an interior point of F_m.

$\Rightarrow$ There exist a closed sphere $S(x_0, r_0)$ such that

$$S(x_0, r_0) \subset F_m, x_0 \in F_m \quad \text{...(3)}$$

$$x_0 \in F_m \Rightarrow \quad |f_i(x_0)| \leq m \ \forall \ i \quad \text{...(4)}$$

Let $x \in B$ such that

$$\|x\| < r_0 \Rightarrow \quad x + x_0 \in S(x_0, r_0) \quad \text{...(5)}$$

Using (3) and (5) we have

$$x + x_0 \in F_m \Rightarrow \quad |f_n(x + x_0)| \leq m \ \forall \ n \quad \text{...(6)}$$

$$F_n(x) = f_n((x + x_0) - x_0) = f_n(x + x_0) - f_n(x_0)$$

$\Rightarrow$ $|f_n(x)| \leq |f_n(x + x_0)| + |f_n(x_0)| + |f_n(x_0)|$

$< m + m$ (Using (4) and (6))

$\Rightarrow$ $|f_n(x)| \leq 2m$

$\Rightarrow$ $\|f_n\| = \sup\ [\ |f_n(x) : \|x\| \leq 1]$

$\leq \sup \{2m : \|x\| \leq 1\} = 2m = c$, a constant

$\Rightarrow$ $\|f_n\| \leq c$

$\Rightarrow$ Sequence $< \|f_n\| >$ is bounded.

THEOREM 4. ***A non-empty subset X of a normed linear space N is bounded if and only if f(x) is bounded set of numbers for each f in N^*.***

(KANPUR–2006, 07, 13, CALCUTTA–2012, 14)

PROOF. Let X be a non-empty subset of a normed linear space N and $f \in N^*$ be arbitrary.
Then $f : N \to K$ is a continuous linear map, where K is a field of scalars.
Now since f is continuous $\Leftrightarrow$ f is bounded
$\Leftrightarrow \exists\, a > 0$ such that $|f(x)| \le a\|x\|$ ...(1)
Let us first suppose X is bounded, then there exist $b > 0$ such that

$$\|x\| \le b \ \forall\, x \in X \qquad \text{...(2)}$$

To prove that $f(x)$ is bounded
Clearly from (1) and (2)

$$\|f(X)\| \le ab \ \forall\, x \in X$$

Hence $f(X)$ is bounded.
Conversely, suppose that $f(x)$ is bounded, therefore X is bounded. ...(3)
Now, define a map
Let us write $X = \{x_i\}$
Since $f(x)$ is bounded, therefore $f(x_i)$ is bounded.

$$F_{x_i} : N^* \to K \text{ such that } F_{x_i}(f) = f(x_i)$$

Then F_{x_i} is a continuous linear map.

$$\Rightarrow \qquad F_{x_i} \in (N^*)^* = N$$

Using (3), we have

$$\left\{F_{x_i}(f)\right\}, \text{ i.e., } \left\{f_{x_i}\right\} \text{ is bounded for all } f \in N^* \text{ and } \left\|f_{x_i}\right\| = \|x_i\|$$

Also, N^* is complete

Therefore, by Uniform boundedness theorem we find that $\left\{\left\|f_{x_i}\right\|\right\}$ is a bounded subset of N^{**}.

But $$\left\|f_{x_i}\right\| = \|x_i\|$$

So, $\{\|x_i\|\}$ is a bounded set.
But $\{x_i\} = X$
Hence, X is a bounded set.

THEOREM 5. **(The Banach-Steinhaus Theorem).** ***If $< T_n >$ be a sequence of bounded linear transformations defined on a Banach space X into a normed space Y and***

$$\lim_{n\to\infty} T_n(x) = T(x) \ \forall x \in X$$

Then, T is bounded linear transformation on X to Y.

(MEERUT–2003, GARHWAL–2003, 05, AGRA–2005, KANPUR–2009, 10 ,11)

PROOF. We know that

$$T_n(x+y) = T_n(x) + T_n(y)$$

$$\Rightarrow \quad \lim_{n\to\infty} [T_n(x+y)] = \lim_{n\to\infty} [T_n(x) + T_n(y)]$$

$$= \lim_{n\to\infty} T_n(x) + \lim_{n\to\infty} T_n(y)$$

$$\Rightarrow \qquad T(x+y) = T(x) + T(y)$$

Similarly,

$$T(\alpha x) = \alpha T(x)$$

Hence, T is a linear transformation.

Now, $\lim_{n\to\infty} T_n(x) = T(x) \ \forall x$

$$\Rightarrow \quad \left\|\lim_{n\to\infty} T_n(x)\right\| = \|T(x)\| \ \forall x$$

$$\Rightarrow \quad \lim_{n\to\infty} \|T_n(x)\| = \|T(x)\| \ \forall x$$

$\Rightarrow$ The sequence $< T_n(x) >$ is convergent and hence bounded for each fixed x.

$$\Rightarrow \quad \sup_n . \|T_n(x)\| < \infty \ \forall x$$

$$\Rightarrow \quad \sup_n . \|T_n\| < \infty \qquad \text{(By uniform boundedness principle)}$$

Thus, $\|T(x)\| = \lim_{n\to\infty} \|T_n(x)\| \le \sup_n \|T_n(x)\| \le \left(\sup_n \|T_n\|\right)\|x\|$

Hence, T is bounded.

4.22 APPLICATIONS OF THE UNIFORM BOUNDEDNESS PRINCIPLE

4.22.1 FOURIER SERIES

The fourier series of a given periodic function f of period 2π of the form

$$\frac{1}{2}a_0 + \sum_{m=1}^{\infty} (a_m \cos mt + b_m \sin mt) \qquad \ldots(1)$$

where

$$\left.\begin{aligned} a_0 &= \frac{1}{\pi}\int_0^{2\pi} f(t)dt \\ a_m &= \frac{1}{\pi}\int_0^{2\pi} f(t)\cos mtdt \\ b_m &= \frac{1}{\pi}\int_0^{2\pi} f(t)\sin mtdt \end{aligned}\right] \qquad \ldots(2)$$

THEOREM 1. ***There exists a continuous real valued functions whose Fourier series diverges at zero.***

PROOF. Consider a normed linear space X of real valued continuous functions x of period 2π with the norm defined as below

$$\|x\| = \max_{0\le t\le 2\pi} |x(t)|$$

If $f_n(x)$ is the value of nth partial sum of the Fourier series at $t = 0$. Now, at $t = 0$, the sine terms vanish and the cosine terms each is equal to 1. Therefore

$$f_n(x) = \frac{1}{2}a_0 + \sum_{m=1}^{n} a_m$$

or

$$f_n(x) = \frac{1}{\pi}\int_0^{2\pi} x(t)\left[\frac{1}{2} + \sum_{m=1}^{n} \cos mt\right] dt \qquad \ldots(3)$$

Further, we have

$$2\sin\frac{t}{2}\sum_{m=1}^{n} \cos mt = \sum_{m=1}^{n} 2\sin\frac{t}{2}\cos mt$$

$$= \sum_{m=1}^{n}\left[-\sin\left(m-\frac{1}{2}\right)t+\sin\left(m+\frac{1}{2}\right)t\right]$$

$$= \left(-\sin\frac{t}{2}+\sin\frac{3}{2}t\right)+\left(-\sin\frac{3t}{2}+\sin\frac{5t}{2}\right)+\ldots$$

$$= -\sin\frac{t}{2}+\sin\left(n+\frac{1}{2}\right)t$$

$$\Rightarrow \quad 2\sum_{m=1}^{n}\cos mt = -1+\frac{\sin\left(n+\frac{1}{2}\right)t}{\sin(t/2)}$$

$$\Rightarrow \quad 1+2\sum_{m=1}^{n}\cos mt = \frac{\sin\left(n+\frac{1}{2}\right)t}{\sin(t/2)}$$

$$\Rightarrow \quad \frac{1}{2}+\sum_{m=1}^{n}\cos mt = \frac{\sin\left(n+\frac{1}{2}\right)t}{2\sin(t/2)}$$

Using this in (3), we get

$$f_n(x) = \frac{1}{2\pi}\int_0^{2\pi} x(t)\, D_n^t\, dt \qquad \ldots(4)$$

where, $$D_n^{(t)} = \frac{\sin\left(n+\frac{1}{2}\right)t}{\sin(t/2)} \qquad \ldots(5)$$

Therefore,

$$|f_n(x)| \le \frac{1}{2\pi}\max|x(t)|\int_0^{2\pi}|D_n^{(t)}|\,dt = \frac{\|x\|}{2\pi}\int_0^{2\pi}|D_n^{(t)}|\,dt$$

whence, $< f_n >$ is bounded.

Also, $$\sup_{\|x\|=1}.\ |f_n(x)| \le \frac{1}{2\pi}\int_0^{2\pi}|D_n^{(t)}|\,dt$$

i.e., $$\|f_n\| \le \frac{1}{2\pi}\int_0^{2\pi}|D_n^{(t)}|\,dt \qquad \ldots(6)$$

Now, for each fixed positive integer p, we have

$$\frac{1}{2\pi}\int_0^{2\pi}|D_n^{(t)}|\,dt = \frac{1}{2\pi}\int_0^{2\pi}|D_n^{(t)}|\left\{\frac{1+p|D_n^{(t)}|}{1+p|D_n^{(t)}|}\right\}dt$$

$$= \frac{1}{2\pi}\left\{\int_0^{2\pi}\frac{|D_n^{(t)}|}{1+p|D_n^{(t)}|}\,dt+\int_0^{2\pi}|D_n^{(t)}|\frac{p|D_n^{(t)}|}{1+p|D_n^{(t)}|}\,dt\right\}$$

$$\le \frac{1}{2\pi}\int_0^{2\pi}\left\{\frac{1}{p}+f_n\left(\frac{p|D_n^{(t)}|}{1+p|D_n^{(t)}|}\right)\right\}dt$$

$$\le \frac{1}{p}+\|f_n\|\left\|\frac{p|D_n^{(t)}|}{1+p|D_n^{(t)}|}\right\| \le \frac{1}{p}+\|f_n\| \qquad \ldots(7)$$

Using (6) and (7), we get

$$\| f_n \| = \frac{1}{2\pi} \int_0^{2\pi} | D_n^{(t)} | \, dt \qquad \ldots(8)$$

Further, using (5) and (8), we get

$$\| f_n \| = \frac{1}{2\pi} \int_0^{2\pi} \left| \frac{\sin\left(n+\frac{1}{2}\right)t}{\sin(t/2)} \right| dt$$

$$> \frac{1}{2\pi} \int_0^{2\pi} \left| \frac{\sin\left(n+\frac{1}{2}\right)t}{t/2} \right| dt > \frac{1}{\pi} \int_0^{2\pi} \left| \frac{\sin\left(n+\frac{1}{2}\right)t}{t} \right| dt$$

$$= \frac{1}{\pi} \int_0^{(2n+1)\pi} \frac{|\sin u|}{u} du \qquad \left(\text{Putting } u = \left(n+\frac{1}{2}\right)t\right)$$

$$> \frac{1}{\pi} \sum_{k=0}^{2n} \int_{k\pi}^{(k+1)\pi} \frac{|\sin u|}{u} du$$

$$\geq \frac{1}{\pi} \sum_{k=0}^{2n} \frac{1}{(k+1)\pi} \int_{k\pi}^{(k+1)\pi} |\sin u| \, du = \frac{2}{\pi} \sum_{k=0}^{2n} \frac{1}{k+1}$$

$$\to \infty \text{ as } n \to \infty$$

Hence, $\langle \| f_n \| \rangle$ is unbounded. But X is complete, therefore by the uniform boundedness principle, it follows that $< f_n(x) >$ is unbounded for some x. Hence, Fourier series of x diverges at $t = 0$.

4.22.2 SUMMABILITY

THEOREM. **(The Toeplitz-Silverman Theorem).** ***A matrix*** $A = [a_{nk}]$, $n, k = 1, 2, \ldots$ ***of complex entries transforms the space C [the linear space of all convergent sequences] into C and presence of limits if and only if***

(i) $\sup_n \sum_{k=1}^{\infty} |a_{nk}| < \infty$ **(ii)** $\lim_{n\to\infty} a_{nk} = 0 \ \forall k$

(iii) $\lim_{n\to\infty} \sum_{k=1}^{\infty} a_{nk} = 1$.

PROOF. **Necessary Condition**

Let $y_n = \sum_{k=1}^{\infty} a_{nk} x_k = A_n(x)$, $n = 1, 2, \ldots$

Take $x =< x_k > = \delta^k$. Then, $y_n = a_{nk}$

and $\lim x_k = \lim y_n = 0$. Therefore $\lim_{n\to\infty} a_{nk} = 0$ which prove the result (ii).

Now, take $x = (1, 1, \ldots)$. Then $y_n = \sum_{k=1}^{\infty} a_{nk}$ and since limit of (1, 1, 1, ...) is 1, we should have $\lim y_n = \lim_{n\to\infty} \sum_{k=1}^{\infty} a_{nk} = 1$, which prove result (iii).

To prove result (i), we first show that

$$\|A\| = \sup_n . \sum_{k=1}^{\infty} |a_{nk}|$$

Also, $x \in C$ and $\|x\| = 1$.

Let $A_n(x) = a_{n1}x_1 + a_{n2}x_2 + \dots, n = 1, 2$

Define a functional on C for each fixed n.

Now, $$|A_n(x)| = |a_{n1}| + |a_{n2}| + \dots + |a_{nk}|$$

where, $$\|A_n\| = \sup_{\|x\|=1} . |A_n(x)| \geq |A_n(x)| = \sum_{j=1}^{k} |a_{nj}|$$

Since $k \geq 1$ is arbitrary, so we get

$$\|A_n\| \geq \sum_{k=1}^{\infty} |a_{nk}| \quad \dots(1)$$

Also, $$|A_n(x)| = \left| \sum_{k=1}^{\infty} a_{nk} x_k \right| \leq \sum_{k=1}^{\infty} |a_{nk}| \|x\|$$

So, $$\|A_n\| \geq \sup_{\|x\|=1} . |A_n(x)| \leq \sum_{k=1}^{\infty} |a_{nk}| \quad \dots(2)$$

Using (1) and (2), we get

$$\|A_n\| = \sum_{k=1}^{\infty} |a_{nk}| \quad \dots(3)$$

Also, for all n

$$|A_n(x)| \leq \sum_{k=1}^{\infty} a|a_{nk}||x_k| \leq \left(\sum_{k=1}^{\infty} |a_{nk}| \right) \|x\|,$$

because A_n is a bounded operator.

Now, since $< A_n(x) >$ should belong to C, therefore, $\lim_{n \to \infty} A_n(x)$ exists.

Also, C is a Banach space. Hence, by uniform boundedness principle, $(\|A_n\|)$ is bounded, so that

$$\sup_n \|A_n\| < \infty \text{ when } \sup_n . \sum_{k=1}^{\infty} |a_{nk}| < \infty$$

which is the required necessary condition.

Sufficient Condition

Condition (1) shows the absolute convergence of $\sum_{k=1}^{\infty} a_{nk}$ and $\sum_{k=1}^{\infty} a_{nk}x_k$, if $x = < x_k > \in C$ for each n. Therefore, if $L = \lim_{k \to \infty} x_k$, we have for $n = 1, 2, \dots$

$$\sum_{k=1}^{\infty} a_{nk} x_k = \sum_{k=1}^{\infty} a_{nk}(x_k - L) + L \sum_{k=1}^{\infty} a_{nk}$$

But $\lim_{n \to \infty} \sum_{k=1}^{\infty} a_{nk} = 1$. Hence, $\lim_{n \to \infty} \sum_{k=1}^{\infty} a_{nk} = L$

Let $$M = \sup_n . \sum_{k=1}^{\infty} |a_{nk}|$$

Since $x_k \to L$, given $\varepsilon > 0$, choose m such that $k > m$ implies

$$|x_k - L| < \frac{\varepsilon}{2M}$$

Fix m.

Now, $\lim_{n\to\infty} a_{nk} = 0 \ \forall k$. Therefore, choose N such that

$$n > N \Rightarrow \sum_{k=1}^{m} |a_{nk}||x_k - L| < \frac{\varepsilon}{2}$$

Then, we have

$$\left|\sum_{k=1}^{\infty} a_{nk}(x_k - L)\right| \le \sum_{k=1}^{\infty} |a_{nk}||x_k - L| + \sum_{k=m+1}^{\infty} |a_{nk}||x_k - L| < \frac{\varepsilon}{2} + M.\frac{\varepsilon}{2M} = \varepsilon$$

$$\Rightarrow \quad \lim_{n\to\infty} \sum_{k=1}^{\infty} a_{nk}\, x_k = L$$

Hence, $A : C \to C$ preserving limits.

4.23 THE CONJUGATE OF AN OPERATOR

Let N be a normed linear space and T be a continuous linear operator on N such that $\alpha \in N \Rightarrow T\alpha \in N$.

Define a linear transformation T^* of N^* into itself as follows :

"If $f \in N^*$, then $T^*(f)$ is given by $(T^*(f))(x) = f(T(x))$. Then T^* is said to be conjugate operator.

THEOREM 1. ***Let T be an operator on a normed linear space N. Then, its conjugate T^*, defined by***

$$\boldsymbol{T^* : N^* \to N^* : T^*(f) = f \, o T}$$

and $$\boldsymbol{[T^*(f)](x) = f(T(x))} \qquad \ldots(1)$$

for all $f \in N^*$ and all $x \in N$, is an operator on N^* and the mapping

$$\boldsymbol{g : B(N) \to B(N^*) : g(T) = T^* \quad \forall T \in B(N)}$$

is an isometric isomorphism of B(N) into B(N^*) which reverses products and preserves the identity transformation.

(MEERUT–1980, 2003, KANPUR–1999, AMRITSAR–1982)

PROOF. Firstly, we shall prove that T^* is linear.

Consider

$$[T^*(\alpha f + \beta g)](x) = (\alpha f + \beta g)(T(x))$$
$$= (\alpha f(T(x)) + (\beta g)(T(x)) = \alpha(f(T(x)) + \beta(g(T(x))$$
$$= \alpha[T^*(f)](x) + \beta[T^*(g)](x)$$
$$= [\alpha T^*(f) + \beta T^*(g)](x)$$

$$\Rightarrow \quad T^*(\alpha f + \beta g) = \alpha T^*(f) + \beta T^*(g)$$

$\Rightarrow \quad T^*$ is linear.

Now, to show T^* is bounded.

Consider

$$\left\| T^{*} \right\| = \sup.\{\left\| T^{*}(f) \right\| : \| f \| \leq 1\}$$

$$= \sup.\{| \ [T^{*}(f)](x) : \| f \| \leq 1|, \| x \| \leq 1| \}$$

$$= \sup.\{| \ f \cdot T(x) : \| f \| \leq 1|, \ \| x \| \leq 1|\} \quad \text{[Using (1)]}$$

$$\leq \sup\{\| f \|.\| T \|.\| x \| : \| f \| \leq 1 \ |, \ \| x \| \leq 1 \ | \ \} \leq \| T \|$$

Since, T is bounded, therefore from (2), we conclude that T^{*} is bounded.

Thus, T^{*} is an operator on N^{*}.

For each non-zero vector x in N, there exists a functional $f \in N^{*}$ such that $\| f \| = 1$ and

$$f(T(x) = \| T(x) \| \quad \ldots(3)$$

$$\therefore \quad \| T \| = \sup.\left\{ \frac{\| T(x) \|}{\| x \|}, \ x \neq 0 \right\}$$

$$= \sup.\left\{ \frac{|f(T(x)|}{\|x\|} : \|f\| = 1, x \neq 0 \right\} \quad \text{[Using (3)]}$$

$$= \sup.\left\{ \frac{|[T^{*}(f)](x)|}{\|x\|} : \|f\| = 1, x \neq 0 \right\}$$

$$= \sup.\{\left\| T^{*}(f) \right\| \ : \ \| f \| = 1\} = \left\| T^{*} \right\| \quad \ldots(4)$$

Using (2) and (4), we have

$$\left\| T^{*} \right\| = \|T\| \quad \ldots(5)$$

Now, we have to show that the mapping

$$g : B(N) \to B(N^{*}) : g(T) = T^{*} \ \ \forall \ T \in \beta(N) \quad \ldots(6)$$

is an isometric isomorphism, *i.e.*, g is one-one linear transformation such that $\|g(T)\| = \|T\|$

From (6), we have $\quad \|g(T)\| = \left\| T^{*} \right\| = \|T\| \quad$ [Using (5)]

(i) *g* is linear

Let $T, U \in B(N)$ be arbitrary and α, β be any scalars

Then, $\quad g(\alpha T + \beta U) = (\alpha T + \beta U)^{*} \quad$ [by (6)]

But $\quad [(\alpha T + \beta U)^{*}(f)](x) = f((\alpha T + \beta U)(x)$

$$= f(\alpha T(x)) + f \beta U(x)$$

$$= \alpha f(T(x) + \beta f(U(x) \quad [\because f \text{ is linear}]$$

$$= \alpha[T^{*}(f)](x) + \beta[U^{*}(f)](x)$$

$$= [\alpha(T^{*}(f)] + \beta[U^{*}(f)](x)$$

$$= [\alpha(T^{*}(f)] + \beta[U^{*}(f)](x)$$

$$\therefore \quad (\alpha T + \beta U)^*(f) = \alpha[T^*(f)] + \beta(U^*(f))$$
$$= (\alpha T^* + \beta U^*)(f)$$
$$\Rightarrow \quad (\alpha T + \beta U)^* = \alpha T^* + \beta U^* \qquad \ldots(7)$$
$$\therefore \quad g(\alpha T + \beta U) = (\alpha T + \beta U)^* = \alpha T^* + \beta U^*$$
$$\Rightarrow \quad g(\alpha T + \beta U) = (\alpha T + \beta U)^* = \alpha T^* + \beta U^*$$
$$= \alpha g(T) + \beta g(U)$$

$\Rightarrow$ g is linear

(ii) g is one-one

We have

$$g(T) = g(U) \Rightarrow T^* = U^* \Rightarrow \left\| T^* - U^* \right\| = 0$$
$$\Rightarrow \left\| (T-U)^* \right\| = 0 \Rightarrow \|T - U\| = 0 \Rightarrow T = U$$

Hence, g is an isometric isomorphism

It remains to prove that x reverse products and preserves the identity transformation. We have

$$[(TU)^*(f)](x) = f((TU)(x) \qquad \text{[By (1)]}$$
$$= f(T(U(x)) = [T^*(f)](U(x))$$
$$= (U^*(T^*(f))(x) = [U^*T^*)(f)](x)$$

Therefore $\quad g(TU) = (TU)^* = U^*T^*$

$\Rightarrow$ g reverses products.

Now, let I denotes the identity operator on N, then

$$[I^*(f)](x) = f(I(x)) = f(x) = (If)(x)$$
$$\Rightarrow \quad I^* = I$$
$$\Rightarrow \quad g(I) = I^* = I.$$

$\Rightarrow$ g preserves the identity transformation.

SOLVED EXAMPLES

EXAMPLE 1. *Let T be an operator on a Banach space B. Show that T has inverse T^{-1} if and only if conjugate T^* of T has an inverse $(T^*)^{-1}$ and that $(T^*)^{-1} = (T^{-1})^*$.*

(KANPUR–2000, 14)

SOLUTION. We have

T has an inverse T^{-1}

$$\Leftrightarrow \quad TT^{-1} = T^{-1}T = I$$
$$\Leftrightarrow \quad (TT^{-1})^* = (T^{-1}T)^* = I^*$$
$$\Leftrightarrow \quad (T^{-1})^*T^* = T^*(T^{-1})^* = I$$

$\Leftrightarrow$ Inverse of T^* is $(T^{-1})^*$

EXAMPLE 2. *If $<x_n>$, $<y_n>$ are sequence in N and $<\alpha_n>$, $<\beta_n>$ are sequences of scalars such that $x_n \to x$, $y_n \to y$ and $\alpha_n \to \alpha$, $\beta_n \to \beta$ then $\alpha_n x_n + \beta_n y_n \to \alpha x + \beta y$.*

SOLUTION. We have to prove that

$$\alpha_n x_n + \beta_n y_n \to \alpha x + \beta y$$

For this consider

$$\|(\alpha_n x_n + \beta_n y_n) - (\alpha x + \beta y)\|$$
$$= \|\alpha_n x_n - \alpha_n x + \alpha_n x - \alpha x + \beta_n y_n - \beta_n y + \beta_n y - \beta y\|$$
$$= \|\alpha_n (x_n - x) + x(\alpha_n - \alpha) + \beta_n (y_n - y) + y(\beta_n - \beta)\|$$
$$\leq |\alpha_n| . \|x_n - x\| + |\alpha_n - \alpha| . \|x\| + |\beta_n| \|y_n - y\| + |\beta_n - \beta| \|y\|$$
$$= |\alpha| .0 + 0\|x\| + |\beta_n| (0) + \|y\| .0$$
$$\text{as } n \to \infty \text{ and } x_n \to x, y_n \to y, \alpha_n \to \alpha, \beta_n \to \beta$$
$$= 0$$
$$\Rightarrow \lim_{n \to \infty} (\alpha_n x_n + \beta_n y_n) = \alpha x + \beta y$$

4.24 STRONG AND WEAK TOPOLOGIES

Definition 1. *The set of all scalar valued bounded linear functional defined on N^*, is a normed linear space. Topology obtained from its character as a metric space is called strong topology on N^*.*

Definition 2. *The weak topology on N^* is the weakest topology on N^* with respect to which all the function in N^{**} are continuous.*

Definition 3. *The weakest topology on N^* with respect to which all the functions in N regarded as a subset of N^{**} remain continuous is called weak star topology or weak* topology.*

REMARKS

- The weak topology is weaker than strong topology.
- Weak star topology is weaker than the weak topology.
- Let $f \in N$ and $F_f \in N^{**}$. The weak star topology on N^* is the weakest topology under which all such F_f's are continuous.
- If g_0 is an arbitrary element in N^* and if $\in > 0$ is given, then the set
$$S(f, g_0, \in) = \{g : g \in N^* \text{ and } |F_f(g) - F(g_0)| < \in\}$$
$$= \{g : g \in N^* \text{ and } |g(f) - g_0(f)| < \in\}$$
is an open set in the weak star topology

THEOREM 1. ***If N is a normed linear space, then the closed unit space S^* in N^* is a compact space in the weak star topology.*** (MEERUT–1986, 87)

PROOF. **(i) $g \in \bar{S}^* \Rightarrow g \in S^*$**

With each vector in N, we associate a compact space C_f, where c_f is the closed interval $[-\|f\|, \|f\|]$, or the closed disc $\{z : |z| \leq \|f\|\}$ according as N is real or complex. We know that by Tychonoff's Theorem of topology, the product C of all the C_f's is also a compact space for each f, the values $g(f)$ of all g's in S^* lie in C_f.

We imbed S^* in C by regarding each g in S^* as identical with the array of all its values at the vectors f in N. Therefore, weak star topology on S^* equals its topology as a subspace of C. Now, since C is compact, S^* is closed subspace of C.

(ii) g is linear as a function defined on N.

Let f and g be any two vectors in N. Also, let $\in > 0$

Since, every neighborhood of g intersect S^*, there exists $\phi \in S^*$ such that

$$|g(f) - \phi(f)| < \in/3$$

$$|\phi(h) - g(h)| < \in/3$$

and $|g(f+h) - \phi(f+h)| < \in/3$

But ϕ is linear so $\phi(f+g) - \phi(f) - \phi(h) = 0$

Therefore, we have

$$|g(f+h) - g(f) - g(h)| = |[g(f+h) - \phi(f+h)| - [g(f) - \phi(f)] - [g(h) - \phi(h)]$$

$$\leq |g(f+h) - \phi(f+h)| + |g(f) - \phi(f)| + |g(h) - \phi(h)|$$

$$< \in/3 + \in/3 + \in/3$$

Since, above inequality is true for every $\in > 0$, we have

$$g(f+h) = g(f) + g(h)$$

Similarly, we can prove that

$$g(af) = a\, g(f) \text{ for } a \in R$$

Hence, g is linear.

THEOREM 2. ***Strong convergence implies weak convergence.***

(KANPUR–2001, MEERUT–2004)

PROOF. Let N be a normed linear space and $<s_n>$ be a sequence in N such that $<s_n>$ converges strongly to $s \in N$.

Then $\quad s_n \to s \text{ as } n \to \infty \quad$ ...(i)

$\Rightarrow \quad \|s_n - s\| \to 0 \text{ as } n \to \infty \quad$ ...(ii)

Let us assume that $f \in N^*$. Then

$$|f(s_n) - f(s)| = |f(s_n - s)|$$

$$\leq \|f\| \cdot \|s_n - s\| \to 0 \text{ as } n \to \infty \qquad \text{[From (ii)]}$$

$$\lim_{n\to\infty} |f(s_n) - f(s)| = 0$$

$\Rightarrow \quad f(s_n) \to f(s) \text{ as } n \to \infty$

Hence, $\quad s_n \to s$

$\Rightarrow \quad f(s_n) \to f(s)$

So, strong convergence is weak convergence.

EXERCISE 4.1

1. Let N be a non-zero normed linear space. Show that N is a Banach space if and only if $\{f : \|f\| \leq 1\}$ is complete.
2. If a Banach space B is the direct sum of two its subspaces M and N such that each f of B is uniquely expressible in the form $f = x + y$ with $x \in M$ $y \in N$.
   If we define $\|f\|_1 = \|x\| + \|y\|$, then show that $\|\cdot\|_1$ is a norm on B.
   (CALCUTTA–2010, MEERUT–1988)
3. Let N be a normed linear space. Inscribe the natural imbedding of N in N^* and show that it is an isometric isomorphism.
   (INDORE–2006, HIMACHAL–2006, KANPUR–1982, 84, 90, 92)
4. Show that there exists Banach spaces in which extension of a bounded linear functional without increasing is the norm is not unique.
5. If the conjugate space N^* of a normed linear space N is separable, then show that N is separable. (RAJASTHAN–2006, ALLAHABAD–2011, DELHI–1980)
6. Show that the set of all continuous linear transformation from one normed vector into another can, with appropriate definition, be equipped with a normed linear space structure.
7. If $k(s, t)$ is a continuous function in the square $0 \leq s \leq 1$, $0 \leq t \leq 1$ and $x(t)$ belongs to the space $C[0, 1]$ of continuous function

on [0, 1], prove that the transformation $T(\mu) = \int_0^1 k(s,t)x(t)dt$ is a bounded linear transformation of $C[0, 1]$ into itself.

8. Prove that every normed linear space X is isomorphic to a linear manifold in the second conjugate space X^{**}_i.

9. Prove that the set of bounded linear operator on a normed linear space X is, with a suitable norm, a Banach space.

10. Let N and N' be two normed linear space of the same finite dimension n, with the same scalar field. Show that N and N' are topologically isomorphic.

11. Let T be a linear transformation of a Banach space M into a Banach space N. If fT be continuous for each $f \in N^*$, show that T is continuous.

(GARHWAL–2001, 05, MEERUT–1982)

12. Let T be an operator on a normed linear space N. If N is considered to be part of N^{**} by means of the natural imbedding, show that T^{**} is an extension of T. Observe that if N is reflexive, then $T^* = T$.

13. Let A be a non-empty subset of a normed linear space X such that $f(A)$ is a bounded set at numbers for each $f \in X^*$. Show that A is a bounded set.

(ROHILKHAND–2006, MEERUT–1982)

14. Let N and N' be normed linear space with the same scalars. If N is infinite dimensional and $N' \neq \{0\}$, show that there exists a linear transformation of N onto N' which is not continuous.

15. Prove that a normed linear space is finite dimensional if and only if every closed and bounded set in it is compact. (AVADH–2001)

16. Let N and N' be normed linear spaces such that $N \neq \{0\}$. If $B(N, N')$ is complete, show that N' must be complete. (BUNDELKHAND–2007)

17. If T is bounded linear operator such that its inverse T^{-1} exists prove that T^{-1} is also continuous.

18. Let N and N' be two normed linear spaces and let $T : N \to T'$ be any linear transformation. Then show that T is continuous either at every point of N or at no point of N.

(KANPUR–2008)

19. Let N be a normed linear space and M be a finite dimensional subspace of N. Then show that M is closed. (AGRA–2008)

20. Let B be a Banach space, then show that B is reflexive if and only if B^* is reflexive.

(KANPUR–2003)

21. If N is a normed linear space then show that the conjugate space N^* of N is separable, then so is N. (KANPUR–2004, AGRA–2009, ROHILKHAND–2009)

22. Prove that every weakly convergent sequence in a normed linear space N is bounded.

(AVADH–2009)

23. Prove that in a finite dimensional space, weak and strong convergence are equivalent.

(ROHILKHAND–2009)

Chapter Review: A Competitive Approach

SELECTED TERMS AND RESULTS

▶ TERMS

- **Normed Linear Space :** A linear space with postulates of norm.
- **Banach Space :** A normed linear space which is complete is called a Banach space.
- **Absolutely Summable Series :** The series Σs_n is said to be absolutely summable if $\sum_{i=1}^{\infty} \|s_i\| < \infty$.
- **Linear Subspace :** A non-empty subset M of a linear space L is a linear subspace if and only if

  (i) $x \in M, y \in M \Rightarrow x + y \in M$

  (ii) $\alpha \in \mathbf{R}, x \in M \Rightarrow \alpha x \in M$
- **Canonical Mapping :** A mapping $f : L \to L/M$ defined by $f(x) = x + M$ is called the canonical mapping of L onto L/M.
- **Continuous Transformation :** Let N and N' be two normed linear spaces with the same scalar and $T : N \to N'$ be the linear transformation, then the linear transformation T is said to be continuous if $x_n \to x$ in N such that $T(x_n) \to T(x)$ in N.
- **Bounded Linear Transformation :** The linear transformation $T : N \to N'$ is said to be bounded if there exists a real number $k \geq 0$ such that

  $$\|Tx\| \leq k\|x\| \quad \forall x \in N$$
- **Isometric Isomorphism :** Let N and N' be two normed linear spaces. An isometric isomorphism of N into N' is a one-one linear transformation T of N into N' such that $\|T(x)\| = \|x\| \; \forall \; x \in N$.
- **Topological Isomorphism :** A topological isomorphism of N into N' is a one-one linear transformation of N into N' such that T and T^{-1} are continuous.
- **Equivalent Norms :** Let L be a normed linear space which is converted into a normed linear space in two ways *i.e.* $\|x\|_1$ and $\|x\|_2$ $\forall x \in L$. Then, these norms are said to be equivalent if and only if they generate the same topology on L.
- **Functional :** A linear bounded or continuous transformation from any arbitrary normed linear space N into **R** or **C** is called a functional.
- **Projection :** Let V be a linear space, which is the direct sum of two subspace M and N, *i.e.*, $V = M \oplus N$, $M \cap N = \{0\}$ such that each element $z \in V$ can be expressed uniquely as $x + y : x \in M, y \in N$, then the projection of M along N is a linear operator E defined by $E(z) = x$.
- **Graph of a Linear Transformation :** Let X and Y be two sets and D be a subset of X and $T : D \to Y$, then the set T_G of ordered pairs (x, Tx) such that $T_G = \{(x, Tx) : x \in D\}$ is called the graph of T.

▶ RESULTS

- Let N be a normed linear space and let d be a function from $N \times N$ into **R** defined by $d(x, y) = \|x - y\|$, then d is a metric on N.
- In a normed linear space, every convergent sequence is a Cauchy sequence.
- A normed linear space which is complete as a metric space is called a Banach space.
- A normed linear space N is a Banach space iff every absolutely summable series in N is summable.
- Let N be a normed linear space. Then the mapping $f : N \to \mathbf{R}$ such that $f(x) = \|x\|$ is continuous, *i.e.*, the norm $\| \; \|$ on N is a continuous function.
- A non-zero normed linear space N is a Banach space if and only if $S = \{x : \|x\| = 1\}$ is complete.
- L_p space are normal linear space
- For $i \leq p < \infty$, L_p space is complete.
- Let N be a non-zero normed linear space and

let $S = \{x \in N : \|x\| \leq 1\}$ be a linear subspace of N. Then N is Banach space if and only if S is complete.

- Let N be a normed linear space and M be a subspace of N. Then, the closure $\bar{M}$ of M is also a subspace of N.
- In a normed linear space, the smallest closed subspace containing a given set of vectors is just the closure of the subspace spanned by that set.
- Let N and N' be normed linear spaces over the same scalar field and T be a linear transformation of N into N'. Then T is bounded if and only if it is continuous.
- Let T be a linear transformation of a normed linear space N into another normed linear space N'. Then, T is bounded if and only if T maps bounded sets in N into bounded sets in N'.
- Let M be a closed linear subspace of a normed linear space N and let ϕ be a natural mapping defined by $\phi(x) = x + M$. Then ϕ is a continuous linear transformation for which $\|\phi\| \leq 1$.
- Let N and N' be normed linear operators and $T : N \to N'$ be a linear transformation. Then Ker(T) is a linear manifold and that Ker(T) is closed if T is continuous.
- A complex or real Banach space which is also a normed algebra is called a Banach Algebra.
- All norms are equivalent on a finite dimensional space.
- Every complete subspace of a normed linear space is closed.
- Every finite dimensional normed linear space is complete.
- Let M be a linear subspace of a normed linear space N and let f be a functional defined on M. Then, f can be extended to a functional f_0 on the whole space N such that $\|f\| = \|f_0\|$.
- Let B and B' be two Banach spaces and T is a continuous linear transformation from B onto B', then the image of every open sphere centered at origin in B contain an open sphere centered at origin in B'.
- Let B and B' be two Banach spaces and if T is continuous linear transformation of B onto B', then T is an open mapping.
- Let B and B' be Banach spaces and let T be a one-one continuous linear transformation of B onto B'. Then, T is a homomorphism. In particular , T^{-1} is automatically continuous.
- Let P be a projection on a Banach space and if M and N are its range and null spaces, then M and N are closed linear subspaces of B such that $B = M \oplus N$.
- T is a closed linear transformation if and only if its graph T_G is a closed subspace.
- Let B and B' be two Banach spaces and if T is a linear transformation of B into B', then T is continuous if and only if its graph is closed.
- A non-empty subset X of a normed linear space N is bounded if and only if $f(X)$ is bounded set of numbers for each $f \in N^*$.
- If N is a normed linear space, then the closed unit space S^* in N^* is a compact space in the weak star topology.

REVIEW QUESTIONS

▶ DESCRIPTIVE QUESTIONS

1. In a normed linear space, prove that a convergent sequence has a unique limit.
2. Show that two norms are equivalent if and only if they define the same bounded sets.
3. Prove that a subset S of a normed linear space is bounded if and only if $\|\lambda_n x_n\| \to 0$ for any $x \in S$ and any scalar $\lambda_n \to 0$.
4. Let N be a Banach space and $B(N)$ be a Banach space of operators on N. Prove that
$$\|TT'\| \leq \|T\|.\|T'\| \quad \forall\, T, T' \in B(N)$$
5. Let M be a closed linear subspace of normed linear space N and $T : N \to N|M$ be defined by $T(x) = x + M \;\forall\, x \in N$. Show that $\|T\| \leq 1$.
6. Let T be a bounded linear transformation from a normed linear space N to N'. If $S = \{x \in N : \|x\| = 1\}$ is a closed unit sphere in N, then prove that its image $T(S)$ is a bounded set in N'.
7. Prove that a normed linear space N is separable if its conjugate space N^* is separable.
8. If n is a normed linear space and x_0 is a non-zero vector in N then prove that T is an open mapping.

▶ MULTIPLE CHOICE QUESTIONS

Choose the most appropriate one.

1. A mapping $f: X \to Y$ is said to be homeomorphism if :
   (a) f is one-one and onto
   (b) f is continuous
   (c) f^{-1} is continuous
   (d) all are true

2. Every complete metric space is of :
(a) first category (b) second category
(c) either (a) and (b) (d) none of these

3. If N and N' be normed linear space and T a linear transformation of N into N'. Then T is continuous if :
(a) T is continuous at the origin
(b) T is bounded
(c) for $k \geq 0$, $\|f(x)\| \leq k\|x\|$
(d) all are true

4. Let M be a closed linear subspace of a normed linear space N if $\|x+m\| = \inf.\{\|x+m\| : m \in M\}$ Then m is a :
(a) normed linear space
(b) complete space
(c) Banach space
(d) all are true

5. Let N be a normed linear space, then N is a Banach space if and only if :
(a) $\{x : \|x\| = 1\}$ is complete
(b) $\{x : \|x\| \leq 1\}$ is complete
(c) $\{x : \|x\| = 1\}$ is bounded
(d) none of these

6. The inequality
$$\sum_{i=1}^{n} |x_i y_i| \leq \left(\Sigma |x_i|^2\right)^{1/2} \left(\Sigma |y_i|^2\right)^{1/2}$$
is called :
(a) Bessel's Inequality
(b) Cauchy's Inequality
(c) Minkowski's Inequality
(d) none of these

7. The inequality
$$\sum_{i=1}^{n} |x_i y_i| \leq \left(\Sigma |x_i|^p\right)^{1/p} . \left(\Sigma |y_i|^q\right)^{1/q}$$
is called :
(a) Holder's Inequality
(b) Cauchy's Inequality
(c) Minkowski's Inequality
(d) none of these

8. The inequality $\|x+y\|_p \leq \|x\|_p + \|y\|_p$ is called :
(a) Holder's Inequality
(b) Cauchy's Inequality
(c) Minkowski's Inequality
(d) none of these

9. Banach space is a :
(a) normed vector space
(b) complete normed vector space
(c) complete vector space
(d) none of these

10. If $\|T\|$ is a norm of T then :
(a) T is bounded
(b) $\|T\| = \inf\{M : \|T_x\| \leq M\|x\|, x \in X\}$
(c) both (a) and (b) are true
(d) none of these

11. A linear transformation T of X into Y is continuous, then :
(a) T is bounded (b) T is unbounded
(c) T is constant (d) none of these

12. A linear transformation T is bounded if there is an integer M such that :
(a) $\|T_x\| = M\|x\| \ \forall \ x \in X$
(b) $\|T_x\| \leq M\|x\| \ \forall \ x \in X$
(c) $\|T_x\| < M\|x\| \ \forall \ x \in X$
(d) none of these

13. A linear transformation T is bounded if :
(a) T is discontinuous
(b) T is continuous
(c) T is discrete
(d) none of these

14. $\|T\|$ is a norm of T then :
(a) T is a linear transformation
(b) $\|T_x\| = \|T\|.\|x\| \ \forall \ x \in X$
(c) both (a) and (b) are true
(d) none of these

15. A normed vector space is finite dimensional if :
(a) the closed bounded sets are compact
(b) open bounded sets are compact
(c) both (a) and (b) are true
(d) none of these

16. If $l \leq p \leq \infty$ and q is conjugate to p, then :
(a) $(l_p)^1 > l_q^*$ (b) $(l_p)^1 < l_q^*$
(c) $(l_p)^1 = l_q^*$ (d) none of thhese

17. The dual of a normed vector space is :
(a) Bounded space
(b) Banach space
(c) both (a) and (b) are true
(d) none of these

18. Cauchy-Schwarz inequality states, for every $x, y \in X$:
(a) $|(x, y)| \leq \|x\|.\|y\|$
(b) $|(x, y)| > \|x\|.\|y\|$
(c) $|(x, y)| = \|x\|.\|y\|$
(d) none of these

19. A linear space together with an inner product in X is called :
(a) closed space
(b) innerproduct space
(c) product space
(d) none of these

20. A Cauchy-Schwarz inequality states that inner product space is :
(a) continuous function on $X \times X$ into F
(b) a constant function
(c) discontinuous function
(d) none of these

21. For $x \in X$, the norm in the space X is :
(a) $\|x\| = (x, x)^2$
(b) $\|x\| = \sqrt{(x,x)} = (x,x)^{1/2}$
(c) both (a) and (b) are true
(d) none of these

22. If X is a normed linear space and f is continuous linear functional on Y, a subspace of X, then :
(a) f can be extended to a linear functional F on X such that $\|f\| = \|F\|$
(b) f cannot be extended to a linear functional F on X such that $\|f\| = \|F\|$
(c) f can be extended to F such that $F = f$
(d) none of these

23. If X is a normed vector space, Y is a subspace of X and f is bounded linear functional on Y with bound $\|f\|$ relation to y, then f has a continuous linear extension $x' \in X'$ with $\|x'\| = \|f\|$. This statement is known as :
(a) hahn Banach theorem
(b) cauchy's theorem
(c) uniform bounded principle
(d) none of these

24. Let X be a Banach space, $<f_n>$ is a sequence of continuous linear functional on X for every $x \in X$, the sequence $<|f_n(x)|>$ is bounded, then the sequence $<\|f_n\|>$ of norms is :
(a) closed (b) unbounded
(c) bounded (d) none of these

25. T is a closed linear transformation if graph $G(T)$ is :
(a) closed linear subspace
(b) open linear subspace
(c) null linear subspace
(d) none of these

26. A linear transformation T from a Banach space X into a Banach space Y with $\text{dom}(T) < X$ is closed if :
(a) $x_n \in \text{dom}(T)$
(b) $\lim_{n \to \infty} x_n = x$
(c) both (a) and (b) are true
(d) none of these

27. For every normed linear space X, there is a set A such that X is isomorphic with a subspace of the Banach space of bounded function f on A with :
(a) $\|f\| = \text{sup.}\{|f(t)| : t \in A\}$
(b) $\|f\| = \{|f(t)| : t \in A\}$
(c) $\|f\| = \text{inf.}\{|f(t)| : t \in A\}$
(d) none of these

28. A graph $G(T)$ of a mapping $T : X \to Y$ is the set of points $(x, T_x) \in X \times Y$ with :
(a) $x \in \text{dom}(T)$
(b) $T_x \in \text{dom}(T)$
(c) both (a) and (b) are true
(d) none of these

29. If $\text{dom}(T)$ is closed in Banach space X and linear transformation T is bounded, then :
(a) T is open (b) T is constant
(c) T is closed (d) none of these

30. If X is a Banach space, Y is a normed linear space and $<T_n>$ is a sequence of continuous linear transformation on X into Y such that for every $x \in X$, the sequence $<\|T_n(x)\|>$ is bounded then the sequence $<\|T_n\|>$ of norm is :
(a) open (b) bounded
(c) unbounded (d) none of these

31. "If X and Y are Banach spaces and T is a linear transformation from X to Y, then $\text{dom}(T)$ is closed and graph $G(T)$ closed $\Rightarrow T$ is bounded". This statement is known as :
(a) closed graph theorem
(b) open mapping theorem
(c) uniform bounded principle
(d) none of these

32. "If X and Y are Banach spaces and T is bounded linear transformation which maps X onto Y. Then T is an open mapping". This statement is called :
(a) closed graph theorem
(b) open mapping theorem
(c) uniform bounded principle
(d) none of these

33. If $C[-1, 1]$ denotes a Banach space of all real valued function $x(t)$ on $[1, -1]$ with norm given by $\|x\| = \max_{t\in[-1,1]} |x(t)|$ then norm of the linear functional f defined on $C[-1, 1]$ by

$$f(x) = \int_{-1}^{0} x(t)dt - \int_{0}^{1} x(t)dt \text{ is :}$$

(GATE–2000)

(a) 0 (b) 1
(c) 2 (3) 3

34. All norms on a normed linear space X are equivalent provided : (GATE–2001)

(a) X is reflexive
(b) X is complete
(c) X is finite dimensional
(d) X is an inner product space

35. Which of the following Banach space is not separable : (GATE–2002)

(a) $L^1[0, 1]$ (b) $L^2[0, 1]$
(c) $C[0, 1]$ (d) $L^\infty[0, 1]$

36. Consider the Banach space $C[0, \pi]$ with the supremum norm. The norm of the linear functional $l : C[0, \pi] \to R$ given by

$$l(t) = \int_0^{\pi} f(x)\sin^2 x dx \text{ is :}$$ (GATE–2002)

(a) π (b) $\pi/2$
(c) 1 (d) 0

37. Let T be a Banach space (not finite dimensional) and $T : B \to B$ be a continuous operator such that the range of T is B and $T(x) = 0 \Rightarrow x = 0$, then : (GATE–2003)

(a) T maps compact sets to opensets
(b) T maps bounded set to compact sets
(c) both (a) and (b) are true
(d) none of these

38. Let X and Y be normed linear spaces and let $T : X \to Y$ be a linear map. Then T is continuous if : (GATE–2007)

(a) X is a finite dimensional
(b) Y is finite dimensional
(c) both (a) and (b) are true
(d) none of these

39. Let X be a normed linear space and $E_1, E_2 \subseteq X$. Define $E_1 + E_2 = [x + y : x \in E_1, y \in E_2]$

Then $E_1 + E_2$ is open if : (GATE–2007)

(a) either E_1 or E_2 is open
(b) both E_1 and E_2 are open
(c) can't say
(d) none of these

40. Let X be a real normed linear space of all real sequences with finitely many non-zero terms with supremum norms and $T : X \to X$ be a one-to-one and onto linear operator defined by

$$T(x_1, x_2, ...) = \left(x_1, \frac{x_2}{2^2}, \frac{x_3}{3^2}, ...\right)$$

Then which of the following is true :

(GATE–2011)

(a) T is bounded
(b) T^{-1} is bounded
(c) both (a) and (b) are true
(d) none of these

41. Given a non-trivial normed linear space, the non-triviality of its dual space is assured by :

(GATE–2001)

(a) the Hahn Banach theorem
(b) uniform bounded principle
(c) closed graph theorem
(d) none of these

42. Let $C[0, 1]$ be the space of all Continuous real valued functions on $[0, 1]$. The identity map $I : (C[0, 1], \|.\|) \to (C[0, 1], \|.\|)$ is :

(GATE–2005)

(a) continuous but not open
(b) continuous and open
(c) open but not continuous
(d) none of these

43. Let X and Y be normed linear space and $<T_n>$ be a sequence of bounded linear operators from X and Y. Consider the statements :

$P : \{\|T_n(x)\| : n \in N\}$ is bounded $\forall\, x \in X$

$Q : \{\|T_n\| : n \in N\}$ is bounded

Which of the following is correct?

(GATE–2010)

(a) if P implies Q, then both X and Y are Banach spaces
(b) if X is Banach then P implies Q
(c) if Y is Banach then P implies Q
(d) none of these

44. Consider the statements

P : If X is normed linear space and $M \subseteq X$ is a subspace then closure of M is also a subspace of X

Q : If X is a Banach space and Σx_n is an absolutely convergent series in X then Σx_n is convergent

R : Let M_1 and M_2 be subspace of an innerproduct space that $M_1 \cup M_2 = \{0\}$ then for all $m_1 \in M_1$, $m_2 \in M_2$,

$$\|m_1 + m_2\|^2 = \|m_1\|^2 + \|m_2\|^2$$

The correct statement amongest the above are : (GATE–2012)

(a) P and Q (b) P and R

(c) Q and R (d) R and S

45. Let $C[0, 1]$ be the space of real valued continuous functions on $[0, 1]$ with the norm $\|f\|_\infty = \{|f(x)| : x \in [0, 1]\}$. Consider the subspace $P_n([0, 1])$ of all polynomials of degree less than or equal to n and the subspace $P([0, 1])$ of all polynomials on $[0, 1]$, then : (GATE–2006)

(a) $P_n([0, 1])$ is closed in $C([0, 1])$ but not in $P([0, 1])$

(b) $P([0, 1])$ is closed in $C([0, 1])$ but not in $P_n([0, 1])$

(c) both (a) and (b) are true

(d) none of these

46. Let $f : (C_{00}, \|.\|_1) \to C$ be a non-zero continuous linear functional then, the number of Hahn Banach extensions of f to $(C^1, \|.\|_1)$ is : (GATE–2009)

(a) 1 (b) 2

(c) 3 (4) ∞

47. Which of the following inequalities is known as Minkowski's inequality for $0 < p < 1$ in L^P-space :

(a) $\|f + g\|_p \leq \|f\|_p + \|g\|_p$

(b) $\|f + g\|_p \geq \|f\|_p + \|g\|_p$

(c) $\|f + g\|_p < \|f\|_p + \|g\|_p$

(d) none of these

48. For two non-negative real numbers a, b and p, q such that $\frac{1}{p} + \frac{1}{q} = 1, p > 1$ which is true :

(a) $\frac{1}{a^p} + \frac{1}{b^q} \leq \frac{a}{p} + \frac{b}{q}$

(b) $ab \leq \frac{a^p}{p} + \frac{b^q}{q}$

(c) both (a) and (b) are true

(d) none of these

49. If S_r denotes the open sphere with radius r centred at origin in Banach space B then for linear transformation T on B : $T(S_r) =$

(a) $rT(S_1)$ (b) $r + T(S_1)$

(c) S_r (d) none of these

50. The natural imbedding of N into N^{**}, where N is a normed linear space and N^{**}, (the second conjugate of N) is :

(a) a linear and norm preserving mapping of N into N^{**}

(b) not a linear mapping of N into N^{**}

(c) not a continuous mapping of N into N^{**}

(d) none of these

ANSWERS

1. (d)	**2.** (b)	**3.** (d)	**4.** (d)	**5.** (a)	**6.** (b)	**7.** (a)	**8.** (c)	**9.** (b)
10. (c)	**11.** (a)	**12.** (b)	**13.** (b)	**14.** (c)	**15.** (a)	**16.** (c)	**17.** (b)	**18.** (a)
19. (b)	**20.** (a)	**21.** (b)	**22.** (a)	**23.** (a)	**24.** (c)	**25.** (a)	**26.** (c)	**27.** (a)
28. (a)	**29.** (c)	**30.** (b)	**31.** (a)	**32.** (b)	**33.** (a)	**34.** (d)	**35.** (b)	**36.** (c)
37. (a)	**38.** (a)	**39.** (b)	**40.** (c)	**41.** (a)	**42.** (a)	**43.** (b)	**44.** (a)	**45.** (a)
46. (b)	**47.** (a)	**48.** (b)	**49.** (a)	**50.** (a)				

1. Let X be a Banach space such that X^* is reflexive. Show that X is reflexive.
2. Let X, Y be Banach spaces and let $B(X, Y)$ denote the set of bounded linear maps from X to Y. Show that $B(X, Y)$ is a Banach space with the operator norm.
3. Let M be a closed linear subspaces of a normed linear space N and let d be the distance from M of a vector f in N which is not in M. Show that there exists a functional ψ on N such that $\psi(m) = 0$, $\psi(f) = 1$ and $\|\psi\| = \frac{1}{d}$.
4. Let A be the set of all non-negative sequences in the closed unit ball of C_0, show that $\sup_{a \in A} . \|x - a\| =$ diameter of A for all $x \in C_0$
5. Prove that a normed space is harmonic to its open unit ball.
6. Prove that the space l_∞ of all bounded sequence has no Schaunder basis.
7. Show that a linear functional f on a normal space X is continuous if and only if its Kernel, $\text{Ker}(f) = [x \in X : f(x) = 0]$ is a closed set in X.
8. Let f be a discontinuous linear functional on a normed space X. Show that $\text{Ker}(f)$ is dense in X.
9. Let X be an infinite-dimensional normed space. Prove that there exists a discontinuous linear functional on X.
10. Prove that every Banach limit has value zero at $(-1, 1, -1, 1, \ldots)$.
11. Prove that every reflexive subspace of l_1 is finite dimensional.
12. Let X be a reflexive Banach space. Prove that X is separable if and only if X^* is separable.
13. Prove that a linear mapping from a normed space into a normed space is continuous if and only if it maps bounded sets.
14. Let E_1 and E_2 be normed spaces. Show that a function $f : E_1 \to E_2$ is continuous if and only if for every $x \in E_1$ and $\varepsilon > 0 \ \exists \ \delta > 0$ such that $\|f(x) - f(y)\| < \varepsilon$ whenever $\|x - y\| < \delta$.
15. Let E_1, E_2 be normed spaces and let $f : E_1 \to E_2$ be a continuous function. Prove that if $S < E_1$ is compact, then $f(S)$ is compact in E_2.

●●●●●●

Chapter 5 Hilbert Spaces

5.1 INTRODUCTION

In this chapter, we first introduce the basic concept of an inner product space and then proceed to discuss certain properties and examples in the theory of Hilbert spaces. The modern developments in Hilbert spaces are concerned largely with the theory of operators on the spaces. In this chapter, we shall also discuss approximate definition and important properties of spaces. Some basic concepts in the theory of operators in Hilbert spaces will also be discussed in this chapter.

5.2 INNER PRODUCT SPACES

(DELHI–2007, 11, CALCUTTA–2005, MEERUT–2006, 08, 11, KANPUR–2001, 09)

Let $L(F)$ be a linear space where F is either the field of real numbers or the field of complex numbers. By an inner product on L, we mean a mapping from $L \times L$ into F which assigns to each ordered pairs of vectors x, y in L, a scalar (α, β) in F such that

(i) $\overline{(x, y)} = (y, x)$ for $x, y \in L$ [*Hermitian conjugate symmetric*]

(ii) $(\alpha x + \beta y, z) = \alpha(x, z) + \beta(y, z)$ for $x, y, z \in L, \ \alpha, \beta \in F$

[*Homogeneity and additivity or linearity*]

(iii) $(x, x) \geq 0$ and $(x, x) = 0 \Leftrightarrow x = 0$ [*Positivity*]

In (i), the bar denotes the complex conjugation.

The linear space L is then said to be an inner product space with respect to the inner product defined on it.

REMARKS

- In the above definition (x, y) does not denote the ordered pair of the vectors x and y, but it denotes the inner product of the vectors x and y.
- If $F = R$, then (x, y) is a real number and if $F = C$ then (x, y) is a complex number.
- If $F = R$, then $\overline{(x, y)} = (y, x) \Rightarrow (x, y) = (y, x)$.
- An inner product space is also called a pre-Hilbert space.

5.3 HILBERT SPACES

A complex Banach space H is called a Hilbert space, if a complex number (x, y) (called the inner product of x and y) is associated to each of the two vectors x and y such that

(*i*) $\overline{(x, y)} = (y, x)$ (*ii*) $(\alpha x + \beta y, z) = \alpha(x, z) + \beta(y, z)$

(*iii*) $(x, x) = \|x\|^2$

For all $x, y, z \in H$ and for all scalars α and β.

(DELHI–2011, 12, MEERUT–2006, 06BP, 07, 08, 09BP, GARHWAL–2003, KANPUR–1990, 2006, 07, 08, 09, 11, AGRA–2009, ROHILKHAND–2008)

ILLUSTRATIONS

(1) Let l_2^n be a Banach space consisting of all n-tuples of complex numbers with the norm of a vector $x = (x_1, ..., x_n)$ defined by

$$\|x\| = \left(\sum_{i=1}^{n} |x_i|^2\right)^{1/2}$$

Then, l_2^n is a Hilbert space.

(2) Let l_2 be a Banach space consisting of all infinite sequence $x = < x_n >$ of complex numbers such that

$$\sum_{n=1}^{\infty} |x_n|^2 < \infty$$

with the norm of a vector $x = < x_n >$ defined by

$$\|x\| = \left(\sum_{n=1}^{\infty} |x_n|^2\right)^{1/2}$$

If the inner product of two vectors $x = < x_n >$ and $y = < y_n >$ is defined by

$$(x, y) = \sum_{n=1}^{\infty} x_n \, \bar{y}_n$$

Then l_2 is a Hilbert space.

THEOREM 1. ***An inner product space is a normed linear space.***

PROOF. Let V be an inner product vector space over K. Let $u, v \in V$ and $a \in K$ be arbitrary.

Define $\|u\| = \sqrt{< u,u >}$

But $<u, u>$ is defined on V, therefore, $\|u\|$ is defined on V.

(i) Since $<u, u> \geq 0$ and $<u, u> = 0$ if and only if $u = 0$, because norm of any vector is greater than or equal to zero.

$\therefore \quad \|u\| \geq 0$ and $\|u\| = 0$ if $u = 0$

(ii)
$$\|au\|^2 = <au, au> = a<u, au>$$
$$= a\bar{a} < u,u > = |a|^2 \|u\|^2$$
$$\Rightarrow \quad \|au\| = |a| . \|u\|$$

(iii) Consider
$$\|u + v\|^2 = <u + v, u + v> = <u, u + v> + <u, u + v>$$
$$= <u, u> + <u, v> + <v, u> + <v, v>$$
$$= \|u\|^2 + \|v\|^2 + < u,v > + \overline{< u,v >}$$
$$= \|u\|^2 + \|v\|^2 + 2\text{Re}<u, v>$$
$$\leq \|u\|^2 + \|v\|^2 + 2|<u, v>|$$
$$\leq \|u\|^2 + \|v\|^2 + 2\|u\|.\|v\| = (\|u\| + \|v\|)^2$$
$$\Rightarrow \quad \|u + v\| \leq \|u\| + \|v\|$$

Thus, the norm defined on an inner product space satisfies all the conditions of a normed vector space and hence it is a normed vector space.

5.4 PROPERTIES OF HILBERT SPACES

THEOREM 1. ***In a Hilbert space H***

(i) $(\alpha x - \beta y, z) = \alpha(x, z) - \beta(y, z)$ (DELHI–2006, MEERUT–2001, 04, 06BP, 07, 09, 10)

(ii) $(x, \beta y + \gamma z) = \bar{\beta}\,(x, y) + \bar{\gamma}(x, z)$ (MEERUT–2001, 04, 06BP, 07, 09)

(iii) $(x, \beta y - \gamma z) = \bar{\beta}(x, y) - \bar{\gamma}(x, z)$

(iv) $(x, 0) = 0 \ \forall\, x \in H$ ***and*** $(0, x) = 0 \ \forall\, x \in H$ (MEERUT–2001, 04, 06BP)

PROOF. Let H be a Hilbert space.

(i) We have, for all $x, y, z \in H$ and for scalars α, β

$$(\alpha x - \beta y, z) = (\alpha x + (-\beta) y, z)$$
$$= \alpha(x, z) + (-\beta)(y, z) = \alpha(x, z) - \beta(y, z)$$

(ii) We have

$$(x, \beta y + \gamma z) = \overline{(\beta y + \gamma z, x)}$$
$$= \overline{\beta(y, x) + \gamma(z, x)} = \overline{\beta(y, x)} + \overline{\gamma(z, x)} = \bar{\beta}\,\overline{(y, x)} + \bar{\gamma}\,\overline{(z, x)}$$
$$= \bar{\beta}(x, y) + \bar{\gamma}(x, z)$$

(iii) We have

$$(x, \beta y - \gamma z) = (x, \beta y + (-\gamma) z)$$
$$= \bar{\beta}(x, y) + (-\bar{\gamma})(x, z) = \bar{\beta}(x, y) + (-1)\bar{\gamma}(x, z)$$
$$= \bar{\beta}(x, y) - \bar{\gamma}(x, z)$$

(iv) We have $(0, x) = (00, x) = 0(0, x) = 0$

Further $(x, 0) = \overline{(0, x)} = 0$

REMARK

- The above results remains true even if x, y and z are vectors belonging to an arbitrary inner product space, which is not necessarily a Hilbert space.

$$(x, y + z) = (x, y) + (x, z).$$

THEOREM 2. **(Schwarz's Inequality).** ***Let H be a Hilbert space. Then for any two vectors x, y in H,*** $|(x, y)| \le \|x\| . \|y\|$. (DELHI–2009, RAJASTHAN–2009, SAGAR–2007, MEERUT–1992, 94, 98, 2000, 02, 03BP, 06, 06BP, 08, 09BP, 11, 13, 15, AGRA–2005, 08, AVADH–2009, GARHWAL–2003, 04, KANPUR–1990, 2005, 06, BUNDELKHAND–1985, ROHILKHAND–2008)

PROOF. **Case – I:** If $y = 0$, then clearly $\|y\| = 0$ which implies $|(x, y)| = 0$.

In this case, result is trivially true.

Case – II: Let $y \neq 0$, then for any scalar λ.

Consider

$$(x + \lambda y,\ x + \lambda y) \ge 0 \qquad (\because (x, x) = \|x\|^2 \ge 0 \ \forall\, x \in H)$$

$$\Rightarrow \qquad (x, x + \lambda y) + \lambda(y, x + \lambda y) \ge 0$$

$$\Rightarrow \quad (x, x) + \bar{\lambda}(x, y) + \lambda(y, x) + \lambda\bar{\lambda}(y, y) \ge 0$$

$$\Rightarrow \|x\|^2 + \bar{\lambda}(x, y) + \lambda(y, x) + |\lambda|^2 \|y\|^2 \ge 0 \qquad \ldots(1)$$

Putting $$\lambda = -\frac{(x, y)}{\|y\|^2}$$

Then, from (1), we have

$$\|x\|^2 - \frac{\overline{(x, y)}}{\|y\|^2}(x, y) - \frac{(x, y)}{\|y\|^2}(y, x) + \frac{|(x, y)|^2}{(\|y\|^2)^2} . \|y\|^2 \ge 0$$

$$\Rightarrow \quad \|x\|^2 - \frac{|(x, y)|^2}{\|y\|^2} - \frac{(x, y)\,\overline{(x, y)}}{\|y\|^2} + \frac{|(x, y)|^2}{\|y\|^2} \geq 0$$

$$\Rightarrow \quad \|x\|^2 - \frac{|(x, y)|^2}{\|y\|^2} - \frac{|(x, y)|^2}{\|y\|^2} + \frac{|(x, y)|^2}{\|y\|^2} \geq 0$$

$$\Rightarrow \quad \|x\|^2 - \frac{|(x, y)|^2}{\|y\|^2} \geq 0$$

i.e.,
$$\|x\|^2 . \|y\|^2 - |(x, y)|^2 \geq 0$$

Hence, $|(x,y)| \leq \|x\| . \|y\|$

REMARK

- In case of arbitrary inner product space, the above result can be written as
$$|(x,y)| \leq \sqrt{(x, x)} . \sqrt{(y, y)}$$

THEOREM 3. ***Let H be a Hilbert space. Then the inner product is jointly continuous, i.e.,***
$$x_n \to x,\ y_n \to y \Rightarrow (x_n, y_n) \to (x,y)$$
(MEERUT–2001, GARHWAL–2003, KANPUR–1990, 2002, 03, 06, 07, ROHILKHAND–2008)

PROOF. Consider

$$\begin{aligned} |(x_n, y_n) - (x, y)| &= |(x_n, y_n) - (x_n, y) + (x_n, y) - (x, y)| \\ &= |(x_n, y_n - y) + (x_n - x, y)| \\ &\leq |(x_n, y_n - y)| + |(x_n - x, y)| \quad \text{[By triangle inequality]} \\ &\leq \|x_n\| . \|y_n - y\| + \|x_n - x\| . \|y\| \end{aligned}$$

[By Schwarz inequality]

As per given, we have
$$x_n \to x \quad \text{and} \quad y_n \to y$$
Therefore, by definition
$$\|y_n - y\| \to 0 \quad \text{and} \quad \|x_n - x\| \to 0 \text{ as } n \to \infty.$$
Thus, $|(x_n, y_n) - (x, y)| \to 0$ as $n \to \infty$.

Hence, $(x_n, y_n) \to (x, y)$, *i.e.*, inner product is jointly continuous.

THEOREM 4. ***Let H be a Hilbert space and x, y be any two vectors of H, then***

(a) Parallelogram law :
$$\| x+y \|^2 + \| x-y \|^2 = 2\| x \|^2 + 2\| y \|^2$$
(MEERUT–2004, 06BP, 07, 07BP, 09BP, 10, 11, KANPUR–1993,95,2000,01)

(b) Polarisation identity :
$$4(x, y) = \| x+y \|^2 - \| x-y \|^2 + i \| x+iy \|^2 - i \| x-iy \|^2$$
(MEERUT–1988, 2002, 03, 04, 09BP, 10, 11, 13, KANPUR–2006, 08, AGRA–1988, 2001)

PROOF. (a) Consider

$$\begin{aligned} \| x+y \|^2 &= (x+y, x+y) = (x, x+y) + (y, x+y) \\ &= (x, x) + (x, y) + (y, x) + (y, y) \\ &= \|x\|^2 + (x, y) + (y, x) + \|y\|^2 \end{aligned} \quad \ldots(1)$$

Also, $\|x-y\|^2 = (x-y, x-y) = (x, x-y) - (y, x-y)$

$$= (x, x) - (x, y) - (y, x) + (y, y)$$

$$= \|x\|^2 - (x, y) - (y, x) + \|y\|^2 \qquad \ldots(2)$$

On adding (1) and (2), we get

$$\|x+y\|^2 + \|x-y\|^2 = 2\|x\|^2 + 2\|y\|^2$$

(b) Using (1) and (2), we have

$$\|x+y\|^2 - \|x-y\|^2 = 2(x, y) + 2(y, x) \qquad \ldots(3)$$

Put iy for y in (3), we get

$$\|x+iy\|^2 - \|x-iy\|^2 = 2(x, iy) + 2(iy, x)$$

$$= 2\bar{i}\,(x, y) + 2i(y, x) = -2i(x, y) + 2i(y, x) \qquad \ldots(4)$$

$$\Rightarrow \quad i\|x+iy\|^2 - i\|x-iy\|^2 = -2i^2(x, y) + 2.i^2(y, x)$$

$$= 2(x, y) - 2(y, x) \qquad \ldots(5)$$

Now, adding (3) and (5), we get

$$\|x+y\|^2 - \|x-y\|^2 + i\|x+iy\|^2 - i\|x-iy\|^2 = 4(x, y)$$

THEOREM 5. ***Let B be a complex Banach space whose norm obeys the parallelogram law and polarisation identity, i.e.,***

$$\mathbf{4(x,y) = \|x+y\|^2 - \|x-y\|^2 + i\|x+iy\|^2 - i\|x-iy\|^2} \qquad \textbf{...(A)}$$

Then, B is a Hilbert space. (DELHI–2006, MEERUT–1997, 2004, 05, 06, 09, KANPUR–2006, AMRITSAR–1982, ROHTAK–2007, PATNA–2004)

PROOF. To show that the inner product defined above has the three properties required by the definition of Hilbert space given by

(1) $(x, y) = \overline{(y, x)}$

(2) $(x, x) = \|x\|^2$

(3) $(\alpha x + \beta y, z) = \alpha(x, z) + \beta(y, z)$

(1) Since $4(x, y) = \|x+y\|^2 - \|x-y\|^2 + i\|x+iy\|^2 - i\|x-iy\|^2$

$$4(y, x) = \|y+x\|^2 - \|y-x\|^2 + i\|y+ix\|^2 - i\|y-ix\|^2$$

Now, $y + ix = i(x - iy)$

$$\therefore \quad \|y+ix\| = \|i(x-iy)\| = |i|.\|x-iy\| = \|x-iy\|$$

Similarly $y - ix = -i(x+iy)$

$$\therefore \quad \|y-ix\| = \|-i(x+iy)\| = |-i|.\|x+iy\| = \|x+iy\|$$

and $\|y-x\| = \|-(x-y)\| = -1|\|x-y\| = \|x-y\|$

$$\therefore \quad 4(y, x) = \|x+y\|^2 - \|x-y\|^2 + i\|x-iy\|^2 - i\|x+iy\|^2$$

$$4\overline{(y, x)} = \|x+y\|^2 - \|x-y\|^2 - i\|x-iy\|^2 + i\|x+iy\|^2$$

$$= 4(x, y)$$

$\|x+iy\|^2$ is a positive real number and so it does not change on taking complex conjugate.

Hence, $(x, y) = \overline{(y, x)}$

(2) Putting $y = x$ in (A), we get

$$4(x, x) = \|x + x\|^2 - \|x - x\|^2 + i\|x + ix\|^2 - i\|x - ix\|^2$$
$$= \|2x\|^2 - \|0\|^2 + i\|(1+i)x\|^2 - i\|(1-i)x\|^2$$
$$= \|2x\|^2 - \|0\|^2 + i\|(1+i)x\|^2 - i\|(1-i)x\|^2$$
$$= 4\|x\|^2 + i\,|(1+i)|^2\,\|x\|^2 - i\,|(1-i)|^2\,\|x\|^2$$

Put $\quad |(1+i)| = \sqrt{(1+1)} = \sqrt{2} = |(1-i)|$

$\therefore \quad 4(x, x) = 4\|x\|^2 + 2i\|x\|^2 - 2i\|x\|^2 = 4\|x\|^2$

$\therefore \quad (x, x) = \|x\|^2$

(3) Here, we shall prove the following two results :

(i) $\quad (x + y, z) = (x, z) + (y, z) \ \forall \ x, y, z \in B$

and (ii) $\quad (\alpha x, y) = \alpha(x, y) \ \forall \ x, y \in B, \alpha \in F$

(i)
$$4(x + y, z) = \|(x + y) + z\|^2 - \|(x + y) - z\|^2$$
$$+ i\|(x + y) + iz\|^2 - i\|(x + y) - iz\|^2 \quad \text{[By definition]}$$

We have already proved that

$\|x + y\|^2 + \|x - y\|^2 = 2\|x\|^2 + 2\|y\|^2$

So that $\|x + y + z\|^2 = \|(x + z) + y\|^2$

$$= 2\|x + z\|^2 + 2\|y\|^2 - \|(x + z) - y\|^2 \quad \ldots(1)$$

Now $\quad \|(x + z) - y\|^2 = \|(z - y) + x\|^2$

$$= 2\|z - y\|^2 + 2\|x\|^2 - \|(z - y) - x\|^2$$
$$= 2\|z - y\|^2 + 2\|x\|^2 - \|x + y - z\|^2$$
$$(\because \|-x\|^2 = \|x\|^2) \quad \ldots(2)$$

From (1) and (2), we get

$$\|x + y + z\|^2 = 2\|x + z\|^2 + 2\|y\|^2 - [2\|z - y\|^2 + 2\|x\|^2 - \|x + y - z\|^2]$$
$$= 2\|x + z\|^2 - 2\|z - y\|^2 + 2\|y\|^2 - 2\|x\|^2 + \|x + y - z\|^2$$

$$\therefore \quad \|x + y + z\|^2 - \|x + y - z\|^2 = 2\|x + z\|^2 - 2\|y - z\|^2 + 2\|y\|^2 - 2\|x\|^2 \quad \ldots(3)$$

Interchanging x and y in (3), we get

$$\|x + y + z\|^2 - \|x + y - z\|^2 = 2\|y + z\|^2 - 2\|x - z\|^2 + 2\|x\|^2 - 2\|y\|^2 \quad \ldots(4)$$

Adding (3) and (4)

$$2\|x + y + z\|^2 - 2\|x + y - z\|^2$$
$$= 2\|x + z\|^2 - 2\|y - z\|^2 + 2\|y + z\|^2 - 2\|x - z\|^2$$

or $\quad \|x + y + z\|^2 - \|x + y - z\|^2$

$$= \|x + z\|^2 - \|y - z\|^2 + \|y + z\|^2 - \|x - z\|^2 \quad \ldots(5)$$

Replacing z by iz in (5) and multiplying both the sides by i, we get

$$i\|x+y+iz\|^2 + i\|x+y-iz\|^2 = i\|x+iz\|^2$$
$$-i\|y-iz\|^2 + i\|y+iz\|^2 - i\|x-iz\|^2 \quad \ldots(6)$$

Adding (5) and (6), we get

$$\text{L.H.S.} = 4\,(x+y, z) \quad \text{[By definition]}$$
$$\text{R.H.S.} = \|x+z\|^2 - \|x-z\|^2 + i\|x+iz\|^2 - i\|x-iz\|^2$$
$$+\|y+z\|^2 - \|y-z\|^2 + i\|y+iz\|^2 - i\|y-iz\|^2$$
$$= 4(x, z) + 4(y, z) \quad \text{[By definition]}$$

or $\quad 4(x+y, z) = 4\,(x, z) + 4(y, z)$

or $\quad (x+y, z) = (x, z) + (y, z)$

(ii) Using (1), we get

$$(x+y, z) = (x, z) + (y, z) \ \forall\ x, y, z \in B$$

So, we can write

$$(x+z, y) = (x, y) + (z, y)$$

Put $\quad z = x$

$\therefore \quad (2x, y) = (x+x, y) = (x, y) + (x, y) = 2(x, y)$

i.e., $\quad (2x, y) = 2(x, y).$

So, the result is true for $\alpha = 2$. Let it be true for $\alpha = n$, a positive integer and we shall show it is true for $\alpha = n+1$.

Consider, $((n+1)x, y) = (nx + x, y) = (nx, y) + (x, y)$

$$= n\,(x, y) + (x, y) = (n+1)\,(x, y)$$

Hence, it is true for every positive integral values of α.

(a) If α is positive.

$$(-x, y) = \frac{1}{4}[\|-x+y\|^2 - \|-x-y\|^2 + i\|-x+iy\|^2 - i\|-x-iy\|^2]$$
$$= \frac{1}{4}[\|x-y\|^2 - \|x-y\|^2 + i\|x-iy\|^2 - i\|x+iy\|^2]$$
$$= -\frac{1}{4}[\|x+y\|^2 - \|x-y\|^2 + i\|x+iy\|^2 - i\|x-iy\|^2]$$
$$= -\frac{1}{4}[4\,(x, y)] \quad \text{[By definition]}$$

$\therefore \quad (-x, y) = -(x, y)$

(b) If α is rational.

Let $\alpha = \dfrac{p}{q}$ where p and q are integers and $q \neq 0$

and let $z = \dfrac{x}{q}$, where z is a vector. $\quad \ldots(7)$

We have to show $\left(\dfrac{p}{q}x, y\right) = \dfrac{p}{q}(x, y)$

Now, $\quad (qz, y) = q\,(z, y)$

$\therefore \quad (z, y) = \dfrac{1}{q}(qz, y)$

and $$(pz, y) = p(z, y) = p \cdot \frac{1}{q}(qz, y)$$

$\therefore$ $$(pz, y) = \frac{p}{q}(qz, y)$$

$$\left(p\frac{x}{q}, y\right) = \frac{p}{q}(x, y) \qquad \text{[By (7)]}$$

i.e., $$\left(\frac{p}{q}x, y\right) = \frac{p}{q}(x, y)$$

(c) If α is complex.

In this case, we have to prove that $(ix, y) = i(x, y)$

$$\begin{aligned}
4(ix, y) &= \|ix + y\|^2 - \|ix - y\|^2 + i\|ix + iy\|^2 - i\|ix - iy\|^2 \\
&= \left\|ix - i^2 y\right\|^2 - \left\|ix + i^2 y\right\|^2 + i\|i(x + y)\|^2 - i\|i(x - y)\|^2 \\
&= |i|^2\|x - iy\|^2 - |i|^2\|x + iy\|^2 + i|i|^2\|x + y\|^2 - i|i|^2\|x - y\|^2 \\
&= i[\|x + y\|^2 - \|x - y\|^2 + i\|x + iy\|^2 - i\|x - iy\|^2] = i[4(x, y)]
\end{aligned}$$

(Using $|i|^2 = 1$ and taking i common and replace $\frac{1}{i}$ by $-i$)

Hence, $(ix, y) = i(x, y)$.

THEOREM 6. ***Let M be a closed linear subspace of a Hilbert space H and x be any vector not in M. If d be the distance from x to M, then there exists a unique vector y_0 in M such that***

$$\|x - y_0\| = d$$

(DELHI–2004, HIMACHAL–2002, PUNJAB–2007, MEERUT–1980, KANPUR–1999, ROHILKHAND–2008)

PROOF. Let $x \notin M$. Also, let the distance of x from M is equal to d.

$\therefore$ $$d(x, M) = d \quad \text{or} \quad \inf\{\|x - y\| : y \in M\} = d \qquad \ldots(1)$$

Let $< y_n >$ be a sequence of points in M such that

$$\lim \|x - y_n\| = d \qquad \ldots(2)$$

Now, for any positive integers m, n, we have

$$\begin{aligned}
y_m - y_n &= (x - y_n) - (x - y_m) \\
&= A - B \text{ (say)}
\end{aligned}$$

$$\Rightarrow \quad \|y_m - y_n\|^2 = \|A - B\|^2$$

Using parallelogram law, we have

$$\|A + B\|^2 + \|A - B\|^2 = 2\|A\|^2 + 2\|B\|^2$$

$$\Rightarrow \quad \|A - B\|^2 = 2\|A\|^2 + 2\|B\|^2 - \|A + B\|^2$$

$$\begin{aligned}
\Rightarrow \quad \|y_m - y_n\|^2 &= 2\|x - y_n\|^2 + 2\|x - y_m\|^2 - \|2x - (y_m + y_n)\|^2 \\
&= 2\|x - y_n\|^2 + 2\|x - y_m\|^2 - 4\left\|x - \frac{y_m + y_n}{2}\right\|^2 \\
&\le 2\|x - y_n\|^2 + 2\|x - y_m\|^2 - 4d^2 \quad \left(\because \frac{y_m + y_n}{2} \in M\right)
\end{aligned}$$

Thus $\quad \| y_m - y_n \|^2 \to 2d^2 + 2d^2 - 4d^2 \to 0$

It follows that $< y_n >$ is a Cauchy sequence in M, which is closed (linear subspace of H). Also, since H is complete, therefore M is also complete. Therefore, the Cauchy sequence $< y_n >$ in M converges to some point $y_0 \in M$.

i.e., $\quad y_n \to y_0$

$\therefore \quad d = \lim \|x - y_n\| = \|x - \lim y_n\| = \|x - y_0\|$

Now, it remains to prove that y_0 is unique.

Let, if possible, y_1 be any other vector in M such that $\| x - y_1 \| = d$

Consider $\quad \| y_0 - y_1 \| = \| (x - y_1) - (x - y_0) \| = \| A - B \|$

$\Rightarrow \quad \| y_0 - y_1 \|^2 = \| A - B \|^2 = 2\| A \|^2 + 2\| B \|^2 - \| A + B \|^2$

Thus, $\quad \| y_0 - y_1 \|^2 = 2\| x - y_1 \|^2 + 2\| x - y_0 \|^2 - \| 2x - (y_0 + y_1) \|^2$

$\leq 2d^2 + 2d^2 - 4d^2 \qquad \left(\because \dfrac{y_0 + y_1}{2} \in M \right)$

$\Rightarrow \quad \| y_0 - y_1 \|^2 \leq 0$

But, by definition of norm

$\| y_0 - y_1 \|^2 \geq 0$

which gives $\quad \| y_0 - y_1 \|^2 = 0 \quad \Rightarrow \quad \|y_0 - y_1\| = 0$

Hence, $y_0 = y_1$, *i.e.,* y_0 is unique.

THEOREM 7. ***Let H be a Hilbert space and x, y be any two vectors of H, then***

(i) $\| x + y \|^2 - \| x - y \|^2 = 4 \operatorname{Re}(x, y)$ (PUNJAB–2009, KANPUR–2004, AGRA–2004)

(ii) $(x, y) = \operatorname{Re}(x, y) + i \operatorname{Re}(x, iy)$ (KANPUR–2002, 07)

PROOF. Let $\alpha = \beta + i\gamma$ is a complex numbers, where $\beta, \gamma \in R$.

Then, $\operatorname{Re}(\alpha) = \beta$ and $\operatorname{Imag}(\alpha) = \gamma$.

(i) We know that

$$\| x + y \|^2 = \| x \|^2 + \| y \|^2 + (x, y) + (y, x) \qquad \ldots(1)$$

and

$$\| x - y \|^2 = \| x \|^2 + \| y \|^2 - (x, y) - (y, x) \qquad \ldots(2)$$

From (1) and (2), we have

$$\| x + y \|^2 - \| x - y \|^2 = 2(x, y) + 2(y, x)$$

$$= 2[(x, y) + \overline{(x, y)}] = 2[2\operatorname{Re}(x, y)] = 4 \operatorname{Re}(x, y)$$

(ii) Here, we have

$$(x, y) = \operatorname{Re}(x, y) + i \operatorname{Imag}(x, y)$$

If $\alpha = \beta + i\gamma$ is a complex number, then

$$\gamma = \operatorname{Imag}(\alpha) = \operatorname{Re}\{-i(\beta + i\gamma)\} = \operatorname{Re}(-i\alpha)$$

$\Rightarrow \quad \operatorname{Imag}(x, y) = \operatorname{Re}\{-i(x, y)\} = \operatorname{Re}(x, iy)$

Hence, $\quad (x, y) = \operatorname{Re}(x, y) + i \operatorname{Re}(x, iy)$

RECAPITULATIONS

➡ An inner product space is a normed linear space.	➡ $\lvert(x, y)\rvert \le \|x\|\cdot\|y\|$
➡ $\|x + y\|^2 + \|x - y\|^2 = 2\|x\|^2 + 2\|y\|^2$	➡ $4(x, y) = \|x + y\|^2 - \|x - y\|^2 + i\|x + iy\|^2 - i\|x - iy\|^2$
➡ $\|x + y\|^2 - \|x - y\|^2 = 4\text{Re}(x, y)$	➡ $(x, y) = \text{Re}(x, y) + i\text{Re}(x, iy)$

5.5 INVARIANT AND REDUCIBILITY

Definition 1. *Let T be a linear operator on a Hilbert space H. A subspace M of H is said to be invariant under T or T-invariant if for all $x \in M$, $T(x) \in M$.*

Definition 2. *Let T be a linear operator on a Hilbert space H. If M is a subspace of H. Then, T is said to be reduced by M, if M and $M^\perp$ both are invariant under T.*

Definition 3. *A function f defined on an open interval $]a, b[$ is said to be convex if*

$$f(\lambda x + (1 - \lambda)y) \le \lambda f(x) + (1 - \lambda)f(y)\ \forall\ x, y \in\ ]a, b[,\ 0 \le \lambda \le 1$$

Definition 4. *Let H(K) be a Hilbert space. A non-empty subset S of H is said to be convex if for all $x, y \in S$*

$$\alpha x + \beta y \in S \text{ where } \alpha + \beta = 1$$

or
$$\forall\ x, y \in S, \qquad \alpha x + (1 - \alpha)y \in S,\ 0 \le \alpha \le 1$$

Definition 5. *A normed linear space N is said to be uniformly convex if for given $\varepsilon > 0\ \exists\ \delta(\varepsilon) > 0$ independent of x and y such that*

$$\|x\| \le 1,\ \|y\| \le 1,\ \|x - y\| \le \varepsilon$$

$$\Rightarrow \qquad \left\|\frac{1}{2}(x + y)\right\| \le 1 - \delta$$

REMARKS

- Finite intersection of convex sets is again convex.
- If S is a convex set in a real linear space $V(R)$ then
  (i) $x + S$ is convex $\forall\ x \in V$
  (ii) aS is convex $\forall\ a \in R$
  (iii) $aS_1 + bS_2$ is convex $\forall\ a, b \in R$, S_1 and S_2 both are convex
- Every subspace of a real linear space is convex.

THEOREM 1. ***A closed convex subset C of a Hilbert space H contains a unique vector of smallest norm.*** (DELHI–2002, MEERUT–1994, 96, 98, 2000, 01, 02, 04, 14, AGRA–2001, 04, SAGAR–2007, KANPUR–1993, 94, 96, 99, 2000, 08, GARHWAL–2003, ROHILKHAND–2007)

PROOF. Define
$$d = \inf\{\|x\| : x \in C\}$$
Then, using above theorem, there must exist a sequence $< x_n >$ of vectors of C such that $\| x_n \| \to d$.

Let x_m and x_n be two vectors belonging to the sequence $< x_n >$. Now, since C is convex, therefore, by definition of convex set

$$x_m, x_n \in C \ \Rightarrow\ \frac{x_m + x_n}{2} \in C$$

Therefore, by definition of d, we have

$$\|(x_m + x_n)/2\| \geq d \text{ so that } \|x_m + x_n\| \geq 2d \qquad \ldots(1)$$

Using parallelogram law, we have

$$\|x_m + x_n\|^2 + \|x_m - x_n\|^2 = 2\|x_m\|^2 + 2\|x_n\|^2$$

$$\Rightarrow \quad \|x_m - x_n\|^2 = 2\|x_m\|^2 + 2\|x_n\|^2 - \|x_m + x_n\|^2$$

$$\leq 2\|x_m\|^2 + 2\|x_n\|^2 - 4d^2$$

Since $\|x_m\| \to d$ and $\|x_n\| \to d$, therefore,

$$2\|x_m\|^2 + 2\|x_n\|^2 - 4d^2 \to 2d^2 + 2d^2 - 4d^2 = 0$$

$$\Rightarrow \quad \|x_m - x_n\|^2 \to 0 \text{ as } m, n \to \infty$$

$\Rightarrow$ $< x_n >$ is a Cauchy sequence in C.

But since H is complete and C is a closed subset of H, thus C is also complete.

$\Rightarrow$ Cauchy sequence $< x_n >$ in C is convergent in C.

$\Rightarrow$ $\exists$ a vector x in C such that $x_n \to x$.

Now, $\quad \|x\| = \|\lim x_n\| = \lim \|x_n\| = d$

Hence, we conclude that x is a vector in C with smallest norm d.

Finally, it remains to prove that x is unique.

Let, if possible, x and x' be two vectors in C such that $\|x\| = d$ and $\|x'\| = d$.

To show $x = x'$.

Since C is convex, so $x, x' \in C \Rightarrow \dfrac{x + x'}{2} \in C$.

Therefore, by definition of d, we have

$$\|(x + x')/2\| \geq d$$

i.e., $\quad \|x + x'\|^2 \geq 2d = 4d^2$

Now, by parallelogram law

$$\|x - x'\|^2 = 2\|x\|^2 + 2\|x'\|^2 - \|x + x'\|^2$$

$$\leq 2\|x\|^2 + 2\|x'\|^2 - 4d^2 = 2d^2 + 2d^2 - 4d^2 = 0$$

$$\Rightarrow \quad \|x - x'\|^2 \leq 0$$

But by definition of norm, $\|x - x'\|^2 \geq 0$.

Therefore, $\quad \|x - x'\|^2 = 0$

$$\Rightarrow \quad x - x' = 0$$

$$\Rightarrow \quad x = x'$$

Hence, x is unique.

THEOREM 2. ***Every Hilbert space is uniformly covex.***

PROOF. Suppose for a given $\varepsilon > 0$, $\|x - y\| \geq \varepsilon$, $\|x\| \leq 1$, $\|y\| \leq 1$, where $x, y \in H$ and H is a Hilbert space.

Then by parallelogram law

$$\|x + y\|^2 + \|x - y\|^2 = 2(\|x\|^2 + \|y\|^2) \leq 2(1 + 1) = 4$$

$$\Rightarrow \qquad \left\|\frac{1}{2}(x+y)\right\|^2 \le 1-\left\|\frac{1}{2}(x-y)\right\|^2 \le 1-\frac{1}{4}\varepsilon^2$$

Let us choose $\delta > 0$ such that $(1-\delta)^2 = 1-\frac{1}{4}\varepsilon^2$. Then we get

$$\left\|\frac{1}{2}(x+y)\right\|^2 \le 1-\delta$$

$\Rightarrow$ H is uniformly convex.

SOLVED EXAMPLES

EXAMPLE 1. *Show that the closed unit sphere $S = \{x : \|x\| \le 1\}$ in a Banach space is always a convex set.* (KANPUR–2009, 11)

SOLUTION. Let B be a Banach space and

$$S = \{x \in B : \|x\| \le 1\}$$

Let $x, y \in S$ and $\alpha + \beta = 1$, $\alpha, \beta > 0$ then

$$\|x\| \le 1, \|y\| \le 1, x, y \in B$$

Now, $x, y \in S \subseteq B \Rightarrow x, y \in B \Rightarrow \alpha x + \beta y \in B$ (By def. of linear space)

$$\therefore \quad \|\alpha x + \beta y\| \le |\alpha|.\|x\| + |\beta|.\|y\| = \alpha\|x\| + \beta\|y\|$$
$$\le \alpha.1 + \beta.1 = \alpha + \beta = 1$$

$\Rightarrow \quad \alpha x + \beta y \in S$

Hence, S is a convex set.

EXAMPLE 2. *Prove that the intersection of two convex sets is a convex set.*

(ROHILKHAND–2007, MADRAS–2009, JAIPUR–2008, UDAYPUR–2012, PUNJAB–2010, CALCUTTA–2011)

SOLUTION. Let A_1 and A_2 be two convex sets.

We have to prove that $A_1 \cap A_2$ is convex.

Let $x, y \in A_1 \cap A_2$ and α, β be scalars such that $\alpha + \beta = 1$

Then, $x, y \in A_1$ and $\alpha + \beta = 1 \Rightarrow \alpha x + \beta y \in A_1$ ($\because A_1$ is convex)

$x, y \in A_2$ and $\alpha + \beta = 1 \Rightarrow \alpha x + \beta y \in A_2$ ($\because A_2$ is convex)

$\Rightarrow \quad \alpha x + \beta y \in A_1 \cap A_2$

Hence, $A_1 \cap A_2$ is a convex set.

EXAMPLE 3. *Let $x = \{\xi, \eta\} \in R^2$ then show that $\|x\| = |\xi| + |\eta|$ does not satisfy parallelogram law.* (KANPUR–2003)

SOLUTION. Let us write

$$x = (m_1, m_2), y = (n_1, n_2) \in R^2$$

Then

$$m_1^2 = |m_1|^2, \|x\| = |m_1| + |m_2|$$

$$\Rightarrow \qquad \|x\|^2 = |m_1|^2 + |m_2|^2 + 2|m_1|\cdot|m_2|$$

$$\Rightarrow \qquad \|x\|^2 = m_1^2 + m_2^2 + 2|m_1 m_2| \qquad \ldots(A)$$

Similarly, $\|y\|^2 = n_1^2 + n_2^2 + 2|n_1 n_2|$ ...(B)

Now adding equation (A) and (B), we get

$$\|x\|^2 + \|y\|^2 = (m_1^2 + m_2^2) + (n_1^2 + n_2^2) + 2(|m_1 m_2| + |n_1 n_2|) \qquad \ldots(1)$$

Here $\quad x + y = (m_1 + n_1, m_2 + n_2)$

and $\quad x - y = (m_1 - n_1, m_2 - n_2)$

Then we have

$$\|x + y\| = |m_1 + n_1| + |m_2 + n_2| \qquad \ldots(2)$$

$$\|x - y\| = |m_1 - n_1| + |m_2 - n_2| \qquad \ldots(3)$$

Now squaring and adding eq^n (2) and (3), we get

$$\begin{aligned}\|x + y\|^2 + \|x - y\|^2 &= |m_1 + n_1|^2 + |m_2 + n_2|^2 + |m_1 - n_1|^2 + |m_2 - n_2|^2 + c \\ &= |m_1^2 + n_1^2 + 2m_1 n_1| + |m_1^2 + n_1^2 - 2m_1 n_1| \\ &\quad + |m_2^2 + n_2^2 + 2m_2 n_2| + |m_2^2 + n_2^2 - 2m_2 n_2| + c\end{aligned}$$

But, we know that

$$|m + n| \le |m| + |n|$$

Therefore

$$\begin{aligned}\|x + y\|^2 + \|x - y\|^2 &\le (m_1^2 + n_1^2 + 2|m_1 n_1|) + (m_1^2 + n_1^2 + 2|m_1 n_1|) \\ &\quad + (m_2^2 + n_2^2 + 2|m_2 n_2|) + (m_2^2 + n_2^2 + 2|m_2 n_2|) + c \\ &= 2(|m_1|^2 + |m_2|^2) + 2(|n_1|^2 + |n_2|^2) + 4(|m_1 n_1| + |m_2 n_2|) + c \\ &\ne 2(\|x\|^2 + \|y\|^2) \qquad \text{[From (1)]}\end{aligned}$$

where $\quad c = 2|m_1 + n_1| \cdot |m_2 + n_2| + 2|m_1 - n_1| \cdot |m_2 - n_2|$

Hence, $\quad \|x\| = |\xi| + |\eta|$ does not satisfy parallelogram law.

5.6 ORTHOGONALITY AND ORTHOGONAL COMPLIMENTS

Definition 1. *Let H be a Hilbert space and x, y be any vectors of H. Then, x is said to be orthogonal to y(written as $x \perp y$) if $(x, y) = 0$.*

REMARKS

- If $x \perp y$, then $y \perp x$, *i.e.*, the relation of orthogonality in a Hilbert space is symmetric.
- If x is orthogonal to y, then every scalar multiple of x is orthogonal to y, *i.e.*, $x \perp y \Rightarrow \alpha x \perp y$ for any scalar α.
- The zero vector is orthogonal to every vector.
- The zero vector is the only vector which is orthogonal to itself.

Definition 2. *Let H be a Hilbert space. A vector x is said to be orthogonal to a non-empty subset S of H ($x \perp S$), if $x \perp y \quad \forall\, y \in S$*

Definition 3. *Two non-empty subset S_1 and S_2 of a Hilbert space H are said to be orthogonal ($S_1 \perp S_2$) if $x \perp y \ \ \forall\, x \in S_1$ and $\forall\, y \in S_2$.*

Definition 4. *Let H be a Hilbert space and S be a non-empty subset of H.*

The orthogonal compliment of S written as $S^{\perp}$ (S-perpendicular) is defined by

$$S^{\perp} = \{ x \in H : x \perp y \ \ \forall\, y \in S \}$$

i.e., $S^{\perp}$ is the set of all those vectors in H which are orthogonal to every vector in S.

Definition 5. *Let H be a Hilbert space and S be a non-empty subset of H. Then, using previous definition $S^{\perp}$ is a subspace of H. Then, the orthogonal compliment of an orthogonal compliment written as $S^{\perp\perp}$ is defined by*

$$S^{\perp\perp} = \{ x \in H : (x, y) = 0 \ \forall \ y \in S^{\perp} \}$$

THEOREM 1. **(Pythogorean Theorem).** ***Let H be a Hilbert space and x, y be any two orthogonal vectors of H, then***

$$\| x + y \|^2 = \| x - y \|^2 = \| x \|^2 + \| y \|^2$$

(AVADH–2009, MEERUT–2006, 08, KANPUR–2005)

PROOF. Since x and y are orthogonal, therefore, we can write $(x, y) = 0$.

Also, since $x \perp y$, then $y \perp x$, *i.e.*, $(y, x) = 0$.

Now, $\| x + y \|^2 = (x + y, x + y) = (x, x) + (x, y) + (y, x) + (y, y)$

$$= \| x \|^2 + 0 + 0 + \| y \|^2 = \| x \|^2 + \| y \|^2$$

Similarly, $\| x - y \|^2 = (x - y, x - y) = (x, x) - (x, y) - (y, x) + (y, y)$

$$= \| x \|^2 - 0 - 0 + \| y \|^2 = \| x \|^2 + \| y \|^2$$

Hence, $\| x + y \|^2 = \| x - y \|^2 = \| x \|^2 + \| y \|^2$

THEOREM 2. ***Let S be a non-empty subset of a Hilbert space H. Then, $S^{\perp}$ is a closed linear subspace of H.***

(MEERUT–1997, 2001, 08, 09, 10, 11, KANPUR–2000, AGRA–2001, 02, ROHILKHAND–2009)

PROOF. By definition of orthogonal compliment, we have

$$S^{\perp} = \{ x \in H : (x, y) = 0 \ \forall \ y \in S \}$$

Now, since $(0, y) = 0 \quad \forall \ y \in S$

$\Rightarrow \quad 0 \in S^{\perp}$, *i.e.*, $S^{\perp} \neq \phi$

Further, let $x_1, x_2 \in S^{\perp}$ and α, β be any scalars, then

$$(x_1, y) = 0 \ \forall \ y \in S \quad \text{and} \quad (x_2, y) = 0 \ \forall \ y \in S$$

Now, for every $y \in S$

$$(\alpha x_1 + \beta x_2, y) = \alpha(x_1, y) + \beta(x_2, y) = \alpha.0 + \beta.0 = 0$$

$\Rightarrow \quad \alpha x_1 + \beta x_2 \in S^{\perp}$

$\Rightarrow \quad S^{\perp}$ is a subspace of H.

It remains to prove that $S^{\perp}$ is a closed subset of H. We know that a set is said to be closed if it contains all its limit points.

Let x be any limit point of $S^{\perp}$. To show $x \in S^{\perp}$

(*i.e.*, $x \in S^{\perp}$, if $(x, y) = 0 \ \forall \ y \in S$).

Now, since x is a limit point of $S^{\perp}$, there exists a sequence $< x_n >$ of points of $S^{\perp}$ such that $x_n \to x$.

Further, let $y \in S$. Since $x_n \in S^{\perp}$ for every n

$\therefore \quad (x_n, y) = 0 \quad \forall \ n$

$\Rightarrow \quad \lim(x_n, y) = 0 \quad \Rightarrow \quad (\lim x_n, y) = 0$

$\Rightarrow \quad (x, y) = 0 \quad \Rightarrow \quad x \in S^{\perp}$

Since x is arbitrary, thus we conclude that $S^{\perp}$ contains all its limit points. Hence, $S^{\perp}$ is a closed subset of H.

THEOREM 3. ***The orthogonal compliment of a subset of a Hilbert space is complete.***

PROOF. Let H be a Hilbert space and S be a subset of H. To show $S^{\perp}$ is complete. Using above theorem, we conclude that $S^{\perp}$ is a closed linear subspace of H. Since H is complete and every closed subset of a complete metric space is complete. Hence, $S^{\perp}$ is complete.

THEOREM 4. ***Let H be a Hilbert space and S, S_1, S_2 are non-empty subset of H. Then***

(i) $\{0\}^{\perp} = H$ (DELHI–2007, KANPUR–2005)

(ii) $H^{\perp} = \{0\}$ (DELHI–2007, KANPUR–2005)

(iii) $S \cap S^{\perp} \subset \{0\}$ (KANPUR–2000)

(iv) $S_1 \subset S_2 \Rightarrow S_2^{\perp} \subset S_1^{\perp}$ (KANPUR–2000)

(v) $S \subset S^{\perp\perp}$ (KANPUR–2003)

(vi) $S^{\perp\perp\perp} = S^{\perp}$ (AGRA–2008, MEERUT–2008)

PROOF. (i) Since, we know that $\{0\}^{\perp} \subset H$. Therefore, it remains to prove that $H \subset \{0\}^{\perp}$.

Let $x \in H$. Then $(x, 0) = 0 \quad \Rightarrow \quad x \in \{0\}^{\perp}$

i.e., $x \in H \Rightarrow x \in \{0\}^{\perp} \quad \Rightarrow \quad H \subset \{0\}^{\perp}$

Hence, $\{0\}^{\perp} = H$.

(ii) Let $x \in H^{\perp}$. Then, by definition, we have

$$(x, y) = 0 \ \forall \ y \in H$$

Putting $y = x$, we get

$$(x, x) = 0 \qquad \Rightarrow \|x\|^2 = 0 \Rightarrow x = 0$$

i.e., $x \in H^{\perp} \Rightarrow x = 0$. Hence, $H^{\perp} = \{0\}$

(iii) Now, to show $S \cap S^{\perp} \subset \{0\}$

Let $x \in S \cap S^{\perp}$

Then, $x \in S$ and $x \in S^{\perp}$

Since, we know that 0 is the only vector which belong to both S and $S^{\perp}$. Therefore

$$S \cap S^{\perp} \subset \{0\}.$$

(iv) It is given that $S_1 \subset S_2$.

Now, $x \in S_2^{\perp} \Rightarrow x$ is orthogonal to each vector in S_2.

$\Rightarrow$ x is orthogonal to each vector in $S_1 \subset S_2$.

$\Rightarrow$ $x \in S_1^{\perp}$

Therefore, $S_1 \subset S_2 \Rightarrow S_2^{\perp} \subset S_1^{\perp}$

(v) Let $x \in S$, then we have

$$(x, y) = 0 \ \forall \ y \in S^{\perp}$$

Thus, by definition of $(S^{\perp})^{\perp}$, $x \in (S^{\perp})^{\perp}$

i.e., $x \in S \Rightarrow x \in S^{\perp\perp}$

Hence, $S \subset S^{\perp\perp}$

(vi) We know that

$$S \subset S^{\perp\perp}$$

Therefore,

$$(S^{\perp\perp})^{\perp} \subset S^{\perp} \quad \Rightarrow \quad S^{\perp\perp\perp} \subset S^{\perp} \qquad \ldots(1)$$

Also,

$$S^{\perp} \subset (S^{\perp})^{\perp\perp} \quad \Rightarrow \quad S^{\perp} \subset S^{\perp\perp\perp} \qquad \ldots(2)$$

Hence, from (1) and (2), we conclude that

$$S^{\perp} = S^{\perp\perp\perp}$$

THEOREM 5. ***Let M be a proper closed linear subspace of a Hilbert space H, then there exists a non-zero vector z_0 in H such that $z_0 \perp M$.***

(MEERUT–1999, 2000, 03, 05BP, KANPUR–1995, AGRA–2008, ROHILKHAND–2008, 09)

PROOF. Let H be a Hilbert space and M be a proper subspace of H. Then there exists a vector x in H which is not in M.

Let d be the distance of x from M.

Therefore,

$$d = \inf \{ \| x - y \| : y \in M \}$$

Again, y can not be equal to x because $x \notin M$ and $y \in M$, so that $d > 0$.

Also, since M is closed subspace of H, therefore, there exists a unique vector y_0 in M such that

$$\| x - y_0 \| = d, \quad x \notin M$$

To prove that $z_0 \perp M,\ z_0 \neq 0$. For this, we shall prove that $(z_0, y) = 0 \ \forall \ y \in M$.

Let $\qquad z_0 = x - y_0$

$$\Rightarrow \quad \|z_0\| = \| x - y_0 \| = d > 0 \qquad \ldots(1)$$

$$\Rightarrow \quad \|z_0\| \neq 0, \quad \text{i.e.,} \quad z_0 \neq 0$$

Select a point $y \in M$ and let α be any scalar, then

$$z_0 - \alpha y = x - y_0 - \alpha y = x - (y_0 + \alpha y), \text{ where } y_0 + \alpha y \in M.$$

Therefore, $\qquad y, y_0 \in M \Rightarrow y_0 + \alpha y \in M$

$$\therefore \quad \| z_0 - \alpha y \| = \| x - (y_0 + \alpha y) \| \geq d = \| z_0 \| \qquad \text{[Using (1)]}$$

$$\Rightarrow \quad \| z_0 - \alpha y \|^2 \geq \| z_0 \|^2$$

$$\Rightarrow \quad (z_0 - \alpha y, z_0 - \alpha y) - (z_0, z_0) \geq 0$$

$$\Rightarrow \quad (z_0, z_0) - (z_0, \alpha y) - (\alpha y, z_0) + (\alpha y, \alpha y) - (z_0, z_0) \geq 0$$

$$\Rightarrow \quad -\bar{\alpha}(z_0, y) - \alpha(y, z_0) + \alpha\bar{\alpha}(y, y) \geq 0$$

$$\Rightarrow \quad -\bar{\alpha}(z_0, y) - \alpha\overline{(z_0, y)} + \alpha\bar{\alpha}(y, y) \geq 0 \qquad \ldots(2)$$

Since, relation (2) is true for all scalars α, putting $\alpha = \beta\,(z_0, y)$, where β is any arbitrary real number (i.e., $\beta = \bar{\beta}$).

$$\Rightarrow \quad \bar{\alpha} = \beta\,\overline{(z_0, y)}$$

Putting in (2), we get

$$-\beta\overline{(z_0, y)}\,(z_0, y) - \beta\overline{(z_0, y)}\,(z_0, y) + \beta^2 (z_0, y)\,\overline{(z_0, y)}\, \| y \|^2 \geq 0$$

$$\Rightarrow \quad -2\beta\, |(z_0, y)|^2 + \beta^2\, |(z_0, y)|^2 \| y \|^2 \geq 0$$

$$\Rightarrow \quad \beta\, |(z_0, y)|^2\, [\beta \| y \|^2 - 2] \geq 0 \qquad \ldots(3)$$

If $(z_0, y) \neq 0$, then (3) holds good for all β.

But, if $\beta > 0$ and $\beta \| y \|^2 < 2$, then (3) does not hold good for all β.

Thus, we must have

$$(z_0, y) = 0 \quad \forall\ y \in M.$$

Hence, $\quad z_0 \perp M.$

THEOREM 6. ***Let H be a Hilbert space and M be a linear subspace of H. Then, M is closed if and only if $M = M^{\perp\perp}$.*** (MEERUT–1998, 2001, 02, KANPUR–2007)

PROOF. Let M be a closed subspace of H. We have already proved that $M \subset M^{\perp\perp}$.

Now, suppose that $M \neq M^{\perp\perp}$.

Therefore, M is a proper closed subspace of $M^{\perp\perp}$, which is also a Hilbert space.

Thus, by previous theorem, there exists a non-zero vector $z_0 \in M^{\perp\perp}$ such that $z_0 \perp M$, *i.e.*, $z_0 \in M^{\perp}$.

Further, $\quad z_0 \in M^{\perp\perp} \quad$ and $\quad z_0 \in M^{\perp}$

$$\therefore \quad z_0 \in M^{\perp} \cap M^{\perp\perp} \qquad \ldots(1)$$

Also, we know that

$$S \cap S^{\perp} = \{0\} \qquad \ldots(2)$$

From (1) and (2), we conclude that $z_0 = 0$, which is a contradiction, because $z_0 \neq 0$.

Hence, $M = M^{\perp\perp}$.

Conversely, let $M = M^{\perp\perp} = S^{\perp}$ (say), where $S = M^{\perp}$.

We know that $S^{\perp}$ is a closed subspace of H.

Hence, M is also a closed subspace of H.

THEOREM 7. ***Let M and N be closed linear subspaces of a Hilbert space H such that $M \perp N$. Then, the linear subspace M + N is also closed.***

(DELHI–2007, MEERUT–1982, 2002, GARHWAL–2004, KANPUR-1991,2007, BHOPAL–2010, RAVISHANKAR–2011, MADRAS–2010)

PROOF. It is given that $M \perp N$ and therefore $M \cap N = \{0\}$.

i.e., M and N are disjoint and complimentary subspaces.

To show $M + N$ is closed.

Let $< z_n >$ be a sequence of points of $M + N$ such that $z_n \to z$. To show $z \in M + N$ ($\because$ A set is said to be closed if it contains all its limit points).

Now, $z_n \in M + N$ implies that there exists x_n in M and y_n in N such that $z_n = x_n + y_n$ uniquely. Also, since every convergent sequence is Cauchy, therefore $< z_n >$ is a Cauchy sequence, therefore, for $\in > 0$, $\exists\ n_0 > 0$ such that

$$\| z_m - z_n \| < \in \quad \forall\ m, n \geq n_0$$

$$\Rightarrow \quad \| (x_m + y_m) - (x_n + y_n) \| < \in \quad \forall\ m, n \geq n_0$$

$$\Rightarrow \quad \| (x_m - x_n) + (y_m - y_n) \|^2 < \in^2 \quad \forall\ m, n \geq n_0 \qquad \ldots(1)$$

Let us write $x_m - x_n = A \in M$ and $y_m - y_n = B \in N$.

Further, since $M \perp N$, so $(A, B) = 0$ and $(B, A) = \overline{(A, B)} = 0$

and $$\| A+B \|^2 = (A+B, A+B)$$
$$= (A, A) + (A, B) + (B, A) + (B, B) = \| A \|^2 + \| B \|^2$$

Therefore, from (1), we get

$$\| x_m - x_n \|^2 + \| y_m - y_n \|^2 < \in^2$$

$\Rightarrow$ $\| x_m - x_n \| < \in$ and $\| y_m - y_n \| < \in$.

Thus, $< x_n >$ and $< y_n >$ are Cauchy sequences in M and N respectively, which are closed.

We know that every closed linear subspace of complete metric space H is complete, therefore M and N are complete.

$\Rightarrow$ $< x_n >$ and $< y_n >$ in M and N respectively are Cauchy.

$\Rightarrow$ $x_n \to x$ and $y_n \to y$.

Finally, $$z = \lim z_n = \lim (x_n + y_n) = \lim x_n + \lim y_n$$
$$= x + y \in M + N$$

Since z is arbitrary, we conclude that $M + N$ contains all its limit points.

Hence, $M + N$ is closed.

THEOREM 8. **(Projection Theorem).** ***If M is a closed linear subspace of a Hilbert space H, then*** $H = M \oplus M^\perp$.

(DELHI–2011, MEERUT–1980, 99, 2003, 05BP, 06, 06BP, 09BP, KANPUR–2000, 06, 07, AGRA–2002, 03)

PROOF. Let M be a closed linear subspace of H. Thus $M^\perp$ is a closed linear subspace. Further, since M and $M^\perp$ both are closed linear subspace, therefore, by previous theorem, we have $M + M^\perp$ is also a closed linear subspace.

Now to show $H = M \oplus M^\perp$.

Since $M \cap M^\perp = \{0\}$. Therefore, here it is sufficient to prove that

$$H = M + M^\perp$$

Let if possible, $H \neq M + M^\perp$, then $M + M^\perp$ is a proper closed linear subspace of H. Therefore, there exists a non-zero vector z_0 in H such that

$$z_0 \perp (M + M^\perp)$$

$\Rightarrow$ $(z_0, (x+y)) = 0$, where $x \in M$ and $y \in M^\perp$

$\Rightarrow$ $$(z_0, x) + (z_0, y) = 0$$

i.e., $$(z_0, x) = 0, \ (z_0, y) = 0$$

$\Rightarrow$ $$z_0 \in M^\perp, \ z_0 \in (M^\perp)^\perp = M^{\perp\perp}$$

Thus, $$z_0 \in M^\perp \cap (M^{\perp\perp})$$

$\Rightarrow$ $$z_0 \in \{0\}$$

which is a contradiction because z_0 is a non-zero vector. Hence, we conclude that $M + M^\perp$ is not a proper closed linear subspace of H.

Thus, $$H = M + M^\perp$$

Also, since $$M \cap M^\perp = \{0\}$$

Hence, $$H = M \oplus M^\perp.$$

THEOREM 9. ***Let M be a non-empty subset of a Hilbert space H. Then $M^\perp$ is a closed linear subspace of H.*** (KANPUR–2000, AGRA–2001, 02, MEERUT–2006, 09, 10, 11)

PROOF. Let M be a non-empty subset of a Hilbert space H. Then by definition

$$M^\perp = \{x \in H : <x, y> = 0 \text{ for all } y \in M\}$$

Let $x, y \in M^\perp$ be any two scalars.

Now, let $u \in M$ be arbitrary.

Then $\quad <x, u> = 0 = <y, u>$

$$<\alpha x + \beta y, u> = \alpha<x, u> + \beta<y, u> = \alpha\cdot 0 + \beta\cdot 0 = 0$$

Then $\alpha x + \beta y \in M^\perp \Rightarrow M^\perp$ is a subspace of H.

Now, it remains to prove that $M^\perp$ is a closed subset of H.

Now we shall show that x_0 is any limiting point of $M^\perp$.

$\Rightarrow \quad x_0 \in M^\perp$

If x_0 is any limit point of $M^\perp$. Then there exist a sequence $<s_n>$ in $M^\perp$ such that

$$\lim s_n = s_0$$

$u \in M, s_n \in M^\perp \quad \Rightarrow \quad <s_n, u> = 0$ for u

$\lim\limits_{n\to\infty} < s_n, u >= 0 \quad \Rightarrow \quad <\lim s_n, u> = 0$

$\Rightarrow \quad <s_0, u> = 0$ and also $u \in M$

$\Rightarrow \quad s_0 \in M^\perp$

Hence, $M^\perp$ is closed.

THEOREM 10. ***Let S be a non-empty subset of a Hilbert space H, then the set [S] of all linear combinations of vectors in S is dense in H if and only if $S^\perp = \{0\}$.***

PROOF. For simplicity, let us write $M = [S]$. We know that a subset M of a metric space H is said to be dense in H if $\bar{M} = H$. Let us first suppose that M is dense in H, *i.e.*, $\bar{M} = H$. To show $S^\perp = \{0\}$. Firstly, we shall prove that $x \perp M$ implies $x \perp \bar{M}$.

Let z be a limit point of M. Then, by definition there exists a sequence $< z_n >$ of points of M such that $z_n \to z$. Since, each $z_n \in M$, thus $x \perp M$ implies $x \perp z_n$ for all n. Therefore, $(x, z_n) = 0$.

$\Rightarrow \quad \lim (x, z_n) = 0$

$\Rightarrow \quad (x, \lim z_n) = 0 \quad$ *i.e.*, $\quad (x, z) = 0 \Rightarrow x \perp z$

i.e., $x \perp M \Rightarrow x \perp z$, if z is a limit point of M.

Thus, we conclude that $x \perp M \Rightarrow x \perp \bar{M}$.

Further, let $x \in S^\perp$. Then $x \in S$. Also, we know that if x is orthogonal to S, then x is orthogonal to every vector in $[S] = M$.

Therefore, $\quad x \in S^\perp \Rightarrow x \perp S \quad x \perp [S]$

$\Rightarrow \quad x \perp M \Rightarrow x \perp \bar{M} \Rightarrow x \perp H \Rightarrow x \perp x \Rightarrow x = 0$

Thus, $\quad x \in S^\perp \Rightarrow x = 0$ and hence $S^\perp = \{0\}$.

Conversely, suppose that $S^\perp = \{0\}$. To prove that $H = \bar{M}$. Let if possible, $H \neq \bar{M}$. Then, there exists a vector x in H such that $x \notin \bar{M}$. Now, $\bar{M}$ is a closed subspace of H. Thus $H = M \oplus M^\perp$.

Thus, we can write $x = y + z, y \in \bar{M}$ and $z \in \bar{M}^\perp$.

Clearly, z can not be a zero vector, otherwise we have $x = y \in \bar{M}$, which is not possible.

Therefore, there exists a non-zero vector z such that

$$z \in \bar{M}^{\perp} \Rightarrow z \perp \bar{M} \Rightarrow z \perp M \qquad [\because M \subset \bar{M}]$$

$$\Rightarrow z \in M^{\perp}$$

$$\Rightarrow z \in S^{\perp} \qquad [\because S \subset [S] \Rightarrow [S]^{\perp} \subset S^{\perp} \Rightarrow M^{\perp} \subset S^{\perp}]$$

So, there exists a non-zero vector z such that $z \in S^{\perp}$, thus $S^{\perp} \neq \{0\}$, which is a contradiction. Hence, we must have $H = \bar{M}$.

THEOREM 11. ***An orthogonal set of non-zero vectors is linearly independent.***

(AGRA–2004, 11, GARHWAL–2006, BHOPAL–2014, MADRAS–2009, ROHILKHAND–2009, KANPUR–2007, 09, 10, PATNA–2005, ROHTAK–2006)

PROOF. Let $S = \{x_1, x_2, ..., x_n\}$ be an orthogonal set of non-zero vectors in Hilbert space H.

We have to show that S is linearly independent.

Let us suppose $\sum_{i=1}^{n} a_i x_i = 0$ where a_i's are scalars. ...(1)

$$\Rightarrow \quad < \sum_{i=1}^{n} a_i x_i, x_i > = < 0, x_r >$$

$$\Rightarrow \quad \sum_{i=1}^{n} a_i < x_i, x_r > = 0$$

By assumption $<x_i, x_r> = 0 \; \forall \; i$ except $i = r$

Therefore, $a_r(x_r, x_r) = 0$ or $a_r = 0$ for $r = 1, 2, ..., n$

Hence, from (1) S is linearly independent.

RECAPITULATIONS

- A closed Convex subset C of a Hilbert space contains a unique norm.
- Every Hilbert space is uniformly convex.
- The Orthonormal Compliment of a subset of a Hilbert space is complete.
- An Orthonormal set of non-zero vectors is linearly independent.

SOLVED EXAMPLES

EXAMPLE 1. *Let L be an inner product space, show that $\sqrt{(x, x)}$ has the properties of a norm.*

SOLUTION. Let us write

$$\| x \| = \sqrt{(x, x)}$$

(i) Since $\| x \| = \sqrt{(x, x)}$

$$\Rightarrow \quad \| x \|^2 = (x, x)$$

By definition of inner product space, we have

$$(x, x) \geq 0 \text{ and } (x, x) = 0 \Leftrightarrow x = 0$$

$$\| x \| \geq 0 \text{ and } \| x \| = 0 \Leftrightarrow x = 0$$

(ii) Consider

$$\| \alpha x \|^2 = (\alpha x, \alpha x) = \alpha\,(x, \alpha x) = \alpha\,\bar{\alpha}\,(x, x) = |\,\alpha\,|^2 . \| x \|^2$$

$$\Rightarrow \quad \| \alpha x \| = |\,\alpha\,| . \| x \|$$

(iii) Consider

$$\|x+y\|^2 = (x+y, x+y) = (x, x) + (x, y) + (y, x) + (y, y)$$
$$= \|x\|^2 + (x, y) + \overline{(x, y)} + \|y\|^2$$
$$= \|x\|^2 + 2\,\text{Re}\,(x, y) + \|y\|^2 \le \|x\|^2 + 2|(x, y)| + \|y\|^2$$
$$\le \|x\|^2 + 2\sqrt{(x, x)} \, . \sqrt{(y, y)} + \|y\|^2$$

[By Schwarz's inequality]

$$= \|x\|^2 + 2.\|x\|.\|y\| + \|y\|^2 = (\|x\| + \|y\|)^2$$

Therefore,

$$\|x+y\|^2 \le (\|x\| + \|y\|)^2$$

i.e., $\|x+y\| \le \|x\| + \|y\|$

EXAMPLE 2. *For the Hilbert space l_2^n, use Cauchy's inequality to prove Schwarz's inequality.*

SOLUTION. Let l_2^n be given Hilbert space and $x = (x_1, x_2, ..., x_n)$, $y = (y_1, y_2, ..., y_n)$ be any two vectors of l_2^n.

Then, by Cauchy's inequality, we have

$$\sum_{i=1}^{n} |x_i \, y_i| \le \left[\sum_{i=1}^{n} |x_i|^2\right]^{1/2} . \left[\sum_{i=1}^{n} |y_i|^2\right]^{1/2}$$

Again, by definition of inner product of l_2^n, we have

$$(x, y) = \sum_{i=1}^{n} x_i \, \bar{y}_i$$

Hence,
$$|(x, y)| = \left|\sum_{i=1}^{n} x_i \, \bar{y}_i\right| = |x_1\bar{y}_1 + x_2\bar{y}_2 + ... + x_n\bar{y}_n|$$
$$\le |x_1\bar{y}_1| + |x_2\bar{y}_2| + ... + |x_n\bar{y}_n|$$
$$= |x_1 \, y_1| + |x_2 \, y_2| + ... + |x_n \, y_n| = \sum_{i=1}^{n} |x_i \, y_i|$$
$$\le \left[\sum_{i=1}^{n} |x_i|^2\right]^{1/2} . \left[\sum_{i=1}^{n} |y_i|^2\right]^{1/2}$$
$$= \sqrt{(x, x)} \, . \sqrt{(y, y)} = \|x\| . \|y\|$$

Hence, $|(x, y)| \le \|x\| . \|y\|$ [Schwarz's inequality]

EXAMPLE 3. *Let S be a non-empty subset of a Hilbert space H, then show that $S^{\perp} = S^{\perp\perp\perp}$.*

(AGRA–2008)

SOLUTION. Let *H* be a Hilbert space and if *M* is a closed subspace of *H*, then $M = M^{\perp\perp}$.

Also, we know that $S^{\perp}$ is a closed subspace of *H*.

Putting $M = S^{\perp}$ in $M = M^{\perp\perp}$, we get

$$S^{\perp} = (S^{\perp})^{\perp\perp} = S^{\perp\perp\perp}$$

EXAMPLE 4. *Let S be a non-empty subset of a Hilbert space H. Show that $S^{\perp\perp} = [\bar{S}]$, i.e., $S^{\perp\perp}$ is the closure of the set of all linear combination of vectors in S.*

SOLUTION. We have already proved that

$$S \subset S^{\perp\perp}$$

Also, $S^{\perp\perp}$ is a closed subspace of H.

$\Rightarrow$ $S^{\perp\perp}$ is a closed subspace of H containing S.

But, by definition of closure, $[\bar{S}]$ is the smallest closed subspace of H containing S. Therefore, we must have

$$[\bar{S}] \subset S^{\perp\perp} \qquad \ldots(1)$$

Also, $S \subset [S]$ and $[S] \subset [\bar{S}] \Rightarrow S \subset [\bar{S}]$

$\Rightarrow$ $[\bar{S}]^{\perp} \subset S^{\perp}$ $\Rightarrow$ $(S^{\perp})^{\perp} \subset [\bar{S}]^{\perp\perp}$

$\Rightarrow$ $S^{\perp\perp} \subset [\bar{S}]^{\perp\perp}$ ...(2)

Further, since $[\bar{S}]$ is a closed subspace of H, therefore

$$[\bar{S}] = [\bar{S}]^{\perp\perp}$$

Then, from (2), we have

$$S^{\perp\perp} \subset [\bar{S}] \qquad \ldots(3)$$

Finally, using (1) and (3), we conclude that

$$S^{\perp\perp} = [\bar{S}]$$

EXAMPLE 5. *Find a non-zero vector w that is orthogonal to $\mu_1 = (1, 2, 1)$ and $\mu_2 = (2, 5, 4)$ in R^3.* (KANPUR–2009, 11, BHOPAL–2010)

SOLUTION. Let $v = (a_1, a_2, a_3) \in R^3$ be the required vector.

We know that

$$< v,u > = v_1\bar{u}_1 + v_2\bar{u}_2 + v_3\bar{u}_3 = v_1u_1 + v_2u_2 + v_3u_3 \text{ for } R$$

Then $v \perp \mu_1 \Rightarrow (v, \mu_1) = 0 \Rightarrow 1.a_1 + 2a_2 + 1a_3 = 0$ ...(1)

and $v \perp \mu_2 \Rightarrow (v, \mu_2) = 0 \Rightarrow 2a_1 + 5a_2 + 4a_3 = 0$ ...(2)

On solving (1) and (2) by cross multiplication method, we have

$$\frac{a_1}{8-5} = \frac{-a_2}{4-2} = \frac{a_3}{5-4} \quad \Rightarrow \quad \frac{a_1}{3} = \frac{a_2}{-2} = \frac{a_3}{1} = k$$

Then $v = (a_1, a_2, a_3) = (3k, -2k, k) = k(3, 2, -1)$

Taking $k = 1$, $v = (3, 2, -1)$.

EXAMPLE 6. *Consider $f(t) = t + 2$ and $g(t) = 3t - 2$ in the polynomial space with inner product $< f,g > = \int_0^1 f(t)g(t)dt$.*

Find (i) $<f, g>$ (ii) $\|f\|$, $\|g\|$ (iii) normalize f and g. (KANPUR–2010)

SOLUTION. (i) $< f,g > = \int_0^1 f(t)g(t)dt = \int_0^1 (t+2)(3t-2)dt$

$$= \int_0^1 (3t^2 + 4t - 4)dt = \left(t^3 + 2t^2 - 4t\right)_{t=0}^{t=1} = 1 + 2 - 4 = -1$$

$\Rightarrow$ $<f, g> = -1$

(ii) We have $\|f\|^2 = <(f,f)> = \int_0^1 f(t)f(t)dt = \int_0^1 (t+2)^2 dt$

$$= \frac{1}{3}\left\{(t+2)^3\right\}_{t=0}^1 = \frac{1}{3}\left[(3)^3 - (2)^3\right] = \frac{19}{3}$$

$\Rightarrow$ $\|f\| = \left(\frac{19}{3}\right)^{1/2}$

Now $\|g\|^2 = \int_0^1 g(t)g(t)dt = \int_0^1 (3t-2)^2 dt$

$$= \frac{1}{3}\cdot\frac{1}{3}\left[(3t-2)^3\right]_{t=0}^1 = \frac{1}{9}\left[1^3 - (-2)^3\right] = 1$$

$\Rightarrow$ $\|g\| = 1$

(iii) Normalization of $f = \frac{f}{\|f\|} = \left(\frac{3}{19}\right)^{1/2}(t+2)$

and normalization of $g = \frac{g}{\|g\|} = \frac{g}{1} = g = (3t-2)$

EXAMPLE 7. *In a Hilbert space H, show that* $\|x + \lambda y\| = \|x - \lambda y\|$, *for all scalars* λ *iff* $x \perp y$.

(KANPUR–2003)

SOLUTION. It is given that

$$\|x + \lambda y\| = \|x - \lambda y\|$$

$\Rightarrow$ $\|x + \lambda y\|^2 = \|x - \lambda y\|^2$ ...(1)

We know that

$$\|f + g\|^2 = \|f\|^2 + \|g\|^2 + <f, g> + <g, f>$$

In view of this

$$\|x + \lambda y\|^2 = \|x\|^2 + \|\lambda y\|^2 + <x, \lambda y> + <\lambda y, x>$$

or $\|x + \lambda y\|^2 = \|x\|^2 + |\lambda|^2\|y\|^2 + \bar{\lambda} <x, y> + \lambda<y, x>$ ...(2)

and $\|x - \lambda y\|^2 = \|x\|^2 + |\lambda|^2\|y\|^2 - \bar{\lambda} <x, y> - \lambda<y, x>$ ...(3)

Using (2) and (3) in (1), we get

$$\bar{\lambda} <x, y> + \lambda<y, x> = -\bar{\lambda} <x, y> - \lambda<y, x>$$

$\Rightarrow$ $2\bar{\lambda} <x, y> + 2\lambda<y, x> = 0$ $\Rightarrow$ $\bar{\lambda} <x, y> + \lambda<y, x> = 0$

which is true for all λ.

Hence, $<x, y> = 0 = <y, x>$ $\Rightarrow$ $x \perp y$

EXAMPLE 8. *If x, y are any two vectors in a Hilbert space H, then show that*

$<x, y> = \text{Re}<x, y> + i\text{Re}<x, iy>$ (ROHILKHAND–2009, KANPUR–2002, 07)

SOLUTION. Let $<x, y> = a + ib$ ...(1)

where $a = \text{Re}<x, y>, b = \text{Im}<x, y>$ ...(2)

Then, $i<x, y> = i(a + ib) = ia - b$

$\Rightarrow$ $i<x, y> = -b + ia$

$\Rightarrow$ $-b = \text{Real part of } i<x, y>$

$\Rightarrow \qquad b = \text{Real part of } -i<x, y>$

$= \text{Real part of } <x, iy> = \text{Re}<x, iy> \qquad ...(3)$

Finally using (2) and (3) in (1), we get

$$<x, y> = \text{Re}(x, y) + i\text{Re}<x, iy>$$

EXAMPLE 9. *Let V(C) be a Hilbert space of all continuous complex valued functions over the unit interval $0 \le t \le 1$. If $f(t), g(t) \in V$, then $< f(t), g(t) > = \int_0^1 f(t)\overline{g(t)}\,dt$. Show that V is an inner product space.*

SOLUTION. Let $f(t), g(t), h(t) \in V, \quad a, b \in C$

(i) $\boldsymbol{<f(t), g(t)> \ge 0}$

$$< f(t), g(t) > = \int_0^1 f(t)\overline{f(t)}\,dt$$

$$= \int_0^1 |f(t)|^2\,dt \ge 0 \qquad (\because |f(t)|^2 \ge 0 \ \forall\ t \in [0, 1])$$

(ii) $\boldsymbol{<f(t), f(t)> = 0}$ **if** $\boldsymbol{f(t) = 0}$

We have

$$< f(t), f(t) > = 0 \Leftrightarrow \int_0^1 |f(t)|^2\,dt = 0 \quad \Leftrightarrow \quad |f(t)|^2 = 0 \ \forall t \in [0,1]$$

$$\Leftrightarrow \quad f(t) = 0 \ \forall t \in [0,1]$$

(iii) $\boldsymbol{<f(t), g(t)> = \overline{< g(t), f(t) >}}$

We have

$$\overline{< g(t), f(t) >} = \left[\overline{\int_0^1 g(t)\overline{f(t)}dt}\right] = \int_0^1 \overline{g(t)\overline{f(t)}}dt$$

$$= \int_0^1 \overline{g(t)}f(t)dt = \int_0^1 f(t)\overline{g(t)}dt = <f(t), g(t)>$$

(iv) $\boldsymbol{<af + bg, h> \le a<f, g> + b<g, h>}$

Consider

$$< af + bg, h > = \int_0^1 (af + bg)\bar{h}dt = \int_0^1 (af\bar{h} + bg\bar{h})dt$$

$$= \int_0^1 af\bar{h}dt + \int_0^1 bg\bar{h}dt = a\int_0^1 f(t)\overline{h(t)}\,dt + b\int_0^1 g(t)\overline{h(t)}\,dt$$

$$= a<f, h> + b<g, h>$$

Hence, V is an inner product space.

EXAMPLE 10. *Let V(R) be a Hilbert space of polynomials with inner product defined by*

$$< f, g > = \int_0^1 f(t)g(t)dt$$

If $f(x) = x^2 + x - 4$, $g(x) = x - 1$, then find $<f, g>$ and $\|g\|$.

SOLUTION. We have $\quad < f, g > = \int_0^1 f(t)g(t)dt = \int_0^1 (x^2 + x - 4)(x - 1)dx$

$$= \int_0^1 (x^3 - x^2 + x^2 - x - 4x + 4)dx$$

$$= \int_0^1 (x^3 - 5x + 4)dx = \left.\frac{x^4}{4} - \frac{5x^2}{2} + 4x\right|_{x=0}^{1}$$

$$= \frac{1}{4} - \frac{5}{2} + 4 = \frac{7}{4}$$

Now, $\|g\| = <g, g> \int_0^1 g(t)g(t)dt = \int_0^1 (t-1)^2 dt$

$$= \frac{(t-1)^3}{3}\Bigg|_{x=0}^{1} = \frac{1}{3}[0-(0-1)^3] = \frac{1}{3}$$

$$\Rightarrow \qquad \|g\| = \frac{1}{\sqrt{3}}$$

5.7 ORTHONORMAL SET

Definition 1. *Let H be a Hilbert space and* $x \in H$ *such that* $\| x \| = 1$, *i.e.,* $(x, x) = 1$. *Then, x is said to be a unit vector or normal vector.*

Definition 2. *Let H be a Hilbert space. A non-empty subset* $\{e_i\}$ *of H is said to be an orthonormal set if*

(i) $\| e_i \| = 1$ or $(e_i, e_i) = 1$, *i.e.,* e_i *is unit vector.*

(ii) $(e_i, e_j) = 0, \quad i \neq j$ *(For orthogonality)*

Both these conditions can be combined as follows :

"A set $[e_i]$ *is said to be orthonormal if* $(e_i, e_j) = \delta_{ij}$ *where* δ_{ij} *(Kroncker delta) is defined by*

$$\delta_{ij} = \begin{cases} 0 & \text{if} \quad i \neq j \\ 1 & \text{if} \quad i = j \end{cases}$$

REMARKS

- An orthonormal set can not contain zero vector $(\because \| 0 \| = 0)$.
- A non-empty subset of a Hilbert space *H* is said to be orthonormal if it consists mutually orthogonal unit vectors.
- The set $\{e\}$ consisting only one vector is necessarily an orthonormal set.

Existence. If $\{x_i\}$ is a non-empty set of mutually orthogonal non-zero vectors in *H* and if this set x_i is replaced by the corresponding unit vector $e_i = \dfrac{x_i}{\| x_i \|}$, then the set $\{e_i\}$ is an orthonormal set.

THEOREM 1. **(Bessel's Inequality for Finite Orthonormal Set).** ***Let*** $\{e_1, e_2, \ldots, e_n\}$ ***be a finite orthonormal set of a Hilbert space H and if x is any vector in H. Then***

$$\sum_{i=1}^{n} |(x, e_i)|^2 \leq \| x \|^2 \qquad \ldots(1)$$

Also,

$$x - \sum_{i=1}^{n} (x, e_i)\, e_i \perp e_j \text{ for each } j \qquad \ldots(2)$$

(MEERUT–1980, 98, 2009BP, 11, KANPUR–1991, 99, 2001, 02, 05, AGRA–2011, DELHI–2008, RAJASTHAN–2006, HIMACHAL–2011, NAGARJUNA–2006, PUNJAB-2006)

PROOF. Let *H* be a Hilbert space

Let
$$y = x - \sum_{i=1}^{n} (x, e_i)\, e_i$$

Then,
$$\| y \|^2 = (y, y) = \left(x - \sum_{i=1}^{n} (x, e_i)\, e_i,\ x - \sum_{j=1}^{n} (x, e_j)\, e_j \right)$$

$$= (x,x) - \sum_{i=1}^{n} (x,e_i)(e_i,x) - \sum_{j=1}^{n} \overline{(x,e_j)}(x,e_j)$$
$$+ \sum_{i=1}^{n} \sum_{j=1}^{n} (x,e_i)\overline{(x,e_j)}(e_i,e_j)$$

Summing the above result w.r.t. j and using $(e_i, e_j) = 1$, when $i = j$ and $(e_i, e_j) = 0$, when $i \neq j$, we get

$$\|y\|^2 = \|x\|^2 - \sum_{i=1}^{n} |(x, e_i)|^2 - \sum_{i=1}^{n} |(x, e_i)|^2 + \sum_{i=1}^{n} |(x, e_i)|^2$$
$$= \|x\|^2 - \sum_{i=1}^{n} |(x, e_i)|^2$$

Further, $\|y\|^2 \geq 0$ implies

$$\|x\|^2 - \sum_{i=1}^{n} |(x, e_i)|^2 \geq 0$$

Hence, $\displaystyle\sum_{i=1}^{n} |(x, e_i)|^2 \leq \|x\|^2$

Also, for j, where $1 \leq j \leq n$, we have

$$\left(x - \sum_{i=1}^{n} (x, e_i)\, e_i, e_j\right) = (x, e_j) - \left(\sum_{i=1}^{n} (x, e_i)\, e_i, e_j\right)$$
$$= (x, e_j) - \sum_{i=1}^{n} (x, e_i)\,(e_i, e_j) = (x, e_j) - (x, e_j)$$
$$= 0 \qquad [\because (e_i, e_j) = 1 \text{ for } i = j,\ (e_i, e_j) = 0 \text{ for } i \neq j]$$

$$\Rightarrow \quad x - \sum_{i=1}^{n} (x, e_i)\, e_i \perp e_j \text{ for each } j.$$

THEOREM 2. ***Let $\{e_i\}$ is an orthonormal set in a Hilbert space H and if x is any vector in H, then the set***

$$\boldsymbol{S = \{ e_i : (x, e_i) \neq 0 \}}$$

is either empty or countable.

PROOF. Define a set

$$S_n = \left\{ e_i : |(x, e_i)|^2 \geq \frac{\|x^2\|}{n} \right\} \text{ for each positive integer } n.$$

The set S_n contains at most $(n-1)$ vectors because if it contains n-vectors, then

$$|(x, e_i)|^2 \geq n \,.\, \frac{\|x^2\|}{n} = \|x^2\|$$

which contradicts the Bessel's inequality just proved in previous theorem. So, the set S_n contains at most $(n-1)$ vectors such that for each n, the set S_n is finite.

Now, let $e_i \in S$, so that $(x, e_i) \neq 0$.

For small value of $|(x, e_i)|^2$, choose n so large such that

$$|(x, e_i)|^2 > \frac{\|x^2\|}{n}$$

and, therefore, e_i must belong to some S_n.

Thus, we can write

$$S = \bigcup_{n=1}^{\infty} S_n \text{ , where } S_n \text{ is a finite set.}$$

Therefore, S has been expressed as a countable union of finite sets, so by definition S is a countable.

Further, if $x \perp e_i \quad \forall\, i$, i.e., $(x, e_i) = 0 \quad \forall\, i$

$\Rightarrow$ S is empty.

Hence, the set S is either empty or countable.

THEOREM 3. **(Bessel's Inequality).** ***Let*** $\{e_i\}$ ***be an orthonormal set on a Hilbert space H, then*** $\Sigma\,|(x, e_i)|^2 \le \|x\|^2$ ***for every vector x in H.***

(DELHI–2011, MEERUT–1991, 92, 93, 94, 95, 2006BP, 08, AMRITSAR–1982, KANPUR–1991)

PROOF. Define a set

$$S = \{\, e_i : (x, e_i) \neq 0 \,\}$$

Then, using previous theorem, either S is empty or countable.

Case – I : If *S* is empty :

If S is empty, then $x \perp e_i \quad \forall\, i$, i.e. $(x, e_i) = 0 \quad \forall\, i$.

Thus, $\quad \Sigma\,|(x, e_i)|^2 = 0$

Also, $\quad \|x\|^2 \ge 0 \;\Rightarrow\; \Sigma\,|(x, e_i)|^2 \le \|x\|^2$

Case-II : If *S* is non-empty

If S is non-empty, then it is either countable finite or countably infinite.

(a) **If *S* is finite :** If S is finite, then

$$\Sigma\,|(x, e_i)|^2 = \sum_{i=1}^{n} |(x, e_i)|^2$$

To show

$$\sum_{i=1}^{n} |(x, e_i)|^2 \le \|x\|^2$$

This has been already proved in theorem 1 (Bessel's inequality for finite orthonormal sets).

(b) Consider the case when S is countably infinite

Then, let $\qquad S = \{e_1, e_2, \ldots, e_n, \ldots\}$

Define $\Sigma\,|(x,e_i)|^2 = \sum_{n=1}^{\infty} |(x,e_n)|^2$

Now, by Bessel's inequality for finite case, we have

$$\left|\sum_{n=1}^{\infty} (x, e_i)\right|^2 \le \|x\|^2$$

This relation holds good for every positive integer n and therefore as such it must also be true in limit.

Therefore

$$\sum_{n=1}^{\infty} |(x, e_i)|^2 \le \|x\|^2$$

THEOREM 4. ***Let $\{e_i\}$ be an orthonormal set in a Hilbert space H and if x is an arbitrary vector in H, then***

$$x - \Sigma (x, e_i)\, e_i \perp e_j \text{ for each } j$$

(MEERUT–1998, KANPUR–2000, CALCUTTA–2002, PATNA–2004, JAIPUR–2007)

PROOF. Define a set

$$S = \{ e_i : (x, e_i) \neq 0 \}$$

Then, we have the following cases :

Case – I : If $S = \phi$, *i.e.*, $(x, e_i) = 0 \ \forall i$ ...(1)

Then, $z = x - \Sigma (x, e_i)\, e_i = x$

To show,

$$(z, e_i) = 0 \quad \forall i \quad \Rightarrow \quad (x, e_j) = 0 \quad \forall j$$

which is true by (1).

Case – II : If $S = \phi$ and finite. Then,

$$\Sigma(x, e_i)\, e_i = \sum_{i=1}^{n} (x, e_i)\, e_i$$

To show $x - \sum_{i=1}^{n} (x, e_i)\, e_i \perp e_j$

which has been already proved in theorem 1.

Case-III : If s is countably infinite

Let $s = \{e_1, e_2, ..., e_n, ...\}$

Select $s_n = \sum_{i=1}^{n} (x, e_i) e_i$

Then, for $m > n$

$$\| s_m - s_n \| = \left\| \sum_{i=n+1}^{m} (x, e_i)\, e_i \right\|^2 = \left(\sum_{i=n+1}^{m} (x, e_i)\, e_i, \ \sum_{j=n+1}^{m} (x, e_j)\, e_j \right)$$

$$= \sum_i \sum_j (x, e_i)\, \overline{(x, e_j)}\, (e_i, e_j)$$

Summing the above result w.r.t. j and using $(e_i, e_j) = 0$ if $i \neq j$ and $(e_i, e_j) = 1$ when $i = j$, we get

$$\| s_m - s_n \| = \sum_{i=n+1}^{m} |(x, e_i)|^2$$

By Bessel's inequality, we have the series $\sum_{n=1}^{\infty} |(x, e_i)|^2$ is convergent and $m, n \to \infty$, we have

$$\sum_{i=1}^{m} |(x, e_i)|^2 \to 0$$

$$\Rightarrow \quad \| s_m - s_n \|^2 \to 0$$

$\Rightarrow$ $< s_n >$ is a Cauchy sequence in H, which is complete.

$\Rightarrow$ $< s_n >$ must converge to some vectors in H.

write $s = \sum_{n=1}^{\infty} (x, e_n)\, e_n$

Also, define

$$\Sigma (x, e_i)\, e_i = \sum_{n=1}^{\infty} (x, e_n)\, e_n = s$$

$$(x - \Sigma(x, e_i)\, e_i, e_j) = (x - s, e_j) = (x, e_j) - (s, e_j)$$

$$= (x, e_j) - (\lim s_n, e_j) \qquad \ldots(1)$$

If $e_j \notin S$, so that by definition of $S, (x, e_j) = 0$. ...(2)

Thus, $$(s_n, e_j) = \left(\sum_{j=1}^{n} (x, e_i)\, e_i, e_j \right) = 0$$

$\Rightarrow$ $\lim (s_n, e_j) = 0$

Therefore, from (1), we have

$$(x - \Sigma(x, e_i)\, e_i,\ e_j) = (x, e_j) - 0 = 0 \qquad \text{[Using (2)]}$$

Thus, $x - \Sigma(x, e_i)\, e_i \perp e_j\ ,\ e_j \notin S$

If, $e_j \in S$, *i.e.*, $(x, e_j) \neq 0$, then

$$(s_n, e_j) = \left(\sum_{i=1}^{n} (x, e_i)\, e_i, e_j \right) = (x, e_j)\,(e_j, e_j) \qquad \text{[Summing w.r.t. } i\text{]}$$

$$= (x, e_j)$$

$\Rightarrow$ $\lim (s_n, e_j) = (x, e_j)$

Hence, from (1), we have

$$(x - \Sigma(x, e_i)\, e_i,\ e_j) = (x, e_j) - (x, e_j) = 0 \quad \forall j$$

Hence, $x - \Sigma(x, e_i)\, e_j \perp e_j \quad \forall j$.

THEOREM 5. ***Let $\{e_1, e_2, \ldots, e_n\}$ be an orthonormal set in a Hilbert space H and $u \in H$. If $a_1, a_2, \ldots, a_n$ are arbitrary scalars, then***

$$\left\| u - \sum_{i=1}^{n} a_i e_i \right\| \textit{ attains its minimum value if } a_i = \langle u, e_i \rangle \qquad \text{(AVADH–2008)}$$

PROOF. Let $\{e_1, e_2, \ldots, e_n\}$ be an orthonormal set and $u \in H$, where H is a Hilbert space. Then consider

$$\left\| u - \sum_{i=1}^{n} a_i e_i \right\|^2$$

$$= \langle u - \sum_{i=1}^{n} a_i e_i, u - \sum_{k=1}^{n} a_k e_k \rangle$$

$$= \langle u, u - \sum_{k=1}^{n} a_k e_k \rangle - \sum_{i=1}^{n} a_i \langle e_i, u - \sum_{k=1}^{n} a_k e_k \rangle$$

$$= \langle u, u \rangle - \langle u, \sum_{k=1}^{n} a_k e_k \rangle - \sum_{i=1}^{n} a_i \langle e_i, u \rangle$$

$$+ \sum_{i=1}^{n} a_i \sum_{k=1}^{n} \bar{a}_k \langle e_i, e_k \rangle$$

But we know that

$$< e_i, e_k > = \begin{cases} i & \text{for } i = k \\ 0 & \text{for } i \neq k \end{cases}$$

$$= \|u\|^2 - \sum_{k=1}^{n} \bar{a}_k < u, e_k > - \sum_{i=1}^{n} a_i \overline{< u, e_i >} + \sum_{i=1}^{n} a_i \cdot \bar{a}_i$$

and $\omega\bar{\omega} = |\omega|^2$

Therefore,

$$\left\| u - \sum_{i=1}^{n} a_i e_i \right\|^2 = \|u\|^2 + \sum_{i=1}^{n} |a_i|^2 - \sum_{i=1}^{n} \{\bar{a}_i < u, e_i > + a_i \overline{< u, e_i >}\} \qquad \ldots(1)$$

Now $|a_i - < u, e_i >|^2 - |< u, e_i >|^2$

$$= (a_i - < u, e_i >)(\bar{a}_i - \overline{< u, e_i >}) - |< u, e_i >|^2$$

$$= a_i\bar{a}_i - a_i \overline{< u, e_i >} - \bar{a}_i < u, e_i > + < u, e_i > \overline{< u, e_i >} - |< u, e_i >|^2$$

$$= |a_i|^2 - (a_i \overline{< u, e_i >} + \bar{a}_i < u, e_i >) + |< u, e_i >|^2 - |< u, e_i >|^2$$

$$= |a_i|^2 - (a_i \overline{< u, e_i >} - \bar{a}_i < u, e_i >)$$

Putting the value in eq^n (1), we get

$$\left\| u - \sum_{i=1}^{n} a_i e_i \right\|^2 = \|u\|^2 - \sum_{i=1}^{n} |a_i - < u, e_i >|^2 - \sum_{i=1}^{n} |< u, e_i >|^2 \qquad \ldots(2)$$

If $a_i - <u, e_i> = 0$ or $a_i = <u, e_i>$

the R.H.S. of eq^n (2) beomes minimum.

5.8 COMPLETE ORTHONORMAL SET

An orthonormal set is said to be complete if it is not contained in any larger orthonormal set.

or

An orthonormal set is said to be complete if it is impossible to adjoin with this set a non-zero vector of unit norm so that the resulting set is an orthonormal set.

THEOREM 1. ***Let H be a Hilbert space and let $\{e_i\}$ be an orthonormal set in H. Then, the following conditions are all equivalent to each other.***

(i) ***$\{e_i\}$ is complete***

(ii) ***$x \perp \{e_i\} \Rightarrow x = 0$***

(iii) ***If x is an arbitrary vector in H, then $x = \Sigma(x, e_i)\, e_i$.*** (KANPUR–2001,04)

(iv) **Parseval's Identity :** ***If x is an arbitrary vector in H, then***

$$\|x\|^2 = \Sigma |(x, e_i)|^2$$

(MEERUT–1996, 97, 2000, 03, 09BP, KANPUR–1990, 94, 2008, DELHI–1980, 2010, AVADH–2008)

PROOF. **(i) ⇒ (ii)**

Let $\{e_i\}$ is complete, *i.e.*, it is not possible to have any other non-zero vector e of unit norm such that $e \perp e_i$, *i.e.*, $(e, e_i) = 0 \ \forall i$.

If x is $\perp$ to $\{e_i\}$ and if $x \neq 0$. Then choose

$$e = \frac{x}{\|x\|}$$

$$\Rightarrow \qquad \|e\| = 1 \quad \text{and} \quad e \perp \{e_i\}$$

Therefore $\{e, e_i\}$ is orthonormal, which is not possible, because $\{e_i\}$ is complete. Hence, $x = 0$.

(ii) $\Rightarrow$ (iii) It is given that

$$x \perp \{e_i\} \Rightarrow x = 0$$

To prove $x = (x, e_i)\, e_i$

Consider

$$\left(x - \Sigma(x, e_i)\, e_i, e_j\right) = (x, e_j) - \left(\sum_i (x, e_i)\, e_i, e_j\right) = (x, e_j) - \sum_i (x, e_i)\,(e_i, e_j)$$

Summing this result w.r.t. i and using $(e_i, e_j) = 0,\ i \neq j$, we get

$$[x - \Sigma(x, e_i)\, e_i, e_j) = (x, e_j) - (x, e_j)\,(e_j, e_j)$$

$$= (x, e_j) - (x, e_j) = 0 \qquad [\because (e_j, e_j) = 1]$$

$$\Rightarrow \qquad x - \sum_i (x, e_i)\, e_i \perp e_j \text{ for each } j.$$

$$\Rightarrow \qquad x - \sum_i (x, e_i)\, e_i \perp e_i$$

$$\Rightarrow \quad x - \Sigma(x, e_i)\, e_i = 0 \Rightarrow \quad x = \Sigma(x, e_i)\, e_i$$

(iii) $\Rightarrow$ (iv)

As per given, we have $x = \Sigma\,(x, e_i)\, e_i$.

To show

$$\|x\|^2 = \Sigma\,|(x, e_i)|^2$$

Now,

$$\|x\|^2 = (x, x) = \left(\sum_i (x, e_i)\, e_i,\ \sum_j (x, e_j)\, e_j\right)$$

$$= \sum_i \sum_j (x, e_i)\,\overline{(x, e_j)}\,(e_i, e_j)$$

Summing this result w.r.t. j and using $(e_i, e_j) = 0$ for $i \neq j$, we get

$$\|x\|^2 = \sum_i (x, e_i)\,\overline{(x, e_i)}\,(e_i, e_i)$$

$$= \Sigma\,|(x, e_i)|^2 \qquad [\text{Using } (e_i, e_i) = \|e_i\|^2 = 1]$$

(iv) $\Rightarrow$ (i) It is given that $\|x\|^2 = \Sigma\,|(x, e_i)|^2$.

To prove that $\{e_i\}$ is complete.

Let if possible $\{e_i\}$ is not complete. Then, $\{e, \{e_i\}\}$, where $e \neq 0$ is complete.

$\Rightarrow$ e is orthogonal to $\{e_i\}$.

$$\Rightarrow \qquad (e, e_i) = 0 \quad \forall i \qquad \ldots(1)$$

$$\therefore \qquad \|x\|^2 = \sum_i |(x, e_i)|^2$$

$$\therefore \qquad \|e\|^2 = \sum_i |(e, e_i)|^2$$

$\Rightarrow \qquad \|e\|^2 = 0$ (Using (1))

i.e., $\qquad e = 0$

Hence, $\{e_i\}$ is complete.

REMARK

- The scalars (x, e_i) using above, are called Fourier coefficients and the expression $\Sigma(x, e_i)\, e_i$ is called Fourier expression of x.

SOLVED EXAMPLES

EXAMPLE 1. *Let H be a Hilbert space and $S = [e_1, e_2,..., e_n, ...]$ be a countably infinite orthonormal set in H. Then show that a series of the form $\sum_{n=1}^{\infty} \alpha_n e_n$ is convergent if and only if $\sum_{n=1}^{\infty} |\alpha_n|^2 < \infty$. Further, if $\sum_{n=1}^{\infty} \alpha_n e_n$ converges to x then show that $\alpha_n = (x, e_n)$.* (KANPUR–2003)

SOLUTION. Let $s_n = \sum_{i=1}^{n} \alpha_i e_i$, $m > n$ be the partial sum of the given series. Then

$$\|s_m - s_n\|^2 = \left\| \sum_{i=n+1}^{m} \alpha_i e_i \right\|^2 = \sum_{i=n+1}^{m} |\alpha_i|^2$$

Now, if the series $\sum_{n=1}^{\infty} \alpha_n e_n$ is convergent, then the sequence $<s_n>$ of partial sums is convergent and every convergent sequence is a Cauchy sequence. So, as $m, n \to \infty$.

$$\|s_m - s_n\|^2 \to 0 \qquad \Rightarrow \qquad \sum_{i=n+1}^{\infty} |\alpha_i|^2 \to 0$$

$\Rightarrow$ The series $\sum_{n=1}^{\infty} |\alpha_n|^2$ is convergent.

Conversely, let us suppose that the series $\sum_{n=1}^{\infty} |\alpha_n|^2$ is convergent, *i.e.,* $\sum_{n=1}^{\infty} |\alpha_n|^2 < \infty$. Then

$$\sum_{i=n+1}^{\infty} |\alpha_i|^2 \to 0 \qquad \Rightarrow \|s_m - s_n\|^2 \to 0 \text{ as } m, n \to \infty$$

$\Rightarrow$ the sequence $<s_n>$ is a Cauchy sequence in H, which is complete.

$\Rightarrow$ $<s_n>$ is a convergent in H.

$\Rightarrow$ $\sum_{n=1}^{\infty} \alpha_n e_n$ is convergent.

Further, suppose that the series $\sum_{n=1}^{\infty} \alpha_n e_n$ is convergent and let $x = \sum_{n=1}^{\infty} \alpha_n e_n$. If $s_n = \sum_{i=1}^{n} \alpha_i e_i$ then for $n \geq j$, we have

$$(s_n, e_j) = \left(\sum_{i=1}^{\infty} \alpha_i e_i, e_j \right) = \alpha_j$$

$\Rightarrow \qquad \alpha_j = \lim(s_n, e_j) = (\lim s_n, e_j)$

$\Rightarrow \qquad \alpha_j = (x, e_i)$ for each j.

EXAMPLE 2. *If $\{u_1, u_2, ..., u_n\}$ is an orthonormal set in V and $u \in V$, then show that $u - \sum_{i=1}^{n} (u,u_i)u_i$ is orthogonal to each u_i.* (ROHILKHAND–2008, AGRA–2001, KANPUR–2001, 08, 13)

SOLUTION. Let $\{u_1, u_2, ..., u_n\}$ be an orthonormal set in V and $u \in V$ be arbitrary so that

$$(u_i, u_j) = \delta_{ij} = \begin{cases} 1 & \text{if } i = j \\ 0 & \text{if } i \neq j \end{cases}$$

Let
$$v = u - \sum_{i=1}^{n} < u, u_i > u_i \quad ...(1)$$

We have to prove that v is orthogonal to each u_i.

Consider
$$(v, u_j) = (u, u_j) - \left(\sum_{i=1}^{n} (u, u_i) u_i, u_j \right) \quad \text{(By (1))}$$

$$= (u, u_j) - \sum_{i=1}^{n} (u, u_i)(u_i, u_j)$$

$$= (u, u_j) - \sum_{i=1}^{n} (u, u_i)\delta_{ij} = (u, u_j) - (u, u_j) = 0$$

Hence, V is orthogonal to each u_i.

EXAMPLE 3. *Find the norm of the vector $V = (1, -2, 5)$. Also, normalize the vector.*

SOLUTION. Let $V = (1, -2, 5)$

Then, $\|V\|^2 = (v, v) = 1^2 + (-2)^2 + 5^2 = 30$

$\Rightarrow \|V\| = \sqrt{30}$

Hence, required normalized vector $= \dfrac{V}{\|V\|} = \left(\dfrac{1}{\sqrt{30}}, \dfrac{-2}{\sqrt{30}}, \dfrac{5}{\sqrt{30}} \right)$

5.9 THE GRAM SCHMIDT ORTHONORMALIZATION PROCESS

(KANPUR–2000, AVADH–2009)

THEOREM 1. ***Let $< x_n >$ be a sequence of linearly independent vectors in a Hilbert space H, then there exists an orthonormal sequence $< e_n >$ such that for every n, the linear subspace of H spanned by $e_1, e_2, ..., e_n$ is the same as that spanned by $x_1, x_2, ..., x_n$.***

PROOF. As per given, the set $(x_1, x_2, ..., x_n)$ is linearly independent. Therefore, none of the x_i's is a zero vector and none of the x_i's is expressible as a linear combination of preceding ones.

Also, $x_1 \neq 0 \Rightarrow \| x_1 \| \neq 0$

Choose $e_1 = \dfrac{x_1}{\| x_1 \|}$, then $e_1 \neq 0$ as $x_1 \neq 0$. ..(1)

and
$$\| e_1 \| = \left\| \frac{x_1}{\| x \|} \right\| = \frac{1}{\| x_1 \|} \cdot \| x_1 \| = 1$$

$\Rightarrow$ The set $\{e_1\}$ is an orthonormal set.

Further, consider

$$y_2 = x_2 - (x_2, e_1)\, e_1$$

Clearly, $y_2 \neq 0$, because if it is zero, then $x_2 = (x_2, e_1)\, e_1$ and e_1 is expressible in terms of x_1, which is not possible because the set x_i's is a linearly independent vector.

Further $\quad (y_2, e_1) = [\, x_2 - (x_2, e_1)\, e_1, e_1\,] = (x_2, e_1) - [(x_2, e_1)\, e_1, e_1]$

$$= (x_2, e_1) - (x_2, e_1)(e_1, e_1) = (x_2, e_1) - (x_2, e_1) \,.\, 1$$

$$= 0 \qquad [\text{Using } (e_1, e_1) = \| e_1 \|^2 = 1]$$

$\Rightarrow \quad y_2$ is orthogonal to e_1.

Again, choose $\quad e_2 = \dfrac{y_2}{\| y_2 \|} = \dfrac{x_2 - (x_2, e_1)\, e_1}{\| x_2 - (x_2, e_1)\, e_1 \|}$ ...(2)

$$\Rightarrow \quad \| e_2 \| = \left\| \frac{y_2}{\| y_2 \|} \right\| = 1$$

Now, since y_2 is orthogonal to e_1, therefore, e_2 is orthogonal to e_1.

$\Rightarrow \quad \{e_2, e_1\}$ is an orthonormal set.

Using (2), we have e_2 is a linear combination of x_2, e_1 or linear combination of x_2, x_1 (By (1)). From (2), we get x_2 is a linear combination of e_1, e_2.

Thus, we conclude that the linear space spanned by $[x_1, x_2]$ is same as that spanned by orthonormal set $\{e_1, e_2\}$.

Consider $\quad y_3 = x_3 - (x_3, e_1)\, e_1 - (x_3, e_2)\, e_2$

Clearly $y_3 \neq 0$, because if it is zero, then x_3 would be a linear combination of e_1, e_2 which means a linear combination of x_1, x_2, which is not possible because of the linear independence of the x_i's.

Also, $\quad (y_3, e_1) = (x_3 - (x_3, e_1)\, e_1 - (x_3, e_2) e_2, e_1)$

$$= (x_3, e_1) - (x_3, e_1)(e_1, e_1) - (x_3, e_2)(e_2, e_1)$$

$$= (x_3, e_1) - (x_3, e_1) \,.\, 1 - 0 = 0$$

Similarly, we can show that

$$(y_3, e_2) = 0$$

Further, choose

$$e_3 = \frac{y_3}{\| y_3 \|} = \frac{x_3 - (x_3, e_1)\, e_1 - (x_3, e_2)\, e_2}{\| x_3 - (x_3, e_1)\, e_1 - (x_3, e_2)\, e_2 \|} \qquad \text{...(3)}$$

Thus, $\quad \|e_3\| = 1$.

$\Rightarrow \quad y_3$ is orthogonal to both e_1 and e_2 and $\{e_1, e_2\}$ is an orthonormal set. Since y_3 is orthogonal to both e_1 and e_2, therefore e_3 is orthogonal to both e_1 and e_2.

$\Rightarrow \quad \{e_1, e_2, e_3\}$ is an orthonormal set.

Further, e_3 is a linear combination of x_3, e_1, e_2 or linear combination of x_3, x_1, x_2 by (1) and (2). Thus, from (3), we get that x_3 is a linear combinations of e_1, e_2 and e_3.

Therefore, we can say that the linear space spanned by $\{x_1, x_2, x_3\}$ is the same as that spanned by orthonormal set $\{e_1, e_2, e_3\}$.

Continuing this process, we can construct an orthonormal set of non-zero vectors

$\{e_1, e_2, ...e_n\}$ such that e_j in the above set is given by

$$e_j = \frac{x_j - (x_j, e_1)\, e_1 - (x_j, e_2) e_2 ... - (x_j, e_{j-1})\, e_{j-1}}{\left\| x_j - (x_j, e_1)\, e_1 - (x_j, e_2) e_2 ... - (x_j, e_{j-1})\, e_{j-1} \right\|}$$

Consider $\quad y_{r+1} = x_{r+1} - \sum_{i=1}^{r} (x_{r+1}, e_i)\, e_i, e_j)$

$$= (x_{r+1}, e_j) - \sum_{i=1}^{r} (x_{r+1}, e_i)\,(e_i, e_j)$$

Summing the above result w.r.t. i and using $(e_i, e_j) = 0$ for $i \neq j$ and $(e_i, e_j) = 1$ for $i = j$, we get

$$(y_{r+1}, e_j) = (x_{r+1}, e_j) - (x_{r+1}, e_j) \,.\, 1 = 0$$

Therefore, y_{r+1} is orthogonal to each $e_1, e_2, ..., e_r$.
Finally, choose

$$e_{r+1} = \frac{y_{r+1}}{\| y_{r+1} \|} = \frac{x_{r+1} - \sum_{i=1}^{r} (x_{r+1}, e_i)\, e_i}{\left\| x_{r+1} - \sum_{i=1}^{r} (x_{r+1}, e_i)\, e_i \right\|} \qquad ...(4)$$

Again, e_{r+1} is a linear combination of $x_{r+1}, e_1, e_2, ..., e_r$ or a linear combination of $x_{r+1}, x_1, x_2, ..., x_r$.

Thus, from (4), we get x_{r+1} is a linear combination of $e_1, e_2, ..., e_{r+1}$. Therefore, we can say that the linear space spanned by $\{x_1, x_2, ..., x_{r+1}\}$ is the same as that spanned by orthonormal set $\{e_1, e_2, ..., e_{r+1}\}$.

Hence, by principle of induction, we prove the existence of an orthonormal set.

THEOREM 2. ***An orthonormal set in a Hilbert space is linearly independent.***

(DELHI–2007, KANPUR–2007, 09, 10, ROHILKHAND–2009, AGRA–2011)

PROOF. Let S be any orthonormal set of vectors in a Hilbert space H.
Let $S_1 = \{e_1, e_2, ..., e_n\}$ be a finite subset of S.

Define $\quad \sum_{j=1}^{n} \alpha_j\, e_j = \alpha_1\, e_1 + \alpha_2 e_2 + ... + \alpha_n e_n = 0 \qquad ...(1)$

Now, for each k, where $1 \leq k \leq n$, we have

$$\left(\sum_{j=1}^{n} \alpha_j\, e_j, e_k \right) = \sum_{j=1}^{n} \alpha_j\, (e_j, e_k)$$

On summing with respect to j and using $(e_j, e_k) = 1$ when $j = k$ and $(e_j, e_k) = 0$ when $j \neq k$, we get

$$\left(\sum_{j=1}^{n} \alpha_j\, e_j, e_k \right) = \alpha_k$$

Also, from (1)

$$\sum_{j=1}^{n} \alpha_j\, e_j = 0 \Rightarrow \left(\sum_{j=1}^{n} \alpha_j\, e_j,\; e_k \right) = (0, e_k) = 0$$

Then, from (1), $\alpha_k = 0$ for each $1 \leq k \leq n$.

$\Rightarrow$ S_1 is linearly independent.

$\Rightarrow$ Every finite subset of S is linearly independent.

Hence, S is linearly independent.

THEOREM 3. ***In a Hilbert space H, an orthonormal set S is complete if and only if*** $x \perp S \Rightarrow x = 0$. (MEERUT–2009)

PROOF. Let H be a Hilbert space. Let us first suppose that S is complete and $x \perp S$.

To show $x = 0$.

Let if possible $x \neq 0$.

Now, $x \neq 0 \Rightarrow \|x\| \neq 0$.

Choose $e = \dfrac{x}{\|x\|} \Rightarrow \|e\| = 1$ and $e \perp S$.

Therefore, $\{e, S\}$ is an orthonormal set which properly contains S, which is a contradiction, because S is complete orthonormal set.

Hence, $x = 0$.

Conversely, let $x \perp S \Rightarrow x = 0$. To show that S is complete.

Let if possible S is not complete. Then, by definition there exists a unit vector e such that $\{e, S\}$ is an orthonormal set properly containing S. Therefore, by definition $e \perp S$.

Then, by our hypothesis $e = 0$, which is a contradiction, because e is a unit vector. Hence, S is complete.

THEOREM 4. ***In a Hilbert space*** l_2^n, ***the set*** $\{e_1, e_2, \ldots, e_n\}$, ***where*** e_i ***is the n-tuple with 1 in the*** i^{th} ***place and zeros elsewhere is a complete orthonormal set.***

PROOF. Let $e_i = [0, 0, \ldots, 1, 0, 0, \ldots, 0]$.

Also, by definition of $x \in l_2^n = (x_1, x_2, \ldots, x_n)$.

Then, $\|x\| = \left[\sum_{i=1}^{n} |x_i|^2\right]^{1/2}$

Therefore, $\|e_i\| = 1$.

Again $y \in l_2^n = [y_1, y_2, \ldots, y_n]$

Then, $(x, y) = \sum_{i=1}^{n} x_i \bar{y}_i$

Also $(e_i, e_j) = 0$

Therefore, the set $S = \{e_1, e_2, \ldots, e_n\}$ is an orthonormal set.

Now to show S is complete. For this we shall prove that $x \perp S \Rightarrow x = 0$ (Using previous theorem).

Let $x \perp S \Rightarrow x \perp e_i \;\; \forall i \qquad \Rightarrow (x, e_i) = 0 \;\; \forall i$

$\Rightarrow ((x_1, x_2, \ldots, x_i, \ldots, x_n), (0, 0, \ldots, 1, 0, 0, \ldots, 0)) = 0$

$\Rightarrow x_1.0 + x_2.0 + \ldots + x_i.1 + \ldots + x_n.0 = 0, \;\; \forall i$

$\Rightarrow x_i = 0 \;\; \forall i \qquad \Rightarrow x = 0$.

Hence, S is complete.

THEOREM 5. ***If $\{e_i\}$ is an orthonormal set in a Hilbert space H and if x, y are arbitrary vectors in H. Then***

$$\Sigma\,|(x, e_i)\overline{(y, e_i)}| \le \|x\|.\|y\|$$

PROOF. Let $S = \{e_i : (x, e_i)\overline{(y, e_i)} \neq 0\}$ such that S is either empty or countable.

Case-I : $S = \phi$

If $S = \phi$, then we have

$$(x, e_i)\overline{(y, e_i)} = 0 \quad \forall i$$

so that $\Sigma\,|(x_i, e_i)\overline{(y, e_i)}| = 0$

Also, we have $\quad 0 \le \|x\|.\|y\|$

Therefore,

$$\Sigma\,|(x, e_i)\overline{(y, e_i)}| \le \|x\|.\|y\|$$

Case-II : $S \neq \phi$. In this case, S is countable, *i.e.*, either S is finite or countably infinite.

(a) **Let S be finite**, then

$$S = \{e_1, e_2, ..., e_n\} \text{ for some positive integer } n.$$

Then define $\Sigma\,|(x, e_i)\overline{(y, e_i)}| = \sum_{i=1}^{n} |(x, e_i)\overline{(y, e_i)}|$

By Cauchy inequality, we have

$$\sum_{i=1}^{n} |(x, e_i)\overline{(y, e_i)}| \le \left(\sum_{i=1}^{n} |(x, e_i)|\right)^{1/2} \left(\sum_{i=1}^{n} |(y, e_i)|^2\right)^{1/2} \le \|x\|.\|y\| \qquad ...(1)$$

(By Bessel's inequality for finite case)

(b) **Let S be countably infinite**

In this case, give the same proof of Bessel's inequality.

Let the vectors in S be arranged in a definite order such that

$$S = \{e_1, e_2, ..., e_n, ...\}$$

Now, we define

$$\Sigma\,|(x, e_i)\overline{(y, e_i)}| = \sum_{n=1}^{\infty} (x, e_n)\overline{(y, e_n)}$$

Now, it remains to prove that the series

$$\sum_{n=1}^{\infty} (x, e_n)\overline{(y, e_n)}$$

is convergent and its sum does not change by rearrangement of its terms.

Since inequality (1) is true for every positive integer n and as such it must be true in limit.

$$\therefore \quad \sum_{n=1}^{\infty} |(x, e_n)\overline{(y, e_n)}| \le \|x\|.\|y\| \qquad ...(2)$$

which shows that the series $\sum_{n=1}^{\infty} |(x, e_n)\overline{(y, e_n)}|$ is convergent, also its all terms are positive, therefore it is absolutely convergent so that its sum does not change by rearrangement of its terms.

Thus, $$\Sigma |(x, e_i) \overline{(y, e_i)}| = \sum_{n=1}^{\infty} |(x, e_n) \overline{(y, e_n)}| \leq \|x\| . \|y\|$$

THEOREM 6. ***A Hilbert space is finite dimensional if and only if every complete orthonormal set is a basis.***

PROOF. Let H be a finite dimensional Hilbert space of dimension n. Also, let $S = \{e_i\}$ be a complete orthonormal set in H. To show that S is a basis for H.

Now, since S is an orthonormal set, therefore, it is linearly independent. Also, since S contains atmost n vectors, therefore S must be finite. Let $x \in H$ and S is a complete orthonormal set. Therefore, we have already proved that

$$x = \sum_{e_i \in S} (x, e_i)\, e_i$$

$\Rightarrow$ Each vector x in H can be written as a linear combination of vectors in S.

$\Rightarrow$ S generates H.

$\Rightarrow$ S is a basis for H.

Conversely, let every complete orthonormal set in H is a basis for H. To show H is finite dimensional.

Let S be a complete orthonormal set in H. By our hypothesis, S is a basis for H. We shall prove that S must be finite. Let, if possible S is infinite.

Then, we can find a denumerable sequence $e_1, e_2, ..., e_n, ...$ of distinct points of S. Consider the series

$$\sum_{n=1}^{\infty} \frac{1}{n^2} . e_n$$

Since, the series $\sum_{n=1}^{\infty} \frac{1}{n^4}$ is convergent, therefore, the series $\sum_{n=1}^{\infty} \frac{1}{n^2} . e_n$ must converge to some vector x in H. Also, since S is a basis for H. Thus, we can write x as a finite linear combination of vectors in S.

Let $$x = e_\lambda\, \alpha_\lambda + ... + e_k\, \alpha_k$$

where $e_\lambda, ..., e_k \in S$ and $\alpha_\lambda ... \alpha_k$ are scalars. Let j be any positive integer having value different from the indices $\lambda ... k$.

We have

$$(x, e_j) = (e_\lambda \alpha_\lambda + ... + \alpha_k e_k, e_j) = \alpha_\lambda (e_\lambda, e_j) + ... + \alpha_k (e_k, e_j) = 0$$

and $$(x, e_j) = \left(\sum_{n=1}^{\infty} \frac{1}{n^2} e_n, e_j \right) = \frac{1}{n^2} \qquad \left(\because x = \sum_{n=1}^{\infty} \frac{1}{n^2} . e_n \right)$$

$\Rightarrow$ $\frac{1}{n^2} = 0$, which is not possible.

Therefore, S must be finite and hence H is finite dimensional.

REMARK

- In a Hilbert space H, an infinite complete orthonormal set can not be a basis.

Theorem 7. ***Any two complete orthonormal sets in a Hilbert space H have the same cardinal number.***

Proof. Let H be a Hilbert space and S_1, S_2 be two complete orthonormal sets in H. Let us suppose one of these sets is finite say S_1.

Let $$S_1 = \{e_1, e_2, \ldots e_n\}$$

Now, since S_1 is an orthonormal set, thus it is linearly independent. Further, since S_1 is complete, therefore if $x \in H$, then we have

$$x = \sum_{i=1}^{n} (x, e_i) e_i$$

$\Rightarrow$ S generates H.

$\Rightarrow$ S is a basis for H

$\Rightarrow$ H is finite dimensional and $\dim H = n$

Also, since S_2 also a complete orthonormal set in H, therefore by previous theorem, S_2 must also be a basis for H.

Since S_1 and S_2 are both the basis for H, thus they must have the same number of elements.

Let us suppose that S_1 and S_2 are infinite. Let $x \in S_1$ and let

$$S_2(x) = \{y : y \in S_2 : (y, x) \neq 0\}$$

$$\Rightarrow \qquad S_2(x) \subset S_2$$

$\Rightarrow$ $S_2(x)$ is countable.

Let z be any arbitrary member of S_2. Now, since, S_1 is complete orthonormal set, then by Parseval's identity, we have

$$\|z\|^2 = \Sigma\ |(z, x)|^2$$

But $z \in S_2 \Rightarrow z$ is a unit vector

Thus, we have

$$1 = \sum_{x \in S_1} |(z, x)|^2$$

$\Rightarrow$ There must exist some vector $x \in S_1$ such that $(z, x) \neq 0$.

Then, by definition of $S_2(x)$, we have $z \in S_2(x)$

$$\Rightarrow \qquad z \in S_2 \Rightarrow z \in S_2(x) \text{ for some } x \in S_1$$

$$\therefore \qquad S_2 = \bigcup_{x \in S_1} S_2(x) \qquad \ldots(1)$$

Let n_1, n_2 be the cardinal numbers of S_1 and S_2 respectively. Since the cardinal number of the union of an arbitrary collection of sets can not exceed the cardinal number of the index set. Thus $n_2 \leq n_1$.

If we interchange the role of S_1 and S_2, then we have $n_1 \leq n_2$.

Hence, $$n_1 = n_2.$$

Theorem 8. ***Every non-zero Hilbert space contains a complete orthonormal set.***

Proof. Let H be a non-zero Hilbert space, then there exists a non-zero vector $x \in H$.

Normalize the vector x by writing $x_1 = \dfrac{x}{\|x\|}$. Then $\{x_i\}$ is clearly an orthonormal set in H. So, every non-zero Hilbert space contains orthonormal set. Consider the collection of all possible orthonormal sets in H. Then by Zorn's lemma this collection has a maximal set M. We claim that M is complete. Let $y \neq 0$, such that $y \perp M$.

Set $$y_1 = \frac{y}{\|y\|}$$

Then $M \cup \{y_i\}$ is also an orthonormal set such that

$$M \cup \{y_i\} \supset M$$

which contradict the maximality of M ($\because$ M is complete)

Finally, $y \perp M \quad \Rightarrow \quad y = 0$

Hence, M is complete orthonormal set.

5.10 THE CONJUGATE SPACE H*

Let H be a Hilbert space, then the set of all linear continuous functionals on H is also a linear space, to be denoted by H^*, is called Conjugate space of H.

THEOREM 1. ***Let y be a fixed vector in a Hilbert space H and let f_y be a scalar valued function on H defined by $f_y(x) = (x, y)$, $\forall x \in H$.***

Show that f_y is a functional in H^*. Also show that

$$\|y\| = \|f_y\|$$

(MEERUT–2000BP, 08)

PROOF. By definition, we have

$$f_y(x) = (x, y) \;\; \forall x \in H$$

Since (x, y) is a scalar, *i.e.*, a complex number for all $x \in H$, thus f_x is a mapping from H to C.

To show f_y is a functional on H. For this we shall prove that f_y is linear and continuous (bounded).

(i) f_y is linear

Let $x_1, x_2 \in H$ and α, β be any scalars.

Consider

$$f_y(\alpha x_1 + \beta x_2) = (\alpha x_1 + \beta x_2, y) = \alpha(x_1, y) + \beta(x_2, y)$$
$$= \alpha f_y(x_1) + \beta f_y(x_2)$$

$\Rightarrow \quad f_y$ is linear.

(ii) f_y is continuous

Consider $|f_y(x)| = |(x, y)|$ (By definition)

$$\leq \|x\| . \|y\| \text{ (By Cauchy Schwarz inequality)} \qquad \ldots(1)$$

Since y is a fixed vector and let $\|y\| = k$

$$\therefore \quad |f_y(x)| \leq k\|x\|$$

$\Rightarrow \quad f_y$ is bounded.

$\Rightarrow \quad f_y$ is continuous.

Hence, f_y is a functional on H, so $f_y \in H^*$.

Now, we shall prove that $\|f_y\| = \|y\|$.

By definition of norm of a continuous linear transformation, we have

$$\| T \| = \sup(\| Tx \| : \| x \| \leq 1\} = \sup\{\| Tx \| : \| x \| = 1\}$$

$$= \sup\left\{\frac{\|Tx\|}{\|x\|} : x \neq 0\right\}$$

Then, from (1) we have

$$\frac{|f_y(x)|}{\| x \|} \leq \| y \|, \; x \neq 0$$

So, $\sup\left\{\dfrac{|f_y(x)|}{\| x \|} : x \neq 0\right\} \leq \| y \|$

$\Rightarrow$ $\| f_y \| \leq \| y \|$ ...(2)

Now, we shall prove that the above relation takes the form of equality.

If $y = 0$, then $f_y = (x, y) = (x, 0) = 0, \; \forall x$.

$\Rightarrow$ f_y is zero functional.

Also $\| y \| = 0$ when $y = 0$

$\therefore$ $\|f_y\| = \|y\| = 0$

If $y \neq 0$, then $\| y \| \neq 0$

Then $\| f_y \| = \sup\{| f_y(x) | : \| x \| = 1\} \geq | f_y(x) | : \| x \| = 1$

Choose $x = \dfrac{y}{\| y \|} \Rightarrow \| x \| = 1$

$$\therefore \quad \| f_y \| \geq \left| f_y\left(\frac{y}{\| y \|}\right)\right| = \left|\left(\frac{y}{\| y \|}\right), y\right|$$

$$= \left|\left(\frac{1}{\| y \|}\right), (y, y)\right| = \left|\frac{1}{\| y \|} . \| y \|^2\right| = \| y \|$$

$\Rightarrow$ $\| f_y \| \geq \| y \|$ ...(3)

From (2) and (3), we conclude that

$$\| f_y \| = \| y \|$$

REMARK

- The mapping $g : H \to H^*$ such that $g(y) = f_y$ is a norm preserving mapping.

THEOREM 2. **(Riesz Representation Theorem)** ***Let H be a Hilbert space and f be an arbitrary functional in H^* then, there exists a unique vector y in H such that***

$$f(x) = (x, y) \;\; \forall x \in H$$

(MEERUT–1980, 2000, 04, 09BP, KANPUR–2005, 06, 08, AGRA–2005, 06, 09, AVADH–2009)

PROOF. Let us assume that there exists y in H such that

$$f(x) = (x, y) \ \forall x \in H$$

To show y is unique, let if possible, y is not unique and let y_1, y_2 be the vectors in H corresponding to an arbitrary functional f in H^* such that

$$f(x) = (x, y_1) \ \forall x \in H$$
$$f(x) = (x, y_2) \ \forall x \in H$$

Thus $\quad (x, y_1) = (x, y_2)$

$\Rightarrow \quad (xy_1) - (xy_2) = 0$

i.e., $\quad (x, y_1 - y_2) = 0$

$\Rightarrow \quad y_1 - y_2 = 0 \Rightarrow y_1 = y_2$

i.e., y is unique.

Now, we shall show the existence of y.

Case (i) If $f = 0$, *i.e.*, a zero functional, then

$$f(x) = 0(x) = 0 \ \forall x \in H$$

and $\quad f(x) = (x, y) \ \forall x \in H$

$\Rightarrow \quad (x, y) = 0 \ \forall x$ and hence $y = 0$.

Case (ii) Consider the case when $f \neq 0$.

Therefore $f(x) \neq 0$ for some $x \in H$. Because, if $f(x) = 0, \ \forall x \in H$, then $f = 0$.

Next, consider the set $M = \{x : f(x) = 0, x \in H\}$.

Since at least $0 \in M$, this M is clearly non-empty. We know that null space is a subspace of H, which is proper, because $f(x) \neq 0$, for some $x \in H$ and all such x do not belong to M.

Also, we know that the null space of continuous linear transformation is closed. Therefore, M is a proper closed linear subspace of H, then there exists a non-zero vector $y_0 \in H$ such that $y_0 \perp M$ or $y_0 \in M^{\perp}$ and $(y_0, x) = 0 \ \forall x \in M$.

Next, we have to prove that there exists a vector $y \in H$ such that

$$f(x) = (x, y) \ \forall x \in H$$

But $x \in M$ be such that

(i) $x \in M$

(ii) $x = y_0$

(iii) x may be any other vector in H.

We shall prove that any suitable scalar multiple of y_0 will be our required y such that $f(x) = (x, y) \ \forall x \in H$

(i) $\quad x \in M \Rightarrow f(x) = 0$

But $\quad f(x) = (x, y)$

$= (x, \alpha y_0)$ $\quad$ Choosing $y = \alpha y_0$

$= \bar{\alpha}(x, y_0)$ $\quad [\because x \in M$ and $y_0 \perp M]$

$= 0$

$\Rightarrow \quad f(x) = 0$ and $(x, y) = 0$ where $y = \alpha y_0$

$\therefore \quad f(x) = (x, y)$

(ii) when $x = y_0$, then

$$f(x) = (x, y) = (x, \alpha y_0)$$

$$\Rightarrow \quad f(y_0) = (y_0, \alpha y_0) = \bar{\alpha}(y_0, y_0) = \bar{\alpha}\|y_0\|^2$$

$$\Rightarrow \quad \bar{\alpha} = \frac{f(y_0)}{\|y_0\|^2} \quad \text{or} \quad \alpha = \frac{\overline{f(y_0)}}{\|y_0\|^2}$$

Therefore, for $x = y_0$ there exists a vector $y = \alpha y_0$ where $\alpha = \frac{\overline{f(y_0)}}{\|y_0\|^2}$.

and $\quad f(x) = (x, y)$.

(iii) When $\quad x \in H \quad (x \notin M$ and $x \neq y_0)$

We have already proved that

$$H = M \oplus M^{\perp}$$

Thus, any vector $x \in H$ can by uniquely expressed as the sum of a vector $m \in M$ and a vector $\beta y_0 \in M^{\perp}$, where β is any scalar.

Therefore, $\quad x = m + \beta y_0$

But $\quad f(m) = 0 \quad$ (By definition of M)

$\Rightarrow \quad f(x - \beta y_0) = 0 \quad$ or $\quad f(x) - \beta f(y_0) = 0$

$$\therefore \quad \beta = \frac{f(x)}{f(y_0)}$$

And $\quad f(x) = f(m + \beta y_0) = f(m) + \beta f(y_0)$

$= (m, y) + \beta(y_0, y)$ [By Case (i) & (ii)]

$= (m, y) + (\beta y_0, y) = (m + \beta y_0, y) = (x, y)$

Therefore, using all above cases, we can say that there exists a unique vector y such that $f(x) = (x, y) \ \forall x \in H$, when y is a scalar multiple of y_0 such that $y_0 \perp M$, where, M is the null space of functional f.

THEOREM 3. ***The mapping $\phi : H \to H^*$ is Conjugate isomorphism, i.e., this is one-one, onto, additive but non-homogeneous. Further, show that the mapping ϕ is an isometry.***

PROOF. **Case (i)** Let $\phi : H \to H^* : \phi(y) = f_y \in H^*$

where $\quad f_y(x) = (x, y) \ \forall x \in H$

(i) ϕ is one-one.

Let $\quad y_1, y_2 \in H$

$\Rightarrow \quad \phi(y_1) = fy_1$ and $\phi(y_2) = fy_2$

Further $\quad \phi(y_1) = \phi(y_2)$

$\Rightarrow \quad fy_1 = fy_2$

$\Rightarrow \quad fy_1(x) = fy_2(x) \ \forall x \in H$

$\Rightarrow \quad (x, y_1) = (x, y_2)$

$\Rightarrow (x, y_1) - (x, y_2) = 0$, *i.e.*, $(x, y_1 - y_2) = 0 \ \forall x$

$\Rightarrow \quad y_1 - y_2 = 0$

i.e., $y_1 = y_2$.

Therefore, ϕ is one-one.

(ii) ϕ is onto.

Let f be an arbitrary functional in H^*. Then by Riesz representation theorem, there exists a unique vector y such that $f(x) = (x, y), \ \forall y \in H$.

By definition of f_y, we have $f = f_y$ and $\phi(y) = f_y = f$.

Thus, ϕ is onto.

(iii) ϕ is additive.

Let $\phi(y_1) = fy_1$ and $\phi(y_2) = fy_2$

Then, $\phi(y_1 + y_2) = f_{y_1+y_2}$

Now, $f_{y_1+y_2}(x) = (x, y_1 + y_2) = (x, y_1) + (x, y_2)$

$$= f_{y_1}(x) + f_{y_2}(x) = (f_{y_1} + f_{y_2})(x) \text{ for all } x$$

Thus, $f_{y_1+y_2} = f_{y_1} + f_{y_2}$

$\Rightarrow \quad \phi(y_1 + y_2) = \phi(y_1) + \phi(y_2)$

Hence, ϕ is additive.

(iv) ϕ is non-homogeneous.

Let $\phi(y) = f_y$ then $\phi(ay) = f_{ay}$

But $f_{ay}(x) = (x, ay) = \bar{a}(x, y) = \bar{a} f_y(x)$

$\therefore \quad f_{ay} = \bar{a} f_y$

or $\phi(ay) = \bar{a}\phi(y)$

$\Rightarrow$ ϕ does not preserves scalar multiplication and hence ϕ is called conjugate isomorphism from H onto H^*, *i.e.*, ϕ is one-one, onto preserves addition but does not preserve scalar multiplication.

(v) ϕ is an isometry.

For $y_1, y_2 \in H$, let $\phi(y_1) = f_{y_1}$ and $\phi(y_2) = f_{y_2}$

Therefore,

$$\|\phi(y_1) - \phi(y_2)\| = \|f_{y_1} - f_{y_2}\| = \|f_{y_1} + (-1) f_{y_2}\|$$

$$= \|f_{y_1} + f_{(-1)y_2}\| = \|f_{y_1} + f_{-y_2}\| = \|f_{y_1 - y_2}\| = \|y_1 - y_2\|$$

$\Rightarrow \quad \phi$ is an isometry.

(vi) ϕ is norm preserving.

Let $y \in H$ and $\phi(y) = f_y$ such that $f_y(x) = (x, y) \ \forall y \in H$.

Now $\|\phi(y)\| = \|f_y\| = \|y\|$.

$\Rightarrow \quad \phi$ preserving the norm.

THEOREM 4. ***Let H be a Hilbert space and H^* be the conjugate space of H whose elements are functional on H. Then H^* is Hilbert space w.r.t. the product defined by $(f_x, f_y) = (y, x)$.*** (KANPUR–2003, 04, 05, 07)

PROOF. We know that every Hilbert space is a normed linear space, *i.e.*, H is a normed linear space. Thus, H^* is a Banach space.

To show H^* is a Hilbert space w.r.t. the inner product defined by $(f_x, f_y) = (y, x)$.
For this, we shall prove the following three conditions

(i) $(\alpha f_x + \beta f_y), f_z) = \alpha(f_x, f_z) + \beta(f_y, f_z)$

(ii) $\overline{(f_x, f_y)} = (f_y, f_x)$

(iii) $(f_x, f_x) = \|f_x\|^2$

(i) We have already proved that $f_{\alpha y} = \bar{\alpha} f_y$

Thus, $\alpha f_y = f_{\bar{\alpha} y}$

Now,

$$\begin{aligned}(\alpha f_x + \beta f_y, f_z) &= (f_{\bar{\alpha}x} + f_{\bar{\beta}y}, f_z)\\ &= (f_{\bar{\alpha}x + \bar{\beta}y}, f_z) = (z, \bar{\alpha}x + \bar{\beta}y)\\ &= (\bar{\bar{\alpha}})(z, x) + (\bar{\bar{\beta}})(z, y) = \alpha(z, x) + \beta(z, y)\\ &= \alpha(f_x, f_z) + \beta(f_y, f_z) = \alpha(f_x, f_z) + \beta(f_y, f_z)\end{aligned}$$

(ii)
$$\begin{aligned}\overline{(f_x, f_y)} &= \overline{(y, x)}\\ &= (x, y) = (f_y, f_x)\end{aligned}$$

(iii) $(fx, fx) = (x, x) = \|x\|^2 = \|fx\|^2$

Hence, H^* is a Hilbert space w.r.t. the inner product $(f_x, f_y) = (y, x)$.

THEOREM 5. ***Let H be a Hilbert space, then for each $x \in H$, the scalar valued function F_n defined by $F_x(f) = f(x)$ $\forall f \in H^*$ is a functional and the mapping $x \to F_n$ is an isometric isomorphism.***

PROOF. Let H be a Hilbert space keep x fixed and let f vary in H^*. Therefore, the expression $f(x)$ is a functional $f(x)$ defined on H^* for fixed x and variable f. Thus, we can write

$$f(x) = F_x(f)$$

Now, consider

$$\begin{aligned}F_x(\lambda_1 f_1 + \lambda_2 f_2) &= (\lambda_1 f_1 + \lambda_2 f_2)x = (\lambda_1 f_1)(x) + (\lambda_2 f_2)x\\ &= \lambda_1 f_1(x) + \lambda_2 f_2(x) = \lambda_1 F_x(f_1) + \lambda_2 F_x(f_2)\end{aligned}$$

$\Rightarrow$ F_x is linear.

Further $\quad |F_x(f)| = |f(x)| \leq \|f\| . \|x\| \leq \|x\| \quad$ if $\|f\| \leq 1$

and $\quad \|F_x\| = \sup\{|F_x(f)| : \|f\| \leq 1\}$

$\Rightarrow \quad \|F_x\| \leq \|x\| \quad$ ...(1)

Then by Hahn Banach theorem, for every $x \in H$, there exists a linear functional f_0 such that $f_0(x) = \|x\|$ and $\|f_0\| = 1$.

Thus, for such a functional

$$|F_x(f_0)| = |f_0(x)| = \|x\|$$

$$\Rightarrow \quad |F_x(f_0)| = \|f_0\| . \|x\|$$

$\Rightarrow \quad \| f_0 \| \| x \| = | F_x(f_0) | \le \| f_x \| \cdot \| f_0 \|$

Thus $\quad \| x \| \le \| F_x \|$

i.e., $\quad \| F_x \| \ge \| x \| \quad \ldots(2)$

Using (1) and (2), we conclude that

$$\| F_x \| = \| x \|$$

Further $\quad F_{x_1+x_2}(f) = f(x_1 + x_2) = f(x_1) + f(x_2)$

$$F_{x_1}(f) + F_{x_2}(f) = (F_{x_1} + F_{x_2})f$$

$\therefore \quad F_{x_1+x_2} = F_{x_1} + F_{x_2}$

$$F_{\lambda x}(x) = f(\lambda x) = \lambda f(x) = \lambda F_x(f)$$

Now $\quad \phi : H \to H^{**}$, i.e., $\phi : x \to F$ such that $\phi(x) = F_x$

Then $\quad \|\phi(x_1) - \phi(x_2)\| = \left\|F_{x_1} - F_{x_2}\right\|$

$$= \left\|F_{x_1} + F_{-x_2}\right\| = \left\| F_{x_1 - x_2}\right\| = \|x_1 - x_2\|$$

$\Rightarrow \quad \phi$ is an isometry.

Hence, for every $x \in H$, there corresponds a well defined functional $F_x \in H^{**}$ and the correspondence is isometric isomorphism.

THEOREM 6. ***If H is a Hilbert space, then H is reflexive.***

(MEERUT–1999, 2000, KANPUR–2001, 08, 10)

PROOF. In a linear space, there exists a natural isomorphism between a vector space V and its second dual V^{**} and the natural isomorphism is called reflexivity between V and V^{**}. Thus, in case of Hilbert space, we have seen that there exists a conjugate isomorphism from H to H^* and a conjugate isomorphism from H^* to H^{**}.

Define the following three mapping as follows :

(i) $T_2 : H \to H^*$ or $T_2 : x \to f_x$ such that $T_2(x) = f_x$

where $f_x(y) = (y, x) \; \forall y \in H \quad \ldots(1)$

(ii) $T_1 : H^* \to H^{**}$ or $T_1 : f_x \to F_{fx}$ such that $T_1(f_x) = F_{fx}$

where $F_{f_x}(f) = (f, f_x) \; \forall f \in H^* \quad \ldots(2)$

(iii) $T : H \to H^{**}$ or $T : x \to F_x$ such that $T(x) = F_x$

where $F_x(f) = f(x) \; \forall f \in H^*$

Now, $T_2 : H \to H^*, T_1 : H^* \to H^{**}$

Thus $T_1T_2 : H \to H^{**}$

Now, to show $\quad T_1T_2 = T$

Consider

$$T_1T_2(x) = T_1[T_2(x)] = T_1(f_x) = F_{f_x}$$

and $\quad T(x) = F_x$

Thus, $T_1T_2(x) = T(x)$ if $F_{f_x} = F_x$ ...(3)

Now choose f to be any arbitrary functional belonging to H^* whose corresponding vector in H be y so that

$$f(x) = (x, y) \quad \forall x \in H$$

Therefore

$$F_{f_x}(f) = (f, f_x) = (x, y)$$

$$\Rightarrow \quad F_{f_x}(f) = f(x) \qquad (\because f(x) = (x, y))$$

$$\Rightarrow \quad F_{f_x}(f) = F_x(f)$$

Now, since f is arbitrary, thus we conclude that

$$F_{f_x} = F_x \qquad \text{...(5)}$$

Using (4) and (5), we have

$$T_1T_2 = T$$

$\Rightarrow$ T is a mapping of H onto H^{**}.

Hence, H is reflexive.

Remark

- H and H^{**} are congruent, *i.e.*, they are equivalent numarically as well as algebraically.

Theorem 7. ***Every finite dimensional linear inner product space is complete.***

Proof. Let S be finite dimensional inner product space of n dimension and $\{e_1, e_2, ..., e_n\}$ be a certain orthonormal set of n vectors in S. Therefore, every f in S can be expressed as

$$f = \sum_{k=1}^{n} (f, e_k)e_k$$

Let $<f_n>$ be a Cauchy sequence of vectors of S. Since the inner product is a continuous function of its arguments (f_1, e_k), (f_2, e_k) ... is also a Cauchy sequence of complex numbers, which therefore converges to certain complex number a_k.

We want to prove that the vector $f = \sum_{k=1}^{n} a_k e_k$ of S is the limit of the segment f_1, f_2, ...

For $\lim_{m\to\infty} (f_m, e_k) = a_k$ for $k = 1, 2, ..., n$

So that for given $\varepsilon > 0$, $N_1(\varepsilon)$, $N_2(\varepsilon)$, ..., $N_n(\varepsilon)$ can be indicated such that

$$|(f_m, e_k) - a_k| < \in / \sqrt{n} \text{ whenever } n \geq N_k(\varepsilon), k = 1, 2, ..., n$$

So, if $\quad m \geq \max(N_1(\varepsilon), ..., N_n(\varepsilon))$

$$\|f - f_m\|^2 = \left\| \sum_{k=1}^{n} a_k - (f_m, e_n)e_k \right\|^2$$

$$= \sum_{k=1}^{n} |a_k - (f_m, e_n)|^2 \leq \varepsilon^2$$

$\Rightarrow$ Every Cauchy sequence has a limit and hence convergent.

$\Rightarrow$ S is complete.

THEOREM 8. ***Every finite dimensional linear inner product space is separable.***

(KANPUR–2011)

PROOF. Let H be a n-dimensional inner product space over the field F (real or complex). Since dim.$H = n$, therefore there exist n orthonormal vectors $x_1, x_2, ..., x_n \in H$ which form basis of H.

$\Rightarrow$ Every vector $x \in H$ can be expressed as $x = \sum_{k=1}^{n} a_k x_k$, $a_k \in F$

It can be shown that $a_n = (x, x_n)$, therefore $x = \sum_{n=1}^{n} (x, x_n) x_n$.

Now consider the countable set S of all linear combinations of the form $\sum_{k=1}^{n} a_k x_k, a_k \in F$. Since, the set S is dense in H. For any vector $x \in S$ and for any $\varepsilon > 0$, we can find scalars $r_1, r_2, ..., r_n$ such that

$$|(x, x_k) - r_k| < \frac{\varepsilon}{\sqrt{n}} \quad \text{for } k = 1, 2, ..., n.$$

Then
$$\left\| x - \sum_{k=1}^{n} r_k x_k \right\|^2 = \left\| \sum_{k=1}^{n} (x, x_k) x_k - \sum_{k=1}^{n} r_k x_k \right\|^2$$

$$\leq \sum_{k=1}^{n} \|((x, x_k) - r_k) x_k\|^2$$

$$= \sum_{k=1}^{n} |(x, x_k) - r_k|^2 \qquad (\because \|x_k\| = 1)$$

$$< \sum_{k=1}^{n} \left[\frac{\varepsilon}{\sqrt{n}} \right]^2 = \varepsilon^2$$

$$\Rightarrow \left\| x - \sum_{k=1}^{n} r_k x_k \right\|^2 < \varepsilon^2$$

Hence, H is separable.

THEOREM 9. ***If a Hilbert space H is separable, then every orthonormal set in H is countable.*** (ROHILKHAND–2009, KANPUR–2011)

PROOF. Let H be a separable Hilbert space. Then by definition there exist countable set S in H such that

$$\bar{S} = H \qquad (\because S \text{ is dense in } H)$$

Now, let B be an orthonormal basis for H.

We have to show that B is a countable set.

Let $x, y \in B$ such that $x \neq y$. Then $\|x\| = 1 = \|y\|$ and $(x, y) = 0 = (y, x)$

Therefore, $\|x - y\|^2 = \|x\|^2 + \|y\|^2 = 1 + 1 = 2$

which implies the open spheres $S\left(x, \frac{1}{2}\right) = \left\{z \in H : \|z - x\| < \frac{1}{2}\right\}$ are all disjoint.

Now since S is dense and so S must contain a point in each $S\left(x, \frac{1}{2}\right)$. Thus, if B is uncountable then S must be uncountable and so H is not separable, which contradict our hypothesis, therefore, B must be countable.

Hence, every orthonormal set is countable.

RECAPITULATIONS

➠ A Hilbert space is finite dimensional if and only if every complete orthonormal set is a basis.	➠ Any two complete orthonormal sets in a Hilbert space have the same cardinal number.
➠ Every non-zero Hilbert space contains a Complete orthonormal set.	➠ If H is a Hilbert space then it is reflexive.
➠ Every finite dimensional linear inner product space is complete.	➠ Every finite dimensional inner product space is separable.
➠ A Hilbert space H is separable then every orthonormal set in H is countable.	

5.11 ADJOINT OF AN OPERATOR

(DELHI–2009, 11, MEERUT–2006, AGRA–2008, KANPUR–2010, PATNA–2006, BILASPUR–2004, RAIPUR–2009)

Let H be a Hilbert space. We know that T is a continuous linear transformation from H into H, then T is called an operator on H. We denote the set of all operator on H by $B(H)$. We have already proved that $B(H)$ is a complex Banach space and $B(H)$ is a complex algebra.

If T is an operator on H, then exists a unique operator T^* on H^* such that

$$(T^* f)(x) = f(Tx) \;\; \forall f \in H^* \quad \text{and} \quad \forall \;\; x \in H$$

This operator T^* on H^* is called the conjugate of T on H. As T gives rise to T^*, in the same way the operator T^* on H^* gives rise the operator T^{**} on H^{**}. Now, since H is reflexive, therefore

$$T^{**} = T$$

Here, we want to associate with each operator T on H a unique operator T^* on H such that $(Tx, y) = (x, T^*y)$ and this operator will be called the adjoint of T.

THEOREM 1. ***Let T be an operator on a Hilbert space H. Then, there exists a unique operator on T^* on H such that $(Tx, y) = (x, T^* y) \;\; \forall x, y \in H$ ...(1)***
(The operator T^* is called the adjoint of T).

PROOF. First of all we shall prove the existence of adjoint operator let y be a vector in H and f_y its corresponding functional in H^*. Let us suppose $T^* f_y = f_z$, where T^* is the conjugate of T.

Now, we have formed the product of the following there mappings :

$$H \to H^* \qquad y \to f_y$$

$$H^* \to H^* \qquad f_y \to T^* f_y = f_z$$

$$H^* \to H \qquad f_z \to z$$

If the product of these mappings by T^*, then T^* is a mapping of H into itself such that $T^* y = z$.

This new mapping T^* of H into itself is called the adjoint of T.

Now, we shall show that the mapping T^* satisfy the property (1).

For every x in H, we have

$$(T^* f_y)x = f_y(T_x) = (Tx, y)$$

Also, $$(T^* f_y)x = f_z(x) = (x, z) = (x, T^* y)$$

Therefore, we have

$$(Tx, y) = (x, T^* y) \ \forall x, y \in H$$

We shall show that T^* is linear and continuous.

(i) T^* is linear.

Let $y_1, y_2 \in H$ and α, β be any two scalars. Then, for any vector x in H, we have

$$(x, T^* (\alpha y_1, \beta y_2) = (Tx, \alpha y_1 + \beta y_2) = \overline{\alpha}(Tx, y_1) + \overline{\beta}(Tx, y_2)$$

$$= \overline{\alpha}(x, T^* y_1) + \overline{\beta}(x, T^* y_2)$$

So, $[x, T^* (\alpha y_1 + \beta y_2)] = (x, \alpha T^* y_1 + \beta T^* y_2) \ \forall \ x \in H$

But, we have know that

$$(x, y) = (x, z) \ \forall x \in H \quad \Rightarrow \quad y = z$$

Thus, we have

$$T^* (\alpha y_1, \beta y_2) = \alpha T^* y_1 + \beta T^* y_2 .$$

$\Rightarrow \quad T^*$ is linear.

(ii) T^* is continuous.

Let $y \in H$ be any vector. Then

$$\left\| T^* y \right\|^2 = (T^* y, T^* y) = (TT^* y, y) = \left| (TT^* y, y) \right|$$

$$\leq \left\| TT^* y \right\| . \| y \|$$

$$\leq \| T \| . \left\| T^* y \right\| . \| y \| \qquad [\text{Using} \| Tx \| \leq \| T \| . \| x \|]$$

$$\Rightarrow \quad \left\| T^* y \right\|^2 \leq \| T \| . \left\| T^* y \right\| . \| y \| \quad \forall y \in H \qquad \ldots(2)$$

If $\left\| T^* y \right\| = 0$ then

$$\left\| T^* y \right\| \leq \| T \| . \| y \|, \text{ because} \| T \| . \| y \| \geq 0 .$$

If $\left\| T^* y \right\| \neq 0$ then, from (2), we have

$$\left\| T^* y \right\| \leq \| T \| . \| y \|$$

let $\| T \| = k$, then $k \geq 0$, we have

$$\left\| T^* y \right\| \leq k . \| y \| \quad \forall y \in H$$

$\Rightarrow \quad T^*$ is bounded.

$\Rightarrow \quad T^*$ is continuous.

Hence, T^* is a continuous linear transformation form H into itself. Thus, T^* is an operator on H.

Now, it remain to prove that T^* is unique. Let if possible there is any mapping T' of H into itself such that

$$(Tx, y) = (x, T'y)$$

We have to show that

$$T' = T^*$$

Consider $\quad (Tx, y) = (x, T'y)$ and $(Tx, y) = (x, T^* y) \ \forall x, y \in H$

$$\Rightarrow (x, T'y) = (x, T^* y) \forall x, y \in H$$

$$\Rightarrow T'y = T^* y \ \ \forall y \in H \Rightarrow T' = T^*$$

Hence, T^* is unique.

THEOREM 2. ***The adjoint operator $T \to T^*$ on $B(H)$ has the following properties:***

(i) $(T_1 + T_2)^* = T_1^* + T_2^*$ **(ii)** $(\alpha T)^* = \bar{\alpha} T^*$

(iii) $(T_1 T_2)^* = T_2^* T_1^*$ **(iv)** $T^{**} = T$

(v) $\|T^*\| = \|T\|$ **(vi)** $\|T^* T\| = \|T\|^2$

(MEERUT–1980, 95, 98, 99, 2000, 01, 02, 06, 06BP, 07, 09BP, 11, AMRITSAR–1982, KANPUR–1990, 99, 2000, 01, 08, BUNDELKHAND–1985, ROHILKHAND–2009, AGRA–2001, 03, 04)

PROOF. (i) Consider

$$(x, (T_1 + T_2)^* y) = ((T_1 + T_2)x, y)$$

$$= (T_1 x + T_2 x, y) = (T_1 x, y) + (T_2 x, y)$$

$$= (x, T_1^* y) + (x, T_2^* y) = (x, T_1^* y + T_2^* y)$$

$$= (x, (T_1^* + T_2^*)y)$$

Thus, from the uniqueness of adjoint operator, we have

$$(T_1 + T_2)^* = (T_1^* + T_2^*)$$

(ii) $\quad (x, (\alpha T)^* y) = ((\alpha T)x, y)$

$$= (\alpha(Tx), y) = \alpha(Tx, y) = (x, \bar{\alpha} T^* y)$$

Thus, from the uniqueness of adjoint operator, we have

$$(\alpha T)^* = \bar{\alpha} T^*$$

(iii) Consider

$$(x, (T_1 T_2)^* y) = ((T_1 T_2)x, y)$$

$$= (T_1(T_2 x), y) = (T_2 x, T_1^* y)$$

$$= (x, T_2^*(T_1^* y)) = (x, (T_2^* T_1^*)y)$$

Thus, from the uniqueness of adjoint operator we have

$$(T_1 T_2)^* = T_2^* T_1^*$$

(iv) We have

$$(x, T^{**} y) = (x, (T^*)^* y) = (T^* x, y) = \overline{(y, T^*(x))}$$
$$= \overline{(Ty, x)} = (x, Ty)$$

Hence, from the uniqueness of adjoint operator, we have

$$T^{**} = T$$

(v) We have

$$\left\|T^* x\right\|^2 = (T^* x, T^* x) = (TT^* x, x)$$

Thus, $\left\|T^* x\right\|^2 = \left|(TT^* x, x)\right|$

$$\leq \left\|TT^* x\right\| . \|x\| \leq \|T\| . \left\|T^* x\right\| \|x\|$$

Therefore

$$\left\|T^* x\right\| \leq \|T\| . \|x\| \quad ...(1)$$

$$\Rightarrow \sup\left\{\frac{\left\|T^* x\right\|}{\|x\|} : x \neq 0\right\} \leq \|T\|$$

$$\Rightarrow \left\|T^*\right\| \leq \|T\| \quad ...(2)$$

If $\|x\| \leq 1$, then from (1), $\left\|T^* x\right\| \leq \|T\|$

Thus, $\sup\{\left\|T^* x\right\| : \|x\| \leq 1\} \leq \|T\|$

$$\Rightarrow \left\|T^*\right\| \leq \|T\|$$

Replacing T by T^*, we get

$$\left\|T^{**}\right\| \leq \left\|T^*\right\| \Rightarrow \|T\| \leq \left\|T^*\right\| \quad ...(3)$$

From (2) and (3), we get $\|T\| = \left\|T^*\right\|$.

(vi) we have

$$\|Tx\|^2 = |(Tx, Tx)| = \left|(T^* Tx, x)\right|$$
$$\leq \left\|T^* Tx\right\| . \|x\| \leq \left\|T^* T\right\| . \|x\| . \|x\| \quad ...(4)$$

If $\|x\| \leq 1$, then form (4), we get

$$\|Tx\|^2 \leq \left\|T^* T\right\|$$

$$\Rightarrow \sup\{\|Tx\|^2 : \|x\| \leq 1\} \leq \left\|T^* T\right\|$$

$$\Rightarrow \sup\{\|Tx\| : \|x\| \leq 1\}^2 \leq \left\|T^* T\right\| \quad ...(5)$$

$$\Rightarrow \|T\|^2 \leq \left\|T^* T\right\|$$

Further

$$\|T^*T\| \le \|T^*\|.\|T\| = \|T\|.\|T\| = \|T\|^2$$

$$\Rightarrow \qquad \|T^*T\| \le \|T\|^2 \qquad \ldots(6)$$

Using (5) and (6), we have

$$\|T^*T\| = \|T\|^2$$

Remark

- The adjoint operator is one-one onto as a mapping of $B(H)$ onto itself.

Theorem 3. ***If 0 and I be respectively the zero and identity operators on a Hilbert space H, then $0^* = 0$ and $I^* = I$. Hence, if T is non-singular operator on H. Then T^* is also non-singular and then $(T^*)^{-1} = (T^{-1})^*$.***

Proof. Consider

$$(x, 0^* y) = (0x, y) = (0, y) = 0 = (x, 0y)$$

Now since adjoint is unique $\Rightarrow 0^* = 0$.

Also, $\qquad (x, I^* y) = (Ix, y) = (x, y) = (x, Iy)$

$\therefore \qquad I^* = I$

Now, since T is non-singular.

$\Rightarrow T$ is invertible

$$\Rightarrow \qquad TT^{-1} = T^{-1}T = I$$

$$\Rightarrow \qquad (TT^{-1})^* = (T^{-1}T)^* = I^*$$

$$\Rightarrow \qquad (T^{-1})^* T^* = T^* (T^{-1})^* = I^* = I \qquad [\because (T_1T_2)^* = T_2^* T_1^*]$$

which shows that T^* is invertibale and hence, non-singular and $(T^*)^{-1} = (T^{-1})^*$.

Hence, Inverse of adjoint of T = Adjoint of inverse of T.

Theorem 4. ***The adjoint operator is one-one onto as a mapping of B(H) into itself.***

Proof. Define a mapping $f : B(H) \to B(H)$, such that $f(T) \to T^* \quad \forall \ T \in B(H)$.

To show that f is one-one and onto.

(i) *f* is one-one: Let $T_1, T_2 \in B(H)$

Then, we have

$$f(T_1) = f(T_2) \Rightarrow T_1^* = T_2^*$$

$$\Rightarrow (T_1^*)^* = (T_2^*)^*$$

$$\Rightarrow T_1^{**} = T_2^{**} \Rightarrow T_1 = T_2$$

$\Rightarrow \qquad f$ is one-one.

(ii) *f* is onto: Let $T \in B(H)$ be arbitrary

Then, $\qquad T^* \in B(H)$

Also, we have $f(T^*) = (T^*)^* = T^{**} = T$

$\Rightarrow \qquad f$ is onto.

5.12 SELF-ADJOINT OPERATOR

Definition. *A linear operator T on a Hilbert space H is said to be self adjoint if* $T^* = T$, *i.e.* $(Tx, y) = (x, T^* y) = (x, Ty)$.

THEOREM 1. ***If T_1 and T_2 be self adjoint operators, then T_1T_2 or T_2T_1 is self adjoint iff $T_1T_2 = T_2T_1$, i.e., the product of two self adjoint operators is self adjoint if and only if they commute.***

(AGRA–2008, 11, MEERUT–2009, 11, AVADH–2009, KANPUR–2008)

PROOF. As per given, T_1 and T_2 are self adjoint, *i.e.*,

$$T_1^* = T_1 \text{ and } T_2^* = T_2 \qquad \ldots(1)$$

Let us first suppose T_1T_2 is self adjoint. To show that $T_1T_2 = T_2T_1$.

Since T_1T_2 is self adjoint, therefore

$$(T_1T_2)^* = T_1T_2$$

$$\Rightarrow \quad T_2^*T_1^* = T_1T_2$$

$$\Rightarrow \quad T_2T_1 = T_1T_2 \qquad \text{[Using (1)]}$$

Conversely, let $T_1T_2 = T_2T_1$. To show that T_1T_2 is self adjoint.

Consider

$$(T_1T_2)^* = T_2^*T_1^* = T_2T_1 \qquad \text{[Using (1)]}$$

$$= T_1T_2 \qquad \text{[By our assumption]}$$

$\Rightarrow$ T_1T_2 is self-adjoint.

THEOREM 2. ***If T be an invertible operator then either both or none of T and T^{-1} are self-adjoint.***

PROOF. As per given, T is invertible, therefore, $TT^{-1} = T^{-1}T = I$

Thus, $\quad (TT^{-1})^* = (T^{-1}T)^* = I^*$

or $\quad (T^{-1})^*T^* = T^*(T^{-1})^* = I$

$$\Rightarrow \quad (T^*)^{-1} = (T^{-1})^* \qquad \ldots(1)$$

If T is self-adjoint, *i.e.*, $T^* = T$, then (1) gives

$$T^{-1} = (T^{-1})^*$$

$\Rightarrow$ T^{-1} is also self-adjoint

$\Rightarrow$ T and T^{-1} both are self-adjoint.

In case T be not self-adjoint, then $T^* \neq T$, then from (1) $(T^{-1})^* \neq T^{-1}$, *i.e.*, T^{-1} will not be self-adjoint. Hence, either both T and T^{-1} are self-adjoint or none of them is self-adjoint.

THEOREM 3. ***If T is self-adjoint operator on a Hilbert space H then αT is self-adjoint if and only if α is real.***

PROOF. Let us first suppose that αT is self-adjoint

Then $\quad (\alpha T)^* = \alpha T$

or $\quad \bar{\alpha}T^* = \alpha T \Rightarrow \bar{\alpha}T = \alpha T \qquad [\because T^* = T]$

$$\Rightarrow \quad \bar{\alpha} = \alpha$$

i.e., α is real

Conversely, let us suppose that α is real. To show (αT) is self-adjoint.

Since α is real.

$\Rightarrow \quad \bar{\alpha} = \alpha$

Then $\quad (\alpha T)^* = \bar{\alpha} T^* = \alpha T$

$\Rightarrow \quad (\alpha T)$ is self-adjoint.

THEOREM 4. ***The self adjoint operator in B(H) form a closed linear subspace and therefore a real Banach space which contains the identity transformation.***

(AGRA–2001, ROHILKHAND–2009)

PROOF. Let T_1, T_2 be self-adjoint operators on a Hilbert space H

$\Rightarrow \quad T_1^* = T_1$ and $T_2^* = T_2$

If α, β be any two real numbers, then

$$(\alpha T_1 + \beta T_2)^* = (\alpha T_1)^* + (\beta T_2)^*$$

$$= \bar{\alpha} T_1^* + \bar{\beta} T_2^* = \alpha T_1 + \beta T_2 \qquad [\because \alpha, \beta \text{ are real}]$$

$\Rightarrow \quad (\alpha T_1 + \beta T_2)$ is self-adjoint on H, when T_1 and T_2 are self-adjoint.

$\Rightarrow \quad$ The set of all self-adjoint operator form a real subspace.

Now, it remains to show that it is closed subset of the Banach space $B(H)$.

Let $< T_n >$ be a sequence of self-adjoint operators converging to an operator T. To show that T is self-adjoint, consider

$$\left\|T - T^*\right\| = \left\|T - T_n + T_n - T^*\right\| \le \left\|T - T_n\right\| + \left\|T_n - T^*\right\|$$

$$= \left\|T - T_n\right\| + \left\|T_n - T_n^* + T_n^* - T^*\right\|$$

$$\le \left\|T - T_n\right\| + \left\|T_n - T_n^*\right\| + \left\|T_n^* - T^*\right\|$$

$$= \left\|-(T_n - T)\right\| + \left\|0\right\| + \left\|(T_n - T)^*\right\|$$

$$= \left\|(T_n - T)\right\| + \left\|(T_n - T)\right\| \qquad [\because \left\|T^*\right\| = \left\|T\right\|]$$

$$= 2\left\|T_n - T\right\| \to 0 \text{ as } T_n \to T$$

Therefore

$$\left\|T_n - T^*\right\| \to 0 \text{ , i.e., } T - T^* = 0$$

$\Rightarrow \quad T = T^*$, *i.e.*, T is self-adjoint.

Also, we know that $I^* = I$, *i.e.*, identity operator is self-adjoint. Hence, the set of all self adjoint operators in $B(H)$ forms a closed linear subspace and therefore, a real Banach space which contains the identity operator.

THEOREM 5. ***If T be an arbitrary operator on a Hilbert space H. Then T = 0 if and only if*** $\boldsymbol{(Tx, y) = 0; \ \forall x, y \in H}$.

(MEERUT–1985, KANPUR–1986, 91, 2008, BUNDELKHAND–1983, ROHILKHAND–2008)

PROOF. Let us first suppose $T = 0$, *i.e.*, zero operator.

Then $\quad Tx = 0$.

Therefore, $\quad (Tx, y) = (0, y) = 0, \ \forall x, y \in H$

Conversely,

If $(Tx, y) = 0 \ \forall x, y$, then, choose $y = Tx$

$\therefore \quad (Tx, y) = 0 \Rightarrow (Tx, Tx) = 0 \ \forall x$

$$\Rightarrow Tx = 0 \Rightarrow T = 0$$

THEOREM 6. ***If T be a linear operator on a Hilbert space H, then***

$$(Tx, x) = 0 \text{ if and only if } T = 0 \quad \forall x \in H$$

(MEERUT–2000, 02, 05, 06BP, 07, 09, 11, KANPUR–1990, 91, 99, 2006, ROHILKHAND–2009)

PROOF. Let us first suppose $T = 0$, then $Tx = 0$.

Therefore, $(Tx,x) = (0,x) = 0 \quad \forall x \in H$

Conversely, let $(Tx,x) = 0$. To show $T = 0$.

Now, $(T(\alpha x + \beta y), \alpha x + \beta y) = 0$

$\Rightarrow \quad (\alpha Tx + \beta Ty, \alpha x + \beta y) = 0$

$\Rightarrow \quad \alpha\bar{\alpha}(Tx,x) + \alpha\bar{\beta}(Tx,y) + \beta\bar{\alpha}(Ty,x) + \beta\bar{\beta}(Ty,y) = 0$

$\Rightarrow \quad \alpha\bar{\beta}(Tx,y) + \beta\bar{\alpha}(Ty,x) = 0 \quad (\because (Tx,x) = 0)$ ...(1)

In particular choosing, $\alpha = 1, \beta = 1 \Rightarrow \bar{\alpha} = \bar{\beta} = 1$

Then from (1)

$(Tx,y) + (Ty,x) = 0$ ...(2)

Further choosing $\alpha = i, \beta = 1$, *i.e.*, $\bar{\alpha} = -i$ and $\bar{\beta} = 1$, then (1) gives

$i(Tx,y) - i(Ty,x) = 0$

$\Rightarrow \quad (Tx,y) - (Ty,x) = 0$ ...(3)

Adding (2) and (3), we get

$2(Tx,y) = 0$

$\Rightarrow \quad (Tx,y) = 0 \quad \forall x, y \in H$

Hence, $T = 0$. [Using above theorem]

THEOREM 7. ***An operator T on a Hilbert space H is self-adjoint if and only if (Tx, x) is real for all x.***

(MEERUT–1999, 2001, 04, 05BP, 08, 10, KANPUR–1990, 2000, 04, 05, AGRA–1986)

PROOF. Let us first suppose T be self-adjoint, *i.e.*, $T^* = T$ ...(1)

To show that (Tx,x) is real for all x.

Consider

$$(Tx,x) = (x,T^*x) = (x,Tx) = \overline{(Tx,x)} \quad \text{[Using (1)]}$$

Thus (Tx,x) is real.

Conversely, suppose that (Tx,x) is real,

Then $(Tx,x) = \overline{(Tx,x)} = (x,T^*x) = (T^*x,x) \ \forall x$

$\Rightarrow \quad (Tx,x) = (T^*x,x) = 0$

$\Rightarrow \quad (T - T^*(x,x)) = 0$

$\Rightarrow \quad T - T^* = 0 \quad \Rightarrow \quad T = T^*$

Hence, T is self-adjoint.

5.13 ORDERED RELATION

Let S be the set of all self-adjoint operators on a Hilbert space H. Then we may define the relation $\leq$ as follows:

"If $T_1, T_2 \in S$, then we write

$$T_1 \leq T_2 \text{ if } (T_1x,x) \leq (T_2x,x) \quad \forall x \in H$$

THEOREM 1. ***The real Banach space of all self-adjoint operators on a Hilbert space H is a partially ordered set whose linear structure and order structure are related by the following properties.***

(i) If $T_1 \leq T_2$ Then $T_1 + T \leq T_2 + T$, for every T

(ii) If $T_1 \leq T_2$ and $\alpha \geq 0$, then $\alpha T_1 \leq \alpha T_2$

PROOF. Let H be a Hilbert space and S be the set of all self adjoint operators on H.

We define a relation $\leq$ as follows :

"If $T_1, T_2 \in S$, we write $T_1 \leq T_2$ if $(T_1x, x) \leq (T_2x, x) \quad \forall x \in H$

To show '$\leq$' is a partial ordered relation.

(a) **$\leq$ is Reflexive:**

Let $T \in S$, Then, we have

$$(Tx, x) = (Tx, x) \quad \forall x \in H$$
$$\Rightarrow (Tx, x) \leq (Tx, x) \quad \forall x \in H$$
$$\Rightarrow T \leq T$$

$\Rightarrow$ Relation $\leq$ is Reflexive.

(b) **$\leq$ is Transitive:** Let $T_1 \leq T_2$ and $T_2 \leq T_3$.

Then $(T_1x, x) \leq (T_2x, x)$ and $(T_2x, x) \leq (T_3x, x)$

On combining above two results, we get

$$(T_1x, x) \leq (T_2x, x) \leq (T_3x, x)$$
$$\Rightarrow (T_1x, x) \leq (T_3x, x)$$
$$\Rightarrow T_1 \leq T_3$$

Thus, the relation $\leq$ is Transitive.

(c) **$\leq$ is anti symmetric:** Let $T_1 \leq T_2$ and $T_2 \leq T_1$

To show $T_1 = T_2$,

Now $T_1 \leq T_2 \Rightarrow (T_1x, x) \leq (T_2x, x) \quad \forall x \in H$

and $T_2 \leq T_1 \Rightarrow (T_2x, x) \leq (T_1x, x) \quad \forall x \in H$

which implies

$$(T_1x, x) = (T_2x, x)$$
$$\Rightarrow T_1 = T_2$$

$\Rightarrow$ Relations $\leq$ is anti symmetric.

Next, we shall prove the remaining part of the theorem.

(i) We have

$$\begin{aligned} T_1 \leq T_2 &\Rightarrow (T_1x, x) \leq (T_2x, x), \quad \forall x \in H \\ &\Rightarrow (T_1x, x) + (Tx, x) \leq (T_2x, x) + (Tx, x), \quad \forall x \in H \\ &\Rightarrow ((T_1 + T)x, x) \leq ((T_2 + T)x, x), \quad \forall x \in H \\ &\Rightarrow T_1 + T \leq T_2 + T \end{aligned}$$

(ii) We have

$$\begin{aligned} T_1 \leq T_2 &\Rightarrow (T_1x, x) \leq (T_2x, x), \quad \forall x \in H \\ &\Rightarrow \alpha(T_1x, x) \leq \alpha(T_2x, x), \quad \forall x \in H \\ &\Rightarrow (\alpha T_1x, x) \leq (\alpha T_2x, x), \quad \forall x \in H \\ &\Rightarrow ((\alpha T_1)x, x) \leq ((\alpha T_2)x, x), \quad \forall x \in H \\ &\Rightarrow \alpha T_1 \leq \alpha T_2 \end{aligned}$$

RECAPITULATIONS

- The adjoint operator is one-one onto as a mapping.
- If T is an invertible operator then either both or none of T and T^{-1} are self-adjoint.
- If T is self-adjoint operator on a Hilbert space H then αT is self-adjoint iff α is real.
- If T be an arbitrary operator on a Hilbert space H then $T = 0$ iff $(Tx, y) = 0 \; \forall \; x, y \in H$.
- If T be a linear operator on a Hilbert space H $(Tx, x) = 0$ iff $T = 0 \; \forall \; x \in H$.
- An operator T on a Hilbert space H is self adjoint iff (Tx, x) is real for all x.

5.14 POSITIVE, NORMAL AND UNITARY OPERATORS

(i) Positive operators. Let H be a Hilbert space. A self adjoint operator T on H is said to be positive operator if $T \geq 0$, *i.e.*, if $(Tx, x) \geq 0, \;\; \forall x \in H$.

(ii) Normal operator. An operator T on a Hilbert space H is said to be normal if $TT^* = T^*T$, *i.e.*, it commutes with its adjoint. Clearly, every self-adjoint operator is normal. (MEERUT–2007, 07BP, KANPUR–1990)

(iii) Unitary operator. An operator U on a Hilbert space H is said to be unitary if $UU^* = U^*U = I$. (MEERUT–2007, 09BP, KANPUR–1990)

Clearly, every unitary operator is normal. It must be noted that an operator U is unitary if and only if it is invertable and its inverse is equal to its adjoint.

REMARKS

- Identity and zero operators are both positive operators.
- For any arbitrary operator T on H, both TT^* and T^*T are positive operators.

THEOREM 1. ***Let H be a Hilbert space and T be a positive operator on H then, I + T is non-singular.*** (MEERUT–2005BP, KANPUR–2006)

PROOF. We want to show that $I + T$ is non-singular. For this we shall proved that $I + T$ is one-one and onto.

(i) $I + T$ is one-one. First of all we shall prove that $(I+T)(x) = 0 \Rightarrow x = 0$.

Now $(I+T)x = 0 \;\Rightarrow\; Ix + Tx = 0 \;\Rightarrow\; x + Tx = 0 \Rightarrow Tx = -x$

$$\Rightarrow (Tx, x) = (-x, x) = -\|x\|^2$$

$$\Rightarrow -\|x\|^2 > 0$$

$$\Rightarrow \|x\|^2 \leq 0$$

but $\quad \|x\|^2 \geq 0$

$\therefore \quad \|x\|^2 = 0$, *i.e.*, $\|x\| = 0$

$\Rightarrow \quad x = 0$

$$(I+T)x = 0 \;\Rightarrow\; x = 0$$

Now let

$$(I+T)x = (I+T)y$$

$\Rightarrow \quad (I+T)(x-y) = 0$

$\Rightarrow \quad x - y = 0$, *i.e.*, $x = y$

Therefore $I + T$ is one-one.

(ii) $I + T$ is onto. Let R be the range of $I+T$.

We know that $I+T$ will be onto if we show $R = H$. Firstly, we shall prove that R is closed. Let $x \in H$, then

$$\|(I+T)x\|^2 = \|x+Tx\|^2 = (x+Tx, x+Tx)$$
$$= (x,x)+(x,Tx)+(Tx,x)+(Tx,Tx)$$
$$= \|x\|^2 + \|Tx\|^2 + \overline{(Tx,x)} + (Tx,x)$$
$$= \|x\|^2 + \|Tx\|^2 + 2(Tx,x)$$

[$\because$ T is positive $\Rightarrow$ T is self-adjoint therefore (Tx, x) is real]

$$\geq \|x\|^2$$

$\Rightarrow$ $\|x\| \leq \|(I+T)x\|, \quad \forall x \in H$

Now, let $<(I+T)x_n>$ be a Cauchy sequence in R. Then by definition for positive integers m, n, we have

$$\|x_m - x_n\| \leq \|(I+T)(x_m - x_n)\| = \|(I+T)x_m - (I+T)x_n\| \to 0$$

[$\because <(I+T)x_n>$ is Cauchy]

$\Rightarrow$ $\|x_m - x_n\| \to 0$

$\Rightarrow$ $<x_n>$ is Cauchy in H, but H is complete.

$\Rightarrow$ $<x_n>$ is convergent in H such that $x_n \to x$.

Now, $\lim\{(I+T)x_n\} = (I+T)(\lim x_n) = (I+T)x \in R$

$\Rightarrow$ Cauchy sequence $<(I+T)x_n>$ in R converges to a vector $(I+T)x$ in R.

$\Rightarrow$ Every Cauchy sequence in R, converges in R.

$\Rightarrow$ R is complete.

But every complete subspace of a complete space is closed.

$\Rightarrow$ R is closed.

Now, to show $R = H$. Let if possible $R \neq H$, then R is a proper closed subspace of H. Then, there exists a non-zero vector $x_0 \in H$ such that $x_0 \perp R$.

Now, since $\quad (I+T)x_0 \in R$

$\therefore$ $\quad x_0 \perp R \Rightarrow ((I+T)x_0, x_0) = 0$

$\Rightarrow$ $\quad (x_0 + Tx_0, x_0) = 0$

$\Rightarrow$ $\quad (x_0, x_0) + (Tx_0, x_0) = 0$, *i.e.*, $\|x\|^2 + (Tx_0, x_0) = 0$

$\Rightarrow$ $\quad -\|x_0\|^2 = (Tx_0, x_0)$

$\Rightarrow$ $\quad -\|x_0\|^2 \geq 0$ $\qquad$ [$\because$ T positive $\Rightarrow (Tx_0, x_0) \geq 0$]

$\Rightarrow$ $\quad \|x_0\|^2 \leq 0$

But $\quad \|x_0\|^2 \geq 0$

Thus $\|x_0\|^2 = 0 \Rightarrow \|x_0\| = 0 \Rightarrow x_0 = 0$, which is a contradiction, because $x_0 \neq 0$.

$\Rightarrow$ Therefore, we must have $R = H$.

$\Rightarrow$ $I+T$ is onto. Hence, $I+T$ is non-singular.

THEOREM 2. ***If T is an arbitrary operator on H, then the operator $I + TT^*$ and $I + T^*T$ are non-singular.*** (MEERUT–1982, KANPUR–1991)

PROOF. Let H be a Hilbert space and T be any arbitrary operator on H. Since, we know that TT^* and T^*T are both positive operator then, using above theorem, we conclude that $I+TT^*$ and $I+T^*T$ are non-singular.

REMARK

- The set of all normed operators on a Hilbert space H is a closed subspace of $B(H)$ which contains the set of all self-adjoint operators and is closed under scalar multiplication.

THEOREM 3. ***If N_1 and N_2 are normal operators on a Hilbert space H with the property that either commutes with the adjoint of the other, then $N_1 + N_2$ and $N_1 . N_2$ are also normal.***

(MEERUT–2000, 03, 05, 09BP, KANPUR–1990, 2006, BUNDELKHAND–1983)

PROOF. As per given, N_1, N_2 are normal operator. Therefore, by definition

$$N_1N_1^* = N_1^*N_1 \quad \text{and} \quad N_2N_2^* = N_2^*N_2 \qquad \ldots(1)$$

Further, it is also given that

$$N_1N_2^* = N_2^*N_1 \quad \text{and} \quad N_2N_1^* = N_1^*N_2 \qquad \ldots(2)$$

To show $N_1 + N_2$ is normal.

Consider $(N_1+N_2)(N_1+N_2)^*$

$$\begin{aligned}
&= (N_1+N_2)(N_1^*+N_2^*) \\
&= N_1N_1^* + N_1N_2^* + N_2N_1^* + N_2N_2^* \\
&= N_1^*N_1 + N_2^*N_1 + N_1^*N_2 + N_2^*N_2 \qquad \text{[Using (1) and (2)]} \\
&= N_1^*(N_1+N_2) + N_2^*(N_1+N_2) \\
&= (N_1^*+N_2^*)(N_1+N_2) = (N_1+N_2)^*(N_1+N_2)
\end{aligned}$$

Thus, N_1+N_2 is normal.

Further, we shall prove that N_1N_2 is normal.

Consider

$$\begin{aligned}
(N_1N_2)(N_1N_2)^* &= (N_1N_2)(N_2^*N_1^*) = N_1(N_2N_2^*)N_1^* \\
&= N_1(N_2^*N_2)N_1^* = (N_1N_2^*)(N_2N_1^*) = (N_2N_1^*)(N_1N_2^*) \\
&= N_2^*(N_1N_1^*)N_2 = N_2^*(N_1^*N_1)N_2 \\
&= (N_2^*N_1^*)(N_1N_2) = (N_1N_2)^*(N_1N_2)
\end{aligned}$$

Hence, N_1N_2 is normal.

THEOREM 4. ***Let H be a Hilbert space, an operator T on H is said to be normal if and only if $\|T^*x\| = \|Tx\|$.***

(DELHI–2005, MEERUT–1980, 85, 2004, 07BP, 09BP, KANPUR–1990, 94, 96, 2001, 03, 06, 10, BUNDELKHAND–1982)

PROOF. Let H be a Hilbert space.

T is normal $\Leftrightarrow TT^* = T^*T \Leftrightarrow TT^* - T^*T = 0 \Leftrightarrow ((TT^* - T^*T)x, x) = 0, \ \forall x \in H$

$$\Leftrightarrow (TT^* x, x) = 0 + (T^* Tx, x), \ \forall x \in H$$

$$\Leftrightarrow (T^* x, T^* x) = (Tx, T^{**} x), \ \forall x \in H$$

$$\Leftrightarrow \left\| T^* x \right\|^2 = \| Tx \|^2, \forall x \in H \qquad [\because T^{**} = T]$$

$$\Leftrightarrow \left\| T^* x \right\| = \| Tx \|$$

THEOREM 5. ***If N is a normal operator on a Hilbert space H. Then,*** $\left\| N^2 \right\| = \| N \|^2$

(MEERUT–1989, 2002, 05BP, 07, KANPUR–1990, 02, 05, BUNDELKHAND–1985, AGRA–2005)

PROOF. Since, we know that if T is a normal operator on H. Then,

$$\|Tx\| = \left\|T^* x\right\|, \ \forall x \qquad \ldots(1)$$

Replacing $T = N$ and $x = Nx$ in (1), we get

$$\|NNx\| = \left\|N^* Nx\right\|, \ \forall x \in H$$

$$\Rightarrow \qquad \left\|N^2 x\right\| = \left\|N^* Nx\right\|, \ \forall x \in H$$

Now, $\left\|N^2\right\| = \sup\{\left\|N^2 x\right\| : \|x\| \le 1\}$

$$= \sup\{\left\|N^* Nx\right\| : \|x\| \le 1\}$$

$$= \left\|N^* N\right\| = \|N\|^2$$

THEOREM 6. ***Any arbitrary operator T on a Hilbert space H can be uniquely expressed as*** $T = T_1 + iT_2$***, where*** T_1 ***and*** T_2 ***are self-adjoint operator.***

(MEERUT–2003, 06BP)

PROOF. Let us write

$$T_1 = \frac{T + T^*}{2} \quad \text{and} \quad T_2 = \frac{1}{2i}(T - T^*)$$

Then $\qquad T = T_1 + iT_2 \qquad \ldots(1)$

We have $\qquad T_1^* = \left[\frac{1}{2}\left(T + T^*\right)\right]^* = \frac{1}{2}(T + T^*)^* = \frac{1}{2}[T^* + T^{**}]$

$$= \frac{1}{2}(T^* + T) = \frac{1}{2}(T + T^*) = T_1$$

$\Rightarrow \quad T_1$ is self-adjoint.

Also, $\qquad T_2^* = \left[\frac{1}{2i}(T - T^*)\right]^* = \left(\overline{\frac{1}{2i}}\right)(T - T^*)^*$

$$= -\frac{1}{2i}(T^* - T^{**}) = -\frac{1}{2i}(T^* - T) = \frac{1}{2i}(T - T^*) = T_2$$

$\Rightarrow \quad T_2$ is self-adjoint.

Therefore, T can be expressed as

$$T = T_1 + iT_2$$

such that T_1, T_2 both are self-adjoint.

Further, let $T = U_1 + iU_2$ be another representation, where U_1 and U_2 are both self-adjoint.

We have

$$T^* = (U_1 + iU_2)^* = U_1^* + (iU_2)^* = U_1^* + \bar{i}U_2^*$$

$$= U_1^* - iU_2^* = U_1 - iU_2$$

[$\because U_1$ and U_2 both are self-adjoint]

$$\Rightarrow \quad T + T^* = (U_1 + iU_2) + (U_1 - iU_2) = 2U_1$$

$$\Rightarrow \quad U_1 = \frac{1}{2}(T + T^*) = T_1$$

Also, $$T - T^* = (U_1 + iU_2) - (U_1 - iU_2) = 2iU_2$$

$$\Rightarrow \quad U_2 = \frac{1}{2i}(T - T^*) = T_2$$

Hence, the expression (1) is unique.

THEOREM 7. ***If T is an operator on a Hilbert space H, then T is normal if and only if its real and imaginary part commutes.***

(MEERUT–2006BP, KANPUR–1990, 91, ROHILKHAND–2009, 14)

PROOF. Let T_1 and T_2 respectively be the real and imaginary parts of T

i.e., $$T = T_1 + iT_2$$

Then using previous theorem we can say that, T_1 and T_2 are self-adjoint.

Now $$T^* = (T_1 + iT_2)^* = (T_1)^* + (iT_2)^* = T_1^* + \bar{i}T_2^*$$

$$= T_1^* - iT_2^* = T_1 - iT_2 \qquad (\because T_1 \text{ and } T_2 \text{ are self-adjoint})$$

and $$TT^* = (T_1 + iT_2)(T_1 - iT_2) = T_1^2 + T_2^2 + i(T_2T_1 - T_1T_2) \qquad \ldots(1)$$

$$T^*T = (T_1 - iT_2)(T_1 + iT_2) = T_1^2 + T_2^2 + i(T_1T_2 - T_2T_1) \qquad \ldots(2)$$

Let us first suppose that T is normal

Therefore $$TT^* = T^*T$$

$$\Rightarrow \quad T_1^2 + T_2^2 + i(T_2T_1 - T_1T_2) = T_1^2 + T_2^2 + i(T_1T_2 - T_2T_1)$$

$$\Rightarrow \quad 2T_2T_1 = 2T_1T_2$$

$$\Rightarrow \quad T_2T_1 = T_1T_2$$

$\Rightarrow$ T_1 and T_2 commutes each other.

Conversely, suppose that T_1 and T_2 commutes, *i.e.*, $T_1T_2 = T_2T_1$

Then, using (1)and (2), we conclude that

$$TT^* = T^*T$$

Hence, T is normal.

THEOREM 8. ***If T is an operator on a Hilbert space H, then following conditions are equivalent.***

(i) $T^*T = I$

(ii) $(Tx, Ty) = (x, y), \quad \forall x, y$

(iii) $\|Tx\| = \|x\|, \forall x$

(MEERUT–1998, KANPUR–2008, 14, ALLAHABAD–2004, BANARAS–2006)

PROOF. **(i)** $\Rightarrow$ **(ii)** Let $T^*T = I$

To show

$$(Tx, Ty) = (x, y), \forall x, y \in H$$

Consider

$$(Tx, Ty) = (x, T^*Ty) = (x, Iy) = (x, y)$$

(ii) $\Rightarrow$ **(iii)** Given that $(Tx, Ty) = (x, y)$.

To show that $\|Tx\| = \|x\|, \forall x$

Consider

$$(Tx, Tx) = (x, x) \qquad \text{(By our hypothesis)}$$

$$\Rightarrow \quad \|Tx\|^2 = \|x\|^2$$

i.e., $$\|Tx\| = \|x\|, \forall x \in H$$

(iii) $\Rightarrow$ **(i)** Here we have

$$\|Tx\| = \|x\|$$

To show $T^*T = I$

Since $\|Tx\| = \|x\|$

$$\Rightarrow \quad \|Tx\|^2 = \|x\|^2$$

$$\Rightarrow \quad (Tx, Tx) = (x, x)$$

$$\Rightarrow \quad (T^*Tx, x) = (x, x)$$

$$\Rightarrow \quad ((T^*T - I)x, x) = 0, \ \forall x$$

$$\Rightarrow \quad T^*T - I = 0$$

Therefore, $T^*T = I$

THEOREM 9. ***An operator T on a Hilbert space H is unitary if and only if it is an isometric isomorphism of H onto itself.***

(MEERUT–2000, 05, KANPUR–1991, AGRA–2005, 06, 09)

PROOF. Let H be a Hilbert space. Let us first suppose that T is unitary. To show that T is an isometric isomorphism of H onto itself.

Since T is unitary $\Rightarrow$ T is invertiable. $\Rightarrow$ T is onto.

Also, we have $T^*T = I$

Thus, using previous theorem ,we have

$$\|Tx\| = \|x\|, \forall x \in H.$$

$\Rightarrow$ T presences norms and therefore T is an isometric isomorphism of H onto itself.

Conversely, let T be an isometric isomorphism of H onto itself.

Then, T is one-one and onto.

$\Rightarrow$ T^{-1} exists

Since T is isometric isomorphism, therefore

$$\|Tx\| = \|x\|$$

$\Rightarrow$ $$T^*T = I$$ [Using previous Theorem-7]

$\Rightarrow$ $$(T^*T)T^{-1} = IT^{-1}$$

$\Rightarrow$ $$T^*(TT^{-1}) = T^{-1}$$

$\Rightarrow$ $$T^*I = T^{-1} \Rightarrow T^* = T^{-1}$$

Thus, we conclude by using $TT^{-1} = T^{-1}T = I$, that

$$TT^* = I = T^*T$$

Hence, T is unitary.

THEOREM 10. ***If T is an arbitrary operator on Hilbert space H and α, β are scalars such that $|\alpha| = |\beta|$. Then $\alpha T + \beta T^*$ is normal.***

(MEERUT–1994, 97, 99, 2001, 06, 06BP, 08, KANPUR–1991, 2007, AGRA–1988, 2001, 04, ROHILKHAND–2008)

PROOF. As per given

$$|\alpha| = |\beta| \Rightarrow |\alpha|^2 = |\beta|^2$$

i.e., $$\alpha\bar{\alpha} = \beta\bar{\beta} \quad \ldots(1)$$

Consider

$$(\alpha T + \beta T^*)(\alpha T + \beta T^*)^* = (\alpha T + \beta T^*)(\bar{\alpha}T^* + \bar{\beta}T) \quad [\because T^{**} = T]$$

$$= \alpha\bar{\alpha}TT^* + \alpha\bar{\beta}TT + \beta\bar{\alpha}T^*T^* + \beta\bar{\beta}T^*T \quad \ldots(2)$$

Also $$(\alpha T + \beta T^*)^*(\alpha T + \beta T^*)$$

$$= (\bar{\alpha}T^* + \bar{\beta}T)(\alpha T + \beta T^*)$$

$$= (\bar{\alpha}\alpha T^*T + \bar{\alpha}\beta T^*T^* + \bar{\beta}\alpha TT + \beta\bar{\beta}TT^*)$$

$$= \beta\bar{\beta}T^*T + \bar{\alpha}\beta T^*T^* + \bar{\beta}\alpha TT + \alpha\bar{\alpha}TT^* \quad \text{(Using (1))} \quad \ldots(3)$$

Therefore, from (2) and (3), we conclude that

$$(\alpha T + \beta T^*)(\alpha T + \beta T^*)^* = (\alpha T + \beta T^*)^*(\alpha T + \beta T^*)$$

$\Rightarrow$ $\alpha T + \beta T^*$ is normal.

THEOREM 11. ***If T be a normal operator on a Hilbert space H and λ be any scalar then $T - \lambda I$ is normal.*** (MEERUT–2006, 06BP, 10)

PROOF. As per given, we have T is normal.

Therefore, $$TT^* = T^*T \quad \ldots(1)$$

Consider

$$(T - \lambda I)(T - \lambda I)^* = (T - \lambda I)(T^* - \bar{\lambda}I) \quad [\because I^* = I]$$

$$= TT^* - \bar{\lambda}T - \lambda T^* + \lambda\bar{\lambda}I \quad (\because IT = TI = I) \quad \ldots(2)$$

and $$(T - \lambda I)^*(T - \lambda I) = (T^* - \bar{\lambda}I)(T - \lambda I)$$

$$= T^*T - \bar{\lambda}T - \lambda T^* + \lambda\bar{\lambda}I \quad \ldots(3)$$

Using (2) and (3), we conclude that

$$(T - \lambda I)(T - \lambda I)^* = (T - \lambda I)^*(T - \lambda I)$$

Hence, $(T - \lambda I)$ is normal.

Theorem 12. ***If T_1 and T_2 are positive operators on a Hilbert space H, then $T_1 + T_2$ is a positive operator.*** (MEERUT–2006)

Proof. Since, T_1 and T_2 are positive operators on a Hilbert space H then for $x \in H$

$$T_1^* = T_1 \quad \ldots(1)$$

and

$$T_2^* = T_2 \quad \ldots(2)$$

Also,

$$(T_1x, x) \geq 0 \quad \ldots(3)$$

and

$$(T_2x, x) \geq 0 \quad \ldots(4)$$

On adding (1) and (2), we get

$$T_1^* + T_2^* = T_1 + T_2 \quad \Rightarrow (T_1 + T_2)^* = T_1 + T_2$$

Similarly, from (3) and (4), we get

$$(T_1x, x) + (T_2x, x) \geq 0$$

$$\Rightarrow \quad ((T_1 + T_2)x, x) \geq 0$$

Hence, $(T_1 + T_2)$ is a positive operator.

5.15 COMPACT OPERATOR

An operator A on a Hilbert space H is said to be compact if for every bounded sequence $<x_n>$ in H, the sequence $<Ax_n>$ contains a convergent sequence.

Illustrations

(1) Every operator on a finite dimensional Hilbert space is compact.

(2) If y and z are fixed elements of a Hilbert space H such that $Tx = (x, y)z$. Then T is a compact operator.

(3) The integral operator T on $L^2[a, b]$ defined by

$$(Tx)s = \int_a^b K(s,t)x(t)dt$$

where a and b are finite and K is continuous, is compact operator.

(4) The projection operator P_s on a finite dimensional subspace S of a Hilbert space H is compact.

Remarks

- The identity operator on an infinite dimensional Hilbert space H is not compact, although it is bounded.
- The collection of all compact operators on a Hilbert space H is a linear (vector) space.

Theorem 1. ***Compact operators are bounded.***

Proof. Let A be an operator which is not bounded. Then there exist a sequence $<x_n>$ such that $\|x_n\| = 1 \ \forall \ n \in N$ and $\|Ax_n\| \to \infty$. Then $<Ax_n>$ does not contain a convergent subsequence which implies that A is not compact.

Hence, contrapositively, compact operators are bounded.

Theorem 2. ***Let A be a compact operator on a Hilbert space H and let B be a bounded operator on H. Then AB and BA are compact.***

Proof. Let $<x_n>$ be a bounded sequence in H. Since B is bounded therefore, $<Bx_n>$ is bounded. Further, since A is compact, the sequence $<ABx_n>$ contains a convergent subsequence, which show that AB is compact. Similarly, we may show that BA is compact.

THEOREM 3. ***Finite dimensional bounded operators are compact.***

PROOF. Let A be a finite dimensional bounded operator and let $\{z_1, z_2, ..., z_k\}$ be an orthonormal basis of the range of A

Let $T_n x = (Ax, z_n)z_n$ for $n = 1, 2, ..., k$

Since $T_n x = (Ax, z_n)z_n = (x, A^* z_n)z_n$

$\Rightarrow$ T_n are compact.

THEOREM 4. ***The limit of a uniformly convergent sequence of compact operators is compact, i.e., if $T_1, T_2, ...$ are compact operators on a Hilbert space H and $\|T_n - T\| \to 0$ as $n \to \infty$ for some operator T on H, then T is compact.***

PROOF. Let H be a Hilbert space and $<x_n>$ be a bounded sequence in H. Since, T_1 is compact, there exists a subsequence $<x_{1n}>$ of $<x_n>$ such that $<T_1 x_{1n}>$ is convergent. Similarly, the sequence $<T_2 x_{1n}>$ contains a convergent subsequence $<T_2 x_{2n}>$ and so on.

In general, for $k \geq 2$, let $<x_{kn}>$ be a subsequence of $<x_{k-1.n}>$ such that $<T_k x_{kn}>$ is convergent.

Now, consider the sequence $<x_{nn}>$. Since it is a subsequence of $<x_n>$, put $x_{p_n} = x_{nn}$ where $<p_n>$ is an increasing sequence of positive integers. Clearly, the sequence $<T_k x_{p_n}>$ converges for every $k \in N$. We have to show that the sequence $<T x_{p_n}>$ also converges.

Let $\varepsilon > 0$ be given; since $\|T_n - T\| \to 0$, $\exists\ k \in N$ such that $\|T_k - T\| < \frac{\varepsilon}{2M}$, where M is a constant such that $\|x_n\| \leq M\ \forall\ n \in N$.

Further, let $k_1 \in N$ be such that

$$\|T_k x_{pn} - T_k x_{p_m}\| < \varepsilon/3\ \forall\ n, m > k_1$$

Then,
$$\|T x_{p_n} - T x_{p_m}\| \leq \|T x_{p_n} - T_k x_{p_n}\| + \|T_k x_{p_n} - T_k x_{p_m}\| + \|T_k x_{p_m} - T x_{p_m}\|$$
$$< \varepsilon/3 + \varepsilon/3 + \varepsilon/3 = \varepsilon, \text{ for sufficiently large } m \text{ and } n$$

$\Rightarrow$ $<T x_{p_n}>$ is a cauchy sequence in H, which is complete

$\Rightarrow$ $<T x_{p_m}>$ is convergent.

REMARK

- The limit of a Convergent sequence of finite dimensional operator S is a Compact operator.

THEOREM 5. ***The adjoint of a compact operator is compact.***

PROOF. Let H be a Hilbert space and T be a compact operator on H and let $<x_n>$ be a bounded sequence in H, so $\|x_n\| \leq M$ for some M and for all $n \in N$.

Let us define $y_n = T^* x_n : n = 1, 2, ...$

Since, T^* is bounded, the sequence $<y_n>$ is bounded.

$\Rightarrow$ $\exists$ a subsequence $<y_{kn}>$ of $<y_n>$ such that $<Ty_{kn}>$ converges in H.

Now, for any $m, n \in N$, we have

$$\| y_{k_m} - y_{k_n} \|^2 = \|T^* x_{k_m} - T^* x_{k_n}\|^2$$
$$= (T^*(x_{k_m} - x_{k_n}), T^*(x_{k_m}, x_{k_n}))$$
$$= (TT^*(x_{k_m} - x_{k_n}), (x_{k_m}, x_{k_n}))$$

$$\leq \|TT^*(x_{k_m} - x_{k_n})\|.\|x_{k_m} - x_{k_n}\|$$

$$\leq 2M \|Ty_{k_m} - Ty_{k_n}\| \to 0 \text{ as } m, n \to \infty$$

$\Rightarrow$ $< y_{k_n} >$ is a cauchy sequence in H, but H is complete therefore $< y_{k_n} >$ is convergent.

Hence, T^* is a compact operator.

SOLVED EXAMPLES

EXAMPLE 1. *If T is a normal operator and f be a polynomial with complex coefficients, then the operator f(T) is also normal.* (MEERUT–1996, KANPUR–1987)

SOLUTION. Let $f(x) = a_0x^n + a_1x^{n-1} + ... + a_n$ be a polynomial.

Then, $f(T) = a_0T^n + a_1T^{n-1} + ... + a_nI$

$\Rightarrow$ $[f(T)]^* = \bar{a}_0T^{*n} + \bar{a}_1T^{*n-1} + ... + \bar{a}_n I$

To show

$$[f(T)][f(T)]^* = [f(T)]^*[f(T)]$$

Since T is normal, therefore $TT^* = T^*T$ ...(1)

$\Rightarrow$ $(T^*)^2T = T^*(TT^*) = T^*(T^*T) = (T^*T)T^*$

$= (TT^*)T^* = T(T^*)^2$

In a similar way, we can show that

$$T^{*3}T = TT^{*3}$$

We can easily proved, by using mathematical induction that the result $T^{*p}T = TT^{*p}$ hold for all p.

Thus, we have

$$[f(T)]^*[f(T)] = [f(T)][f(T)]^*$$

Hence, $f(T)$ is normal.

EXAMPLE 2. *Show that the set of all unitary operators on Hilbert space H form a multiplicative group.*

(JABALPUR–2007, BHOPAL–2008, MEERUT–2002, 05BP, KANPUR–2002)

SOLUTION. Let G be the set of all unitary operators on H.

(i) **Closure property:** Let $T_1, T_2 \in G$ Then T_1 and T_2 both are invertiable and preserve inner product and norms.

Also $T_1^{-1} = T_1^*, T_2^{-1} = T_2^*$

$\Rightarrow$ T_1T_2 or T_2T_1 is also invertible

Also $(T_1T_2)^{-1} = T_2^{-1}T_1^{-1} = T_2^*T_1^* = (T_1T_2)^*$

and $\|T_1T_2(x)\| = \|T_2x\| = \|x\|$ [$\because$ T_1 and T_2 presences norms]

$\Rightarrow$ T_1T_2 preserves norms.

$\Rightarrow$ T_1T_2 is unitary.

$\Rightarrow$ $T_1T_2 \in G$.

$\Rightarrow$ Closure property is satisfied.

(ii) **Associativity:** Since, the multiplication of linear operators is always associative. Hence, multiplication of unitary operators is always associative.

(iii) **Existence of Identity:**

Since, we know that the identity operator I is always invertiable.

and $\| I(x) \| = \| x \|, \forall x$

$\Rightarrow$ Identity element exists and unitary.

(iv) **Existence of inverse:**

We know that if T_1 is unitary then T_1^{-1} exists.

Also $\|T_1 x\| = \|x\|$

Choose $T_1 x = y \quad \Rightarrow x = T_1^{-1} y$

Now $\left\|T_1 T_1^{-1} y\right\| = \left\|T_1^{-1} y\right\|$ [$\because T_1$ is unitary]

$\Rightarrow \quad \|y\| = \left\|T_1^{-1}\right\|$ [$\because T_1 T_1^{-1} = I$]

$\Rightarrow$ T_1^{-1} preserves norms.

$\Rightarrow$ T_1^{-1} is unitary.

$\Rightarrow$ T_1^{-1} exists and unitary.

Hence, the set of unitary operators forms a multiplicative group.

EXAMPLE 3. *Prove that an operator T on a Hilbert space H is unitary if and only if $T\{e_i\}$ is a complete orthonormal set whenever $\{e_i\}$.* (MEERUT–1980,96)

SOLUTION. Let us first suppose that T be unitary, *i.e.*, $TT^* = I$.

Let $\{e_i\}$ is complete so that $\left(e, e_j\right) = \delta_{ij}$.

To show that $T(e_i)$ is complete orthonormal set.

Since, T is unitary so that

$$\left(Te_i, Te_j\right) = \left(e_i, e_j\right) = \delta_{ij}$$

$\Rightarrow$ $T\{e_i\}$ is orthonormal.

Let if possible $T\{e_i\}$ be not complete, then there exists $x \neq 0$ such that

$$x \perp T\{e_i\}$$

$$\Rightarrow \quad (x, Te_i) = 0, \forall e_i$$

$$\Rightarrow \quad \left(T^* x, e_i\right) = 0, \forall e_i$$

$\Rightarrow$ $T^* x \perp \{e_i\}$, where $\{e_i\}$ is complete.

$$\Rightarrow \quad T^* x = 0.$$

$\Rightarrow$ $TT^* x = T.0$, *i.e.*, $Ix = 0 \Rightarrow x = 0$.

which is a contradiction as $x \neq 0$. Hence, $T\{e_i\}$ is complete.

Conversely, let $T\{e_i\}$ is complete orthonormal whenever $\{e_i\}$ is complete orthonormal. To show T is unitary. For this we shall prove that it is an isometric isomorphism of H onto itself.

(i) ***T* is an isometry** (*i.e.*, *T* preserves norms)

If $x = 0$, then $Tx = 0 \Rightarrow \| Tx \| = \| x \|$.

If $x \neq 0$, then $\frac{x}{\|x\|}$ is a unit vector and the set $\left\{\frac{x}{\|x\|}\right\}$ is an orthonormal set in *H*.

Then there exists a complete orthonormd set in *H* containing this singleton orthonormal set. By given condition *T*-image of this complete orthonormal set is also a complete orthonormal set.

Thus, we conclude that

$$T\left(\frac{x}{\|x\|}\right) \text{ is a unit vector.}$$

$$\Rightarrow \quad \left\|T\left(\frac{x}{\|x\|}\right)\right\| = 1 \quad \text{or} \quad \left\|\frac{1}{\|x\|}T(x)\right\| = 1$$

$$\Rightarrow \quad \frac{1}{\|x\|}.\| Tx \| = 1$$

$$\Rightarrow \quad \| Tx \| = \| x \|$$

i.e., *T* preserves norms, There it is an isometry and therefore it is one-one also.

(ii) ***T* is onto :** Now, it remains to prove that $R(T) = H$.

Clearly, $R(T)$ is a subspace of *H*. To show it is closed. Let *y* be a limit point of $R(T)$. To show $y \in R(T)$. By definition of limit point, there exists sequence $< Tx_n >$ of distinct points of $R(T)$ such that $Tx_n \to y$

Now, consider

$$\|x_m - x_n\|^2 = \|T(x_m - x_n)\|^2 \qquad [\because T \text{ preserve norm}]$$

$$= \|Tx_m - Tx_n\|^2 \qquad [\because T \text{ is linear}]$$

$$\to 0 \qquad [\because < Tx_n > \text{ is convergent in } H]$$

$\Rightarrow < Tx_n >$ is a Cauchy sequence in *H*.

$\Rightarrow < x_n >$ is a Cauchy sequence in *H*, which is complete and therefore, the sequence is convergent in *H*.

$\Rightarrow \exists\, x \in H$ such that $x_n \to x$

and $$y = \lim(Tx_n) = T(\lim x_n) = T_x \qquad [\because T \text{ is continuous}]$$

$\Rightarrow y \in R(T)$. Therefore, $R(T)$ be a closed subspace of *H*.

Further, if $R(T) \neq H$, then $R(T)$ is a proper closed linear subspace of *H* such that there exists a non-zero vector *z* in *H* such that $z \perp R(T)$.

But, $R(T)$, being closed linear subspace of Hilbert space *H* must itself be a Hilbert space. Hence, there exists a complete orthonormal set in $R(T)$.

And, $z \perp R(T)$, which is orthonormal therefore $z = 0$, which is contradiction because $z \neq 0$. Therefore, we must have $R(T) = H$.

$\Rightarrow$ *T* is onto.

Hence, *T* is an isometric isomorphism of *H* onto itself and as such it is unitary.

EXAMPLE 4. *A linear operator on R^2 is defined by $T[x, y] = (x + 2y, x - y)$. Find the adjoint T^* of T, if the inner product is standard one. If $\alpha = (1, 3)$, find $T^*(\alpha)$.*

SOLUTION. Let $e_1 = (1, 0), e_2 = (0, 1)$ and $B = (e_1, e_2)$ be the basis of the inner product space $R^2(R)$.

Let $T : R^2 \to R^2$ be a linear map such that $T(x, y) = (x + 2y, x - y)$

$$T(e_1) = T(1, 0) = (1, 1) = 1e_1 + 1e_2$$
$$T(e_2) = T(0, 1) = (2, -1) = 2e_1 - e_2$$

Coefficient matrix $= \begin{bmatrix} 1 & 1 \\ 2 & -1 \end{bmatrix}$ and its transpose is $\begin{bmatrix} 1 & 2 \\ 1 & -1 \end{bmatrix} = [T]_B$

$[T^*]B$ = conjugate transpose of $[T]_B = \begin{bmatrix} 1 & 1 \\ 2 & -1 \end{bmatrix}$

The coordinate matrix of $T^*(x, y)$ in the basis B is

$$\begin{bmatrix} 1 & 1 \\ 2 & -1 \end{bmatrix}\begin{bmatrix} x \\ y \end{bmatrix} = \begin{bmatrix} x+y \\ 2x-y \end{bmatrix}$$

$\Rightarrow \qquad T^*(x, y) = \langle(x + y), (2x - y)\rangle$

Now, $\qquad T^*(\alpha) = T^*(1, 3) = (1 + 3, 2 - 3) = (4, -1)$

EXAMPLE 5. *Find the adjoint of linear map $F : R^3 \to R^3$ defined by*

$$F(x, y, z) = (3x + 4y - 5z, 2x - 6y + 7z, 5x - 9y + z)$$ (KANPUR–2009, 11)

SOLUTION. Let $e_1 = (1, 0, 0), e_2 = (0, 1, 0), e_3 = (0, 0, 1)$

Then, $\{e_1, e_2, e_3\} = B$ is a basis of R^3

$$F(e_1) = F(1, 0, 0) = (3, 2, 5) = 3e_1 + 2e_2 + 5e_3$$
$$F(e_2) = F(0, 1, 0) = (4, -6, -9) = 4e_1 - 6e_2 - 9e_3$$
$$F(e_3) = F(0, 0, 1) = (-5, 7, 1) = -5e_1 + 7e_2 + 1\cdot e_3$$

$$\text{Coefficient matrix} = \begin{bmatrix} 3 & 2 & 5 \\ 4 & -6 & -9 \\ -5 & 7 & 1 \end{bmatrix}$$

Its transpose is

$$\begin{bmatrix} 3 & 4 & -5 \\ 2 & -6 & 7 \\ 5 & -9 & 1 \end{bmatrix} = [F]_B$$

Now, $\qquad [F^*]_B$ = conjugate transpose of $[F]_B$

$$= \begin{bmatrix} 3 & 2 & 5 \\ 4 & -6 & -9 \\ -5 & 7 & 1 \end{bmatrix}$$

Coordinate matrix of $F^*(x, y, z)$ in the basis B is

$$\begin{bmatrix} 3 & 2 & 5 \\ 4 & -6 & -9 \\ -5 & 7 & 1 \end{bmatrix}\begin{bmatrix} x \\ y \\ z \end{bmatrix} = \begin{bmatrix} 3x+2y+5z \\ 4x-6y-9z \\ -5x+7y+z \end{bmatrix}$$

$\Rightarrow \qquad F^*(x, y, z) = (3x + 2y + 5z, 4x - 6y - 9z, -5x + 7y + z)$

EXAMPLE 6. *Find the adjoint of the linear map $T : C^3 \to C^3$ defined by*

$$T(x, y, z) = (2x + (1 - i)y, (3 + 2i)x - 4iz, 2ix + (4 - 3i)y - 3z)$$

where $x = (x_1, x_2, x_3), y = (y_1, y_2, y_3)$ (KANPUR–2010)

Solution. By definiton of adjoint operator, we have

$$(T^*x, y) = (x, T(y))$$
$$= x_1\overline{\{2y_1 + (1-i)y_2\}} + x_2\overline{\{(3+2i)y_1 - 4iy_3\}}$$
$$+ x_3\overline{\{2iy_1 + (4-3i)y_2 - 3y_3\}}$$
$$= x_1\{2\bar{y}_1 + (1+i)\bar{y}_2\} + x_2\{(3-2i)\bar{y}_1 + 4i\,\bar{y}_3\}$$
$$+ x_3\{-2i\,\bar{y}_1 + (4+3i)\bar{y}_2 - 3\bar{y}_3\}$$
$$= \{2x_1 + (3-2i)x_2 - 2ix_3\}\bar{y}_1 + \{(1+i)x_1$$
$$+ (4+3i)x_3\}\bar{y}_2 + \{4ix_2 - 3x_3\}\bar{y}_3$$

$$\Rightarrow \quad T^*(x) = T^*(x_1, x_2, x_3)$$
$$= (2x_1 + (3-2i)x_2 - 2ix_3,\ (1+i)x_1 + (4+3i)x_3,\ (4ix_2 - 3x_3)$$

$$\Rightarrow \quad G^*(x, y, z) = (2x + (3-2i)y - 2iz,\ (1+i)x + (4+3i)z,\ (4iy - 3z))$$

Example 7. *Let a linear map $T : C^2 \to C^2$ be defined by*

$$T(x) = (x_1 + ix_2,\ x_1 - ix_2)\ \forall\ x = (x_1, x_2) \in C^2$$

Find the adjoint T^* and show that $TT^* = T^*T = 2I$

Solution. Let $x = (x_1, x_2)$, $y = (y_1, y_2)$

We know that

$$< x, y > = x_1\bar{y}_1 + x_2\bar{y}_2 \qquad \ldots(1)$$
$$T(x) = (x_1 + ix_2,\ x_1 - ix_2) \qquad \ldots(2)$$

By definition of adjoint operator $<T^*x, y> = <x, T(y)>$; $x = (x_1, x_2)$

and $\quad T(y) = (y_1 + iy_2,\ y_1 - iy_2)$

$$= x_1\overline{(y_1 + iy_2)} + x_2\overline{(y_1 - iy_2)} \qquad \text{(Using (1))}$$
$$= x_1(\bar{y}_1 - i\bar{y}_2) + x_2(\bar{y}_1 + i\bar{y}_2)$$
$$= (x_1 + x_2)\bar{y}_1 + (-ix_1 + ix_2)\bar{y}_2$$

From definition of inner product

$$T^*(x) = T^*(x_1, x_2) = (x_1 + x_2,\ i(-x_1 + x_2)) \qquad \ldots(3)$$

Now, $\quad (TT^*)(x) = TT^*(x) = T(x_1 + x_2,\ i(-x_1 + x_2))$

$$= ((x_1 + x_2) + i(-ix_1 + ix_2),\ (x_1 + x_2) - i(-ix_1 + ix_2)) \qquad \text{(From (2))}$$
$$= (x_1 + x_2 + x_1 - x_2,\ x_1 + x_2 - x_1 + x_2)$$
$$= (2x_1, 2x_2) = 2(x_1, x_2) = 2I(x)$$

Similarly, we may prove that

$$T^*T = 2I$$

Hence, $\quad T^*T = TT^* = 2I$

Example 8. *Find the matrix A that represent usual inner product on R^2 relative to the basis $B = \{(1, 2), (4, -2)\}$.* (KANPUR–2011)

Solution. Let $\quad u = (a_1, a_2),\ v = (b_1, b_2) \in R^2$

Then $(u, v) = a_1b_1 + a_2b_2$ (By definition of inner product)

Let $A = [a_{ij}]_{2\times 2}, e_1 = (1, 2), e_2 = (4, -2)$

$$a_{11} = <e_1, e_1> = 1.1 + 2.2 = 5$$
$$a_{22} = <e_2, e_2> = 4^2 + (-2)^2 = 20$$
$$a_{12} = a_{21} = <e_1, e_2> = 1.4 + 2(-2) = 0$$

Hence, the required matrix is given by

$$A = \begin{bmatrix} a_{11} & a_{12} \\ a_{21} & a_{22} \end{bmatrix} = \begin{bmatrix} 5 & 0 \\ 0 & 20 \end{bmatrix}$$

EXAMPLE 9. *Find matrix A that represents usual inner product on R^2 relative to the basis $B = \{(1, 4), (2, -3)\}$* (KANPUR–2011)

SOLUTION. Let $u = (a_1, a_2), v = (b_1, b_2) \in R^2$

Then $(u, v) = a_1b_1 + a_2b_2$ (By definition of usual inner product space)

Let $e_1 = (1, 4), e_2 = (2, -3)$

Also, let $A = [a_{ij}]_{2\times 2}$, then

$$a_{11} = <e_1, e_1> = 1^2 + 4^2 = 17$$
$$a_{12} = a_{21} = <e_1, e_2> = 1.2 + 4(-3) = -10$$
$$a_{22} = <e_2, e_2> = 2^2 + (-3)^2 = 13$$

Hence, the required matrix is given by

$$A = \begin{bmatrix} a_{11} & a_{12} \\ a_{21} & a_{22} \end{bmatrix} = \begin{bmatrix} 17 & -10 \\ -10 & 13 \end{bmatrix}$$

EXAMPLE 10. *Find the matrix P that represents the usual inner product on R^3 relative to the basis $B = \{1, 1 + i, 1 - 2i\}$.* (KANPUR–2010)

SOLUTION. We know that in inner product on $C^3\,(u, v) = u\bar{v}$

Suppose $e_1 = 1, e_2 = 1 + i, e_3 = 1 - 2i$

Also, let $P = [p_{ij}]_{3\times 3}$, then

$$p_{11} = <e_1, e_1> = e_1\bar{e}_1 = 1$$
$$p_{12} = <e_1, e_2> = e_1\bar{e}_2 = 1(1-i) = 1-i$$
$$p_{13} = <e_1, e_3> = e_1\bar{e}_3 = 1(1+2i) = 1+2i$$
$$p_{21} = <e_2, e_1> = e_2\bar{e}_1 = (1+i)$$
$$p_{22} = <e_2, e_2> = e_2\bar{e}_2 = (1+i)(1-i) = 2$$
$$p_{23} = <e_2, e_3> = e_2\bar{e}_3 = (1+i)(1+2i) = -1+3i$$
$$p_{31} = <e_3, e_1> = e_3\bar{e}_1 = (1-2i).1 = 1-2i$$
$$p_{32} = <e_3, e_2> = e_3\bar{e}_2 = (1-2i)(1-i) = -1-3i$$
$$p_{33} = <e_3, e_3> = e_3\bar{e}_3 = (1-2i)(1+2i) = 5$$

Hence, the required matrix is given by

$$P = \begin{bmatrix} p_{11} & p_{12} & p_{13} \\ p_{21} & p_{22} & p_{23} \\ p_{31} & p_{32} & p_{33} \end{bmatrix} = \begin{bmatrix} 1 & 1-i & 1+2i \\ 1+i & 2 & -1+3i \\ 1-2i & -1-3i & 5 \end{bmatrix}$$

EXAMPLE 11. *Show that the determinant of a unitary operator has absolute value 1.*

SOLUTION. Let T be a unitary operator on a finite dimensional inner product space $V(K)$ and B be an orthonormal basis for V.

If A denote the matrix of T relative to B, then

$$\det(T) = \det(A) = \det(T)_B$$

Since T is unitary, therefore $T^*T = I$

$\Rightarrow$ $[T^*T]_B = [I]_B$

$\Rightarrow$ $[T^*]_B[T]_B = [I]_B$

$\Rightarrow$ $\det[T^*]_B.\det[T]_B = I$

$\Rightarrow$ $\det A^*.\det A = I$

$\Rightarrow$ $(\overline{\det A}).(\det A) = I$

$\Rightarrow$ $|\det A|^2 = I$

$\Rightarrow$ $\det A = I$

$\Rightarrow$ det. A has absolute value 1.

$\Rightarrow$ det. T has absolute value 1.

Example 12. *It T is a self-adjoint linear operator on a finite dimensional inner product space, then show that det.T is real.*

Solution. Let B be an orthonormal basis for an inner product space V such that

$$\dim. V = \text{finite}$$

Then $[T^*]_B = [T]_B^*$

But $T = T^*$

Therefore, $[T]_B^* = [T]_B$

Let $A = [T]_B$, then $A = A^*$

$\Rightarrow$ $\det. A = \det. A^* = \det(\overline{A})$

$\Rightarrow$ $\det. A = \det.\overline{A}$

$\Rightarrow$ det. A is real.

$\Rightarrow$ det. T is real.

Example 13. *Show that the set of all unitary operators on a Hilbert space H is group under the composition of operators.* (GARHWAL–2002, KANPUR–2002, MEERUT–2002)

Solution. Let G be the set of all unitary operators on a Hilbert space H. Let $T, S, F \in G$ are arbitrary, where T, S, F are unitary operators on Hilbert space H. Then

$$TT^* = T^*T = I \text{ and } SS^* = S^*S = I$$

Now, we have to show that TS is unitary operators.

$$(TS)(TS)^* = (TS)(S^*T^*) = T(SS^*)T^*$$
$$= TIT^* = TT^* = I$$
$$(TS)^*(TS) = (S^*T^*)(TS) = S^*(T^*T)S$$
$$= S^*IS = S^*S = I$$
$$(TS)^*(TS) = (TS)(TS)^* = I$$

Hence, TS is a unitary operator.

Therefore $TS \in G$

(i) Closure Property. If $T, S \in G$, then also, $TS \in G$

(ii) Associativity. For associativity we know that

$$(TS)F = T(SF)$$

Now, $[(TS)F](x) = (TS)[F(x)], x \in H$

$$= T[S(F(x))] = T[(SF)(x)]$$
$$[(TS)F](x) = [T(SF)](x) \ \forall \ x \in H$$
$$(TS)F = T(SF)$$

(iii) Existence of identity. There exist an identity element $I \in G$ such that

$$TI = IT = T$$

$$\Rightarrow \quad II^* = I = I^*I$$

(iv) Existence of inverse. If $T \in G$ then $T^{-1} \in G$

We know that $TT^* = T^*T = I$

$$\Rightarrow \quad T^* = T^{-1}$$

and $\quad T^{-1}(T^{-1})^* = T^*(T^*)^{-1} = I$

also $\quad (T^{-1})^*(T^{-1}) = I$

$$\Rightarrow \quad (T^{-1})(T^{-1})^* = I = (T^{-1})^*(T^{-1})$$

Hence, T^{-1} is a unitary operator. If $T \in G$ then also $T^{-1} \in G$.

Hence, we conclude that G is group w.r.t. operation of product of operators.

5.16 PERPENDICULAR PROJECTION

(MEERUT–2006)

Let V be a linear space and M, N be two subspaces of V such that $V = M \oplus N$, *i.e.*, V is direct sum of M and N. Then, $z \in V$ can be expressed as $z = x + y$, $x \in M, y \in N$ uniquely. Then, a projection E on M along N is linear operator if and only if

(i) E is idempotent, *i.e.*, $E^2 = E$.

(ii) $E(z) = x$ and $E(y) = 0$, *i.e.*, Range space of E is M and null space of E is N

$$(R(E) = M, N(E) = N)$$

Further, we have proved that Hilbert space H can be expressed as the direct sum of a subspace M and its orthogonal compliment $M^{\perp}$, *i.e.*, $H = M \oplus M^{\perp}$.

Therefore, $z = x + y, x \in M, y \in N$ so that $(x, y) = 0$

If E is a projection on M along $M^{\perp}$ such that $E(z) = x, E(y) = 0$ and null space of E, *i.e.*, $N(E) = M^{\perp}$. In this case range and null space of E are orthogonal to each other.

Definition. *Let H be a Hilbert space. A projection P on H is said to be a perpendicular projection if its range and null spaces are orthogonal.*

In other words, we can say that a linear operator P is called a perpendicular projection on H if

(i) $P^2 = P$ *(Idempotent)*

(ii) *Range and null spaces of P are orthogonal.*

REMARKS

- Since, in case of perpendicular projection, range and null spaces are orthogonal, so it will sufficient to specify only the range spaces and speak as P is perpendicular projection on M, (It being understood along $M^{\perp}$).
- If P is a projection on M along N, then $I - P$ is a projection on N along M.

THEOREM 1. ***If P is a projection on a Hilbert space H with range M and null space N, then M ⊥ N if and only if P is self-adjoint and in this case $N = M^1$.***

(KANPUR–1982, 84)

PROOF. Here, P is a projection on H, then by definition $P^2 = P$ with range M and null space N.

Therefore, $H = M \oplus N$ such that a vector z in H uniquely expressed as

$$z = x + y; x \in M, y \in N.$$

Also, $Pz = x$ and $Py = 0$

If $M \perp N$, then $(x, y) = 0 = (y, x)\ \forall x \in M$ and $\forall y \in N$.

To show P is self-adjoint, *i.e.*, $p^* = P$.

Consider,

$$\left(P^* z, z\right) = (z, Pz) = (z, x) = (x + y, x)$$
$$= (x, x) + (y, x) = (x, x) + 0 = (x, x) \qquad \ldots(1)$$

and $\quad (Pz, z) = (x, z) = (x, x + y) = (x, x) + (x, y) = (x, x) \qquad \ldots(2)$

Using (1) and (2), we have

$$\left(P^* z, z\right) = (Pz, z)$$

Thus $\quad P^* = P$

$\Rightarrow \quad P$ is self-adjoint.

Conversely, let $P^* = P$ To show that $M \perp N$, *i.e.*, $(x, y) = 0$.

Consider

$$(x, y) = (Px, y) = \left(x, P^* y\right) = (x, Py)$$
$$= (x, 0) = 0$$

$\Rightarrow \quad M \perp N$.

Now, it remains to prove that $N = M^{\perp}$

Let $y \in N$, therefore $(y, x) = 0\ \forall x \in M$

$\Rightarrow \quad y \in M^{\perp}$

$\Rightarrow \quad N \subset M^{\perp}$

Further, if $N \neq M^{\perp}$, then it is a proper subset of $M^{\perp}$ and therefore a closed linear subspace of $M^{\perp}$. Thus, there exists a non-zero vector $z_0 \in M^{\perp}$ such that $z_0 \perp N$.

Now, $z_0 \in M^{\perp} \Rightarrow z_0 \perp M$, we have shown that $z_0 \perp N$.

Therefore, $z_0 \perp M \oplus N$ or $z_0 \perp H$. But, we know that only zero vector is orthogonal to whole of H.

Since $H^{\perp} = \{0\}$, therefore $z_0 = 0$, which is a contradiction because $z_0 \neq 0$.

Therefore, N is a proper subset of $M^{\perp}$ is not possible, which implies $N = M^{\perp}$, *i.e.*, null space of P is orthogonal to range to P so that P is a perpendicular projection.

Remark

- The above theorem can be restated as follows :
  "A linear operator P on a Hilbert space H is a perpendicular projection if and only if $E = E^2 = E^*$

THEOREM 2. ***P is a projection on a closed linear subspace M of H if and only if I – P is a projection on $M^{\perp}$.*** (MEERUT–1998, 99)

PROOF. Let us first suppose P be a projection on H, then be definition

$$P^2 = P \quad \text{and} \quad P^* = P \qquad \ldots(1)$$

Consider

$$(I - P)^* = I^* - P^* = I - P$$

$\Rightarrow$ $I - P$ is self-adjoint.

Also, $(I-P)^2 = (I-P)(I-P) = I - P - P + P^2 = I - P - P + P$
$= I - P$

Thus, $(I - P)$ is also a projection on H. Now, let M be the range of P. To show that range of $I - P$ is $M^{\perp}$.

Let N be the range of $I - P$, then $x \in N$

$\Rightarrow$ $(I-P)x = x$

$\Rightarrow$ $x - Px = x$, *i.e.*, $x \in$ Null space of P

$\Rightarrow$ $Px = 0$

$\Rightarrow$ $x \in M^{\perp}$

$\Rightarrow$ $N \subset M^{\perp}$ (since x is arbitrary) ...(2)

Further, let $x \in M^{\perp}$, which is a null space of P, so that

$Px = 0$

$\Rightarrow$ $x - px = x \Rightarrow (I-P)x = x$

$\Rightarrow$ $x \in$ Range of $(I-P)$

i.e., $x \in N$

Thus, $M^{\perp} \subset N$ ...(3)

From (2) and (3), we have

$N = M^{\perp}$

$\Rightarrow$ Range of $(I-P)$ is $M^{\perp}$.

Conversely, let $(I-P)$ be a projection on $M^{\perp}$, then using first part (just proved above)

$I - (I-P)$ is a projection on $(M^{\perp})^{\perp}$.

Hence, P is a projection on M. Hence, P is a projection on M.

THEOREM 3. ***If P is a perpendicular projection on a closed subspace M of H, then*** $\boldsymbol{x \in M \Leftrightarrow Px = x \Rightarrow \|Px\| = \|x\|}$ (MEERUT–1999)

PROOF. Let P is a perpendicular projection on a closed linear subspace M of H.

Let us first suppose $x \in M$. Let $Px = z$, therefore $x \in R(P)$ which is M and therefore $x \in M$

$\therefore$ $Px = x \Rightarrow x \in M$

Now, let $x \in M$. To show that $Px = x$.

Suppose $Px = y$. So $P(Px) = Py$.

$\Rightarrow$ $P^2x = Py \Rightarrow Px = Py$ $[\because P^2 = P]$

$\Rightarrow$ $P(x-y) = 0 \Rightarrow x - y \in$ Null space of P

$\Rightarrow$ $x - y \in M^{\perp} \Rightarrow x - y = z,\ z \in M^{\perp}$

$\Rightarrow$ $x = y + z, y \in R(P)$, *i.e.*, $Px = y$ and $z \in M^{\perp}$

Since $x = x + 0,\ x \in M$ and $0 \in M^{\perp}$.

Therefore, there are two representations for x, *i.e.*, $x = y + z$ and $x = x + 0$ as the sum of a vectors in M and $M^{\perp}$.

But $H = M \oplus M^{\perp}$.

Thus, above representation must be unique.

Hence, we must have $y = x$ and $z = 0$.

Thus, $Px = y \Rightarrow Px = x$

Also, $Px = x \Leftrightarrow \| Px \| = \| x \|$

Let $Px = x$, then clearly $\| Px \| = \| x \|$

Conversely, let $\| Px \| = \| x \|$. To show that $Px = x$.

Now, $x = Px + (I - P)x = \alpha + \beta$ (say)

where, $\alpha = Px \in R(P) = M$

and $\beta = (I - Px \in R(I - P)$, i.e., $M^{\perp}$

But $\alpha \in M$ and $\beta \in M^{\perp}$

$\Rightarrow$ $(\alpha, \beta) = 0 \Rightarrow (\beta, \alpha) = 0$

Therefore

$$\begin{aligned} \| x \|^2 &= \| \alpha + \beta \|^2 = (\alpha + \beta, \alpha + \beta) \\ &= (\alpha, \alpha) + (\beta, \beta) + (\alpha, \beta) + (\beta, \alpha) \\ &= \| \alpha \|^2 + \| \beta \|^2 + 0 + 0 \\ &= \| Px \|^2 + \| (I - P)x \|^2 \end{aligned}$$

$\Rightarrow$ $0 = \| (I - P)x \|^2$ $[\because \| Px \| = \| x \|]$

Hence $(I - P)x = 0$

$\Rightarrow$ $Ix - Px = 0$

$\Rightarrow$ $x = Px$

Remark

- Perpendicular projections on a Hilbert space H are positive linear operator such that $0 \le P \le 1$ and having the property that

$$\| Pz \| \le \| z \| \ \forall z \in H \text{ and } \| P \| \le 1$$

Theorem 4. ***If P and Q are projection on closed linear subspace M and N of a Hilbert space H, then following are equivalent*** :

(i) $P \le Q$ ***(ii)*** $\|Px\| \le \|Qx\| \ \forall x \in H$

(iii) $M \subset N$ ***(iv)*** $QP = P$

Solution. We have $(Px, x) = (P^2x, x) = (PPx, x)$

$= (Px, P^*x) = (Px, Px)$

$\Rightarrow$ $(Px, x) = \|Px\|^2 \ \forall x \in H$ ...(1)

(i) $\Rightarrow$ (ii)

$$\begin{aligned} P \le Q \quad &\Rightarrow \quad Px \le Qx \\ &\Rightarrow \quad (Px, x) \le (Qx, x) \\ &\Rightarrow \quad \|Px\|^2 \le \|Qx\|^2 \quad \text{(By (1))} \\ &\Rightarrow \quad \|Px\| \le \|Qx\| \end{aligned}$$

(ii) $\Rightarrow$ (iii) For all $x \in M$

$Px = x \Rightarrow \|Px\|^2 = \|x\|^2$

$\Rightarrow \qquad \|x\| = \|Px\| \leq \|Qx\|$

$\Rightarrow \qquad \|x\| \leq \|Qx\|$ ...(2)

But we have $\|Qx\| \leq \|x\|$ ...(3)

Now (2) and (3) implies

$$\|Qx\| = \|x\|$$

$$\Rightarrow \quad Qx = x_1 \quad \Rightarrow \quad x \in N$$

So for any $x \in M \Rightarrow x \in N \Rightarrow M \subset N$

(iii) $\Rightarrow$ (iv) Let $M \subset N$

$Px \in M$ (= range space of P) but $M \subset N$

$\therefore \qquad Px \in$ N and so $Q(Px) \in N$

$\Rightarrow \qquad Q(Px) = Px$

For $\qquad y \in N \Rightarrow Qy = y$

or $\qquad (QP)x = Px \ \forall \ x \in H$

Therefore, $\qquad QP = P$.

(iv) $\Rightarrow$ (v) It is given that $QP = P$

$\Rightarrow \qquad (QP)^* = P^*$

$\Rightarrow \qquad P^*Q^* = P^*$

$\Rightarrow \qquad PQ = P \qquad (\because P$ & Q are projections $\Rightarrow P^* = P$ and $Q^* = Q)$

(v) $\Rightarrow$ (i) It is given that $PQ = P$

Let $x \in H$ then from (1)

$$(Px, x) = \|Px\|^2 = \|PQ(x)\|^2 \qquad (\because PQ = P)$$
$$= \|P(Qx)\|^2$$
$$\leq \|Qx\|^2 = (Qx, x) \qquad \text{(From (1))}$$

$\Rightarrow \qquad (Px, x) \leq (Qx, x)$

$\Rightarrow \qquad P \leq Q$

5.17 INVARIANCE AND REDUCIBILITY

Definition 1. *Let M be a closed linear subspace of a Hilbert space H and T be a linear operator on H, then M is said to be invariant under T if $x \in M$ implies $T(x) \in M$.*

Definition 2. *The operator T on H induces an operator T_M on M defined by*

$$T_M(x) = T(x) \quad \forall x \in M$$

Then, operator T_M is called the restriction of T to M.

Definition 3. *Let H be a Hilbert space and M be a closed linear subspace of H. We know that*

$$H = M \oplus M^{\perp}$$

such that M and $M^{\perp}$ are invariant under T. Then we say that M reduces T, i.e., T is said to be reduced by M if both M and $M^{\perp}$ are invariant under T.

REMARKS

- *M* is invariant under zero operator.
- Every closed linear subspace is invariant under identity operator *I*.

THEOREM 1. ***A closed linear subspace M of a Hilbert space H is invariant under an operator T if and only if $M^{\perp}$ is invariant under T^*.***

(MEERUT–1988, 90, 98, KANPUR–2006)

PROOF. Let us first suppose that *M* is invariant under *T*.

i.e., $x \in M \Rightarrow T(x) \in M$

To show that $M^{\perp}$ is invariant under T^*. For this, we shall prove that

$$y \in M^{\perp} \Rightarrow T^*(y) \in M^{\perp}$$

Let $x \in M$ and $y \in M^{\perp}$, then $(x, y) = 0$.

Now $(x, T^* y) = (Tx, y) = 0$.

Since M is invariant under T, therefore $Tx \in M$ and $y \in M^{\perp}$.

Therefore $(Tx, y) = 0$

$\therefore$ $(x, T^* y) = 0$.

This is true for every $x \in M$ and therefore $T^* y \in M^{\perp}$, $y \in M^{\perp}$.

$\Rightarrow$ $M^{\perp}$ is invariant under T^*.

Conversely, let $M^{\perp}$ is invariant under T^*.

$\Rightarrow$ $(M^{\perp})^{\perp}$ is invariant under $(T^*)^*$ [Using previous case just proved]

But $M^{\perp\perp} = M$ and $T^{**} = T$.

Thus, we conclude that M is invariant under T.

THEOREM 2. ***Let H be a Hilbert space. A closed linear subspace M of H reduces an operator T if and only if M is invariant under both T and T^*.***

PROOF. Let us first suppose that M reduces an operator T. To show that M is invariant under T and T^*.

Since $M^{\perp}$ is invariant under T, then using previous theorem,
$(M^{\perp})^{\perp}$ is invariant under T^*.

$\Rightarrow$ M is invariant under T^*.

Hence, M is invariant under T and T^* both.

Conversely, let M is invariant under T and T^* both, i.e., M is invariant under T.

Now, M is invariant under T^*.

$\Rightarrow$ $M^{\perp}$ is invariant under $(T^*)^* = T$.

$\Rightarrow$ Both M and $M^{\perp}$ are invariant under T. Hence, M reduces an operator T.

THEOREM 3. ***If P is a projection on a closed linear subspace M of a Hilbert space H and T be a linear operator on H, then M is invariant under T if and only if TP = PTP.*** (KANPUR–1993, 96, MEERUT–1987)

PROOF. Let us first suppose that M is invariant under T so that

$$x \in M \Rightarrow Tx \in M \qquad \ldots(1)$$

Now $PTP(z) = PT(x) = Tx$ ...(2)

and $TP(z) = Tx$ ...(3)

Using (2) and (3), we get

$$PTP(z) = TP(z) \quad \forall z \in H$$

Thus, $PTP = TP$ ...(4)

Conversely, let $PTP = TP$

If $x \in M$ then $Px = x$ and

$TP(x) = PT(Px)$ [Using (4)]

$\Rightarrow$ $Tx = PTx$ or $P(Tx) = Tx$

$\Rightarrow$ $Tx \in M$ [$\because$ P is a projection on M]

Since $x \in M \Rightarrow Tx \in M$, Hence, M is invariant under T.

THEOREM 4. ***If P is a projection on a closed linear subspace M of a Hilbert space H, then M reduces an operator T if and only if TP = PT.***

(KANPUR–1992, 93, MEERUT–1986, 87)

PROOF. Let us first suppose that M reduces operator T, which implies M is invariant under both T and T^*.

$$\Rightarrow \quad TP = PTP \quad \text{and} \quad T^* P = PT^* P$$

$$\Rightarrow \quad TP = PTP \quad \text{and} \quad (T^* P)^* = (PT^* P)^*$$

$$\Rightarrow \quad TP = PTP \quad \text{and} \quad P^* T^{**} = P^* T^{**} P^*$$

$$\Rightarrow \quad TP = PTP \quad \text{and} \quad PT = PTP \qquad [\because P^* = P \text{ and } T^{**} = T]$$

$$\Rightarrow \quad TP = PT$$

Conversely, let $TP = PT$

$$\Rightarrow \quad TP^2 = PTP \quad \text{and} \quad PTP = P^2T$$

$$\therefore \quad TP = PTP \quad \text{and} \quad PTP = PT$$

$$\Rightarrow \quad (PTP)^* = (PT)^*$$

$$\Rightarrow \quad P^* T^* P^* = T^* P^*$$

$$\Rightarrow \quad PT^* P = T^* P \qquad [\because P^* = P]$$

$\Rightarrow$ M is invariant under both T and T^*. Hence, M reduces operator T.

THEOREM 5. ***If P_1 and P_2 are perpendicular projection on a closed linear subspace M_1 and M_2 respectively of a Hilbert space H, then M_1 and M_2 are orthogonal if and only if $P_1P_2 = 0 \Leftrightarrow P_2P_1 = 0$***

(MEERUT–1980, 2000, KANPUR–1991, 96)

PROOF. We know that if $P_1P_2 = 0$

Then, $(P_1P_2)^* = 0^* \Rightarrow P_2^* P_1^* = 0 \Rightarrow P_2P_1 = 0$.

Therefore, if is sufficient to prove that $P_1P_2 = 0$. Let $z = x_1 + y_1$, be any point corresponding to P_2, a projection on M_2 so that $P_2z = x_2$

Consider

$$(P_1P_2z, z) = (P_2z, P_1^* z)$$

$$= (P_2z, P_1z) \qquad [\because P_1^* = P_1]$$

$$= (x_2, x_1), \qquad [x_2 \in M_2, x_1 \in M_1]$$

$$= 0 \qquad [M_1 \perp M_2]$$

$$\Rightarrow \quad (P_1P_2z, z) = 0$$

$\therefore$ $P_1P_2 = 0$ because $(Tx, x) = 0 \forall x \Rightarrow T = 0$.

Conversely, let $P_1P_2 = 0$. To show that M_1 and M_2 are orthogonal, let $x_1 \in M_1$ and $x_2 \in M_2$ so that $P_1x_1 = x_1$ and $P_2x_2 = x_2$

Therefore,

$$(x_1, x_2) = (P_1x_1, P_2x_2) = (x_1, P_1^* P_2 x)$$

$$= (x_1, P_1P_2x_2) = (x_1, 0.x_2) = (x_1, 0) = 0$$

Hence, M_1 and M_2 are orthogonal.

5.18 ORTHOGONAL PROJECTIONS

Definition. *Two projections P_1 and P_2 are said to be orthogonal if $P_1P_2 = P_2P_1 = 0$.*

THEOREM 1. ***If $P_1, P_2, ..., P_n$ are perpendicular projections on closed linear subspace $M_1, M_2, ..., M_n$ of Hilbert space H, then their sum $P = P_1 + P_2 + ... + P_n$ is a perpendicular projection if and only if $P_i^{'s}$ are pair wise orthogonal, i.e., iff $P_iP_j = 0 \cdot i \pm j$ and in the case P is a projection on M, where***

$$M = M_1 + M_2 + ... + M_n \qquad \text{(KANPUR–1996)}$$

PROOF. Let us first suppose that P_i 's are perpendicular projections which are pair wise orthogonal.

i.e., $\quad P_i^2 = P_i = P_i^*$ and $P_iP_j = 0, \quad i \neq j$

To show $P = P_1 + P_2 + ... + P_n$ is a perpendicular projection

$$P^* = (P_1 + P_2 + ... + P_n)^* = P_1^* + P_2^* + ... + P_n^*$$

$$P^* = P_1 + P_2 + ... + P_n = P \qquad [\because P_i^* = P_i, \ \forall i]$$

$\Rightarrow \quad P$ is self-adjoint.

Also, $\quad P^2 = (P_1 + P_2 + ... + P_n)^2$

$$= P_1^2 + P_2^2 + ... + P_n^2 + 2\Sigma P_i P_j$$

$$= P_1 + P_2 + ... + P_n + 0 \qquad \text{[Using given conditions]}$$

$\therefore \quad P^2 = P$.

Hence, P is idempotent and self-adjoint.

$\Rightarrow \quad P$ is a perpendicular projection.

Conversely, let P_i's are perpendicular projections and $P = P_1 + P_2 + ... + P_n$ is also a perpendicular projections.

To show, P_i's are pairwise orthogonal, *i.e.*, $P_iP_j = 0, i \neq j$.

We have already proved the following results.

(i) $\| Pz \| \leq \| z \|, \forall z$

(ii) $\| Px \| = \| x \|$, if x belongs to range of P.

(iii) If P_1 and P_2 be perpendicular projections on M_1 and M_2, then

$$M_1 \perp M_2 \Leftrightarrow P_1P_2 = 0 = P_2P_1$$

(iv) $\quad (Pz, z) = \| Pz \|^2 \ \forall z$

$$\because \quad (Pz, z) = (P^2 z, z) = (Pz, P^* z) = \| Pz \|^2$$

Now, let x be any vector belong to M_t, the range of P_t.

Therefore, $\quad \| x \|^2 = \| P_i x \|^2 \qquad$ (By (ii))

Now, $\quad \| x \|^2 = \| P_i x \|^2 \leq \sum_{j=1}^{n} \| P_j x \|^2 = \sum_{j=1}^{n} (P_j x, x)$

$$= (P_1 x, x) + (P_2 x, x) + ...$$

$$= ((P_1 + P_2 + ...)x, x) = (Px, x) = \| Px \|^2$$

But by (i) $\| P_i x \|^2 \le \| x \|^2$, x does not belong to the range of P.

Therefore, $\| x \|^2 \le \| P_i x \|^2 \le \| x \|^2$.

This result is true if and only if the sign of equality hold throughout.

Therefore, $\left\| P_j x \right\|^2 = \sum_{j=1}^{n} \left\| P_j x \right\|^2$. This will hold good if and only if

$$\left\| P_j x \right\| = 0 \text{ for } j \ne i$$

$$P_j x = 0, \ \forall j \ne i$$

$\Rightarrow$ x belongs to the null space of P_j whose range space is M_j and therefore, null space is $M_j^{\perp}$.

$\therefore$ $x \in M_i \Rightarrow x \in M_j^{\perp}, j \ne i$

$\Rightarrow$ $M_i \perp M_j \ \ \forall j \ne i$.

Now, P_1 and P_2 are perpendicular projections whose ranges M_i and M_j are orthogonal and therefore, these projections are also orthogonal. Thus, using result (iii), we have $P_i P_j = 0$, $i \pm j$, *i.e.*, they are pair wise orthogonal.

Further, we shall prove that range of P, *i.e.*, $R(P)$ is M, where $M = M_1 + M_2 + \ldots + M_n$.

For this, first we shall prove that

$$R(P) \subset M$$

Let $x \in R(P)$

$\Rightarrow$ $x = Px$

$\Rightarrow$ $x = (P_1 + P_2 + \ldots + P_n)x$

$= P_1 x + P_2 x + \ldots + P_n x$

$\Rightarrow$ $P_i x \in$ Range of P_t, *i.e.*, M_t

$\Rightarrow$ $x \in R(P) \Rightarrow x \in M_1 + M_2 + \ldots M_n \Rightarrow x \in M$.

$\Rightarrow$ $R(P) \subset M$ (Since x is arbitrary)

Now, we shall prove that $M \subseteq R(P)$. Let $x \in$ Range of P_t, *i.e.*, M_t.

Then $P_i x = x$.

$\Rightarrow$ $\|P_i x\| = \|x\|$

Again, $P_i x = 0$, $j \ne i$, because $x \in M_t \Rightarrow x \in M_j^{\perp}$.

Since $M_i \perp M_j$ and $M_j^{\perp}$ is null space of P_j.

Therefore $P_j x = 0$

$\Rightarrow$ $Px = (P_1 + P_2 + \ldots + P_n)x = P_1 x + P_2 x + \ldots + P_n x = P_t x$

Also, $\| P_i x \| = \| x \|$ or $\| Px \| = \| x \| \quad \forall x \in M_t$

$\Rightarrow$ $x \in R(P)$

$\therefore$ $x \in M_t \Rightarrow x \in R(P)$

$M_t \subseteq R(P)$

Hence, $M =$ Range of P.

Theorem 2. ***If P_1 and P_2 are the projections on closed linear subspaces M_1 and M_2, then P_1P_2 is a projection, if and only if $P_1P_2 = P_2P_1$. In this case P_1P_2 is a projection on $M_1 \cap M_2$.*** (MEERUT–1982)

Proof. Let H be a Hilbert space and P_1, P_2 are the projections on H, therefore, by definition

$$\left.\begin{aligned} P_1^2 = P_1, P_1^* = P_1 \\ P_2^2 = P_2, P_2^* = P_2 \end{aligned}\right\} \quad \ldots(1)$$

It is also given that M_1 is the range of P_1 and M_2 is the range of P_2.

Let us suppose that P_1P_2 is a projection on H. To show $P_1P_2 = P_2P_1$

Since, P_1P_2 is a projection on H

$$(P_1P_2)^* = P_1P_2$$

$$\therefore \quad P_2^* P_1^* = P_1P_2 \Rightarrow P_2P_1 = P_1P_2 \quad \text{[Using (1)]}$$

Conversely, let us suppose that

$$P_1P_2 = P_2P_1$$

To show P_1P_2 is a projection on H

Consider

$$(P_1P_2)^* = P_2^* P_1^* = P_2P_1 = P_1P_2$$

and

$$(P_1P_2)^2 = (P_1P_2)(P_1P_2) = (P_1P_2)(P_2P_1)$$
$$= P_1P_2^2P_1 = P_1P_2P_1 = P_2P_1P_1 = P_2P_1^2 = P_2P_1 = P_1P_2$$

Therefore, P_1P_2 is a projection on H.

Now, it remains to prove that P_1P_2 is a projection on $M_1 \cap M_2$. For this, we shall prove that the range of P_1P_2 is $M_1 \cap M_2$.

Let $R(P_1P_2)$ denote the range of P_1P_2.

Let $x \in M_1 \cap M_2$ Then $x \in M_1$ and $x \in M_2$

We have

$$(P_1P_2)x = P_1(P_2x) = P_1x$$

$$[\because M_2 \text{ is the range of } P_2 \text{ and } x \in M_2 \Rightarrow P_2x = x]$$

$$= x$$

$$\therefore \quad (P_1P_2)x = x \quad \Rightarrow \quad x \in R(P_1P_2)$$

i.e.,

$$x \in M_1 \cap M_2 \quad \Rightarrow \quad x \in R(P_1P_2)$$

$$\Rightarrow \quad M_1 \cap M_2 \subset R(P_1P_2) \quad \Rightarrow \quad P_1[(P_1P_2)x] = P_1x$$

$$\Rightarrow \quad [P_1(P_1P_2)]x = P_1x \quad \Rightarrow \quad [P_1(P_1P_2)]x = P_1x$$

$$\Rightarrow \quad (P_1^2P_2)x = P_1x$$

$$\Rightarrow \quad (P_1P_2)x = P_1x \quad \text{but} \quad (P_1P_2)x = x$$

Thus, we have $\quad P_1x = x$

$$\Rightarrow \quad x \in M_1, \text{ i.e., } R(P_1)$$

Again $\quad P_1P_2 = P_2P_1$

$$\Rightarrow \quad x \in R(P_1P_2) \quad \Rightarrow \quad (P_1P_2)x = x \Rightarrow (P_1P_2)x = x$$

$$\Rightarrow \quad P_2[(P_2P_1)x] = P_2x \quad \Rightarrow (P_2^2P_1)x = P_2x$$

$$\Rightarrow \quad (P_1P_2)x = P_2x$$

But we have $(P_2P_1)x = x$

Thus, we have

$$P_2x = x \Rightarrow x \in M_2$$

$$\therefore \quad x \in R(P_1P_2) \Rightarrow x \in M_1, x \in M_2 \Rightarrow x \in M_1 \cap M_2$$

$$\Rightarrow \quad R(P_1P_2) \subset M_1 \cap M_2$$

Hence, $\quad R(P_1P_2) = M_1 \cap M_2$.

SOLVED EXAMPLES

EXAMPLE 1. *Let W be a subspace of an inner product space V spanned by* {(0, 1, 1, 0), (0, 5, –3, –2), (–3, –3, 5, –7)}. *Find an orthonormal basis for W.*

SOLUTION. Let $\alpha_1 = (0, 1, 1, 0)$; $\alpha_2 = (0, 5, -3, -2)$; $\alpha_3 = (-3, -3, 5, -7)$

Let us suppose $\quad \beta_1 = \alpha = (0, 1, 1, 0)$

Then $\quad \|\beta_1\|^2 = 0^2 + 1^2 + 1^2 + 0^2 = 2$

Further let
$$\beta_2 = \alpha_2 - \frac{\langle \alpha_2, \beta_1 \rangle}{\|\beta_1\|^2}\beta_1$$

$$= (0,5,-3,-2) - \frac{\langle (0,5,-3,-2),(0,1,1,0) \rangle}{2}(0,1,1,0)$$

$$= (0,5,-3,-2) - \frac{(0+5-3-0)}{2}(0,1,1,0)$$

$$= (0,5,-3,-2) - (0,1,1,0) = (0,4,-4,-2)$$

Then $\quad \|\beta_2\|^2 = 0^2 + 4^2 + (-4)^2 + (-2)^2 = 36$

Also, let
$$\beta_3 = \alpha_3 - \frac{\langle \alpha_3, \beta_1 \rangle}{\|\beta_1\|^2}\beta_1 - \frac{\langle \alpha_3, \beta_2 \rangle}{\|\beta_2\|^2}\beta_2$$

$$= (-3,-3,5,-7) - \frac{\langle (-3,-3,5,-7),(0,1,1,0) \rangle}{2}(0,1,1,0) - \frac{\langle (-3,-3,5,-7),(0,4,-4,-2) \rangle}{36}(0,4,-4,-2)$$

$$= (-3,-3,5,-7) - \frac{(-0-3+5+0)}{2}(0,1,1,0) - \frac{(0-12-20+14)}{36}(0,4,-4,-2)$$

$$= (-3,-3,5,-7) - (0,1,1,0) + \frac{1}{2}(0,4,-4,-2)$$

$$= (-3,-3,5,-7) - (0,1,1,0) + (0,2,-2,-1) = (-3,-2,2,-8)$$

Then $\quad \|\beta_3\|^2 = (-3)^2 + (-2)^2 + 2^2 + (-8)^2 = 81$

Hence, the required orthonormal basis for W is

$$B = \left\{ \frac{\beta_1}{\|\beta_1\|}, \frac{\beta_2}{\|\beta_2\|}, \frac{\beta_3}{\|\beta_3\|} \right\}$$

$$= \left\{ \frac{(0,1,1,0)}{\sqrt{2}}, \frac{(0,4,-4,-2)}{\sqrt{36}}, \frac{(-3,-2,2,-8)}{\sqrt{81}} \right\}$$

$$= \left\{ \frac{(0,1,1,0)}{\sqrt{2}}, \frac{(0,4,-4,-2)}{6}, \frac{(-3,-2,2,-8)}{9} \right\}$$

$$= \left\{ \frac{1}{\sqrt{2}}(0,1,1,0), \frac{1}{3}(0,2,-2,-1), \frac{1}{9}(-3,-2,2,-8) \right\}$$

Example 2. *Convert the linear basis {(1, 0, 0), (1, 1, 0), (1, 1, 1)} of Hilbert space R^3 into an orthonormal basis.* (KANPUR–2003)

Solution. Using Gram Schmidt Orthogeneralization process define inner product of two vectors

$$<u, v> = <v, u> = a_1b_1 + a_2b_2 + a_3b_3$$

$$< u, v >= \|u\|^2 = a_1^2 + a_2^2 + a_3^2$$

Let $u_1 = (1, 0, 0)$ $\quad \|u_1\|^2 = 1^2 + 0^2 + 0^2 = 1$

$u_2 = (1, 1, 0)$ $\quad \|u_2\|^2 = 1^2 + 1^2 + 0^2 = 2$

$u_3 = (1, 1, 1)$ $\quad \|u_3\|^2 = 1^2 + 1^2 + 1^2 = 3$

$$<u_2, u_1> = (1)(1) + (0)(1) + (0)(0) = 1$$

$$<u_3, u_1> = (1)(1) + (0)(1) + (0)(1) = 1$$

$$<u_2, u_3> = (1)(1) + (1)(1) + (0)(1) = 2$$

If $$v_1 = u_1, v_2 = u_2 - \frac{< u_2, v_1 > v_1}{\|v_1\|^2} \quad \ldots(1)$$

$$v_3 = u_3 - \frac{< u_3, v_1 > v_1}{\|v_1\|^2} - \frac{< u_3, v_2 > v_2}{\|v_2\|^2} \quad \ldots(2)$$

$$v_2 = u_2 - \frac{(1)u_1}{1} = u_2 - u_1 = (1,1,0) - (1,0,0)$$

$\Rightarrow$ $\quad v_2 = (0, 1, 0)$ and $\|v_2\| = 1$

$$<u_3, v_1> = <u_3, u_1> = 1,$$

$$<u_3, v_2> = (1)(0) + (1)(1) + (1)(0) = 1$$

Putting values in (2)

$$v_3 = u_3 - u_1 - \frac{v_2}{1} = u_3 - u_1 - (0,1,0) = (0,0,1)$$

So $\quad v_1 = (1, 0, 0), v_2 = (0, 1, 0), v_3 = (0, 0, 1)$

$\Rightarrow$ $\quad \|v_1\| = \|v_2\| = \|v_3\| = 1$

Let us write $$w_1 = \frac{v_1}{\|v_1\|}, w_2 = \frac{v_2}{\|v_2\|}, w_3 = \frac{v_3}{\|v_3\|}$$

Required set is $\{w_1, w_2, w_3\} = \{(1, 0, 0), (0, 1, 0), (0, 0, 1)\}$

Example 3. *Let P_2 be a family of polynomials of degree 2 at most. Define an inner product on P_2 as*

$$\langle f(x), g(x) \rangle = \int_0^1 f(x)g(x)dx$$

Let $\{1, x, x^2\}$ be a basis of the inner product space P_2. Find out an orthonormal basis from this basis.

SOLUTION. Let $\{u_1, u_2, u_3\}$ be a basis of P_2, where $u_1 = 1, u_2 = x, u_3 = x^2$. From this, we have

$$\|u_1\|^2 = < u_1, u_1 > = \int_0^1 1 \times 1 dx = 1,$$
$$\|u_2\|^2 = < u_2, u_2 > = \int_0^1 x\, x dx = 1/3,$$
$$\|u_3\|^2 = < u_3, u_3 > = \int_0^1 x^2\, x^2 dx = 1/5,$$
$$< u_2, u_1 > = \int_0^1 x.1 dx = 1/2,$$
$$< u_3, u_1 > = \int_0^1 x^2.1 dx = 1/3,$$

and $$< u_3, u_2 > = \int_0^1 x^2.x dx = 1/4.$$

Write $$u_1 = v_1, v_2 = u_2 - \frac{\langle u_2, v_1\rangle v_1}{\|v_1\|^2} = x - \frac{1\times 1}{2\times 1} = x - \frac{1}{2}$$

$$v_3 = u_3 - \frac{\langle u_3, v_1\rangle v_1}{\|v_1\|^2} - \frac{\langle u_3, v_2\rangle v_2}{\|v_2\|^2}$$
$$= x^2 - \frac{1\times 1}{3\times 1} - \left[\left(x - \frac{1}{2}\right)\int_0^1 x^2\left(x - \frac{1}{2}\right)dx\right] \Big/ \int_0^1 \left(x - \frac{1}{2}\right)^2 dx$$
$$= x^2 - \frac{1}{3} - \left(x - \frac{1}{2}\right)\frac{1}{12} \Big/ \frac{1}{12}$$
$$= x^2 - \frac{1}{3} - \left(x - \frac{1}{2}\right) = x^2 - x + \frac{1}{6}$$

$$\|v_1\| = \|u_1\| = 1$$
$$\|v_2\|^2 = \int_0^1 \left(x - \frac{1}{2}\right)\left(x - \frac{1}{2}\right) dx = \frac{1}{12}$$
$$\|v_3\|^2 = \int_0^1 \left(x^2 - x + \frac{1}{6}\right)\left(x^2 - x + \frac{1}{6}\right) dx$$
$$= \left(x^4 + x^2 - 2x^3 + \frac{1}{36} + \frac{x^2}{3} - \frac{x}{3}\right)_{x=0}^{x=1}$$
$$= \frac{1}{5} + \frac{1}{3} - \frac{2}{4} + \frac{1}{36} + \frac{1}{9} - \frac{1}{6} = \frac{1}{180}$$

So, $$w_1 = \frac{v_1}{\|v_1\|}, w_2 = \frac{v_2}{\|v_2\|}, w_3 = \frac{v_3}{\|v_3\|}$$

Then $$\{w_1, w_2, w_3\} = \left\{1, \sqrt{(12)}\left(x - \frac{1}{2}\right), \left(x^2 - x + \frac{1}{6}\right)\sqrt{(180)}\right\}$$
$$= \left\{1, (2x-1)\sqrt{3}, \left(x^2 - x + \frac{1}{6}\right)3\sqrt{(20)}\right\}$$

is the required set.

EXAMPLE 4. *Apply the Gram-schmidt process to the vectors $u_1 = (1, 0, 1)$, $u_2 = (1, 0, -1)$, $u_3 = (0, 3, 4)$ to obtain an orthonormal basis for $R^3(R)$ with the standard inner product.*

SOLUTION. If $u = (a_1, a_2, a_3), (b_1, b_2, b_3) \in R^3$, then by definition of inner product say

$$<u, v> = a_1b_1 + a_2b_2 + a_3b_3$$

$$\|u_1\|^2 = <u_1, u_1> = a_1^2 + a_2^2 + a_3^2$$

$$\|u_1\|^2 = <u_1, u_1> = 1 \times 1 + 0 \times 0 + 1 \times 1 = 2$$

$$\|u_2\|^2 = <u_2, u_2> = 1 \times 1 + 0 \times 0 + (-1) \times (-1) = 1 + 1 = 2$$

$$\|u_3\|^2 = 0 \times 0 + 3 \times 3 + 4 \times 4 = 25$$

$$<u_2, u_1> = 1 \times 1 + 0 \times 0 + (-1) \times 1 = 0$$

$$<u_3, u_1> = 0 \times 1 + 3 \times 0 + 4 \times 1 = 4$$

and $$<u_3, u_2> = 0 \times 1 + 3 \times 0 + 4 \times (-1) = -4$$

Set $$v_1 = u_1, v_2 = u_2 - \frac{<u_2, v_1> v_1}{\|v_1\|^2}$$

$$v_3 = u_3 - \frac{<u_3, v_1> v_1}{\|v_1\|^2} - \frac{<u_3, v_2> v_2}{\|v_2\|^2}$$

Then $$v_2 = u_2 - \frac{<u_2, u_1> u_1}{\|u_1\|^2} = u_2 - \frac{0u_1}{2} = u_2$$

$\therefore$ $$v_2 = u_2$$

$$v_3 = u_3 - \frac{<u_3, u_1> u_1}{\|u_1\|^2} - \frac{<u_3, u_2> u_2}{\|u_2\|^2}$$

$$= u_3 - \frac{4u_1}{2} - \frac{(-4)u_2}{2} = u_3 - 2u_1 + 2u_2$$

$$= (0, 3, 4) - 2(1, 0, 1) + 2(1, 0, -1)$$

$$= (0 - 2 + 2, 3 - 0 + 0, 4 - 2 - 2) = (0, 3, 0)$$

Thus, $$v_1 = u_1, v_2 = u_2, v_3 = (0, 3, 0), \|v_3\| = 3$$

Set $$w_1 = \frac{v_1}{\|v_1\|}, w_2 = \frac{v_2}{\|v_2\|}, w_3 = \frac{v_3}{\|v_3\|}$$

Then $\{w_1, w_2, w_3\} = \left\{\left(\frac{1}{\sqrt{2}}, 0, \frac{1}{\sqrt{2}}\right), \left(\frac{1}{\sqrt{2}}, 0, -\frac{1}{\sqrt{2}}\right), (0, 1, 0)\right\}$ is the required set.

EXAMPLE 5. *Orthonormalize the set of linearly independent vectors.*

$$\{(1, 0, 1, 1), (-1, 0, -1, 1), (0, -1, 1, 1)\} \text{ of } V_4.$$

SOLUTION. Given $\alpha_1 = (1, 0, 1, 1), \alpha_2 = (-1, 0, -1, 1), \alpha_3 = (0, -1, 1, 1)$

Let $\beta_1 = \alpha_1 = (1, 0, 1, 1)$, then $\|\beta_1\|^2 = 1^2 + 0^2 + 1^2 + 1^2 = 3$

$$\beta_2 = \alpha_2 - \frac{\langle \alpha_2, \beta_1 \rangle}{\|\beta_1\|^2} \beta_1$$

$$= (-1,0,-1,1) - \frac{\langle(-1,0,-1,1),(1,0,1,1)\rangle}{3}(1,0,1,1)$$

$$= (-1,0,-1,1) - \frac{(-1+0-1+1)}{3}(1,0,1,1)$$

$$= (-1,0,-1,1) - \left(\frac{1}{3},\frac{0}{3},\frac{1}{3},\frac{1}{3}\right)$$

$$= \left(-1+\frac{1}{3},0+\frac{0}{3},-1+\frac{1}{3},1+\frac{1}{3}\right) = \left(-\frac{2}{3},0,-\frac{2}{3},\frac{4}{3}\right)$$

$$\|\beta_2\|^2 = \frac{4}{9}+0+\frac{4}{9}+\frac{16}{9} = \frac{24}{9} = \frac{8}{3}$$

$$\beta_3 = \alpha_3 - \frac{\langle\alpha_3,\beta_1\rangle}{\|\beta_1\|^2}\beta_1 - \frac{\langle\alpha_3,\beta_2\rangle}{\|\beta_2\|^2}\beta_2$$

$$= (0,-1,1,1) - \frac{\langle(0,-1,1,1),(1,0,1,1)\rangle}{3}(1,0,1,1)$$

$$- \frac{\left\langle(0,-1,1,1),\left(-\frac{2}{3},0,\frac{-2}{3},\frac{4}{3}\right)\right\rangle}{8/3}\left(-\frac{2}{3},0,\frac{-2}{3},\frac{4}{3}\right)$$

$$= (0,-1,1,1) - \frac{0+0+1+1}{3}(1,0,1,1)$$

$$- \frac{0+0+\frac{-2}{3}+\frac{4}{3}}{8/3}\left(-\frac{2}{3},0,\frac{-2}{3},\frac{4}{3}\right)$$

$$= (0,-1,1,1) - \left(\frac{2}{3},0,\frac{2}{3},\frac{2}{3}\right) - \frac{2}{3}\times\frac{3}{8}\left(-\frac{2}{3},0,-\frac{2}{3},\frac{4}{3}\right)$$

$$= (0,-1,1,1) - \left(\frac{2}{3},0,\frac{2}{3},\frac{2}{3}\right) - \left(-\frac{1}{6},0,-\frac{1}{6},\frac{1}{3}\right)$$

$$= \left(-\frac{3}{6},-1,-\frac{3}{6},0\right) = \left(-\frac{1}{2},-1,-\frac{1}{2},0\right)$$

Thus, $\{\beta_i\}$, where

$$\beta_1 = (1,0,1,1), \|\beta_1\| = \sqrt{3}$$

$$\beta_2 = \left(-\frac{2}{3},0,\frac{-2}{3},\frac{4}{3}\right), \|\beta_2\| = \frac{2\sqrt{2}}{\sqrt{3}}$$

$$\beta_3 = \left(-\frac{1}{2},-1,-\frac{1}{2},0\right), \|\beta_3\| = \frac{\sqrt{6}}{2}$$

is an orthonormal set.

Hence, the required orthonormal set $\{\gamma_i\}$ is given by

$$\gamma_1 = \frac{\beta_1}{\|\beta_1\|} = \frac{1}{\sqrt{3}}(1,0,1,1) = \left(\frac{1}{\sqrt{3}},0,\frac{1}{\sqrt{3}},\frac{1}{\sqrt{3}}\right)$$

$$\gamma_2 = \frac{\beta_2}{\|\beta_2\|} = \frac{3\sqrt{2}}{2\sqrt{2}}\left(-\frac{2}{3}, 0, -\frac{2}{3}, \frac{4}{3}\right) = \left(-\frac{1}{\sqrt{6}}, 0, -\frac{1}{\sqrt{6}}, \frac{2}{\sqrt{6}}\right)$$

$$\gamma_3 = \frac{\beta_3}{\|\beta_3\|} = \frac{2}{\sqrt{6}}\left(-\frac{1}{2}, -1, \frac{1}{2}, 0\right) = \left(-\frac{1}{\sqrt{6}}, -\frac{2}{\sqrt{6}}, \frac{1}{\sqrt{6}}, 0\right)$$

5.19 WEAK AND STRONG CONVERGENCE

(KANPUR–1990, 92, 94,2001, 07, AVADH–2009, MEERUT–2003, 04)

Definition 1. *Let N be a normed linear space. The sequence $<f_n>$ of N is said to converge weakly to $f \in N$, if for every $\phi \in N^*$ (dual space of N)*

$$\phi(fn) \to \phi(f) \text{ as } n \to \infty$$

It is written as $f_n \xrightarrow{w} f$.

Also, f is said to be the weak limit of the sequence $< f_n >$.

Definition 2. *The convergence with respect to the norm is called strong convergence. Therefore, the sequence $< f_n >$ converges strongly to f if*

$$\| f_n - f \| \to 0 \text{ as } n \to \infty \qquad \text{(KANPUR–1990, 91, 94)}$$

THEOREM 1. ***Weak limit of a sequence is unique.***

PROOF. Let $< f_n >$ be the given sequence. Let, if possible, f_0 and f be two weak limit of $< f_n >$.

i.e., $f_n \xrightarrow{w} f_0$ and $f_n \xrightarrow{w} f$

$\Rightarrow \quad \phi(f_n) \to \phi(f_0)$ and $\phi(f_n) \to \phi(f)$

where ϕ is an arbitrary operator in N^*,

Then, $\quad \phi(f_0) = \phi(f)$, *i.e.*, $\phi(f_0 - f) = 0$

If we select a ϕ_0 with $\| \phi_0 \| = 1$ and $\phi_0(f_0 - f) = \| f_0 - f \|$

Then, we have $\quad \| f - f_0 \| = 0$

$\Rightarrow \quad f = f_0$.

Hence, weak limit of the sequence is always unique.

REMARKS

- If $f_n \xrightarrow{w} f_0$, then every arbitrary subsequence $< f_{n_k} >$ converges weakly to f_0.
- If $f_n \xrightarrow{w} f$ and $g_n \xrightarrow{w} g$, then $f_n + g_n \xrightarrow{w} f + g$, and
- If c is any scalar, then $cf_n \xrightarrow{w} cf$

THEOREM 2. ***Strong convergence implies weak convergence.*** (KANPUR–1991, 93, 95)

PROOF. Let us suppose the sequence $< f_n >$ converges strongly to *f*, *i.e.*,

$$\| f_n - f \| \to 0 \text{ as } n \to \infty$$

To show $\quad f_n \xrightarrow{w} f$

Consider $\| \phi(f_n) \to \phi(f) \| \le \| \phi \| . \| f_n - f \| \to 0$ as $n \to \infty$.

Therefore, $\quad \phi(f_n) \to \phi(f)$

$\Rightarrow \quad f_n \xrightarrow{w} f$

REMARK

- Converse of the above theorem is not necessarily true.

THEOREM 3. ***In a finite dimensional normed linear space, the notions of weak and strong convergence are equal.*** (KANPUR–1992, 94, 95)

PROOF. We have already proved that strong convergence of a sequence in a finite dimensional space implies weak convergence. Therefore, here it remains to prove that weak convergence implies strong convergence to the same element.

Let N be a finite dimensional space and $< f_k >$ is a sequence such that $f_k \xrightarrow{w} f$. Since N is finite dimensional, then there exists a finite system of linearly independent element $g_1, g_2, ..., g_n$ such that every $f \in N$ can be written as

$$f = a_1 g_1 + a_2 g_2 + ... + a_n g_n = \sum_{1}^{n} a_k g_k \text{ for scalers } a_k, k = 1, 2, ... n$$

Then, we can write

$$f_k = a_{k_1} g_1 + a_{k_2} g_2 + ... + a_{k_n} g_n$$

and $$f = a_1 g_1 + a_2 g_2 + ... + a_n g_n$$

Now, consider the functional $\phi_i \in N^*$ such that $\phi_i(g_i) = 1$ and $\phi_i(g_k) = 0$ for $k \neq i$

Then $\phi_i(f_k) = a_{k_i}$ and $\phi_i(f) = a_i$

Since, the sequence $\phi(f_k) \to \phi(f)$ for every linear functional ϕ. Thus

$$\phi_i(f_k) \to \phi_i(f) \quad \Rightarrow \quad a_{k_i} \to a_i \text{ for } i \in N$$

Let $$M = \max \|g_k\|$$

Therefore, for any given $\in > 0$, there exists n_0 such that

$$\left| a_{k_i} - a \right| < \frac{\in}{M.n} \quad \forall i \in N \text{ and } k \geq n_0 .$$

So,
$$\| f_k - f \| = \left\| \sum_{i=1}^{n} (a_{k_i} - a_i) g_i \right\|$$
$$\leq \sum_{i=1}^{n} \left\| a_{k_i} - a_i \right\| . \| g_i \| < \in$$

Hence, the sequence $< f_k >$ converges strongly to f.

REMARK

- If a sequence $<f_n>$ converges weakly to f, then there exists a sequence $\left\langle \sum_{i=1}^{m} a_i f_i \right\rangle$ of linear combination, which converges strongly to f.

5.20 CONVERGENCE OF SEQUENCE OF OPERATORS

(1) Strong Convergence (KANPUR–2007)

The sequence $< T_n > \in B(N, N')$ is said to converge strongly to T if for any $f \in N$

$$T_n(f) \to T(f)$$

It is written as $T_n \xrightarrow{s} T$

(2) Uniform Convergence (KANPUR–1995, 2001)

A sequence $< T_n >$ in $B(N, N')$ is said to converge uniformly to T if for any given $\in > 0$, there exists some integer $N(\in)$ such that

$$\|T_n - T\| < \in, \forall\ n \geq n(\in)$$

(3) Weak Convergence (KANPUR–2007)

The sequence of bounded linear transformation $< T_n(f) >$ is said to converge weakly to T if for any $f \in N$, the sequence $< T_n(f) >$ of elements of N' converges weakly to $T(f)$,

i.e., $T_n \xrightarrow{w} T$ if for every $f \in N, \phi \in N^*$

$$\phi(T_n(f)) \to \phi(T(f))$$

REMARKS

- Uniform convergence implies strong convergence.
- Strong convergence implies weak convergence.
- Strong and weak convergence are equivalent in $B(N, F)$.
- Uniform convergence ⇒ Strong convergence ⇒ Weak convergence.

5.21 BANACH-STEINHAUS PRINCIPLE OF UNIFORM BOUNDEDNESS

THEOREM 1. ***If a sequence of linear operators $<T_n>$ from $B(N, N')$ is a Cauchy sequence in the sense of weak convergence at every point $f \in N$, where N is complete normed space, then the sequence $< \|T_n\| >$ is bounded uniformly.***

(MEERUT–1986, 87, KANPUR–1992, 2004, DELHI–1981)

PROOF. Let if possible, the sequence $< \|T_n\| >$ is not bounded. Then, we prove that the set $< \|T_n f\| >$ is also not bounded on every closed sphere $\| f - f_0 \| < \in$.

If the set $< \|T_n f\| >$ is bounded on every closed sphere $\| f - f_0 \| < \in$.

Then $\|T_n - f\| \leq k \quad \forall n$ and $\forall f$ satisfying $\| f - f_0 \| < \in$.

Choose the element

$$g' = \frac{\in}{\| g \|} g + f_0$$

Then g' would belong to the above sphere for every g

$$\therefore \qquad \|T_n g'\| \leq k$$

Also, $\dfrac{\in}{\| g \|} \| T_n g \| - \| T_n f_0 \| \leq \left\| \dfrac{\in}{\| g_\in \|} T_n g + T_n f_0 \right\| \leq k$

Therefore,

$$\| T_n g \| \leq \frac{k + \| T_n f_0 \|}{\in} \| g \| = k_1 \| g \|$$

$$\Rightarrow \qquad \| T_n g \| \leq k_1 \| g \|$$

and $\qquad \|T_n\| \leq k_1 \ \forall n$

which is a contradiction.

Therefore, we have proved that if $< \|T_n\| >$ is not bounded then the set $< \|T_n f\| >$ is also not bounded on any closed sphere $\|f - f_0\| < \in$

Further let $\overline{S}$ be any closed sphere of N.

Now, since $< \|T_n f\| >$ is not bounded, $\exists$ a T_{n_1} and $f_1 \in \overline{S}$ such that

$$\left\|T_{n_1} f\right\| > 1 \quad \ldots(1)$$

Since, T_{n_1} is continuous, the inequality (1) holds on a closed sphere $\overline{S}_{\epsilon_1}(f_1)$ where $\overline{S}_{\epsilon_1}(f_1)$ is a closed sphere with radius ϵ_1 and centre $f_1 \in S$

The sequence $< \|T_n f\| >$ is again not bounded on $\overline{S}_{\epsilon_1}(f_1)$ and therefore, there exists $T_{n_2}, n_2 > n_1$ and $f_2 \in \overline{S}_{\epsilon_1}(f_1)$ such that

$$\left\|T_{n_2} f_2\right\| > 2$$

Continue this process and let $\epsilon_n \to 0$ as $n \to \infty$, there exists a point $\hat{f}$ such that

$$\left\|T_{n_k} \hat{f}\right\| \geq k$$

which is not possible, because the sequence $\|T_n f\|$ is bounded as $< \|T_n f\| >$ converges weakly for every f. Hence, the sequence $< \|T_n\| >$ is bounded.

REMARK

- The above principle can also be restated as follows:
  "If a sequence of linear operator $< T_n >$ of $B(N, N')$ is vector wise bounded
  i.e., $\sup\|T_n\| < \infty$ for all $f \in N$, then $\sup\|T_n\| < \infty$

THEOREM 2. ***Let Δ be an index set of arbitrary cardinality and suppose $\{< T_\alpha >: \alpha \in \Delta\}$ is a collection of members of $B(N, N^*)$, where N is a Banach space, if $\sup_\alpha \|T_\alpha\| < \infty$ $\forall f \in N$ Then $\sup_\alpha \|T_\alpha f\| < \infty$*** (KANPUR–1982, 88)

PROOF. Let us suppose $< T_\alpha >: \alpha \in \Delta$ satisfies the hypothesis and suppose at the same time that

$$\sup\|T_\alpha\| = \infty \quad \ldots(1)$$

Here (1) implies that there exists $T_1, T_2, \ldots, T_n \ldots$ such that

$$\|T_1\| > 1, \|T_2\| > 2 \quad \ldots(2)$$

Clearly $\{T_1, T_2, \ldots, T_n \ldots\}$ satisfies the hypothesis of above theorem.

Thus, using above theorem,

$$\sup_\alpha \|T_n\| < \infty$$

which is not true.

Hence, $\quad \sup\|T_\alpha f\| < \infty$.

Particular Cases

Case-I. If a sequence $< T_n >$ of linear operator in B^* converges pointwise to T such that $Tf = \lim_{n \to \infty} T_n(f)$, then $T \in B^*$ (KANPUR-1996)

Case-II. If a sequence $< \phi >$ of linear functional form $B(N, F)$ converges weakly to the functional ϕ_0, then the norms of ϕ_n are bounded uniformly.

THEOREM 3. ***A sequence $< g_n >$ of linear functional converges weakly to a linear functional g_0 if and only if***

(i) the sequence $\|g_n\|$ is bounded and

(ii) $g_n(f) \to g_0(f)$***, for arbitrary f belonging to the set G for which the linear combinations of its elements are everywhere dense in N.***

PROOF. **(i) Necessary Condition**

Let us first suppose the sequence $< g_n >$ of linear functional converges weakly to a linear functional g_0.

Then, by using particular case-II of above theorem, we can easily proved the required conditions.

(ii) Sufficient Conditions

Define $$M = \sup_n \|g_n\|$$

and f be an arbitrary element in N.

Further, let f_0, which is a linear combination of elements of G be choosen such that

$$\|f - f_0\| < \frac{\in}{4M}$$

Consider

$$\begin{aligned}\|g_n(f) - g_0(f)\| &\le |g_n(f) - g_n(f_0)| + |g_n(f_0) - g_0(f_0)| + |g_0(f_0) - g_0(f)| \\ &\le |g_n(f_0) - g_0(f_0)| + (\|g_n\| + \|g_o\|)\|f - f_0\| \\ &< |g_n(f_0) - g_0(f_0)| + \in/2\end{aligned}$$

Now, by given condition (ii) $g_n(f_0) \to g_0(f_0)$, so that there exists an N_0 such that

$$|g_n(f_0) - g_0(f_0)| < \in/2, \ \forall\, n \ge N_0$$

$$\Rightarrow \quad |g_n(f) - g_0(f)| < \in, \ \forall\, n \ge N_0.$$

Since f is an arbitrary element of N.

Hence, the sequence $< g_n >$ of linear functional converges weakly to a linear functional g_0.

SOME IMPORTANT RESULTS

(1) The sequence $< f_n >$, belonging to N, converges weakly to $f_0 \in N$ if and only if

(i) the sequence $< \|f_n\| >$ is bounded, and

(ii) $g(f_n) \to g(f_0)$, for every g of the set of linear functional whose linear combinations are everywhere in N^*.

(2) A sequence $< f_n >$, $f_n = \{f_n(k)\}_{k=1}^{\infty} \in l^p$ converges weakly to

$f_0 = \{f_0(k)\}_{k=1}^{\infty} \in l^p$ if and only if

(i) the sequence $\|f_n\|$ is bounded, and

(ii) $f_n(k) \to f_0(k)$ as $n \to \infty \ \forall k$

(3) The necessary and sufficient condition for weak convergence of the sequence $< f_n(x) >, f_n(x) \in L^p(0,1)$ to $f_0(x) \in L^p(0,1)$ is that

(i) the sequence $< \|f_n\| >$ is bounded, and

(ii) $\int_o^T f_n(x)dx \to \int_o^T f_o(x)dx$ for arbitrary $T \in [0,1]$

(4) If a sequence $< f_n >: f_n \in N$ converges weakly to f_0, then the norms of the elements are bounded.

5.22 WEAK AND STRONG CONVERGENCE IN HILBERT SPACE

Definition 1. *Let H be a Hilbert space. The sequence $<f_n>$ of H converges weakly to f if $<(f_n, h)>$ converges to (f, h) for every $h \in H$.*

Therefore, f is said to be the weak limit of $<f_n>$ if

$$(f,h) = \lim_{n\to\infty} (f_n h), \forall\ h \in H$$

If $f_n \xrightarrow{w} f_0$, then clearly, every arbitrary subsequence $< f_k >$ also converges weakly to f_0.

Also, if $f_n \xrightarrow{w} f, g_n \xrightarrow{w} g$, then $f_n + g_n \xrightarrow{w} f + g$

and $a\, f_n \xrightarrow{w} af$ for any scalar G.

Definition 2. *The sequence $< f_n >, f_n \in H$ converges strongly to $f \in H$ if $\| f_n - f \| \to 0$ as $n \to \infty$.*

REMARK

- The strong convergence always implies weak convergence. Converse is not necessarily true.

THEOREM 1. ***If $f_n \xrightarrow{w} f$, where $f_n, f \in H$ and $\| f_n \| \to \| f \|$ then f_n converges strongly to f.***

PROOF. We have

$$\|f_n - f\|^2 = (f_n - f, f_n - f) = \|f_n\|^2 - \|f\|^2 - (f_n, f) - (f, f_n)$$

$$\to \|f\|^2 - \|f\|^2 - (f,f) - (f,f)$$

[As per given $f_n \xrightarrow{w} f$ and $\| f_n \| \to \| f \|$]

Thus, we have

$$\|f_n - f\|^2 \to 0 \text{ as } n \to \infty$$

Hence, $f_n \to f$ strongly.

THEOREM 2. ***Let H be a Hilbert space, which is finite dimensional, then weak convergence, implies strong convergence.***

PROOF. Since H is finite dimensional, therefore, we can find m linearly independent vectors $e_1, e_2, ..., e_m$ such that any vector $h \in H$ can be written as

$$h = a_n e_n + a_{n+1} e_{n+1} + ... + a_{n+m} e_{n+m}, \text{ when } a_n^{'s} \text{ as scalar}$$

$$= \sum_{n=1}^{m} a_n e_n$$

Now, consider the sequence $< f_n >$ where.

$$f_n = a_{n_1} e_1 + a_{n_2} e_2 + ... + a_{n_m} e_m$$

and let $f = a_1 e_1 + a_2 e_2 + ... + a_m e_m$

It is given that $< f_n >$ converges weakly to f, we have

$$\lim_{n\to\infty} (f_n, h) = (f, h)\ \ \forall\ \ h \in H$$

$$\Rightarrow \quad \lim_{n\to\infty} (f_n - f, h) = 0$$

$$\Rightarrow \quad \lim_{n\to\infty} \left(\sum_{k=1}^{m} (a_{n_k} - a_k) e_k, h \right) = 0\ \ \forall\ h \in H$$

Setting $e_k = h$, we get

$$\lim_{n\to\infty}\left(\sum_{k=1}^{m}(a_{n_k}-a_k)e_k, e_k\right)=0$$

$$\Rightarrow \quad \lim_{n\to\infty}\left(\sum_{k=1}^{m}(a_{n_k}-a_k)\right)=0$$

$$\Rightarrow \quad \lim_{n\to\infty} a_{n_k} = a_k \text{ for all } k=1,2,\ldots,m.$$

Finally, consider

$$\|f_n - f\|^2 = (f_n - f, f_n - f)$$

$$= \left(\sum_{k=1}^{m}(a_{n_k}-a_k)e_k, \sum_{k=1}^{m}(a_{n_k}-a_k)e_k\right)$$

$$= \sum_{k=1}^{m}\left|a_{n_k}-a_k\right|^2$$

Therefore

$$\lim_{n\to\infty}\|f_n - f\|^2 = \lim_{n\to\infty}\sum_{k=1}^{m}\left|a_{n_k}-a_k\right|^2 = 0$$

$$\Rightarrow \quad \|f_n - f\| \to 0 \text{ as } n\to\infty.$$

Hence, $< f_n >$ converges strongly to f.

THEOREM 3. **(Banach-Steinhaus uniform boundedness principle for functionals)** ***Let $g_1, g_2\ldots$ are bounded linear functional on H and $g_1(h), g_2(h)\ldots$ are vectorwise bounded for every $h\in H$. Then $<\|g_n\|>$ is also a uniformly bounded sequence.*** (KANPUR–1982)

PROOF. Define a functional

$$p_n(h) = |g_n(h)|$$

We shall prove that $p_n(h)$ is a convex continuous functional on H.

(i) **p_n is convex.**

Let $h_1, h_2 \in H$,

then
$$p_n(h_1+h_2) = |g_n(h_1+h_2)|$$
$$= |g_n(h_1)+g_n(h_2)| \le |g_n(h_1)| + |g_n(h_2)|$$
$$= p_n(h_1)+p_n(h_2)$$

and
$$p_n(ah) = |g_n(ah)| = |a(g_n(h)| = |a|.|g_n(h)|$$
$$= ap_n(h), \text{ for all scalars } a$$

Therefore, p_n is a convex functional.

(ii) **p_n is continuous.**

Let $h, h_0 \in H$. Now, since g_n is a bounded linear functional then it is continuous also. Thus, by definition, corresponding to given $\in > 0$ we can find an integer n_0 such that

$|g_n(h) - g_n(h_0)| < \in \ \forall n \ge n_0$ and $\forall h$ such that $\| h - h_0 \| < \delta$.

If $n \geq n_0$, then

$$p_n(h) - p_n(h_0) = |g_n(h)| - |g_n(h_0)| \leq |g_n(h) - g_n(h_0)|$$
$$< \in \text{for} \| h - h_0 \| < \delta$$

Therefore, p_n is a convex continuous functional and $p_1(h), p_2(h)...$ are bounded for every $h \in H$.

Now, take $p(h) = \sup_n p_n(h)$

Then, $p(h)$ is also a continuous convex functional and hence bounded.

Therefore, there exists $k > 0$ such that $p(h) \leq k$ for every h in the unit sphere of H.

$$\Rightarrow \quad \sup_n |g_n(h)| \leq k$$

Thus, $|g_n(h)| \leq k$ for all n and for every h in the unit sphere of H.

Here, $< \|g_n\| >$ is bounded uniformly.

REMARK

- Every weakly convergent sequence in H is bounded.

THEOREM 4. ***The Hilbert space is weakly complete.***

PROOF. Let H be a Hilbert space. Suppose that $< f_n >$ be a weakly Cauchy sequence, *i.e.*, $\lim_{n \to \infty} \{(f_n - f_{n+m}, h\} = 0$

Since $< f_n >$ converges weakly, $\| f_1 \|, \| f_2 \|, ...$ is bounded, say by k.

If we take $\quad g(h) = \lim_{n \to \infty} (h, f_n)$

then, for any vector h of the unit sphere of H,

$$|g(h)| = \left| \lim_{n \to \infty} (h, f_n) \right| \leq \sup_n |(h, f_n)|$$
$$\leq \sup_n \|f_n\| . \|h\| \leq k.1 = k$$

Therefore, g is a bounded linear functional defined on H.

Now, by Riesz Representation theorem, we can find a vector $h \in H$ such that

$$g(h) = (h, f) \forall h \in H$$

Thus $\quad g(h) = (h, f) = \lim_{n \to \infty} (h, f_n)$

Hence, $f_n \xrightarrow{w} f$ and $f \in H$, *i.e.*, H is weakly complete.

THEOREM 5. ***A non-empty subset S of a normed linear space N is bounded if and only if g(S) is a bounded set of numbers for each $g \in N^*$***

(MEERUT–1985, 88)

PROOF. **(i) Necessary Condition**

Let S be a bounded subset of N, then by definition there exists a positive constant k_1 such that

$$\| f \| \leq k_1 \ \forall f \in S \quad ...(1)$$

Now, to show that $g(S)$ is bounded, we have

$$g \in N^* \Rightarrow g \text{ is bounded} \Rightarrow \exists \ k_2 > 0 \text{ such that}$$
$$|g(f)| \leq k_2 \| f \| \quad \forall f \in N \quad ...(2)$$

Form (1) and (2), we conclude that

$$|g(f)| \le k_1 k_2 \quad \forall f \in S$$

$\Rightarrow$ $g(S)$ is a bounded set of numbers for each $g \in N^*$.

(ii) Sufficient condition

Let $g(S)$ be bounded set of numbers for each $g \in N^*$ and let S be the set such that

$$S = \{f_1, f_2 ...\}$$

To show that S is bounded, consider the natural imbedding $f_i \to f_{f_i}$

To pass from S to the corresponding subset $\{F_{f_i}\}$ of N^* defined by

$$F_{f_i}(g) = g(f_i)$$

Now, our assumption that $g(S) = \{g(f_i)\}$ is bounded for each g is equivalent to the statement $\{F_{f_i}(g)\}$ is a bounded set for each $g \in N^*$ and since, N^* is complete, then by uniform boundedness theorem we have $\{F_{f_i}\}$ is a bounded set of numbers.

Also, since we know that the natural imbedding preserves norm, thus, we have $\| F_{f_i} \| = \| f_i \|$ for each $f_i \in S$.

Hence, $\{\|f_i\|\}$ is a bounded set of numbers, *i.e.*, $S = \{f_i\}$ is a bounded subset of N.

REMARK

- A sequence $g_1, g_2, ...$ of vectors of Hilbert space H is weakly convergent if and only if
  (i) $\|g_k\|$ is bounded for each k, and
  (ii) (g_k, f) converges for all f in a certain dense subset M of H.

5.23 WEAK COMPACTNESS IN HILBERT SPACE

(i) Strongly Compact : Let H be a Hilbert space. A subset S of H is said to be strongly compact if for every sequence of elements of S, we can extract a certain strongly convergent subsequence.

(ii) Weakly Compact : A subset S of a Hilbert space H is said to be weakly compact if from every sequence of S we can extract a certain weakly convergent subsequence.

(iii) Sequentially Compact : A subset of a Hilbert space H is said to be sequentially compact if every sequence S contains a convergent subsequence.

REMARKS

- The closed unit sphere $\overline{S_1(0)}$ of an infinite-dimensional Hilbert space is closed and bounded but not strongly compact.
- Every closed bounded subset of H is weakly compact.

EXERCISE 5.1

1. Show that if $\|Tf\| = \|Tf^*\|$ for all f, belonging to the finite dimensional inner product space H, then the linear transformation T is normal.
2. Prove that the linear transformation T is normal if and only if $T(M) \subset M \Rightarrow T(M^{\perp}) \subset M^{\perp}$, when M is a subspace of a Hilbert space H.
3. If A is a self-adjoint linear transformation on the finite dimensional inner product space H, show that $T^n = I$ for some positive integer n, show that $T^2 = I$.
4. If T_1 and T_2 are self-adjoint operators on a Hilbert space H, show that $T_1T_2 + T_2T_1$ is self-adjoint.

5. If T_1 and T_2 are positive linear operators on a Hilbert space H, show that $T_1 + T_2$ is also positive.
6. Show that unitary operators on H form a group. (KASHMIR–2006, KANPUR–1982, 85)
7. Let y and z be closed linear subspace of a Banach spaces X such that $X = y \oplus z$. If $x = y + z$ is the unique representation in y and z and $P(x) = y$, show that P is continuous. (MEERUT–1986)
8. Let y be a closed linear subspace of a normed linear space X and suppose that x_0 is a vector not in Y. Show that there exists a continuous linear functional f on X such that $f(x) = 1$ and $f(y) = 0$. (ANDHRA–2008, MEERUT–1986, 87)
9. Let T be a positive operator on a finite dimensional Hilbert space H. Show that there exists a positive operator S on H such that $S^2 = T$. (DELHI–2002, MEERUT–1987)
10. If P and Q are the projections on closed linear subspaces M and of H, then following statement are equivalent :

    (i) $P \le Q$ (ii) $\| Px \| \le \| Qx \| \ \forall x$

    (iii) $M \subset N$ (iv) $PQ = P$

    (v) $QP = P$

    (ANDHRA–2003, MEERUT–1980, KANPUR–1991)
11. If P and Q are the projections on closed linear subspaces M and N of H, show that $Q - P$ is a projection if and only if $P \le Q$. Hence, show that $Q - P$ is the projection on $N \cap M^{\perp}$.
12. Show that an idempotent operator on a Hilbert space H is a projection if and only if it is normal.
13. Show that parallelogram law is not true in the Banach space $l_1^n (n > 1)$. (AMRITSAR–1982)
14. Let T be a normal operator on a Hilbert space H and f be a polynomial with complex coefficients. Show that $f(T)$ is normal. (BANGLORE–2003)
15. If P_1 and P_2 are non-zero orthogonal projections and $P_1P_2 = 0$, show that $\|P_1 + P_2\| \le \|P_1\| + \|P_2\|$.
16. Prove that every finite dimensional normed linear space is Banach space. (GARHWAL–2002)
17. Let H be a Hilbert space. Then show that H contains a denumerable basis if H is separable. (ROHILKHAND–2009, KANPUR–2011)
18. Let $\{e_1, e_2, \ldots, e_n\}$ be a finite orthonormal set of a Hilbert space H and if x is any vector in H. Then show that

$$\left\| x - \sum_{i=1}^{n} \alpha_i e_i \right\|$$

attains its minimum value $\alpha_i = \langle x, e_i \rangle$ where $\alpha_1, \alpha_2, \ldots, \alpha_n$ are arbitrary scalars.

Chapter Review: A Competitive Approach

SELECTED TERMS AND RESULTS

▶ TERMS

- **Hilbert Space :** A complex Banach space is called a Hilbert space.
- **Orthogonal Vectors :** Two vectors x and y in a Hilbert space H are said to be orthogonal if $(x, y) = 0$.
- **Orthonormal Set :** A non-empty subset $\{e_i\}$ of a Hilbert space H is said to be an orthonormal set if
  (i) $\|e_i\| = 1$ i.e., $(e_i, e_i) = 1$
  (ii) $(e_i, e_j) = 0, \quad i \neq j$
- **Complete Orthonormal Set :** An orthonormal set is said to be complete if it is not contained in any larger orthonormal set.
- **Self-adjoint operator :** If $T^* = T$
- **Positive operator :** If $(Tx, x) \geq 0 \ \forall \ x \in H$
- **Normal operator :** If $TT^* = T^*T$
- **Unitary operator :** If $TT^* = T^*T = I$
- **Compact operator :** An operator A on a Hilbert space H is said to be compact if for every bounded sequence $<x_n>$ in H, the sequence $<Ax_n>$ contains a convergent subsequence.

▶ RESULTS

- If H be a Hilbert space, then for any two vectors x, y in $H, |(x, y)| \leq \|x\| + \|y\|$. This is called Schwarz inequality.
- Let H be a Hilbert space. Then the inner product is jointly continuous, *i.e.*,
  $x_n \to x, \ y_n \to y \Rightarrow (x_n, y_n) \to (x, y)$
- Let M be a closed linear subspace of a Hilbert space H and x be any vector not in M. If d be the distance from x to M, then there exists a unique vector y_0 in M such that $\|x - y_0\| = d$
- A closed convex subset C of a Hilbert space H contains a unique vector of smallest norm.
- Let H be a Hilbert space and x, y be any vectors of H. Then, x is said to be orthogonal to y(written as $x \perp y$) if $(x, y) = 0$.
- Let S be a non-empty subset of a Hilbert space H. Then, $S^\perp$ is a closed linear subspace of H.
- The orthogonal compliment of a subset of a Hilbert space is complete.
- Let M be a proper closed linear subspace of a Hilbert space H, then there exists a non-zero vector z_0 in H such that $z_0 \perp M$.
- Let H be a Hilbert space and M be a linear subspace of H. Then, M is closed if and only if $M = M^{\perp\perp}$.
- Let M and N be closed linear subspaces of a Hilbert space H such that $M \perp N$. Then, the linear subspace $M + N$ is also closed.
- If M is a closed linear subspace of a Hilbert space H, then $H = M \oplus M^\perp$. This is known as projection theorem.
- Let S be a non-empty subset of a Hilbert space H, then the set $[S]$ of all linear combinations of vectors in S is dense in H if and only if $S^\perp = \{0\}$.
- Let H be a Hilbert space and $x \in H$ such that $\|x\| = 1$, *i.e.*, $(x, x) = 1$. Then, x is said to be a unit vector or normal vector.
- Let $\{e_i\}$ be an orthonormal set on a Hilbert space H, then $\Sigma |(x, e_i)|^2 \leq \|x\|^2$ for every vector x in H. This is called Bessel's inequality.
- An orthonormal set is said to be complete if it is not contained in any larger orthonormal set.
- An orthonormal set in a Hilbert space is linearly independent.
- In a Hilbert space H, an orthonormal set S is complete if and only if $x \perp S \Rightarrow x = 0$.
- A Hilbert space is finite dimensional if and only if every complete orthonormal set is a basis.

- Any two complete orthonormal sets in a Hilbert space H have the same cardinal number.
- Let H be a Hilbert space and f be an arbitrary functional in H^* then, there exists a unique vector y in H such that $f(x) = (x, y)\ \forall x \in H$. This is known as Riesz representation theorem.
- The mapping $\phi : H \to H^*$ is Conjugate isomorphism, *i.e.*, this is one-one, onto, additive but non-homogeneous. Further, the mapping ϕ is an isometry.
- If H is a Hilbert space, then H is reflexive.
- The adjoint operation is one-one onto as a mapping of $B(H)$ onto itself.
- The adjoint operator is one-one onto as a mapping of $B(H)$ into itself.
- If T be an invertible operator then either both or none of T and T^{-1} are self-adjoint.
- If T is self-adjoint operator on a Hilbert space H then αT is self-adjoint if and only if α is real.
- The self-adjoint operator in $B(H)$ form a closed linear subspace and therefore a real Banach space which contains the identity transformation.
- If T be an arbitrary operator on a Hilbert space H. Then $T = 0$ if and only if $(Tx, y) = 0, \forall x, y \in H$.
- An operator T on a Hilbert space H is self-adjoint if and only if (Tx, x) is real for all x.
- If T is an arbitrary operator on H, then the operator $I + TT^*$ and $I + T^*T$ are non-singular.
- If N_1 and N_2 are normal operators on a Hilbert space H with the property that either commutes with the adjoint of the other, then $N_1 + N_2$ and $N_1 . N_2$ are also normal.
- If N is a normal operator on a Hilbert space H. Then, $\|N^2\| = \|N\|^2$.
- An operator T on a Hilbert space H is unitary if and only if it is an isometric isomorphism of H onto itself.
- If T be a normal operator on a Hilbert space H and λ be any scalar then $T - \lambda I$ is normal.
- A closed linear subspace M of a Hilbert space H is invariant under an operator T if and only if $M^\perp$ is invariant under T^*.
- If P is a projection on a closed linear subspace M of a Hilbert space H, then M reduces an operator T if and only if $TP = PT$.

REVIEW QUESTIONS

▶ DESCRIPTIVE QUESTIONS

1. Prove that every non-zero Hilbert space contains an orthonormal basis.
2. Prove that a Hilbert space has a countable orthonormal basis if and only if it is separable.
3. Prove that two Hilbert spaces are isomorphic if and only if their dimensions are equal.
4. Prove that every bounded sequence in a Hilbert space contains a subsequence which converges weakly.
5. Prove that every finite dimensional inner product space is necessarily a Hilbert space.
6. Prove that every one dimensional normed space is an inner product space.
7. Let H be a Hilbert space such that every bounded sequence in H has a convergent subsequence. Prove that H is finite dimensional.
8. If $\|x\| = \|y\| = 1$ in an inner product space, show that $(x + y)$ is orthogonal to $(x - y)$.
9. Let S and T be bounded linear transformation on a Hilbert space H to H. If $(Sx, x) = (Tx, x)$ for every $x \in H$. Show that $S = T$.
10. Let $T : H \to H$ be a linear homomorphism. Show that
$$(T^{-1})^* = (T^*)^{-1}$$

▶ MULTIPLE CHOICE QUESTIONS

Choose the most appropriate one.

1. For a linear operator T if $(Tx, x) = 0$, then :
   (a) $T = 0$ (b) $T = 1$
   (c) $T \neq 0$ (d) none of these
2. Every inner product space is a :
   (a) Banach space
   (b) Normed linear space
   (c) Hilbert space
   (d) none of these
3. For adjoint operator, which is not true ?
   (a) $(T_1 + T_2)^* = T_1^* + T_2^*$
   (b) $(\alpha T)^* = \bar{\alpha} T^*$
   (c) $(T_1 T_2)^* = T_1^* T_2^*$
   (d) none of these
4. If $(x, z) = (y, z)\ \forall\ z \in H$ then :
   (a) $x > y$ (b) $x < y$
   (c) $x \neq y$ (d) $x = y$
5. If T is self-adjoint then αT is self-adjoint iff :
   (a) α is real (b) α is complex
   (c) $\alpha = 0$ (d) none of these

6. The value of $\|x + y\|^2 - \|x - y\|^2 =$
(a) $4xy$ (b) $4(y, x)$
(c) $4\text{Re}(x, y)$ (d) none of these

7. The value of $\|T.T^*\| =$
(a) T^2 (b) TT^*
(c) $\|T\|^2$ (d) none of these

8. If inner product of two linear spaces is zero, then they called :
(a) orthogonal (b) orthonormal
(c) equal (d) none of these

9. The value of $\|x + y\|^2 - \|x - y\|^2 + i\|x + iy\|^2 - i\|x - iy\|^2 =$
(a) $4(x, y)$ (b) $4\text{Re}(x, y)$
(c) $4(y, x)$ (d) none of these

10. In a Hilbert space H, if $x_n \to x$, $y_n \to y$ then $(x_n, y_n) \to$
(a) x (b) y
(c) (x, y) (d) none of these

11. H is called Hilbert space, then :
(a) If H is complete under the norm obtained from its inner product
(b) If H is complete space
(c) If H is not complete
(d) none of these

12. Banach space is Hilbert space if :
(a) Parallelogram law holds
(b) Pythogorean theorem hold
(c) both (a) and (b) are true
(d) none of these

13. $x \perp y \Rightarrow \|x\|^2 + \|y\|^2 = \|x + y\|^2$ is valid in inner product space, is called :
(a) Pythogorean theorem
(b) Parallelogram law
(c) both (a) and (b) are true
(d) none of these

14. The law $\|x + y\|^2 + \|x - y\|^2 = 2\{\|x\|^2 + \|y\|^2\}$ is called :
(a) Pythogorean theorem
(b) Parallelogram law
(c) both (a) and (b) are true
(d) none of these

15. If $\{x_1, x_2, ..., x_n\}$ is an orthonormal set, then for any $x \in H$
(a) $\sum_{i=1}^{n} |(x, x_i)|^2 \leq |x|^2$
(b) $\sum_{i=1}^{n} |(x, x_i)|^2 \leq \|x\|^2$
(c) both (a) and (b) are true
(d) none of these

16. Let H be an Hilbert space, $S = \{x_\alpha : \alpha \in \Delta\}$ is an orthonormal set in H and $x \in H$ then $\sum_{\alpha \in A} |(x, x_\alpha)|^2 \leq \|x\|^2$ is called :
(a) Bessel's inequality
(b) Cauchy-Schwarz inequality
(c) Parseval's formula
(d) none of these

17. Let H be a Hilbert space, $S = \{x_\alpha : \alpha \in \Delta\}$ is an orthonormal set. Then S is a basis if and only if equality holds in Bessel's inequality is called:
(a) Parseval's formula
(b) Minkowski inequality
(c) both (a) and (b) are true
(d) none of these

18. If the Hilbert space H is separable then :
(a) there is a countable basis for H
(b) there is a non-countable basis for H
(c) there is a null basis for H
(d) none of these

19. Two Hilbert spaces are isomorphic if they have :
(a) same dimensions
(b) different dimensions
(c) can't say
(d) none of these

20. If T is a bounded linear operator on Hilbert space H, then :
(a) T is normal iff $\|Tx\| = \|T^*x\|$ for every $x \in H$
(b) If T is normal and $Tx = \alpha x$ for $x \in H$, $\alpha \in C$ then $T^* x = \bar{\alpha} x$
(c) both (a) and (b) are true
(d) none of these

21. If U is an unitary operator on H then :
(a) $(Ux, Uy) = (x, y)$
(b) $\|Ux\| = \|x\| \ \forall \ x \in H$
(c) both (a) and (b) are true
(d) none of these

22. A set S of elements of an inner product space X is an orthogonal set if :
(a) $x \perp y$ whenever $x, y \in S$, $x \neq y$
(b) $x \perp y$ whenever $x, y \in S$, $x = y$
(c) both (a) and (b) are true
(d) none of these

23. Let T be a bounded linear operator of Hilbert space H then there exists a unique operator on H called adjoint of T, i.e., T^* such that :
(a) $(Tx, y) = (x, T^*y) \ \forall \ x, y \in H$

(b) $\|T^*\| = \|T\|$
(c) both (a) and (b) are true
(d) none of these

24. A bounded linear operator T on a Hilbert space H is said to be self-adjoint if :
(a) $T^* = T$ (b) $T^* = -T$
(c) $\|T\|^2 = T^*$ (d) none of these

25. A bounded linear operator T is called normal if :
(a) $T^* = T$ (b) $TT^* = T^*T$
(c) $T^*T = TT^* = I$ (d) none of these

26. For inner product space, which of the following is true ?
(a) $(x, \alpha y) = \bar{\alpha}(x, y)$ (b) $(x, \alpha y) = (x, y)$
(c) $(x, \alpha y) = \alpha(x, y)$ (d) none of these

27. For adjoint operator, which one of following is true ?
(a) $(\alpha T)^* = \bar{\alpha} T$
(b) $(\alpha T)^* = \bar{\alpha} T^*$
(c) both (a) and (b) are true
(d) none of these

28. Let H be a complex Hilbert space, $T : H \to H$ be a bounded linear operator and let T^* denote the adjoint of T. Then which of the following statement is always true? **(GATE–2012)**
P : For all $x, y \in H$, $<Tx, y> = <x, T^*y>$
Q : For all $x, y \in H$, $<x, Ty> = <T^*x, y>$
R : For all $x, y \in H$, $<x, Ty> = <x, T^*y>$
S : For all $x, y \in H$, $<Tx, Ty> = <T^*x, T^*y>$
(a) P and Q (b) P and R
(c) Q and S (d) none of these

29. Let H be a complex Hilbert space and H^* its dual. The mapping $\phi : H \to H^*$ defined by $\phi(y) = f_y$ where $f_y(x) = (x, y)$ is : **(GATE–2011)**
(a) not linear but onto
(b) both linear and onto
(c) linear but not onto
(d) none of these

30. Let $<e_n>$ be an orthonormal sequence in a Hilbert space H and let $x \neq 0$, then : **(GATE–2009)**
(a) $\lim_{n\to\infty} (x, e_n)$ does not exist
(b) $\lim_{n\to\infty} (x, e_n) = \|x\|$
(c) $\lim_{n\to\infty} (x, e_n) = 1$
(d) none of these

31. Let $C[0, 1]$ be the space of all continuous real valued functions on $[0, 1]$. The identity map $I : (C[0, 1], \|.\|_\infty) \to (C[0, 1], \|.\|_1)$ is : **(GATE–2005)**
(a) continuous but not open
(b) open but not continuous
(c) both continuous and open
(d) none of these

32. Let X be a real normed linear space of all real sequences with finitely many non-zero terms with supremum norm and $T : X \to X$ be a one-to-one and onto linear operator defined by :

$$T(x_1, x_2, \ldots) = \left(x_1, \frac{x_2}{2^2}, \frac{x_3}{3^2}, \ldots\right)$$

then, which of the following is true ? **(GATE–2011)**
(a) T is bounded but T^{-1} is not bounded
(b) T is not bounded but T^{-1} is bounded
(c) Both T and T^{-1} are bounded
(d) none of these

33. Let the continuous linear operator $T : l^2 \to l^2$ be defined by

$$T(x_1, x_2, \ldots) = (0, x_1, 0, x_3, 0, x_5, 0, \ldots)$$

then : **(GATE–2006)**
(a) T is compact but not T^2
(b) T^2 is compact but not T
(c) both T and T^2 are compact
(d) none of these

34. Let X be the space of bounded real sequences with supremum norm. Define a linear operator $T : X \to X$ by

$$T(x) = \left(\frac{x_1}{1}, \frac{x_2}{2}, \ldots\right) \text{ for } x = (x_1, x_2, \ldots) \in X$$

then : **(GATE–2004)**
(a) T is bounded but not one-to-one
(b) T is one-one but is not bounded
(c) T is bounded and its inverse exists and is not bounded
(d) T is bounded and its inverse exists and is bounded

35. Let the sequence $<e_n>$ be a complex orthonormal set in a Hilbert space H, then : **(GATE–2003)**
(a) for all bounded linear operator T on H, the sequence $<Te_n>$ is convergent in H
(b) for all identity operator I on H, the sequence $<Ie_n>$ is convergent in H
(c) for all bounded linear functional f on H, the sequence $<fe_n>$ is convergent in R
(d) none of these

36. Let $A : H \to H$ be any bounded linear operator on a complex Hilbert space H such that $\|Ax\| = \|A^*x\|$ for all x in H where A^* is the adjoint of A. If there is a non-zero x in H such that $A^*x = (2 + 3i)x$ then A is : **(GATE–2003)**
(a) unitary operator on H
(b) self-adjoint but not unitary
(c) self-adjoint but not normal
(d) a normal operator

37. All norms on a normed vector space X are equivalent provided : **(GATE–2001)**
(a) X is reflexive
(b) X is complete
(c) X is finite dimensional
(d) none of these

38. We can define a mappping of a Hilbert space H onto its conjugate space H^* by the formula $y \to f_x$ where $f_y(x) = (x, y)$ which is :
(a) one-to-one and linear
(b) one-to-one and norm preserving
(c) many one and non-linear
(d) none of these

39. An operator T on a Hilbert space H is self-adjoint if and only if :
(a) $(Tx, x) = 0 \; \forall \, x \in H$
(b) (Tx, x) is real $\forall \, x \in H$
(c) $(Tx, x) = 1 \; \forall \, x \in H$
(d) none of these

40. A linear operator P on a Hilbert space H is a projection if and only if :
(a) $P^2 = P$ (b) $P^2 = P = P^*$
(c) $P^2 = P \neq P^*$ (d) none of these

41. Angle θ between non-zero vectors x, y of a Hilbert space H is given by :
(a) $\cos\theta = \dfrac{(x, y)}{\|x\|.\|y\|}$
(b) $\cos\theta = \dfrac{|(x, y)|}{\|x\|.\|y\|}$
(c) both (a) and (b) are true
(d) none of these

42. Let S_1, S_2 be subspaces of a Hilbert space H, then :
(a) $\{0\}^\perp = H$
(b) $H^\perp = \{0\}$
(c) both (a) and (b) are true
(d) none of these

43. If P and Q are projections on closed linear subspace M and N of a Hilbert space H, then $M \perp N$ if and only if :
(a) $PQ = 0$ (b) $QP = 0$
(c) $PQ = 0 = QP$ (d) none of these

44. If T is a linear operator, then :
(a) $\|T^2\| = \|T\|^2$
(b) $\|Tx\| = \|T^*x\|$
(c) both (a) and (b) are true
(d) none of these

45. If $<T_n>$ is a sequence of self-adjoint operators which converges to an operator T then T is :
(a) self-adjoint (b) normal
(c) projection (d) none of these

ANSWERS

1. (a)	**2.** (b)	**3.** (c)	**4.** (d)	**5.** (a)	**6.** (c)	**7.** (c)	**8.** (a)	**9.** (a)
10. (c)	**11.** (a)	**12.** (a)	**13.** (a)	**14.** (b)	**15.** (b)	**16.** (a)	**17.** (a)	**18.** (a)
19. (a)	**20.** (c)	**21.** (d)	**22.** (a)	**23.** (a)	**24.** (a)	**25.** (b)	**26.** (a)	**27.** (b)
28. (a)	**29.** (b)	**30.** (c)	**31.** (a)	**32.** (c)	**33.** (a)	**34.** (d)	**35.** (a)	**36.** (d)
37. (a)	**38.** (a)	**39.** (b)	**40.** (b)	**41.** (a)	**42.** (c)	**43.** (c)	**44.** (c)	**45.** (a)

1. Let S and T be positive operators. Show that $S + T$ is positive.
2. If T is an unitary operator on an inner product space X then show that $\|T\| = 1$.
3. Let T be a normal operator, show that $\lambda I - T$ is normal for all complex numbers λ.
4. Let T be a self-adjoint operator on a Hilbert space H and let $T^2x = 0$ for some x in H. Prove that $T(x) = 0$. Also, deduce that $N \cap M = \{0\}$, where N is the null space of T and M is the range of T.
5. If S and T are self-adjoint operator, prove that $ST + TS$ is self-adjoint.
6. Prove that the set of all projections on a Hilbert space forms a complete lattice in their natural ordering as self-adjoint operator.
7. Let X be a Banach space. If $T : X \to X$ is compact operator, show that the equation $x - Tx = 0$ has only finitely many linearly independent solutions.
8. Let X be a normed space and let $T : X \to X$ be a compact operator. Prove that for every $\lambda \neq 0$, the range of $T - \lambda I$ is closed.
9. Let $T : X \to X$ be a compact linear operator and $S : X \to X$ be a bounded linear operator on a normed space X. Then show that TS and ST are compact.
10. Let T be a bounded linear operator on H. Then, show that T is normal if and only if its real and imaginary parts commutes.
11. Find a necessary and sufficient condition for equality in Schwarz inequality.
12. Let H be a separable Hilbert space. Prove that a bounded linear transformation $T : H \to H$ can be expressed by a matrix.
13. Show that unitary operators on a Hilbert space H form a group.
14. Show that compact operators map orthonormal sequences into sequences strongly convergent to 0.
15. Let $T : R^2 \to R^2$ be defined by $T(x, y) = (x + 3y, 2x + y)$. Show that $T^* \neq T$.
16. If A is self-adjoint operator and B is a bounded operator. Show that B^*AB is self-adjoint.
17. If $A^*A + B^*B = 0$, show that $A = B = 0$.
18. If T is self-adjoint and $T \neq 0$, show that $T^n \neq 0$ $\forall\, n \in N$.
19. If A and B are positive operators and $A + B = 0$, show that $A = B = 0$.
20. Show that for any self-adjoint operator A, there exists positive operators S and T such that $A = S - T$ and $ST = 0$.

Chapter 6 Finite Dimensional Spectral Theory

6.1 INTRODUCTION

In this chapter we shall discuss the concepts of eigen values and eigen vectors on a linear operator T in a Hilbert space H. We shall also discuss the spectrum of the eigen values.

6.2 FINITE DIMENSIONAL SPECTRAL THEORY

Definition 1. *Let T be an operator on a Hibert space H. A scalar λ is said to be an eigen value of T if there exists a non-zero vector x in H such that $Tx = \lambda x$. If λ is an eigen value of T, then any non-zero vector x in H such that $Tx = \lambda x$ is called an eigen vector corresponding to the eigen value λ.*

Definition 2. *Let A and B be sequence matrices of order n over the field of complex numbers. Then B is said to be similar to A if there exists an $n \times n$ non-singular matrix C over the field of complex numbers such that*

$$B = \bar{C}AC$$

6.2.1 IMPORTANT RESULTS (TO BE USED DIRECTLY)

1. If x is an eigen vector of T corresponding to the eigen value λ, then λx is also an eigen vector of T corresponding to the same eigen value λ.
2. If x is an eigen vector of T, then x can not correspond to more than one eigen value of T.
3. Similar matrices have the same determinant.

Definition 3. *The set of eigen values of an operator T is called the spectrum of T. It is denoted by $\sigma(T)$.*

Definition 4. *A set $M = [v \in H : Tv = \lambda v]$ is called the eigen space of λ.*

Definition 5. *Let T be a linear operator on a Hilbert space H. If there exists distinct complex numbers $\lambda_1, \lambda_2, \ldots, \lambda_n$ and non-zero pairwise orthognal projections $P_1, P_2, \ldots, P_n$ such that $T = \sum_{i=1}^{n} \lambda_i P_i$. Then this expression is called spectral solution for T.*

THEOREM 1. ***Let λ be an eigen value of an operator T on a Hilbert space H. Let M be the eigen space of λ then M is closed subspace of H.***

PROOF. Let $M = \{x \in H : Tx = \lambda x\}$

If $x, y \in M \Rightarrow Tx = \lambda x, Ty = \lambda y$

$$T(ax + by) = aTx + bTy; \ a, b \text{ are scalars}$$

$$= a\lambda x + b\lambda y = \lambda(ax + by) \qquad (\because T \text{ is linear})$$

$$\Rightarrow T(ax + by) = \lambda(ax + by),\ ax + by \in H$$

$$\Rightarrow ax + by \in M$$

$\Rightarrow$ M is a subspace of H.

It remains to prove that M is closed

Consider, $M = \{x \in H : Tx = \lambda x = \lambda Ix\}$

$= \{x \in H : (T - \lambda I)x = 0\}$

$=$ null space of linear map $T - \lambda I$

$=$ closed space ($\because$ Null space of linear map is always closed)

Hence, M is a closed subspace of H.

THEOREM 2. ***The eigen space of a linear operator T on a Hilbert space H is invariant under T.***

PROOF. Let M be the eigen space of T corresponding to eigen value λ. Then

$$M = \{x \in H : Tx = \lambda x\}$$

Now, $\lambda \in K, x \in M \Rightarrow \lambda x \in M$ (By definition of subspace)

$Tx \in M$ (By definition of M)

Therefore, $x \in M \Rightarrow Tx \in M$

Hence, M is invariant under T.

THEOREM 3. ***The eigen values of a self adjoint operator are all real.***

(LUCKNOW–2006, KANPUR–2005, BHOPAL–2008)

PROOF. Let T be a self adjoint operator on a Hilbert space H over K and $Tx = \lambda x$, then clearly λ is an eigen value of T relative to eigen vector x.

We have to prove that λ is real.

Consider $\lambda(x, x) = (\lambda x, x) = (Tx, x) = (x, T^* x)$

$= (x, Tx)$ ($\because T = T^*$)

$= (x, \lambda x) = \bar{\lambda}(x, x)$

$\Rightarrow$ $(\lambda - \bar{\lambda})(x, x) = 0$

$\Rightarrow$ $\lambda - \bar{\lambda} = 0$ ($\because (x, x) > 0$)

$\Rightarrow$ $\lambda = \bar{\lambda}$

$\Rightarrow$ λ is real.

THEOREM 4. ***If x, y are eigen vectors of a self adjoint operator T on a Hilbert space H relative to different eigen values of T, then*** $x \perp y$.

PROOF. Let H be a Hilbert space over K and T be a self adjoint operator on H.

Suppose

$$Tx = \lambda x, \; Ty = \mu y \qquad (\lambda \neq \mu)$$

We have to prove that $x \perp y$

Since, T is self adjoint therefore $T^* = T$

$\Rightarrow$ $\bar{\lambda} = \lambda$ and $\bar{\mu} = \mu$

($\because$ The eigen values of a self adjoint operator are real)

Consider $\lambda(x, y) = (\lambda x, y) = (Tx, y) = (x, T^* y) = (x, Ty)$

$= (x, \mu y) = \bar{\mu}(x, y) = \mu(x, y)$

$\Rightarrow$ $(\lambda - \mu)(x, y) = 0$

$\Rightarrow$ $(x, y) = 0$ ($\because \lambda \neq \mu$)

Hence, $x \perp y$.

THEOREM 5. ***An operator T on a Hilbert space H is singular if and only if there exists a non-zero vector x in H such that $Tx = 0$*** (MEERUT–2000)

PROOF. Let H be a Hilbert space. Let us first suppose that there exists a non-zero vector x in H such that

$$Tx = 0$$

Then, we have

$$Tx = 0 = T.0$$

Now, since $x \neq 0$, thus x and 0 are two distinct elements in H and they have the same image under T. Thus, the mapping T is not one-one. Therefore, T is not non-singular. Hence, T is singular. Conversely, let us suppose that T is singular. Let if possible there exists no non-zero vector x such that $Tx = 0$.
i.e., suppose that $Tx = 0 \Rightarrow x = 0$

Then, T must be one-one, because for $T_y = T_z$

$\Rightarrow \quad T(y - z) = 0$

$\Rightarrow \quad y - z = 0$, *i.e.*, $y = z$.

Since H is finite dimensional, therefore, T is one-one implies, T is onto therefore, T is non-singular, which is a contradiction because T is singular. Hence, there must exists a non-zero vector x such that $Tx = 0$.

THEOREM 6. ***If T is a normal operator on a Hilbert space H. Then x is an eigen vector of T with eigen value λ if and only if x is an eigen vector of T^* with value $\bar{\lambda}$.*** (MEERUT–2001, KANPUR–2000)

PROOF. We know that if T is a normal operator on H, then $T - \lambda I$ is also a normal operator on H.

Now, $\quad (T - \lambda I)^* = T^* - \bar{\lambda} I^* = T^* - \bar{\lambda} I$.

Since $(T - \lambda I)$ is normal, therefore,

$$\|(T - \lambda I)x\| = \left\|(T - \lambda I)^* x\right\|, \ \forall x \in H$$

$$\Leftrightarrow \quad \|(T - \lambda I)x\| = \left\|(T^* - \bar{\lambda} I)x\right\|, \forall x \in H$$

$$\Leftrightarrow \quad \|T - \lambda x\| = \left\|T^* - \bar{\lambda} x\right\|, \forall x \in H$$

Thus, $Tx - \lambda x = 0$ if and only if $T^* x - \bar{\lambda} x = 0$.

Hence, x is an eigen vector of T with eigen value λ if and only if it is an eigen vector of T^* with eigen value $\bar{\lambda}$.

THEOREM 7. ***If T is a normal operator on a Hilbert space H, then the eigen spaces of T are pairwise orthogonal.*** (MEERUT–2000, KANPUR–2004)

PROOF. Let us suppose that M_1, M_2 be eigen space of a normal operator T on H corresponding to the distinct eigen values λ_1 and λ_2 respectively. To show $M_1 \perp M_2$,

Let x_1 and x_2 be arbitrary vectors in M_1 and M_2 respectively. Then, we have

$$Tx_1 = \lambda_1 x_1 \qquad \text{and} \qquad Tx_2 = \lambda_2 x_2$$

Now $\lambda_1.(x_1,x_2) = (\lambda_1 x_1, x_2) = (Tx_1, x_2)$

$$= (x_1, T^* x_2) = (x_1, \bar{\lambda}_2 x_2) \qquad [\because T^* x_1 = \bar{\lambda}_2 x_2]$$

$$= \lambda_2 (x_1, x_2)$$

$\Rightarrow \quad (\lambda_1 - \lambda_2)(x_1, x_2) = 0$

$\Rightarrow \quad (x_1, x_2) = 0 \qquad [\because \lambda_1 \neq \lambda_2]$

$\Rightarrow \quad x_1, x_2$

Since, x_1, x_2 are arbitrary thus $x_1 \perp x_2 \quad \forall x_1 \in M_1$ and $x_2 \in M_2$.

Hence, $M_1 \perp M_2$.

6.3 SPECTRAL THEOREM

THEOREM 8. ***Let T be an operator on a finite dimensional Hilbert space H. Let $\lambda_1, \lambda_2, ..., \lambda_m$ be the distinct eigen values of T. Let $M_1, M_2, ..., M_m$ be their corresponding eigen space and let, $P_1, P_2, ..., P_m$ be the projections on these eigen spaces. Then, following statements are equivalent.***

(i) The M_i's are pairwise orthogonal and span H.

(ii) The P_i's are pairwise orthogonal.

$P_1 + P_2 + ... + P_m = I$ and $T = \lambda_1 P_1 + \lambda_2 P_2 + ... + \lambda_m P_m$ (KANPUR–2003)

(iii) T is normal. (MEERUT–1980, 82, 2001)

PROOF. (i) $\Rightarrow$ (ii) Suppose that M_i 's are orthogonal in a space H. Then each vector $x \in H$ can be uniquely expressed in the form of

$$x = x_1 + x_2 + ... + x_n, x_i \in M, \ \forall i = 1, 2, ..., m \qquad ...(1)$$

As per given M_i 's are pairwise orthogonal and P_i's are pairwise orthogonal

i.e., $\quad P_i P_j = 0$ for $i \neq j$

Let x be any vector in H, then using (1), for each we have

$$P_i x = P_i(x_1 + x_2 + ... + x_m)$$

$$= P_i x_1 + P_i x_2 + ... + P_i x_m \qquad ...(2)$$

Now, $x_i \in M_i$, which is the range of P_i. Thus, $P_i x_i = x_i$.

Also, let $j \neq i$, then $M_j \perp M_i$.

Since $x_j \in M_j$ for each j. Thus, if $j \neq i$, then $x_j \perp M_i$, *i.e.*, $x_j \in M_i^{\perp}$.

But $M_i^{\perp}$ is the null space of P_i. Thus, $x_j \in M_i^{\perp} \Rightarrow P_i x_j = 0$

$\Rightarrow \quad P_i x_j = 0$ if $j \neq i$.

and $\quad P_i x_i = x$

Then, from (2) we have

$$P_i x = x_i \qquad ...(3)$$

Now for all $x \in H$, we can write

$$Ix = x = x_1 + x_2 + ... + x_m \qquad \text{(Using (1))}$$

$$= P_1x + P_2x + \ldots + P_mx \qquad \text{(Using (3))}$$
$$= (P_1 + P_2 + \ldots + P_m)x$$
$$I = P_1 + P_2 + \ldots + P_m$$

Further for all $x \in H$, we have

$$Tx = T(x_1 + x_2 + \ldots + x_m) = Tx_1 + Tx_2 + \ldots + Tx_m$$
$$= \lambda_1 x_1 + \lambda_2 x_2 + \ldots + \lambda_m x_m = \lambda_1 P_1 x + \lambda_2 P_2 x + \ldots + \lambda_m P_m x$$
$$= (\lambda_1 P_1 + \lambda_2 P_2 + \ldots + \lambda_m P_m)x$$
$$\Rightarrow \quad T = \lambda_1 P_1 + \lambda_2 P_2 + \ldots + \lambda_m P_m$$

(ii) ⇒ (iii) Let us suppose that P_i 's are orthogonal with

$$P_1 + P_2 + \ldots + P_m = I$$

and $\quad T = \lambda_1 P_1 + \lambda_2 P_2 + \ldots + \lambda_m P_m$

To show, T is normal.

Since, each P_i is a projection, thus $P_i^* = P_i$ and $P_i^2 = P_i$.

Also, P_i 's are pairwise orthogonal. Therefore $i \neq j \Rightarrow P_i P_j = 0$

We have

$$T^* = (\lambda_1 P_1 + \lambda_2 P_2 + \ldots + \lambda_m P_m)^*$$
$$= \bar{\lambda}_1 P_1^* + \bar{\lambda}_2 P_2^* + \ldots + \bar{\lambda}_m P_m^* = \bar{\lambda}_1 P_1 + \bar{\lambda}_2 P_2 + \ldots + \bar{\lambda}_m P_m$$

Therefore

$$TT^* = (\lambda_1 P_1 + \lambda_2 P_2 + \ldots + \lambda_m P_m)(\bar{\lambda}_1 P_1 + \bar{\lambda}_2 P_2 + \ldots + \bar{\lambda}_m P_m)$$
$$= |\lambda_1|^2 P_1^2 + |\lambda_2|^2 P_2^2 + \ldots + |\lambda_m|^2 P_m^2 \qquad [\because P_i P_j = 0 \text{ for } i \neq j]$$
$$= |\lambda_1|^2 P_1 + |\lambda_2|^2 P_2 + \ldots + |\lambda_m|^2 P_m \qquad [\because P_i^2 = P_i]$$

Similarly, we can find

$$T^*T = |\lambda_1|^2 P_1 + \ldots + |\lambda_m|^2 P_m$$

Thus, we conclude that

$$T^*T = TT^*$$

Hence, T is normal.

(iii) ⇒ (i) Let T be a normal operator on H. To show that M_i 's are pairwise orthogonal and M_i 's span H.

Since we assume that T is normal and M_i 's are eigen spaces of T, then clearly M_i 's are pairwise orthogonal.

Now to show that M_i 's span H.

Since M_i 's are pairwise orthogonal and P_i 's are projection on M_i 's.

Then, P_i 's are pairwise orthogonal. Now, let

$$M = M_1 + M_2 + \ldots + M_m$$

Then, clearly M is a closed linear subspace of H and its associated projection is

$$P = P_1 + P_2 + \ldots + P_m$$

Since T is normal operator on H, thus each eigen spaces M_i of T reduces T. Again, P_i is the projection on the closed linear subspace M_i of H, thus M_i reduces T implies that $P_iT = TP_i$

$\Rightarrow$ $P_iT = TP_i$ for each P_i.

Therefore,

$$TP = T(P_1 + P_2 + ... + P_m) = TP_1 + TP_2 + ... + TP_m$$
$$= P_1T + P_2T + ... + P_mT = (P_1 + P_2 + ... + P_m)T = PT$$

Further $TP = PT$ and P is the projection on M. Then M reduces T and therefore $M^\perp$ is invariant under T.

Let U be the restriction of T to $M^\perp$, then clearly U is an operator on a finite dimensional Hilbert space $M^\perp$ and $Ux = Tx$ for all $x \in M^\perp$.

If x is an eigen vector for U corresponding to the eigen value λ, then $x \in M^\perp$ and $Ux = \lambda x$. Thus, $Tx = \lambda x$.

$\Rightarrow$ x is an eigen vector for T also.

Thus, each eigen vector for U is also an eigen vectors for T. But T has no eigen vector in $M^\perp$ since all the eigen vectoers for T are in M and $M \cap M^\perp = \{0\}$, therefore U is an eigen vector and so an eigen value.

$\Rightarrow$ $M^\perp = \{0\}$ because if $M^\perp \neq \{0\}$, then every operator on a non-zero finite dimensional Hilbert space must have an eigen value

$$M^\perp = \{0\}$$

which implies $M = H$. Hence, $M_1 + M_2 + ... + M_m = H$ and so $M_i^{'s}$ span H.

THEOREM 9. ***If T is a normal operator on a Hilbert space H then eigen vectors of T belonging to distinct eigen values are orthogonal.***

SOLUTION. Let H be a Hilbert space over K and T be a linear operator such that

$$Tx = \lambda x,\ Ty = \mu y \qquad (\mu \neq \lambda)$$

We have to prove that $x \perp y$

Since $Tx = \lambda x$ and $Ty = \mu y$, then

$$T^*x = \bar{\lambda}x, T^*y = \bar{\mu}y$$

Now, consider

$$\lambda(x, y) = (\lambda x, y) = (Tx, y)$$
$$= (x, T^*y) = (x, \bar{\mu}y) = \mu(x, y)$$

$\Rightarrow$ $(\lambda - \mu)(x, y) = 0; \qquad (\lambda \neq \mu)$

$\Rightarrow$ $(x, y) = 0$

Hence, $x \perp y$.

THEOREM 10. ***The spectral resolution of a normal operator on a finite dimensional non-zero Hilbert space is unique.***

PROOF. Let T be a normal operator on a finite dimensional non-zero Hilbert space H. Further let

$$T = \sum_{i=1}^{n} \lambda_i P_i \qquad ...(1)$$

be a spectral resolution of T where P_i's are pairwise non-zero orthogonal projections such that

$$\sum_{i=1}^{n} P_i = I \qquad \ldots(2)$$

λ_i's are distinct complex numbers.

Firstly, we shall show that λ_i's are eigen value of T. Then, there exists a non-zero vector x in the range of P_i such that $P_i x = x$

and $$T_x = \left(\sum_{j=1}^{n} \lambda_j P_j\right) x = \sum_{j=1}^{n} \lambda_j P_j P_i(x)$$

$$\Rightarrow \quad T_x = \lambda_i P_i^2(x) \qquad (\because P_i P_j = 0 \text{ if } i \neq j)$$

$$= \lambda_i P_i x \qquad (\because P_i^2 = P_i)$$

$$= \lambda_i x$$

$\Rightarrow$ λ_i is an eigen value of T $\forall$ i

Now, write $A = [\lambda_1, \lambda_2, \ldots, \lambda_n]$

We will prove that eigen value of T are precisely members of A.

Let if possible, λ be an arbitrary eigen value of T such that $Tx = \lambda x$.

Then $$Tx = \lambda I x$$

$$\Rightarrow \quad \left(\sum_{i=1}^{n} \lambda_i P_i\right) x = \lambda \left(\sum_{i=1}^{n} P_i\right) x$$

$$\Rightarrow \quad \sum_{i=1}^{n} \lambda_i P_i(x) = \lambda \sum_{i=1}^{n} P_i(x)$$

$$\Rightarrow \quad \sum_{i=1}^{n} \lambda_i P_j \mathrm{P}_i(x) = \lambda \sum_{i=1}^{n} P_j P_i(x)$$

$$\Rightarrow \quad \lambda_j P_j^2(x) = \lambda P_j^2(x)$$

$$\Rightarrow \quad \lambda_j P_j x = \lambda P_j(x) \qquad (\because P_j^2 = P_j)$$

$$\Rightarrow \quad \lambda_j x = \lambda x$$

$$\Rightarrow \quad (\lambda_j - \lambda) x = 0$$

$$\Rightarrow \quad \lambda_j = \lambda \qquad (\because x \neq 0)$$

$\Rightarrow$ Every eigen value of T is a member of A.

$\Rightarrow$ In the spectral resolution of (1) of T, the scalars λ_i's are distinct eigen value of T.

Now, it remains to prove the uniqueness of T.

Let if possible $$T = \sum_{i=1}^{n} \alpha_i Q_i \qquad \ldots(3)$$

be another spectral resolution of T, where α_i's are distinct eigen value of T and Q_i's are pairwise orthogonal projections.

Now, from (3)

$$T = \sum_{i=1}^{n} \lambda_i \theta_i \qquad \ldots(4)$$

$$\Rightarrow \quad T^2 = \left(\sum_i \lambda_i P_i\right)\left(\sum_j \lambda_j P_j\right) = \sum_{i,j} \lambda_i \lambda_j P_i P_j$$

$$= \sum \lambda_i^2 P_i^2 \qquad (\because P_i P_j = 0 \text{ if } i \neq j)$$

$$= \sum \lambda_i^2 P_i$$

$$\Rightarrow \quad T^2 = \sum_{i=1}^{n} \lambda_i^2 P_i \qquad \ldots(5)$$

Let $g(t)$ be any polynomial with complex coefficient in the complex variable t. Taking linear combination

$$g(T) = \sum_{i=1}^{n} g(\lambda_i) P_i$$

Let p_i be a polynomial such that $p_i(\lambda_i) = \begin{cases} 0 & \text{if } \quad i = j \\ 1 & \text{if } \quad i \neq j \end{cases}$ ...(6)

Taking p_i in place of g

$$P_j(T) = \sum_{i=1}^{n} p_j(\lambda_i) P_i = P_j$$

$$\Rightarrow \quad p_j(T) = P_j$$

If we take $\quad p_i(t) = \dfrac{(t-\lambda_1)(t-\lambda_2)\ldots(t-\lambda_n)}{(\lambda_i - \lambda_1)(\lambda_i - \lambda_2)\ldots(\lambda_i - \lambda_n)}$

$\Rightarrow \quad p_i$ satisfies equation (6)

Apply the same discussion for Q_i's then we shall get

$$Q_i = p_i(T) \text{ for each } i$$

$$\Rightarrow \quad P_i = Q_i \text{ for each } i$$

Hence, spectral resolution given by (1) is unique.

RECAPITULATIONS

- The eigen space of a linear operator T on a Hilbert space H is invariant under T.
- The eigen values of a self-adjoint operator are real.
- An operator T on a Hilbert space H is singular if and only if there exists a non-zero vector x in H such that $Tx = 0$.
- If T is a normal operator on a Hilbert space H. Then x is an eigen vector of T with eigen value λ if and only if x is an eigen vector of T^* with value $\overline{\lambda}$.
- If T is a normal operator on a Hilbert space H, then the eigen spaces of T are pairwise orthogonal.
- If T is a normal operator on a Hilbert space H then eigen vectors of T belonging to distinct eigen values are orthogonal.
- The spectral resolution of a normal operator on a finite dimensional non-zero Hilbert space is unique.

SOLVED EXAMPLES

EXAMPLE 1. *Let $B = \{e_1, e_2, ..., e_n\}$ be an orthogonal basis for a finite dimensional Hilbert space H. If T is an operator on H whose metric to B is $[\alpha_{ij}]$, show that the matrix T^* relative to B is the conjugate transpose of the matrix $[\alpha_{ij}]$.* (MEERUT–2000)

SOLUTION. As per given, we have the matrix T relative to the basis B is $[\alpha_{ij}]$.

Therefore, $$T_{e_j} = \sum_{i=1} \alpha_{ij} e_i \ ; \ j = 1, 2, ..., n$$

Now $$(Te_j, e_i) = \left(\sum_{i=1}^{n} \alpha_{ij} e_i, e_i \right) = (\alpha_{ij} e_1 + \alpha_{2j} e_2 + ... + \alpha_{nj} e_n, e_i)$$
$$= \alpha_{ij}(e_1 e_i) + \alpha_{2j}(e_2 e_i) + ... + \alpha_{nj}(e_n, e_i) = \alpha_{ij}$$

Therefore, if $[\alpha_{ij}]$ is the matrix of T relative to the basis B.

Then, $$\alpha_{ij} = (Te_j, e_i) \text{ for } i \in N, \ j \in N \qquad ...(1)$$

Let $[B_{ij}]$ be the matrix of T^* relative to the basis B then using (1), we have
$$B_{ij} = (T^* e_j, e_i) = \overline{(e_i, T^* e_j)} = \overline{(T^* e_i, e_j)} = \overline{\alpha}_{ji}$$

Hence, the matrix $[B_{ij}]$ is the conjugate transpose of the matrix $[\alpha_{ij}]$.

EXAMPLE 2. *Show that an operator T on a finite dimensional Hilbert space H is normal if and only if its adjoint T^* is a polynomial in T.* (MEERUT–1982, 2001, 02)

SOLUTION. Let us first suppose that T^* is a polynomial in T.

Let $$T^* = \alpha_0 I + \alpha_1 T + \alpha_2 T^2 + ... + \alpha_k T^k.$$

Then, we can easily proved that $T^* T = TT^*$.

$\Rightarrow$ T is normal.

Conversely, let T be normal and let
$$T = \lambda_1 P_1 + \lambda_2 P_2 + ... + \lambda_m P_m$$
be the spectral resolution of T, then
$$T^* = (\lambda_1 P_1 + \lambda_2 P_2 + ... + \lambda_m P_m)^*$$
$$= \overline{\lambda}_1 P_1^* + \overline{\lambda}_2 P_2^* + ... + \overline{\lambda}_m P_m^*$$
$$= \overline{\lambda}_1 P_1 + \overline{\lambda}_2 P_2 + ... + \overline{\lambda}_m P_m \qquad [\because P_i^* = P_i]$$

$\Rightarrow$ P is a polynomial in T for each i

Hence, P^* is also a polynomial in T.

EXAMPLE 3. *Let T be a normal operator on a Hilbert space H and let M be an eigen space of λ, where λ is an eigen value of T. Then show that M and $M^\perp$ both are invariant under T.*

SOLUTION. By definition of eigen space, we can write
$$M = \{x \in H : Tx = \lambda x\}$$

Now $\lambda \in K, x \in M \Rightarrow Tx = \lambda x, \lambda x \in M$ ($\because$ M is a subspace)

$\Rightarrow Tx \in M$

$\Rightarrow$ M is invariant under T.

Since T is normal, therefore $TT^* = T^* T$

$\Rightarrow$ $$T(T^* x) = [TT^*](x) = (T^* T)x$$

$$= T^*(Tx) = T^*(\lambda x) = \lambda T^* x$$

$\Rightarrow \quad T(T^*x) = \lambda(T^*x)$

$\Rightarrow \quad T^*x \in M$ (By definiton of M)

Therefore, for all $x \in M$, $T^*x \in M$

$\Rightarrow$ T^* is invariant under T^*

Findly, M is invariant under T^*

$\Rightarrow$ $M^\perp$ is invariant under $(T^*)^* = T$

$\Rightarrow$ $M^\perp$ is invariant under T

EXAMPLE 4. *Let T be a normal operator on a finite dimensional Hilbert space and let p be a polynomial. Show that if λ is an eigen value of T, then $p(\lambda)$ is an eigen value of $p(T)$.*

SOLUTION. Let T be a normal operator on a finite dimensional Hilbert space H. Also, let λ is an eigen value of T, *i.e.*, $T\mu = \lambda\mu$

We have already known that $p(T)$ is a normal operator. So it remains to prove that $p(\lambda)$ is an eigen value of $p(T)$.

Now, $\quad p(x) = a_0 + a_1x + a_2x^2 + ... + a_mx^m$

$\Rightarrow \quad p(T) = a_0I + a_1T + a_2T^2 + ... + a_mT^m \quad ...(1)$

$\therefore \quad Tu = \lambda u \Rightarrow \quad T^2u = T.Tu = T(\lambda u) = \lambda(Tu) = \lambda.\lambda u = \lambda^2 u$

$\Rightarrow \quad T^2u = \lambda^2 u$

... ...

In general $\quad T^n u = \lambda^n u$

Using all these in (1), we get

$$p(T)u = a_0Iu + a_1Tu + a_2T^2u + ... + a_mT^mu$$
$$= a_0u + a_1\lambda u + a_2\lambda^2 u + ... + a_m\lambda^m u$$
$$= (a_0 + a_1\lambda + a_2\lambda^2 + ... + a_m\lambda^m)u = p(\lambda)u$$

$\Rightarrow \quad p(T)u = p(\lambda)u$

Hence, $p(\lambda)$ is an eigen value of $p(T)$.

EXAMPLE 5. *Let T be a linear operator on a finite dimensional complex Hilbert space $H(K)$ then T is nomal if and only if the adjoint T^* is a polynomial in T.*

SOLUTION. Let us first suppose T is normal. We have to prove that T^* is a polynomial in T. Since, T is normal and so, T is expressible as the spectral solution given by

$$T = \sum_{i=1}^{n} c_iE_i, c_i \in K$$

and E_i is pairwise orthogonal projections so that

$$E_i^* = E_i = E_i^2$$

Now, T is normal therefore $TT^* = T^*T$

$$T^* = (\sum_{i=1}^{n} c_iE_i)^* = \sum_{i=1}^{n} \overline{c_i}E_i^* = \sum_{i=1}^{n} \overline{c_i}E_i$$

$$\Rightarrow \quad T^* = \sum_{i=1}^{n} \overline{c_i}E_i$$

$\Rightarrow$ T^* is a polynomial in T ($\because$ Each E_i is a polynomial in T)

Conversely, let T^* be a polynomial in T, so that $T^* = f(T)$

We have to prove that T is normal.

Consider $TT^* = Tf(T)$ and $T^*T = f(T).T = Tf(T) = TT^*$

$\Rightarrow$ $$TT^* = T^*T$$

$\Rightarrow$ T is normal.

EXAMPLE 6. *If T is unitary then show its eigen values has absolute value one.*

SOLUTION. Let T be unitary then by definition

$$T^*T = TT^* = I \qquad \ldots(1)$$

Also, let $$Tu = \lambda u \qquad \ldots(2)$$

We have to prove that $|\lambda| = 1$

Now, $$\|T(u)\|^2 = (Tu, Tu) = (\lambda u, \lambda u)$$
$$= \lambda\bar{\lambda}(u,u) = |\lambda|^2.\|u\|^2$$

$\Rightarrow$ $$\|T(u)\|^2 = |\lambda|^2.\|u\|^2 \qquad \ldots(3)$$

Again, $$\|T(u)\|^2 = (Tu, Tu) = (T^*Tu, u)$$
$$= (Iu, u) = (u, u) = \|u\|^2 \qquad \ldots(4)$$

Using (3) and (4), we have

$$|\lambda|^2 \|u\|^2 = \|u\|^2$$

$\Rightarrow$ $$|\lambda|^2 = 1$$

$\Rightarrow$ $$|\lambda| = 1$$

EXAMPLE 7. *Let T be any operator on a finite dimensional Hilbert space H and N a normal operator on H. Show that if T commutes with N, then T also commutes with N^*.*

SOLUTION. Let T commutes with normal operator N

Then $$NN^* = N^*N \qquad \ldots(1)$$

and $$TN = NT \qquad \ldots(2)$$

We have to prove that $TN^* = N^*T$

We claim that $$TN^k = N^kT \qquad \ldots(3)$$

The result is true for $k = 1$, by (2)

Let (3) be true for $k = m$, then

$$TN^m = N^mT \qquad \ldots(4)$$

Now, $$TN^{m+1} = (TN^m)N = (N^mT)N$$
$$= N^m(TN) = N^m(NT) = N^{m+1}T$$

or $$TN^{m+1} = N^{m+1}T$$

$\Rightarrow$ Result is true for $n = m + 1$, if it is true for $n = m$.

Hence, by principle of mathematical induction result (3) is true for all n.

Let $p(t) = a_0 + a_1t + a_2t^2 + \ldots + a_nt^n$ be any polynomial with complex coefficients.

Then, $$T_p(N) = T[a_0I + a_1N + a_2N^2 + \ldots + a_nN^n]$$
$$= a_0T + a_1TN + a_2TN^2 + \ldots + a_nTN^n$$

$$= a_0IT + a_1NT + a_2N^2T + ... + a_nN^nT \quad \text{(By 3)}$$
$$= (a_0I + a_1N + a_2N^2 + ... + a_nN^n)T$$

$$\Rightarrow \quad T_p(N) = p(N).T \quad ...(5)$$

Now, let $N = \sum_{i=1}^{n} \lambda_i P_i$ be the spectral resolution of the normal operator N. Here P_i is projection and therefore

$$P_i^2 = P_i^* = P_i$$

Then $N^* = (\Sigma\lambda_i P_i)^* = \Sigma\bar{\lambda}_i P_i^* = \sum_i \bar{\lambda}_i P_i$

$\Rightarrow \quad N^* = \sum_i \bar{\lambda}_i P_i$

But each operator P_i is a polynomial in N.
Hence, from (5), N^* commutes with T.

EXAMPLE 8. *If T is an operator on a finite dimensional Hilbert space H, then prove the following :*

(i) T is singular if and only if 0 is spectrum of T.

(ii) If T is non-singular then $\lambda \in \sigma(T) \Rightarrow \lambda^{-1} \in \sigma(T^{-1})$

(iii) If A is non-singular then $\sigma(ATA^{-1}) = \sigma(T)$

(iv) If $\lambda \in \sigma(T)$ and if p is any polynomial, then $p(\lambda) \in \sigma[p(T)]$

(MEERUT–2003, 09)

SOLUTION. (i) T is singular $\Leftrightarrow \exists\, x \in H$ such that $Tx = 0, x \neq 0$

$\Leftrightarrow Tx = 0x$

$\Leftrightarrow$ 0 is an eigen value of T

$\Leftrightarrow 0 \in \sigma(T)$, *i.e.*, 0 is the spectrum of T

(ii) T is non-singular, $\lambda \in \sigma(T)$

$\Rightarrow$ λ is an eigen value of non-singular operator T

$\Rightarrow$ $\lambda \neq 0$ is eigen value of non-singular operator T

$\Rightarrow$ $\exists\, x \in H$ s.t. $x \neq 0$ and $Tx = \lambda x$

$\Rightarrow$ $x = T^{-1}(\lambda x) = \lambda T^{-1}(x)$

$\Rightarrow$ $\lambda^{-1}x = T^{-1}x$

$\Rightarrow$ $\lambda^{-1} \in \sigma(T^{-1})$

(iii) Let $S = ATA^{-1}$, then

$$S - \lambda I = ATA^{-1} - \lambda IAA^{-1} = AA^{-1} - A(\lambda IA^{-1})$$
$$= A[T - \lambda I]A^{-1}$$

$$\Rightarrow \det(S - \lambda I) = \det A.\det(T - \lambda I).\det(A^{-1})$$
$$= (\det A.\det A^{-1}) \det(T - \lambda I)$$
$$= (\det AA^{-1}) \det(T - \lambda I)$$
$$= I.\det(T - \lambda I) = \det(T - \lambda I)$$

$\Rightarrow \det(S - \lambda I) = \det(T - \lambda I)$

$\therefore \det(S - \lambda I) = 0 \quad \Leftrightarrow \quad \det(T - \lambda I) = 0$

But we have

$\det(S - \lambda I) = 0 \quad \Leftrightarrow \quad \lambda \in \sigma(S)$

Therefore, $\lambda \in \sigma(S) \quad \Leftrightarrow \quad \lambda \in \sigma(T)$

or $\sigma(S) = \sigma(T)$

or $\sigma(ATA^{-1}) = \sigma(T)$

(iv) Let $\lambda \in \sigma(T)$ then there exist non-zero vector x such that

$$Tx = \lambda x$$

$$\Rightarrow \quad T^2x = T(\lambda x) = \lambda Tx = \lambda.\lambda x = \lambda^2 x$$

$$\Rightarrow \quad T(T^2x) = T(\lambda^2 x) = \lambda^2 Tx$$

$$\Rightarrow \quad T^3x = \lambda^2\lambda x = \lambda^3 x$$

i.e., $Tx = \lambda x,\ T^2x = \lambda^2 x,\ T^3x = \lambda^3 x$

In general, $T^n x = \lambda^n x$

Let $p(t) = a_0 + a_1 t + a_2 t^2 + \ldots + a_n t^n$ be a polynomial in t of degree n where $a_n \neq 0$

Then
$$\begin{aligned} p(T)x &= (a_0 I + a_1 T + a_2 T^2 + \ldots + a_n T^n)x \\ &= a_0 x + a_1 \lambda x + a_2 \lambda^2 x + \ldots + a_n \lambda^n x \\ &= (a_0 + a_1 \lambda + a_2 \lambda^2 + \ldots + a_n \lambda^n)x = p(\lambda)x \end{aligned}$$

$\Rightarrow$ $p(T)x = p(\lambda)x$ with $Tx = \lambda x$

Hence $\lambda \in \sigma(T) \Rightarrow p(\lambda) \in \sigma[p(T)]$

EXAMPLE 9. *Let T be a linear operator on a finite dimensional Hilbert space H with spectrum $\lambda_1, \lambda_2, \ldots, \lambda_m$. Then prove the following*

(i) T is self adjoint $\Leftrightarrow$ each λ_i is real

(ii) T is positive $\Leftrightarrow$ each $\lambda_i \geq 0$

(iii) T is unitary $\Leftrightarrow$ $|\lambda_i| = 1\ \forall\ i$

(MEERUT–2003, 04, BHOPAL–2011)

SOLUTION. Let
$$T = \sum_{i=1}^{n} \lambda_i P_i \quad \ldots(1)$$

be the spectral resolution of T, where $\lambda_1, \ldots, \lambda_n$ are eigen values of T and P_1, $P_2, \ldots, P_n$ are non-zero pairwise orthogonal projection, therefore

and
$$\left.\begin{aligned} P_i^2 = P_i^* = P_i \quad & \forall i \\ P_i P_j = 0 \quad & \text{for } i \neq j \end{aligned}\right] \quad \ldots(2)$$

Also we have
$$\sum_{i=1}^{n} P_i = I$$

Now
$$T^* = (\Sigma\lambda_i P_i)^* = \sum_i \bar{\lambda}_i P^* = \Sigma\bar{\lambda}_i P_i$$

$$\Rightarrow \quad T^* = \Sigma\bar{\lambda}_i P_i \quad \ldots(3)$$

(i) Let T be self-adjoint, then $T = T^*$ ...(4)

Using (1) and (3) in (4), we get

$$\sum_i \bar{\lambda}_i P_i = \Sigma\lambda_i P_i$$

$$\Rightarrow \quad \sum_i (\bar{\lambda}_i - \lambda_i) P_i = 0$$

Multiplying both sides by P_j and noting that $P_i P_j = 0\ \forall\ i$ except $i = j$

We get $(\bar{\lambda}_j - \lambda_j)P_j^2 = 0$

or $(\bar{\lambda}_j - \lambda_j)P_j = 0$

$\Rightarrow$ $\bar{\lambda}_j - \lambda_j = 0$

$\Rightarrow$ $\bar{\lambda}_j = \lambda_j$

$\Rightarrow$ λ_j is real

Conversely, if λ_i is real for i then $\lambda_i = \bar{\lambda}_i$ ...(5)

To prove T is self-adjoint

Using (3) and (5) we get

$$T^* = \Sigma\lambda_i P_i \quad \text{...(6)}$$

From (1) and (6), we have

$$T^* = T$$

$\Rightarrow$ T is self-adjoint.

(ii) Consider

$$(Tx, x) = (Tx, Ix) = (\Sigma\lambda_i P_i x, \Sigma P_j x)$$

$$= \sum_i \sum_j \lambda_i (P_i x, P_j x) = \sum_{i,j} \lambda_i (P_j^* P_i x, x) \quad \text{...(7)}$$

But $$P_j^* P_i = P_j P_i = \begin{cases} P_i^2 & \text{for } i = j \\ 0 & \text{for } i \neq j \end{cases}$$

$$P_j^* = P_j$$

Using this in (7), we get

$$(Tx, x) = \sum_i \lambda_i (P_i^2 x, x) \quad \text{...(8)}$$

But $(P_i^2 x, x) = (P_i x, P_i^* x) = (P_i x, P_i x) = \|P_i x\|^2$

Using this in (8), we get

$$(Tx, x) = \sum_i \lambda_i \|P_i x\|^2 \quad \text{...(9)}$$

Let T be positive, then $(Tx, x) \geq 0$ ...(10)

We want to prove that $\lambda_i \geq 0 \ \forall \ i$

Using (9) and (10), we have

$$\sum_i \lambda_i \|P_i x\|^2 \geq 0 \quad \text{...(11)}$$

Now, let us suppose that x is in the range of P_1, then

$$P_1 x = x,\ P_i x = 0 \ \forall \ i \text{ except } i = 1$$

Then from (11), $\lambda_1 \geq 0$

Similarly, we may prove that $\lambda_2 \geq 0$

... ...

In general, $\lambda i \geq 0 \ \forall \ i$

Conversely, suppose that $\lambda_i \geq 0 \ \forall \ i$ ...(12)

To prove that, T is positive

Using (9) and (12) we have

$$(Tx, x) \geq 0$$

$\Rightarrow$ T is positive

(iii) Consider $TT^* = (\Sigma\lambda_i P_i)(\Sigma\lambda_j P_j)$

$$= \sum_i \sum_j \lambda_i \bar{\lambda}_j P_i P_j = \sum \lambda_i \bar{\lambda}_i P_i^2 = \sum |\lambda_i|^2 P_i \quad \ldots(13)$$

Now, T is unitary $\Leftrightarrow TT^* = I$

$$\Leftrightarrow \sum_i |\lambda_i|^2 P_i = \sum P_i \quad \text{(By (2) and (13))}$$

$$\Leftrightarrow \sum_i (|\lambda_i|^2 - 1) P_i = 0$$

Multiplying both sides by P_j and noting that $P_i P_j = 0 \ \forall \ i$ except $i = j$
We get

$$(|\lambda_j|^2 - 1) P_j = 0$$

$$\Leftrightarrow |\lambda_j|^2 - 1 = 0$$

$$\Leftrightarrow |\lambda_j| = 1 \ \forall \ j$$

EXAMPLE 10. *If T is a bounded linear operator on a Banach space E and $\|T\| < |\lambda|$ then show that $T_\lambda = (T - \lambda I)^{-1}$ is a bounded operator*

$$T_\lambda = -\sum_{n=0}^{\infty} \frac{T^n}{\lambda^{n+1}} \quad \ldots(1)$$

and

$$\|T_\lambda\| \le \frac{1}{|\lambda| - \|T\|} \quad \ldots(2)$$

SOLUTION. Since, we have

$$\|T/\lambda\| < 1$$

$$\therefore \quad \sum_{n=0}^{\infty} \left\| \frac{T^n}{\lambda^n} \right\| \le \sum_{n=0}^{\infty} \left\| \frac{T}{\lambda} \right\|^n \to \infty$$

So, there exists a bounded linear operator B on E such that

$$B = \sum_{n=0}^{\infty} \frac{T^n}{\lambda^n}$$

Further,

$$|T - \lambda I| B = (T - \lambda I)\left(\sum_{n=0}^{\infty} \frac{T^n}{\lambda^n} \right) = \sum_{n=0}^{\infty} (T - \lambda I) \frac{T^n}{\lambda^n}$$

$$= \sum_{n=0}^{\infty} \frac{T^{n+1} - \lambda T^n}{\lambda^n} = \lambda \sum_{n=0}^{\infty} \left(\frac{T^{n+1}}{\lambda^{n+1}} - \frac{T^n}{\lambda^n} \right) = -\lambda I$$

Similarly, $B(T - \lambda I) = -\lambda I$, therefore

$$T_\lambda = (T - \lambda I)^{-1} = -\frac{B}{\lambda} = -\sum_{n=0}^{\infty} \frac{T^n}{\lambda^{n+1}}$$

Further,

$$\|T_\lambda\| \le \frac{1}{|\lambda|} \sum_{n=0}^{\infty} \left\| \frac{T}{\lambda} \right\|^n = \frac{1}{\lambda} \left[\frac{1}{1 - \|T/\lambda\|} \right] = \frac{1}{|\lambda| - \|T\|}$$

REMARK

- The above reperesentation (2) usually called as 'Neumann series'.

EXAMPLE 11. *Let T be an invertible operator on a linear space E and let A be an operator on E. Then show that the operator A and TAT^{-1} have the same eigen values.*

SOLUTION. Let λ be an eigen value of A. Then by definition, there exists a non-zero vector u, such that $Au = \lambda u$.

Now, since T is invertible, $Tu \neq 0$ and

$$TAT^{-1}(Tu) = TAu = T(\lambda u) = \lambda Tu$$

$\Rightarrow$ λ is an eigen value of TAT^{-1}

Further, assume that λ is an eigen value of TAT^{-1}, *i.e.*, $TAT^{-1} = u = \lambda u$ for some non-zero vector $u = Tu$

Now, since $AT^{-1}u = \lambda T^{-1}u$ and $T^{-1}u \neq 0$

$\Rightarrow$ λ is an eigen value of A

Hence, we conclude that A and TAT^{-1} both have the same eigen values.

EXAMPLE 12. *Show that the eigen spaces corresponding to non-zero eigen values of a compact self-adjoint operator are finite dimensional.*

SOLTION. Let $\lambda \neq 0$ be an eigen values of a compact self-adjoint operator T and let $\in_\lambda$ be the eigen space corresponding to eigen value λ. If $\in_\lambda$ is of infinite dimesion. Now, let $[x_1, x_2, ...]$ be an orthogonal basis of $\in_\lambda$. Then $Tx_n \to 0$ as $n \to \infty$ (because $<x_n>$ is weakly convergent to zero). This is not possible because $Tx_n = \lambda x_n$ for all $n \in N$ and $\lambda \neq 0$. Hence, the eigen space corresponding to non-zero eigen values of a compact self-adjoint operator are finite dimensional.

EXAMPLE 13. *Show that the set of distinct non-zero eigen value (λ_n) of a compact self-adjoint operator is either finite or countable with* $\lim_{n\to\infty} \lambda_n = 0$.

SOLUTION. Let T be a compact self adjoint operator that has infinitely many eigen values λ_n, $n \in N$. Further, let u_n be an eigen vector corresponding to λ_n such that $\|u_n\| = 1$. Then $<u_n>$ is an orthogonal sequence converges weakly to 0. Hence, $<Tu_n>$ converges strongly to 0.

Thus, $$|\lambda_n| = \|\lambda_n u_n\| = \|Tu_n\| \to 0 \text{ as } n \to \infty.$$

EXAMPLE 14. *If T is a compact operator, then show that every non-zero approximate eigen value of T is an eigen value.*

SOLUTION. Let $<x_n>$ be a sequence of vectors such that

$$\|x_n\| = 1 \ \forall \ n \in N$$

and $\|Tx_n - \lambda x_n\| \to 0$ as $n \to \infty$, for some $\lambda \neq 0$

Now, since, T is compact, there exists a subsequence $< x_{p_n} >$ of $<x_n>$ such that $Tx_{p_n} \to y$ as $n \to \infty$ for some y.

Then $\left\|y - \lambda x_{p_n}\right\| \leq \left\|y - Tx_{p_n}\right\| + \left\|Tx_{p_n} - \lambda x_{p_n}\right\| \to 0$ as $n \to \infty$

Further, since $\lambda \neq 0$, $\quad x_{p_n} \to \dfrac{y}{\lambda}$

Let $\dfrac{y}{\lambda} = u$, then $\qquad \|u\| = 1$

and $\|Tu - \lambda u\| \leq \left\|Tu - Tx_{p_n}\right\| + \left\|Tx_{p_n} - y\right\| \to 0$ as $n \to \infty$

$\Rightarrow$ $$Tu = \lambda u$$

EXAMPLE 15. *If T is compact, self-adjoint operator on a Hilbert space, show that at least one of the numbers $\|T\|$ or $-\|T\|$ is an eigen value of T.*

SOLUTION. If $T = 0$, then result is trivally true. Assume that T is a non-zero compact self-adjoint operator on a Hilbert space H. Then there exists a sequence $<x_n>$ in H such that $\|x_n\| = 1$.

and $$Tx_n - \lambda x_n \to 0 \text{ as } n \to \infty$$

By definition of compactness we can say that the sequence $<x_n>$ has a subsequence $< x_{p_n} >$ such that the sequence $< Tx_{p_n} >$ converges. Now, since $T \neq 0$, therefore, $x_{p_n} \to u$ for some $u \in H$. Because $\|u\| = 0$ as $\left\|x_{p_n}\right\| = 1 \ \forall n \in N$.

Hence, by continuity of T, we get

$$Tu = \lambda u.$$

RECAPITULATIONS

- If T is unitary then its eigen values has absolute value 1.
- The eigenspace corresponding to non-zero eigen values of a compact self-adjoint operator are finite dimensional.
- If T is a compact operator then every non-zero approximate eigen value of T is an eigen value.
- Eigen values of a self-adjoint operator are real.

EXERCISE 6.1

1. Define eigen value, eigen vector, eigen space and spectrum of an operator T on a Hilbert space H.
2. Show that the eigen values of a self-adjoint operator are real.
3. Let $T : l_2 \to l_2$ be defined by

$$T(x_1, x_2, \ldots)\left(\frac{x_1}{1}, \frac{x_2}{2}, \frac{x_3}{3}, \ldots\right)$$

   (i) Find $\sigma(T)$.
   (ii) Find the eigen values of T.
   (iii) Show that T is a compact operator.
4. Let $T : R^2 \to R^2$ be defined by $T(x, y) = (x + 2y, 3x + 2y)$. Find the eigen values and eigen vectors of T.
5. Prove that if T is normal operator then x is an eigen vector of T with eigen value λ if and only if x is an eigen vector of T^* with eigen value $\bar{\lambda}$.

Chapter Review: A Competitive Approach

SELECTED TERMS AND RESULTS

▶ TERMS

- **Eigen Values and Eigen Vectors :** Let T be a linear operator on a Hilbert space H. Any scalar $\lambda \in K$ is called an eigen value of T if there exist a non-zero vector $v \in H$ such that $Tv = \lambda v$ such that vector v is called eigen vector of T corresponding to the eigen value λ.
- **Spectrum :** The set of all eigen values of an operator T is called the spectrum of T.
- **Eigen space :** The set $M = \{x \in H : Tx = \lambda x\}$ is called the eigen space of λ.
- **Spectral solution :** Let T be a linear operator on a Hilbert space H. If there exists distinct complex numbers $\lambda_1, \lambda_2, \ldots, \lambda_n$ and non-zero pairwise orthogonal projections $P_1, P_2, \ldots, P_n$ such that $T = \sum_{i=1}^{n} \lambda_i P_i$ and $I = \sum_{i=1}^{n} P_i$. Then expression is called spectral solution for T.

▶ RESULTS

- Let λ be an eigen value of an operator T on a Hilbert space H. Let M be the eigen space of λ then M is closed subspace of H.
- The eigen space of a linear operator T on a Hilbert space H is invariant under T.
- The eigen values of a self adjoint operator are all real.
- If x, y are eigen vectors of a self adjoint operator T on a Hilbert space H relative to different eigen values of T, then $x \perp y$.
- An operator T on a Hilbert space H is singular if and only if there exists no non-zero vector x in H such that $Tx = 0$.
- If T is a normal operator on a Hilbert space H. Then x is an eigen vector of T with eigen value λ if and only if x is an eigen vector of T^* with value $\overline{\lambda}$.
- If T is a normal operator on a Hilbert space H, then the eigen spaces of T are pairwise orthogonal.
- Let T be an operator on a finite dimensional Hilbert space H. Let $\lambda_1, \lambda_2, \ldots, \lambda_m$ be the distinct eigen values of T. Let $M_1, M_2, \ldots, M_m$ be their corresponding eigen space and let, $P_1, P_2, \ldots, P_m$ be the projections on these eigen spaces. Then, following statements are equivalent.
  (i) The M_i's are pairwise orthogonal and span H.
  (ii) The P_i's are pairwise orthogonal.
  $P_1 + P_2 + \ldots + P_m = I$
  and $T = \lambda_1 P_1 + \lambda_2 P_2 + \ldots + \lambda_m P_m$
  (iii) T is normal.
- If T is a normal operator on a Hilbert space H then eigen vectors of T belonging to distinct eigen values are orthogonal.
- The spectral resolution of a normal operator on a finite dimensinal non-zero Hilbert space is unique.

REVIEW QUESTIONS

▶ DESCRIPTIVE QUESTIONS

1. If T is a linear operator on a n-dimnesional Hilbert space H, then show that T cannot have more than n distinct characteristic values.
2. Prove that if T is positive or strictly positive then the characteristic values are positive or strictly positive respectively.
3. Show that every characteristic vector of a

normal operator T is also a characteristic vector for T^*.

4. Show that the eigen spaces of a normal operator T are pairwise orthogonal.

5. Let T be a linear operator on a finite dimensional Hilbert space H then prove the following :

(i)Each characteristic value of T is non-negaitve $\Rightarrow T$ is non-negative.

(ii)Each characteristic value of T is different from zero $\Rightarrow T$ is invertible.

(iii)Each characteristic value of T is zero or one $\Rightarrow T$ is idempotent.

▶ MULTIPLE CHOICE QUESTIONS

Choose the most appropriate one.

1. Let T be a linear operator on a finite dimensional complex Hilbert space, then each characteristic value of T is real if and only if T is :

(a) self adjoint (b) 0
(c) adjoint (d) none of these

2. If T is a linear operator on a finite dimensional Hilbert space H, then each characteristic value of T is zero or one if and only if T is :

(a) invertible (b) idempotent
(c) negative (d) none of these

3. The characteristic vector of T and T^* are :

(a) equal (b) unequal
(c) zero (d) none of these

4. The non-zero characteristic vectors $x_1, x_2, ..., x_n$ corresponding to characteristic values $\lambda_1,\lambda_2,...,\lambda_n$ of linear operator T on H are :

(a) linearly dependent
(b) linearly independent
(c) equal
(d) none of these

5. A non-zero vectro $x \in H$ is called characteristic vector of T if there exists a scalars $\lambda \in F$ such that :

(a) $Tx = \lambda x$ (b) $Tx = \lambda$
(c) $Tx = x$ (d) none of these

6. If T is unitary, then eigen values has absolute values :

(a) 0 (b) 1
(c) 2 (d) none of these

7. The eigen values of a self-adjoint operator are all :

(a) zero (b) real
(c) complex (d) none of these

8. If x, y are eigen values of self-adjoint operator T on a Hilbert space H relatve to different eigen values of T then :

(a) $x = y$ (b) $x \neq y$
(c) $x \perp y$ (d) none of these

ANSWERS

1. (a) **2.** (b) **3.** (a) **4.** (b) **5.** (a) **6.** (b) **7.** (b) **8.** (c)

1. If x is a characteristic vector of T associated with characteristic value λ, then show that every scalar multiple of x is also a characteristic vector for the same characteristic value λ.
2. Show that the characteristic vectors corresponding to distinct characteristic values of a self-adjoint operator are orthogonal.
3. Show that the eigen spaces of a normal operator are pairwise orthogonal.
4. If T be a linear operator then show that each eigen space M_i reduces T.
5. State and prove spectral theorem of Hilbert space H for any operator T on H.
6. Let T be a linear operator on a finite dimensional complex Hilbert space, then show that T is normal iff the adjoint T^* is a polynomial in T.
7. Let X be a normed linear space and let $T : X \to X$ be a compact operator. Prove that for every $\lambda \neq 0$, the range of $T - \lambda I$ is closed.
8. Let $T : l_2 \to l_2$ be defined by $T(x) : (0, x_1, x_2, ...)$ for $x = (x_n) \in l_2$. Find $\sigma(T)$.
9. If x is an eigen vector of T, show that x cannot correspond to more than one eigen value of T.
10. Let T be a normal operator on a Hilbert space H, then show that the eigen spaces of T are pairwise orthogonal.

Chapter 7 Banach Algebra

7.1 INTRODUCTION

In this chapter we shall discuss the theory and application of Banach algebra. The theroy of Banach algebra is a large area in functional analysis.

7.2 SOME BASIC DEFINITIONS

(1) Algebra : A linear vector space V over F is called an algebra if for a composition (.) defined on V following conditions are satisfied :

(i) $(u \cdot v)w = u \cdot v + u \cdot w$

(ii) $u \cdot (v + w) = u \cdot v + u \cdot w$

$(v + w) \cdot u = v \cdot u + w \cdot u$

(iii) $\lambda(u \cdot v) = (\lambda u) \cdot v = u \cdot (\lambda v)\ \forall\ u, v \in V$ and $\lambda \in F$

or in other words we can say that an algebra V over any field F is a vector space over F with a mapping.

$$(.) : v \times v \to V \text{ such that } (x, y) \to x \cdot y$$

in which the scalar multiplication is related to the following property.

$$\lambda(x \cdot y) = (\lambda x)y = x(\lambda y)\ \forall\ x, y \in V, \lambda \in F$$

For example : R(R), C(C) are the example of algebra where R and C are respectively the set of real and complex numbers.

Remark

- An algebra V over F is called Complex algebra if F = C and if F = R then it is called real algebra.

(2) Division Algebra : An algebra with identity is called division algebra.

(3) Normed Algebra : An algebra B over F (= R or C) with multiplicative identity 1 which has a norm making it into a normed linear space satisfying

$$\|xy\| \leq \|x\| \cdot \|y\|\ \forall\ x, y \in B$$

(4) Banach Algebra : An algebra B over a field F is called Banach algebra if following conditions are satisfied :

(i) $\|xy\| \leq \|x\| \cdot \|y\|\ \forall\ x, y \in B$

(ii) $\exists\ e \in B$ such that $ex = xe = x$ and $\|e\| = 1\ \forall\ x, y \in B$

(iii) B is complete.

Remarks

- A normed algebra B is called Banach algebra if it is complete as a normed linear space.
- A Banach algebra admits vector addition, vector multiplication and scalar multiplication.
- If B has an identity e then the mapping $\alpha \to \alpha e$ is an isomorphism of F into B and $\|\alpha e\| = |\alpha|$. So, it will be assumed that $F \subseteq B$.
- A Banach algebra B is called a unital Banach algebra if B has the identity e with $\|e\| = 1$.

(5) Regular or invertible element : Let B be a Banach algebra. If corresponding to an element $f \in B$ there exist an element $g \in B$ such that $fg = gf = e$. Then g is called inverse of f (denoted by f^{-1}).

i.e., $g = f^{-1}$ and $ff^{-1} = f^{-1}f = e$

If an element $f \in B$ has an inverse then it is called regular element.

REMARKS

- The identity element e is always regular in B.
- The set of all regular elements of B forms a group.

(6) Singular Element : An element f of a Banach algebra B is said to be singular if it is not regular.

REMARK

- The zero vector **0** is always singular element of A.

ILLUSTRATIONS

(1) C, is a commutative Banach algebra with identity $e = 1$.

(2) The set $B(H, H)$ of all bounded linear transformation on a Hilbert space H with the operator norm is a Banach algebra.

(3) Let X be a Compact T_2-space and let $C(X)$ denotes the set of all continuous complex valued functions defined on X then $C(X)$ is a Banach algebra with respect to pointwise linear operator and the norm defined by $\|f\| = \sup_{x \in X} . |f(x)|$.

(4) Let set $D = \{z \in C : |z| < 1\}$ and $S = \{z \in C : |z| \le 1\}$. If B is the set of all complex functions f analytic on D and continuous on S. Then B is a Banach algebra with respect to pointwise algebraic operations of the functions on B.

(5) If (X, Ω, μ) is a σ-finite measurable space and $B = L^{\infty}(X, \Omega, \mu)$ then B is an abelian Banach algebra if the operations are defined pointwise.

7.3 SPECTRUM

Let B be a Banach algebra, then for any fixed x in X, the set

$\sigma_B(x) = \{\lambda \in \mathbf{C} : x - \lambda I\}$ is not invertible in X is called the spectrum of x.

Further, its compliment $\mathbf{C}$-$\sigma(x)$ is called resolvent set of x.

Also, the spectrum $\sigma(T)$ of an operator T on a Hilbert space H is defined by

$\sigma(T) = \{\lambda : T - \lambda I$ is singular$\}$, where λ is an eigen value of T.

REMARK

- The notation $\sigma_B(x)$ has been used to show the fact that the spectrum of x depends on B and x both.

7.3.1 RESOLVENT EQUATION

We know that $x - \lambda I$ is a continuous function of λ with values in B. Also, the set of singular elements of B is closed and therefore $\sigma(x)$ is also closed, *i.e.*,

$$\sigma(x) = \{z \in C : |z| \le \|x\|\}$$

The resolvent set of $x \in A$ is denoted by $\rho(x)$, which is the compliment of $\sigma(x)$. Hence,

$\rho(x) =$ An open subset of **C** such that

$$\rho(x) = [z \in C : |z| > \|x\|]$$

The resolvent of x is denoted by $x(\lambda)$ and is defined as the function with values in x such that

$$x(\lambda) = (x - \lambda I)^{-1} \to 0 \text{ as } \lambda \to \infty$$

If λ and μ both are in $\rho(x)$, then

$$x(\lambda) - x(\mu) = (\lambda - \mu)x(\lambda)\cdot x(\mu)$$

which is called resolvent equation.

7.3.2 RADICAL OF A BANACH ALGEBRA

Let B be a Banach algebra and M be its maximal left ideal radical of B denoted by R is the intersection of all its maximal left ideal M in B and so

$$R = \cap M(I_l) \text{ where } I_l = \text{left ideal}$$

7.3.3 SPECTRAL RADIUS

Let B be a complex Banach space and f be any element of B. Then we have

$$\sigma(f) \subset \{z : \|z\| \leq \|f\|\} = S[\|f\|, 0]$$

$\Rightarrow$ $\sigma(f)$ is wholly contained in the closed space of radius $\|f\|$ with the centre at origin. If it is possible that $\sigma(f)$ might be contained in the smaller closed circle about the origin then radius of this smallest circle containing $\sigma(f)$ is called spectral radius of f, *i.e.*, spectral radius

$$r(f) = \sup\{|\lambda| : \lambda \in \sigma_B(f)\}$$

REMARKS

- $0 \leq r(f) \leq \|f\|$
- $r(f) = \lim \|f^n\|^{1/n}$
- $r(f^n) = [r(f)]^n$

7.3.4 DIVISION ALGEBRA

A unital Banach algebra B is said to be division algebra if x^{-1} exists for each non-zero x in B.

7.4 COMMUTATIVE BANACH ALGEBRA

A Banach algebra B is said to be commutative if it is commutative as a ring.

7.5 B* ALGEBRA AND BANACH ALGEBRA WITH INVOLUTION

An involution in a Banach algebra B is a mapping $x \to x^*$ of x into itself with the following properties :

(i) $(x + y)^* = x^* + y^*$ (ii) $(\alpha x)^* = \bar{\alpha} x^*$

(iii) $(xy)^* = y^* x^*$ (iv) $x^{**} = x$

Further, a Banach algebra with an evolution defined on it is called Banach algebra with involution.

Definition. *A Banach algebra with involution is called a B^*-algebra if it has the property that*

$$\|x^* x\| = \|x\|^2$$

REMARK

- An element x in B^*-algebra is said to be
  (i) self adjoint if $x = x^*$
  (ii) normal if $xx^* = x^* x$

7.6 TOPOLOGICAL DIVISORS OF ZERO

Let B be a Banach algebra. An element x in B is called topological divisor of zero if there exist a sequence $<s_n>$ in B such that

$$\|s_n\| = 1 \text{ and } s\cdot s_n \to 0$$

The set of all topological divisors of zero be denoted by Z.

7.7 SOME IMPORTANT RESULTS

THEOREM 1. ***Let B be a Banach algebra then every element f for which $\|f - e\| < 1$ is regular and the inverse of f is given by***

$$f^{-1} = e + \sum_{n=1}^{\infty} (e - f)^n$$

(KANPUR–2005, 09, MADRAS–2007)

PROOF. Let B be a Banach algebra. Also, let

$$f \in A \text{ such that } \|f - e\| < 1 \qquad \ldots(1)$$

Here e is the identity element therefore

$$ef = fe = f \;\forall\, f \in B \qquad \ldots(2)$$

We have to prove that $f^{-1} = e + \sum_{n=1}^{\infty} (e - f)^n$

Let $r = \|f - e\|$ then clearly $r < 1$ [By (1)]

and $\|(e - f)^n\| \le \|e - f\|^n = r^n$

$\Rightarrow$ $\|(e - f)^n\| \le r^n \to 0$ as $n \to \infty$ $(\because r < 1)$

$\Rightarrow$ the partial sum of the series $\sum_{n=1}^{\infty} (e - f)^n$ form a Cauchy sequence in B, but B is complete therefore, by definition these partial sums will converges to an element in B. Let this element be

$$\sum_{n=1}^{\infty} (e - f)^n$$

Define an element g by

$$g = e + \sum_{n=1}^{\infty} (e - f)^n \qquad \ldots(3)$$

Consider,

$$g - fg = eg - fg = (e - f)g = (e - f)\left[e + \sum_{n=1}^{\infty} (e - f)^n\right]$$

$$= (e - f)e + \sum_{n=1}^{\infty} (e - f)^{n+1} = (e - f) + \sum_{n=1}^{\infty} (e - f)^{n+1}$$

$$= (e - f) + (e - f)^2 + (e - f)^3 + \ldots = \sum_{n=1}^{\infty} (e - f)^n$$

$$= g - e \qquad \text{[By (3)]}$$

$\Rightarrow$ $g - fg = g - e$

$\Rightarrow$ $fg = e$

In a similar way, we may prove that

$$gf = e$$

So, $fg = gf = e$

$\Rightarrow$ $g = f^{-1}$

Putting this value in (3), we get

$$f^{-1} = e + \sum_{n=1}^{\infty} (e - f)^n$$

REMARK

- If x is regular in B such that $\|x - 1\| < 1$. Then above result can be written as

$$x^{-1} = 1 + \sum_{n=1}^{\infty} (1 - x)^n$$

THEOREM 2. ***Let B be a Banach algebra such that $|\lambda| > \|f\|$ then $\lambda e - f$ is a unit, and***

$$(\lambda e - f)^{-1} = \sum_{n=1}^{\infty} \lambda^{-n} f^{n-1}$$ (KANPUR–2005)

PROOF. Let B be a Banach algebra, $f \in B$ and λ be a scalar such that

$$|\lambda| \geq \|f\| \quad \ldots(1)$$

We have to prove that $\lambda e - f$ is a unit and

$$(\lambda e - f)^{-1} = \sum_{n=1}^{\infty} \lambda^{-n} f^{n-1}$$

Now from (1), we have

$$\frac{\|f\|}{|\lambda|} < 1 \quad \Rightarrow \quad \left\|\frac{f}{\lambda}\right\| < 1 \quad \ldots(2)$$

$\Rightarrow \quad e - \dfrac{f}{\lambda}$ is a unit.

Since, we have if x is regular such that $\|x - e\| < 1$ ...(3)

$$\Rightarrow \quad x^{-1} = e + \sum_{n=1}^{\infty} (e - x)^n \quad \text{[From theorem (1)]} \quad \ldots(4)$$

Let us take $x = e - \dfrac{f}{\lambda}$ then we have

$$\left\|\left(e - \frac{f}{\lambda}\right) - e\right\| = \left\|\frac{f}{\lambda}\right\| < 1 \quad \text{[Using (2)]}$$

$\Rightarrow$ (3) is satisfied.

Now, according to (4), we get

$$\left(e - \frac{f}{\lambda}\right)^{-1} = e + \sum_{n=1}^{\infty}\left[e - \left(e - \frac{f}{\lambda}\right)\right]^n$$

$$\Rightarrow \quad \lambda(\lambda e - f)^{-1} = e + \sum_{n=1}^{\infty} \lambda^{-n} f^n$$

$$\Rightarrow \quad (\lambda e - f)^{-1} = \lambda^{-1} f^0 + \sum_{n=1}^{\infty} \lambda^{-(n+1)} f^n \quad (\because f^0 = e)$$

$$\Rightarrow \quad (\lambda e - f)^{-1} = \lambda^{-1} f^0 + \lambda^{-2} f + \lambda^{-3} f^2 + \ldots = \sum_{n=1}^{\infty} \lambda^{-n} f^{n-1}$$

THEOREM 3. ***The set of regular elements of a Banach algebra B is an open set.***

PROOF. Let B be a Banach algerbra and G be the set of regular elements, *i.e.*,

$$G = \{x \in B : x^{-1} \in B\}, e \in G$$

We have to show that G is an open set.

Consider a unit sphere $S(e, 1)$ with centre e and of unit radius such that

$$S[e, 1] = [x \in B : \|x - e\| < 1]$$

Clearly $S(e, 1) \subset G$ ...(1)

Now let $x \in B$ be any regular element then $x^{-1} \in B$ and

$$xx^{-1} = e \in S(e, 1)$$

Now for $\varepsilon > 0$, we have

$$S(x, \varepsilon).x^{-1} \in S(e, 1) \subset G \quad (\because \text{Ring operator is commutative}) \quad \ldots(2)$$

where $S(x, \varepsilon)x^{-1} = [yx^{-1} : y \in S(x, \varepsilon)] \; \forall \, y \in S(x, \varepsilon)$

$$\Rightarrow \quad yx^{-1} \in S(x, \varepsilon)x^{-1}$$

$$\Rightarrow \quad yx^{-1} \in G. \quad \text{[Using (2)]}$$

Now, yx^{-1} is a unit and therefore $\exists\, z \in B$ such that

$$(yx^{-1})z = z(yx^{-1}) = e$$

$\Rightarrow \quad y(x^{-1}z) = e$

$\Rightarrow \quad y^{-1} = x^{-1}z$

$\Rightarrow$ y is a unit and so $y \in G$

Thus, we conclude that

For all $\quad y \in G \quad \Rightarrow \quad \exists$ a nbd $S(x, \varepsilon)x^{-1}$ such that $y \in S(x, \varepsilon)x^{-1} \subset G$

But y is arbitrary. Hence, G is open.

THEOREM 4. ***Let B be a Banach algebra and $x \in B$. Then $\sigma(x)$ is Compact and $r(x) \le \|x\|$.***

PROOF. Let us define a mapping

$$f : C \to B \text{ such that } f(\lambda) = x - \lambda$$

Then clearly f is continuous. If G is the set of all invertible elements then $\rho(x) = f^{-1}(G)$.

Since G is open and inverse image of an open set is open therefore $\rho(x)$ is open. Consequently, $\sigma(x)$ is closed.

Now, if $\quad |\lambda| > |x|$, then

$$\left\|1-\left(1-\frac{x}{\lambda}\right)\right\| = \left\|\frac{x}{\lambda}\right\| = \frac{\|x\|}{|\lambda|} < 1$$

$\Rightarrow \quad \left(1-\dfrac{x}{\lambda}\right)$ is invertible.

$\Rightarrow \quad (x - \lambda)$ is invertible.

$\Rightarrow \quad \lambda \notin \sigma(x) \qquad$ (By definiton $\sigma(x) = \{\lambda \in C : x - \lambda\}$ is not invertible)

Therefore, $\sigma(x)$ is bounded and $\rho(x) \le \|x\|$ and hence, $\sigma(x)$ being closed and bounded is compact.

THEOREM 5. ***For a Banach algebra B, $\sigma(x) \neq \phi, x \in B$.***

PROOF. Let $f : \rho(x) \to x$ be a mapping defined by

$$f(\lambda) = (x - \lambda)^{-1}$$

Now, for $\lambda_0 \in \rho(x)$

$$\lim_{\lambda\to\lambda_0} \frac{f(\lambda) - f(\lambda_0)}{\lambda - \lambda_0} = \lim_{\lambda\to\lambda_0} \frac{(x-\lambda_0)^{-1}[(x-\lambda_0)-(x-\lambda)](x-\lambda)^{-1}}{\lambda-\lambda_0}$$

$$= \lim_{\lambda\to\lambda_0} (x-\lambda_0)^{-1}(x-\lambda)^{-1} = (x-\lambda_0)^{-2}$$

Therefore, for $\phi \in X^*$, the function $\phi(f)$ is a complex analytic function on $\rho(x)$

Also, for $|\lambda| > \|x\|, 1 - \dfrac{x}{\lambda}$ is invertible, therefore,

$$\left\|\left(1-\frac{x}{\lambda}\right)^{-1}\right\| \le \frac{1}{1-\left\|\dfrac{x}{\lambda}\right\|}$$

Therefore, $\quad \displaystyle\lim_{\lambda\to\infty} \|f(\lambda)\| = \lim_{\lambda\to 0}\left\|\frac{1}{\lambda}\left(\frac{x}{\lambda}-1\right)^{-1}\right\|$

$$\le \lim_{\lambda\to 0} \sup.\frac{1}{|\lambda|}.\frac{1}{1-\left\|\dfrac{x}{\lambda}\right\|}$$

$\Rightarrow$ for $\phi \in x^*$, $\lim\limits_{\lambda \to \infty} \phi(f(\lambda)) = 0$

Further, let if possible $\sigma(x) = \phi$ so that $\rho(x) = C$

Therefore, for $\phi \in x^*$, $\phi(f)$ is an entire function and vanishes at infinity.

$\Rightarrow$ $\phi(f) = 0$ (By Liouville's theorem for analytic function)

Now, since for a fixed $\lambda \in C$, $\phi(f(\lambda)) = 0$ for each $\phi \in x^*$

$\Rightarrow$ $f(\lambda) = 0$, which is a contradiction since $f(\lambda)$ is an invertible element of x.

Hence, $\sigma(x) \neq \phi$.

THEOREM 6. (Gelfand-Mazur Theorem).

Let B be a division Banach algebra then there is an isometric isomorphism of B onto C.

PROOF. Recall that a Banach division algebra is a Banach algebra in which every non-zero element is invertible.

Now, let $x \in B$ then clearly $\sigma(x) \neq 0$.

Let $\lambda x \in \sigma(x)$ then by definition of spectrum we have $x - \lambda x$ is not invertible.

Since, B is a division algebra, we must have $x - \lambda_x = 0$.

If $\lambda \neq \lambda_x$ then $x - \lambda = \lambda_x - \lambda$ is invertible in B. Therefore, for each $x \in B$ there is exactly one complex number λ_x in $\sigma(x)$. Then the mapping $\phi : B \to \mathbf{C}$ such that $\phi(x) = \lambda_x$ is an isometric isomoprphism of B onto C.

THEOREM 7. (Spectral Radius Formula). ***Let B be a Banach algebra and $x \in B$ then spectral radius is given by***

$$r(x) = \|x^n\|^{1/n}$$

(KANPUR–2002, 11)

PROOF. We know that $\sigma(x^n) = \sigma(x)^n$ and by definition of spectral radius, we have

$$r(x) = \sup. \{|\lambda| : \lambda \in \sigma(x)\} \quad \ldots(1)$$

$$\Rightarrow \quad r(x^n) = \sup. \{|\lambda| : \lambda \in \sigma(x^n)\} \quad \ldots(2)$$

$$= \sup. \{|\lambda| : \lambda \in \sigma(x)^n\} = r(x)^n \quad \ldots(3)$$

Again, by definition of norm

$$0 \leq r(x) \leq \|x\|$$

$$\Rightarrow \quad 0 \leq r(x^n) \leq \|x^n\|$$

$$\Rightarrow \quad 0 \leq r(x)^n \leq \|x^n\| \quad \text{(Using (3))}$$

$$\Rightarrow \quad r(x) \leq \|x^n\|^{1/n} \text{ and } r(x) \text{ is positive} \quad \ldots(4)$$

Further, assume that a is any real number such that

$$r(x) < a$$

We have to prove that $\|x^n\|^{1/n} \leq a$ for all finite numbers of n

Let us choose $|\lambda| > \|x\|$ then

$$x(\lambda) = (x - \lambda I)^{-1}\left(\frac{x}{\lambda} - 1\right)^{-1} = -\lambda^{-1}\left(1 - \frac{x}{\lambda}\right)^{-1}$$

$$= -\lambda^{-1}\left(1 + \sum_{n=1}^{\infty} \frac{x^n}{\lambda^n}\right) \quad \ldots(5)$$

Now, for any functional f on B, equation (5) becomes

$$f(x) - \lambda = -\lambda^{-1}\left[f(1) + \sum_{n=1}^{\infty} f\left(\frac{x^n}{\lambda^n}\right)\right]$$

$$= -\lambda^{-1}\left[f(1) + \sum_{n=1}^{\infty} f(x^n)\lambda^{-n} \right] \quad \ldots(6)$$

Since $f(\lambda) = f(x(\lambda))$ is differentiable, therefore $f(x(\lambda))$ is analytic within the region $|\lambda| > r(x)$. Therefore, we can say that the expansion given by (6) is the Laurent series expansion of $f(x(\lambda))$ for all $|\lambda| > \|x\|$ valid in the region $|\lambda| > r(x)$.

Also, if b is any another real number such that $r(x) < b < a$

Then the series $\sum_{n=1}^{\infty} f\left(\dfrac{x^n}{b^n}\right)$ is convergent so that its terms form a bounded sequence for all $f \in B^*$, where B^* is the conjugate space of B.

$\Rightarrow$ the elements $\dfrac{x^n}{b^n}$ form a bounded sequence in B so that

$\left\|\dfrac{x^n}{b^n}\right\| \leq k \Rightarrow \left\|x^n\right\|^{1/n} \leq bk^{1/n}$, for some positive integer k and for all n.

$\Rightarrow \quad \left\|x^n\right\|^{1/n} \leq a$ for sufficiently large n when $b < a \Rightarrow bk^{1/n} < a$

$$\Rightarrow \quad \left\|x^n\right\|^{1/n} \leq a \ \forall n$$

Hence, we conclude that

$$r(x) \leq \left\|x^n\right\|^{1/n} \leq a$$

$$\Rightarrow \quad r(x) = \lim \left\|x^n\right\|^{1/n}$$

THEOREM 8. ***σ(f) is non-empty.*** (KANPUR–2002, 05, 11, DELHI–2011, CALCUTTA–2014)

PROOF. Let if possible $\sigma(f) = \phi$.

We know that the function $f(\lambda) = (f - \lambda e)^{-1}$ is an analytic function over the resolvent set $\rho(f)$. If $\sigma(f) = \phi$ then $f(\lambda)$ is analytic everywhere and hence it is an entire function.

Clearly, $$e - \frac{f}{\lambda} \to e^{-1} = e$$

$$\Rightarrow \quad \left\|(f - \lambda e)^{-1}\right\| = \left|\lambda^{-1}\right| . \left\|(e - f/\lambda)^{-1}\right\| \to \frac{\|e\|}{|\lambda|} \to 0 \text{ as } |\lambda| \to \infty \quad \ldots(1)$$

So, for any given $m_1 > 0$ there exist $r > 0$ such that

$$\left\|(f - \lambda e)^{-1}\right\| < m_1 \text{ for } |\lambda| > \text{r}$$

Now, let $$m_2 = \max_{|\lambda| \leq r} \|f(\lambda)\|$$

We observe that $f(\lambda) = (f - \lambda e)^{-1}$ is bounded everywhere by larger of m_1 and m_2. Then by Liouville's theorem, $f(\lambda)$ is constant.

($\because$ Every entire function which is bounded must reduces to a constant)

From (1), the constant must be zero which is not possible because if $(f - \lambda e)^{-1} = 0$ then $(f - e\lambda)(f - \lambda e)^{-1} \neq e$

Thus our assumption is wrong.

Hence, $\sigma(f)$ must be non-empty.

Theorem 9. ***Every maximal ideal in Banach algebra B is closed and hence the radical is a proper closed ideal.***

Proof. Recall that the closure of any ideal is again an ideal and the closure of any proper ideal is a proper ideal. Let, there exists a maximal ideal M which is not closed then closure of M, *i.e.*, $\bar{M}$ is a proper ideal which properly contains M (By definition of closure, $\bar{M}$ is the smallest closed set containing M) but M is maximal so it is not contain in any other ideal. Hence, M must be closed. Further, since by definition of radical, the redical R is a proper ideal and it is the intersection of closed sets. Therefore, it is a proper closed ideal.

Theorem 10. ***Let B be a Banach algebra and M is a proper closed ideal in B then the quotient algebra B|M is a Banach algebra.***

Proof. We have already proved in the chapter of Banach space that $B|M$ is a banach space with the norm defined by

$$\|x + M\| = \inf\{\|x + m\| : m \in M\}$$

Here $B|M$ is an algebra with identity $1 + M$

Further, $\|1 + M\| = \inf\{\|1 + m\| : m \in M\} \leq \|1\| = 1$

$$\begin{aligned}\text{Now, } \|(x + M)|(y + M)\| &= \|xy + M\| = \inf\{\|xy + m\| : m \in M\} \\ &\leq \inf\{\|(x + m_1)(y + m_2)\| : m_1, m_2 \in M\} \\ &\leq \inf\{\|(x + m_1)\|.\|(y + m_2)\| : m_1, m_2 \in M\} \\ &= [\inf\{\|(x + m_1)\| : m_1 \in M].[\|(y + m_2)\| : m_2 \in M\}] \\ &= \|x + M\|.\|y + M\|\end{aligned}$$

Now, we shall prove that $\|1 + M\| = 1$

We have $\|1 + M\| \leq \|(1 + M)^2\| \leq \|1 + M\|^2$

$\Rightarrow$ $\|1 + M\| \geq 1$

But we have already proved that $\|1 + M\| \leq 1$

$\Rightarrow$ $\|1 + M\| = 1$

Hence, $B|M$ is a Banach algebra.

Remark

- The maximal ideal space is a Compact T_2-space.

Theorem 11. ***Let B be a Banach algebra and S be the set of singular elements of B and Z be a the set of all topological divisors of zero, then***

(i) Z is a subset of S

(ii) the elements of S is a subset of Z.

(KANPUR–2003, REEVA–2006, SAGAR–2009)

Proof. (i) Let $z \in Z$ and $<z_n>$ be a sequence and $\|z_n\| = 1$ such that $z \cdot z_n \to 0$. Let G be the set of all regular elements of B and $z \in G$, then

$$z^{-1}(zz_n) = (z^{-1}z)z_n = 1.z_n = z_n \to 0 \text{ as } zz_n \to 0$$

which is a conradiction, because we have assume that $\|z_n\| = 1$

Hence, $Z \subseteq S$.

(ii) Since, S is closed therefore, its boundary consists of all points in S which are limits of convergent sequence in G.

We have to prove that $z \in S \Rightarrow \exists$ a sequence $<r_n>$ in G such that

$$r_n \to z \Rightarrow z \in Z$$

Consider $\quad r_n^{-1}.z - 1 = r_n^{-1}z - r_n^{-1}r_n = r_n^{-1}(z - r_n)$

$\Rightarrow \quad \|r_n^{-1}z - 1\| \leq \|r_n^{-1}\|.\|z - r_n\|$

$\Rightarrow \; < r_n^{-1} >$ is unbounded

therefore, let us assume that $\|r_n^{-1}\| \to \infty$

set $\quad z_n = \dfrac{r_n^{-1}}{\left\|r_n^{-1}\right\|}$

$$\Rightarrow \quad \|z_n\| = \left\|\frac{r_n^{-1}}{\left\|r_n^{-1}\right\|}\right\| = \left\|r_n^{-1}\right\|.\frac{1}{\left\|r_n^{-1}\right\|} = 1$$

Now, $\quad zz_n = \dfrac{zr_n^{-1}}{\left\|r_n^{-1}\right\|} = \dfrac{1 + zr_n^{-1} - r_n r_n^{-1}}{\left\|r_n^{-1}\right\|} \quad$ (as $r_n r_n^{-1} = 1$)

$$= \frac{1 + (z - r_n)r_n^{-1}}{\left\|r_n^{-1}\right\|} = \frac{1}{\left\|r_n^{-1}\right\|} + (z - r_n)z_n \to 0$$

($\because r_n \to z$ and $\left\|r_n^{-1}\right\| \to \infty$)

$\Rightarrow$ Topological divisors of Banach algebra B are permanently singular and hence have the permanent boundary.

$\Rightarrow$ The boundary of S is a subset of Z.

THEOREM 12. ***If B is a B^*-algebra and $x \in B$ then $\left\|x^*\right\| = x$ and therefore $\left\|x^* x\right\| = \|x\|$. Further if B is normal $\left\|x^2\right\| = \|x\|^2$.***

PROOF. Let B be a B^*-algebra. Then we have

$$\|x\|^2 = \|x^*x\| \leq \|x^*\|.\|x\|$$

$\Rightarrow \quad \|x\| \leq \|x^*\|$

$\Rightarrow \quad \|x^*\| \leq \|x^{**}\| = \|x\|$

Therefore, $\quad \|x^*\| = \|x\| \quad \Rightarrow \quad \|x^*x\| = \|x^*\|.\|x\|$

Now, let us assume that x is normal

Also, $\quad \|x^2\| \leq \|x\|^2 \quad$ ...(1)

Therefore, $\quad \|x^*\|^2 \cdot \|x\|^2 = (\|x^*\|\,\|x\|)^2 = \|x^*x\|^2$

$$= \|(x^*x)x^*x\| = \|x^*xx^*x\|$$

$$= \|x^*x^*xx\| = \|(x^*)^2x^2\| = \|(x^2)^*\| \cdot \|x^2\|$$

$$= \|(x^2)^*\| \cdot \|x^2\| \leq \|x^*\|^2 \cdot \|x^2\|$$

$\Rightarrow \quad \|x^*\|^2 \leq \|x\|^2 \quad$ ...(2)

From (1) and (2) we conclude that

$$\|x^2\| = \|x\|^2$$

THEOREM 13. ***Let B be a Banach algebra and $x \in B$ then $r(x) = \|x\|$ if and only if $\|x^2\| = \|x\|^2$. Particularly, if x is a normal element in a B^*-algebra, $r(x) = \|x\|$.***

PROOF. Let B be a Banach algebra and $x \in B$.

Let us first suppose $r(x) = \|x\|$ then we have

$$\|x^2\| = r(x^2) = \|x\|^2$$

Conversely, let $\|x^2\| = \|x\|^2$

then for every positive integer k, we have

$$\left\|x^{2^k}\right\| = \|x\|^{2^k}$$

Then by spectral-radius formula, we have

$$r(x) = \lim \|x^n\|^{1/n} = \lim \left\|x^{2^k}\right\|^{1/2^k}$$

$$= \lim \|x\| = \|x\|$$

THEOREM 14. ***The mapping $f : x \to x^{-1}$ of G into G is continuous and is therefore a homomorphism of onto itself.***

PROOF. Let B be a Banach algebra and G be the set of regular elements of B.

Define a mapping $f : G \to G$ such that $f(x) = x^{-1} \ \forall \ x \in G$.

We have to prove that f is a homomorphism.

Since G is an open set, therefore f is open.

Let $x_1, x_2 \in G$ then $f(x_1) = f(x_2)$

$\Rightarrow \quad x_1^{-1} = x_2^{-1}$

$\Rightarrow \quad (x_1)^{-1} = (x_2)^{-1}$

$\Rightarrow \quad x_1 = x_2$

$\Rightarrow$ f is one-one

Now we shall show that f is onto also.

For given $y \in G \ \exists \ y^{-1} \in G$ such that

$$f(y^{-1}) = (y^{-1})^{-1} = y$$

$\Rightarrow$ f is onto

Now, it remains to prove that f is continuous.

Consider a convergent sequence $<x_n>$ in G such that $x_n \to x$ in G. We will prove that

$$f(x_n) = f(x), \text{ i.e., } x_n^{-1} = x^{-1}$$

$$x_n \to x \quad \Rightarrow \quad x^{-1}x_n \to x^{-1}x = e$$

$\Rightarrow \quad x^{-1}x_n \to e$

So, for given $\varepsilon > 0 \ \exists$ positive integer n_0 such that for all $n \geq n_0$ such that

$$\|x^{-1}x_n - e\| < \varepsilon$$

Let us choose n_0' such that

$$\|x^{-1}x_n - e\| < \varepsilon \ \forall \ n \geq n_0'$$

Further, consider the series

$$e + \sum_{K=1}^{\infty} (e - x^{-1}x_n)^k \ \forall n \geq n_0' \text{ such that } \|e - x^{-1}x_n\| < 1$$

So this series converges to $(x^{-1}x_n)^{-1} = x_n^{-1}x$

Hence, $$x_n^{-1}x = e + \sum_{K=1}^{\infty} (e - x^{-1}x_n)^k$$

or $$x_n^{-1}x - e = \sum_{K=1}^{\infty} (e - x^{-1}x_n)^k$$

$\Rightarrow$ $$\left\|x_n^{-1}x - e\right\| \leq \sum_{K=1}^{\infty} \left\|(e - x^{-1}x_n)^k\right\| \to 0 \text{ as } n \to \infty$$

because $e - x^{-1}x_n = e - x^{-1}x = e - e = 0$ as $x \to \infty$

Hence, $$x_n^{-1}x - e \to 0$$

or $x_n^{-1}x \to e$ or $x_n^{-1} \to ex^{-1}$ or $x_n^{-1} \to x^{-1}$

Finally, $x_n \to x \Rightarrow x_n^{-1} \to x^{-1} \Rightarrow f(x_n) \to f(x)$

$\Rightarrow$ $x_n \to x \Rightarrow f(x_n) \to f(x)$

$\Rightarrow$ f is continuous.

Hence, we conclude that f is a homomorphism.

RECAPITULATIONS

- An algebra with identity is called division algebra.
- A normed algebra is called Banach algebra if it is complete as normed linear space.
- A Banach algebra is said to be Commutative if it is Commutative as a ring.
- The set of regular elements of a Banach algebra is an open set.
- Every maximal ideal in Banach algebra B is closed and hence the radical is a proper closed ideal.
- Let B be a Banach algebra and M is a proper closed ideal in B then the quotient algebra $B|M$ is a Banach algebra.

SOLVED EXAMPLES

EXAMPLE 1. *If $\lambda_n \to \lambda$ and $x_n \to x$, $\lambda_n \in \sigma(x_n)$, then show that $\lambda \in \sigma(x)$ in a Banach algebra B.*

SOLUTION. Let B be a Banach algebra.

It is given that $$\lambda_n \in \sigma(x_n)$$

$\Rightarrow$ $$x_n - \lambda_n e \notin U \text{ where } U = [x \in B : x^{-1} \in B]$$

$\Rightarrow$ $$x_n - \lambda_n e \in B - U$$

$\Rightarrow$ $$\lim_{n\to\infty} (x_n - \lambda_n e) \in (B - U) \qquad (\because B - U \text{ is closed})$$

$\Rightarrow$ $$(x - \lambda e) \notin B$$

$\Rightarrow$ $$\lambda \in \sigma(x)$$

EXAMPLE 2. *Prove that $\sigma(e) = \{1\}$.*

SOLUTION. If $\lambda = 1$ then we have

$$e - \lambda e = 0$$

Therefore, $e - \lambda e$ is not invertible.

$\Rightarrow$ $$e - \lambda e \notin U$$

Hence, $1 \in \sigma(e)$

If $\lambda \neq 1$ then $e - \lambda e$ cannot be zero therefore, in this case $e - \lambda e$ is invertible and inverse of $e - \lambda e$ is given by

$$(e - \lambda e)^{-1}(x) = \left(\frac{1}{1-\lambda}\right)x$$

$$\Rightarrow \qquad \lambda \notin \sigma(e)$$

Hence, $\sigma(e) = \{1\}$.

EXAMPLE 3. *Let B be a Banach algebra. If $xy = yx$, then show that $r(xy) \leq r(x).r(y)$.*

SOLUTION. Consider
$$(xy)^2 = xy.xy = x(yx)y = x(xy)y \qquad (\because xy = yx)$$
$$= x^2y^2$$

In general
$$(xy)^n = x^ny^n$$

Therefore,
$$\|(xy)^n\| = \|x^ny^n\| \leq \|x^n\|.\|y^n\|$$

$$\Rightarrow \qquad \|(xy)^n\|^{1/n} \leq \|x^n\|^{1/n}.\|y^n\|^{1/n} \qquad \text{(By Schwarz's inequality)}$$

Taking limit of both sides, we get
$$\lim_{n\to\infty}\left\|(xy)^n\right\|^{1/n} \leq \lim_{n\to\infty}\left\|x^n\right\|^{1/n}.\lim_{n\to\infty}\left\|y^n\right\|^{1/n}$$

Hence, $r(x, y) \leq r(x)r(y)$.

EXAMPLE 4. *In a B^*-algebra if $xyy^* = 0$, show that $xy = 0$.*

SOLUTION. Consider
$$\|xy\|^2 = \|(xy)(xy)^*\| = \|xyy^*x^*\| = \|0.x^*\| \qquad (\because xyy^* = 0)$$
$$= 0$$

Hence, $\|xy\| = 0$, *i.e.*, $xy = 0$.

EXAMPLE 5. *Show that $\lambda \in \sigma(x) \Rightarrow \bar{\lambda} \in \sigma(x^*)$.*

SOLUTION. Let $\lambda \in \sigma(x)$ then we have
$$x - \lambda e \in U \qquad (U = x \in B : x^{-1} \in B)$$
$$\Rightarrow \qquad (x - \lambda e)^* \in U$$
$$\Rightarrow \qquad x^* - \bar{\lambda} e \in U$$
$$\Rightarrow \qquad \bar{\lambda} \notin \sigma(x^*)$$

Hence, $\bar{\lambda} \in \sigma(x^*) \Rightarrow \lambda \in \sigma(x)$

EXAMPLE 6. *Let B be a unital Banach algebra. If $x \in B$ is invertible and $xy = yx$ for $y \in B$, show that $x^{-1}y = yx^{-1}$.*

SOLUTION. Let B be a unital Banach algebra and e be the identity in B, then we have
$$x^{-1}y = x^{-1}ye = x^{-1}yxx^{-1}$$
$$= x^{-1}(yx)x^{-1} = x^{-1}(xy)x^{-1} \qquad (\because xy = yx)$$
$$= (x^{-1}x)yx^{-1} = eyx^{-1} = yx^{-1}$$

EXAMPLE 7. *Show that multiplication in a Banach algebra is jointly continuous.*

SOLUTION. Let B be a Banach algebra.

Let us suppose that $x_n \to x$ and $y_n \to y$

We have to prove that $x_ny_n \to xy$

Since $x_n \to x$ and $y_n \to y$ therefore,
$$\|x_n - x\| \to 0 \text{ and } \|y_n - y\| \to 0 \text{ as } n \to \infty \qquad \ldots(1)$$

Also, we know that every convergent sequence is bounded, therefore, $\|x_n\| \leq M$ $\forall\, n$ and constant $M > 0$.

Now consider
$$\begin{aligned}\|x_n y_n - xy\| &= \|x_n y_n - yx_n + yx_n - xy\| \\ &= \|x_n(y_n - y) + (x_n - x)y\| \\ &\le \|x_n(y_n - y) + (x_n - x)y\| \\ &\le \|x_n\|.\|y_n - y\| + \|x_n - x\|.\|y\| \\ &\le M\|y_n - y\| + \|x_n - x\|.\|y\| \\ &\to 0 \text{ as } n \to \infty \qquad \text{[Using (1)]}\end{aligned}$$

Hence, $x_n y_n \to xy$.

EXAMPLE 8. *Let X be a Banach space and B(X, X) be the set of all bounded linear transformation on X onto X, then show that B(X, X) is also a Banach algebra.*

SOLUTION. In the chapter of Banach space, we have already proved that B(X, X) is a Banach space with the operator norm.

$$\|T\| = \sup_{\|x\|\le 1} \|Tx\|$$

Now
$$\|TS\| = \sup_{\|x\|\le 1} \|TS(x)\| = \sup_{\|x\|\le 1} \|T[S(x)]\| = \|T\| \sup_{\|x\|\le 1} \|S(x)\|$$
$$= \|T\|\|S\| \sup_{\|x\|\le 1} \|x\| \le \|T\|\|S\|$$

Hence, $B(X, X)$ is a Banach algebra.

EXERCISE 7.1

1. Show that the set of complex numbers **C** is a commutative Banach algebra with identity $e = 1$.
2. Show that in an algebra with identity, a proper ideal contains no invertible elements.
3. Let B be a Banach algebra with identity e. Show that the set of all invertible elements is a multiplicative group.
4. In a Banach algebra, show that xy and yx always have the same spectral radius.
5. Let B be a Banach algebra with identity and $A = \begin{pmatrix} 0 & 1 \\ -1 & 0 \end{pmatrix}$ is a 2×2 matrix. Show that $\sigma(A) = \phi$.
6. Let B be a unital Banach algebra such that $\|x^{-1}\| = \dfrac{1}{\|x\|}$ $\forall x$. Show that B is one-dimensional.
7. Prove that every Banach algebra is isomorphic to a subalgebra of the algebra of the bounded linear transformations on a Banach space into itself.
8. If G and S are respectively the sets of regular and singular elemets in a Banach algebra B, then show that if G is an open set, S is closed.
9. If B is a divison Banach algebra, then show that it equals the set of all scalar multiples of the identity.
10. Show that the set $e(X)$ of all bounded continuous complex functions defined on a topological space X forms a Banach algebra.

Chapter Review: A Competitive Approach

SELECTED TERMS AND RESULTS

Terms

- **Banach Algebra :** An algebra B over a field F is called Banach algebra if following conditions are satisfied :
  (i) $\|xy\| \leq \|x\|.\|y\|$
  (ii) $\exists\ e \in B$ such that $ex = xe = x$ and $\|e\| = 1$ $\forall\ x, y \in B$
- **Division Algebra :** An algebra with identity.
- **Regular Elements :** Let B be a Banach algebra. If corresponding to an element $f \in B$ there exist an element $g \in B$ such that $fg = gf = e$. Then g is called inverse of f. If an element $f \in B$ has an inverse then it is called regular element.
- **Singular Element :** An element of a Banach algebra, which is not regular is called singular element.
- **Spectral Radius :**
  $$\sigma_B(x) = \sigma(x) = [\lambda : x - \lambda I = \text{singular}]$$
- **Topological Divisor of Zero :** An element x of a Banach algebra B is called topological divisor of zero if $\exists$ a sequence $<x_n>$ in B such that $\|x_n\| = 1$ and $x.x_n \to 0$

Results

- Every algebra is a ring.
- Multiplication of vectors is jointly continuous in a Banach algebra.
- The set of all regular elements of a Banach algebra forms a multiplicative group.
- The zero vector is always singular element of B.
- Units of Banach algebra form an open set.
- Spectral radius $r(x)$ is given by
  $$r(x) = \|x^n\|^{1/n}$$
- $\sigma(f)$ is always non-empty.
- A division algebra is equals the set of all scalar multiples of the identtiy.
- The mapping $f : G \to G$, *i.e.*, $f : x \to x^{-1}$ is continuous and is therefore a homomorphism of G onto itself.
- In a Banach algebra, the maximal ideal space is a Compact Housedroff space.
- Every maximal ideal in a Banach algebra B is closed.

REVIEW QUESTIONS

1. Let B be a commutative Banach algebra with identity. Show that $x \in B$ belongs the radical if and only if the sequence $<(\lambda x)^n>$ converges to zero for every $\lambda \in C$.
2. Let B be a Banach algebra. If there exist a constant $m > 0$ such that $\|xy\| \geq m\|x\|.\|y\|$ for all $x, y \in B$ then show that B is isomorphic to C.
3. Show that in a Banach algebra, closure of an ideal is again an ideal.
4. Prove that Banach algebra B is commutative if $\|x^2\| = \|x\|^2\ \forall\ x \in B$.
5. Show that the spectral radius is not necessarily continuous.
6. Let B be a complex Banach algebra such that 0 is the only topological divisor of zero in B. Prove that B is isomorphic to C.
7. Let B be a commutative Banach algebra with identity and let R be the radical of B. Show that $X|B$ is semi-simple.
8. Let B be a commutative Banach algebra with an involution satisfying $\|x\|^2 \leq k\|xx^*\|\ \forall$ $x \in B$, for some constant $k > 0$. Show that $(r(x))^2 = r(xx^*)$.
9. If $\|x\| \leq 1$, show that $x^n \to 0$ as $n \to \infty$ is a normal algebra.
10. In a Banach algebra B, show that $\|x^n\| \leq \|x\|^n$ $\forall\ x \in B$.

Self Assessment Test

1. Show that the set of complex numbers C is a commutative Banach algebra with identity 1.
2. Show that the set of all bounded linear transformations on a Hilbert space H with the operator norm is a Banach algebra.
3. If $|\lambda| > \|x\|$ then show that $\lambda \in \rho(x)$ and $\sigma(x)$ is compact.
4. Show that the surjection $f : U \to U$ defined by $f(x) = x^{-1}$ is continous where U is the set of all invertible elements.
5. Let $x_n \in U$ for $n = 1, 2, \ldots$ and $x_n \to x$ with $x \notin U$ then show that $\|x_n^{-1}\| \to \infty$ as $n \to \infty$.
6. Show that every B^*-algebra has continuous involution.
7. Give an example to show that the spectral radius $r(x) = 0$ with $x \neq 0$.
8. Let B be a unital Banach algebra and $x \in B$. If $\lambda \in \rho(x)$, show that
   distance $(x, \sigma(x)) \geq \|(\lambda - x)^{-1}\|^{-1}$
9. Let B be a unital Banach algebra such that zero is the only topological divisor of zero in B. Show that B is isomorphic to C.
10. Prove that the algebra-isomorphism between semisimple Banach algebra is a homomorphism.
11. For a Banach algebra B and $x \in B$, prove the following :
    (i) e is Hermitian
    (ii) x^*x is Hermitian
    (iii) xx^* is Hermitian

●●●●●●

Bibliography

1.	**Banach, S.**	*Theorie des Operations Lineaires,* Warszawa, (1932)
2.	**Cheney, E. W.**	*Introduction to approximation theory,* McGraw-Hill, New York (1966)
3.	**Hewitt, E. and K. Stromberg**	*Real and abstract analysts,* Springer Verlag, Berlin, Heidelberg, New York (1969)
4.	**Kreyszig, E.**	*Introduction to functional analysis with applications,* John Wiley & Sons, New York.
5.	**Limaye, B. V.**	*Functional analysis,* Wiley Easters Ltd. (1981)
6.	**Maddox, I. J.**	*Elements of functional analysis,* Cambridge University Press (1970)
7.	**Rall, L. B.**	*Computational solution of non-linear operator equations,* John Wiley & Sons (1969).
8.	**Rall, L. B. (ed.)**	*Non-linear functional analysis and applications,* Academic Press (1971)
9.	**Rudlin, W.**	*Real and complex analysis,* Tata McGraw Hill Publishing Co. Ltd., New Delhi (1966)
10.	**Rudin, W.**	*Functional analysis,* Tata McGraw-Hill Publishing Co. Ltd., New Delhi (1973)
11.	**Simmons, G. F.**	*Introduction to topology and modern analysis,* McGraw-Hill Book Co. Inc., International Student Edition (1963)
12.	**Smart, D. R.**	*Fixed point theorems,* Cambridge University Press (1974)
13.	**Taylor, A. E.**	*Introduction to functional analysis,* Toppan Co. Ltd. (1958)
14.	**Wilansky, A.**	*Functional analysis,* Blaisdell Publishing Co. (1964).

Index

A

B

C

D

E

F

H

I

L

S

READER'S NOTES

READER'S NOTES